MS&A

Volume 8

For further volumes:
http://www.springer.com/series/8377

Alfio Quarteroni

Numerical Models for Differential Problems

Second Edition

Springer

Alfio Quarteroni
CMCS-MATHICSE
Ecole Polytechnique Fédérale de Lausanne
Switzerland
and
MOX, Department of Mathematics "F. Brioschi"
Politecnico di Milano
Italy

Translated by Silvia Quarteroni from the original Italian edition:
A. Quarteroni, Modellistica Numerica per Problemi Differenziali. 5ª ed.,
Springer-Verlag Italia, Milano 2012

Linguistic copy-editing: Simon Chiossi

ISSN print edition: 2037-5255 ISSN electronic edition: 2037-5263
MS&A – Modeling, Simulation & Applications
ISBN 978-88-470-5883-5 ISBN 978-88-470-5522-3 (eBook)
DOI 10.1007/978-88-470-5522-3
Springer Milan Heidelberg New York Dordrecht London

9 8 7 6 5 4 3 2 1

Cover-Design: Beatrice 𝔅, Milano
The cardiac numerical simulations reported on the front cover are due to Ricardo Ruiz Baier from CMCS-MATHICSE, EPFL, Lausanne
Typesetting with LaTeX: PTP-Berlin, Protago TeX-Production GmbH, Germany (www.ptp-berlin.de)

Springer is a part of Springer Science+Business Media (www.springer.com)

To Fulvia, Silvia and Marzia

Preface to the second edition

Differential equations (DEs) are the foundation on which many mathematical models for real-life applications are built. These equations can seldom be solved in 'closed' form: in fact, the exact solution can rarely be characterized through explicit, and easily computable, mathematical formulae. Almost invariably one has to resort to appropriate numerical methods, whose scope is the approximation (or discretization) of the exact differential model and, hence, of the exact solution.

This is the second edition of a book that first appeared in 2009. It presents in a comprehensive and self-contained way some of the most successful numerical methods for handling DEs, for their analysis and their application to classes of problems that typically show up in the applications.

Although we mostly deal with partial differential equations (PDEs), both for steady problems (in multiple space dimensions) and time-dependent problems (with one or several space variables), part of the material is specifically devoted to ordinary differential equations (ODEs) for one-dimensional boundary-value problems, especially when the discussion is interesting in itself or relevant to the PDE case.

The primary concern is on the finite-element (FE) method, which is the most popular discretization technique for engineering design and analysis. We also address other techniques, albeit to a lesser extent, such as finite differences (FD), finite volumes (FV), and spectral methods, including further *ad-hoc* methods for specific types of problems. The comparative assessment of the performance of different methods is discussed, especially when it sheds light on their mutual interplay.

We also introduce and analyze numerical strategies aimed at reducing the computational complexity of differential problems: these include operator-splitting and fractional-step methods for time discretization, preconditioning, techniques for grid adaptivity, domain decomposition (DD) methods for parallel computing, and reduced-basis (RB) methods for solving parametrized PDEs efficiently.

Besides the classical elliptic, parabolic and hyperbolic linear equations, we treat more involved model problems that arise in a host of applicative fields: linear and nonlinear conservation laws, advection-diffusion equations with dominating advection, Navier-Stokes equations, saddle-point problems and optimal-control problems.

Here is the contents' summary of the various chapters.

Chapter 1 briefly surveys PDEs and their classification, while Chapter 2 introduces the main notions and theoretical results of functional analysis that are extensively used throughout the book.

In Chapter 3 we illustrate boundary-value problems for elliptic equations (in one and several dimensions), present their weak or variational formulation, treat boundary conditions and analyze well-posedness. Several examples of physical interest are introduced.

The book's first cornerstone is Chapter 4, where we formulate Galerkin's method for the numerical discretization of elliptic boundary-value problems and analyze it in an abstract functional setting. We then introduce the Galerkin FE method, first in one dimension, for the reader's convenience, and then in several dimensions. We construct FE spaces and FE interpolation operators, prove stability and convergence results and derive several kinds of error estimates. Eventually, we present grid-adaptive procedures based on either *a priori* or *a posteriori* error estimates.

The numerical approximation of parabolic problems is explained in Chapter 5: we begin with semi-discrete (continuous in time) Galerkin approximations, and then consider fully-discrete approximations based on FD schemes for time discretization. For both approaches stability and convergence are proven.

Chapters 6, 7 and 8 are devoted to the algorithmic features and the practical implementation of FE methods. More specifically, Chapter 6 illustrates the main techniques for grid generation, Chapter 7 surveys the basic algorithms for the solution of ill-conditioned linear algebraic systems that arise from the approximation of PDEs, and Chapter 8 presents the main operational phases of a FE code, together with a complete working example.

The basic principles underlying finite-volume methods for the approximation of diffusion-transport-reaction equations are discussed in Chapter 9. FV methods are commonly used in computational fluid dynamics owing to their intrinsic, built-in conservation properties.

Chapter 10 addresses the multi-faceted aspects of spectral methods (Galerkin, collocation, and the spectral-element method), analyzing thoroughly the reasons for their superior accuracy properties.

Galerkin discretization techniques relying on discontinuous polynomial subspaces are the subject of Chapter 11. We present, more specifically, the discontinuous Galerkin (DG) method and the mortar method, together with their use in the context of finite elements or spectral elements.

Chapter 12 focuses on singularly perturbed elliptic boundary-value problems, in particular diffusion-transport equations and diffusion-reaction equations, with small diffusion. The exact solutions to this type of problems can exhibit steep gradients in tiny subregions of the computational domains, the so-called internal or boundary layers. A great deal of attention is paid to stabilization techniques meant to prevent the on-rise of oscillatory numerical solutions. Upwinding techniques are discussed for FD approximations, and their analogy with FE with artificial diffusion is analyzed. We introduce and discuss other stabilization approaches in the FE context, as well, which lead to the sub-grid generalized Galerkin methods, the Petrov-Galerkin methods and Galerkin's Least-Squares method.

The ensuing three chapters form a thematic unit focusing on the approximation of first-order hyperbolic equations. Chapter 13 addresses classical FD methods. Stability is investigated using both the energy method and the Von Neumann analysis. Using the latter we also analyze the properties of dissipation and dispersion featured by a numerical scheme. Chapter 14 is devoted to spatial approximation by FE methods, including the DG methods and spectral methods. Special emphasis is put on characteristic compatibility conditions for the boundary treatment of hyperbolic systems. A very quick overview of the numerical approximation of nonlinear conservation laws is found in Chapter 15. Due to the relevance of this particular topic the interested reader is advised to consult the specific monographs mentioned in the references.

In Chapter 16 we discuss the Navier-Stokes equations for incompressible flows, plus their numerical approximation by FE, FV and spectral methods. A general stability and convergence theory is developed for spatial approximation of saddle-point problems, which comprises strategies for stabilization. Next we propose and analyze a number of time-discretization approaches, among which finite differences, characteristic methods, fractional-step methods and algebraic factorization techniques. Special attention is devoted to the numerical treatment of interfaces in the case of multiphase flows.

Chapter 17 discusses the issue of optimal control for elliptic PDEs. The problem is first formulated at the continuous level, where conditions of optimality are obtained using two different methods. Then we address the interplay between optimization and numerical approximation. We present several examples, some of them elementary in character, others involving physical processes of applicative relevance.

Chapter 18 regards domain-decomposition methods. These techniques are specifically devised for parallel computing and for the treatment of multiphysics' PDE problems. The families of Schwarz methods (with overlapping subdomains) and Schur methods (with disjoint subdomains) are illustrated, and their convergence properties of optimality (grid invariance) and scalability (subdomain-size invariance) studied. Several examples of domain-decomposition preconditioners are provided and tested numerically.

Finally, in Chapter 19 we introduce the reduced-basis (RB) method for the efficient solution of PDEs. RB methods allow for the rapid and reliable evaluation of input/output relationships in which the output is expressed as a functional of a field variable that is the solution of a parametrized PDE. Parametrized PDEs model several processes relevant in applications such as steady and unsteady transfer of heat or mass, acoustics, solid and fluid mechanics, to mention a few. The input-parameter vector variously characterizes the geometric configuration of the domain, physical properties, boundary conditions or source terms. The combination with an efficient *a posteriori* error estimate, and the splitting between offline and online calculations, are key factors for RB methods to be computationally successful.

Many important topics that would have deserved a proper treatment were touched only partially (in some cases completely ignored). This depends on the desire to offer a reasonably-sized textbook on one side, and our own experience on the other. The list of notable omissions includes, for instance, the approximation of equations for the structural analysis and the propagation of electromagnetic waves. Detailed studies can be found in the references' specialized literature.

This text is intended primarily for graduate students in Mathematics, Engineering, Physics and Computer Science and, more generally, for computational scientists. Each chapter is meant to provide a coherent teaching unit on a specific subject. The first eight chapters, in particular, should be regarded as a comprehensive and self-contained treatise on finite elements for elliptic and parabolic PDEs. Chapters 9–16 represent an advanced course on numerical methods for PDEs, while the last three chapters contain more subtle and sophisticated topics for the numerical solution of complex PDE problems.

This work has been used as a textbook for graduate-level courses at the Politecnico di Milano and the École Polytechnique Fédérale de Lausanne. We would like to thank the many people – students, colleagues and readers – who contributed, at various stages and in many different ways, to its preparation and to the improvement of early drafts. A (far from complete) list includes Paola Antonietti, Luca Dedé, Marco Discacciati, Luca Formaggia, Loredana Gaudio, Paola Gervasio, Andrea Manzoni, Stefano Micheletti, Nicola Parolini, Anthony T. Patera, Luca Pavarino, Simona Perotto, Gianluigi Rozza, Fausto Saleri, Benjamin Stamm, Alberto Valli, Alessandro Veneziani, and Cristoph Winkelmann. Special thanks go to Luca Paglieri for the technical assistance, to Francesca Bonadei of Springer for supporting this project since its very first Italian edition, and, last but not least, to Silvia Quarteroni for the translation from Italian and to Simon G. Chiossi for the linguistic revision of the second edition.

Milan and Lausanne, October 2013 Alfio Quarteroni

3.3.1 The homogeneous Dirichlet problem ... 41
3.3.2 Equivalence, in the sense of distributions, between weak and strong form of the Dirichlet problem ... 43
3.3.3 The problem with mixed, non homogeneous conditions ... 44
3.3.4 Equivalence, in the sense of distributions, between weak and strong form of the Neumann problem ... 47
3.4 More general elliptic problems ... 48
3.4.1 Existence and uniqueness theorem ... 50
3.5 Adjoint operator and adjoint problem ... 51
3.5.1 The nonlinear case ... 55
3.6 Exercises ... 56

The Galerkin finite element method for elliptic problems ... 61
4.1 Approximation via the Galerkin method ... 61
4.2 Analysis of the Galerkin method ... 63
4.2.1 Existence and uniqueness ... 63
4.2.2 Stability ... 64
4.2.3 Convergence ... 64
4.3 The finite element method in the one-dimensional case ... 67
4.3.1 The space X_h^1 ... 67
4.3.2 The space X_h^2 ... 69
4.3.3 The approximation with linear finite elements ... 71
4.3.4 Interpolation operator and interpolation error ... 73
4.3.5 Estimate of the finite element error in the H^1 ... 75
4.4 Finite elements, simplices and barycentric coordinates ... 76
4.4.1 An abstract definition of finite element in the Lagrangian case ... 76
4.4.2 Simplexes ... 78
4.4.3 Barycentric coordinates ... 78
4.5 The finite element method in the multi-dimensional case ... 80
4.5.1 Finite element solution of the Poisson problem ... 82
4.5.2 Conditioning of the stiffness matrix ... 85
4.5.3 Estimate of the approximation error in the energy norm ... 88
4.5.4 Estimate of the approximation error in the L^2 norm ... 95
4.6 Grid adaptivity ... 98
4.6.1 A priori adaptivity based on derivatives reconstruction ... 100
4.6.2 A posteriori adaptivity ... 103
4.6.3 Numerical examples of adaptivity ... 107
4.6.4 A posteriori error estimates in the L^2 norm ... 111
4.6.5 A posteriori estimates of a functional of the error ... 112
4.7 Exercises ... 114

Contents

1 A brief survey of partial differential equations
- 1.1 Definitions and examples
- 1.2 Numerical solution
- 1.3 PDE Classification
  - 1.3.1 Quadratic form associated to a PDE
- 1.4 Exercises

2 Elements of functional analysis
- 2.1 Functionals and bilinear forms
- 2.2 Differentiation in linear spaces
- 2.3 Elements of distributions
  - 2.3.1 Square-integrable functions
  - 2.3.2 Differentiation in the sense of distributions
- 2.4 Sobolev spaces
  - 2.4.1 Regularity of the spaces $H^k(\Omega)$
  - 2.4.2 The space $H^1_0(\Omega)$
  - 2.4.3 Trace operators
- 2.5 The spaces $L^\infty(\Omega)$ and $L^p(\Omega)$, with $1 \leq p < \infty$
- 2.6 Adjoint operators of a linear operator
- 2.7 Spaces of time-dependent functions
- 2.8 Exercises

3 Elliptic equations
- 3.1 An elliptic problem example: the Poisson equation
- 3.2 The Poisson problem in the one-dimensional case
  - 3.2.1 Homogeneous Dirichlet problem
  - 3.2.2 Non-homogeneous Dirichlet problem
  - 3.2.3 Neumann Problem
  - 3.2.4 Mixed homogeneous problem
  - 3.2.5 Mixed (or Robin) boundary conditions
- 3.3 The Poisson problem in the two-dimensional case

5 **Parabolic equations** ... 121
5.1 Weak formulation and its approximation ... 122
5.2 A priori estimates ... 125
5.3 Convergence analysis of the semi-discrete problem ... 128
5.4 Stability analysis of the θ-method ... 132
5.5 Convergence analysis of the θ-method ... 135
5.6 Exercises ... 138

6 **Generation of 1D and 2D grids** ... 141
6.1 Grid generation in 1D ... 141
6.2 Grid of a polygonal domain ... 144
6.3 Generation of structured grids ... 146
6.4 Generation of non-structured grids ... 149
6.4.1 Delaunay triangulation ... 149
6.4.2 Advancing front technique ... 153
6.5 Regularization techniques ... 155
6.5.1 Diagonal swap ... 156
6.5.2 Node displacement ... 157

7 **Algorithms for the solution of linear systems** ... 161
7.1 Direct methods ... 161
7.2 Iterative methods ... 164
7.2.1 Classical iterative methods ... 164
7.2.2 Gradient and conjugate gradient methods ... 167
7.2.3 Krylov subspace methods ... 169
7.2.4 The Multigrid method ... 175

8 **Elements of finite element programming** ... 179
8.1 Working steps of a finite element code ... 179
8.1.1 The code in a nutshell ... 182
8.2 Numerical computation of integrals ... 183
8.2.1 Numerical integration using barycentric coordinates ... 185
8.3 Storage of sparse matrices ... 187
8.4 Assembly ... 189
8.4.1 Coding geometrical information ... 191
8.4.2 Coding of functional information ... 192
8.4.3 Mapping between reference and physical element ... 193
8.4.4 Construction of local and global systems ... 198
8.4.5 Boundary conditions prescription ... 201
8.5 Integration in time ... 203
8.6 A complete example ... 206

9 **The finite volume method** ... 213
9.1 Some basic principles ... 214
9.2 Construction of control volumes for vertex-centered schemes ... 216

9.3 Discretization of a diffusion-transport-reaction problem 219
9.4 Analysis of the finite volume approximation 221
9.5 Implementation of boundary conditions 222

10 Spectral methods 225
10.1 The spectral Galerkin method for elliptic problems 225
10.2 Orthogonal polynomials and Gaussian numerical integration 229
10.2.1 Orthogonal Legendre polynomials 229
10.2.2 Gaussian integration 232
10.2.3 Gauss-Legendre-Lobatto formulae 233
10.3 G-NI methods in one dimension 236
10.3.1 Algebraic interpretation of the G-NI method 237
10.3.2 Conditioning of the stiffness matrix in the G-NI method . . . 239
10.3.3 Equivalence between G-NI and collocation methods 240
10.3.4 G-NI for parabolic equations 243
10.4 Generalization to the two-dimensional case 245
10.4.1 Convergence of the G-NI method 246
10.5 G-NI and SEM-NI methods for a one-dimensional model problem . . 254
10.5.1 The G-NI method 255
10.5.2 The SEM-NI method 258
10.6 Spectral methods on triangles and tetrahedra 261
10.7 Exercises 265

11 Discontinuous element methods (DG and mortar) 267
11.1 The discontinuous Galerkin method (DG) for the Poisson problem . . 267
11.1.1 Numerical results for the DG approximation of Poisson problem 272
11.2 The mortar method 273
11.2.1 Characterization of the space of constraints by spectral elements 275
11.2.2 Characterization of the space of constraints by finite elements 276
11.3 Mortar formulation for the Poisson problem 277
11.4 Choosing basis functions 278
11.5 Choosing quadrature formulae for spectral elements 280
11.6 Choosing quadrature formulae for finite elements 281
11.7 Solving the linear system of the mortar method 282
11.8 The mortar method for combined finite and spectral elements 283
11.9 Generalization of the mortar method to multi-domain decompositions 285
11.10 Numerical results for the mortar method 286

12 Diffusion-transport-reaction equations 291
12.1 Weak problem formulation 291
12.2 Analysis of a one-dimensional diffusion-transport problem 294
12.3 Analysis of a one-dimensional diffusion-reaction problem 299

12.4 Finite elements and finite differences (FD) . . . 301
12.5 The mass-lumping technique . . . 302
12.6 Decentred FD schemes and artificial diffusion . . . 304
12.7 Eigenvalues of the diffusion-transport equation . . . 307
12.8 Stabilization methods . . . 309
12.8.1 Artificial diffusion and decentred finite element schemes . . . 310
12.8.2 The Petrov-Galerkin method . . . 312
12.8.3 The artificial diffusion and streamline-diffusion methods in the two-dimensional case . . . 313
12.8.4 Consistency and truncation error for the Galerkin and generalized Galerkin methods . . . 314
12.8.5 Symmetric and skew-symmetric part of an operator . . . 315
12.8.6 Strongly consistent methods (GLS, SUPG) . . . 316
12.8.7 On the choice of the stabilization parameter τ_K . . . 319
12.8.8 Analysis of the GLS method . . . 321
12.8.9 Stabilization through bubble functions . . . 327
12.9 DG methods for diffusion-transport equations . . . 329
12.10 Mortar methods for the diffusion-transport equations . . . 330
12.11 Some numerical tests . . . 332
12.12 An example of goal-oriented adaptivity . . . 334
12.13 Exercises . . . 336

13 Finite differences for hyperbolic equations . . . 339
13.1 A scalar transport problem . . . 339
13.1.1 An a priori estimate . . . 341
13.2 Systems of linear hyperbolic equations . . . 343
13.2.1 The wave equation . . . 345
13.3 The finite difference method . . . 346
13.3.1 Discretization of the scalar equation . . . 347
13.3.2 Discretization of linear hyperbolic systems . . . 349
13.3.3 Boundary treatment . . . 350
13.4 Analysis of the finite difference methods . . . 350
13.4.1 Consistency and convergence . . . 350
13.4.2 Stability . . . 351
13.4.3 Von Neumann analysis and amplification coefficients . . . 356
13.4.4 Dissipation and dispersion . . . 360
13.5 Equivalent equations . . . 364
13.5.1 The upwind scheme case . . . 364
13.5.2 The Lax-Friedrichs and Lax-Wendroff case . . . 367
13.5.3 On the meaning of coefficients in equivalent equations . . . 367
13.5.4 Equivalent equations and error analysis . . . 368
13.6 Exercises . . . 369

14 Finite elements and spectral methods for hyperbolic equations . . . 371
14.1 Temporal discretization . . . 371

14.1.1 The forward and backward Euler schemes ... 371
14.1.2 The upwind, Lax-Friedrichs and Lax-Wendroff schemes ... 373
14.2 Taylor-Galerkin schemes ... 378
14.3 The multi-dimensional case ... 382
14.3.1 Semi-discretization: strong and weak treatment of the boundary conditions ... 382
14.3.2 Temporal discretization ... 385
14.4 Discontinuous finite elements ... 388
14.4.1 The one-dimensional upwind DG method ... 388
14.4.2 The multi-dimensional case ... 393
14.4.3 DG method with jump stabilization ... 395
14.5 Approximation using spectral methods ... 396
14.5.1 The G-NI method in a single interval ... 397
14.5.2 The DG-SEM-NI method ... 401
14.6 Numerical treatment of boundary conditions for hyperbolic systems ... 402
14.6.1 Weak treatment of boundary conditions ... 406
14.7 Exercises ... 408

15 Nonlinear hyperbolic problems ... 409
15.1 Scalar equations ... 409
15.2 Finite difference approximation ... 414
15.3 Approximation by discontinuous finite elements ... 415
15.3.1 Temporal discretization of DG methods ... 418
15.4 Nonlinear hyperbolic systems ... 424

16 Navier-Stokes equations ... 429
16.1 Weak formulation of Navier-Stokes equations ... 431
16.2 Stokes equations and their approximation ... 435
16.3 Saddle-point problems ... 439
16.3.1 Problem formulation ... 439
16.3.2 Analysis of the problem ... 440
16.3.3 Galerkin approximation, stability and convergence analysis 444
16.4 Algebraic formulation of the Stokes problem ... 447
16.5 An example of stabilized problem ... 451
16.6 A numerical example ... 453
16.7 Time discretization of Navier-Stokes equations ... 455
16.7.1 Finite difference methods ... 456
16.7.2 Characteristics (or Lagrangian) methods ... 458
16.7.3 Fractional step methods ... 459
16.8 Algebraic factorization methods and preconditioners for saddle-point systems ... 462
16.9 Free surface flow problems ... 468
16.9.1 Navier-Stokes equations with variable density and viscosity 469
16.9.2 Boundary conditions ... 470
16.9.3 Application to free surface flows ... 471

16.10 Interface evolution modelling ... 473
16.10.1 Explicit interface descriptions ... 473
16.10.2 Implicit interface descriptions ... 473
16.11 Finite volume approximation ... 478
16.12 Exercises ... 481

17 Optimal control of partial differential equations ... 483
17.1 Definition of optimal control problems ... 483
17.2 A control problem for linear systems ... 485
17.3 Some examples of optimal control problems for the Laplace equation 486
17.4 On the minimization of linear functionals ... 487
17.5 The theory of optimal control for elliptic problems ... 489
17.6 Some examples of optimal control problems ... 494
17.6.1 A Dirichlet problem with distributed control ... 494
17.6.2 A Neumann problem with distributed control ... 495
17.6.3 A Neumann problem with boundary control ... 495
17.7 Numerical tests ... 496
17.8 Lagrangian formulation of control problems ... 502
17.8.1 Constrained optimization in $\mathbb{R}^n$ ... 502
17.8.2 The solution approach based on the Lagrangian ... 503
17.9 Iterative solution of the optimal control problem ... 505
17.10 Numerical examples ... 510
17.10.1 Heat dissipation by a thermal fin ... 510
17.10.2 Thermal pollution in a river ... 512
17.11 A few considerations about observability and controllability ... 514
17.12 Two alternative paradigms for numerical approximation ... 516
17.13 A numerical approximation of an optimal control problem for advection–diffusion equations ... 517
17.13.1 The strategies "optimize–then–discretize" and "discretize–then–optimize" ... 519
17.13.2 A posteriori error estimates ... 520
17.13.3 A test problem on control of pollutant emission ... 523
17.14 Exercises ... 525

18 Domain decomposition methods ... 527
18.1 Some classical iterative DD methods ... 528
18.1.1 Schwarz method ... 528
18.1.2 Dirichlet-Neumann method ... 530
18.1.3 Neumann-Neumann algorithm ... 532
18.1.4 Robin-Robin algorithm ... 532
18.2 Multi-domain formulation ... 533
18.2.1 The Steklov-Poincaré operator ... 533
18.2.2 Equivalence between Dirichlet-Neumann and Richardson methods ... 535
18.3 Finite element approximation ... 538

18.3.1 The Schur complement . . . 541
18.3.2 The discrete Steklov-Poincaré operator . . . 543
18.3.3 Equivalence between Dirichlet-Neumann and Richardson methods in the discrete case . . . 545
18.4 Generalization to the case of many subdomains . . . 547
18.4.1 Some numerical results . . . 550
18.5 DD preconditioners in case of many subdomains . . . 551
18.5.1 Jacobi preconditioner . . . 552
18.5.2 Bramble-Pasciak-Schatz preconditioner . . . 554
18.5.3 Neumann-Neumann preconditioner . . . 555
18.5.4 FETI (Finite Element Tearing & Interconnecting) methods . 559
18.5.5 FETI-DP (Dual Primal FETI) methods . . . 563
18.5.6 BDDC (Balancing Domain Decomposition with Constraints) methods . . . 566
18.6 Schwarz iterative methods . . . 566
18.6.1 Algebraic form of Schwarz method for finite element discretizations . . . 567
18.6.2 Schwarz preconditioners . . . 569
18.6.3 Two-level Schwarz preconditioners . . . 573
18.7 An abstract convergence result . . . 576
18.8 Interface conditions for other differential problems . . . 577
18.9 Exercises . . . 580

19 Reduced basis approximation for parametrized partial differential equations . . . 585
19.1 Elliptic coercive parametric PDEs . . . 587
19.1.1 Two simple examples . . . 588
19.2 Main components of computational reduction techniques . . . 590
19.3 The reduced basis method . . . 593
19.3.1 RB Spaces . . . 594
19.3.2 Galerkin projection . . . 594
19.3.3 Offline-Online computational procedure . . . 596
19.4 Algebraic and geometric interpretations of the RB problem . . . 597
19.4.1 Algebraic interpretation of the (G-RB) problem . . . 598
19.4.2 Geometric interpretation of the (G-RB) problem . . . 600
19.4.3 Alternative formulations: Least-Squares and Petrov-Galerkin RB problems . . . 602
19.5 Construction of reduced spaces . . . 605
19.5.1 Greedy algorithm . . . 605
19.5.2 Proper Orthogonal Decomposition . . . 608
19.6 Convergence of RB approximations . . . 611
19.6.1 A priori convergence theory: a simple case . . . 611
19.6.2 A priori convergence theory: greedy algorithms . . . 612
19.7 A posteriori error estimation . . . 615
19.7.1 Some preliminary estimates . . . 615

19.7.2 Error bounds ... 616
19.8 Non-compliant problems ... 617
19.9 Parametrized geometries and operators ... 619
19.9.1 Physical parameters ... 620
19.9.2 Geometrical parameters ... 622
19.10 A working example: a diffusion-convection problem ... 626

References ... 635

Index ... 649

1

A brief survey of partial differential equations

The purpose of this chapter is to recall the basic concepts related to partial differential equations (PDEs, in short). For a wider coverage see [RR04, Eva98, LM68, Sal08].

1.1 Definitions and examples

Partial differential equations are differential equations containing derivatives of the unknown function with respect to several variables (temporal or spatial). In particular, if we denote by u the unknown function in the $d+1$ independent variables $\mathbf{x} = (x_1, \ldots, x_d)^T$ and t, we denote by

$$\mathcal{P}(u,g) = F\left(\mathbf{x}, t, u, \frac{\partial u}{\partial t}, \frac{\partial u}{\partial x_1}, \ldots, \frac{\partial u}{\partial x_d}, \ldots, \frac{\partial^{p_1+\cdots+p_d+p_t} u}{\partial x_1^{p_1} \ldots \partial x_d^{p_d} \partial t^{p_t}}, g\right) = 0 \quad (1.1)$$

a generic PDE, g being the set of data on which the PDE depends, while $p_1, \ldots, p_d$, $p_t \in \mathbb{N}$.

We say that (1.1) is of *order* q if q is the maximum order of the partial derivatives appearing in the equation, i.e. the maximum value taken by the integer $p_1 + p_2 + \ldots + p_d + p_t$.

If (1.1) depends linearly on the unknown u and on its derivatives, the equation is said to be *linear*. In the particular case where the derivatives having maximal order only appear linearly (with coefficients which may depend on lower-order derivatives), the equation is said to be *quasi-linear*. It is said to be *semi-linear* when it is quasi-linear and the coefficients of the maximal order derivatives only depend on $\mathbf{x}$ and t, and not on the solution u. Finally, if the equation contains no terms which are independent of the unknown function u, the PDE is said to be *homogeneous*.

We list below some examples of PDEs frequently encountered in the applied sciences.

Example 1.1. A first-order linear equation is the *transport* (or *advection*) *equation*

$$\frac{\partial u}{\partial t} + \nabla \cdot (\beta u) = 0, \quad (1.2)$$

A. Quarteroni: *Numerical Models for Differential Problems*, 2nd Ed.
MS&A – Modeling, Simulation & Applications 8
DOI 10.1007/978-88-470-5522-3_1,

having denoted by

$$\nabla \cdot \mathbf{v} = \mathrm{div}(\mathbf{v}) = \sum_{i=1}^{d} \frac{\partial v_i}{\partial x_i}, \quad \mathbf{v} = (v_1, \ldots, v_d)^T,$$

the *divergence operator*. Integrated on a region $\Omega \subset \mathbb{R}^d$, (1.2) expresses the mass conservation of a material system (a continuous media) occupying the region Ω. The u variable is the system's density, while $\beta(\mathbf{x}, \mathbf{t})$ is the velocity of a particle in the system that occupies position $\mathbf{x}$ at time t. ■

Example 1.2. Linear second-order equations include:
the *potential equation*

$$-\Delta u = f, \tag{1.3}$$

that describes the diffusion of a fluid in a homogeneous and isotropic region $\Omega \subset \mathbb{R}^d$, but also the vertical displacement of an elastic membrane;
the *heat* (or *diffusion*) *equation*

$$\frac{\partial u}{\partial t} - \Delta u = f; \tag{1.4}$$

the *wave equation*

$$\frac{\partial^2 u}{\partial t^2} - \Delta u = 0. \tag{1.5}$$

We have denoted by

$$\Delta u = \sum_{i=1}^{d} \frac{\partial^2 u}{\partial x_i^2} \tag{1.6}$$

the *Laplace operator* (*Laplacian*). ■

Example 1.3. An example of a quasi-linear first-order equation is *Burgers' equation*

$$\frac{\partial u}{\partial t} + u \frac{\partial u}{\partial x_1} = 0,$$

while its variant obtained by adding a second-order perturbation

$$\frac{\partial u}{\partial t} + u \frac{\partial u}{\partial x_1} = \varepsilon \frac{\partial^2 u}{\partial x_1^2}, \quad \varepsilon > 0,$$

is an example of a semi-linear equation.
Another second-order, non-linear equation, is

$$\left(\frac{\partial^2 u}{\partial x_1^2}\right)^2 + \left(\frac{\partial^2 u}{\partial x_2^2}\right)^2 = f.$$

■

A function $u = u(x_1,\ldots,x_d,t)$ is said to be a *solution* (or a *particular integral*) of (1.1) if it makes (1.1) an identity once it is replaced in (1.1) together with all of its derivatives. The set of all solutions of (1.1) is called the *general integral* of (1.1).

Example 1.4. The transport equation in the one-dimensional case,

$$\frac{\partial u}{\partial t} - \frac{\partial u}{\partial x_1} = 0, \tag{1.7}$$

admits a general integral of the form $u = w(x_1 + t)$, w being a sufficiently regular arbitrary function (see Exercise 2). Similarly, the one-dimensional wave equation

$$\frac{\partial^2 u}{\partial t^2} - \frac{\partial^2 u}{\partial x_1^2} = 0 \tag{1.8}$$

admits as a general integral

$$u(x_1,t) = w_1(x_1+t) + w_2(x_1-t),$$

w_1 and w_2 being two sufficiently regular arbitrary functions (see Exercise 3). ■

Example 1.5. Let us consider the one-dimensional heat equation

$$\frac{\partial u}{\partial t} - \frac{\partial^2 u}{\partial x_1^2} = 0,$$

for $0 < x < 1$ and $t > 0$, with boundary conditions

$$u(0,t) = u(1,t) = 0, \quad t > 0$$

and initial condition $u|_{t=0} = u_0$. Its solution is

$$u(x_1,t) = \sum_{j=1}^{\infty} u_{0,j} e^{-(j\pi)^2 t} \sin(j\pi x_1),$$

where $u_0 = u_{|t=0}$ is the initial datum and

$$u_{0,j} = 2\int_0^1 u_0(x_1)\sin(j\pi x_1)\, dx_1, \qquad j = 1,2,\ldots$$ ■

1.2 Numerical solution

In general, it is not possible to obtain a solution of (1.1) in closed (explicit) form. Indeed, the available analytical integration methods (such as the technique of separation of variables) are of limited applicability. On the other hand, even in the case where a general integral is known, it is not guaranteed that a particular integral may be determined. Indeed, in order to obtain the latter, it will be necessary to assign appropriate conditions on u (and/or its derivatives) at the boundary of the domain Ω.

Besides, from the examples provided it is evident that the general integral depends on a number of *arbitrary functions* (and not on arbitrary *constants*, as it happens for ordinary differential equations), so that the imposition of the boundary conditions will result in the solution of mathematical problems that are generally rather involved.

Thus, from a theoretical point of view, the analysis of a given PDE is often bound to investigating *existence*, *uniqueness*, and, possibly, *regularity* of its solutions, but lacks practical tools for their actual determination.

It follows that it is extremely important to have *numerical methods* at one's disposal, that allow to construct an approximation u_N of the exact solution u and to evaluate (in some suitable norm) the error $u_N - u$ when substituting to the exact solution u the approximate solution u_N. In general, $N \geq 1$ is a positive integer that denotes the (finite) dimension of the approximate problem. Schematically, we will obtain the following situation:

$$
\begin{array}{ll}
\mathcal{P}(u,g) = 0 & \text{Exact PDE} \\
\downarrow & \text{[numerical methods]} \\
\mathcal{P}_N(u_N, g_N) = 0 & \text{Approximate PDE.}
\end{array}
$$

We have denoted by g_N an approximation of the set of data g on which the PDE depends, and with $\mathcal{P}_N$ the new functional relation characterizing the approximated problem. For simplicity, one writes $u = u(g)$ and $u_N = u_N(g_N)$.

We will present several numerical methods starting from Chap. 4. Here, we only recall their main features. A numerical method is *convergent* if

$$\|u - u_N\| \to 0 \quad \text{as } N \to \infty$$

for a given norm. More precisely, we have convergence if and only if

$$\forall \varepsilon > 0,\ \exists N_0 = N_0(\varepsilon) > 0,\ \exists \delta = \delta(N_0, \varepsilon) : \ \forall N > N_0,\ \forall g_N \text{ such that } \|g - g_N\| < \delta,$$
$$\|u(g) - u_N(g_N)\| \leq \varepsilon.$$

(The norm used for the data is not necessarily the same as that used for the solutions.) A direct verification of the convergence of a numerical method may not be easy. A verification of its consistency and stability properties is recommendable, instead. A numerical method is said to be *consistent* if

$$\mathcal{P}_N(u,g) \to 0 \text{ as } N \to \infty, \tag{1.9}$$

and *strongly consistent* (or *fully consistent*) if

$$\mathcal{P}_N(u,g) = 0 \quad \forall N \geq 1. \tag{1.10}$$

Notice that (1.9) can be equivalently formulated as

$$\mathcal{P}_N(u,g) - \mathcal{P}(u,g) \to 0 \text{ as } N \to \infty.$$

This expresses the property that $\mathcal{P}_N$ (the approximated PDE) "tends" to $\mathcal{P}$ (the exact one) as $N \longrightarrow \infty$. Instead, we say that a numerical method is *stable* if to small perturbations to the data correspond small perturbations to the solution. More precisely,

$$\forall \varepsilon > 0,\ \exists \delta = \delta(\varepsilon) > 0:\ \forall \delta g_N:\ \|\delta g_N\| < \delta \Rightarrow \|\delta u_N\| \leq \varepsilon,\ \forall N \geq 1.$$

$u_N + \delta u_N$ being the solution of the *perturbed problem*

$$\mathcal{P}_N(u_N + \delta u_N, g_N + \delta g_N) = 0.$$

(See also [QSS07, Chap. 2] for an in-depth coverage.)

The fundamental result, known as the Lax-Richtmyer *equivalence theorem*, finally guarantees that

Theorem 1.1. *If a method is consistent, then it is convergent if and only if it is stable.*

Other important properties will obviously influence the choice of a numerical method, such as its *convergence rate* (i.e. the order with respect to $1/N$ with which the error tends to zero) and its *computational cost*, that is the computation time and memory required to implement such method on the computer.

1.3 PDE Classification

Partial differential equations can be classified into three different families: *elliptic*, *parabolic* and *hyperbolic* equations, for each of which appropriate specific numerical methods will be considered. For the sake of brevity, here we will limit ourselves to the case of a linear second-order PDE, with constant coefficients, of the form $Lu = G$,

$$Lu = A\frac{\partial^2 u}{\partial x_1^2} + B\frac{\partial^2 u}{\partial x_1 \partial x_2} + C\frac{\partial^2 u}{\partial x_2^2} + D\frac{\partial u}{\partial x_1} + E\frac{\partial u}{\partial x_2} + Fu, \tag{1.11}$$

with assigned function G and $A, B, C, D, E, F \in \mathbb{R}$. (Notice that any of the x_i variables could represent the temporal variable.) In that case, the classification is carried out based on the sign of the *discriminant*, $\triangle = B^2 - 4AC$. In particular:

if $\triangle < 0$ the equation is said to be *elliptic*,
if $\triangle = 0$ the equation is said to be *parabolic*,
if $\triangle > 0$ the equation is said to be *hyperbolic*.

Example 1.6. The wave equation (1.8) is hyperbolic, while the potential equation (1.3) is elliptic. An example of a parabolic problem is given by the heat equation (1.4), but also by the following *diffusion-transport equation*

$$\frac{\partial u}{\partial t} - \mu \Delta u + \nabla \cdot (\boldsymbol{\beta} u) = 0$$

where the constant $\mu > 0$ and the vector field β are given. ■

The criterion introduced above makes the classification depend on the sole coefficients of the highest derivatives and is justified via the following argument. As the reader will recall, the quadratic algebraic equation

$$Ax_1^2 + Bx_1x_2 + Cx_2^2 + Dx_1 + Ex_2 + F = G,$$

represents a hyperbola, a parabola or an ellipse in the Cartesian plane (x_1, x_2) depending whether $\triangle$ is positive, null or negative. This parallel motivates the name assigned to the three classes of partial derivative operators.

Let us investigate the difference between the three classes more attentively. Let us suppose, without this being restrictive, that D, E, F and G be null. We look for a change of variables of the form

$$\xi = \alpha x_2 + \beta x_1, \quad \eta = \gamma x_2 + \delta x_1, \tag{1.12}$$

with α, β, γ and δ to be chosen so that Lu becomes a multiple of $\partial^2 u/\partial\xi\partial\eta$. Since

$$\begin{aligned} Lu = {} & (A\beta^2 + B\alpha\beta + C\alpha^2)\frac{\partial^2 u}{\partial\xi^2} \\ & + (2A\beta\delta + B(\alpha\delta + \beta\gamma) + 2C\alpha\gamma)\frac{\partial^2 u}{\partial\xi\partial\eta} + (A\delta^2 + B\gamma\delta + C\gamma^2)\frac{\partial^2 u}{\partial\eta^2}, \end{aligned} \tag{1.13}$$

we need to require that

$$A\beta^2 + B\alpha\beta + C\alpha^2 = 0, \quad A\delta^2 + B\gamma\delta + C\gamma^2 = 0. \tag{1.14}$$

If $A = C = 0$, the trivial trasformation $\xi = x_2$, $\eta = x_1$ (for instance) provides Lu in the desired form.

Let us then suppose that A or C be not null. It is not restrictive to suppose $A \neq 0$. Then, if $\alpha \neq 0$ and $\gamma \neq 0$, we can divide the first equation of (1.14) by α^2 and the second one by γ^2. We find two identical quadratic equations for the ratios β/α and δ/γ. By solving them, we have

$$\frac{\beta}{\alpha} = \frac{1}{2A}\left[-B \pm \sqrt{\triangle}\right], \quad \frac{\delta}{\gamma} = \frac{1}{2A}\left[-B \pm \sqrt{\triangle}\right].$$

In order for the transformation (1.12) to be non-singular, the quotients β/α and δ/γ must be different. We must therefore take the positive sign in one case, and the negative sign in the other. Moreover, we must assume $\triangle > 0$. If $\triangle$ were indeed null, the two fractions would still be coincident, while if $\triangle$ were negative none of the two fractions could be real. To conclude, we can take the following values as coefficients of transformation (1.12):

$$\alpha = \gamma = 2A, \quad \beta = -B + \sqrt{\triangle}, \quad \delta = -B - \sqrt{\triangle}.$$

Correspondingly, (1.12) becomes

$$\xi = 2Ax_2 + \left[-B + \sqrt{\triangle}\right] x_1, \quad \eta = 2Ax_2 + \left[-B - \sqrt{\triangle}\right] x_1,$$

and, after the transformation, the original differential problem $Lu = 0$ becomes

$$Lu = -4A\triangle \frac{\partial^2 u}{\partial\xi\partial\eta} = 0. \tag{1.15}$$

(For ease of notation, we still denote by u the transformed solution and by L the transformed differential operator.) The case $A = 0$ and $C \neq 0$ can be treated in a similar way by taking $\xi = x_1$, $\eta = x_2 - (C/B)x_1$.

To conclude, the original term Lu can become a multiple of $\partial^2 u/\partial\xi\partial\eta$ based on the transformation (1.12) if and only if $\triangle > 0$, and in such case, as we have anticipated, the problem is said to be *hyperbolic*. It is easy to verify that the general solution of problem (1.15) is

$$u = p(\xi) + q(\eta),$$

p and q being arbitrary differentiable functions in one variable. The lines $\xi = constant$ and $\eta = constant$ are said to be the *characteristics* of L and are characterized by the fact that on these lines, the functions p and q, respectively, remain constant. In particular, possible discontinuities of the solution u propagate along the characteristic lines (this will be shown in more detail in Chap. 13). Indeed, if $A \neq 0$, by identifying x_1 with t and x_2 with x, the transformation

$$x' = x - \frac{B}{2A}t, \quad t' = t,$$

transforms the hyperbolic operator L such that

$$Lu = A\frac{\partial^2 u}{\partial t^2} + B\frac{\partial^2 u}{\partial t\partial x} + C\frac{\partial^2 u}{\partial x^2}$$

in a multiple of the wave operator L such that

$$Lu = \frac{\partial^2 u}{\partial t^2} - c^2\frac{\partial^2 u}{\partial x^2}, \text{ with } c^2 = \triangle/4A^2.$$

The latter is the wave operator in a coordinate system moving with velocity $-B/2A$. The characteristic lines of the wave operator are the lines verifying

$$\left(\frac{dt}{dx}\right)^2 = \frac{1}{c^2},$$

that is

$$\frac{dt}{dx} = \frac{1}{c} \text{ and } \frac{dt}{dx} = -\frac{1}{c}.$$

When $\triangle = 0$, as previously stated L is *parabolic*. In this case there exists only one value of β/α in corrispondence of which the coefficient of $\partial^2 u/\partial\xi^2$ in (1.13) becomes zero:

precisely, $\beta/\alpha = -B/(2A)$. On the other hand, since $B/(2A) = 2C/B$, this choice also implies that the coefficient of $\partial^2 u/\partial\xi\partial\eta$ becomes zero. Hence, the change of variables

$$\xi = 2Ax_2 - Bx_1, \quad \eta = x_1,$$

transforms the original problem $Lu = 0$ into the following

$$Lu = A\frac{\partial^2 u}{\partial \eta^2} = 0,$$

the general solution of which has the form

$$u = p(\xi) + \eta q(\xi).$$

A parabolic operator therefore has only one family of characteristics, precisely $\xi = constant$. The discontinuities in the derivatives of u propagate along such characteristic lines.

Finally, if $\triangle < 0$ (*elliptic* operators) there does not exist any choice of β/α or δ/γ that makes the coefficients $\partial^2 u/\partial\xi^2$ and $\partial^2 u/\partial\eta^2$ null. However, the transformation

$$\xi = \frac{2Ax_2 - Bx_1}{\sqrt{-\triangle}}, \quad \eta = x_1,$$

transforms $Lu= 0$ into

$$Lu = A\left(\frac{\partial^2 u}{\partial \xi^2} + \frac{\partial^2 u}{\partial \eta^2}\right) = 0,$$

i.e. a multiple of the potential equation. The latter has therefore no family of characteristic lines.

1.3.1 Quadratic form associated to a PDE

We can associate to equation (1.11) the so-called principal symbol S^p defined by

$$S^p(\mathbf{x},\mathbf{q}) = -A(\mathbf{x})q_1^2 - B(\mathbf{x})q_1 q_2 - C(\mathbf{x})q_2^2.$$

This quadratic form can be represented in matrix form as follows:

$$S^p(\mathbf{x},\mathbf{q}) = \mathbf{q}^T \begin{bmatrix} -A(\mathbf{x}) & -\frac{1}{2}B(\mathbf{x}) \\ -\frac{1}{2}B(\mathbf{x}) & -C(\mathbf{x}) \end{bmatrix} \mathbf{q}. \tag{1.16}$$

A quadratic form is said to be *definite* if all of the eigenvalues of its associated matrix have the same sign (either positive or negative); it is *indefinite* if the matrix has eigenvalues of both signs; it is *degenerate* if the matrix is singular.

It can then be said that equation (1.11) is elliptic if its quadratic form (1.16) is definite (positive or negative), hyperbolic if it is indefinite, and parabolic if it is degenerate.

The matrices associated to the potential equation (1.3), the (one-dimensional) heat equation (1.4) and the wave equation (1.5) are given respectively by

$$\begin{bmatrix} 1 & 0 \\ 0 & 1 \end{bmatrix}, \begin{bmatrix} 0 & 0 \\ 0 & 1 \end{bmatrix} \text{ and } \begin{bmatrix} -1 & 0 \\ 0 & 1 \end{bmatrix}$$

and are positive definite in the first case, singular in the second case, and indefinite in the third case.

1.4 Exercises

1. Classify the following equations based on their order and linearity:

$$(a) \quad \left[1+\left(\frac{\partial u}{\partial x_1}\right)^2\right]\frac{\partial^2 u}{\partial x_2^2} - 2\frac{\partial u}{\partial x_1}\frac{\partial u}{\partial x_2}\frac{\partial^2 u}{\partial x_1 \partial x_2} + \left[1+\left(\frac{\partial u}{\partial x_2}\right)^2\right]\frac{\partial^2 u}{\partial x_1^2} = 0,$$

$$(b) \quad \rho\frac{\partial^2 u}{\partial t^2} + K\frac{\partial^4 u}{\partial x_1^4} = f,$$

$$(c) \quad \left(\frac{\partial u}{\partial x_1}\right)^2 + \left(\frac{\partial u}{\partial x_2}\right)^2 = f.$$

[*Solution*: (a) quasi-linear, second-order; it is Plateau's equation which governs, under appropriate hypotheses, the plane motion of a fluid. The u appearing in the equation is the so-called *kinetic potential*; (b) linear, fourth-order. It is the *vibrating rod* equation, ρ is the rod's density, while K is a positive quantity that depends on the geometrical properties of the rod itself; (c) non-linear, first-order.]

2. Reduce the one-dimensional transport equation (1.7) to an equation of the form $\partial w/\partial y = 0$, having set $y = x_1 - t$, and obtain that $u = w(x_1 + t)$ is a solution of the original equation.
[*Solution*: operate the substitution of variables $z = x_1 + t$, $y = x_1 - t$, $u(x_1,t) = w(y,z)$. In such way $\partial u/\partial x_1 = \partial w/\partial z + \partial w/\partial y$, where $\partial u/\partial t = \partial w/\partial z - \partial w/\partial y$, and thus $-2\partial w/\partial y = 0$. Note at this point that the equation obtained thereby admits a solution $w(y,z)$ that does not depend on y and, using the original variables, we get $u = w(x_1 + t)$.]

3. Prove that the wave equation

$$\frac{\partial^2 u}{\partial t^2} - c^2\frac{\partial^2 u}{\partial x_1^2} = 0,$$

with constant c, admits as a solution $u(x_1,t) = w_1(x_1 + ct) + w_2(x_1 - ct)$, w_1 and w_2 being two sufficiently regular arbitrary functions.

[*Solution*: proceed as in Exercise 2, by applying the substitution of variables $y = x_1 + ct$, $z = x_1 - ct$ and setting $u(x_1,t) = w(y,z)$.]

4. Verify that the Korteveg-de-Vries equation

$$\frac{\partial u}{\partial t} + \beta \frac{\partial u}{\partial x_1} + \alpha \frac{\partial^3 u}{\partial x_1^3} = 0,$$

admits a general integral of the form $u = a\cos(kx_1 - \omega t)$ with an appropriate ω to be determined, and a, β and α being assigned constants. This equation describes the position u of a fluid with respect to a reference position, in the presence of long wave propagation.
[*Solution*: the given u satisfies the equation only if $\omega = k\beta - \alpha k^3$.]

5. Consider the equation

$$x_1^2 \frac{\partial^2 u}{\partial x_1^2} - x_2^2 \frac{\partial^2 u}{\partial x_2^2} = 0$$

with $x_1 x_2 \neq 0$. Classify it and determine its characteristic lines.

6. Consider the generic second-order semi-linear differential equation

$$a(x_1,x_2)\frac{\partial^2 u}{\partial x_1^2} + 2b(x_1,x_2)\frac{\partial^2 u}{\partial x_1 \partial x_2} + c(x_1,x_2)\frac{\partial^2 u}{\partial x_2^2} + f(u,\nabla u) = 0,$$

where $\nabla u = \left(\frac{\partial u}{\partial x_1}, \frac{\partial u}{\partial x_2}\right)^T$ is the gradient of u. Write the equation of its characteristic lines and deduce from it the classification of the proposed equation, by distinguishing the different cases.

7. Set $r(\mathbf{x}) = |\mathbf{x}| = (x_1^2 + x_2^2)^{1/2}$ and define $u(\mathbf{x}) = \ln(r(\mathbf{x}))$, $\mathbf{x} \in \mathbb{R}^2 \setminus \{\mathbf{0}\}$. Verify that

$$\Delta u(\mathbf{x}) = 0, \quad \mathbf{x} \in \Omega,$$

where Ω is any given open set such that $\bar{\Omega} \subset \mathbb{R}^2 \setminus \{\mathbf{0}\}$.
[*Solution*: observe that

$$\frac{\partial^2 u}{\partial x_i^2} = \frac{1}{r^2}\left(1 - \frac{2x_i^2}{r^2}\right), \qquad i = 1,2.]$$

2

Elements of functional analysis

In this chapter we recall a number of concepts used extensively in this textbook: functionals and bilinear forms, distributions, Sobolev spaces, L^p spaces. For a more in-depth reading, the reader can refer to e.g. [Sal08],[Yos74], [Bre86], [LM68], [Ada75].

2.1 Functionals and bilinear forms

Definition 2.1. *Given a function space V, we call* functional *on V an operator associating a real number to each element of V*

$$F : V \mapsto \mathbb{R}.$$

The functional is often denoted as $F(v) = \langle F, v\rangle$, an expression called *duality* or *crochet*. A functional is said to be *linear* if it is linear with respect to its argument, that is if

$$F(\lambda v + \mu w) = \lambda F(v) + \mu F(w) \quad \forall \lambda, \mu \in \mathbb{R},\ \forall v, w \in V.$$

A linear functional is *bounded* if there is a constant $C > 0$ such that

$$|F(v)| \leq C\|v\|_V \quad \forall v \in V. \tag{2.1}$$

A linear and bounded functional on a Banach space (i.e. a normed and complete space) is also continuous. We then define the space V', called *dual* of V, as the set of linear and bounded functionals on V, that is

$$V' = \{F : V \mapsto \mathbb{R} \text{ such that } F \text{ is linear and bounded } \},$$

and we equip it with the norm $\|\cdot\|_{V'}$ defined as

$$\|F\|_{V'} = \sup_{v \in V \setminus \{0\}} \frac{|F(v)|}{\|v\|_V}. \tag{2.2}$$

The constant C appearing in (2.1) is greater or equal to $\|F\|_{V'}$.

A. Quarteroni: *Numerical Models for Differential Problems*, 2nd Ed.
MS&A – Modeling, Simulation & Applications 8
DOI 10.1007/978-88-470-5522-3_2, © Springer-Verlag Italia 2014

The following theorem, called identification or representation theorem ([Yos74]), holds.

Theorem 2.1 (Riesz representation theorem). *Let H be a Hilbert space, that is a Banach space whose norm is induced by a scalar product $(\cdot,\cdot)_H$. For each linear and bounded functional f on H there exists a unique element $x_f \in H$ such that*

$$f(y) = (y, x_f)_H \quad \forall y \in H, \quad \text{and} \quad \|f\|_{H'} = \|x_f\|_H. \tag{2.3}$$

Conversely, each element $x \in H$ identifies a linear and bounded functional f_x on H such that

$$f_x(y) = (y, x)_H \quad \forall y \in H \quad \text{and} \quad \|f_x\|_{H'} = \|x\|_H. \tag{2.4}$$

If H is a Hilbert space, its dual space H' of linear and bounded functionals on H is a Hilbert space too. Moreover, thanks to Theorem 2.1, there exists a bijective and isometric (i.e. norm-preserving) transformation $f \leftrightarrow x_f$ between H' and H thanks to which H' and H can be identified. We can denote this transformation as follows:

$$\begin{array}{ll} \Lambda_H : H \to H', & x \to f_x = \Lambda_H x, \\ \Lambda_H^{-1} : H' \to H, & f \to x_f = \Lambda_H^{-1} x. \end{array} \tag{2.5}$$

We now introduce the notion of fom.

Definition 2.2. *Given a normed functional space V we call* form *an application which associates to each pair of elements of V a real number*

$$a : V \times V \mapsto \mathbb{R}.$$

A form is called:

bilinear if it is linear with respect to both its arguments, i.e. if

$$\begin{array}{ll} a(\lambda u + \mu w, v) = \lambda a(u,v) + \mu a(w,v) & \forall \lambda, \mu \in \mathbb{R}, \forall u, v, w \in V, \\ a(u, \lambda w + \mu v) = \lambda a(u,v) + \mu a(u,w) & \forall \lambda, \mu \in \mathbb{R}, \forall u, v, w \in V; \end{array}$$

continuous if there exists a constant $M > 0$ such that

$$|a(u,v)| \leq M \|u\|_V \|v\|_V \quad \forall u, v \in V; \tag{2.6}$$

symmetric if

$$a(u,v) = a(v,u) \quad \forall u, v \in V; \tag{2.7}$$

positive (or positive definite) if

$$a(v,v) > 0 \quad \forall v \in V; \tag{2.8}$$

coercive if there exists a constant $\alpha > 0$ such that

$$a(v,v) \geq \alpha \|v\|_V^2 \quad \forall v \in V. \tag{2.9}$$

Definition 2.3. *Let X and Y be two Hilbert spaces. We say that X is contained in Y with* continuous injection *if there exists a constant C such that $\|w\|_Y \leq C\|w\|_X$ $\forall w \in X$. Moreover X is* dense *in Y if each element belonging to Y can be obtained as the limit, in the $\|\cdot\|_Y$ norm, of a sequence of elements of X.*

Given two Hilbert spaces V and H, such that $V \subset H$, the injection of V in H is continuous and moreover V is dense in H, we have that H is a subspace of V', the dual of V, and we have

$$V \subset H \simeq H' \subset V'. \tag{2.10}$$

For elliptic problems, the spaces V and H will typically be chosen respectively as $H^1(\Omega)$ (or one of its subspaces, $H_0^1(\Omega)$ or $H_{\Gamma_D}^1(\Omega)$) and $L^2(\Omega)$, see Chap. 3.

For control problems, the space H typically represents the functional space of the forcing term f of the state equation (17.1) and, sometimes, the one where we seek the control u, see Chap. 17.

Definition 2.4. *A linear and bounded (hence continuous) operator T between two functional spaces X and Y is an* isomorphism *if it maps bijectively the elements of the spaces X and Y and its inverse T^{-1} exists. If also $X \subset Y$ holds, such isomorphism is called* canonical.

2.2 Differentiation in linear spaces

In this section, we briefly report the notions of differentiability and differentiation for applications on linear functional spaces; for a further analysis of this topic, as well as an extension of such notions to more general cases, see [KF89].

Let us begin by considering the notion of *strong* (or *Fréchet*) *differential*:

Definition 2.5. *Let X and Y be two normed linear spaces and F an application of X in Y, defined on an open set $E \subset X$; such application is called* differentiable *at $x \in E$ if there exists a linear and bounded operator $L_x : X \to Y$ such that:*

$$\forall \varepsilon > 0,\ \exists \delta > 0 \ : \ ||F(x+h) - F(x) - L_x h||_Y \leq \varepsilon\ ||h||_X \ \forall h \in X \text{ with } ||h||_X < \delta.$$

We call the expression $L_x h$ (or $L_x[h]$), which generates an element in Y for each $h \in X$, strong differential *(or* Fréchet differential*) of the application F at $x \in E$; the operator L_x is called* strong derivative *of the application F at x and is generally denoted as $F'(x)$, that is $F'(x) = L_x$.*

From the definition, we deduce that a differentiable application in x is also continuous in x. We list below some properties deriving from this definition:

- if $F(x) = constant$, then $F'(x)$ is the null operator, that is $L_x[h] = 0,\ \forall h \in X$;
- the strong derivative of a continuous linear application $F(x)$ is the application itself, that is $F'(x) = F(x)$;
- given two continuous applications F and G of X in Y, if these are differentiable at x_0, so are the applications $F+G$ and αF, for all $\alpha \in \mathbb{R}$, and we have:

$$(F+G)'(x_0) = F'(x_0) + G'(x_0),$$
$$(\alpha F)'(x_0) = \alpha F'(x_0).$$

Consider now the following definition of *weak* (or *Gâteaux*) *differential*:

Definition 2.6. *Let F be an application of X in Y; we call* weak (or *Gâteaux*) differential *of the application F at x the limit:*

$$DF(x,h) = \lim_{t \to 0} \frac{F(x+th) - F(x)}{t} \qquad \forall h \in X,$$

where $t \in \mathbb{R}$ and the convergence of the limit must be intended with respect to the norm of the space Y. If the weak differential $DF(x,h)$ is linear (in general it is not), it can be expressed as

$$DF(x,h) = F'_G(x)h \qquad \forall h \in X.$$

The linear and bounded operator $F'_G(x)$ is called weak derivative *(or* Gâteaux derivative*) of F. Moreover, we have*

$$F(x+th) - F(x) = tF'_G(x)h + o(t) \qquad \forall h \in X,$$

which implies

$$||F(x+th) - F(x) - tF'_G(x)h|| = o(t) \qquad \forall h \in X.$$

Note that if an application F has a strong derivative, then it also admits a weak derivative, coinciding with the strong one; the converse instead is not generally true. However, the following theorem holds (see [KF89]):

Theorem 2.2. *If on a neighbourhood $U(x_0)$ of x_0 there exists a weak derivative $F'_G(x)$ of the application F and such derivative is a function of x on such neighborhood, continuous at x_0, then the strong derivative $F'(x_0)$ at x_0 exists, too, and coincides with the weak one, that is $F'(x_0) = F'_G(x_0)$.*

2.3 Elements of distributions

In this section we want to recall the main definitions regarding the theory of distributions and Sobolev spaces, useful for a better comprehension of the subjects introduced in the textbook. For a more in-depth treatment, see, e.g., the monographs [Bre86], [Ada75] and [LM68].
Let Ω be an open set of $\mathbb{R}^n$ and $f : \Omega \mapsto \mathbb{R}$.

Definition 2.7. *By* support *of a function f we mean the closure of the set where the function itself takes values different from zero*

$$\text{supp} f = \overline{\{\mathbf{x} : \ f(\mathbf{x}) \neq 0\}}.$$

A function $f : \Omega \mapsto \mathbb{R}$ is said to have a *compact support* in Ω if there exists a compact set [1] $K \subset \Omega$ such that $\text{supp} f \subset K$.

At this point, we can provide the following definition:

Definition 2.8. *$\mathcal{D}(\Omega)$ is the space of infinitely differentiable functions with compact support in Ω, that is*

$$\mathcal{D}(\Omega) = \{f \in C^\infty(\Omega) : \ \exists K \subset \Omega, \ compact : \ \text{supp} f \subset K\}.$$

We introduce the multi-index notation for the derivatives. Let $\alpha = (\alpha_1, \alpha_2, \ldots, \alpha_n)$ be an n-tuple of non-negative integers (called *multi-index*) and let $f : \Omega \mapsto \mathbb{R}$ be a function defined on $\Omega \subset \mathbb{R}^n$. We will use the following notation

$$D^\alpha f(\mathbf{x}) = \frac{\partial^{|\alpha|} f(\mathbf{x})}{\partial x_1^{\alpha_1} \partial x_2^{\alpha_2} \ldots \partial x_n^{\alpha_n}},$$

[1] With $\Omega \subset \mathbb{R}^n$, a compact set is a closed and bounded set.

$|\alpha| = \alpha_1 + \alpha_2 + \ldots + \alpha_n$ being the length of the multi-index coinciding with the order of differentiation of f.

In the space $\mathcal{D}(\Omega)$ we can introduce the following notion of convergence:

Definition 2.9. *Given a sequence $\{\phi_k\}$ of functions of $\mathcal{D}(\Omega)$ we say that these converge in $\mathcal{D}(\Omega)$ to a function ϕ, and we will write $\phi_k \underset{\mathcal{D}(\Omega)}{\longrightarrow} \phi$, if:*

1. *the supports of the functions ϕ_k are all contained in a fixed compact set K of Ω;*
2. *we have uniform convergence of the derivatives of all orders, that is*

$$D^\alpha \phi_k \longrightarrow D^\alpha \phi \quad \forall \alpha \in \mathbb{N}^n.$$

We are now able to define the space of distributions on Ω:

Definition 2.10. *Let T be a linear transformation from $\mathcal{D}(\Omega)$ into $\mathbb{R}$ and let us denote by $\langle T, \varphi \rangle$ the value taken by T on the element $\varphi \in \mathcal{D}(\Omega)$. We say that T is continuous if*

$$\lim_{k \to \infty} \langle T, \varphi_k \rangle = \langle T, \varphi \rangle$$

where $\{\varphi_k\}_{k=1}^{\infty}$ is an arbitrary sequence of $\mathcal{D}(\Omega)$ that converges toward $\varphi \in D(\Omega)$. We call distribution *on Ω any linear and continuous transformation T from $\mathcal{D}(\Omega)$ into $\mathbb{R}$. The space of distributions on Ω is therefore given by the dual space $\mathcal{D}'(\Omega)$ of $\mathcal{D}(\Omega)$.*

The action of a distribution $T \in \mathcal{D}'(\Omega)$ on a function $\phi \in \mathcal{D}(\Omega)$ will always be denoted via the identity pairing $\langle T, \phi \rangle$.

Example 2.1. Let $\mathbf{a}$ be a point of the set Ω. The *Dirac delta* relative to point $\mathbf{a}$ is the distribution $\delta_{\mathbf{a}}$ defined by the following relation

$$\langle \delta_{\mathbf{a}}, \phi \rangle = \phi(\mathbf{a}) \quad \forall \phi \in \mathcal{D}(\Omega). \qquad \blacksquare$$

For another example, see Exercise 4. Also in $\mathcal{D}'(\Omega)$ we introduce a notion of convergence:

Definition 2.11. *A sequence of distributions $\{T_n\}$ converges to a distribution T in $\mathcal{D}'(\Omega)$ if we have*

$$\lim_{n \to \infty} \langle T_n, \phi \rangle = \langle T, \phi \rangle \qquad \forall \phi \in \mathcal{D}(\Omega).$$

2.3.1 Square-integrable functions

We consider the space of square-integrable functions on $\Omega \subset \mathbb{R}^n$,

$$\mathrm{L}^2(\Omega) = \{f : \Omega \mapsto \mathbb{R} \text{ such that } \int_\Omega (f(\mathbf{x}))^2 \, d\Omega < +\infty\}.$$

More precisely, $\mathrm{L}^2(\Omega)$ is a space of *equivalence classes* of measurable functions, the equivalence relation to be intended as follows: v is equivalent to w if and only if v and w are equal almost everywhere, i.e. they differ at most on a subset of Ω with zero measure. The expression "almost everywhere in Ω" (in short, a.e. in Ω) means exactly "for all the $\mathbf{x} \in \Omega$, except for a zero-measure set, at most".
The space $\mathrm{L}^2(\Omega)$ is a Hilbert space whose scalar product is

$$(f,g)_{\mathrm{L}^2(\Omega)} = \int_\Omega f(\mathbf{x})g(\mathbf{x}) \, d\Omega.$$

The norm in $\mathrm{L}^2(\Omega)$ is the one induced by this scalar product, i.e.

$$\|f\|_{\mathrm{L}^2(\Omega)} = \sqrt{(f,f)_{\mathrm{L}^2(\Omega)}}.$$

To each function $f \in \mathrm{L}^2(\Omega)$ we associate a distribution $T_f \in \mathcal{D}'(\Omega)$ defined in the following way

$$\langle T_f, \phi \rangle = \int_\Omega f(\mathbf{x})\phi(\mathbf{x}) \, d\Omega \qquad \forall \phi \in \mathcal{D}(\Omega).$$

The following result holds:

Lemma 2.1. *The space $\mathcal{D}(\Omega)$ is dense in $L^2(\Omega)$.*

Due to the latter, it is possible to prove that the correspondence $f \rightarrow T_f$ is injective, thus we can identify $\mathrm{L}^2(\Omega)$ with a subset of $\mathcal{D}'(\Omega)$, writing:

$$\mathrm{L}^2(\Omega) \subset \mathcal{D}'(\Omega).$$

Example 2.2. Let $\Omega = \mathbb{R}$ and let us denote by $\chi_{[a,b]}(x)$ the *characteristic function* of the interval $[a,b]$, defined as

$$\chi_{[a,b]}(x) = \begin{cases} 1 & \text{if } x \in [a,b], \\ 0 & \text{otherwise.} \end{cases}$$

Let us then consider the sequence of functions $f_n(x) = \frac{n}{2}\chi_{[-1/n,1/n]}(x)$ (see Fig. 2.1). We want to verify that the sequence $\{T_{f_n}\}$ of the distributions associated to the former converges to the distribution δ_0, i.e. the Dirac delta relative to the origin. As a matter

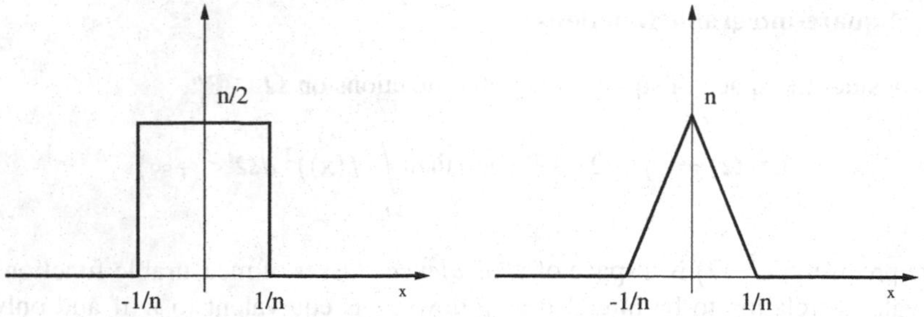

Fig. 2.1. The characteristic function of the interval $[-1/n, 1/n]$ (left) and the triangular function f_n (right)

of fact, for each function $\phi \in \mathcal{D}(\Omega)$, we have

$$\langle T_{f_n}, \phi \rangle = \int_{\mathbb{R}} f_n(x)\phi(x)\,dx = \frac{n}{2}\int_{-1/n}^{1/n} \phi(x)\,dx = \frac{n}{2}[\Phi(1/n) - \Phi(-1/n)],$$

Φ being a primitive of ϕ. If we now set $h = 1/n$, we can write

$$\langle T_{f_n}, \phi \rangle = \frac{\Phi(h) - \Phi(-h)}{2h}.$$

When $n \to \infty$, $h \to 0$ and thus, following the definition of derivative, we have

$$\frac{\Phi(h) - \Phi(-h)}{2h} \to \Phi'(0).$$

By construction $\Phi' = \phi$, and therefore

$$\langle T_{f_n}, \phi \rangle \to \phi(0) = \langle \delta_0, \phi \rangle,$$

having used the definition of δ_0 (see Example 2.1).

The same limit can be obtained by taking a sequence of triangular functions (see Fig. 2.1) or Gaussian functions, instead of rectangular ones (provided that they still have unit integral).

Finally, we point out that in the usual metrics, such sequences converge to a function which is null almost everywhere. ■

2.3.2 Differentiation in the sense of distributions

Let $\Omega \subset \mathbb{R}^n$ and $T \in \mathcal{D}'(\Omega)$. Its derivatives $\frac{\partial T}{\partial x_i}$ in the *sense of distributions* are distributions defined in the following way

$$\langle \frac{\partial T}{\partial x_i}, \phi \rangle = -\langle T, \frac{\partial \phi}{\partial x_i} \rangle \qquad \forall \phi \in \mathcal{D}(\Omega), \quad i = 1, \dots, n.$$

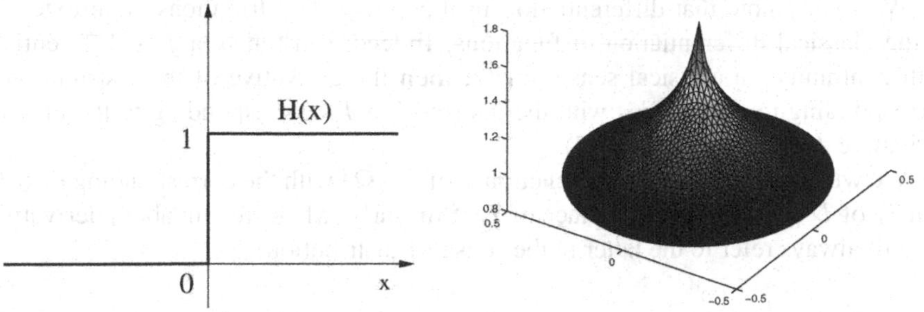

Fig. 2.2. The Heaviside function (left). On the right, the function of Example 2.6 with $k = 1/3$. Note that this function tends to infinity at the origin

In a similar way, we define derivatives of arbitrary order. Precisely, for each multi-index $\alpha = (\alpha_1, \alpha_2, \ldots, \alpha_n)$, we have that $D^\alpha T$ is a new distribution defined as

$$\langle D^\alpha T, \phi \rangle = (-1)^{|\alpha|} \langle T, D^\alpha \phi \rangle \qquad \forall \phi \in \mathcal{D}(\Omega).$$

Example 2.3. The *Heaviside function* on $\mathbb{R}$ (see Fig. 2.2) is defined as

$$H(x) = \begin{cases} 1 & \text{if } x > 0, \\ 0 & \text{if } x \leq 0. \end{cases}$$

The derivative of the distribution T_H associated to the latter is the Dirac distribution relative to the origin (see Example 2.1); upon identifying the function H with the associated distribution T_H, we will then write

$$\frac{dH}{dx} = \delta_0.$$ ■

Differentiation in the context of distributions enjoys some important properties that do not hold in the more restricted context of differentiation for functions in classical terms.

Property 2.1. *The set $\mathcal{D}'(\Omega)$ is closed with respect to the differentiation operation (in the sense of distributions), that is each distribution is differentiable infinitely many times and its distributional derivatives are themselves distributions.*

Property 2.2. *Differentiation in $\mathcal{D}'(\Omega)$ is a continuous operation, in the sense that if $T_n \underset{\mathcal{D}'(\Omega)}{\longrightarrow} T$ for $n \to \infty$, then it also results that $D^\alpha T_n \underset{\mathcal{D}'(\Omega)}{\longrightarrow} D^\alpha T$ for $n \to \infty$, for each multi-index α.*

We finally note that differentiation in the sense of distributions is an extension of the classical differentiation of functions. Indeed, if a function f is differentiable with continuity (in classical sense) on Ω, then the derivative of the distribution T_f corresponding to f coincides with the distribution $T_{f'}$ corresponding to the classical derivative f' of f (see Exercise 7).

We will invariably identify a function f of $\mathrm{L}^2(\Omega)$ with the corresponding distribution T_f of $\mathcal{D}'(\Omega)$, writing f in place of T_f. Similarly, when we talk about derivatives, we will always refer to the latter in the sense of distributions.

2.4 Sobolev spaces

In Sect. 2.3.1 we have noted that the functions of $\mathrm{L}^2(\Omega)$ are particular distributions. However, this does not guarantee that their derivatives (in the sense of distributions) are still functions of $\mathrm{L}^2(\Omega)$, as shown in the following example.

Example 2.4. Let $\Omega \subset \mathbb{R}$ and let $[a,b] \subset \Omega$. Then, the *characteristic function* of the interval $[a,b]$ (see Example 2.2) belongs to $\mathrm{L}^2(\Omega)$, while its distributional derivative $d\chi_{[a,b]}/dx = \delta_a - \delta_b$ (see Example 2.3) does not. ■

It is therefore reasonable to introduce the following spaces:

Definition 2.12. *Let Ω be an open set of $\mathbb{R}^n$ and k be a positive integer. We call Sobolev space of order k on Ω the space formed by the totality of functions of $\mathrm{L}^2(\Omega)$ whose (distributional) derivatives up to order k belong to $\mathrm{L}^2(\Omega)$:*

$$\mathrm{H}^k(\Omega) = \{f \in L^2(\Omega) : D^\alpha f \in L^2(\Omega) \quad \forall \alpha \, : \, |\alpha| \leq k\}.$$

It follows, obviously, that $\mathrm{H}^{k+1}(\Omega) \subset \mathrm{H}^k(\Omega)$ for each $k \geq 0$ and this inclusion is continuous. The space $\mathrm{L}^2(\Omega)$ is sometimes denoted by $\mathrm{H}^0(\Omega)$.

The Sobolev spaces $\mathrm{H}^k(\Omega)$ are Hilbert spaces with respect to the following scalar product

$$(f,g)_k = \sum_{|\alpha|\leq k} \int_\Omega (D^\alpha f)(D^\alpha g)\, d\Omega,$$

from which descend the norms

$$\|f\|_k = \|f\|_{\mathrm{H}^k(\Omega)} = \sqrt{(f,f)_k} = \sqrt{\sum_{|\alpha|\leq k} \int_\Omega (D^\alpha f)^2\, d\Omega}. \tag{2.11}$$

Finally, we define the seminorms

$$|f|_k = |f|_{\mathrm{H}^k(\Omega)} = \sqrt{\sum_{|\alpha|=k} \int_\Omega (D^\alpha f)^2\, d\Omega},$$

so that (2.11) becomes

$$\|f\|_{\mathrm{H}^k(\Omega)} = \sqrt{\sum_{m=0}^{k} |f|^2_{\mathrm{H}^m(\Omega)}}.$$

Example 2.5. If $n = 1$ and $k = 1$ we have:

$$(f,g)_1 = (f,g)_{\mathrm{H}^1(\Omega)} = \int_\Omega fg \, d\Omega + \int_\Omega f'g' \, d\Omega;$$

$$\|f\|_1 = \|f\|_{\mathrm{H}^1(\Omega)} = \sqrt{\int_\Omega f^2 \, d\Omega + \int_\Omega f'^2 \, d\Omega} = \sqrt{\|f\|^2_{\mathrm{L}^2(\Omega)} + \|f'\|^2_{\mathrm{L}^2(\Omega)}};$$

$$|f|_1 = |f|_{\mathrm{H}^1(\Omega)} = \sqrt{\int_\Omega (f')^2 \, d\Omega} = \|f'\|_{\mathrm{L}^2(\Omega)}.$$

■

2.4.1 Regularity of the spaces $\mathrm{H}^k(\Omega)$

We now want to relate the fact that a function belongs to a space $\mathrm{H}^k(\Omega)$ with its continuity properties.

Example 2.6. Let $\Omega = B(0,1) \subset \mathbb{R}^2$ be the ball centered at the origin and of radius 1. Then the function

$$f(x_1,x_2) = \left| \ln \frac{1}{\sqrt{x_1^2 + x_2^2}} \right|^k \tag{2.12}$$

belongs to $\mathrm{H}^1(\Omega)$ when $0 < k < 1/2$, see Fig. 2.2 (right). It develops a singularity at the origin and therefore it is neither continuous nor bounded. A similar conclusion can be drawn for

$$f(x_1,x_2) = \ln(-\ln(x_1^2 + x_2^2)),$$

this time with $\Omega = B(0,1/2) \subset \mathbb{R}^2$. ■

Not all of the functions of $\mathrm{H}^1(\Omega)$ are therefore continuous if Ω is an open set of $\mathbb{R}^2$ (or $\mathbb{R}^3$). In general, the following result holds:

Property 2.3. *If Ω is an open set of $\mathbb{R}^n$, $n \geq 1$, provided with a "sufficiently regular" boundary, then*

$$\mathrm{H}^k(\Omega) \subset C^m(\overline{\Omega}) \quad \textit{if } k > m + \frac{n}{2}.$$

In particular, in one spatial dimension ($n = 1$), the functions of $\mathrm{H}^1(\Omega)$ are continuous (they are indeed *absolutely continuous*, see [Sal08] and [Bre86]), while in two or three dimensions they are not necessarily so. Instead, the functions of $\mathrm{H}^2(\Omega)$ are always continuous for $n = 1,2,3$.

2.4.2 The space $\mathrm{H}_0^1(\Omega)$

If Ω is bounded, the space $\mathcal{D}(\Omega)$ is not dense in $\mathrm{H}^1(\Omega)$. We can then give the following definition:

Definition 2.13. *We denote by* $\mathrm{H}_0^1(\Omega)$ *the closure of* $\mathcal{D}(\Omega)$ *in* $\mathrm{H}^1(\Omega)$.

The functions of $\mathrm{H}_0^1(\Omega)$ enjoy the following properties:

Property 2.4 (Poincaré inequality). *Let* Ω *be a* bounded *set in* $\mathbb{R}^n$*; then there exists a constant* C_Ω *such that:*

$$\|v\|_{\mathrm{L}^2(\Omega)} \leq C_\Omega |v|_{\mathrm{H}^1(\Omega)} \qquad \forall v \in \mathrm{H}_0^1(\Omega). \tag{2.13}$$

Proof. Ω being bounded, we can always find a sphere $S_D = \{\mathbf{x} : |\mathbf{x}-\mathbf{g}| < D\}$ with centre $\mathbf{g}$ and radius $D > 0$, containing Ω. Since $\mathcal{D}(\Omega)$ is dense in $\mathrm{H}_0^1(\Omega)$ it is sufficient to prove the inequality for a function $u \in \mathcal{D}(\Omega)$. (In the general case where $v \in \mathrm{H}_0^1(\Omega)$ it will suffice to build a sequence $u_i \in \mathcal{D}(\Omega)$, $i = 1,2,\ldots$ converging to v in the norm of $\mathrm{H}^1(\Omega)$, apply the inequality to the terms of the sequence and pass to the limit.) Integrating by parts and exploiting the fact that $\mathrm{div}(\mathbf{x}-\mathbf{g}) = n$,

$$\begin{aligned}
\|u\|^2_{\mathrm{L}^2(\Omega)} &= n^{-1}\int_\Omega n\cdot|u(\mathbf{x})|^2\,d\Omega = -n^{-1}\int_\Omega(\mathbf{x}-\mathbf{g})\cdot\nabla(|u(\mathbf{x})|^2)\,d\Omega \\
&= -2n^{-1}\int_\Omega(\mathbf{x}-\mathbf{g})\cdot[u(\mathbf{x})\nabla u(\mathbf{x})]\,d\Omega \leq 2n^{-1}\|\mathbf{x}-\mathbf{g}\|_{\mathrm{L}^\infty(\Omega)}\|u\|_{\mathrm{L}^2(\Omega)}\|u\|_{\mathrm{H}^1(\Omega)} \\
&\leq 2n^{-1}D\|u\|_{\mathrm{L}^2(\Omega)}\|u\|_{\mathrm{H}^1(\Omega)}.
\end{aligned}$$
◇

As an immediate consequence, we have that:

Property 2.5. *The seminorm* $|v|_{H^1(\Omega)}$ *is a norm on the space* $\mathrm{H}_0^1(\Omega)$ *that turns out to be equivalent to the norm* $\|v\|_{H^1(\Omega)}$.

Proof. We recall that two norms, $\|\cdot\|$ and $|||\cdot|||$, are said to be *equivalent* if there exist two positive constants c_1 and c_2, such that

$$c_1|||v||| \leq \|v\| \leq c_2|||v||| \quad \forall v \in V.$$

As $\|v\|_1 = \sqrt{|v|_1^2 + \|v\|_0^2}$ it is evident that $|v|_1 \leq \|v\|_1$. Conversely, exploiting Property 2.4,

$$\|v\|_1 = \sqrt{|v|_1^2 + \|v\|_0^2} \leq \sqrt{|v|_1^2 + C_\Omega^2 |v|_1^2} \leq C_\Omega^* |v|_1,$$

from which we deduce the equivalence of the two norms. ◇

In a similar way, we define the spaces $\mathrm{H}_0^k(\Omega)$ as the closure of $\mathcal{D}(\Omega)$ in $\mathrm{H}^k(\Omega)$.

2.4.3 Trace operators

Let Ω be a domain of $\mathbb{R}^n$. By that we mean:

- an open bounded interval if $n = 1$;
- an open bounded connected set, with a sufficiently regular boundary $\partial\Omega$. For instance, a polygon if $n = 2$ (i.e. a domain whose boundary is a finite union of segments), or a polyhedron if $n = 3$ (i.e. a domain whose boundary is a finite union of polygons).

Let v be an element of $\mathrm{H}^1(\Omega)$: the remarks formulated in Sect. 2.4.1 show that it is not simple to define the "value" of v on the boundary of Ω, a value that we will call the *trace* of v on $\partial\Omega$. We exploit the following result:

Theorem 2.3. *Let Ω be a domain of $\mathbb{R}^n$ provided with a "sufficiently regular" boundary $\partial\Omega$, and let $k \geq 1$. There exists one and only one linear and continuous application*

$$\gamma_0 : \mathrm{H}^k(\Omega) \mapsto \mathrm{L}^2(\partial\Omega),$$

such that $\gamma_0 v = v|_{\partial\Omega}$, $\forall v \in H^k \cap C^0(\overline{\Omega})$; $\gamma_0 v$ is called trace *of v on $\partial\Omega$. The continuity of γ_0 implies that there exists a constant $C > 0$ such that*

$$\|\gamma_0 v\|_{\mathrm{L}^2(\Gamma)} \leq C\|v\|_{\mathrm{H}^k(\Omega)}.$$

The result still holds if we consider the trace operator $\gamma_\Gamma : \mathrm{H}^k(\Omega) \mapsto \mathrm{L}^2(\Gamma)$ where Γ is a sufficiently regular portion of the boundary of Ω with positive measure.

Owing to this result, Dirichlet boundary conditions make sense when seeking solutions v in $\mathrm{H}^k(\Omega)$, with $k \geq 1$, provided we interpret the boundary value in the sense of the trace.

Remark 2.1. The trace operator γ_Γ is not surjective on $\mathrm{L}^2(\Gamma)$. In particular, the set of functions of $\mathrm{L}^2(\Gamma)$ which are traces of functions of $\mathrm{H}^1(\Omega)$ constitutes a subspace of $\mathrm{L}^2(\Gamma)$ denoted by $\mathrm{H}^{1/2}(\Gamma)$ and characterized by intermediate regularity properties between those of $\mathrm{L}^2(\Gamma)$ and those of $\mathrm{H}^1(\Gamma)$. More generally, for every $k \geq 1$ there exists a unique linear and continuous application $\gamma_0 : \mathrm{H}^k(\Omega) \mapsto \mathrm{H}^{k-1/2}(\Gamma)$ such that $\gamma_0 v = v_{|\Gamma}$ for each $v \in \mathrm{H}^k(\Omega) \cap C^0(\overline{\Omega})$. •

The trace operators allow for an interesting characterization of the previously defined space $\mathrm{H}_0^1(\Omega)$. Indeed, we have the following property:

Property 2.6. *Let Ω be a domain of $\mathbb{R}^n$ provided with a sufficiently regular boundary $\partial\Omega$ and let γ_0 be the trace operator from $\mathrm{H}^1(\Omega)$ in $\mathrm{L}^2(\partial\Omega)$. We then have*

$$\mathrm{H}_0^1(\Omega) = \mathrm{Ker}(\gamma_0) = \{v \in \mathrm{H}^1(\Omega) : \gamma_0 v = 0\}$$

In other words, $\mathrm{H}_0^1(\Omega)$ is formed by the functions of $\mathrm{H}^1(\Omega)$ having null trace on the boundary. Analogously, we define $\mathrm{H}_0^2(\Omega)$ as the subspace of functions of $\mathrm{H}^2(\Omega)$ whose traces, together with the traces of the normal derivative, vanish at the boundary.

2.5 The spaces $\mathrm{L}^\infty(\Omega)$ and $\mathrm{L}^p(\Omega)$, with $1 \le p < \infty$

The space $\mathrm{L}^2(\Omega)$ can be generalized in the following way: for each real number p with $1 \le p < \infty$ we can define the following space of equivalence classes of measurable functions

$$\mathrm{L}^p(\Omega) = \{v : \Omega \mapsto \mathbb{R} \text{ suchthat } \int_\Omega |v(\mathbf{x})|^p \, d\Omega < \infty\},$$

whose norm is given by

$$\|v\|_{\mathrm{L}^p(\Omega)} = \left(\int_\Omega |v(\mathbf{x})|^p d\Omega \right)^{1/p}.$$

Furthermore, we define the space

$$\mathrm{L}^1_{loc}(\Omega) = \{f : \Omega \to \mathbb{R} \textit{ suchthat } f|_K \in \mathrm{L}^1(K) \text{ for each compact set } K \subset \Omega\}.$$

If $1 \le p < \infty$, then $\mathcal{D}(\Omega)$ is dense in $\mathrm{L}^p(\Omega)$.
In the case where $p = \infty$, we define $\mathrm{L}^\infty(\Omega)$ to be the space of functions that are bounded a.e. in Ω. Its norm is defined as follows

$$\|v\|_{\mathrm{L}^\infty(\Omega)} = \inf\{C \in \mathbb{R} : |v(x)| \le C, \text{ a.e. in } \Omega\} \tag{2.14}$$

$$= \sup\{|v(x)|, \text{ a.e. in } \Omega\}. \tag{2.15}$$

For $1 \le p \le \infty$, the spaces $\mathrm{L}^p(\Omega)$, provided with the norm $\|\cdot\|_{\mathrm{L}^p(\Omega)}$, are Banach spaces. We recall the *Hölder inequality*: given $v \in \mathrm{L}^p(\Omega)$ and $w \in \mathrm{L}^{p'}(\Omega)$ with $1 \le p \le \infty$ and

$\frac{1}{p} + \frac{1}{p'} = 1$, then $vw \in L^1(\Omega)$ and

$$\int_\Omega |v(\mathbf{x})\, w(\mathbf{x})| d\Omega \le \|v\|_{L^p(\Omega)} \|w\|_{L^{p'}(\Omega)}. \tag{2.16}$$

The index p' is called conjugate of p.
If $1 < p < \infty$, then $L^p(\Omega)$ is a reflexive space: this means that any linear and continuous form $\varphi : L^p(\Omega) \to \mathbb{R}$ can be identified to an element of $L^{p'}(\Omega)$, i.e. there exists a unique $g \in L^{p'}(\Omega)$ such that

$$\varphi(f) = \int_\Omega f(\mathbf{x}) g(\mathbf{x})\, d\Omega \quad \forall\, f \in L^p(\Omega).$$

If $p = 2$, then $p' = 2$, so the Hölder inequality becomes

$$(v, w)_{L^2(\Omega)} \le \|v\|_{L^2(\Omega)} \|w\|_{L^2(\Omega)} \quad \forall\, v,\, w \in L^2(\Omega)\,. \tag{2.17}$$

As such, it is known as Cauchy-Schwarz inequality. Moreover, the following inequality holds

$$\|vw\|_{L^2(\Omega)} \le \|v\|_{L^4(\Omega)} \|w\|_{L^4(\Omega)} \quad \forall v, w \in L^4(\Omega). \tag{2.18}$$

If $\Omega \subset \mathbb{R}^n$ is a domain, for $1 \le p \le q \le \infty$ we have

$$L^q(\Omega) \subset L^p(\Omega) \subset L^1(\Omega) \subset L^1_{loc}(\Omega)\,.$$

If Ω is unbounded, we always have

$$L^p(\Omega) \subset L^1_{loc}(\Omega) \quad \forall\, p \ge 1\,.$$

Moreover, if $\Omega \subset \mathbb{R}^n$ and for $n > 1$ the boundary $\partial\Omega$ is polygonal (more generally, it is Lipschitz continuous), we have the following continuous inclusions:

$$\begin{aligned}
&\text{if } 0 < 2s < n && \text{then } H^s(\Omega) \subset L^q(\Omega)\ \forall q \text{ such that } 1 \le q \le q^* \text{ with } q^* = 2n/(n-2s);\\
&\text{if } 2s = n && \text{then } H^s(\Omega) \subset L^q(\Omega)\ \forall q \text{ such that } 1 \le q < \infty;\\
&\text{if } 2s > n && \text{then } H^s(\Omega) \subset C^0(\overline{\Omega}).
\end{aligned} \tag{2.19}$$

Finally, we introduce the Sobolev space $W^{k,p}(\Omega)$, with k a non-negative integer and $1 \le p \le \infty$, as the space of functions $v \in L^p(\Omega)$ such that all the distributional derivatives of v of order up to k are in $L^p(\Omega)$.

$$W^{k,p}(\Omega) = \{v \in L^p(\Omega) : D^\alpha v \in L^p(\Omega)$$
$$\text{for each non-negative multi-index } \alpha \text{ such that } |\alpha| \le k\}.$$

For $1 \le p < \infty$ this is a Banach space with norm

$$\|v\|_{W^{k,p}(\Omega)} = \left(\sum_{|\alpha|\le k} \|D^\alpha v\|^p_{L^p(\Omega)} \right)^{1/p}.$$

Its seminorm $|v|_{W^{k,p}(\Omega)}$ is defined similarly, provided we sum over multi-integers α such that $|\alpha| = k$.
Note that, for $k = 0$, $W^{k,p}(\Omega) = L^p(\Omega)$, and that, for $p = 2$, $W^{k,2}(\Omega) = H^k(\Omega)$.

2.6 Adjoint operators of a linear operator

Let X and Y be two Banach spaces and $\mathcal{L}(X,Y)$ be the space of linear and bounded operators from X to Y. Given $L \in \mathcal{L}(X,Y)$, the *adjoint* (or *coniugate*) *operator* of L is another operator $L' : Y' \to X'$ defined by

$$_{X'}\langle L'f, x\rangle_X = {}_{Y'}\langle f, Lx\rangle_Y \quad \forall f \in Y', x \in X. \tag{2.20}$$

L' is a linear and bounded operator between Y' and X', that is $L' \in \mathcal{L}(Y', X')$, moreover $\|L'\|_{\mathcal{L}(Y',X')} = \|L\|_{\mathcal{L}(X,Y)}$, where we have set

$$\|L\|_{\mathcal{L}(X,Y)} = \sup_{\substack{x\in X\\ x\neq 0}} \frac{\|Lx\|_Y}{\|x\|_X}. \tag{2.21}$$

In the case where X and Y are two Hilbert spaces, an additional adjoint operator, $L^T : Y \to X$, called *transpose* of L, can be introduced. It is defined by

$$(L^T y, x)_X = (y, Lx)_Y \quad \forall x \in X, y \in Y. \tag{2.22}$$

Here, $(\cdot,\cdot)_X$ denotes the scalar product of X, while $(\cdot,\cdot)_Y$ denotes the scalar product of Y. The above definition can be explained as follows: for any given element $y \in Y$, the real-valued function $x \to (y, Lx)_Y$ is linear and continuous, hence it defines an element of X'. By Riesz's theorem (Theorem 2.1) there exists an element x of X, which we name $L^T y$, that satisfies (2.22). Such operator belongs to $\mathcal{L}(Y,X)$ (that is, it is linear and bounded from Y to X) and moreover

$$\|L^T\|_{\mathcal{L}(Y,X)} = \|L\|_{\mathcal{L}(X,Y)}. \tag{2.23}$$

Thus, in the case where X and Y are two Hilbert spaces, we have two notions of adjoint operator, L' and L^T. The relationship between the two operators is

$$\Lambda_X L^T = L' \Lambda_Y, \tag{2.24}$$

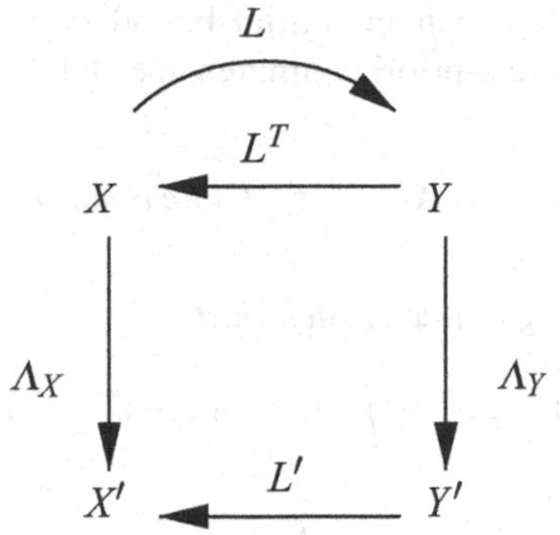

Fig. 2.3. The adjoint operators L^T and L' of the operator L

Λ_X and Λ_Y being Riesz's canonical isomorphisms from X to X' and from Y to Y', respectively (see (2.5)). Indeed, $\forall x \in X, y \in Y$,

$$\begin{aligned} {}_{X'}\langle \Lambda_X L^T y, x \rangle_X &= (L^T y, x)_X = (y, Lx)_Y = {}_{Y'}\langle \Lambda_Y y, Lx \rangle_Y \\ &= {}_{X'}\langle L' \Lambda_Y y, x \rangle_X . \end{aligned}$$

Identity (2.24) can be equivalently expressed by stating that the diagramme in Fig. 2.3 is commutative.

2.7 Spaces of time-dependent functions

When considering space-time functions $v(\mathbf{x},t)$, $\mathbf{x} \in \Omega \subset \mathbb{R}^n$, $n \geq 1, t \in (0,T)$, $T > 0$, it is natural to introduce the functional space

$$L^q(0,T; W^{k,p}(\Omega)) = \left\{ v : (0,T) \to W^{k,p}(\Omega) \text{ such that } v \text{ is measurable and } \int_0^T \|v(t)\|^q_{W^{k,p}(\Omega)} dt < \infty \right\}, \tag{2.25}$$

where $k \geq 0$ is a non-negative integer, $1 \leq q < \infty$, $1 \leq p \leq \infty$, endowed with the norm

$$\|v\|_{L^q(0,T;W^{k,p}(\Omega))} = \left(\int_0^T \|v(t)\|^q_{W^{k,p}(\Omega)} dt \right)^{1/q} . \tag{2.26}$$

For every $t \in (0,T)$ we have used the shorthand notation $v(t)$ to indicate the function:

$$v(t) : \Omega \to \mathbb{R}, \; v(t)(\mathbf{x}) = v(\mathbf{x},t) \quad \forall \mathbf{x} \in \Omega. \tag{2.27}$$

The spaces $L^\infty(0,T : W^{k,p}(\Omega))$ and $C^0([0,T]; W^{k,p}(\Omega))$ are defined in a similar way.

When dealing with time-dependent initial-boundary value problems, the following result can be useful to derive a-priori estimates and stability inequalities.

Lemma 2.2 (Gronwall). *Let* $A \in \mathrm{L}^1(t_0, T)$ *be a non-negative function,* φ *a continuous function on* $[t_0, T]$.

i) *If* g *is non-decreasing and* φ *is such that*

$$\varphi(t) \leq g(t) + \int_{t_0}^{t} A(\tau)\varphi(\tau)d\tau \qquad \forall t \in [t_0, T], \tag{2.28}$$

then

$$\varphi(t) \leq g(t) exp\left(\int_{t_0}^{t} A(\tau)d\tau\right) \qquad \forall t \in [t_0, T]. \tag{2.29}$$

ii) *If* g *is a non-negative constant and* φ *a non-negative function such that*

$$\varphi^2(t) \leq g + \int_{t_0}^{t} A(\tau)\varphi(\tau)d\tau \qquad \forall t \in [t_0, T], \tag{2.30}$$

then

$$\varphi(t) \leq \sqrt{g} + \frac{1}{2}\int_{t_0}^{t} A(\tau)d\tau \qquad \forall t \in [t_0, T]. \tag{2.31}$$

A discrete counterpart of this lemma, that will be useful when dealing with fully discrete (in space and time) approximations of initial-boundary value problems, is the following

Lemma 2.3 (discrete Gronwall lemma). *Assume that* k_n *is a non-negative sequence, and that the sequence* φ_n *satisfies*

$$\varphi_0 \leq g_0, \quad \varphi_n \leq g_0 + \sum_{m=0}^{n-1} p_m + \sum_{m=0}^{n-1} k_m \varphi_m, \; n \geq 1. \tag{2.32}$$

If $g_0 \geq 0$ *and* $p_m \geq 0$ *for* $m \geq 0$, *then*

$$\varphi_n \leq (g_0 + \sum_{m=0}^{n-1} p_m) exp(\sum_{m=0}^{n-1} k_m), \; n \geq 1. \tag{2.33}$$

For the proof of these two lemmas, see, e.g., [QV94, Chap. 1].

2.8 Exercises

1. Let $\Omega = (0,1)$ and, for $\alpha > 0$, $f(x) = x^{-\alpha}$. For which α do we have $f \in \mathrm{L}^p(\Omega)$, $1 \le p < \infty$? Is there an $\alpha > 0$ for which $f \in \mathrm{L}^\infty(\Omega)$?
2. Let $\Omega = (0, \frac{1}{2})$ and $f(x) = \frac{1}{x(\ln x)^2}$. Show that $f \in \mathrm{L}^1(\Omega)$.
3. Prove for which $\alpha \in \mathbb{R}$ we have that $f \in \mathrm{L}^1_{loc}(0,1)$, with $f(x) = x^{-\alpha}$.
4. Let $u \in \mathrm{L}^1_{loc}(\Omega)$. Define $T_u \in \mathcal{D}'(\Omega)$ as follows

$$\langle T_u, \varphi \rangle = \int_\Omega \varphi(\mathbf{x}) u(\mathbf{x})\, d\Omega \quad \forall \varphi \in \mathcal{D}(\Omega).$$

   Verify that T_u is effectively a distribution, and that the application $u \rightarrow T_u$ is injective. We can therefore identify u with T_u, and conclude by observing that $\mathrm{L}^1_{loc}(\Omega) \subset \mathcal{D}'(\Omega)$.
5. Show that the function defined as follows:

$$f(x) = e^{1/(x^2-1)} \text{ if } x \in (-1,1)$$
$$f(x) = 0 \text{ if } x \in]-\infty, -1] \cup [1, +\infty[$$

   belongs to $\mathcal{D}(\mathbb{R})$.
6. Prove that for the function f defined in (2.12) we have

$$\|f\|^2_{\mathrm{H}^1(\Omega)} = 2\pi \int_0^r |\log s|^{2k} s\, ds + 2\pi k^2 \int_0^r \frac{1}{s} |\log s|^{2k-2}\, ds,$$

   hence f belongs to $\mathrm{H}^1(\Omega)$ for every $0 < k < \frac{1}{2}$.
7. Let $\varphi \in C^1(-1,1)$. Show that the derivative $\frac{d\varphi}{dx}$ computed in the classical sense is equal to $\frac{d\varphi}{dx}$ computed in the sense of distributions, after observing that $C^0(-1,1) \subset \mathrm{L}^1_{loc}(-1,1) \subset \mathcal{D}'(-1,1)$.
8. Prove that if $\Omega = (a,b)$ the Poincaré inequality (2.13) holds with $C_\Omega = (b-a)/\sqrt{2}$.

   [*Solution*: observe that the Cauchy-Schwarz inequality implies

$$v(x) = \int_a^x v'(t)dt \le \left(\int_a^x [v'(t)]^2 dt \right)^{1/2} \left(\int_a^x 1 dt \right)^{1/2} \le \sqrt{x-a} \|v'\|_{\mathrm{L}^2(a,b)},$$

   whence

$$\|v\|^2_{L^2(a,b)} \le \|v'\|^2_{L^2(a,b)} \int_a^b (x-a)dx].$$

3

Elliptic equations

This chapter is devoted to the introduction of elliptic problems and to their weak formulation. Although our introduction is quite basic, the complete novice to functional analysis is invited to consult Chapter 2 before reading it.

For the sake of simplicity, we will focus primarily on one-dimensional and two-dimensional problems. However, the generalization to three-dimensional problems is (almost always) straightforward.

3.1 An elliptic problem example: the Poisson equation

Consider a domain $\Omega \subset \mathbb{R}^2$, i.e. an open bounded and connected set, and let $\partial\Omega$ be its boundary. We denote by $\mathbf{x}$ the spatial variable pair (x_1,x_2). The problem under examination is

$$-\Delta u = f \quad \text{in } \Omega, \tag{3.1}$$

where $f = f(\mathbf{x})$ is a given function and the symbol Δ denotes the Laplacian operator (1.6) in two dimensions. (3.1) is an elliptic, linear, non-homogeneous (if $f \neq 0$) second-order equation. We call (3.1) the *strong formulation* of the Poisson equation. We also recall that, in the case where $f = 0$, equation (3.1) is known as the Laplace equation.

Physically, u can represent the vertical displacement of an elastic membrane due to the application of a force with intensity equal to f, or the electric potential distribution due to an electric charge with density f.

To obtain a unique solution, suitable boundary conditions must be added to (3.1), that is we need information about the behaviour of the solution u at the domain boundary $\partial\Omega$. For instance, the value of the displacement u on the boundary can be assigned

$$u = g \quad \text{on } \partial\Omega, \tag{3.2}$$

where g is a given function, and in such case we will talk about a *Dirichlet problem*. The case where $g = 0$ is said to be *homogeneous*.

A. Quarteroni: *Numerical Models for Differential Problems*, 2nd Ed.
MS&A – Modeling, Simulation & Applications 8
DOI 10.1007/978-88-470-5522-3_3, © Springer-Verlag Italia 2014

Alternatively, the value of the *normal derivative* of u can be imposed

$$\nabla u \cdot \mathbf{n} = \frac{\partial u}{\partial n} = h \quad \text{on } \partial\Omega,$$

$\mathbf{n}$ being the outward unit normal vector on $\partial\Omega$ and h an assigned function. The associated problem is called a *Neumann problem* and corresponds, in the case of the membrane problem, to imposing the traction at the boundary of the membrane itself. Once again, the case $h = 0$ is said to be *homogeneous*.

Finally, different types of conditions can be assigned to different portions of the boundary of the computational domain Ω. For instance, supposing that $\partial\Omega = \Gamma_D \cup \Gamma_N$ with $\overset{\circ}{\Gamma}_D \cap \overset{\circ}{\Gamma}_N = \emptyset$, the following conditions can be imposed:

$$\begin{cases} u = g & \text{on } \Gamma_D, \\ \dfrac{\partial u}{\partial n} = h & \text{on } \Gamma_N. \end{cases}$$

The notation $\overset{\circ}{\Gamma}$ has been used to indicate the interior of Γ. In such a case, the associated problem is said to be *mixed*.

Also in the case of homogeneous Dirichlet problems where f is a continuous function in $\overline{\Omega}$ (the closure of Ω), it is not guaranteed that problem (3.1), (3.2) admits a regular solution. For instance, if $\Omega = (0,1) \times (0,1)$ and $f = 1$, u may not belong to the space $C^2(\overline{\Omega})$. Indeed, if it were so, we would have

$$-\Delta u(0,0) = -\frac{\partial^2 u}{\partial x_1^2}(0,0) - \frac{\partial^2 u}{\partial x_2^2}(0,0) = 0,$$

as the boundary conditions would imply that $u(x_1,0) = u(0,x_2) = 0$ for all x_1, x_2 belonging to $[0,1]$. Hence u could not satisfy equation (3.1), that is

$$-\Delta u = 1 \quad \text{in } (0,1) \times (0,1).$$

What can be learned from this counterexample is that, even if $f \in C^0(\overline{\Omega})$, it makes no sense in general to look for a solution $u \in C^2(\overline{\Omega})$ to problem (3.1), (3.2), while one has greater probabilities to find a solution $u \in C^2(\Omega) \cap C^0(\overline{\Omega})$ (a larger space than $C^2(\overline{\Omega})$!).

We are therefore interested in finding an alternative formulation to the strong one, also because, as we will see in the following section, the latter does not allow the treatment of some physically significant cases. For instance, it is not guaranteed that, in the presence of non-smooth data, the physical solution lies in the space $C^2(\Omega) \cap C^0(\overline{\Omega})$, and not even that it lies in $C^1(\Omega) \cap C^0(\overline{\Omega})$.

3.2 The Poisson problem in the one-dimensional case

Our first step is the introduction of the weak formulation of a simple boundary-value problem in one dimension.

3.2.1 Homogeneous Dirichlet problem

Let us consider the homogeneous Dirichlet problem in the one-dimensional interval $\Omega = (0,1)$

$$\begin{cases} -u''(x) = f(x), & 0 < x < 1, \\ u(0) = 0, & u(1) = 0. \end{cases} \tag{3.3}$$

This problem governs, for instance, the equilibrium configuration of an elastic string with tension equal to one, fixed at the endpoints, in a small displacement configuration and subject to a transversal force with intensity f. The overall force acting on the section $(0,x)$ of the string is

$$F(x) = \int_0^x f(t)dt.$$

The function u describes the vertical displacement of the string relative to the resting position $u = 0$.

The strong formulation (3.3) is in general inadequate. If we consider, for instance, the case where the elastic string is subject to a charge concentrated in one or more points (in such case f can be represented via Dirac delta functions), the physical solution exists and is continuous, but not differentiable. Fig. 3.1 shows the case of a unit charge concentrated only in the point $x = 0.5$ (left) and in the two points $x = 0.4$ and $x = 0.6$ (right). These functions cannot be solutions of (3.3), as the latter would require the solution to have a continuous second derivative. Similar considerations hold in the case where f is a piecewise constant function. For instance, in the case represented in Fig. 3.2 of a null load, except for the interval $[0.4, 0.6]$ where it is equal to -1, the analytical solution is only of class $C^1([0,1])$, since it is given by

$$u(x) = \begin{cases} -\dfrac{1}{10}x & \text{for } x \in [0, 0.4], \\ \dfrac{1}{2}x^2 - \dfrac{1}{2}x + \dfrac{2}{25} & \text{for } x \in [0.4, 0.6], \\ -\dfrac{1}{10}(1-x) & \text{for } x \in [0.6, 1]. \end{cases}$$

A formulation of the problem alternative to the strong one is therefore necessary to allow reducing the order of the derivation required for the unknown solution u. We move from a second-order differential problem to a first-order one in integral form, which is called the *weak formulation* of the differential problem.

To this end, we operate a sequence of formal transformations of (3.3), without worrying at this stage whether all the operations appearing in it are allowed. We start by multiplying equation (3.3) by a (so far arbitrary) *test function* v and integrating on the interval $(0,1)$,

$$-u''v = fv \quad \Rightarrow \quad -\int_0^1 u''v\,dx = \int_0^1 fv\,dx.$$

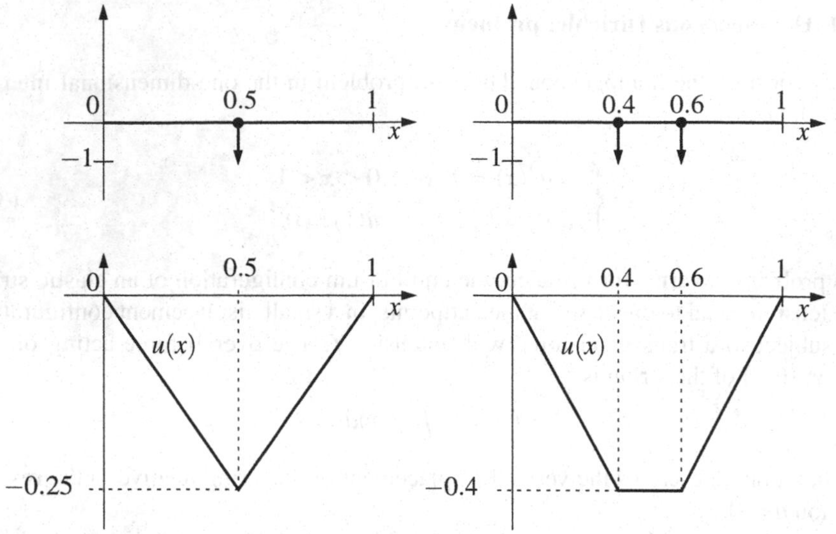

Fig. 3.1. We display on the left the equilibrium configuration of the string corresponding to the unit charge concentrated in $x = 0.5$, represented in the upper part of the figure. On the right we display the one corresponding to two unit charges concentrated in $x = 0.4$ and $x = 0.6$, also represented in the upper part of the figure

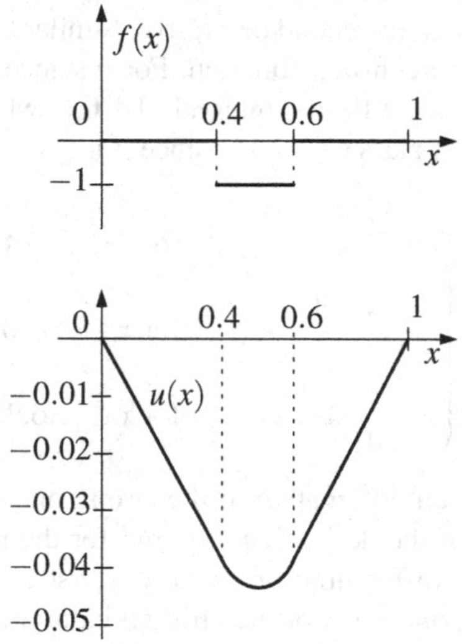

Fig. 3.2. Displacement relative to the discontinuous charge represented in the upper part of the figure

We integrate by parts the first integral, with the purpose of eliminating the second derivative, in order to impose a lower regularity on the solution. We find

$$-\int_0^1 u''v\,dx = \int_0^1 u'v'\,dx - [u'v]_0^1.$$

Since u is known at the boundary, we can consider only test functions which vanish at the endpoints of the interval, hence the contribution of the boundary terms vanishes. In this way, the equation becomes

$$\int_0^1 u'v'\,dx = \int_0^1 fv\,dx. \tag{3.4}$$

The test function space V must therefore be such that if $v \in V$ then $v(0) = v(1) = 0$. Note that the solution u, being null at the boundary and having the same requirements of regularity as the test functions, will also be sought in the same space V.

There remains to specify the regularity requirements which must be satisfied by the space V, so that all the operations introduced make sense. Evidently, if u and v belonged to $C^1([0,1])$, we would have $u', v' \in C^0([0,1])$ and therefore the integral appearing in the left-hand side of (3.4) would make sense. However, the examples in Fig. 3.1 tell us that the physical solutions might not be continuously differentiable: we must therefore require a lower regularity. Moreover, even when $f \in C^0([0,1])$, there is no garantee that the problem admits solutions in the space

$$V = \{v \in C^1([0,1]) : v(0) = v(1) = 0\}. \tag{3.5}$$

This may be attributed to the fact that such vector space, when provided with a scalar product

$$[u,v]_1 = \int_0^1 u'v'dx, \tag{3.6}$$

is not a complete space, that is, not all of the Cauchy sequences with values in V converge to an element of V. (Verify as an exercise that (3.6) is indeed a scalar product.)

Let us then proceed as follows. We recall the definition of the spaces L^p of functions whose p-th power is Lebesgue integrable. For $1 \le p < \infty$, these are defined as follows (see Sect. 2.5):

$$\mathrm{L}^p(0,1) = \{v : (0,1) \mapsto \mathbb{R} \text{ such that } \|v\|_{\mathrm{L}^p(0,1)} = \left(\int_0^1 |v(x)|^p\,dx\right)^{1/p} < +\infty\}.$$

Since we want the integral $\int_0^1 u'v'\,dx$ to be well defined, the minimum requirement on u' and v' is that the product $u'v'$ lies in $\mathrm{L}^1(0,1)$. To this purpose, the following property holds:

Property 3.1. *Given two functions* φ, $\psi : (0,1) \to \mathbb{R}$, *if*

$$\varphi^2,\ \psi^2 \text{ are integrable then } \varphi\psi \text{ is integrable,}$$

that is, equivalently,

$$\varphi,\ \psi \in \mathrm{L}^2(0,1) \quad \Longrightarrow \quad \varphi\psi \in \mathrm{L}^1(0,1).$$

This result is a direct consequence of the *Cauchy-Schwarz inequality*:

$$\left| \int_0^1 \varphi(x)\psi(x)\,dx \right| \leq \|\varphi\|_{\mathrm{L}^2(0,1)} \|\psi\|_{\mathrm{L}^2(0,1)}, \tag{3.7}$$

where

$$\|\varphi\|_{\mathrm{L}^2(0,1)} = \sqrt{\int_\Omega |\varphi(x)|^2 dx} \tag{3.8}$$

is the norm of φ in $\mathrm{L}^2(0,1)$. Since $\|\varphi\|_{\mathrm{L}^2(0,1)}, \|\psi\|_{\mathrm{L}^2(0,1)} < \infty$ by hypothesis, this proves that there also exists a (finite) integral of $\varphi\psi$.

In order for the integrals appearing in (3.4) to make sense, functions, as well as their derivatives, must be square integrable. We therefore define the *Sobolev space*

$$\mathrm{H}^1(0,1) = \{v \in \mathrm{L}^2(0,1) : v' \in \mathrm{L}^2(0,1)\}.$$

The derivative must be interpreted in the sense of distributions (see Sect. 2.3). Let us hence choose as V the following subspace of $\mathrm{H}^1(0,1)$,

$$\mathrm{H}^1_0(0,1) = \{v \in \mathrm{H}^1(0,1) : v(0) = v(1) = 0\},$$

constituted by the functions of $\mathrm{H}^1(0,1)$ that are null at the endpoints of the interval. If we suppose $f \in \mathrm{L}^2(0,1)$, the integral on the right-hand side of (3.4) also makes sense. Problem (3.3) is then reduced to the following integral problem,

$$\text{find } u \in V : \quad \int_0^1 u'v'\,dx = \int_0^1 fv\,dx \quad \forall v \in V, \tag{3.9}$$

with $V = \mathrm{H}^1_0(0,1)$.

Remark 3.1. The space $\mathrm{H}^1_0(0,1)$ is the closure, with respect to the scalar product (3.6), of the space defined in (3.5).
The functions of $\mathrm{H}^1(0,1)$ are not necessarily differentiable in a traditional sense, that is $\mathrm{H}^1(0,1) \not\subset C^1([0,1])$. For instance, functions that are piecewise continuous on a partition of the interval $(0,1)$ with derivatives that do not match at all endpoints of

the partition belong to $H^1(0,1)$ but not to $C^1([0,1])$. Hence, also continuous but not differentiable solutions of the previous examples are considered. •

The weak problem (3.9) turns out to be equivalent to a *variational problem*, due to the following result:

Theorem 3.1. *The problem*

$$\text{find } u \in V: \begin{cases} J(u) = \min_{v\in V} J(v) \quad \text{with} \\ J(v) = \dfrac{1}{2}\displaystyle\int_0^1 (v')^2\,dx - \int_0^1 fv\,dx, \end{cases} \tag{3.10}$$

is equivalent to problem (3.9), *in the sense that u is a solution of* (3.9) *if and only if u is a solution of* (3.10).

Proof. Suppose that u is a solution of the variational problem (3.10). Then, setting $v = u + \delta w$, with $\delta \in \mathbb{R}$, we have that

$$J(u) \leq J(u+\delta w) \quad \forall w \in V.$$

The function $\psi(\delta) = J(u+\delta w)$ is a quadratic function in δ with minimum reached for $\delta = 0$. Thus,

$$\psi'(\delta)\Big|_{\delta=0} = \frac{\partial J(u+\delta w)}{\partial \delta}\Big|_{\delta=0} = 0.$$

From the definition of derivative we have

$$\frac{\partial J(u+\delta w)}{\partial \delta} = \lim_{\delta\to 0} \frac{J(u+\delta w) - J(u)}{\delta} \quad \forall w \in V.$$

Let us consider the term $J(u+\delta w)$:

$$\begin{aligned} J(u+\delta w) &= \frac{1}{2}\int_0^1 [(u+\delta w)']^2\,dx - \int_0^1 f(u+\delta w)\,dx \\ &= \frac{1}{2}\int_0^1 [u'^2 + \delta^2 w'^2 + 2\delta u'w']\,dx - \int_0^1 fu\,dx - \int_0^1 f\delta w\,dx \\ &= J(u) + \frac{1}{2}\int_0^1 [\delta^2 w'^2 + 2\delta u'w']\,dx - \int_0^1 f\delta w\,dx. \end{aligned}$$

Hence,

$$\frac{J(u+\delta w)-J(u)}{\delta}=\frac{1}{2}\int_0^1[\delta w'^2+2u'w']\,dx-\int_0^1 fw\,dx.$$

Passing to the limit for $\delta \to 0$ and setting to 0, we obtain

$$\int_0^1 u'w'\,dx-\int_0^1 fw\,dx=0 \quad \forall w\in V,$$

that is, u satisfies the weak problem (3.9).

Conversely, if u is a solution of (3.9), by setting $v=\delta w$, we have in particular that

$$\int_0^1 u'\delta w'\,dx-\int_0^1 f\delta w\,dx=0,$$

and therefore

$$\begin{aligned}J(u+\delta w)&=\frac{1}{2}\int_0^1[(u+\delta w)']^2\,dx-\int_0^1 f(u+\delta w)\,dx\\&=\frac{1}{2}\int_0^1 u'^2\,dx-\int_0^1 fu\,dx+\int_0^1 u'\delta w'\,dx-\int_0^1 f\delta w\,dx+\frac{1}{2}\int_0^1\delta^2 w'^2\,dx\\&=J(u)+\frac{1}{2}\int_0^1\delta^2 w'^2\,dx.\end{aligned}$$

Since

$$\frac{1}{2}\int_0^1\delta^2 w'^2\,dx\geq 0 \quad \forall\, w\in V,\ \forall\delta\in\mathbb{R},$$

we deduce that

$$J(u)\leq J(v) \quad \forall v\in V,$$

that is u also satisfies the variational problem (3.10). ◇

Remark 3.2 (Principle of virtual work). Let us consider again the problem of studying the configuration assumed by a unit-tension string, fixed at the endpoints and subject to a forcing term f, described by equation (3.3). We indicate with v an admissible displacement of the string (that is a null displacement at the endpoints) from the equilibrium position u. Equation (3.9), expressing the equality between the work performed by the internal forces and by the external forces in correspondence to the displacement v, is nothing but the *principle of virtual work* of mechanics. Moreover, as in our case there exists a potential (indeed, $J(w)$ defined in (3.10) expresses the global potential energy corresponding to the configuration w of the system), the principle of virtual works establishes that any displacement allowed by the equilibrium configuration causes an increment of the system's potential energy. In this sense, Theorem 3.1 states that the weak solution is also the one minimizing the potential energy. •

3.2.2 Non-homogeneous Dirichlet problem

In the non-homogeneous case the boundary conditions in (3.3) are replaced by

$$u(0) = g_0, \quad u(1) = g_1,$$

g_0 and g_1 being two assigned values.

We can reduce to the homogeneous case by noticing that if u is a solution of the non-homogeneous problem, then the function $\overset{\circ}{u}= u - [(1-x)g_0 + xg_1]$ is a solution of the corresponding homogeneous problem (3.3). The function $R_g = (1-x)g_0 + xg_1$ is said *lifting* (or *extension*, or *prolongation*) of the boundary data.

3.2.3 Neumann Problem

Let us now consider the following Neumann problem

$$\begin{cases} -u'' + \sigma u = f, & 0 < x < 1, \\ u'(0) = h_0, & u'(1) = h_1, \end{cases}$$

σ being a positive function and h_0, h_1 two real numbers. We observe that in the case where $\sigma = 0$ the solution of this problem would not be unique, being defined up to an additive constant. By applying the same procedure followed in the case of the Dirichlet problem, that is by multiplying the equation by a test function v, integrating on the interval $(0,1)$ and applying the formula of integration by parts, we get the equation

$$\int_0^1 u'v'\,dx + \int_0^1 \sigma uv\,dx - [u'v]_0^1 = \int_0^1 fv\,dx.$$

Let us suppose $f \in \mathrm{L}^2(0,1)$ and $\sigma \in \mathrm{L}^\infty(0,1)$, that is that σ is a bounded function almost everywhere (a.e.) on $(0,1)$ (see (2.14)). The boundary term is known from the Neumann conditions. On the other hand, the unknown u is not known at the boundary in this case, hence it must not be required that v is null at the boundary. The weak formulation of the Neumann problem is therefore: *find* $u \in \mathrm{H}^1(0,1)$ *such that*

$$\int_0^1 u'v'\,dx + \int_0^1 \sigma uv\,dx = \int_0^1 fv\,dx + h_1 v(1) - h_0 v(0) \quad \forall v \in \mathrm{H}^1(0,1). \qquad (3.11)$$

In the homogeneous case $h_0 = h_1 = 0$, the weak problem is characterized by the same equation as the Dirichlet case, but the space V of test functions is now $\mathrm{H}^1(0,1)$ instead of $\mathrm{H}^1_0(0,1)$.

3.2.4 Mixed homogeneous problem

Analogous considerations hold for the mixed homogeneous problem, that is when we have a homogeneous Dirichlet condition at one endpoint and a homogeneous Neumann condition at the other,

$$\begin{cases} -u'' + \sigma u = f, & 0 < x < 1, \\ u(0) = 0, & u'(1) = 0. \end{cases} \tag{3.12}$$

In such case it must be required that the test functions are null in $x = 0$. Setting $\Gamma_D = \{0\}$ and defining

$$\mathrm{H}^1_{\Gamma_D}(0,1) = \{v \in \mathrm{H}^1(0,1) : v(0) = 0\},$$

the weak formulation of problem (3.12) is: *find* $u \in \mathrm{H}^1_{\Gamma_D}(0,1)$ *such that*

$$\int_0^1 u'v'\,dx + \int_0^1 \sigma uv\,dx = \int_0^1 fv\,dx \quad \forall v \in \mathrm{H}^1_{\Gamma_D}(0,1),$$

with $f \in \mathrm{L}^2(0,1)$ and $\sigma \in \mathrm{L}^\infty(0,1)$. The formulation is once again the same as in the homogeneous Dirichlet problem, however the space where to find the solution changes.

3.2.5 Mixed (or Robin) boundary conditions

Finally, consider the following problem

$$\begin{cases} -u'' + \sigma u = f, & 0 < x < 1, \\ u(0) = 0, & u'(1) + \gamma u(1) = r, \end{cases}$$

where $\gamma > 0$ and r are two assigned constants.
Also in this case, we will use test functions that are null at $x = 0$, the value of u being thereby known. As opposed to the Neumann case, the boundary term for $x = 1$, deriving from the integration by parts, no longer provides a known quantity, but a term proportional to the unknown u. As a matter of fact, we have

$$-[u'v]_0^1 = -rv(1) + \gamma u(1)v(1).$$

The weak formulation is therefore: *find* $u \in \mathrm{H}^1_{\Gamma_D}(0,1)$ *such that*

$$\int_0^1 u'v'\,dx + \int_0^1 \sigma uv\,dx + \gamma u(1)v(1) = \int_0^1 fv\,dx + rv(1) \quad \forall v \in \mathrm{H}^1_{\Gamma_D}(0,1).$$

A boundary condition that is a linear combination between the value of u and the value of its first derivative is called *Robin* (or *Newton*, or *third-type*) *condition*.

3.3 The Poisson problem in the two-dimensional case

In this section, we consider the boundary-value problems associated to the Poisson equation in the two-dimensional case.

3.3.1 The homogeneous Dirichlet problem

The problem consists in finding u such that

$$\begin{cases} -\Delta u = f & \text{in } \Omega, \\ u = 0 & \text{on } \partial\Omega, \end{cases} \tag{3.13}$$

where $\Omega \subset \mathbb{R}^2$ is a bounded domain with boundary $\partial\Omega$. We proceed in a similar way as for the one-dimensional case. By multiplying the differential equation in (3.13) by an arbitrary function v and integrating on Ω, we find

$$-\int_{\Omega} \Delta u v \, d\Omega = \int_{\Omega} f v \, d\Omega.$$

At this point, it is necessary to apply the multi-dimensional analogue of the one-dimensional formula of integration by parts. This can be obtained by applying the divergence (Gauss) theorem by which

$$\int_{\Omega} \mathrm{div}(\mathbf{a}) \, d\Omega = \int_{\partial\Omega} \mathbf{a} \cdot \mathbf{n} \, d\gamma, \tag{3.14}$$

$\mathbf{a}(\mathbf{x}) = (a_1(\mathbf{x}), a_2(\mathbf{x}))^T$ being a sufficiently regular vector-valued function and $\mathbf{n}(\mathbf{x}) = (n_1(\mathbf{x}), n_2(\mathbf{x}))^T$ the outward unit normal vector on $\partial\Omega$. If we apply (3.14) first to the function $\mathbf{a} = (\varphi\psi, 0)^T$ and then to $\mathbf{a} = (0, \varphi\psi)^T$, we get the relations

$$\int_{\Omega} \frac{\partial \varphi}{\partial x_i} \psi \, d\Omega = -\int_{\Omega} \varphi \frac{\partial \psi}{\partial x_i} \, d\Omega + \int_{\partial\Omega} \varphi\psi n_i \, d\gamma, \quad i = 1, 2. \tag{3.15}$$

Note also that if we take $\mathbf{a} = \mathbf{b}\varphi$, where $\mathbf{b}$ and φ are respectively a vector and a scalar field, then (3.14) yields

$$\int_{\Omega} \varphi \mathrm{div} \mathbf{b} \, d\Omega = -\int_{\Omega} \mathbf{b} \cdot \nabla\varphi \, d\Omega + \int_{\partial\Omega} \mathbf{b} \cdot \mathbf{n} \varphi \, d\gamma \tag{3.16}$$

which is called *Green formula* for the divergence operator.

We exploit (3.15) by keeping into account the fact that $\Delta u = \text{div}\nabla u = \sum_{i=1}^{2} \frac{\partial}{\partial x_i}\left(\frac{\partial u}{\partial x_i}\right)$. Supposing that all the integrals make sense, we find

$$\begin{aligned}
-\int_{\Omega} \Delta u v \, d\Omega &= -\sum_{i=1}^{2} \int_{\Omega} \frac{\partial}{\partial x_i}\left(\frac{\partial u}{\partial x_i}\right) v \, d\Omega \\
&= \sum_{i=1}^{2} \int_{\Omega} \frac{\partial u}{\partial x_i}\frac{\partial v}{\partial x_i} \, d\Omega - \sum_{i=1}^{2} \int_{\partial\Omega} \frac{\partial u}{\partial x_i} v n_i \, d\gamma \\
&= \int_{\Omega} \sum_{i=1}^{2} \frac{\partial u}{\partial x_i}\frac{\partial v}{\partial x_i} \, d\Omega - \int_{\partial\Omega} \left(\sum_{i=1}^{2} \frac{\partial u}{\partial x_i} n_i\right) v \, d\gamma.
\end{aligned}$$

We obtain the following relation, called *Green formula* for the Laplacian

$$-\int_{\Omega} \Delta u v \, d\Omega = \int_{\Omega} \nabla u \cdot \nabla v \, d\Omega - \int_{\partial\Omega} \frac{\partial u}{\partial n} v \, d\gamma. \tag{3.17}$$

Similarly to the one-dimensional case, the homogeneous Dirichlet problem will lead us to choose test functions that vanish at the boundary, and, consequently, the boundary term that appears in (3.17) will in turn vanish.

Taking this into account, we get the following weak formulation for problem (3.13)

$$\text{find u } \in \mathrm{H}_0^1(\Omega): \quad \int_{\Omega} \nabla u \cdot \nabla v \, d\Omega = \int_{\Omega} f v \, d\Omega \quad \forall v \in \mathrm{H}_0^1(\Omega), \tag{3.18}$$

f being a function of $\mathrm{L}^2(\Omega)$ and having set

$$\begin{aligned}
&\mathrm{H}^1(\Omega) = \{v : \Omega \to \mathbb{R} \text{ such that } v \in \mathrm{L}^2(\Omega), \frac{\partial v}{\partial x_i} \in \mathrm{L}^2(\Omega),\ i = 1,2\}, \\
&\mathrm{H}_0^1(\Omega) = \{v \in \mathrm{H}^1(\Omega):\ v = 0 \text{ on } \partial\Omega\}.
\end{aligned}$$

The derivatives must be understood in the sense of distributions and the condition $v = 0$ on $\partial\Omega$ in the sense of the traces (see Chap. 2).
In particular, we observe that if $u, v \in \mathrm{H}_0^1(\Omega)$, then $\nabla u, \nabla v \in [\mathrm{L}^2(\Omega)]^2$ and therefore $\nabla u \cdot \nabla v \in \mathrm{L}^1(\Omega)$. The latter property is obtained by applying the following inequality

$$|(\nabla u, \nabla v)| \leq \|\nabla u\|_{\mathrm{L}^2(\Omega)} \|\nabla v\|_{\mathrm{L}^2(\Omega)},$$

a direct consequence of the *Cauchy-Schwarz inequality* (2.17).

Hence, the integral appearing in the left side of (3.18) is perfectly meaningful, and so is the one appearing at the right.

Similarly to the one-dimensional case, it can be shown also in the two-dimensional case that problem (3.18) is equivalent to the following *variational problem*

$$\text{find } u \in V : \begin{cases} J(u) = \inf_{v \in V} J(v), \text{ with} \\ J(v) = \dfrac{1}{2} \int_{\Omega} |\nabla v|^2 \, d\Omega - \int_{\Omega} f v \, d\Omega, \end{cases}$$

having set $V = \mathrm{H}_0^1(\Omega)$.

We can rewrite the weak formulation (3.18) in a more compact way by introducing the following form

$$a : V \times V \to \mathbb{R}, \quad a(u,v) = \int_{\Omega} \nabla u \cdot \nabla v \, d\Omega \tag{3.19}$$

and the following functional

$$F : V \to \mathbb{R}, \quad F(v) = \int_{\Omega} f v \, d\Omega$$

(functionals and forms are introduced in Chap. 2).
Problem (3.18) therefore becomes:

$$\text{find } u \in V : \quad a(u,v) = F(v) \quad \forall v \in V.$$

We notice that $a(\cdot,\cdot)$ is a bilinear form (that is, linear in to both its arguments), while F is a linear functional. Then

$$|F(v)| \le \|f\|_{\mathrm{L}^2(\Omega)} \|v\|_{\mathrm{L}^2(\Omega)} \le \|f\|_{\mathrm{L}^2(\Omega)} \|v\|_{\mathrm{H}^1(\Omega)}.$$

Consequently, F is also bounded. Following definition (2.2), its norm is bounded by $\|F\|_{V'} \le \|f\|_{\mathrm{L}^2(\Omega)}$. Consequently, F belongs to V', the dual space of V, that is the set of linear and continuous functionals defined on V (see Sect. 2.1).

3.3.2 Equivalence, in the sense of distributions, between weak and strong form of the Dirichlet problem

We want to prove that the equations of problem (3.13) are actually satisfied by the weak solution, albeit only in the sense of distributions.
To this end, we consider the weak formulation (3.18). Let $\mathcal{D}(\Omega)$ now be the space of functions that are infinitely differentiable and with compact support in Ω (see Chap. 2). We recall that $\mathcal{D}(\Omega) \subset \mathrm{H}_0^1(\Omega)$. Hence, by choosing $v = \varphi \in \mathcal{D}(\Omega)$ in (3.18), we have

$$\int_{\Omega} \nabla u \cdot \nabla \varphi \, d\Omega = \int_{\Omega} f \varphi \, d\Omega \quad \forall \varphi \in \mathcal{D}(\Omega). \tag{3.20}$$

By applying Green's formula (3.17) to the left-hand side of (3.20), we find

$$-\int_{\Omega} \Delta u \varphi \, d\Omega + \int_{\partial\Omega} \frac{\partial u}{\partial n} \varphi \, d\gamma = \int_{\Omega} f \varphi \, d\Omega \quad \forall \varphi \in \mathcal{D}(\Omega),$$

where the integrals are to be understood via duality, that is:

$$-\int_{\Omega} \Delta u \varphi \, d\Omega = {}_{\mathcal{D}'(\Omega)}\langle -\Delta u, \varphi \rangle_{\mathcal{D}(\Omega)},$$

$$\int_{\partial\Omega} \frac{\partial u}{\partial n} \varphi \, d\gamma = {}_{\mathcal{D}'(\partial\Omega)}\langle \frac{\partial u}{\partial n}, \varphi \rangle_{\mathcal{D}(\partial\Omega)}.$$

Since $\varphi \in \mathcal{D}(\Omega)$, the boundary integral is null, so that

$${}_{\mathcal{D}'(\Omega)}\langle -\Delta u - f, \varphi \rangle_{\mathcal{D}(\Omega)} = 0 \quad \forall \varphi \in \mathcal{D}(\Omega),$$

which corresponds to saying that $-\Delta u - f$ is the null distribution, that is

$$-\Delta u = f \quad \text{in } \mathcal{D}'(\Omega).$$

The differential equation (3.13) is therefore verified, as long as we intend the derivatives in the sense of distributions and we interpret the equality between $-\Delta u$ and f not in a pointwise sense, but in the sense of distributions (and thus almost everywhere in Ω). Finally, the fact that u vanishes on the boundary (in the sense of traces) is a direct consequence of u being in $\mathrm{H}^1_0(\Omega)$.

3.3.3 The problem with mixed, non homogeneous conditions

The problem we want to solve is now the following

$$\begin{cases} -\Delta u = f & \text{in } \Omega, \\ u = g & \text{on } \Gamma_D, \\ \dfrac{\partial u}{\partial n} = \phi & \text{on } \Gamma_N, \end{cases} \tag{3.21}$$

where Γ_D and Γ_N yield a partition of $\partial\Omega$, that is $\Gamma_D \cup \Gamma_N = \partial\Omega$, $\mathring{\Gamma}_D \cap \mathring{\Gamma}_N = \emptyset$ (see Fig. 3.3).

In the case of the Neumann problem, where $\Gamma_D = \emptyset$, the data f and ϕ must verify the following *compatibility condition*

$$-\int_{\partial\Omega} \phi \, d\gamma = \int_{\Omega} f \, d\Omega \tag{3.22}$$

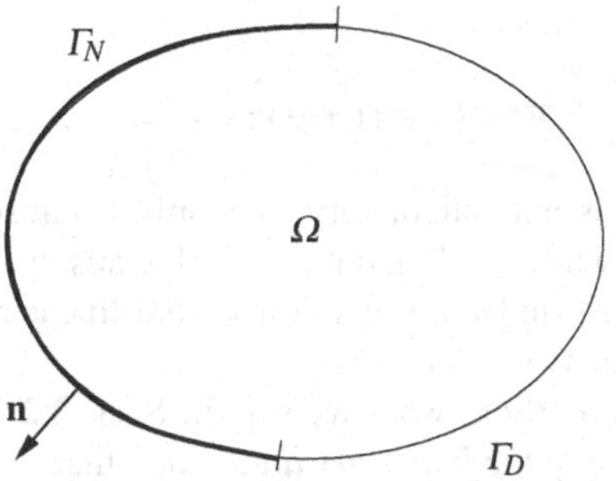

Fig. 3.3. The computational domain Ω

in order for the problem to have a solution. Condition (3.22) is deduced by integrating the differential equation in (3.21) and applying the divergence theorem (3.14)

$$-\int_{\Omega} \Delta u \, d\Omega = -\int_{\Omega} \operatorname{div}(\nabla u) \, d\Omega = -\int_{\partial\Omega} \frac{\partial u}{\partial n} \, d\gamma.$$

Moreover, we observe that also in the case of the Neumann problem, the solution is defined only up to an additive constant. In order to have uniqueness it would be sufficient, for example, to find a function with null average in Ω.

Let us now suppose that $\Gamma_D \neq \emptyset$ in order to ensure the uniqueness of the solution to the strong problem without conditions of compatibility on the data. Let us also suppose that $f \in \mathrm{L}^2(\Omega)$, $g \in \mathrm{H}^{1/2}(\Gamma_D)$ and $\phi \in \mathrm{L}^2(\Gamma_N)$, having denoted by $\mathrm{H}^{1/2}(\Gamma_D)$ the space of functions of $\mathrm{L}^2(\Gamma_D)$ that are traces of functions of $\mathrm{H}^1(\Omega)$ (see Sect. 2.4.3).

By Green's formula (3.17) we obtain from (3.21)

$$\int_{\Omega} \nabla u \cdot \nabla v \, d\Omega - \int_{\partial\Omega} \frac{\partial u}{\partial n} v \, d\gamma = \int_{\Omega} f v \, d\Omega. \tag{3.23}$$

We recall that $\partial u/\partial n = \phi$ on Γ_N, and by exploiting the additivity of integrals, (3.23) becomes

$$\int_{\Omega} \nabla u \cdot \nabla v \, d\Omega - \int_{\Gamma_D} \frac{\partial u}{\partial n} v \, d\gamma - \int_{\Gamma_N} \phi \, v \, d\gamma = \int_{\Omega} f v \, d\Omega. \tag{3.24}$$

By forcing the test function v to vanish on Γ_D, the first boundary integral appearing in (3.24) vanishes. The mixed problem therefore admits the following weak formulation

$$\text{find } u \in V_g : \quad \int_{\Omega} \nabla u \cdot \nabla v \, d\Omega = \int_{\Omega} f v \, d\Omega + \int_{\Gamma_N} \phi \, v \, d\gamma \quad \forall v \in V, \tag{3.25}$$

having denoted by V the space

$$V = \mathrm{H}^1_{\Gamma_D}(\Omega) = \{v \in \mathrm{H}^1(\Omega) : \, v|_{\Gamma_D} = 0\}, \tag{3.26}$$

and having set

$$V_g = \{v \in \mathrm{H}^1(\Omega) : v|_{\Gamma_D} = g\}.$$

The formulation (3.25) is not satisfactory, not only because the choice of spaces is "asymmetrical" ($v \in V$, while $u \in V_g$), but mainly because V_g is an affine manifold, but not a subspace of $\mathrm{H}^1(\Omega)$ (indeed, it is not true that linear combinations of elements of V_g are still elements of V_g).

We then proceed similarly to what we saw in Sect. 3.2.2. We suppose to know a function R_g, called *lifting of the boundary data*, such that

$$R_g \in \mathrm{H}^1(\Omega), \quad R_g|_{\Gamma_D} = g.$$

Furthermore, we suppose that such lifting are continuous, i.e. that

$$\exists C > 0 : \|R_g\|_{\mathrm{H}^1(\Omega)} \leq C\|g\|_{\mathrm{H}^{1/2}(\Gamma_D)} \forall g \in \mathrm{H}^{1/2}(\Gamma_D).$$

We set $\overset{\circ}{u} = u - R_g$ and we begin by observing that $\overset{\circ}{u}|_{\Gamma_D} = u|_{\Gamma_D} - R_g|_{\Gamma_D} = 0$, that is $\overset{\circ}{u} \in \mathrm{H}^1_{\Gamma_D}(\Omega)$. Moreover, since $\nabla u = \nabla\overset{\circ}{u} + \nabla R_g$, problem (3.25) becomes

$$\text{find } \overset{\circ}{u} \in \mathrm{H}^1_{\Gamma_D}(\Omega) : \quad a(\overset{\circ}{u}, v) = F(v) \quad \forall v \in \mathrm{H}^1_{\Gamma_D}(\Omega), \tag{3.27}$$

having defined the bilinear form $a(\cdot,\cdot)$ as in (3.19), while the linear functional F now takes the form

$$F(v) = \int_\Omega f v \, d\Omega + \int_{\Gamma_N} \phi \, v \, d\gamma - \int_\Omega \nabla R_g \cdot \nabla v \, d\Omega.$$

The problem is now symmetrical since the space where the (new) unknown solution is sought coincides with the test function space.

The Dirichlet conditions are said to be *essential* as they are imposed explicitly in the functional space in which the problem is set.

The Neumann conditions are instead said to be *natural*, as they are satisfied implicitly by the solution of the problem (to this end, see Sect. 3.3.4). This difference in treatment has important ripercussions on the approximate problems.

Remark 3.3. The reduction of the problem to a "symmetrical" form allows to obtain a linear system with a symmetric matrix when solving the problem numerically (for instance via the finite elements method). •

Remark 3.4. Building a lifting R_g of a boundary function with an arbitrary form can turn out to be problematic. Such task is simpler in the context of a numerical approximation, where one generally builds a lifting of an approximation of the function g (see Chap. 4). •

3.3.4 Equivalence, in the sense of distributions, between weak and strong form of the Neumann problem

Let us consider the nonhomogeneous Neumann problem

$$\begin{cases} -\Delta u + \sigma u = f & \text{in } \Omega, \\ \dfrac{\partial u}{\partial n} = \phi & \text{on } \partial\Omega, \end{cases} \tag{3.28}$$

where σ is a positive constant or, more generally, a function $\sigma \in \mathrm{L}^\infty(\Omega)$ such that $\sigma(\mathbf{x}) \geq \alpha_0$ a.e. in Ω, for a well-chosen constant $\alpha_0 > 0$. Let us also suppose that $f \in \mathrm{L}^2(\Omega)$ and that $\phi \in \mathrm{L}^2(\partial\Omega)$. By proceeding as in Sect. 3.3.3, the following weak formulation can be derived:

$$\begin{aligned} &\text{find } u \in \mathrm{H}^1(\Omega): \\ &\int_\Omega \nabla u \cdot \nabla v \, d\Omega + \int_\Omega \sigma u v \, d\Omega = \int_\Omega f v \, d\Omega + \int_{\partial\Omega} \phi v \, d\gamma \qquad \forall v \in \mathrm{H}^1(\Omega). \end{aligned} \tag{3.29}$$

By taking $v = \varphi \in \mathcal{D}(\Omega)$ and counterintegrating by parts, we obtain

$$_{\mathcal{D}'(\Omega)}\langle -\Delta u + \sigma u - f, \varphi \rangle_{\mathcal{D}(\Omega)} = 0 \quad \forall \varphi \in \mathcal{D}(\Omega).$$

Hence

$$-\Delta u + \sigma u = f \quad \text{in } \mathcal{D}'(\Omega)$$

i.e.

$$-\Delta u + \sigma u - f = 0 \quad \text{a.e. in } \Omega. \tag{3.30}$$

In the case where $u \in C^2(\Omega)$ the application of Green's formula (3.17) in (3.29) leads to

$$\int_\Omega (-\Delta u + \sigma u - f) v \, d\Omega + \int_{\partial\Omega} \left(\frac{\partial u}{\partial n} - \phi\right) v = 0 \quad \forall v \in \mathrm{H}^1(\Omega),$$

and therefore, by (3.30),

$$\frac{\partial u}{\partial n} = \phi \quad \text{on } \partial\Omega.$$

In the case where the solution u of (3.29) is only in $\mathrm{H}^1(\Omega)$ the generalized Green formula can be used, which states that there exists a unique linear and continuous functional $g \in (\mathrm{H}^{1/2}(\partial\Omega))'$ (called generalized normal derivative), which operates on the space $\mathrm{H}^{1/2}(\partial\Omega)$ and satisfies

$$\int_\Omega \nabla u \cdot \nabla v \, d\Omega = \langle -\Delta u, v \rangle + \ll g, v \gg \quad \forall v \in \mathrm{H}^1(\Omega).$$

We have denoted by $< \cdot, \cdot >$ the pairing between $\mathrm{H}^1(\Omega)$ and its dual, and by $\ll \cdot, \cdot \gg$ the pairing between $\mathrm{H}^{1/2}(\partial\Omega)$ and its dual. Clearly g coincides with the classical normal derivative of u if u has sufficient regularity. For the sake of simplicity we use the notation $\partial u/\partial n$ for the generalized normal derivative in the remainder of this chapter. We therefore obtain that for $v \in \mathrm{H}^1(\Omega)$

$$\langle -\Delta u + \sigma u - f, v\rangle + \ll \partial u/\partial n - \phi, v \gg = 0;$$

using (3.30) we finally conclude that

$$\ll \partial u/\partial n - \phi, v \gg = 0 \quad \forall v \in \mathrm{H}^1(\Omega)$$

and thus that $\partial u/\partial n = \phi$ a.e. on $\partial\Omega$.

3.4 More general elliptic problems

Let us now consider the problem

$$\begin{cases} -\mathrm{div}(\mu\nabla u) + \sigma u = f & \text{in } \Omega, \\ u = g & \text{on } \Gamma_D, \\ \mu\dfrac{\partial u}{\partial n} = \phi & \text{on } \Gamma_N, \end{cases} \tag{3.31}$$

where $\Gamma_D \cup \Gamma_N = \partial\Omega$ with $\mathring{\Gamma}_D \cap \mathring{\Gamma}_N = \emptyset$. We will suppose that $f \in \mathrm{L}^2(\Omega)$, $\mu, \sigma \in \mathrm{L}^\infty(\Omega)$. Furthermore, we suppose that there is a $\mu_0 > 0$ such that $\mu(\mathbf{x}) \geq \mu_0$ and $\sigma(\mathbf{x}) \geq 0$ a.e. in Ω. Only in the case where $\sigma = 0$ we will require that Γ_D is non-empty in order to prevent the solution from losing uniqueness. Finally, we will suppose that g and ϕ are sufficiently regular functions on $\partial\Omega$, for instance $g \in \mathrm{H}^{1/2}(\Gamma_D)$ and $\phi \in \mathrm{L}^2(\Gamma_N)$.

Also in this case, we proceed by multiplying the equation by a test function v and by integrating (once again formally) on the domain Ω:

$$\int_\Omega [-\mathrm{div}(\mu\nabla u) + \sigma u]\, v\, d\Omega = \int_\Omega f v\, d\Omega.$$

By applying Green's formula we obtain

$$\int_\Omega \mu\nabla u \cdot \nabla v\, d\Omega + \int_\Omega \sigma u v\, d\Omega - \int_{\partial\Omega} \mu\frac{\partial u}{\partial n} v\, d\gamma = \int_\Omega f v\, d\Omega,$$

which can also be rewritten as

$$\int_\Omega \mu\nabla u \cdot \nabla v\, d\Omega + \int_\Omega \sigma u v\, d\Omega - \int_{\Gamma_D} \mu\frac{\partial u}{\partial n} v\, d\gamma = \int_\Omega f v\, d\Omega + \int_{\Gamma_N} \mu\frac{\partial u}{\partial n} v\, d\gamma.$$

The function $\mu\partial u/\partial n$ is called *conormal derivative* of u associated to the operator $-\mathrm{div}(\mu\nabla u)$. On Γ_D we impose that the test function v is null, while on Γ_N we impose that the conormal derivative is equal to ϕ. We obtain

$$\int_{\Omega} \mu \nabla u \cdot \nabla v \, d\Omega + \int_{\Omega} \sigma \, u \, v \, d\Omega = \int_{\Omega} f v \, d\Omega + \int_{\Gamma_N} \phi v \, d\gamma.$$

Having denoted by R_g a lifting of g, we set $\overset{\circ}{u} = u - R_g$. The weak formulation of problem (3.31) is therefore

$$\begin{aligned} &\text{find } \overset{\circ}{u} \in \mathrm{H}^1_{\Gamma_D}(\Omega) : \\ &\int_{\Omega} \mu \nabla \overset{\circ}{u} \cdot \nabla v \, d\Omega + \int_{\Omega} \sigma \overset{\circ}{u} v \, d\Omega = \int_{\Omega} f v \, d\Omega \\ &- \int_{\Omega} \mu \nabla R_g \cdot \nabla v \, d\Omega - \int_{\Omega} \sigma R_g v \, d\Omega + \int_{\Gamma_N} \phi v \, d\gamma \qquad \forall v \in \mathrm{H}^1_{\Gamma_D}(\Omega). \end{aligned}$$

We define the bilinear form

$$a : V \times V \to \mathbb{R}, \quad a(u,v) = \int_{\Omega} \mu \nabla u \cdot \nabla v \, d\Omega + \int_{\Omega} \sigma \, u \, v \, d\Omega$$

and the linear and continuous functional

$$F : V \to \mathbb{R}, \; F(v) = -a(R_g, v) + \int_{\Omega} f v \, d\Omega + \int_{\Gamma_N} \phi v \, d\gamma. \tag{3.32}$$

The previous problem can then be rewritten as

$$\text{find } \overset{\circ}{u} \in \mathrm{H}^1_{\Gamma_D}(\Omega) : \quad a(\overset{\circ}{u}, v) = F(v) \quad \forall v \in \mathrm{H}^1_{\Gamma_D}(\Omega). \tag{3.33}$$

A yet more general problem than (3.31) is the following

$$\begin{cases} Lu = f & \text{in } \Omega, \\ u = g & \text{on } \Gamma_D, \\ \dfrac{\partial u}{\partial n_L} = \phi & \text{on } \Gamma_N, \end{cases}$$

where, as usual, $\Gamma_D \cup \Gamma_N = \partial\Omega$, $\overset{\circ}{\Gamma}_D \cap \overset{\circ}{\Gamma}_N = \emptyset$, and having defined

$$Lu = -\sum_{i,j=1}^{2} \frac{\partial}{\partial x_i} \left(a_{ij} \frac{\partial u}{\partial x_j} \right) + \sigma u.$$

The coefficients a_{ij} are functions defined on Ω. The derivative

$$\frac{\partial u}{\partial n_L} = \sum_{i,j=1}^{2} a_{ij} \frac{\partial u}{\partial x_j} n_i \tag{3.34}$$

is called *conormal derivative* of u associated to the operator L (it coincides with the normal derivative when $Lu = -\Delta u$).
Let us suppose that $\sigma(\mathbf{x}) \in L^\infty(\Omega)$ and that there exists an $\alpha_0 > 0$ such that $\sigma(\mathbf{x}) \geq \alpha_0$ a.e. in Ω. Furthermore, let us suppose that the coefficients $a_{ij} : \bar{\Omega} \to \mathbb{R}$ are continuous functions $\forall i, j = 1, 2$, and that there exists a positive constant α such that

$$\forall \boldsymbol{\xi} = (\xi_1, \xi_2)^T \in \mathbb{R}^2 \quad \sum_{i,j=1}^{2} a_{ij}(\mathbf{x})\xi_i\xi_j \geq \alpha \sum_{i=1}^{2} \xi_i^2 \quad \text{a.e. in } \Omega. \tag{3.35}$$

In such case, the weak formulation is still the same as (3.33), the functional F is still the one introduced in (3.32), while

$$a(u,v) = \int_\Omega \left(\sum_{i,j=1}^{2} a_{ij} \frac{\partial u}{\partial x_j} \frac{\partial v}{\partial x_i} + \sigma uv \right) d\Omega. \tag{3.36}$$

It can be shown (see Exercise 2) that under the ellipticity hypothesis on the coefficients (3.35), this bilinear form is continuous and coercive, in the sense of definitions (2.6) and (2.9). These properties will be exploited in the analysis of the well-posedness of problem (3.33) (see Sect. 3.4.1).
Elliptic problems for fourth-order operators are proposed in Exercises 4 and 6, while an elliptic problem deriving from the linear elasticity theory is analyzed in Exercise 7.

Remark 3.5 (Robin conditions). The case where Robin boundary conditions are enforced on the whole boundary, say

$$\mu \frac{\partial u}{\partial n} + \gamma u = 0 \quad \text{on } \partial\Omega,$$

requires more care. The weak form of the problem reads

$$\text{find } u \in H^1(\Omega) : a(u,v) = \int_\Omega f v d\Omega \quad \forall v \in H^1(\Omega),$$

where the bilinear form $a(u,v) = \int_\Omega \mu \nabla u \cdot \nabla v d\Omega + \int_\Omega \gamma u v d\Omega$ this time is not coercive if $\gamma < 0$. The analysis of this problem can be carried out by means of the Peetre-Tartar lemma, see [EG04]. •

3.4.1 Existence and uniqueness theorem

The following fundamental result holds (refer to Sect. 2.1 for definitions):

Lemma 3.1 (Lax-Milgram). *Let V be a Hilbert space, $a(\cdot,\cdot) : V \times V \to \mathbb{R}$ a continuous and coercive bilinear form, $F(\cdot) : V \to \mathbb{R}$ a linear and continuous functional. Then, there exists a unique solution to the problem*

$$\text{find } u \in V : \quad a(u,v) = F(v) \quad \forall v \in V. \tag{3.37}$$

Proof. This is based on two classical results of Functional Analysis: the Riesz representation theorem (see Theorem 2.1, Chap. 2), and the Banach closed range theorem. The interested reader can refer to, e.g., [QV94, Chap. 5]. ◇

The Lax-Milgram lemma thus ensures that the weak formulation of an elliptic problem is well posed, as long as the hypotheses on the form $a(\cdot,\cdot)$ and on the functional $F(\cdot)$ hold. Several consequences derive from this lemma. We report one of the most important in the following corollary.

Corollary 3.1. *The solution of* (3.37) *is bounded by the data, that is*

$$\|u\|_V \leq \frac{1}{\alpha}\|F\|_{V'},$$

where α is the coercivity constant of the bilinear form $a(\cdot,\cdot)$, while $\|F\|_{V'}$ is the norm of the functional F, see (2.2).

Proof. It is sufficient to choose $v = u$ in (3.37) and then to use the coercivity of the bilinear form $a(\cdot,\cdot)$. Indeed, we have

$$\alpha\|u\|_V^2 \leq a(u,u) = F(u).$$

On the other hand, since F is linear and continuous it is also bounded, and the upper bound

$$|F(u)| \leq \|F\|_{V'}\|u\|_V$$

holds, hence the claim follows. ◇

Remark 3.6. If the bilinear form $a(\cdot,\cdot)$ is additionally *symmetric*, that is

$$a(u,v) = a(v,u) \quad \forall u,v \in V,$$

then (3.37) is equivalent to the following variational problem (see Exercise 1)

$$\begin{cases} \text{find } u \in V: \quad J(u) = \min_{v\in V} J(v), \\ \text{with } J(v) = \dfrac{1}{2}a(v,v) - F(v). \end{cases} \tag{3.38}$$

•

3.5 Adjoint operator and adjoint problem

In this section we will introduce the concept of *adjoint* of a given operator in Hilbert spaces, as well as the *adjoint* (or *dual*) problem of a given boundary-value problem. Then we will show how to obtain dual problems, with associated boundary conditions.

The adjoint problem of a given differential problem plays a fundamental role, for instance, when establishing error estimates for Galerkin methods, both a priori and a posteriori (see Sects. 4.5.4 and 4.6.4–4.6.5, respectively), but also for the solution of optimal-control problems, as we will see in Chapter 17.

Let V be a Hilbert space with scalar product $(\cdot,\cdot)_V$ and norm $\|\cdot\|_V$, and let V' be its dual space. Let $a : V \times V \to \mathbb{R}$ be a continuous and coercive bilinear form and let $A : V \to V'$ be its associated elliptic operator, that is $A \in \mathcal{L}(V,V')$,

$$_{V'}\langle Av, w\rangle_V = a(v,w) \quad \forall v,w \in V. \tag{3.39}$$

Let $a^* : V \times V \to \mathbb{R}$ be the bilinear form defined by

$$a^*(w,v) = a(v,w) \quad \forall v,w \in V, \tag{3.40}$$

and consider the operator $A^* : V \to V'$ associated to the form $a^*(\cdot,\cdot)$, that is

$$_{V'}\langle A^*w, v\rangle_V = a^*(w,v) \quad \forall v,w \in V. \tag{3.41}$$

Thanks to (3.40) we have the following relation, known as the *Lagrange identity*

$$_{V'}\langle A^*w, v\rangle_V = {}_{V'}\langle Av, w\rangle_V \quad \forall v,w \in V. \tag{3.42}$$

Note that this is precisely the equation that stands at the base of the definition (2.20) of the adjoint of a given operator A acting between a Hilbert space and its dual. For coherence with (2.20), we should have noted this operator A'. However, we prefer to denote it A^* because the latter notation is more customarily used in the context of elliptic boundary value problems.

If $a(\cdot,\cdot)$ is a symmetric form, $a^*(\cdot,\cdot)$ coincides with $a(\cdot,\cdot)$ and A^* with A. In such case A is said to be *self-adjoint*; A is said to be *normal* if $AA^* = A^*A$.
Naturally, the identity operator I is self-adjoint ($I = I^*$), while if an operator is self-adjoint, then it is also normal.
Some properties of the adjoint operators which are a consequence of the previous definition, are listed below:

- A being linear and continuous, then also A^* is, that is $A^* \in \mathcal{L}(V,V')$;
- $\|A^*\|_{\mathcal{L}(\mathcal{V},\mathcal{V}')} = \|A\|_{\mathcal{L}(\mathcal{V},\mathcal{V}')}$ (these norms are defined in (2.21));
- $(A+B)^* = A^* + B^*$;
- $(AB)^* = B^*A^*$;
- $(A^*)^* = A$;
- $(A^{-1})^* = (A^*)^{-1}$ (if A is invertible);
- $(\alpha A)^* = \alpha A^* \quad \forall \alpha \in \mathbb{R}$.

When we need to find the adjoint (or dual) problem of a given (primal) problem, we will use the Lagrange identity to characterize the differential equation of the dual problem, as well as its boundary conditions.

We provide an example of such a procedure, starting from a simple one-dimensional diffusion transport equation, completed by homogeneous Robin-Dirichlet boundary conditions

$$\begin{cases} Av = -v'' + v' = f, & x \in I = (0,1), \\ v'(0) + \beta v(0) = 0, & v(1) = 0, \end{cases} \tag{3.43}$$

assuming β constant. Note that the weak form of this problem is

$$\text{find } u \in V \text{such that } a(u,v) = \int_0^1 fv dx \quad \forall v \in V, \tag{3.44}$$

where $V = \{v \in H^1(0,1) : v(1) = 0\}$ and

$$a : V \times V \to \mathbb{R}, \; a(u,v) = \int_0^1 (u' - u)v' dx - (\beta + 1)u(0)v(0).$$

By (3.40) we obtain, $\forall v, w \in V$,

$$\begin{aligned} a^*(w,v) = a(v,w) &= \int_0^1 (v' - v)w' dx - (\beta + 1)v(0)w(0) \\ &= -\int_0^1 v(w'' + w')dx + [vw']_0^1 - (\beta + 1)v(0)w(0) \\ &= \int_0^1 (-w'' - w')vdx - [w'(0) + (\beta + 1)w(0)]v(0). \end{aligned}$$

Since definition (3.41) must hold, we will have

$$A^*w = -w'' - w' \quad \text{in } \mathcal{D}'(0,1).$$

Moreover, as $v(0)$ is arbitrary, w will need to satisfy the boundary conditions

$$[w' + (\beta + 1)w](0) = 0, \quad w(1) = 0.$$

We observe that the transport field of the dual problem has an opposite direction with respect to that of the primal problem. Moreover, to homogeneous Robin-Dirichlet boundary conditions for the primal problem (3.43) correspond conditions of exactly the same nature for the dual problem.

The procedure illustrated for problem (3.43) can clearly be extended to the multi-dimensional case. In Table 3.1 we provide a list of several differential operators with boundary conditions, and their corresponding adjoint operators with associated boundary conditions. (On the functions appearing in the table assume all the necessary regularity for the considered differential operators to be well-defined). We note, in particular, that to a given type of primal conditions do not necessarily correspond dual

Table 3.1. Differential operators and boundary conditions (B.C.) for the primal problem and corresponding dual (adjoint) operators (with associated boundary conditions)

Primal operator	*Primal B.C.*	*Dual (adjoint) operator*	*Dual B.C.*
$-\Delta u$	$u=0$ on Γ $\dfrac{\partial u}{\partial n}=0$ on $\partial\Omega\backslash\Gamma$	$-\Delta w$	$w=0$ on Γ, $\dfrac{\partial w}{\partial n}=0$ on $\partial\Omega\backslash\Gamma$
$-\Delta u+\sigma u$	$u=0$ on Γ, $\dfrac{\partial u}{\partial n}+\gamma u=0$ on $\partial\Omega\backslash\Gamma$	$-\Delta w+\sigma w$	$w=0$ on Γ, $\dfrac{\partial w}{\partial n}+\gamma w=0$ on $\partial\Omega\backslash\Gamma$
$-\Delta u+\mathbf{b}\cdot\nabla u+\sigma u$, $\mathrm{div}\,\mathbf{b}=0$	$u=0$ on Γ, $\dfrac{\partial u}{\partial n}+\gamma u=0$ on $\partial\Omega\backslash\Gamma$	$-\Delta w-\mathbf{b}\cdot\nabla w+\sigma w$, $\mathrm{div}\,\mathbf{b}=0$	$w=0$ on Γ, $\dfrac{\partial w}{\partial n}+(\mathbf{b}\cdot\mathbf{n}+\gamma)w=0$ on $\partial\Omega\backslash\Gamma$
$-\Delta u+\mathbf{b}\cdot\nabla u+\sigma u$, $\mathrm{div}\,\mathbf{b}=0$	$u=0$ on Γ, $\dfrac{\partial u}{\partial n}=0$ on $\partial\Omega\backslash\Gamma$	$-\Delta w-\mathbf{b}\cdot\nabla w+\sigma w$, $\mathrm{div}\,\mathbf{b}=0$	$w=0$ on Γ, $\dfrac{\partial w}{\partial n}+\mathbf{b}\cdot\mathbf{n}w=0$ on $\partial\Omega\backslash\Gamma$
$-\mathrm{div}(\mu\nabla u)+\mathrm{div}(\mathbf{b}u)+\sigma u$	$u=0$ on Γ, $\mu\dfrac{\partial u}{\partial n}-\mathbf{b}\cdot\mathbf{n}u=0$ on $\partial\Omega\backslash\Gamma$	$-\mathrm{div}(\mu\nabla w)-\mathbf{b}\cdot\nabla w+\sigma w$	$w=0$ on Γ, $\mu\dfrac{\partial w}{\partial n}=0$ on $\partial\Omega\backslash\Gamma$
$-\mathrm{div}(\mu\nabla u)+\mathbf{b}\cdot\nabla u+\sigma u$, $\mathrm{div}\,\mathbf{b}=0$	$u=0$ on Γ, $\mu\dfrac{\partial u}{\partial n}=0$ on $\partial\Omega\backslash\Gamma$	$-\mathrm{div}(\mu\nabla w)-\mathrm{div}(\mathbf{b}w)+\sigma w$, $\mathrm{div}\,\mathbf{b}=0$	$w=0$ on Γ, $\mu\dfrac{\partial w}{\partial n}+\mathbf{b}\cdot\mathbf{n}w=0$ on $\partial\Omega\backslash\Gamma$

conditions of the same type, and that, for an operator that is not self-adjoint, to a conservative (resp. non-conservative) formulation of the primal problem corresponds a non-conservative (resp. conservative) formulation of the dual one.

3.5.1 The nonlinear case

The extension of the analysis in the previous section to the nonlinear case is not so immediate. For simplicity, we consider the one-dimensional problem

$$\begin{cases} A(v)v = -v'' + vv' = f, & x \in I = (0,1), \\ v(0) = v(1) = 0, \end{cases} \tag{3.45}$$

having denoted by $A(v)$ the operator

$$A(v)\cdot = -\frac{d^2\cdot}{dx^2} + v\frac{d\cdot}{dx}. \tag{3.46}$$

The Lagrange identity (3.42) is now generalized as

$$_{V'}\langle A(v)u, w\rangle_V = {}_V\langle u, A^*(v)w\rangle_{V'} \tag{3.47}$$

for each $u \in D(A)$ and $w \in D(A^*)$, $D(A)$ being the set of functions of class C^2 that are null at $x = 0$ and $x = 1$, and $D(A^*)$ the domain of the adjoint (or dual) operator A^* whose properties will be identified by imposing (3.47). Starting from such identity, let us see which adjoint operator A^* and which dual boundary conditions we get for problem (3.45). By integrating by parts the diffusion term twice and the transport term of order one once, we obtain

$$\begin{aligned} _{V'}\langle A(v)u, w\rangle_V &= -\int_0^1 u'' w\,dx + \int_0^1 vu' w\,dx \\ &= \int_0^1 u' w'\,dx - u' w\Big|_0^1 - \int_0^1 (vw)' u\,dx + vuw\Big|_0^1 \\ &= -\int_0^1 uw''\,dx + uw'\Big|_0^1 - u' w\Big|_0^1 - \int_0^1 (vw)' u\,dx + vuw\Big|_0^1. \end{aligned} \tag{3.48}$$

Let us analyze the boundary terms separately, by makin the contribution at both endpoints explicit. In order to guarantee (3.47), we must have

$$u(1)\,w'(1) - u(0)\,w'(0) - u'(1)\,w(1) + u'(0)\,w(0) + v(1)\,u(1)\,w(1) - v(0)\,u(0)\,w(0) = 0$$

for each u and $v \in D(A)$. We observe that the fact that u belongs to $D(A)$ allows us to ignore, as vanishing, both the first two and the last two summands, so that we end up having

$$-u'(1)\,w(1) + u'(0)\,w(0) = 0.$$

Since such relation must hold for each $u \in D(A)$, we must choose homogeneous Dirichlet conditions for the dual operator, i.e.

$$w(0) = w(1) = 0. \tag{3.49}$$

Reverting to (3.48), we then have

$$\begin{aligned} {}_{V'}\langle A(v)u, w\rangle_V &= -\int_0^1 u'' w \, dx + \int_0^1 v u' w \, dx \\ &= -\int_0^1 u w'' \, dx - \int_0^1 (vw)' u \, dx = {}_V\langle u, A^*(v) w\rangle_{V'}. \end{aligned}$$

The adjoint operator A^* of the primal operator A defined in (3.46) therefore reads

$$A^*(v)\cdot = -\frac{d^2 \cdot}{dx^2} + \frac{d}{dx} v \cdot$$

while the dual boundary conditions are provided by (3.49). To conclude, we note that the dual problem is always linear, even though we started from a nonlinear primal problem.

For more details on the differentiation and on the analysis of the adjoint problems, we refer the reader to, e.g., [Mar95].

3.6 Exercises

1. Prove that the weak problem (3.37) is equivalent to the variational problem (3.38) if the bilinear form is coercive and symmetric.
   [*Solution:* let $u \in V$ be the solution of the weak problem and let w be a generic element of V. Thanks to the bilinearity and to the symmetry of the form, we find

$$\begin{aligned} J(u+w) &= \frac{1}{2}[a(u,u) + 2a(u,w) + a(w,w)] - [F(u) + F(w)] \\ &= J(u) + [a(u,w) - F(w)] + \frac{1}{2}a(w,w) = J(u) + \frac{1}{2}a(w,w). \end{aligned}$$

   Thanks to the coercivity we then obtain that $J(u+w) \geq J(u) + (\alpha/2)\|w\|_V^2$, that is $\forall v \in V$ with $v \neq u$, $J(v) > J(u)$. Conversely, if u is a minimum for J, then by writing the extremality condition $\lim_{\delta \to 0} (J(u+\delta v) - J(u))/\delta = 0$ we find (3.37).]

2. Prove that the bilinear form (3.36) is continuous and coercive under the hypotheses listed in the text on the coefficients.
   [*Solution:* the bilinear form is obviously continuous. Thanks to the hypothesis (3.35) and to the fact that $\sigma \in \mathrm{L}^\infty(\Omega)$ is positive a.e. in Ω, it is also coercive as

$$a(v,v) \geq \alpha |v|^2_{\mathrm{H}^1(\Omega)} + \alpha_0 \|v\|^2_{\mathrm{L}^2(\Omega)} \geq \min(\alpha, \alpha_0)\|v\|^2_V \quad \forall v \in V.$$

We point out that if $V = \mathrm{H}^1(\Omega)$ then the condition $\alpha_0 > 0$ is necessary for the bilinear form to be coercive. In the case where $V = \mathrm{H}^1_0(\Omega)$, it is sufficient that $\alpha_0 > -\alpha/C^2_\Omega$, C_Ω being the constant intervening in the Poincaré inequality. In this case, the equivalence between $\|\cdot\|_{\mathrm{H}^1(\Omega)}$ and $|\cdot|_{\mathrm{H}^1(\Omega)}$ can indeed be exploited. We have denoted by $|v|_{\mathrm{H}^1(\Omega)} = \|\nabla v\|_{\mathrm{L}^2(\Omega)}$ the seminorm of v in $\mathrm{H}^1(\Omega)$ (see Example 2.11 in Chap. 2).]

3. Let $V = \mathrm{H}^1_0(0,1)$, and take $a : V \times V \to \mathbb{R}$ and $F : V \to \mathbb{R}$ defined in the following way:

$$F(v) = \int_0^1 (-1-4x)v(x)\,dx, \quad a(u,v) = \int_0^1 (1+x)u'(x)v'(x)\,dx.$$

Prove that the problem: find $u \in V$ such that $a(u,v) = F(v)\ \forall v \in V$, admits a unique solution. Moreover, verify that this solution coincides with $u(x) = x^2 - x$.
[*Solution:* it can be easily shown that the bilinear form is continuous and coercive in V. Then, since F is a linear and continuous functional, by the Lax-Milgram lemma we can conclude that there exists a unique solution in V. We verify that the latter is indeed $u(x) = x^2 - x$. The latter function belongs for sure to V (since it is continuous and differentiable and such that $u(0) = u(1) = 0$). Moreover, from the relation

$$\int_0^1 (1+x)u'(x)v'(x)\,dx = -\int_0^1 ((1+x)u'(x))'v(x)\,dx = \int_0^1 (-1-4x)v(x)\,dx,$$

valid $\forall v \in V$, we deduce that in order for u to be a solution we must have $((1+x)u'(x))' = 1+4x$ almost everywhere in $(0,1)$. Such property holds for the proposed u.]

4. Find the weak formulation of the problem

$$\begin{cases} \Delta^2 u = f & \text{in } \Omega, \\ u = 0 & \text{on } \partial\Omega, \\ \dfrac{\partial u}{\partial n} = 0 & \text{on } \partial\Omega, \end{cases}$$

$\Omega \subset \mathbb{R}^2$ being a bounded open set with regular boundary $\partial\Omega$, $\Delta^2 \cdot = \Delta\Delta\cdot$ the *bilaplacian* operator and $f \in \mathrm{L}^2(\Omega)$ an assigned function.
[*Solution:* the weak formulation, obtained by applying Green's formula twice to the bilaplacian operator, is

$$\text{find } u \in \mathrm{H}^2_0(\Omega): \quad \int_\Omega \Delta u \Delta v\,d\Omega = \int_\Omega fv\,d\Omega \quad \forall v \in \mathrm{H}^2_0(\Omega), \tag{3.50}$$

where $\mathrm{H}^2_0(\Omega) = \{v \in \mathrm{H}^2(\Omega) : v = 0, \partial v/\partial n = 0 \text{ on } \partial\Omega\}$.]

5. For each function v of the Hilbert space $\mathrm{H}^2_0(\Omega)$, defined in Exercise 4, it can be shown that the seminorm $|\cdot|_{\mathrm{H}^2(\Omega)}$ defined as $|v|_{\mathrm{H}^2(\Omega)} = (\int_\Omega |\Delta v|^2\,d\Omega)^{1/2}$ is in fact

equivalent to the norm $\|\cdot\|_{H^2(\Omega)}$. Using such property, prove that problem (3.50) admits a unique solution.
[*Solution:* let us set $V = H_0^2(\Omega)$. Then

$$a(u,v) = \int_\Omega \Delta u \Delta v \, d\Omega \quad \text{and } F(v) = \int_\Omega f v \, d\Omega$$

are a bilinear form from $V \times V \to \mathbb{R}$ and a linear and continuous functional, respectively. To prove existence and uniqueness it is sufficient to invoke the Lax-Milgram lemma as the bilinear form is coercive and continuous. Indeed, thanks to the equivalence between norm and seminorm, there exist two positive constants α and M such that

$$a(u,u) = |u|_V^2 \geq \alpha \|u\|_V^2, \quad |a(u,v)| \leq M \|u\|_V \|v\|_V.]$$

6. Write the weak formulation of the fourth-order problem

$$\begin{cases} -\text{div}\,(\mu \nabla u) + \Delta^2 u + \sigma u = 0 & \text{in } \Omega, \\ u = 0 & \text{on } \partial\Omega, \\ \dfrac{\partial u}{\partial n} = 0 & \text{on } \partial\Omega, \end{cases}$$

by introducing appropriate functional spaces, knowing that $\Omega \subset \mathbb{R}^2$ is a bounded open set with regular boundary $\partial\Omega$ and that $\mu(\mathbf{x})$ and $\sigma(\mathbf{x})$ are known functions defined on Ω.
[*Solution:* proceed as in the two previous exercises by supposing that the coefficients μ and σ lie in $L^\infty(\Omega)$.]

7. Let $\Omega \subset \mathbb{R}^2$ be a domain with a smooth boundary $\partial\Omega = \Gamma_D \cup \Gamma_N$ and $\mathring{\Gamma}_D \cap \mathring{\Gamma}_N = \emptyset$. By introducing appropriate functional spaces, find the weak formulation of the following linear elasticity problem

$$\begin{cases} -\displaystyle\sum_{j=1}^{2} \frac{\partial}{\partial x_j} \sigma_{ij}(\mathbf{u}) = f_i & \text{in } \Omega, \quad i = 1,2, \\ u_i = 0 & \text{on } \Gamma_D, \quad i = 1,2, \\ \displaystyle\sum_{j=1}^{2} \sigma_{ij}(\mathbf{u}) n_j = g_i & \text{on } \Gamma_N, \quad i = 1,2, \end{cases} \tag{3.51}$$

having denoted as usual by $\mathbf{n} = (n_1, n_2)^T$ the outward unit normal vector to $\partial\Omega$, by $\mathbf{u} = (u_1, u_2)^T$ the unknown vector, and by $\mathbf{f} = (f_1, f_2)^T$ and $\mathbf{g} = (g_1, g_2)^T$ two assigned vector functions. Moreover, it has been set for $i, j = 1,2$,

$$\sigma_{ij}(\mathbf{u}) = \lambda \text{div}(\mathbf{u}) \delta_{ij} + 2\mu \varepsilon_{ij}(\mathbf{u}), \quad \varepsilon_{ij}(\mathbf{u}) = \frac{1}{2}\left(\frac{\partial u_i}{\partial x_j} + \frac{\partial u_j}{\partial x_i} \right),$$

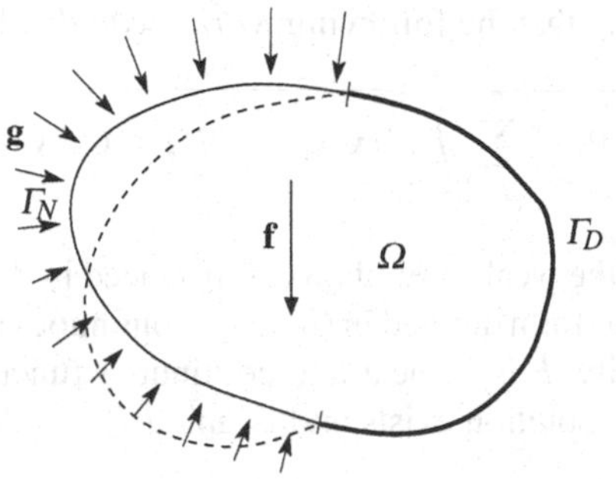

Fig. 3.4. A partially constrained body subject to the action of an external charge

λ and μ being two positive constants and δ_{ij} the Kronecker symbol. The system (3.51) allows to describe the displacement $\mathbf{u}$ of an elastic body, homogeneous and isotropic, that occupies in its equilibrium position the region Ω, under the action of an external body force whose density is $\mathbf{f}$ and of a surface charge distributed on Γ_N with intensity $\mathbf{g}$ (see Fig. 3.4).
[*Solution*: the weak formulation of (3.51) can be found by observing that $\sigma_{ij} = \sigma_{ji}$ and by using the following Green formula

$$\begin{aligned}\sum_{i,j=1}^{2} \int_{\Omega} \sigma_{ij}(\mathbf{u})\varepsilon_{ij}(\mathbf{v})\, d\Omega &= \sum_{i,j=1}^{2} \int_{\partial\Omega} \sigma_{ij}(\mathbf{u}) n_j v_i \, d\gamma \\ &- \sum_{i,j=1}^{2} \int_{\Omega} \frac{\partial \sigma_{ij}(\mathbf{u})}{\partial x_j} v_i \, d\Omega.\end{aligned} \tag{3.52}$$

By assuming $\mathbf{v} \in V = (\mathrm{H}^1_{\Gamma_D}(\Omega))^2$ (the space of vectorial functions that have components $v_i \in \mathrm{H}^1_{\Gamma_D}(\Omega)$ for $i = 1, 2$), the weak formulation reads

$$\text{find } \mathbf{u} \in V \text{ such that } a(\mathbf{u}, \mathbf{v}) = F(\mathbf{v}) \ \forall \mathbf{v} \in V,$$

with

$$a(\mathbf{u}, \mathbf{v}) = \int_{\Omega} \lambda \operatorname{div}(\mathbf{u}) \operatorname{div}(\mathbf{v})\, d\Omega + 2\mu \sum_{i,j=1}^{2} \int_{\Omega} \varepsilon_{ij}(\mathbf{u})\varepsilon_{ij}(\mathbf{v})\, d\Omega,$$

$$F(\mathbf{v}) = \int_{\Omega} \mathbf{f} \cdot \mathbf{v}\, d\Omega + \int_{\Gamma_N} \mathbf{g} \cdot \mathbf{v}\, d\gamma.$$

In order for the integrals to make sense, it will be sufficient to require $\mathbf{f} \in (\mathrm{L}^2(\Omega))^2$ and $\mathbf{g} \in (\mathrm{L}^2(\Gamma_N))^2$.]

8. Prove, by applying the Lax-Milgram Lemma, that the solution of the weak formulation (3.52) exists and is unique under appropriate conditions on the regularity of

the data and knowing that the following *Korn inequality* holds:

$$\exists C_0 > 0 : \sum_{i,j=1}^{2} \int_{\Omega} \varepsilon_{ij}(\mathbf{v})\varepsilon_{ij}(\mathbf{v})\, d\Omega \geq C_0 \|\mathbf{v}\|_V^2 \quad \forall \mathbf{v} \in V.$$

[*Solution:* consider the weak formulation introduced in the solution to the previous exercise. The bilinear form defined in (3.52) is continuous and also coercive because of the Korn inequality. F is a linear and continuous functional; hence, by the Lax-Milgram lemma, the solution exists and is unique.]

4

The Galerkin finite element method for elliptic problems

In this chapter we describe the numerical solution of the elliptic boundary-value problems considered in Chapter 3 by introducing the Galerkin method. We then illustrate the finite element method as a particular case. The latter will be further developed in the following chapters.

4.1 Approximation via the Galerkin method

As seen in Chapter 3.2, the weak formulation of a generic elliptic problem set on a domain $\Omega \subset \mathbb{R}^d$, $d = 1,2,3$, can be written in the following way

$$\text{find } u \in V: \quad a(u,v) = F(v) \quad \forall v \in V, \tag{4.1}$$

V being an appropriate Hilbert space, subspace of $\mathrm{H}^1(\Omega)$, $a(\cdot,\cdot)$ being a continuous and coercive bilinear form from $V \times V$ in $\mathbb{R}$, $F(\cdot)$ being a continuous linear functional from V in $\mathbb{R}$. Under such hypotheses, the Lax-Milgram Lemma of Sect. 3.4.1 ensures existence and uniqueness of the solution.

Let V_h be a family of spaces that depends on a positive parameter h, such that

$$V_h \subset V, \quad \dim V_h = N_h < \infty \quad \forall h > 0.$$

The approximate problem takes the form

$$\text{find } u_h \in V_h: \quad a(u_h, v_h) = F(v_h) \quad \forall v_h \in V_h, \tag{4.2}$$

and is called *Galerkin problem*. Denoting with $\{\varphi_j, j = 1,2,\dots,N_h\}$ a basis of V_h, it suffices that (4.2) be verified for each function of the basis, as all the functions in the space V_h are a linear combination of the φ_j. We will then require that

$$a(u_h, \varphi_i) = F(\varphi_i), \quad i = 1,2,\dots,N_h. \tag{4.3}$$

A. Quarteroni: *Numerical Models for Differential Problems*, 2nd Ed.
MS&A – Modeling, Simulation & Applications 8
DOI 10.1007/978-88-470-5522-3_4, © Springer-Verlag Italia 2014

Obviously, since $u_h \in V_h$,

$$u_h(\mathbf{x}) = \sum_{j=1}^{N_h} u_j \varphi_j(\mathbf{x}),$$

where the $u_j, j = 1,\dots,N_h$, are unknown coefficients. Equations (4.3) then become

$$\sum_{j=1}^{N_h} u_j \, a(\varphi_j, \varphi_i) = F(\varphi_i), \quad i = 1,2,\dots,N_h. \tag{4.4}$$

We denote by A the matrix (called *stiffness* matrix) with elements

$$a_{ij}=a(\varphi_j, \varphi_i)$$

and by $\mathbf{f}$ the vector with components $f_i = F(\varphi_i)$. If we denote by $\mathbf{u}$ the vector having as components the unknown coefficients u_j, (4.4) is equivalent to the linear system

$$\mathrm{A}\mathbf{u} = \mathbf{f}. \tag{4.5}$$

We point out some characteristics of the stiffness matrix that are independent of the basis chosen for V_h, but exclusively depend on the properties of the weak problem that is being approximated. Other properties, instead, such as the condition number or the sparsity structure, depend on the basis under exam and are therefore addressed in the sections dedicated to the specific numerical methods. For instance, bases formed by functions with small support are appealing, as all the elements a_{ij} relating to basis functions having supports with empty intersections will be null. More in general, from a computational viewpoint, the most convenient choices of V_h will be the ones requiring a modest computational effort for the computation of the matrix elements as well as the source term $\mathbf{f}$.

Theorem 4.1. *The matrix* A *associated to the Galerkin discretization of an elliptic problem whose bilinear form is coercive is positive definite.*

Proof. We recall that a matrix $\mathrm{B} \in \mathbb{R}^{n\times n}$ is said to be positive definite if

$$\mathbf{v}^T\mathrm{B}\mathbf{v} \geq 0 \quad \forall \mathbf{v} \in \mathbb{R}^n \quad \text{and also } \mathbf{v}^T\mathrm{B}\mathbf{v} = 0 \Leftrightarrow \quad \mathbf{v} = \mathbf{0}. \tag{4.6}$$

The correspondence

$$\mathbf{v} = (v_i) \in \mathbb{R}^{N_h} \leftrightarrow v_h(x) = \sum_{j=1}^{N_h} v_j \phi_j \in V_h \tag{4.7}$$

defines a bijection between the spaces $\mathbb{R}^{N_h}$ and V_h . Given a generic vector $\mathbf{v} = (v_i)$ of $\mathbb{R}^{N_h}$, thanks to the bilinearity and coercivity of the form $a(\cdot,\cdot)$, we obtain

$$\begin{aligned}
\mathbf{v}^T \mathrm{A}\mathbf{v} &= \sum_{j=1}^{N_h}\sum_{i=1}^{N_h} v_i a_{ij} v_j = \sum_{j=1}^{N_h}\sum_{i=1}^{N_h} v_i a(\varphi_j,\varphi_i) v_j \\
&= \sum_{j=1}^{N_h}\sum_{i=1}^{N_h} a(v_j\varphi_j, v_i\varphi_i) = a\left(\sum_{j=1}^{N_h} v_j\varphi_j, \sum_{i=1}^{N_h} v_i\varphi_i\right) \\
&= a(v_h,v_h) \geq \alpha \|v_h\|_V^2 \geq 0.
\end{aligned}$$

Moreover, if $\mathbf{v}^T \mathrm{A}\mathbf{v} = 0$, then, by what we have just obtained, $\|v_h\|_V^2 = 0$ too, i.e. $v_h = 0$ and so $\mathbf{v} = \mathbf{0}$. Consequently the claim is proved, as the two conditions in (4.6) are fulfilled. ◇

Furthermore, the following property can be proved (see Exercise 4):

Property 4.1. *The matrix* A *is symmetric if and only if the bilinear form* $a(\cdot,\cdot)$ *is symmetric.*

For instance, in the case of the Poisson problem with either Dirichlet (3.18) or mixed (3.27) boundary conditions, the matrix A is symmetric and positive definite. The numerical solution of such a system can be efficiently performed using both direct methods such as the Cholesky factorization, and iterative methods such as the conjugate gradient method (see Chap. 7 and, e.g., [QSS07, Chap. 4]).

4.2 Analysis of the Galerkin method

In this section, we aim at studying the Galerkin method, and in particular at verifying three of its fundamental properties:

- *existence* and *uniqueness* of the discrete solution u_h;
- *stability* of the discrete solution u_h;
- *convergence* of u_h to the exact solution u of problem (4.1), as $h \to 0$.

4.2.1 Existence and uniqueness

The Lax-Milgram Lemma stated in Sect. 3.4.1 holds for any Hilbert space, hence in particular for the space V_h, as the latter is a closed subspace of the Hilbert space V. Furthermore, the bilinear form $a(\cdot,\cdot)$ and the functional $F(\cdot)$ are the same as in the variational problem (4.1). The hypotheses required by the Lemma are therefore fulfilled. The following result then follows:

Corollary 4.1. *The solution of the Galerkin problem* (4.2) *exists and is unique.*

It is nonetheless instructive to provide a constructive proof of this corollary without using the Lax-Milgram Lemma. As we have seen, in fact, the Galerkin problem (4.2) is equivalent to the linear system (4.5). Proving the existence and uniqueness for one means to prove automatically the existence and uniqueness of the other. We therefore focus our attention on the linear system (4.5).

The matrix A is invertible as the unique solution of system A$\mathbf{u} = \mathbf{0}$ is the identically null solution. This immediately descends from the fact that A is positive definite. Consequently, the linear system (4.5) admits a unique solution, hence also its corresponding Galerkin problem admits a unique solution.

4.2.2 Stability

Corollary 3.1 allows us to provide the following stability result.

Corollary 4.2. *The Galerkin method is stable, uniformly with respect to h, by virtue of the following upper bound for the solution*

$$\|u_h\|_V \leq \frac{1}{\alpha}\|F\|_{V'}.$$

The stability of the method guarantees that the norm $\|u_h\|_V$ of the discrete solution remains bounded for h tending to zero, uniformly with respect to h. Equivalently, it guarantees that $\|u_h - w_h\|_V \leq \frac{1}{\alpha}\|F - G\|_{V'}$, u_h and w_h being numerical solutions corresponding to two different data F and G.

4.2.3 Convergence

We now want to prove that the weak solution of the Galerkin problem converges to the solution of the weak problem (4.1) when h tends to zero. Consequently, by taking a sufficiently small h, it will be possible to approximate the exact solution u by the Galerkin solution u_h as accurately as desired.

Let us first prove the following consistency property.

Lemma 4.1 (Galerkin orthogonality). *The solution u_h of the Galerkin method satisfies*

$$a(u - u_h, v_h) = 0 \quad \forall v_h \in V_h. \tag{4.8}$$

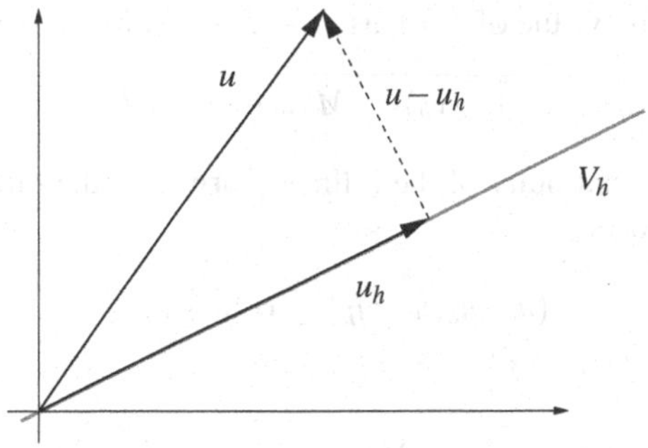

Fig. 4.1. Geometric interpretation of the Galerkin orthogonality

Proof. Since $V_h \subset V$, the exact solution u satisfies the weak problem (4.1) for each element $v = v_h \in V_h$, hence we have

$$a(u, v_h) = F(v_h) \quad \forall v_h \in V_h. \tag{4.9}$$

By subtracting side by side (4.2) from (4.9), we obtain

$$a(u, v_h) - a(u_h, v_h) = 0 \quad \forall v_h \in V_h,$$

from which, thanks to the bilinearity of the form $a(\cdot,\cdot)$, the claim follows. $\diamond$

Let us point out that (4.9) coincides with the definition of *strong consistency* given in (1.10).

Property (4.8) is known as Galerkin orthogonality for the following reason. If $a(\cdot,\cdot)$ is symmetric, it defines a scalar product in V. Then (4.8) is interpreted as the orthogonality condition with respect to the scalar product $a(\cdot,\cdot)$, between the approximation error, $u - u_h$, and the subspace V_h. Borrowing terminology from the Euclidean case, the solution u_h of the Galerkin method is said to be the *orthogonal projection* on V_h of the exact solution u. Among all elements of V_h, v_h is the one minimizing the distance to the exact solution u in the *energy norm*, i.e. in the following norm induced by the scalar product $a(\cdot,\cdot)$:

$$\|u - u_h\|_a = \sqrt{a(u - u_h, u - u_h)}.$$

Remark 4.1. The geometric interpretation of the Galerkin method makes sense only in the case where the form $a(\cdot,\cdot)$ is symmetric. However, this does not impair the generality of the method or its consistency property in the case where the bilinear form is not symmetric. •

Let us now consider the value taken by the bilinear form when both its arguments are equal to $u - u_h$. If v_h is an arbitrary element of V_h we obtain

$$a(u - u_h, u - u_h) = a(u - u_h, u - v_h) + a(u - u_h, v_h - u_h).$$

The last term is null by virtue of (4.8), as $v_h - u_h \in V_h$. Moreover

$$|a(u-u_h, u-v_h)| \leq M\|u-u_h\|_V \|u-v_h\|_V,$$

having exploited the continuity of the bilinear form. On the other hand, by the coercivity of $a(\cdot,\cdot)$ it follows

$$a(u-u_h, u-u_h) \geq \alpha\|u-u_h\|_V^2,$$

hence we have

$$\|u-u_h\|_V \leq \frac{M}{\alpha}\|u-v_h\|_V \quad \forall v_h \in V_h.$$

Such inequality holds for all functions $v_h \in V_h$ and therefore we find

$$\|u-u_h\|_V \leq \frac{M}{\alpha} \inf_{w_h \in V_h} \|u-w_h\|_V. \tag{4.10}$$

This fundamental property of the Galerkin method is known as *Céa Lemma*.
It is then evident that in order for the method to converge, it will be sufficient to require that, for h tending to zero, the space V_h tends to "fill" the entire space V. Precisely, it must turn out that

$$\lim_{h\to 0} \inf_{v_h \in V_h} \|v-v_h\|_V = 0 \quad \forall v \in V. \tag{4.11}$$

In that case, the Galerkin method is convergent and it can be written that

$$\lim_{h\to 0} \|u-u_h\|_V = 0.$$

The space V_h must therefore be carefully chosen in order to guarantee the density property (4.11). Once this requirement is satisfied, convergence will be verified in any case, independently of how u looks like; conversely, the speed with which the discrete solution converges to the exact solution, i.e. the order of decay of the error with respect to h, will depend, in general, on both the choice of V_h and the regularity of u (see Theorem 4.3).

Remark 4.2. Obviously, $\inf_{v_h \in V_h} \|u-v_h\|_V \leq \|u-u_h\|_V$. Consequently, by (4.10), if $\frac{M}{\alpha}$ is has order 1, the error due to the Galerkin method can be identified with the best approximation error for u in V_h. In any case, both errors have the same infinitesimal order with respect to h. •

Remark 4.3. In the case where $a(\cdot,\cdot)$ is a symmetric bilinear form, and also continuous and coercive, then (4.10) can be improved as follows (see Exercise 5)

$$\|u-u_h\|_V \leq \sqrt{\frac{M}{\alpha}} \inf_{w_h \in V_h} \|u-w_h\|_V. \tag{4.12}$$

•

4.3 The finite element method in the one-dimensional case

Let us suppose that Ω is an interval (a,b). The goal of this section is to create approximations of the space $\mathrm{H}^1(a,b)$ that depend on a parameter h. To this end, we introduce a partition $\mathcal{T}_h$ of (a,b) in $N+1$ subintervals $K_j = (x_{j-1}, x_j)$, also called *elements*, having width $h_j = x_j - x_{j-1}$ with

$$a = x_0 < x_1 < \ldots < x_N < x_{N+1} = b, \tag{4.13}$$

and set $h = \max_j h_j$.
Since the functions of $\mathrm{H}^1(a,b)$ are continuous functions on $[a,b]$, we can construct the following family of spaces

$$X_h^r = \left\{ v_h \in C^0\left(\overline{\Omega}\right) : v_h|_{K_j} \in \mathbb{P}_r \, \forall K_j \in \mathcal{T}_h \right\}, \quad r = 1, 2, \ldots \tag{4.14}$$

having denoted by $\mathbb{P}_r$ the space of polynomials with degree lower than or equal to r in the variable x. The spaces X_h^r are all subspaces of $\mathrm{H}^1(a,b)$, as they are constituted by differentiable functions except for at most a finite number of points (the vertices x_i of the partition $\mathcal{T}_h$). They represent possible choices for the space V_h, provided that the boundary conditions are properly incorporated. The fact that the functions of X_h^r are locally (element-wise) polynomials will make the stiffness matrix easy to compute.

We must now choose a basis $\{\varphi_i\}$ for the X_h^r space. It is convenient, by what exposed in Sect. 4.1, that the support of the generic basis function φ_i have non-empty intersection only with the support of a negligible number of other functions of the basis. In such way, many elements of the stiffness matrix will be null. It is also convenient that the basis be *Lagrangian*: in that case, the coefficients of the expansion of a generic function $v_h \in X_h^r$ in the basis itself will be the values taken by v_h at carefully chosen points, which we call *nodes* and which, as we will see, generally form a superset of the vertices of $\mathcal{T}_h$. This does not prevent the use of non-Lagrangian bases, especially in their hierarchical version (as we will see later). We now provide some examples of bases for the spaces X_h^1 and X_h^2.

4.3.1 The space X_h^1

This space is constituted by the piecewise continuous and linear functions on a partition $\mathcal{T}_h$ of (a,b) of the form (4.13). Since only one straight line can pass through two different points and the functions of X_h^1 are continuous, the *degrees of freedom* of the functions of this space, i.e. the values that must be assigned to define uniquely the functions themselves, will be equal to the number $N+2$ of vertices of the partition. In this case, therefore, nodes and vertices coincide. Consequently, having assigned $N+2$ basis functions φ_i, $i = 0, \ldots, N+1$, the whole space X_h^1 will be completely defined. The characteristic Lagrangian basis functions are characterized by the following property

$$\varphi_i \in X_h^1 \quad \text{such that} \quad \varphi_i(x_j) = \delta_{ij}, \quad i, j = 0, 1, \ldots, N+1,$$

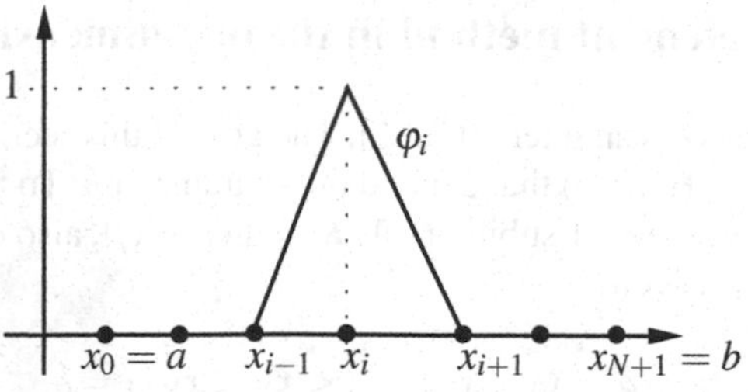

Fig. 4.2. The basis function of X_h^1 associated to node x_i

δ_{ij} being the Kronecker delta. The function φ_i is therefore piecewise linear and equal to one at x_i and zero at the remaining nodes of the partition (see Fig. 4.2). Its expression is given by

$$\varphi_i(x) = \begin{cases} \dfrac{x - x_{i-1}}{x_i - x_{i-1}} & \text{for } x_{i-1} \le x \le x_i, \\ \dfrac{x_{i+1} - x}{x_{i+1} - x_i} & \text{for } x_i \le x \le x_{i+1}, \\ 0 & \text{otherwise.} \end{cases} \tag{4.15}$$

Obviously φ_i has as support the union of the intervals $[x_{i-1}, x_i]$ and $[x_i, x_{i+1}]$ only, if $i \neq 0$ or $i \neq N+1$ (for $i = 0$ or $i = N+1$ the support will be limited to the interval $[x_0, x_1]$ or $[x_N, x_{N+1}]$, respectively). Consequently, the only basis functions whose support overlaps with that of φ_i are φ_{i-1} and φ_{i+1} (and, of course, φ_i). Hence the stiffness matrix is tridiagonal as $a_{ij} = 0$ if $j \notin \{i-1, i, i+1\}$.

As visible in expression (4.15), the two basis functions φ_i and φ_{i+1} defined on each interval $[x_i, x_{i+1}]$ basically repeat themselves with no changes, up to a scaling factor linked to the length of the interval itself. In practice, the two basis functions φ_i and φ_{i+1} can be obtained by transforming two basis functions $\widehat{\varphi}_0$ and $\widehat{\varphi}_1$ built once and for all on a reference interval, typically the $[0,1]$ interval.

To this end, it is sufficient to exploit the fact that the generic interval (x_i, x_{i+1}) of the partition of (a,b) can be obtained starting from the interval $(0,1)$ via the linear transformation $\phi : [0,1] \to [x_i, x_{i+1}]$ defined as

$$x = \phi(\xi) = x_i + \xi(x_{i+1} - x_i). \tag{4.16}$$

If we define the two basis functions $\widehat{\varphi}_0$ and $\widehat{\varphi}_1$ on $[0,1]$ as

$$\widehat{\varphi}_0(\xi) = 1 - \xi, \quad \widehat{\varphi}_1(\xi) = \xi,$$

the basis functions φ_i and φ_{i+1} on $[x_i, x_{i+1}]$ will simply be given by

$$\varphi_i(x) = \widehat{\varphi}_0(\xi(x)), \quad \varphi_{i+1}(x) = \widehat{\varphi}_1(\xi(x))$$

since $\xi(x) = (x - x_i)/(x_{i+1} - x_i)$ (see Figs. 4.3 and 4.4).

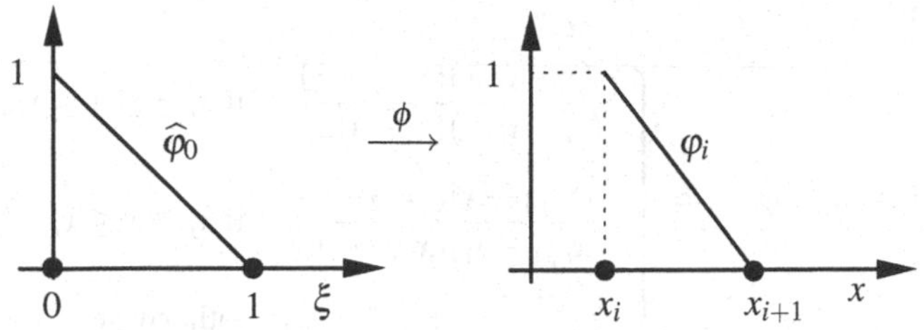

Fig. 4.3. The basis function φ_i in $[x_i, x_{i+1}]$ and the corresponding basis function $\widehat{\varphi}_0$ on the reference element

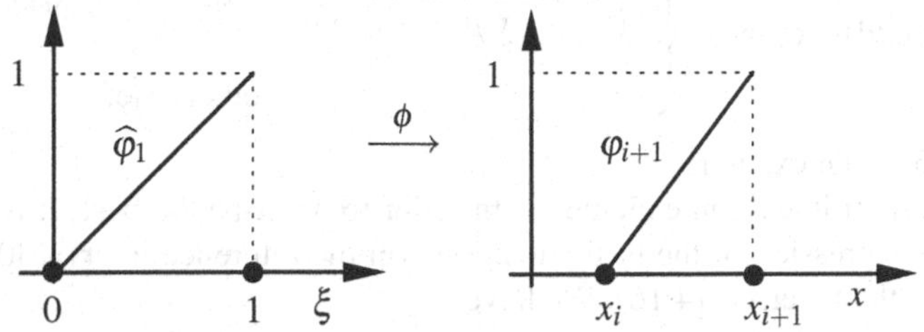

Fig. 4.4. The basis function φ_{i+1} in $[x_i, x_{i+1}]$ and the corresponding basis function $\widehat{\varphi}_1$ on the reference element

This way of proceeding (defining the basis on a reference element and then transforming it on a specific element) will be of fundamental importance when considering problems in several dimensions.

4.3.2 The space X_h^2

The functions of X_h^2 are piecewise polynomials of degree 2 on each interval of $\mathcal{T}_h$ and, consequently, they are determined once the values they take at three distinct points of each interval K_j are assigned. To guarantee the continuity of the functions of X_h^2 two of these points will be the endpoints of the generic interval of $\mathcal{T}_h$, the third will be the midpoint of the latter. The degrees of freedom of the space X_h^2 are therefore the values of v_h taken at the endpoints of the intervals composing the partition $\mathcal{T}_h$ and at their midpoints. We order the nodes starting from $x_0 = a$ to $x_{2N+2} = b$; in such way the midpoints correspond to the nodes with odd indices, and the endpoints to the nodes with even indices (refer to Exercise 6 for alternative numberings).

Exactly as in the previous case the Lagrangian basis for X_h^2 is the one formed by the functions

$$\varphi_i \in X_h^2 \quad \text{such that} \quad \varphi_i(x_j) = \delta_{ij}, \quad i, j = 0, 1, \ldots, 2N+2.$$

These are therefore piecewise quadratic functions that are equal to 1 at the node to which they are associated and are null at the remaining nodes. Here is the explicit expression of the generic basis function associated to the endpoints of the intervals in the partition:

$$(i \text{ even}) \quad \varphi_i(x) = \begin{cases} \dfrac{(x-x_{i-1})(x-x_{i-2})}{(x_i-x_{i-1})(x_i-x_{i-2})} & \text{if } x_{i-2} \le x \le x_i, \\[2ex] \dfrac{(x_{i+1}-x)(x_{i+2}-x)}{(x_{i+1}-x_i)(x_{i+2}-x_i)} & \text{if } x_i \le x \le x_{i+2}, \\[2ex] 0 & \text{otherwise.} \end{cases}$$

For the midpoints of the intervals, we have

$$(i \text{ odd}) \quad \varphi_i(x) = \begin{cases} \dfrac{(x_{i+1}-x)(x-x_{i-1})}{(x_{i+1}-x_i)(x_i-x_{i-1})} & \text{if } x_{i-1} \le x \le x_{i+1}, \\[2ex] 0 & \text{otherwise.} \end{cases}$$

See Fig. 4.5 for an example.

As in the case of linear finite elements, in order to describe the basis it is sufficient to provide the expression of the basis functions on the reference interval $[0,1]$ and then to transform the latter via (4.16). We have

$$\widehat{\varphi}_0(\xi) = (1-\xi)(1-2\xi), \quad \widehat{\varphi}_1(\xi) = 4(1-\xi)\xi, \quad \widehat{\varphi}_2(\xi) = \xi(2\xi-1).$$

We represent these functions in Fig. 4.5. Note that the generic basis function φ_{2i+1} relative to node x_{2i+1} has a support coinciding with the element to which the midpoint belongs. For its peculiar form, it is known as *bubble function*.

As previously anticipated, we can also introduce other non-Lagrangian bases. A particularly interesting one is the one constructed (locally) by the three functions

$$\widehat{\psi}_0(\xi) = 1-\xi, \quad \widehat{\psi}_1(\xi) = \xi, \quad \widehat{\psi}_2(\xi) = (1-\xi)\xi.$$

A basis of this kind is said to be *hierarchical* because, to construct the basis for X_h^2, it exploits the basis functions of the lower-dimension space, X_h^1. It is convenient from a computational viewpoint if one decides, during the approximation of a problem, to increase only locally, i.e. only for such elements, the degree of interpolation (that is if one intends to perform the so-called adaptivity in the degree, or *adaptivity of type p*).

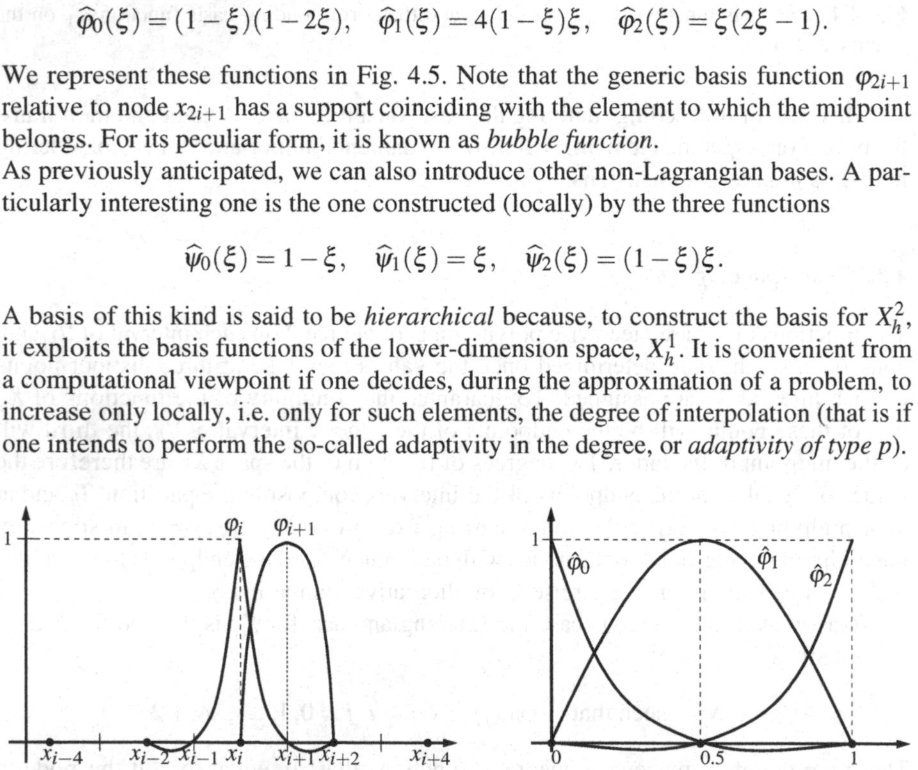

Fig. 4.5. The basis functions of X_h^2 (on the left) and the corresponding functions on the reference interval (on the right)

The Lagrange polynomials are linearly independent by construction. In general, however, such property must be verified to ensure that the set of chosen polynomials is effectively a basis. In the case of the functions $\widehat{\varphi}_0$, $\widehat{\varphi}_1$ and $\widehat{\varphi}_2$ we must verify that

$$\text{if } \alpha_0\widehat{\varphi}_0(\xi)+\alpha_1\widehat{\varphi}_1(\xi)+\alpha_2\widehat{\varphi}_2(\xi)=0\ \forall\xi, \quad \text{then} \quad \alpha_0=\alpha_1=\alpha_2=0.$$

Indeed, the equation

$$\alpha_0\widehat{\varphi}_0(\xi)+\alpha_1\widehat{\varphi}_1(\xi)+\alpha_2\widehat{\varphi}_2(\xi)=\alpha_0+\xi(\alpha_1-\alpha_0+\alpha_2)-\alpha_2\xi^2=0$$

implies $\alpha_0=0$, $\alpha_2=0$ and therefore $\alpha_1=0$. We notice that the stiffness matrix in the case of finite elements of degree 2 will be pentadiagonal.

By proceeding in the same way it will be possible to generate bases for X_h^r with an arbitrary positive integer r: we point out, however, that as the polynomial degree increases, the number of degrees of freedom increases and so does the computational cost of solving the linear system (4.5). Moreover, a well known fact from polynomial interpolation theory, the use of high degrees combined with equispaced node distributions leads to increasingly less stable approximations, in spite of the theoretical increase in accuracy. A successful remedy is provided by the spectral element approximation that, using well-chosen nodes (the ones from the Gaussian quadrature), allows to generate approximations with arbitrarily high accuracy. To this purpose see Chap. 10.

4.3.3 The approximation with linear finite elements

We now examine how to approximate the following problem

$$\begin{cases} -u''+\sigma u=f, & a<x<b,\\ u(a)=0, & u(b)=0, \end{cases}$$

whose weak formulation, as we have seen in the previous chapter, is

$$\text{find } u\in \mathrm{H}_0^1(a,b): \quad \int_a^b u'v'\,dx+\int_a^b \sigma uv\,dx=\int_a^b fv\,dx \quad \forall v\in \mathrm{H}_0^1(a,b).$$

As we did in (4.13), we introduce a decomposition $\mathcal{T}_h$ of $(0,1)$ in $N+1$ subintervals K_j and use linear finite elements. We therefore introduce the space

$$V_h=\{v_h\in X_h^1 :\ v_h(a)=v_h(b)=0\}, \tag{4.17}$$

that is the space of piecewise linear functions that vanish at the boundary (a function of such space has been introduced in Fig. 4.6). This is a subspace of $\mathrm{H}_0^1(a,b)$.
The corresponding finite element problem is therefore given by

$$\text{find } u_h\in V_h: \quad \int_a^b u_h'v_h'\,dx+\int_a^b \sigma u_hv_h\,dx=\int_a^b fv_h\,dx \quad \forall v_h\in V_h. \tag{4.18}$$

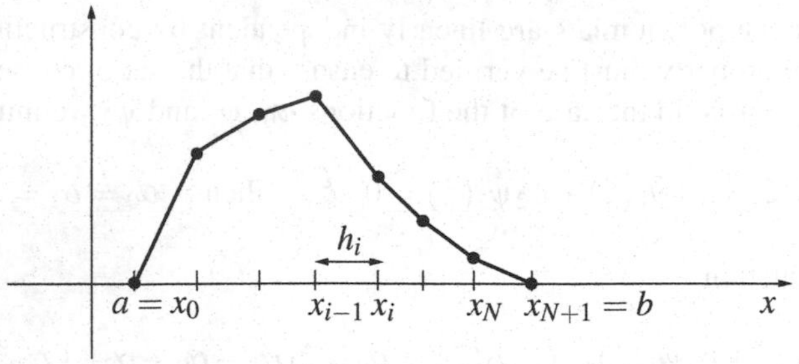

Fig. 4.6. Example of a function of V_h

We use as a basis of X_h^1 the set of hat functions defined in (4.15) by caring to consider only the indices $1 \le i \le N$. By expressing u_h as a linear combination of such functions $u_h(x) = \sum_{i=1}^N u_i \varphi_i(x)$, and imposing that (4.18) is satisfied for each element of the basis of V_h, we obtain a system of N equations

$$\mathbf{A}\mathbf{u} = \mathbf{f}, \tag{4.19}$$

where

$$\mathrm{A} = [a_{ij}], \quad a_{ij} = \int_a^b \varphi_j' \varphi_i' \, dx + \int_a^b \sigma \varphi_j \varphi_i \, dx;$$

$$\mathbf{u} = [u_i]; \quad \mathbf{f} = [f_i], \quad f_i = \int_a^b f \varphi_i \, dx.$$

Note that $u_i = u_h(x_i)$, $1 \le i \le N$, that is the finite element unknowns are the nodal values of the finite element solution u_h.
To find the numerical solution u_h it is now sufficient to solve the linear system (4.19).

In the case of linear finite elements, the stiffness matrix A is not only sparse, but also results to be tridiagonal. To compute its elements, we proceed as follows. As we have seen it is not necessary to operate directly on the basis functions on the single intervals, but it is sufficient to refer to the ones defined on the reference interval: it will then be enough to transform appropriately the integrals that appear in the definition of the coefficients of A.

A generic non-null element of the stiffness matrix is given by

$$a_{ij} = \int_a^b (\varphi_i' \varphi_j' + \sigma \varphi_i \varphi_j) dx = \int_{x_{i-1}}^{x_i} (\varphi_i' \varphi_j' + \sigma \varphi_i \varphi_j) dx + \int_{x_i}^{x_{i+1}} (\varphi_i' \varphi_j' + \sigma \varphi_i \varphi_j) dx.$$

Let us consider the first summand by supposing $j=i-1$. Evidently, via the coordinate transformation (4.16), we can rewrite it as

$$\int_{x_{i-1}}^{x_i} (\varphi_i' \varphi_{i-1}' + \sigma \varphi_i \varphi_{i-1}) dx =$$

$$\int_0^1 [\varphi_i'(x(\xi))\varphi_{i-1}'(x(\xi)) + \sigma(x(\xi))\varphi_i(x(\xi))\varphi_{i-1}(x(\xi))] h_i \, d\xi,$$

having noted that $dx = d(x_{i-1} + \xi h_i) = h_i d\xi$. On the other hand $\varphi_i(x(\xi)) = \widehat{\varphi}_1(\xi)$ and $\varphi_{i-1}(x(\xi)) = \widehat{\varphi}_0(\xi)$. We also note that

$$\frac{d}{dx}\varphi_i(x(\xi)) = \frac{d\xi}{dx}\widehat{\varphi}_1'(\xi) = \frac{1}{h_i}\widehat{\varphi}_1'(\xi).$$

Similarly, we find that $\varphi_{i-1}'(x(\xi)) = (1/h_i)\widehat{\varphi}_0'(\xi)$. Hence, the element $a_{i,i-1}$ becomes

$$a_{i,i-1} = \int_0^1 \left(\frac{1}{h_i}\widehat{\varphi}_1'(\xi)\widehat{\varphi}_0'(\xi) + \sigma \widehat{\varphi}_1(\xi)\widehat{\varphi}_0(\xi) h_i \right) d\xi.$$

The advantage of this expression lies in the fact that in the case of constant coefficients, all the integrals appearing within the matrix A can be computed once and for all. We will see in the multi-dimensional case that this way of proceeding maintains its importance also in the case of variable coefficients.

4.3.4 Interpolation operator and interpolation error

Let us set $I = (a,b)$. For each $v \in C^0(\overline{I})$, we call *interpolant* of v in the space of X_h^1, determined by the partition $\mathcal{T}_h$, the function $\Pi_h^1 v$ such that

$$\Pi_h^1 v(x_i) = v(x_i) \quad \forall x_i \text{ node of the partition, } i = 0, \ldots, N+1.$$

By using the Lagrangian basis $\{\varphi_i\}$ of the space X_h^1, the interpolant can be expressed in the following way

$$\Pi_h^1 v(x) = \sum_{i=0}^{N+1} v(x_i)\varphi_i(x).$$

Hence, when v and a basis of X_h^1 are known, the interpolant of v is easy to compute. The operator $\Pi_h^1 : C^0(\overline{I}) \mapsto X_h^1$ mapping a function v to its interpolant $\Pi_h^1 v$ is called *interpolation operator*.

Analogously, we can define the operators $\Pi_h^r : C^0(\overline{I}) \mapsto X_h^r$, for all $r \geq 1$. Having denoted by $\Pi_{K_j}^r$ the local interpolation operator mapping a function v to the polynomial $\Pi_{K_j}^r v \in \mathbb{P}_r(K_j)$, interpolating v at the $r+1$ nodes of the element $K_j \in \mathcal{T}_h$, we define

$\Pi_h^r v$ as

$$\Pi_h^r v \in X_h^r : \quad \Pi_h^r v\big|_{K_j} = \Pi_K^r (v\big|_{K_j}) \qquad \forall K_j \in \mathcal{T}_h. \tag{4.20}$$

Theorem 4.2. *Let $v \in \mathrm{H}^{r+1}(I)$, for $r \geq 1$, and let $\Pi_h^r v \in X_h^r$ be its interpolating function defined in (4.20). The following estimate of the interpolation error holds*

$$|v - \Pi_h^r v|_{\mathrm{H}^k(I)} \leq C_{k,r} h^{r+1-k} |v|_{\mathrm{H}^{r+1}(I)} \qquad \text{for } k = 0, 1. \tag{4.21}$$

The constants $C_{k,r}$ are independent of v and h. We recall that $\mathrm{H}^0(I) = \mathrm{L}^2(I)$ and that $|\cdot|_{\mathrm{H}^0(I)} = \|\cdot\|_{\mathrm{L}^2(I)}$.

Proof. We prove (4.21) for the case $r = 1$, and refer to [QV94, Chap. 3] or [Cia78] for the more general case. We start by observing that if $v \in H^{r+1}(I)$ then $v \in C^r(I)$. In particular, for $r = 1$, $v \in C^1(I)$. Let us set $e = v - \Pi_h^1 v$. Since $e(x_j) = 0$ for each node x_j, Rolle's theorem allows to conclude that there exist some $\xi_j \in K_j = (x_{j-1}, x_j)$, with $j = 1, \ldots, N+1$, for which we have $e'(\xi_j) = 0$.

$\Pi_h^1 v$ being a linear function in each interval K_j, we obtain that for $x \in K_j$

$$e'(x) = \int_{\xi_j}^{x} e''(s)ds = \int_{\xi_j}^{x} v''(s)ds,$$

from which we deduce that

$$|e'(x)| \leq \int_{x_{j-1}}^{x_j} |v''(s)|ds \qquad \text{for } x \in K_j.$$

Now, by using the Cauchy-Schwarz inequality we obtain

$$|e'(x)| \leq \left(\int_{x_{j-1}}^{x_j} 1^2 ds \right)^{1/2} \left(\int_{x_{j-1}}^{x_j} |v''(s)|^2 ds \right)^{1/2} \leq h^{1/2} \left(\int_{x_{j-1}}^{x_j} |v''(s)|^2 ds \right)^{1/2}. \tag{4.22}$$

Hence,

$$\int_{x_{j-1}}^{x_j} |e'(x)|^2 dx \leq h^2 \int_{x_{j-1}}^{x_j} |v''(s)|^2 ds. \tag{4.23}$$

An upper bound for $e(x)$ can be obtained by noting that, for each $x \in K_j$,

$$e(x) = \int_{x_{j-1}}^{x} e'(s)ds,$$

and therefore, by applying inequality (4.22),

$$|e(x)| \leq \int_{x_{j-1}}^{x_j} |e'(s)|ds \leq h^{3/2} \left(\int_{x_{j-1}}^{x_j} |v''(s)|^2 ds \right)^{1/2}.$$

Hence,

$$\int_{x_{j-1}}^{x_j} |e(x)|^2 dx \leq h^4 \int_{x_{j-1}}^{x_j} |v''(s)|^2 ds. \tag{4.24}$$

By summing over the indices j from 1 to $N+1$ in (4.23) and (4.24) we obtain the inequalities

$$\left(\int_a^b |e'(x)|^2 dx \right)^{1/2} \leq h \left(\int_a^b |v''(x)|^2 dx \right)^{1/2}$$

and

$$\left(\int_a^b |e(x)|^2 dx \right)^{1/2} \leq h^2 \left(\int_a^b |v''(x)|^2 dx \right)^{1/2}$$

respectively, that correspond to the desired estimates (4.21) for $r = 1$, with $C_{k,1} = 1$ and $k = 0, 1$. ◇

4.3.5 Estimate of the finite element error in the H^1 norm

Owing to result (4.21) we can obtain an estimate of the approximation error of the finite element method.

Theorem 4.3. *Let $u \in V$ be the exact solution of the variational problem* (4.1) *(in our case $\Omega = I = (a,b)$) and u_h its approximate solution via the finite element method of degree r, i.e. the solution of problem* (4.2) *where $V_h = X_h^r \cap V$. Moreover, let $u \in \mathrm{H}^{p+1}(I)$, for a suitable p such that $r \leq p$. Then the following inequality, also called a priori error estimate, holds*

$$\|u - u_h\|_V \leq \frac{M}{\alpha} C h^r |u|_{\mathrm{H}^{r+1}(I)}, \tag{4.25}$$

C being a constant independent of u and h.

Proof. From (4.10), by setting $w_h = \Pi_h^r u$, the interpolant of degree r of u in the space V_h, we obtain

$$\|u - u_h\|_V \leq \frac{M}{\alpha} \|u - \Pi_h^r u\|_V.$$

The right-hand side can now be bounded from above via the interpolation error estimate (4.21) for $k = 1$, from which the claim follows. ◇

It follows from the latter theorem that, in order to increase the accuracy, two different strategies can be followed: reducing h, i.e. refining the grid, or increasing r, that is using finite elements of higher degree. However, the latter strategy makes sense only if the solution u is regular enough: as a matter of fact, from (4.25) we immediately

Table 4.1. Order of convergence with respect to h for the finite element method for varying regularity of the solution and degree r of the finite elements. We have highlighted on each column the result corresponding to the "optimal" choice of the polynomial degree

r	$u \in \mathrm{H}^1(I)$	$u \in \mathrm{H}^2(I)$	$u \in \mathrm{H}^3(I)$	$u \in \mathrm{H}^4(I)$	$u \in \mathrm{H}^5(I)$
1	converges	$\boxed{h^1}$	h^1	h^1	h^1
2	converges	h^1	$\boxed{h^2}$	h^2	h^2
3	converges	h^1	h^2	$\boxed{h^3}$	h^3
4	converges	h^1	h^2	h^3	$\boxed{h^4}$

infer that, if $u \in V \cap \mathrm{H}^{p+1}(I)$, the maximum value of r that it makes sense to take is $r = p$. Values higher than p do not ensure a better rate of convergence: therefore if the solution is not very regular it is not convenient to use finite elements of high degree, as the greater computational cost is not compensated by an improvement of the convergence. An interesting case is when the solution only has the minimum regularity ($p = 0$). From the relations (4.10) and (4.11) we obtain that there is anyhow convergence, but estimate (4.25) is no longer valid. It is then impossible to say how the norm V of the error tends to zero when h decreases. We summarize these situations in Table 4.1.

In general, we can state that: if $u \in \mathrm{H}^{p+1}(I)$, for a given $p > 0$, then there exists a constant C independent of u and h, such that

$$\|u - u_h\|_{\mathrm{H}^1(I)} \leq C h^s |u|_{\mathrm{H}^{s+1}(I)}, \quad s = \min\{r, p\}. \tag{4.26}$$

4.4 Finite elements, simplices and barycentric coordinates

Before introducing finite element spaces in 2D and 3D domains we can attempt to provide a formal definition of *finite element*.

4.4.1 An abstract definition of finite element in the Lagrangian case

From the examples we considered we can deduce that there are three ingredients allowing to characterize a finite element in the general case, i.e. independently of the dimension:

- the domain of definition K of the element. In the one-dimensional case it is an interval, in the two-dimensional case it is generally a triangle but it can also be a quadrilateral; in the three-dimensional case it can be a tetrahedron, a prism or a hexahedron;
- a space of polynomials Π_r of dimension N_r defined on K and a basis $\{\varphi_j\}_{j=1}^{N_r}$ of Π_r. In the monodimensional case, Π_r has been introduced in Sect. 4.3 and $N_r = r+1$. For the multidimensional case, see Sect. 4.4.2;

- a set $\Sigma = \{\gamma_i : \Pi_r \to \mathbb{R}\}_{i=1}^{N_r}$ of functionals on Π_r, satisfying $\gamma_i(\varphi_j) = \delta_{ij}$, δ_{ij} being the Kronecker delta. These allow a unique identification of the coefficients $\{\alpha_j\}_{j=1}^{N_r}$ of the expansion of a polynomial $p \in \Pi_r$ with respect to the chosen basis, $p(x) = \sum_{j=1}^{N_r} \alpha_j \varphi_j(x)$. As a matter of fact, we have $\alpha_i = \gamma_i(p)$, $i = 1, \ldots, N_r$. These coefficients are called *degrees of freedom* of the finite element.

In the case of *Lagrange finite elements* the chosen basis is provided by the Lagrange polynomials and the degree of freedom α_i is equal to the value taken by the polynomial p at a point $\mathbf{a}_i$ of K, called *node*, that is we have $\alpha_i = p(\mathbf{a}_i)$, $i = 1, \ldots, N_r$. We can then set, with a slight notation abuse, $\Sigma = \{\mathbf{a}_j\}_{j=1}^{N_r}$, since knowing the position of the nodes allows us to find the degrees of freedom (notice however that this is not true in general, think only of the case of the hierarchical basis introduced previously). In the remainder, we will exclusively refer to the case of Lagrange finite elements.

In the construction of a Lagrange finite element, the choice of nodes is not arbitrary. Indeed, the problem of interpolation on a given set K may be ill posed. For this reason the following definition proves useful:

Definition 4.1. *A set $\Sigma = \{\mathbf{a}_j\}_{j=1}^{N_r}$ of points of K is called unisolvent on Π_r if, given N_r arbitrary scalars α_j, $j = 1, \ldots, N_r$, there exists a unique function $p \in \Pi_r$ such that*

$$p(\mathbf{a}_j) = \alpha_j, \quad j = 1, \ldots, N_r.$$

In such case, the triple (K, Σ, Π_r) is called *Lagrangian finite element*. In the case of Lagrangian finite elements, the element is generally recalled by citing the sole polynomial space: hence the linear finite elements introduced previously are called $\mathbb{P}_1$, the quadratic ones $\mathbb{P}_2$, and so forth.

As we have seen in the 1D case, for the finite elements based on the use of local $\mathbb{P}_1$ and $\mathbb{P}_2$ polynomial spaces, it is convenient to define the finite element starting from a reference element $\widehat{K}$; typically this is the interval $(0,1)$. It will tipically be the right triangle with vertices $(0,0)$, $(1,0)$ and $(0,1)$ in the two-dimensional case (when using triangular elements). (See Sect. 4.4.2 for the case in arbitrary dimensions.) Hence, via a transformation ϕ, we move to the finite element defined on K. The transformation therefore concerns the finite element as a whole. More precisely, we observe that if $(\widehat{K}, \widehat{\Sigma}, \widehat{\Pi}_r)$ is a Lagrangian finite element and $\phi : \widehat{K} \to \mathbb{R}^d$ a continuous and injective map, and we define

$$K = \phi(\widehat{K}), \quad P_r = \{p : K \to \mathbb{R} : \; p \circ \phi \in \widehat{\Pi}_r\}, \quad \Sigma = \phi(\widehat{\Sigma}),$$

then (K, Σ, P_r) is still said to be a Lagrangian finite element. The space of polynomials defined on triangles and tetrahedra can be introduced as follows.

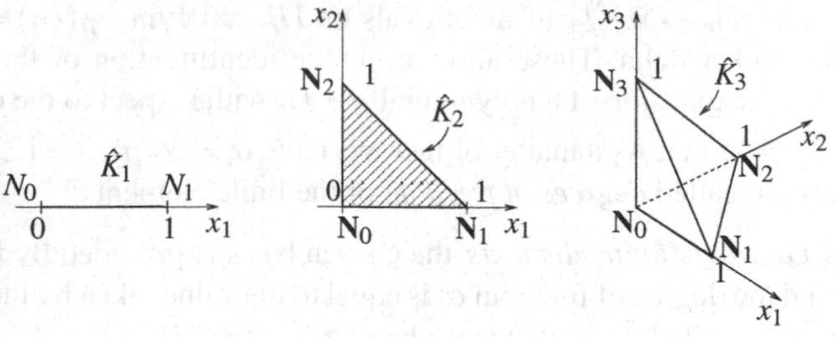

Fig. 4.7. The unitary simplex in $\mathbb{R}^d, d = 1,2,3$

4.4.2 Simplexes

If $\{\mathbf{N}_0, \dots, \mathbf{N}_d\}$ are $d+1$ points in $\mathbb{R}^d$, $d \geq 1$, and the vectors $\{\mathbf{N}_1 - \mathbf{N}_0, \dots, \mathbf{N}_d - \mathbf{N}_0\}$ are linearly independent, then the convex hull of $\{\mathbf{N}_0, \dots, \mathbf{N}_d\}$ is called a *simplex*, and $\{\mathbf{N}_0, \dots, \mathbf{N}_d\}$ area called the *vertices* of the simplex. The *standard simplex* of $\mathbb{R}^d$ is the set

$$\hat{K}_d = \{\mathbf{x} \in \mathbb{R}^d : x_i \geq 0,\ 1 \leq i \leq d,\ \sum_{i=1}^{d} x_i \leq 1\} \tag{4.27}$$

and it is a unit interval in $\mathbb{R}^1$, a unit triangle in $\mathbb{R}^d$, a unit tetrahedron in $\mathbb{R}^d$ (see Fig. 4.7). Its vertices are ordered in such a way that the Cartesian coordinates of $\mathbf{N}_i$ are all null, except the i-th one that is equal to 1. On a d-dimensional simplex, the space of polynomials $\mathbb{P}_r$ is defined as follows

$$\mathbb{P}_r = \{p(\mathbf{x}) = \sum_{\substack{0 \leq i_1, \dots, i_d \\ i_1 + \dots + i_d \leq d}} a_{i_1 \dots i_d} x_1^{i_1} \dots x_d^{i_d};\quad a_{i_1 \dots i_d} \in \mathbb{R}\}. \tag{4.28}$$

Then

$$N_r = \dim \mathbb{P}_r = \binom{r+d}{r} = \frac{1}{d!} \prod_{k=1}^{d} (r+k). \tag{4.29}$$

4.4.3 Barycentric coordinates

For a given simplex K in $\mathbb{R}^d$ (see Sect. 4.5.1) it is sometimes convenient to consider a coordinate frame alternative to the Cartesian one, that of the *barycentric coordinates*. The latter are $d+1$ functions, $\{\lambda_0, \dots, \lambda_d\}$, defined as follows

$$\lambda_i : \mathbb{R}^d \to \mathbb{R},\ \lambda_i(\mathbf{x}) = 1 - \frac{(\mathbf{x} - \mathbf{N}_i) \cdot \mathbf{n}_i}{(\mathbf{N}_j - \mathbf{N}_i) \cdot \mathbf{n}_i}, \qquad 0 \leq i \leq d. \tag{4.30}$$

For every $i = 0, \dots, d$ let F_i denote the *face* of K opposite to $\mathbf{N}_i$; F_i is in fact a vertex if $d = 1$, an edge if $d = 2$, a triangle if $d = 3$. In (4.30), $\mathbf{n}_i$ denotes the outward normal to F_i, while $\mathbf{N}_j$ is an arbitrary vertex belonging to F_i. The definition of λ_i is however independent of which vertex of F_i is chosen.

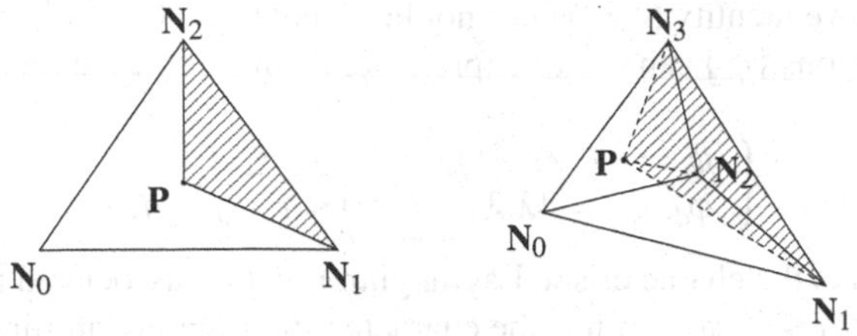

Fig. 4.8. The barycentric coordinate λ_i of the point **P** is the ratio $\frac{|K_i|}{|K|}$ between the measure of the simplex K_i (whose vertices are **P** and $\{\mathbf{N}_j, j \neq i\}$) and that of the given simplex K (a triangle on the left, a tetrahedron on the right). The shadowed simplex is K_0

Barycentric coordinates have a geometrical meaning. Indeed, for every point **P** belonging to K, its barycentric coordinate λ_i, $0 \leq i \leq d$, represents the ratio between the measure of the simplex K_i whose vertices are **P** and the vertices of K sitting on the face F_i opposite to the vertex $\mathbf{N}_i$, and the measure of K. See Fig. 4.8.

Remark 4.4. Let us consider the unitary simplex $\hat{K}_d$, whose vertices $\{\hat{\mathbf{N}}_\mathbf{0}, \ldots, \hat{\mathbf{N}}_\mathbf{d}\}$ are ordered in such a way that all the Cartesian coordinates of $\hat{\mathbf{N}}_\mathbf{i}$ are null, except x_i which is equal to one. Then

$$\lambda_i(\mathbf{x}) = x_i, \quad 1 \leq i \leq d, \quad \lambda_0(\mathbf{x}) = 1 - \sum_{i=1}^{d} \lambda_i(\mathbf{x}). \tag{4.31}$$

The barycentric coordinate λ_i is therefore an affine function that is equal to 1 at $\mathbf{N}_i$ and vanishes on the face F_i opposite to $\mathbf{N}_i$.

On a general simplex K in $\mathbb{R}^d$, the following *partition of unity* property is satisfied

$$0 \leq \lambda_i(\mathbf{x}) \leq 1, \quad \sum_{i=0}^{d} \lambda_i(\mathbf{x}) = 1 \quad \forall \mathbf{x} \in K. \tag{4.32}$$

•

A point **P** belonging to the interior of K has therefore all its barycentric coordinates positive. This property is useful whenever one has to check which triangle in 2D or tetrahedron in 3D a given point belongs to, a situation that occurs when using Lagrangian derivatives (see Sect. 16.7.2) or computing suitable quantities (fluxes, streamlines, etc.) as a post-processing of finite element computations.

A remarkable property is that the center of gravity of K has all its barycentric coordinates equal to $(d+1)^{-1}$. Another remarkable property is that

$$\varphi_i = \lambda_i, \quad 0 \leq i \leq d, \tag{4.33}$$

where $\{\varphi_i, 0 \leq i \leq d\}$ are the characteristic Lagrangian functions on the simplex K of degree $r = 1$, that is

$$\varphi_i \in \mathbb{P}_1(K_d), \quad \varphi_i(\mathbf{N}_j) = \delta_{ij}, \quad 0 \leq j \leq d. \tag{4.34}$$

(See Fig. 4.10, left, for the nodes.)

For $r = 2$ the above identity (4.33) does not hold anymore, however the characteristic Lagrangian functions $\{\varphi_i\}$ can still be expressed in terms of the barycentric coordinates $\{\lambda_i\}$ as follows:

$$\begin{cases} \varphi_i = \lambda_i(2\lambda_i - 1), & 0 \le i \le d, \\ \varphi_{d+i+j} = 4\lambda_i\lambda_j, & 0 \le i < j \le d. \end{cases} \tag{4.35}$$

For $0 \le i \le d$, φ_i is the characteristic Lagrangian function associated to the vertex $\mathbf{N}_i$, while for $0 \le i < j \le d$, φ_{d+i+j} it is the characteristic Lagrangian function associated to the midpoint of the edge whose endpoints are the vertices $\mathbf{N}_i$ and $\mathbf{N}_j$ (see Fig. 4.10, middle).

The previous identities justify the name "coordinates" that is used for the λ_i's. Indeed, if $\mathbf{P}$ is a generic point of the simplex K, its Cartesian coordinates $\{x_j^{(P)}, 1 \le j \le d\}$ can be expressed in terms of the barycentric coordinates $\{\lambda_i^{(P)}, 0 \le i \le d\}$ as follows

$$x_j^{(P)} = \sum_{i=0}^{d} \lambda_i^{(P)} x_j^{(i)}, \, 1 \le j \le d, \tag{4.36}$$

where $\{x_j^{(i)}, 1 \le j \le d\}$ denote the Cartesian coordinates of the i-th vertex $\mathbf{N}_i$ of the simplex K.

4.5 The finite element method in the multi-dimensional case

In this section we extend the finite element method introduced previously for one-dimensional problems to the case of boundary-value problems in multi-dimensional regions. We will also specifically refer to the case of simplexes. Many of the results presented are in any case immediately generalizable to more general finite elements (see, for instance, [QV94]).

For the sake of simplicity, most often we will consider domains $\Omega \subset \mathbb{R}^2$ with polygonal shape and meshes (or grids) $\mathcal{T}_h$ which represent their cover with non-overlapping triangles. For this reason, $\mathcal{T}_h$ is also called a triangulation. We refer to Chapter 6 for a more detailed description of the essential features of a generic grid $\mathcal{T}_h$.
In this way, the discretized domain

$$\Omega_h = int\left(\bigcup_{K \in \mathcal{T}_h} K\right)$$

represented by the internal part of the union of the triangles of $\mathcal{T}_h$ perfectly coincides with Ω. We recall that we denote by $int(A)$ the internal part of the set A, that is the region obtained by excluding the boundary from A. In fact, we will not discuss the issue relating to the approximation of a non-polygonal domain with a finite element grid (see Fig. 4.9). Hence, from now on we will adopt the symbol Ω to denote without distinction both the computational domain and its (optional) approximation.

Also in the multidimensional case, the parameter h is related to the spacing of the grid. Having set $h_K = \text{diam}(K)$, for each $K \in \mathcal{T}_h$, where $\text{diam}(K) = \max_{x,y \in K} |x - y|$ is

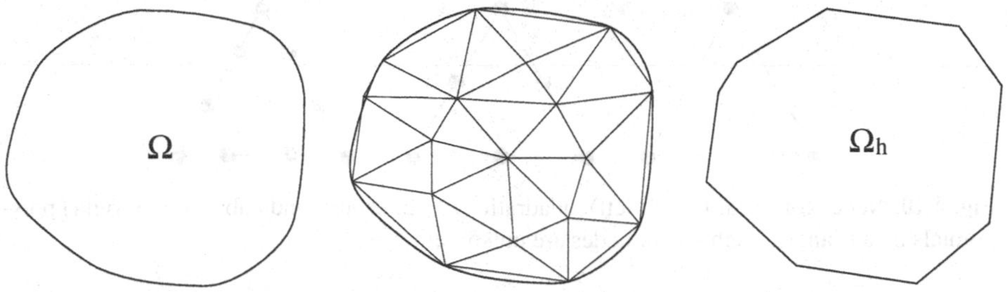

Fig. 4.9. Example of grid of a non-polygonal domain. The grid induces an approximation Ω_h of the domain Ω such that $\lim_{h\to 0} \text{meas}(\Omega - \Omega_h) = 0$. This issue is not addressed in the present text. The interested reader may consult, for instance, [Cia78] or [SF73]

the *diameter* of element K, we define $h = \max_{K\in\mathcal{T}_h} h_K$. Moreover, we will impose that the grid satisfy the following *regularity* condition. Let ρ_K be the diameter of the circle inscribed in the triangle K (also called *sphericity* of K); a family of grids $\{\mathcal{T}_h, h > 0\}$ is said to be *regular* if, for a suitable $\delta > 0$, the condition

$$\frac{h_K}{\rho_K} \leq \delta \quad \forall K \in \mathcal{T}_h \tag{4.37}$$

is verified. We observe that condition (4.37) excludes immediately very deformed (i.e. stretched) triangles, and hence the option of using *anisotropic* computational grids.

On the other hand, anisotropic grids are often used in the context of fluid dynamics problems in the presence of boundary layers. See Remark 4.6, and especially references [AFG$^+$00, DV02, FMP04]. Additional details on the generation of grids on two-dimensional domains are provided in Chapter 6.

We denote by $\mathbb{P}_r$ the space of polynomials of global degree less than or equal to r, for $r = 1, 2, \ldots$. According to the general formula (4.28) we find

$$\begin{aligned}
\mathbb{P}_1 &= \{p(x_1,x_2) = a + bx_1 + cx_2, \text{ with } a,b,c \in \mathbb{R}\},\\
\mathbb{P}_2 &= \{p(x_1,x_2) = a + bx_1 + cx_2 + dx_1x_2 + ex_1^2 + fx_2^2, \text{ with } a,b,c,d,e,f \in \mathbb{R}\},\\
&\ \ \vdots\\
\mathbb{P}_r &= \{p(x_1,x_2) = \sum_{i,j\geq 0, i+j\leq r} a_{ij}x_1^i x_2^j, \text{ with } a_{ij} \in \mathbb{R}\}.
\end{aligned}$$

According to (4.29), the spaces $\mathbb{P}_r$ have dimension $(r+1)(r+2)/2$. For instance, it results that $\dim \mathbb{P}_1 = 3$, $\dim \mathbb{P}_2 = 6$ and $\dim \mathbb{P}_3 = 10$, hence on every element of the grid $\mathcal{T}_h$ the generic function v_h is well defined whenever its value at 3, resp. 6, resp. 10 suitably chosen nodes, is known (see Fig. 4.10). The nodes for linear ($r = 1$), quadratic ($r = 2$), and cubic ($r = 3$) polynomials on a three dimensional simplex are shown in Fig. 4.11.

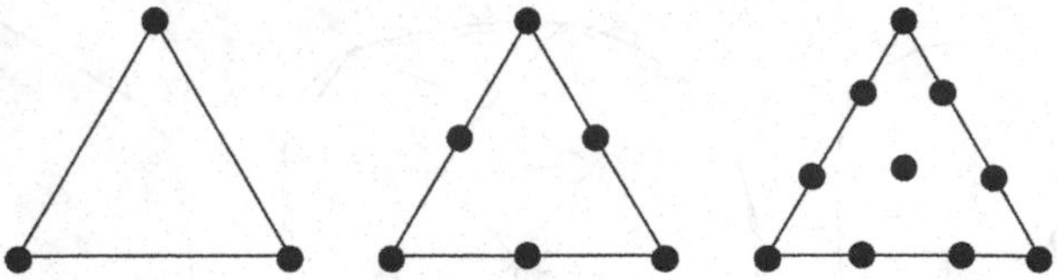

Fig. 4.10. Nodes for linear ($r = 1$, left), quadratic ($r = 2$, center) and cubic ($r = 3$, right) polynomials on a triangle. Such sets of nodes are unisolvent

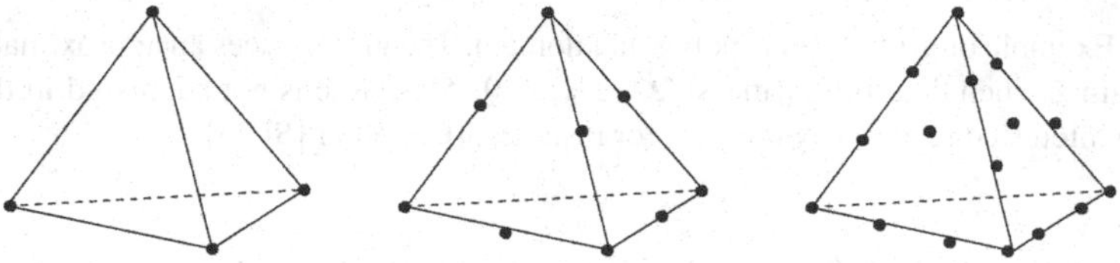

Fig. 4.11. Nodes for linear ($r = 1$, left), quadratic ($r = 2$, center) and cubic ($r = 3$, right) polynomials on a tetrahedron (only those on visible faces are shown)

4.5.1 Finite element solution of the Poisson problem

We introduce the space of finite elements

$$X_h^r = \{v_h \in C^0(\overline{\Omega}) : v_h|_K \in \mathbb{P}_r \forall K \in \mathcal{T}_h\}, \qquad r = 1,2,\ldots \tag{4.38}$$

that is the space of globally continuous functions that are polynomials of degree r on the single triangles (elements) of the triangulation $\mathcal{T}_h$.
Moreover, we define

$$\overset{\circ}{X_h^r} = \{v_h \in X_h^r : v_h|_{\partial\Omega} = 0\}. \tag{4.39}$$

The spaces X_h^r and $\overset{\circ}{X_h^r}$ are suitable for the approximation of $\mathrm{H}^1(\Omega)$, resp. $\mathrm{H}_0^1(\Omega)$, thanks to the following property (for its proof see, e.g., [QV94]):

Property 4.2. *A sufficient condition for a function v to belong to $\mathrm{H}^1(\Omega)$ is that $v \in C^0(\overline{\Omega})$ and $v \in \mathrm{H}^1(K)$ $\forall K \in \mathcal{T}_h$.*

Having set $V_h = \overset{\circ}{X_h^r}$, we can introduce the following finite element problem for the approximation of the Poisson problem (3.1) with Dirichlet boundary condition (3.2), in the homogeneous case (that is with $g = 0$)

$$\text{find } u_h \in V_h : \int_\Omega \nabla u_h \cdot \nabla v_h \, d\Omega = \int_\Omega f v_h \, d\Omega \quad \forall v_h \in V_h. \tag{4.40}$$

As in the one-dimensional case, each function $v_h \in V_h$ is characterized, uniquely, by the values it takes at the nodes $\mathbf{N}_i$, with $i = 1,\ldots,N_h$, of the grid $\mathcal{T}_h$ (excluding the

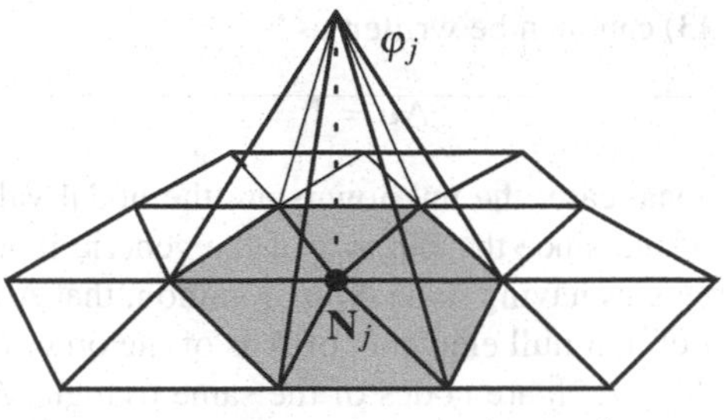

Fig. 4.12. The basis function φ_j of the space X_h^1 and its support

boundary nodes where $v_h = 0$); consequently, a basis in the space V_h can be the set of the characteristic Lagrangian functions $\varphi_j \in V_h$, $j = 1, \ldots, N_h$, such that

$$\varphi_j(\mathbf{N}_i) = \delta_{ij} = \begin{cases} 0 & i \neq j, \\ 1 & i = j, \end{cases} \quad i, j = 1, \ldots, N_h. \tag{4.41}$$

In particular, if $r = 1$, the nodes are vertices of the elements, with the exception of those vertices belonging to the boundary of Ω, while the generic function φ_j is linear on each triangle and is equal to 1 at the node $\mathbf{N}_j$ and 0 at all the other nodes of the triangulation (see Fig. 4.12).
A generic function $v_h \in V_h$ can be expressed through a linear combination of the basis functions of V_h in the following way

$$v_h(\mathbf{x}) = \sum_{i=1}^{N_h} v_i \varphi_i(\mathbf{x}) \quad \forall\, \mathbf{x} \in \Omega, \text{ with } v_i = v_h(\mathbf{N}_i). \tag{4.42}$$

By expressing the discrete solution u_h in terms of the basis $\{\varphi_j\}$ via (4.42), $u_h(\mathbf{x}) = \sum_{j=1}^{N_h} u_j \varphi_j(\mathbf{x})$, with $u_j = u_h(\mathbf{N}_j)$, and imposing that it verifies (4.40) for each function of the basis itself, we find the following linear system of N_h equations in the N_h unknowns u_j, equivalent to problem (4.40),

$$\sum_{j=1}^{N_h} u_j \int_\Omega \nabla\varphi_j \cdot \nabla\varphi_i \, d\Omega = \int_\Omega f \varphi_i \, d\Omega, \quad i = 1, \ldots, N_h. \tag{4.43}$$

The stiffness matrix has dimensions $N_h \times N_h$ and is defined as

$$\mathrm{A} = [a_{ij}] \quad \text{with} \quad a_{ij} = \int_\Omega \nabla\varphi_j \cdot \nabla\varphi_i \, d\Omega. \tag{4.44}$$

Moreover, we introduce the vectors

$$\mathbf{u} = [u_j] \quad \text{with} \quad u_j = u_h(\mathbf{N}_j), \qquad \mathbf{f} = [f_i] \quad \text{with} \quad f_i = \int_\Omega f \varphi_i \, d\Omega. \tag{4.45}$$

The linear system (4.43) can then be written as

$$\mathbf{A}\mathbf{u} = \mathbf{f}. \tag{4.46}$$

As in the one-dimensional case, the unknowns are the nodal values of the finite element solution. It is evident, since the *support* of the generic function with basis φ_i is only formed by the triangles having node $\mathbf{N}_i$ in common, that A is a sparse matrix. In particular, the number of non-null elements of A is of the order of N_h, as a_{ij} is different from zero only if $\mathbf{N}_j$ and $\mathbf{N}_i$ are nodes of the same triangle. A has not necessarily a definite structure (e.g. banded), as that will depend on how the nodes are numbered.

Let us consider now the case of a *non-homogeneous* Dirichlet problem represented by equations (3.1)–(3.2). We have seen in the previous chapter that we can in any case resort to the homogeneous case through a lifting (also called extension, or prolongation) of the boundary datum. In the corresponding discrete problem we build a lifting of a well-chosen approximation of the boundary datum, by proceeding in the following way.
We denote by N_h the internal nodes of the grid $\mathcal{T}_h$ and by N_h^t the total number, including the boundary nodes that for the sake of simplicity we will suppose to be numbered last. The set of boundary nodes will then be formed by $\{\mathbf{N}_i,\, i = N_h+1,\dots,N_h^t\}$. A possible approximation g_h of the boundary datum g can be obtained by interpolating g on the space formed by the trace functions on $\partial\Omega$ of functions of X_h^r. This can be written as a linear combination of the traces of the basis functions of X_h^r associated to the boundary nodes

$$g_h(\mathbf{x}) = \Sigma_{i=N_h+1}^{N_h^t} g(\mathbf{N}_i)\varphi_i(\mathbf{x}) \quad \forall \mathbf{x} \in \partial\Omega. \tag{4.47}$$

Its lifting $R_{g_h} \in X_h^r$ is constructed as follows

$$R_{g_h}(\mathbf{x}) = \Sigma_{i=N_h+1}^{N_h^t} g(\mathbf{N}_i)\varphi_i(\mathbf{x}) \quad \forall \mathbf{x} \in \Omega. \tag{4.48}$$

In Fig. 4.13 we provide an example of a possible lifting of a non-homogeneous Dirichlet boundary datum (3.2), in the case where g has a non-constant value. The finite element formulation of the Poisson problem then becomes:
find $\overset{\circ}{u}_h \in V_h$:

$$\int_\Omega \nabla \overset{\circ}{u}_h \cdot \nabla v_h \, d\Omega = \int_\Omega f v_h \, d\Omega - \int_\Omega \nabla R_{g_h} \cdot \nabla v_h \, d\Omega \quad \forall\, v_h \in V_h. \tag{4.49}$$

The approximate solution will then be provided by $u_h = \overset{\circ}{u}_h + R_{g_h}$.

Notice how the particular lifting we adopted allows for the following algebraic interpretation of (4.49)

$$\mathbf{A}\mathbf{u} = \mathbf{f} - \mathbf{B}\mathbf{g}$$

where A and $\mathbf{f}$ are defined as in (4.44) and (4.45), now with $u_j = \overset{\circ}{u}_h(\mathbf{N}_j)$. Having set $N_h^b = N_h^t - N_h$ (this is the number of boundary nodes), the vector $\mathbf{g} \in \mathbb{R}^{N_h^b}$ and the matrix

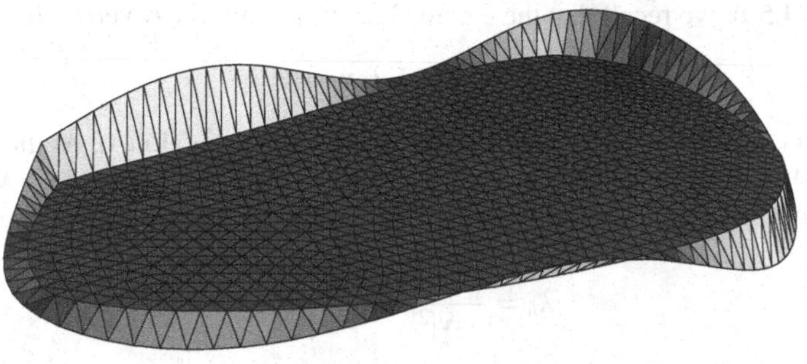

Fig. 4.13. Example of lifting of a non-homogeneous Dirichlet boundary datum $u = g$, g being variable

$B \in \mathbb{R}^{N_h \times N_h^b}$ have, respectively, the components

$$g_i = g(\mathbf{N}_{i+N_h}), \quad i = 1, \ldots, N_h^b,$$

$$b_{ij} = \int_{\Omega} \nabla \varphi_{j+N_h} \cdot \nabla \varphi_i \, d\Omega, \quad i = 1, \ldots, N_h, \, j = 1, \ldots, N_h^b.$$

Remark 4.5. The matrices A and B are both sparse. An efficient program will store exclusively their non-null elements. (See, e.g., [Saa96] for a description of possible storage formats for sparse matrices, and also Chapter 8). In particular, thanks to the special lifting we have adopted, in the matrix B, all the lines corresponding to non-adjacent nodes to a boundary node will be null. (Two grid nodes are said to be adjacent if there exists an element $K \in \mathcal{T}_h$ to which they both belong.) •

4.5.2 Conditioning of the stiffness matrix

We have seen that the stiffness matrix $A = [a(\varphi_j, \varphi_i)]$ associated to the Galerkin problem and therefore, in particular, to the finite element method, is positive definite; moreover A is symmetric if the bilinear form $a(\cdot,\cdot)$ is symmetric.

For a symmetric and positive definite matrix, its condition number with respect to the norm $\|.\|_2$ is given by

$$K_2(A) = \frac{\lambda_{max}(A)}{\lambda_{min}(A)},$$

$\lambda_{max}(A)$ and $\lambda_{min}(A)$ being the maximum and minimum eigenvalues, respectively, of A.

It can be proved that, both in the one-dimensional and the multi-dimensional case, the following relation holds for the stiffness matrix

$$K_2(A) = Ch^{-2}, \tag{4.50}$$

where C is a constant independent of the parameter h, but dependent on the degree of the finite elements being used.

To prove (4.50), we recall that the eigenvalues of the matrix A verify the relation

$$\mathrm{A}\mathbf{v} = \lambda_h \mathbf{v},$$

$\mathbf{v}$ being an eigenvector corresponding to the eigenvalue λ_h. Let v_h be the function of the space V_h whose nodal values are the components v_i of $\mathbf{v}$, see (4.7). We suppose $a(\cdot,\cdot)$ to be symmetric, so A is symmetric and its eigenvalues are real and positive. We then have

$$\lambda_h = \frac{(\mathrm{A}\mathbf{v},\mathbf{v})}{|\mathbf{v}|^2} = \frac{a(v_h,v_h)}{|\mathbf{v}|^2} \tag{4.51}$$

where $|\cdot|$ is the Euclidean vector norm. We suppose that the grid family $\{\mathcal{T}_h, h>0\}$ is regular (i.e. satisfies (4.37)) and moreover *quasi-uniform*, i.e. such that there exists a constant $\tau > 0$ with

$$\min_{K\in\mathcal{T}_h} h_K \geq \tau h \qquad \forall h > 0.$$

We now observe that, under the hypotheses made on $\mathcal{T}_h$, the following *inverse inequality* holds (for the proof, refer to [QV94])

$$\exists C_I > 0 \quad : \quad \forall v_h \in V_h, \quad \|\nabla v_h\|_{\mathrm{L}^2(\Omega)} \leq C_I h^{-1} \|v_h\|_{\mathrm{L}^2(\Omega)}, \tag{4.52}$$

the constant C_I being independent of h. We can now prove that there exist two constants $C_1, C_2 > 0$ such that, for each $v_h \in V_h$ as in (4.7), we have

$$C_1 h^d |\mathbf{v}|^2 \leq \|v_h\|^2_{\mathrm{L}^2(\Omega)} \leq C_2 h^d |\mathbf{v}|^2 \tag{4.53}$$

d being the spatial dimension, with $d = 1,2,3$. For the proof in the general case we refer to Proposition 6.3.1. [QV94]. We here limit ourselves to proving the second inequality in the one-dimensional case ($d = 1$) and for linear finite elements. Indeed, on each element $K_i = [x_{i-1}, x_i]$, we have

$$\int_{K_i} v_h^2(x)\,dx = \int_{K_i} \big(v_{i-1}\varphi_{i-1}(x) + v_i\varphi_i(x)\big)^2 dx,$$

with φ_{i-1} and φ_i defined according to (4.15). Then, a direct computation shows that

$$\int_{K_i} v_h^2(x)\,dx \leq 2\,\Big(v_{i-1}^2 \int_{K_i} \varphi_{i-1}^2(x)\,dx + v_i^2 \int_{K_i} \varphi_i^2(x)\,dx\Big) = \frac{2}{3}\,h_i\,(v_{i-1}^2 + v_i^2)$$

with $h_i = x_i - x_{i-1}$. The inequality

$$\|v_h\|^2_{\mathrm{L}^2(\Omega)} \leq C h |\mathbf{v}|^2$$

with $C = 4/3$, can be found by simply summing the intervals K and observing that each nodal contribution v_i is counted twice.
On the other hand, from (4.51) we obtain, thanks to the continuity and coercivity of the bilinear form $a(\cdot,\cdot)$,

$$\alpha\,\frac{\|v_h\|^2_{\mathrm{H}^1(\Omega)}}{|\mathbf{v}|^2} \leq \lambda_h \leq M\,\frac{\|v_h\|^2_{\mathrm{H}^1(\Omega)}}{|\mathbf{v}|^2},$$

M and α being the continuity and coercivity constants, respectively. Now, $\|v_h\|^2_{H^1(\Omega)} \geq \|v_h\|^2_{L^2(\Omega)}$ by the definition of the norm in $H^1(\Omega)$, while $\|v_h\|_{H^1(\Omega)} \leq C_3 h^{-1} \|v_h\|_{L^2(\Omega)}$ (for a well-chosen constant $C_3 > 0$) thanks to (4.52). Thus, by using inequalities (4.53), we obtain

$$\alpha C_1 h^d \leq \lambda_h \leq M C_3^2 C_2 h^{-2} h^d.$$

We therefore have

$$\frac{\lambda_{max}(A)}{\lambda_{min}(A)} \leq \frac{M C_3^2 C_2}{\alpha C_1} h^{-2},$$

that is (4.50).

When the grid-size h decreases, the condition number of the stiffness matrix increases, and therefore the associated system becomes more and more ill-conditioned. In particular, if the datum $\mathbf{f}$ of the linear system (4.46) is subject to a perturbation $\boldsymbol{\delta}\mathbf{f}$ (i.e. it is affected by error), the latter in turn affects the solution with a perturbation $\boldsymbol{\delta}\mathbf{u}$; then it can be proved that, if there are no perturbations on the matrix A,

$$\frac{|\boldsymbol{\delta}\mathbf{u}|}{|\mathbf{u}|} \leq K_2(A) \frac{|\boldsymbol{\delta}\mathbf{f}|}{|\mathbf{f}|}.$$

It is evident that the higher the conditioning number is, the more the solution is affected by the perturbation on the data. (On the other hand, notice that the latter is always affected by perturbations on the data caused by the inevitable roundoff errors introduced by the computer.)

As a further example we can study how conditioning affects the solution method. Consider, for instance, solving the linear system (4.46) using the conjugate gradient method (see Chap. 7). Then a sequence $\mathbf{u}^{(k)}$ of approximate solutions is iteratively constructed, converging to the exact solution $\mathbf{u}$. In particular, we have

$$\|\mathbf{u}^{(k)} - \mathbf{u}\|_A \leq 2 \left(\frac{\sqrt{K_2(A)} - 1}{\sqrt{K_2(A)} + 1} \right)^k \|\mathbf{u}^{(0)} - \mathbf{u}\|_A,$$

having denoted by $\|\mathbf{v}\|_A = \sqrt{\mathbf{v}^T A \mathbf{v}}$ the so-called "A-norm" of a generic vector $\mathbf{v} \in \mathbb{R}^{N_h}$. If we define

$$\rho = \frac{\sqrt{K_2(A)} - 1}{\sqrt{K_2(A)} + 1},$$

such quantity gives an idea of the convergence rate of the method: the closer ρ is to 0, the faster the method converges, wilst the closer ρ is to 1, the slower the convergence. Indeed, following (4.50), the more accurate one wants to be, by decreasing h, the more ill-conditioned the system will be, and therefore the more "problematic" its solution will turn out to be.

This calls for the system to be preconditioned, i.e. it is necessary to find an invertible matrix P, called *preconditioner*, such that

$$K_2(P^{-1}A) \ll K_2(A),$$

and then apply the iterative method to the system preconditioned with P (see Chapter 7).

4.5.3 Estimate of the approximation error in the energy norm

Analogously to the one-dimensional case, for each $v \in C^0(\overline{\Omega})$ we define *interpolant* of v in the space of X_h^1, determined by the grid $\mathcal{T}_h$, the function $\Pi_h^1 v$ such that

$$\Pi_h^1 v(\mathbf{N}_i) = v(\mathbf{N}_i) \quad \text{for each node}\,\mathbf{N}_i \text{ of } \mathcal{T}_h, \text{ for } i = 1, \ldots, N_h.$$

If $\{\varphi_i\}$ is the Lagrangian basis of the space X_h^1, then

$$\Pi_h^1 v(\mathbf{x}) = \sum_{i=1}^{N_h} v(\mathbf{N}_i)\varphi_i(\mathbf{x}).$$

The operator $\Pi_h^1 : C^0(\overline{\Omega}) \to X_h^1$, associating to a continuous function v its interpolant $\Pi_h^1 v$, is called *interpolation operator*.

Similarly, we can define an operator $\Pi_h^r : C^0(\overline{\Omega}) \to X_h^r$, for each integer $r \geq 1$. Having denoted by Π_K^r the local interpolation operator associated to a continuous function v the polynomial $\Pi_K^r v \in \mathbb{P}_r(K)$, interpolating v in the degrees of freedom of the element $K \in \mathcal{T}_h$, we define

$$\Pi_h^r v \in X_h^r : \quad \Pi_h^r v\big|_K = \Pi_K^r(v\big|_K) \qquad \forall K \in \mathcal{T}_h. \tag{4.54}$$

We will suppose that $\mathcal{T}_h$ belongs to a family of regular grids of Ω.

In order to obtain an estimate for the approximation error $\|u - u_h\|_V$ we follow a similar procedure to the one used in Theorem 4.3 for the one-dimensional case. The first step is to derive a suitable estimate for the interpolation error. To this end, we will obtain useful information starting from the geometric parameters of each triangle K, i.e. its diameter h_K and sphericity ρ_K. Moreover, we will exploit the affine and invertible transformation $F_K : \widehat{K} \to K$ between the reference triangle $\widehat{K}$ and the generic triangle K (see Fig. 4.14). Such map is defined by $F_K(\hat{\mathbf{x}}) = B_K\hat{\mathbf{x}} + \mathbf{b}_K$, with $B_K \in \mathbb{R}^{2\times 2}$ and $\mathbf{b}_K \in \mathbb{R}^2$, and satisfies the relation $F_K(\widehat{K}) = K$. We recall that the choice of the reference triangle $\widehat{K}$ is not unique.
We will need some preliminary results.

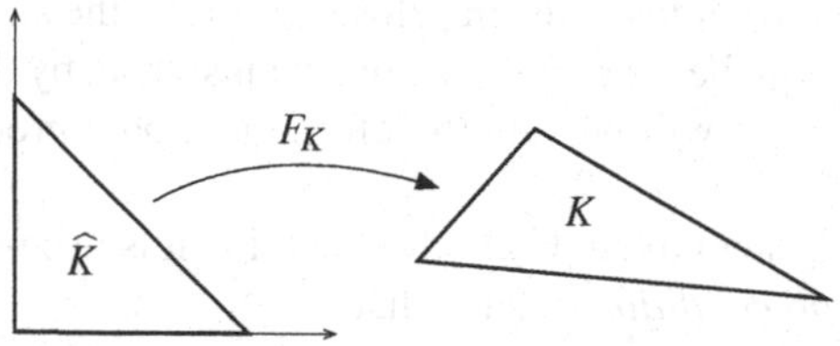

Fig. 4.14. The map F_K between the reference triangle $\widehat{K}$ and the generic triangle K

Lemma 4.2 (Transformation of the seminorms). *For each integer $m \geq 0$ and each $v \in \mathrm{H}^m(K)$, let $\hat{v} : \widehat{K} \to \mathbb{R}$ be the function defined by $\hat{v} = v \circ F_K$. Then $\hat{v} \in \mathrm{H}^m(\widehat{K})$. Moreover, there exists a constant $C = C(m) > 0$ such that:*

$$|\hat{v}|_{\mathrm{H}^m(\widehat{K})} \leq C \|B_K\|^m |\det B_K|^{-\frac{1}{2}} |v|_{\mathrm{H}^m(K)}, \tag{4.55}$$

$$|v|_{\mathrm{H}^m(K)} \leq C \|B_K^{-1}\|^m |\det B_K|^{\frac{1}{2}} |\hat{v}|_{\mathrm{H}^m(\widehat{K})}, \tag{4.56}$$

$\|\cdot\|$ being the matrix norm associated to the Euclidean vector norm $|\cdot|$, i.e.

$$\|B_K\| = \sup_{\boldsymbol{\xi} \in \mathbb{R}^2, \boldsymbol{\xi} \neq \mathbf{0}} \frac{|B_K \boldsymbol{\xi}|}{|\boldsymbol{\xi}|}. \tag{4.57}$$

Proof. Since $C^m(K) \subset \mathrm{H}^m(K)$ densely, for each $m \geq 0$, we can limit ourselves to proving the previous two inequalities for the functions of $C^m(K)$, then extending by density the result to the functions of $\mathrm{H}^m(K)$. The derivatives in the remainders will therefore have to be intended in the classical sense. We recall that

$$|\hat{v}|_{\mathrm{H}^m(\widehat{K})} = \Big(\sum_{|\alpha|=m} \int_{\widehat{K}} |D^\alpha \hat{v}|^2 \, d\hat{\mathbf{x}} \Big)^{1/2},$$

by referring to Chapter 2.3 for the definition of the derivative D^α. By using the chain rule for the differentiation of composite functions, we obtain

$$\|D^\alpha \hat{v}\|_{\mathrm{L}^2(\widehat{K})} \leq C \|B_K\|^m \sum_{|\beta|=m} \|(D^\beta v) \circ F_K\|_{\mathrm{L}^2(\widehat{K})}.$$

Then

$$\|D^\alpha \hat{v}\|_{\mathrm{L}^2(\widehat{K})} \leq C \|B_K\|^m |\det B_K|^{-\frac{1}{2}} \|D^\alpha v\|_{\mathrm{L}^2(K)}.$$

Inequality (4.55) follows after summing on the multi-index α, for $|\alpha| = m$. The result (4.56) can be proved by proceeding in a similar way. ◇

Lemma 4.3 (Estimates for the norms $\|B_K\|$ and $\|B_K^{-1}\|$). *We have the following upper bounds:*

$$\|B_K\| \leq \frac{h_K}{\hat{\rho}}, \tag{4.58}$$

$$\|B_K^{-1}\| \leq \frac{\hat{h}}{\rho_K}, \tag{4.59}$$

$\hat{h}$ and $\hat{\rho}$ being the diameter and the sphericity of the reference triangle $\widehat{K}$.

Proof. Thanks to (4.57) we have

$$\|B_K\| = \frac{1}{\hat{\rho}} \sup_{\xi \in \mathbb{R}^2, |\xi| = \hat{\rho}} |B_K \xi|.$$

For each $\boldsymbol{\xi}$, with $|\boldsymbol{\xi}| = \hat{\rho}$, we can find two points $\hat{\mathbf{x}}$ and $\hat{\mathbf{y}} \in \widehat{K}$ such that $\hat{\mathbf{x}} - \hat{\mathbf{y}} = \boldsymbol{\xi}$. Since $B_K \boldsymbol{\xi} = F_K(\hat{\mathbf{x}}) - F_K(\hat{\mathbf{y}})$, we have $|B_K \boldsymbol{\xi}| \leq h_K$, that is (4.58).
An analogous procedure leads to the result (4.59). ◇

What we now need is an estimate in $\mathrm{H}^m(\widehat{K})$ of the seminorm of $(v - \Pi_K^r v) \circ F_K$, for each function v of $\mathrm{H}^m(K)$. In the remainder, we denote the interpolant $\Pi_K^r v \circ F_K$ with $[\Pi_K^r v]^\wedge$. The nodes of K are $\mathbf{N}_i^K = F_K(\widehat{\mathbf{N}}_i)$, $\widehat{\mathbf{N}}_i$ being the nodes of $\widehat{K}$, and, analogously, the basis functions $\hat{\varphi}_i$ defined on $\widehat{K}$ are determined by the relation $\hat{\varphi}_i = \varphi_i^K \circ F_K$, having denoted by φ_i^K the basis functions associated to the element K. Thus,

$$[\Pi_K^r v]^\wedge = \Pi_K^r v \circ F_K = \sum_{i=1}^{M_K} v(\mathbf{N}_i^K) \varphi_i^K \circ F_K = \sum_{i=1}^{M_K} v(F_K(\widehat{\mathbf{N}}_i)) \hat{\varphi}_i = \Pi_{\widehat{K}}^r \hat{v},$$

M_K being the number of nodes on K determined by the choice made for the degree r. It then follows that

$$|(v - \Pi_K^r v) \circ F_K|_{\mathrm{H}^m(\widehat{K})} = |\hat{v} - \Pi_{\widehat{K}}^r \hat{v}|_{\mathrm{H}^m(\widehat{K})}. \tag{4.60}$$

In order to estimate the right side of the previous equality, we start by proving the following result:

Lemma 4.4 (Bramble-Hilbert Lemma). *Let $\widehat{L} : \mathrm{H}^{r+1}(\widehat{K}) \to \mathrm{H}^m(\widehat{K})$, with $m \geq 0$ and $r \geq 0$, be a linear and continuous transformation such that*

$$\widehat{L}(\hat{p}) = 0 \qquad \forall \hat{p} \in \mathbb{P}_r(\widehat{K}). \tag{4.61}$$

Then, for each $\hat{v} \in \mathrm{H}^{r+1}(\widehat{K})$, we have

$$|\widehat{L}(\hat{v})|_{\mathrm{H}^m(\widehat{K})} \leq \|\widehat{L}\|_{\mathcal{L}(\mathrm{H}^{r+1}(\widehat{K}), \mathrm{H}^m(\widehat{K}))} \inf_{\hat{p} \in \mathbb{P}_r(\widehat{K})} \|\hat{v} + \hat{p}\|_{\mathrm{H}^{r+1}(\widehat{K})}, \tag{4.62}$$

where $\mathcal{L}(\mathrm{H}^{r+1}(\widehat{K}), \mathrm{H}^m(\widehat{K}))$ denotes the space of linear and continuous transformations $l : \mathrm{H}^{r+1}(\widehat{K}) \to \mathrm{H}^m(\widehat{K})$, normed by

$$\|l\|_{\mathcal{L}(\mathrm{H}^{r+1}(\widehat{K}), \mathrm{H}^m(\widehat{K}))} = \sup_{v \in \mathrm{H}^{r+1}(\widehat{K}), v \neq 0} \frac{\|l(v)\|_{\mathrm{H}^m(\widehat{K})}}{\|v\|_{\mathrm{H}^{r+1}(\widehat{K})}}. \tag{4.63}$$

Proof. Let $\hat{v} \in \mathrm{H}^{r+1}(\widehat{K})$. For each $\hat{p} \in \mathbb{P}_r(\widehat{K})$, thanks to (4.61) and to definition (4.63) of the norm, we obtain

$$|\widehat{L}(\hat{v})|_{\mathrm{H}^m(\widehat{K})} = |\widehat{L}(\hat{v} + \hat{p})|_{\mathrm{H}^m(\widehat{K})} \leq \|\widehat{L}\|_{\mathcal{L}(\mathrm{H}^{r+1}(\widehat{K}), \mathrm{H}^m(\widehat{K}))} \|\hat{v} + \hat{p}\|_{\mathrm{H}^{r+1}(\widehat{K})}.$$

Then (4.62) can be deduced thanks to the fact that $\hat{p}$ is arbitrary. ◇

The following result (whose proof is given, e.g., in [QV94, Chap. 3]) provides the last necessary tool to obtain the estimate for the interpolation error that we are seeking.

Lemma 4.5 (Deny-Lions Lemma). *For each $r \geq 0$, there exists a constant $C = C(r, \widehat{K})$ such that*

$$\inf_{\hat{p} \in \mathbb{P}_r} \|\hat{v} + \hat{p}\|_{\mathrm{H}^{r+1}(\widehat{K})} \leq C |\hat{v}|_{\mathrm{H}^{r+1}(\widehat{K})} \qquad \forall \hat{v} \in \mathrm{H}^{r+1}(\widehat{K}). \tag{4.64}$$

As a consequence of the two previous lemmas, we can provide the following

Corollary 4.3. *Let $\widehat{L} : \mathrm{H}^{r+1}(\widehat{K}) \to \mathrm{H}^m(\widehat{K})$, with $m \geq 0$ and $r \geq 0$, be a linear and continuous transformation such that $\widehat{L}(\hat{p}) = 0 \; \forall \hat{p} \in \mathbb{P}_r(\widehat{K})$. Then there exists a constant $C = C(r, \widehat{K})$ such that, for each $\hat{v} \in \mathrm{H}^{r+1}(\widehat{K})$, we have*

$$|\widehat{L}(\hat{v})|_{\mathrm{H}^m(\widehat{K})} \leq C \|\widehat{L}\|_{\mathcal{L}(\mathrm{H}^{r+1}(\widehat{K}), \mathrm{H}^m(\widehat{K}))} |\hat{v}|_{\mathrm{H}^{r+1}(\widehat{K})}. \tag{4.65}$$

We are now able to prove the interpolation error estimate.

Theorem 4.4 (Local estimate of the interpolation error). *Let $r \geq 1$ and $0 \leq m \leq r+1$. Then there exists a constant $C = C(r, m, \widehat{K}) > 0$ such that*

$$|v - \Pi_K^r v|_{\mathrm{H}^m(K)} \leq C \frac{h_K^{r+1}}{\rho_K^m} |v|_{\mathrm{H}^{r+1}(K)} \qquad \forall v \in \mathrm{H}^{r+1}(K). \tag{4.66}$$

Proof. From Property 2.3 we derive first of all that $\mathrm{H}^{r+1}(K) \subset C^0(K)$, for $r \geq 1$. The interpolation operator Π_K^r thus results to be well defined in $\mathrm{H}^{r+1}(K)$. By using in succession (4.56), (4.60), (4.59) and (4.65) we have

$$\begin{aligned}
|v - \Pi_K^r v|_{\mathrm{H}^m(K)} &\leq C_1 \|B_K^{-1}\|^m |\det B_K|^{\frac{1}{2}} |\hat{v} - \Pi_{\widehat{K}}^r \hat{v}|_{\mathrm{H}^m(\widehat{K})} \\
&\leq C_1 \frac{\hat{h}^m}{\rho_K^m} |\det B_K|^{\frac{1}{2}} \underbrace{|\hat{v} - \Pi_{\widehat{K}}^r \hat{v}|_{\mathrm{H}^m(\widehat{K})}}_{\widehat{L}(\hat{v})} \\
&\leq C_2 \frac{\hat{h}^m}{\rho_K^m} |\det B_K|^{\frac{1}{2}} \|\widehat{L}\|_{\mathcal{L}(\mathrm{H}^{r+1}(\widehat{K}), \mathrm{H}^m(\widehat{K}))} |\hat{v}|_{\mathrm{H}^{r+1}(\widehat{K})} \\
&= C_3 \frac{1}{\rho_K^m} |\det B_K|^{\frac{1}{2}} |\hat{v}|_{\mathrm{H}^{r+1}(\widehat{K})},
\end{aligned}$$

where $C_1 = C_1(m)$, $C_2 = C_2(r, m, \widehat{K})$ and $C_3 = C_3(r, m, \widehat{K})$ are suitably chosen constants.

We note that the result (4.65) has been applied when identifying $\widehat{L}$ with the operator $I-\Pi^r_{\widehat{K}}$, with $(I-\Pi^r_{\widehat{K}})\hat{p}=0$, for $\hat{p}\in\mathbb{P}_r(\widehat{K})$. Moreover the quantity $\hat{h}^m$ and the norm of the operator $\widehat{L}$ have been included in the constant C_3.
At this point, by applying (4.55) and (4.58) we obtain (4.66), that is

$$|v-\Pi^r_K v|_{\mathrm{H}^m(K)}\leq C_4\frac{1}{\rho_K^m}\|B_K\|^{r+1}|v|_{\mathrm{H}^{r+1}(K)}\leq C_5\frac{h_K^{r+1}}{\rho_K^m}|v|_{\mathrm{H}^{r+1}(K)},\qquad(4.67)$$

$C_4=C_4(r,m,\widehat{K})$ and $C_5=C_5(r,m,\widehat{K})$ being two well-chosen constants. The quantity $\hat{\rho}^{r+1}$ generated by (4.58) and relating to the sphericity of the reference element has been directly included in the constant C_5. ◇

Finally, we can prove the global estimate for the interpolation error:

Theorem 4.5 (Global estimate for the interpolation error). *Let $\{\mathcal{T}_h\}_{h>0}$ be a family of regular grids of the domain Ω and let $m=0,1$ and $r\geq 1$. Then there exists a constant $C=C(r,m,\widehat{K})>0$ such that*

$$|v-\Pi^r_h v|_{\mathrm{H}^m(\Omega)}\leq C\Big(\sum_{K\in\mathcal{T}_h}h_K^{2(r+1-m)}|v|^2_{\mathrm{H}^{r+1}(K)}\Big)^{1/2}\qquad\forall v\in\mathrm{H}^{r+1}(\Omega).\quad(4.68)$$

In particular, we obtain

$$|v-\Pi^r_h v|_{\mathrm{H}^m(\Omega)}\leq C h^{r+1-m}|v|_{\mathrm{H}^{r+1}(\Omega)}\qquad\forall v\in\mathrm{H}^{r+1}(\Omega).\qquad(4.69)$$

Proof. Thanks to (4.66) and to the regularity condition (4.37), we have

$$\begin{aligned}|v-\Pi^r_h v|^2_{\mathrm{H}^m(\Omega)}&=\sum_{K\in\mathcal{T}_h}|v-\Pi^r_K v|^2_{\mathrm{H}^m(K)}\\&\leq C_1\sum_{K\in\mathcal{T}_h}\left(\frac{h_K^{r+1}}{\rho_K^m}\right)^2|v|^2_{\mathrm{H}^{r+1}(K)}\\&=C_1\sum_{K\in\mathcal{T}_h}\left(\frac{h_K}{\rho_K}\right)^{2m}h_K^{2(r+1-m)}|v|^2_{\mathrm{H}^{r+1}(K)}\\&\leq C_1\,\delta^{2m}\sum_{K\in\mathcal{T}_h}h_K^{2(r+1-m)}|v|^2_{\mathrm{H}^{r+1}(K)},\end{aligned}$$

i.e. (4.68), with $C_1=C_1(r,m,\widehat{K})$ and $C=C_1\,\delta^{2m}$. (4.69) follows thanks to the fact that $h_K\leq h$, for each $K\in\mathcal{T}_h$, and that

$$|v|_{\mathrm{H}^p(\Omega)}=\Big(\sum_{K\in\mathcal{T}_h}|v|^2_{\mathrm{H}^p(K)}\Big)^{1/2},$$

for each integer $p\geq 0$. ◇

In the $m = 0$ case, regularity of the grid is not necessary to obtain the estimate (4.69). This is no longer true for $m = 1$. As a matter of fact, given a triangle K and a function $v \in \mathrm{H}^{r+1}(K)$, with $r \geq 1$, it can be proved that the following inequality holds [QV94],

$$|v - \Pi_h^r v|_{\mathrm{H}^m(K)} \leq \tilde{C} \frac{h_K^{r+1}}{\rho_K^m} |v|_{\mathrm{H}^{r+1}(K)}, \quad m = 0, 1,$$

with $\tilde{C}$ independent of v and $\mathcal{T}_h$. Hence, in the case $m = 1$ for a family of regular grids we obtain (4.69) by setting $C = \delta\tilde{C}$, δ being the constant appearing in (4.37). On the other hand, the need for a regularity condition can be proved by considering the particular case where, for each $C > 0$, a (non-regular) grid can be constructed for which inequality (4.69) is not true, as we are about to prove in the following example which relates to the case $r = 1$.

Example 4.1. Consider the triangle K_l illustrated in Fig. 4.15, with vertices $(0,0)$, $(1,0)$, $(0.5, l)$, with $l \leq \frac{\sqrt{3}}{2}$, and the function $v(x_1, x_2) = x_1^2$. Clearly $v \in \mathrm{H}^2(K_l)$, and its linear interpolant on K_l is given by $\Pi_h^1 v(x_1, x_2) = x_1 - (4l)^{-1} x_2$. Since in this case $h_{K_l} = 1$, inequality (4.69), applied to the single triangle K_l, would yield

$$|v - \Pi_h^1 v|_{\mathrm{H}^1(K_l)} \leq C |v|_{\mathrm{H}^2(K_l)}. \tag{4.70}$$

Let us now consider the behaviour of the relation

$$\eta_l = \frac{|v - \Pi_h^1 v|_{\mathrm{H}^1(K_l)}}{|v|_{\mathrm{H}^2(K_l)}}$$

when l tends to zero, that is when the triangle is squeezed. We note that allowing l to tend to zero is equivalent to violating the regularity condition (4.37), because for small enough values of l, $h_{K_l} = 1$. At the same time, denoting by p_{K_l} the perimeter of K_l and

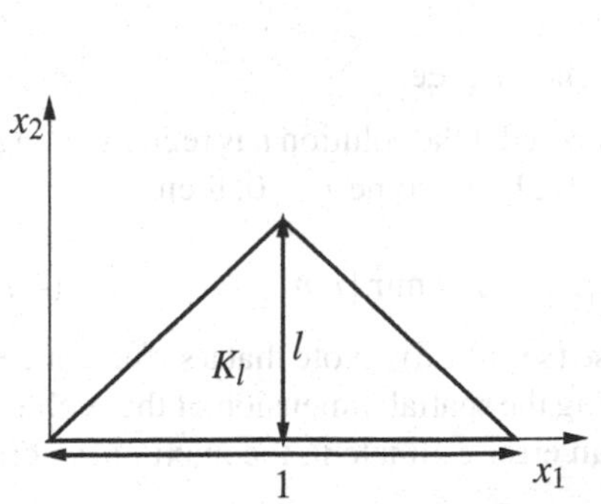

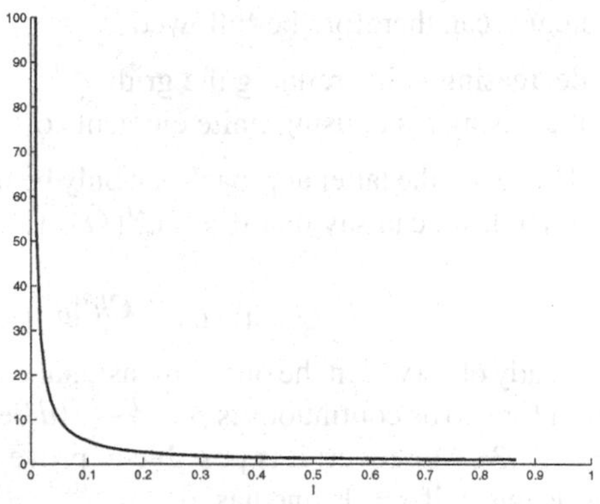

Fig. 4.15. The triangle K_l (left) and the behaviour of the relation $|v - \Pi_h^1 v|_{\mathrm{H}^1(K_l)} / |v|_{\mathrm{H}^2(K_l)}$ as a function of l (right)

by $|K_l|$ we have the surface of the element K_l, the sphericity of K_l

$$\rho_{K_l} = \frac{4|K_l|}{p_{K_l}} = \frac{2l}{1+\sqrt{1+4l^2}}$$

tends to zero. We have

$$\eta_l \geq \frac{\|\partial_{x_2}(v - \Pi_h^1 v)\|_{L^2(K_l)}}{|v|_{H^2(K_l)}} = \left(\frac{\int_{K_l} \left(\frac{1}{4l}\right)^2 d\mathbf{x}}{2l} \right)^{\frac{1}{2}} = \frac{1}{8l}.$$

Hence $\lim_{l\to 0} \eta_l = +\infty$ (see Fig. 4.15). Consequently, there cannot exist a constant C, independent of $\mathcal{T}_h$, for which (4.70) holds. ■

The theorem on the interpolation error estimate immediately provides us with an estimate for the approximation error of the Galerkin method. The proof is analogous to that of Theorem 4.3 for the one-dimensional case. Indeed, it is sufficient to apply (4.10) and Theorem 4.5 (for $m = 1$) to obtain the following error estimate:

Theorem 4.6. *Let $u \in V$ be the exact solution of the variational problem* (4.1) *and u_h its approximate solution using the finite element method of degree r. If $u \in \mathrm{H}^{r+1}(\Omega)$, then the following* a priori *error estimates hold:*

$$\|u - u_h\|_{\mathrm{H}^1(\Omega)} \leq \frac{M}{\alpha} C \Big(\sum_{K \in \mathcal{T}_h} h_K^{2r} |u|^2_{\mathrm{H}^{r+1}(K)} \Big)^{1/2}, \tag{4.71}$$

$$\|u - u_h\|_{\mathrm{H}^1(\Omega)} \leq \frac{M}{\alpha} C h^r |u|_{\mathrm{H}^{r+1}(\Omega)}, \tag{4.72}$$

C being a constant independent of h and u.

Also in the multi-dimensional case, in order to increase the accuracy two different strategies can therefore be followed:

1. decreasing h, i.e. refining the grid;
2. increasing r, i.e. using finite elements of higher degree.

However, the latter approach can only be pursued if the solution u is regular enough. In general, we can say that if $u \in C^0(\bar{\Omega}) \cap \mathrm{H}^{p+1}(\Omega)$ for some $p > 0$, then

$$\|u - u_h\|_{\mathrm{H}^1(\Omega)} \leq C h^s |u|_{\mathrm{H}^{s+1}(\Omega)}, \quad s = \min\{r, p\}, \tag{4.73}$$

as already observed in the one-dimensional case (see (4.26)). Note that a sufficient condition for u to be continuous is $p > \frac{d}{2} - 1$ (d being the spatial dimension of the problem, $d = 1, 2, 3$). Moreover, it is possible to prove an error estimate in the maximum norm. For instance, if $r = 1$, one has

$$\|u - u_h\|_{L^\infty(\Omega)} \leq C h^2 |\log h| |u|_{W^{2,\infty}(\Omega)}$$

where C is a positive constant independent of h and the last term on the right-hand side is the seminorm of u in the Sobolev space $W^{2,\infty}(\Omega)$ (see Sect. 2.5). For the proof of this and other error estimates in $W^{k,\infty}(\Omega)$-norms see, e.g., [Cia78] and [BS94].

Remark 4.6 (Case of anisotropic grids). The interpolation error estimate (4.66) (and the consequent discretization error estimate) can be generalized in the case of *anisotropic grids*. In such case however, the left term of (4.66) takes a more complex expression: these estimates, in fact, because of their *directional* nature, must take into account the information coming from the characteristic directions associated to the single triangles which replace the "global" information concentrated in the seminorm $|v|_{\mathrm{H}^{r+1}(K)}$. The interested reader can consult [Ape99, FP01]. Moreover, we refer to Fig. 4.18 and 12.15 for examples of anisotropic grids. •

4.5.4 Estimate of the approximation error in the L^2 norm

The inequality (4.72) provides an estimate of the approximation error in the energy norm. Analogously, it is possible to obtain an error estimate in the L^2 norm. Since the latter norm is weaker than the former one, one must expect a higher convergence rate with respect to h.

Lemma 4.6 (Elliptic regularity). *Consider the homogeneous Dirichlet problem*

$$\begin{cases} -\Delta w = g & \text{in } \Omega, \\ w = 0 & \text{on } \partial\Omega, \end{cases}$$

with $g \in \mathrm{L}^2(\Omega)$. If $\partial\Omega$ is sufficiently regular (for instance, if $\partial\Omega$ is a curve of class C^2, or else if Ω is a convex polygon), then $w \in \mathrm{H}^2(\Omega)$, and moreover there exists a constant $C > 0$ such that

$$\|w\|_{\mathrm{H}^2(\Omega)} \leq C\|g\|_{\mathrm{L}^2(\Omega)}. \tag{4.74}$$

For the proof see, e.g., [Bre86, Gri11].

Theorem 4.7. *Let $u \in V$ be the exact solution of the variational problem* (4.1) *and u_h its approximate solution obtained with the finite element method of degree r. Moreover, let $u \in \mathrm{C}^0(\bar{\Omega}) \cap \mathrm{H}^{p+1}(\Omega)$ for a given $p > 0$. Then, the following* a priori *error estimate in the norm of* $\mathrm{L}^2(\Omega)$ *holds*

$$\|u - u_h\|_{\mathrm{L}^2(\Omega)} \leq Ch^{s+1}|u|_{\mathrm{H}^{s+1}(\Omega)}, \quad s = \min\{r, p\}, \tag{4.75}$$

C being a constant independent of h and u.

Proof. We will limit ourselves to proving this result for the Poisson problem (3.13), the weak formulation of which is given in (3.18). Let $e_h = u - u_h$ be the approximation error, and consider the following auxiliary Poisson problem (called *adjoint problem*, see Sect. 3.5) with source term given by the error function e_h

$$\begin{cases} -\Delta\phi = e_h & \text{in } \Omega, \\ \phi = 0 & \text{on } \partial\Omega, \end{cases} \tag{4.76}$$

whose weak formulation is

$$\text{find } \phi \in V: \quad a(\phi, v) = \int_\Omega e_h v \, d\Omega \qquad \forall v \in V, \tag{4.77}$$

with $V = \mathrm{H}_0^1(\Omega)$. Taking $v = e_h \, (\in V)$, we have

$$\|e_h\|^2_{\mathrm{L}^2(\Omega)} = a(\phi, e_h).$$

Since the bilinear form is symmetric, by the Galerkin orthogonality (4.8) we have

$$a(e_h, \phi_h) = a(\phi_h, e_h) = 0 \qquad \forall \phi_h \in V_h.$$

It follows that

$$\|e_h\|^2_{\mathrm{L}^2(\Omega)} = a(\phi, e_h) = a(\phi - \phi_h, e_h). \tag{4.78}$$

Now, taking $\phi_h = \Pi_h^1 \phi$, applying the Cauchy-Schwarz inequality to the bilinear form $a(\cdot,\cdot)$ and using the interpolation error estimate (4.69) we obtain

$$\|e_h\|^2_{\mathrm{L}^2(\Omega)} \le |e_h|_{\mathrm{H}^1(\Omega)} |\phi - \phi_h|_{\mathrm{H}^1(\Omega)} \le |e_h|_{\mathrm{H}^1(\Omega)} C h |\phi|_{\mathrm{H}^2(\Omega)}. \tag{4.79}$$

Notice that the interpolation operator Π_h^1 can be applied to ϕ since, $\phi \in \mathrm{H}^2(\Omega)$ thanks to Lemma 4.6 and thus, in particular, $\phi \in C^0(\overline{\Omega})$, thanks to property 2.3 in Chap. 2. By applying Lemma 4.6 to the adjoint problem (4.76) we obtain the inequality

$$|\phi|_{\mathrm{H}^2(\Omega)} \le C \|e_h\|_{\mathrm{L}^2(\Omega)}, \tag{4.80}$$

which, applied to (4.79), eventually provides

$$\|e_h\|_{\mathrm{L}^2(\Omega)} \le C h |e_h|_{\mathrm{H}^1(\Omega)},$$

where C accounts for all the constants that have appeared so far. By now exploiting the error estimate in the energy norm (4.72), we obtain (4.75). ◇

Let us generalize the result we have just proved for the Poisson problem to the case of a generic elliptic boundary-value problem approximated with finite elements and for which an estimate of the approximation error in the energy norm such as (4.72) holds, and an elliptic regularity property analogous to the one expressed in Lemma 4.6 is true.

In particular, let us consider the case where the bilinear form $a(\cdot,\cdot)$ is not necessarily symmetric. Let u be the exact solution of the problem

$$\text{find } u \in V: \quad a(u,v) = (f,v) \quad \forall v \in V, \tag{4.81}$$

and u_h the solution of the Galerkin problem

$$\text{find } u_h \in V_h: \quad a(u_h,v_h) = (f,v_h) \quad \forall v_h \in V_h.$$

Finally, suppose that the error estimate (4.72) holds and let us consider the following problem, which we will call *adjoint problem* of (4.81): for each $g \in \mathrm{L}^2(\Omega)$,

$$\text{find } \phi = \phi(g) \in V: \quad a^*(\phi,v) = (g,v) \quad \forall v \in V, \tag{4.82}$$

where we have defined (see (3.40)) $a^*: V \times V \to \mathbb{R}$.
Obviously if a is symmetric the two problems coincide, as seen for instance in the case of problem (4.77).
Let us suppose that for the solution u of the primal problem (4.81) an elliptic regularity result holds; it can then be verified that the same result is valid for the adjoint problem (4.82), that is

$$\exists C > 0: \qquad \|\phi(g)\|_{\mathrm{H}^2(\Omega)} \le C\|g\|_{\mathrm{L}^2(\Omega)} \quad \forall g \in \mathrm{L}^2(\Omega).$$

In particular, this is true for a generic elliptic problem with Dirichlet or Neumann (but not mixed) data on a polygonal and convex domain Ω [Gri11]. We now choose $g = e_h$ and denote, for simplicity, $\phi = \phi(e_h)$. Furthermore, having chosen $v = e_h$, we have

$$\|e_h\|^2_{\mathrm{L}^2(\Omega)} = a(e_h,\phi).$$

Since by the elliptic regularity of the adjoint problem $\phi \in \mathrm{H}^2(\Omega)$, and $\|\phi\|_{\mathrm{H}^2(\Omega)} \le C\|e_h\|_{\mathrm{L}^2(\Omega)}$ thanks to the Galerkin orthogonality, we have that

$$\begin{aligned}
\|e_h\|^2_{\mathrm{L}^2(\Omega)} &= a(e_h,\phi) = a(e_h,\phi - \Pi_h^1\phi) \\
&\le C_1\|e_h\|_{\mathrm{H}^1(\Omega)}\,\|\phi - \Pi_h^1\phi\|_{\mathrm{H}^1(\Omega)} \\
&\le C_2\|e_h\|_{\mathrm{H}^1(\Omega)}\, h\,\|\phi\|_{\mathrm{H}^2(\Omega)} \\
&\le C_3\|e_h\|_{\mathrm{H}^1(\Omega)}\, h\,\|e_h\|_{\mathrm{L}^2(\Omega)},
\end{aligned}$$

where we have exploited the continuity of the form $a(\cdot,\cdot)$ and the estimate (4.72). Thus

$$\|e_h\|_{\mathrm{L}^2(\Omega)} \le C_3 h\|e_h\|_{\mathrm{H}^1(\Omega)},$$

from which (4.75) follows, using the estimate (4.73) of the error in $\mathrm{H}^1(\Omega)$.

Remark 4.7. The technique illustrated above, depending upon the use of the adjoint problem for the estimate of the L^2-norm of the discretization error, is known in the

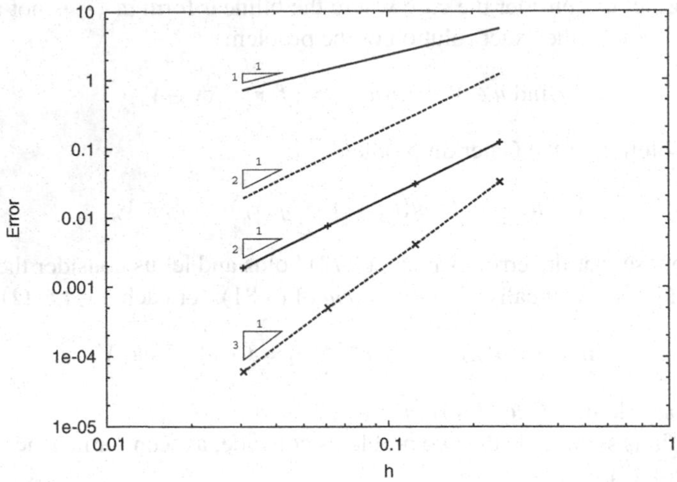

Fig. 4.16. Behaviour with respect to h of the error in $H^1(\Omega)$ norm (lines without crosses) and in $L^2(\Omega)$ norm (lines with crosses) for linear (solid lines) and quadratic (etched lines) finite elements for the solution of the problem reported in Example 4.2

literature as *Aubin-Nitsche trick* [Aub67, Nit68]. Several examples of how to determine the adjoint of a given problem will be presented in Sect. 3.5. •

Example 4.2. We consider the model problem $-\Delta u + u = f$ in $\Omega = (0,1)^2$ with $u = g$ on $\partial\Omega$. Suppose to choose the source term f and the function g so that the exact solution of the problem is $u(x,y) = \sin(2\pi x)\cos(2\pi y)$. We solve such a problem with the Galerkin method with finite elements of degree 1 and 2 on a uniform grid with step-size h. The graph of Fig. 4.16 shows the behaviour of the error when the grid-size h decreases, both in the norm $L^2(\Omega)$ and in that of $H^1(\Omega)$. As shown by inspecting the slope of the lines in the figure, the error's decrease when using L^2 norm (crossed lines) is quadratic if linear finite elements are used (solid line), and cubic when quadratic finite elements are used (etched line).

With respect to the H^1 norm (lines without crosses) instead, there is a linear reduction of the erorr with respect to the linear finite elements (solid line), and quadratic when quadratic finite elements are used (etched line). Fig. 4.17 shows the solution on the grid with grid-size 1/8 obtained with linear (left) and quadratic (right) finite elements. ■

4.6 Grid adaptivity

In Theorems 4.6 and 4.7 we have derived some a priori estimates for the finite element approximation error.
Since the parameter h is the maximal length of the finite element edges, if we referred to (4.72) we could be tempted to refine the grid everywhere in the hope of reducing

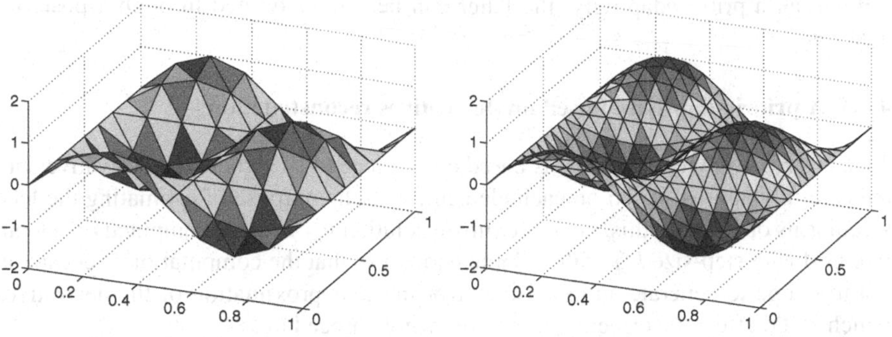

Fig. 4.17. Solutions computed using piecewise linear (left) and piecewise quadratic (right) finite elements on a uniform grid with grid-size 1/8

the error $\|u-u_h\|_{\mathrm{H}^1(\Omega)}$. However, it is more convenient to refer to (4.71) where the upper bound is the sum of elemental contributions involving the solution seminorm $|u|_{\mathrm{H}^{r+1}(K)}$ on each element K and the local grid-size h_K.
Indeed, in order to have an efficient grid that minimizes the number of elements necessary to obtain the desired accuracy, we can *equidistribute* the error on each element $K \in \mathcal{T}_h$. In particular, we would like to obtain

$$h_K^r |u|_{\mathrm{H}^{r+1}(K)} \simeq \eta \qquad \forall K \in \mathcal{T}_h,$$

where η is a well-chosen constant that only depends on the desired accuracy and on the number of elements of the grid.
A larger contribution from $|u|_{\mathrm{H}^{r+1}(K)}$ (due to a more pronounced variability of $u|_K$) will need to be balanced either by a smaller local grid-size h_K or by a higher polynomial degree r. In the first case, we will talk about *h-adaptivity* of the grid, in the second case of *p-adaptivity* (where p stands for "polynomial"). In the remainder of this chapter we will only focus on the first technique. However, we refer to Chap. 10 for the analysis of error estimates which are better suited for polynomial adaptivity.

The remarks made up to now, although correct, turn out to be of little use as the solution u is not known. We can therefore proceed according to different strategies. The first one is to use the a priori error estimate (4.71) by replacing the exact solution u with a well-chosen approximation, easily computable on each single element. In such case, we talk about *a priori adaptivity*.

A second approach is instead based on the use of an *a posteriori error estimate* able to link the approximation error to the behaviour of the approximate numerical solution u_h, known after solving the problem numerically. In such case, the optimal computational grid will be constructed through an iterative process where *solution, error estimate* and *modification of the computational grid* are recomputed until reaching the requested accuracy. In this case, we talk about *a posteriori adaptivity*.

The a priori and a posteriori adaptivity strategies are not mutually exclusive, actually they can coexist. For instance, having generated an appropriate starting grid through an a priori adaptivity, the latter can be further refined through a posteriori analysis.

4.6.1 A priori adaptivity based on derivatives reconstruction

An a priori adaptivity technique is based on estimate (4.71) where the derivatives of u are carefully approximated on each element, with the purpose of estimating the local seminorms of u. To do this, an approximate solution u_{h^*} is used, computed on a tentative grid with step-size h^*, with h^* large enough so that the computation is cheap, but not too large to generate an excessive error in the approximation of the derivatives, which could affect the effectiveness of the whole procedure.

We exemplify the algorithm for linear finite elements, in which case (4.71) takes the form

$$\|u-u_h\|_{\mathrm{H}^1(\Omega)} \leq C\left(\sum_{K\in\mathcal{T}_h} h_K^2 |u|^2_{\mathrm{H}^2(K)}\right)^{\frac{1}{2}} \tag{4.83}$$

(C accounts for the continuity and coercivity constants of the bilinear form). Our aim is eventually to solve our problem on a grid $\mathcal{T}_h$ guaranteeing that the right-hand side of (4.83) stands below a predefined tolerance $\varepsilon > 0$. Let us suppose that we have computed a solution, say u_{h^*}, on a preliminary grid $\mathcal{T}_{h^*}$ with N^* triangles. We use u_{h^*} to approximate the second derivatives of u that intervene in the definition of the seminorm $|u|^2_{\mathrm{H}^2(K)}$. Since u_{h^*} does not have any continuous second derivatives in Ω, it is necessary to proceed with an adequate *reconstruction technique*. For each node $\mathbf{N}_i$ of the grid we consider the set (*patch*) $K_{\mathbf{N}_i}$ of the elements sharing $\mathbf{N}_i$ as a node (that is the set of the elements forming the support of φ_i, see Fig. 4.12). We then find the planes $\pi_i^j(\mathbf{x}) = \mathbf{a}_i^j \cdot \mathbf{x} + b_i^j$ by minimizing

$$\int_{K_{\mathbf{N}_i}} \left|\pi_i^j(\mathbf{x}) - \frac{\partial u_{h^*}}{\partial x_j}(\mathbf{x})\right|^2 d\mathbf{x}, \qquad j=1,2, \tag{4.84}$$

solving a two-equation system for the coefficients $\mathbf{a}_i^j$ and b_i^j. This can be regarded as the local projection phase. We thus build a piecewise linear approximation $\mathbf{g}_{h^*} \in (X_{h^*}^1)^2$ of the gradient ∇u_{h^*} defined as

$$[\mathbf{g}_{h^*}(\mathbf{x})]^j = \sum_i \pi_i^j(\mathbf{x}_i)\varphi_i(\mathbf{x}), \qquad j=1,2, \tag{4.85}$$

where the sum spans over all the nodes $\mathbf{N}_i$ of the grid. Once the gradient is reconstructed we can proceed in two different ways, depending on the type of reconstruction that we want to obtain for the second derivatives. We recall first of all that the Hessian matrix associated to a function u is defined by $\mathbf{D}^2(u) = \nabla(\nabla u)$, that is

$$\left[\mathbf{D}^2(u)\right]_{i,j} = \frac{\partial^2 u}{\partial x_i \partial x_j}, \quad i,j=1,2.$$

A *piecewise constant* approximation of the latter is obtained by setting, for each $K^* \in \mathcal{T}_{h^*}$,

$$\mathbf{D}_h^2\big|_{K^*} = \frac{1}{2}\left(\nabla \mathbf{g}_{h^*} + (\nabla \mathbf{g}_{h^*})^T\right)\big|_{K^*}. \tag{4.86}$$

Notice the use of the symmetric form of the gradient, which is necessary for Hessian symmetry.

Should one be interested in a piecewise linear reconstruction of the Hessian, the same projection technique defined by (4.84) and (4.85) could be directly applied to the reconstructed $\mathbf{g}_{h^*}$, by then symmetrizing the matrix obtained in this way via (4.86). In any case, we are now able to compute an approximation of $|u|_{\mathrm{H}^2(K^*)}$ on a generic triangle K^* of $\mathcal{T}_{h^*}$, an approximation that will obviously be linked to the reconstructed $\mathbf{D}_h^2$.

From (4.83) we deduce that, to obtain the approximate solution u_h with an error smaller than or equal to a predefined tolerance ε, we must construct a new grid $\mathcal{T}_h^{new}$ such that

$$\sum_{K \in \mathcal{T}_h^{new}} h_K^2 |u|_{\mathrm{H}^2(K)}^2 \simeq \sum_{K \in \mathcal{T}_h^{new}} h_K^2 \sum_{i,j=1}^{2} \|[\mathbf{D}_h^2]_{ij}\|_{\mathrm{L}^2(K)}^2 \leq \left(\frac{\varepsilon}{C}\right)^2.$$

Ideally one would wish the error to be equidistributed on each element K of the new grid.

A possible adaptation procedure then consists in generating the new grid by appropriately partitioning all of the N^* triangles K^* of $\mathcal{T}_{h^*}$ for which we have

$$\eta_{K^*}^2 = h_{K^*}^2 \sum_{i,j=1}^{2} \|[\mathbf{D}_h^2]_{ij}\|_{\mathrm{L}^2(K^*)}^2 > \frac{1}{N^*}\left(\frac{\varepsilon}{C}\right)^2. \tag{4.87}$$

This method is said to be a *refinement* as it only aims at creating a *finer* grid than the initial one, but it clearly does not allow to fully satisfy the equidistribution condition.

More sophisticated algorithms also allow to *derefine* the grid in presence of the triangles for which the inequality (4.87) is verified with the sign $\ll$ (i.e. much smaller than) instead of $>$. However, derefinement procedures are of more difficult implementation than refinement ones. Hence, one often prefers to construct the new grid from scratch (a procedure called *remeshing*). For this purpose, on the basis of the error estimate, the following *spacing function* H (constant on each element) is introduced

$$H\big|_{K^*} = \frac{\varepsilon}{C\sqrt{N^*}\left(\sum_{i,j=1}^{2} \|[\mathbf{D}_h^2]_{ij}\|_{\mathrm{L}^2(K)}^2\right)^{1/2} |u_{h^*}|_{\mathrm{H}^2(K^*)}} \quad \forall K^* \in \mathcal{T}_{h^*} \tag{4.88}$$

and is used to construct the adapted grid by applying one of the grid generation algorithms illustrated in Chap. 6. The adaptation algorithm often requires the function H to be continuous and linear on each triangle. In this case we can again resort to a local projection, like that in (4.84).

The adaptation can then be repeated for the solution computed on the new grid, until inequality (4.87) is inverted on all of the elements.

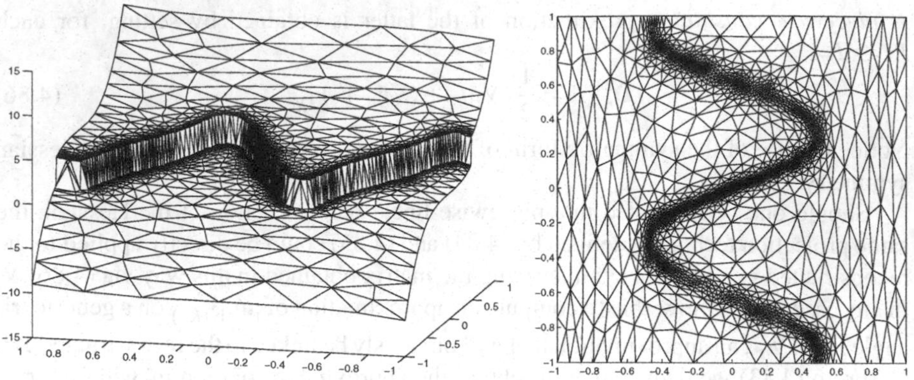

Fig. 4.18. The function u (left) and the third adapted grid (right) for Example 4.3

Remark 4.8. The C constant appearing in inequality (4.83) can be estimated by applying the same inequality to known functions (which makes therefore possible to compute the exact error). An alternative that does not require explicitly knowing C consists in realizing the grid that equally distributes the error for a number N^* of a priori fixed elements. In this case the value of H computed by setting ε and C to one in (4.88) is rescaled, by multiplying it by a constant, so that the new grid has a number N^* of elements fixed a priori. •

Example 4.3. We consider the function $u(x,y) = 10x^3 + y^3 + \tan^{-1}(10^{-4}/(\sin(5y) - 2x))$ on the domain $\Omega = (-1,1)^2$, which features a strong gradient across the curve $x = 0.5\sin(5y)$, as can be observed from Fig. 4.18 on the left. Starting from an initial structured grid constituted by 50 triangles and using an adaptive procedure guided by the Hessian of u, we obtain, after 3 iterations, the grid in Fig. 4.18 (right), made of 3843 elements. Most of the triangles are located in the proximity of the functions' jump: indeed, while few medium-large surface triangles are necessary to describe u in a satisfactory way in the regions located far enough from the jump, the abrupt variation of u in presence of discontinuities requires the use of small triangles, i.e. a reduced discretization grid-size. Furthermore, we note the anisotropic nature of the grid in Fig. 4.18, visible by the presence of elements whose shape is very stretched with respect to that of an equilateral triangle (typical of an isotropic grid). Such grid has been obtained by generalizing the estimator (4.87) to the anisotropic case. The idea is essentially to exploit the information provided by the components $[\mathbf{D}_h^2]_{ij}$ *separately* instead of "mixing" them through the $L^2(K^*)$ norm. By using the same adaptive procedure in the isotropic case (i.e. the estimator in (4.87)), we would have obtained, after 3 iterations, an adapted grid made of 10535 elements. ■

4.6.2 A posteriori adaptivity

The procedures described in the previous section can be unsatisfactory because the recostruction of u's derivatives starting from u_{h^*} is often subject to errors that are not easy to quantify.

A radical alternative consists in adopting *a posteriori estimates* of the error. The latter do not make use of the a priori estimate (4.71) (and consequently of any approximate derivatives of the unknown solution u). Rather, they are obtained as a function of *computable* quantities, normally based on the so-called *residue* of the approximate solution.

Let us consider problem (4.1) together with its Galerkin approximation (4.2). We define the residue $R \in V'$ by

$$\langle R, v\rangle = F(v) - a(u_h, v) \quad \forall v \in V, \tag{4.89}$$

that is

$$\langle R, v\rangle = a(u - u_h, v) \quad \forall v \in V. \tag{4.90}$$

Then

$$\alpha\|u - u_h\|_V \le \|R\|_{V'} \le M\|u - u_h\|_V. \tag{4.91}$$

Indeed, using (4.90) and the continuity of $a(\cdot,\cdot)$,

$$\|R\|_{V'} = \sup_{v\in V} \frac{\langle R, v\rangle}{\|v\|_V} \le M\|u - u_h\|_V.$$

On the other hand, taking $v = u - u_h$ in (4.90) and using the coercitivity of $a(\cdot,\cdot)$,

$$\begin{aligned}\alpha\|u - u_h\|_V^2 &\le a(u - u_h, u - u_h) = \langle R, u - u_h\rangle \\ &\le \|R\|_{V'}\|u - u_h\|_V,\end{aligned}$$

whence the first inequality of (4.91).
Now our goal is to express R in terms of computable quantities on every element K of the finite element triangulation. For the sake of exposition let us consider, as an example, the Poisson problem (3.13). Its weak formulation is given by (3.18), while its approximation using finite elements is described by (4.40), where V_h is the space $\overset{\circ}{X}{}^r_h$ defined in (4.39). In this specific case, $V = H_0^1(\Omega)$, $V' = H^{-1}(\Omega)$, $\alpha = M = 1$. Using the Galerkin orthogonality, together with (4.90) and (4.89), for every $v \in H_0^1(\Omega)$ and every $v_h \in V_h$, we have

$$\begin{aligned}\langle R, v\rangle &= \int_\Omega \nabla(u - u_h)\cdot\nabla v\, d\Omega = \int_\Omega \nabla(u - u_h)\cdot\nabla(v - v_h)\, d\Omega \\ &= \int_\Omega f(v - v_h)\, d\Omega - \int_\Omega \nabla u_h\cdot\nabla(v - v_h)\, d\Omega\end{aligned}$$

$$= \int_\Omega f(v - v_h)\, d\Omega + \sum_{K \in \mathcal{T}_h} \int_K \Delta u_h (v - v_h)\, d\Omega - \sum_{K \in \mathcal{T}_h} \int_{\partial K} \frac{\partial u_h}{\partial n}(v - v_h)\, d\gamma$$

$$= \sum_{K \in \mathcal{T}_h} \int_K (f + \Delta u_h)(v - v_h)\, d\Omega - \sum_{K \in \mathcal{T}_h} \int_{\partial K} \frac{\partial u_h}{\partial n}(v - v_h)\, d\gamma. \tag{4.92}$$

We observe that all the local integrals make sense.
Having denoted by e a side of the generic triangle K, we define the *jump* of the normal derivative of u_h through the internal side e the quantity

$$\left[\frac{\partial u_h}{\partial n}\right]_e = \nabla u_h\big|_{K_1} \cdot \mathbf{n}_1 + \nabla u_h\big|_{K_2} \cdot \mathbf{n}_2 = \left(\nabla u_h\big|_{K_1} - \nabla u_h\big|_{K_2}\right) \cdot \mathbf{n}_1, \tag{4.93}$$

where K_1 and K_2 are the two triangles sharing the side e, whose normal outgoing unit vectors are given by $\mathbf{n}_1$ and $\mathbf{n}_2$ respectively, with $\mathbf{n}_1 = -\mathbf{n}_2$ (see Fig. 4.19). In order to extend such definition also to the boundary sides, we introduce the so-called *generalized jump*, given by

$$\left[\frac{\partial u_h}{\partial n}\right] = \begin{cases} \left[\dfrac{\partial u_h}{\partial n}\right]_e & \text{for } e \in \mathcal{E}_h, \\ 0 & \text{for } e \in \partial\Omega, \end{cases} \tag{4.94}$$

where $\mathcal{E}_h$ indicates the set of inner sides in the grid. We note that, in the case of linear finite elements, (4.94) identifies a piecewise constant function defined on all the sides of the grid $\mathcal{T}_h$. Moreover, the definition (4.94) can be suitably modified in the case where problem (3.13) is completed with boundary conditions that are not necessarily of Dirichlet type.
Thanks to (4.94) we can therefore write that

$$\begin{aligned} -\sum_{K \in \mathcal{T}_h} \int_{\partial K} \frac{\partial u_h}{\partial n}(v - v_h)\, d\gamma &= -\sum_{K \in \mathcal{T}_h} \sum_{e \in \partial K} \int_e \frac{\partial u_h}{\partial n}(v - v_h)\, d\gamma \\ = -\sum_{K \in \mathcal{T}_h} \sum_{e \in \partial K} \frac{1}{2} \int_e \left[\frac{\partial u_h}{\partial n}\right](v - v_h)\, d\gamma &= -\frac{1}{2} \sum_{K \in \mathcal{T}_h} \int_{\partial K} \left[\frac{\partial u_h}{\partial n}\right](v - v_h)\, d\gamma, \end{aligned} \tag{4.95}$$

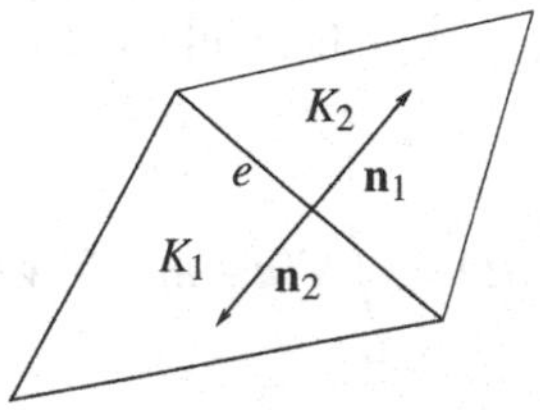

Fig. 4.19. Triangles involved in the definition of the jump of the normal derivative of u_h through an internal side e

where the factor $1/2$ takes into account the fact that each internal side e of the grid is shared by two elements. Moreover, since $v - v_h = 0$ on the boundary, in (4.94) we could assign any value different from zero in presence of $e \in \partial\Omega$, as the terms of (4.95) associated to the boundary sides would be null in any case.
By now inserting (4.95) in (4.92) and applying the Cauchy-Schwarz inequality, we obtain

$$\langle R, v\rangle \leq \sum_{K\in\mathcal{T}_h} \left\{ \|f + \Delta u_h\|_{\mathrm{L}^2(K)} \|v - v_h\|_{\mathrm{L}^2(K)} + \frac{1}{2} \left\| \left[\frac{\partial u_h}{\partial n} \right] \right\|_{\mathrm{L}^2(\partial K)} \|v - v_h\|_{\mathrm{L}^2(\partial K)} \right\}. \tag{4.96}$$

Now we look for $v_h \in V_h$ that allows to express the norms of $v - v_h$ as a function of a well-chosen norm of v. Moreover, we want this norm to be "local", i.e. computed over a region $\widetilde{K}$containing K, but as little as possible. If v were continuous, we could take as v_h the Lagrangian interpolant of v and use the previously cited interpolation error estimates on K. Unfortunately, in our case $v \in \mathrm{H}^1(\Omega)$ is not necessarily continuous. However, if $\mathcal{T}_h$ is a regular grid, we can introduce the so-called *Clément* interpolation operator $\mathcal{R}_h : \mathrm{H}^1(\Omega) \to V_h$ defined, in the case of linear finite elements, as

$$\mathcal{R}_h v(\mathbf{x}) = \sum_{\mathbf{N}_j} (P_j v)(\mathbf{N}_j)\varphi_j(\mathbf{x}) \qquad \forall v \in \mathrm{H}^1(\Omega), \tag{4.97}$$

where $P_j v$ denotes a local L^2 projection of v. More precisely it is a linear function defined on the patch $K_{\mathbf{N}_j}$ of the grid elements that share the node $\mathbf{N}_j$ (see Fig. 4.20), which is determined by the relations

$$\int_{K_{\mathbf{N}_j}} (P_j v - v)\psi\, d\mathbf{x} = 0 \quad \text{for } \psi = 1, x, y.$$

As usual, the φ_j are the characteristic Lagrangian basis functions of the finite element space under exam.

For each $v \in \mathrm{H}^1(\Omega)$ and each $K \in \mathcal{T}_h$, the following inequalities hold (see, e.g., [BG98, BS94, Clé75]):

$$\|v - \mathcal{R}_h v\|_{\mathrm{L}^2(K)} \leq C_1 h_K\, |v|_{\mathrm{H}^1(\widetilde{K})},$$
$$\|v - \mathcal{R}_h v\|_{\mathrm{L}^2(\partial K)} \leq C_2 h_K^{\frac{1}{2}}\, \|v\|_{\mathrm{H}^1(\widetilde{K})},$$

where C_1 and C_2 are two positive constants that depend on the minimal angle of the elements of the triangulation, while $\widetilde{K} = \{K_j \in \mathcal{T}_h : K_j \cap K \neq \emptyset\}$ represents the union of K with all the triangles that share an edge or a vertex with it (see Fig. 4.20).
Alternatively to $\mathcal{R}_h v$ we could use the local Scott-Zhang interpolation operator, see [BS94, Sect. 4.8]. The rest of the proof would proceed similarly.

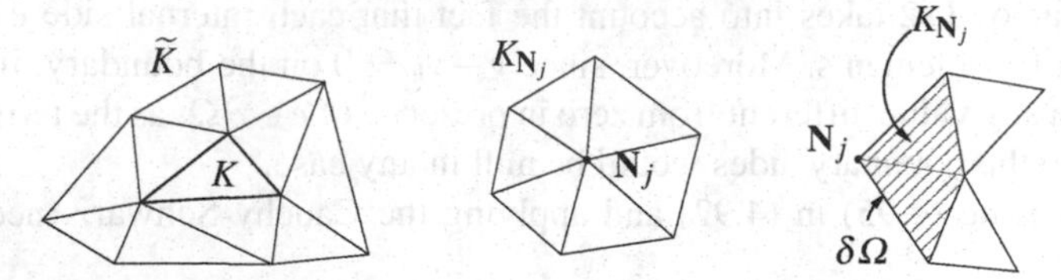

Fig. 4.20. The set $\widetilde{K}$ of elements that have in common with K at least a node of the grid (left), and the set $K_{\mathbf{N}_j}$ of the elements that share node $\mathbf{N}_j$ (middle and right)

By choosing in (4.96) $v_h = \mathcal{R}_h v$, setting $C = \max(C_1, C_2)$ and using the discrete Cauchy-Schwarz inequality, we obtain

$$
\begin{aligned}
\langle R, v\rangle &\leq C \sum_{K\in\mathcal{T}_h} \rho_K(u_h) \|v\|_{\mathrm{H}^1(\widetilde{K})} \\
&\leq C \left(\sum_{K\in\mathcal{T}_h} [\rho_K(u_h)]^2\right)^{\frac{1}{2}} \left(\sum_{K\in\mathcal{T}_h} \|v\|^2_{\mathrm{H}^1(\widetilde{K})}\right)^{\frac{1}{2}}.
\end{aligned}
$$

We have denoted by

$$
\rho_K(u_h) = h_K \, \|f + \Delta u_h\|_{\mathrm{L}^2(K)} + \frac{1}{2} h_K^{\frac{1}{2}} \left\| \left[\frac{\partial u_h}{\partial n}\right] \right\|_{\mathrm{L}^2(\partial K)} \tag{4.98}
$$

the so-called *local residue*, constituted by the internal residue $\|f + \Delta u_h\|_{\mathrm{L}^2(K)}$ and by the boundary residue $\left\| \left[\frac{\partial u_h}{\partial n}\right] \right\|_{\mathrm{L}^2(\partial K)}$.

We now observe that, since $\mathcal{T}_h$ is regular, the number of elements in $\widetilde{K}$ is necessarily bounded by a positive integer independent of h, which we denote by n. Thus,

$$
\|v\|_{H^1(\Omega)} \leq \left(\sum_{K\in\mathcal{T}_h} \|v\|^2_{\mathrm{H}^1(\widetilde{K})}\right)^{\frac{1}{2}} \leq \sqrt{n} \|v\|_{\mathrm{H}^1(\Omega)}.
$$

Because of the Poincaré inequality (2.13),

$$
\|v\|_{H^1(\Omega)} \leq C \|v\|_{H^1_0(\Omega)}, \quad C = \sqrt{1 + C_\Omega^2}
$$

(see the proof of Property 2.5), whence

$$
\|R\|_{H^{-1}(\Omega)} = \sup_{v\in H^1_0(\Omega)} \frac{\langle R, v\rangle}{\|v\|_{H^1_0(\Omega)}} \leq C\sqrt{n} \left(\sum_{k\in\mathcal{T}_h} [\rho_k(u_h)]^2\right)^{\frac{1}{2}}.
$$

Thanks to the first inequality of (4.91) (and the fact that $\alpha = 1$ in the current case), we conclude with the following *residual-based* a posteriori error estimate

$$\|u-u_h\|_{\mathrm{H}^1(\Omega)} \leq C\sqrt{n}\left(\sum_{K\in\mathcal{T}_h}[\rho_K(u_h)]^2\right)^{\frac{1}{2}}. \tag{4.99}$$

Notice that $\rho_K(u_h)$ is an effectively computable quantity, being a function of the datum f, of the geometric parameter h_K and of the computed solution u_h. The most delicate point of this analysis is the not-always-immediate estimate of the constants C and n.

The a posteriori estimate (4.99) can, for instance, be used in order to guarantee that

$$\frac{1}{2}\varepsilon \leq \frac{\|u-u_h\|_{\mathrm{H}^1(\Omega)}}{\|u_h\|_{\mathrm{H}^1(\Omega)}} \leq \frac{3}{2}\varepsilon, \tag{4.100}$$

$\varepsilon > 0$ being a pre-established tolerance. To this end, via an iterative procedure illustrated in Fig. 4.21, we can locally make finer and coarser the grid $\mathcal{T}_h$ until when, for each K, the following *local* inequalities are satisfied

$$\frac{1}{4}\frac{\varepsilon^2}{N}\|u_h\|^2_{\mathrm{H}^1(\Omega)} \leq [\rho_K(u_h)]^2 \leq \frac{9}{4}\frac{\varepsilon^2}{N}\|u_h\|^2_{\mathrm{H}^1(\Omega)}, \tag{4.101}$$

having denoted by N the number of elements of the grid $\mathcal{T}_h$. This ensures that the *global* inequalities (4.100) are satisfied, up to the contribution of the constant $C\sqrt{n}$. Alternatively, we can construct a well-chosen grid spacing function H, analogously to what was done in Sect. 4.6.1.

Naturally, the flow diagram reported in Fig. 4.21 can also be used for boundary-value problems differing from (4.40).

4.6.3 Numerical examples of adaptivity

We illustrate the concept of grid adaptivity on two simple differential problems. For this purpose, we adopt the iterative procedure reported in Fig. 4.21, although we will limit ourselves to the sole refinement phase. The coarsening process turns out to be of more difficult implementation: as a matter of fact, the most commonly used software only allows to refine the initial grid, hence it will be necessary to choose the latter to be suitably coarse.

Finally, for both reported examples, the reference estimator for the discretization error is represented by the right term of (4.99).

First example

Let us consider the problem $-\Delta u = f$ in $\Omega = (-1,1)^2$, with homogeneous Dirichlet conditions on the whole boundary $\partial\Omega$. Moreover, we choose a forcing term f such that the exact solution is $u = \sin(\pi x)\sin(\pi y)\exp(10x)$. We begin the adaptive procedure by starting from a uniform initial grid, made of 324 elements, and with a tolerance $\varepsilon = 0.2$. The iterative procedure converges after 7 iterations. We report in Fig. 4.22 the initial grid together with three of the adapted grids obtained in this way, while Table 4.2 summarizes the number $\mathcal{N}_h$ of elements of the grid $\mathcal{T}_h$, the relative error

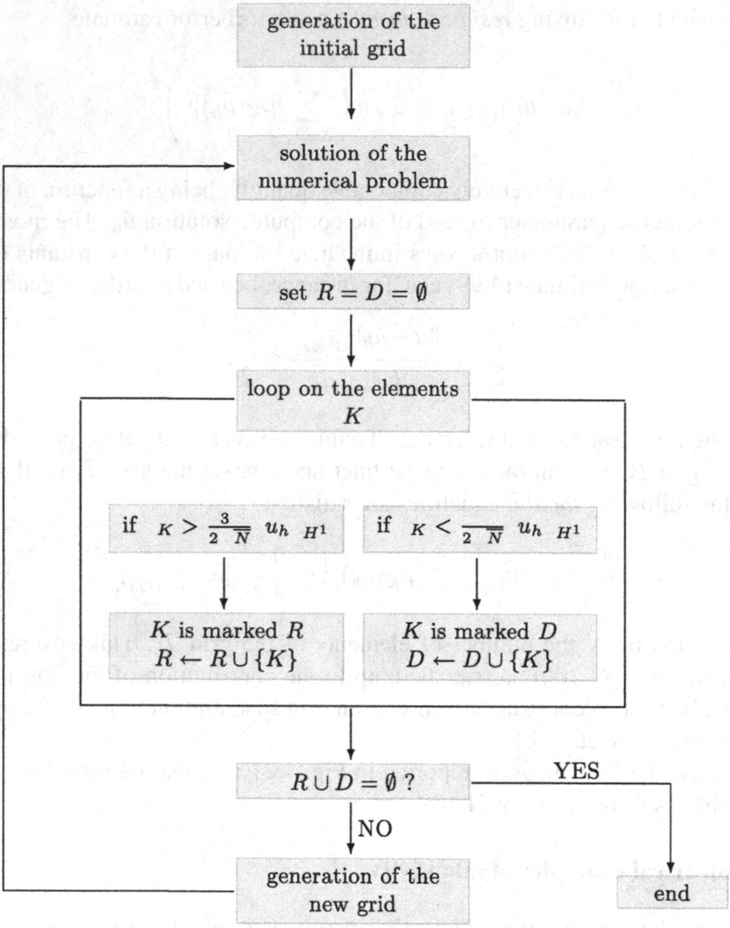

Fig. 4.21. Example of iterative grid adaptation procedure

Table 4.2. Cardinality, relative error and normalized estimator associated with the initial grid and with the first six adaptive grids

iteration	$\mathcal{N}_h$	$\|u-u_h\|_{H^1(\Omega)}/\|u_h\|_{H^1(\Omega)}$	$\eta/\|u_h\|_{H^1(\Omega)}$
0	324	0.7395	5.8333
1	645	0.3229	3.2467
2	1540	0.1538	1.8093
3	3228	0.0771	0.9782
4	7711	0.0400	0.5188
5	17753	0.0232	0.2888
6	35850	0.0163	0.1955

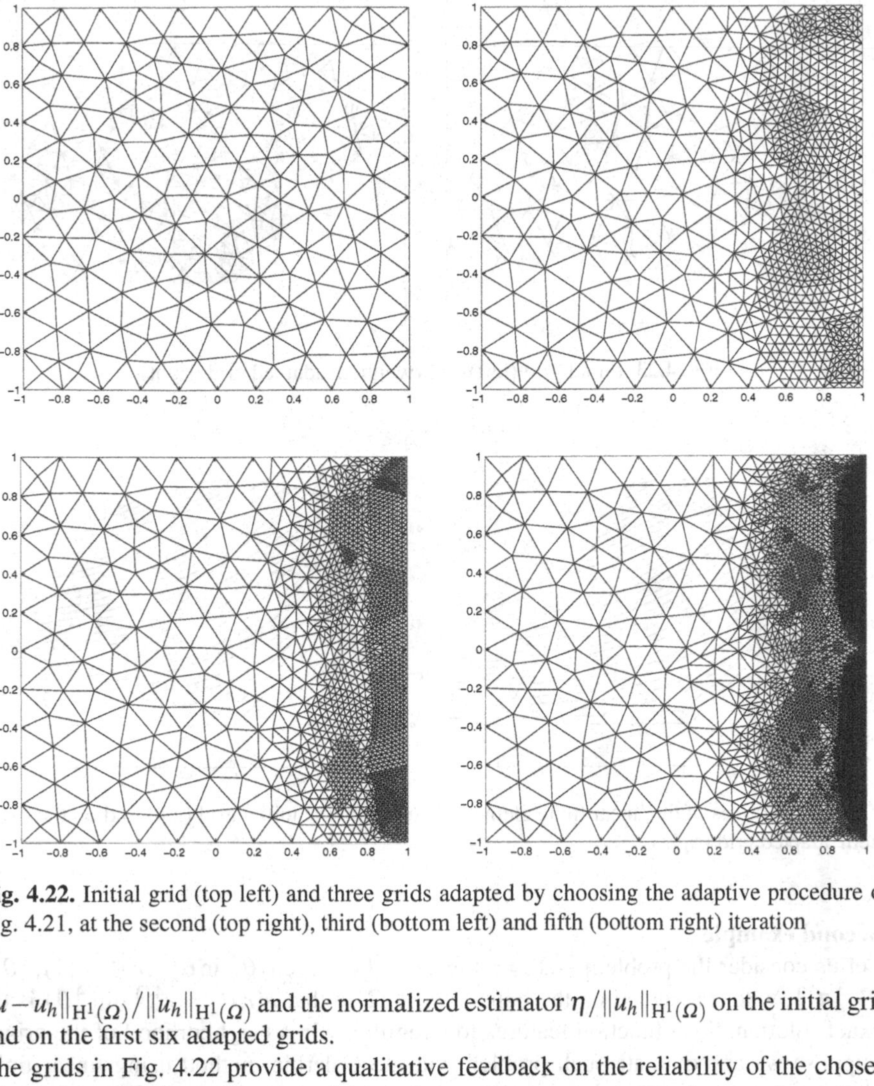

Fig. 4.22. Initial grid (top left) and three grids adapted by choosing the adaptive procedure of Fig. 4.21, at the second (top right), third (bottom left) and fifth (bottom right) iteration

$\|u-u_h\|_{H^1(\Omega)}/\|u_h\|_{H^1(\Omega)}$ and the normalized estimator $\eta/\|u_h\|_{H^1(\Omega)}$ on the initial grid and on the first six adapted grids.

The grids in Fig. 4.22 provide a qualitative feedback on the reliability of the chosen adaptivity procedure: as expected, triangles tend to concentrate in those regions where u attains its extrema. On the other hand, the values in Table 4.2 also allow to perform a quantitative analysis: both the relative error and the normalized estimator progressively decrease, when the iterations increase. However, we can notice an average overestimate of about 10-11 times with respect to the fixed tolerance ε. This is not unusual and can basically be explained by the fact that the constant $C\sqrt{n}$ in the inequalities (4.100) and (4.101) has been neglected (i.e. set to 1). It is clear that such choice actually leads to requiring a tolerance $\widetilde{\varepsilon} = \varepsilon/(C\sqrt{n})$, that will therefore coincide with the original ε only in the case where we have $C\sqrt{n} \sim 1$. More precise procedures, taking the constant $C\sqrt{n}$ into account, are in any case possible by starting, e.g., from the (theoretical and numerical) analysis provided in [BDR92, EJ88].

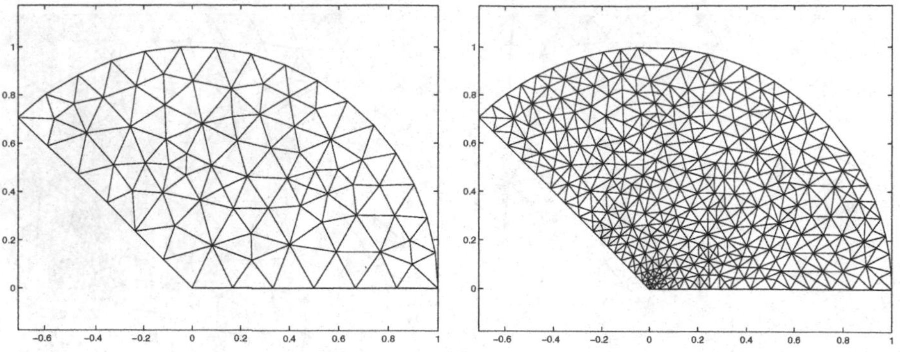

Fig. 4.23. Initial grid (left) and twentieth adapted grid (right)

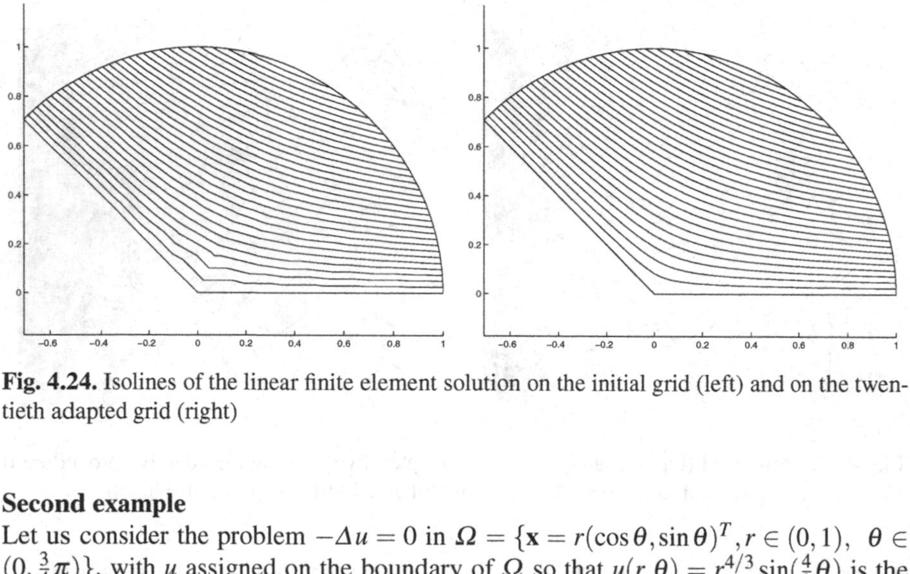

Fig. 4.24. Isolines of the linear finite element solution on the initial grid (left) and on the twentieth adapted grid (right)

Second example

Let us consider the problem $-\Delta u = 0$ in $\Omega = \{\mathbf{x} = r(\cos\theta, \sin\theta)^T, r \in (0,1),\ \ \theta \in (0, \frac{3}{4}\pi)\}$, with u assigned on the boundary of Ω so that $u(r,\theta) = r^{4/3}\sin(\frac{4}{3}\theta)$ is the exact solution. Such function features low regularity in a neighborhood of the origin. Suppose we approximate such problem via the Galerkin method using linear finite elements on the quasi-uniform grid drawn on in the left of Fig. 4.23, and made of 138 triangles. The distortion in the isolines of u_h in the left of Fig. 4.24 shows that the solution obtained in this way is quite inaccurate near the origin. We now use the estimator (4.99) to generate an adapted grid which better suits the approximation of u. By following an adaptive procedure such as the one illustrated in Fig. 4.21 we obtain after 20 steps the grid made of 859 triangles of in Fig. 4.23 on the right. As in Fig. 4.24 on the right, the isolines associated to the corresponding discrete solution denote a higher regularity, an evidence of the improved quality of the solution. As a comparison, in order to obtain a solution characterized by the same accuracy ε with respect to the norm H^1 of the error (required to be equal to 0.01) on a uniform grid, 2208 triangles are necessary.

4.6.4 A posteriori error estimates in the L^2 norm

Besides (4.99) it is possible to derive an a posteriori estimate of the error in L^2 norm. To this end, we will again resort to the duality technique of Aubin-Nitsche used in Sect. 4.5.4, and in particular we will consider the adjoint problem (4.76) associated to the Poisson problem (3.13). Moreover, we will suppose that the domain Ω is sufficiently regular (for instance, a convex polygon) in order to guarantee that the elliptic regularity result (4.74) stated in Lemma 4.6 is true.

Moreover, we will exploit the following local estimates for the interpolation error associated with the operator Π_h^r applied to functions $v \in H^2(\Omega)$

$$\|v - \Pi_h^r v\|_{L^2(\partial K)} \leq \widetilde{C}_1 h_K^{\frac{3}{2}} |v|_{H^2(K)} \tag{4.102}$$

(see [BS94] or [Cia78]), and

$$\|v - \Pi_h^r v\|_{L^2(K)} \leq \widetilde{C}_2 h_K^2 |v|_{H^2(K)}. \tag{4.103}$$

The latter inequality is obtained from (4.67).
Starting from the adjoint problem (4.76) and exploiting the Galerkin orthogonality (4.8), we have, for each $\phi_h \in V_h$,

$$\|e_h\|^2_{L^2(\Omega)} = \sum_{K\in\mathcal{T}_h} \int_K f(\phi - \phi_h)\, d\Omega - \sum_{K\in\mathcal{T}_h} \int_K \nabla u_h \cdot \nabla(\phi - \phi_h)\, d\Omega.$$

Counterintegrating by parts, we obtain

$$\|e_h\|^2_{L^2(\Omega)} = \sum_{K\in\mathcal{T}_h} \int_K (f + \Delta u_h)(\phi - \phi_h)\, d\Omega - \sum_{K\in\mathcal{T}_h} \int_{\partial K} \frac{\partial u_h}{\partial n}(\phi - \phi_h)\, d\gamma.$$

Using the definition (4.94) of generalized jump of the normal derivative of u_h across the triangle edges and setting $\phi_h = \Pi_h^r \phi$, we have

$$\|e_h\|^2_{L^2(\Omega)} = \sum_{K\in\mathcal{T}_h} \Bigg[\int_K (f + \Delta u_h)(\phi - \Pi_h^r\phi)\, d\Omega - \frac{1}{2} \int_{\partial K} \left[\frac{\partial u_h}{\partial n}\right] (\phi - \Pi_h^r\phi)\, d\gamma \Bigg]. \tag{4.104}$$

We estimate the two terms in the right-hand side separately. By using the Cauchy-Schwarz inequality and (4.103), it follows that

$$\begin{aligned} \left| \int_K (f + \Delta u_h)(\phi - \Pi_h^r\phi)\, d\Omega \right| &\leq \|f + \Delta u_h\|_{L^2(K)} \|\phi - \Pi_h^r\phi\|_{L^2(K)} \\ &\leq \widetilde{C}_2 h_K^2 \|f + \Delta u_h\|_{L^2(K)} |\phi|_{H^2(K)}. \end{aligned} \tag{4.105}$$

Moreover, thanks to (4.102) we obtain

$$\begin{aligned}\left|\int\limits_{\partial K}\left[\frac{\partial u_h}{\partial n}\right](\phi-\Pi_h^r\phi)\,d\gamma\right| &\leq \left\|\left[\frac{\partial u_h}{\partial n}\right]\right\|_{L^2(\partial K)}\|\phi-\Pi_h^r\phi\|_{L^2(\partial K)}\\ &\leq \widetilde{C}_1 h_K^{\frac{3}{2}}\left\|\left[\frac{\partial u_h}{\partial n}\right]\right\|_{L^2(\partial K)}|\phi|_{H^2(K)}.\end{aligned}\tag{4.106}$$

By now inserting (4.105) and (4.106) in (4.104) and applying the discrete Cauchy-Schwarz inequality we have

$$\begin{aligned}\|e_h\|^2_{L^2(\Omega)} \leq C\sum_{K\in\mathcal{T}_h} h_K\rho_K(u_h)|\phi|_{H^2(K)} &\leq C\sqrt{\sum_{K\in\mathcal{T}_h}[h_K\rho_K(u_h)]^2}|\phi|_{H^2(\Omega)}\\ &\leq C\sqrt{\sum_{K\in\mathcal{T}_h}[h_K\rho_K(u_h)]^2}\,\|e_h\|_{L^2(\Omega)},\end{aligned}$$

with $C=\max(\widetilde{C}_1,\widetilde{C}_2)$, having introduced the notation (4.98) and having exploited the elliptic regularity property (4.80) in the last inequality. We can then conclude that

$$\|u-u_h\|_{L^2(\Omega)} \leq C\left(\sum_{K\in\mathcal{T}_h} h_K^2[\rho_K(u_h)]^2\right)^{\frac{1}{2}},\tag{4.107}$$

$C>0$ being a constant independent of h.

Remark 4.9. Among the most widespread a posteriori estimates in engineering, we cite for its simplicity and computational effectiveness the estimator proposed by Zienkiewicz and Zhu in the context of a finite element approximation of linear elasticity problems [ZZ87]. The basic idea of this estimator is very simple. Suppose we want to control the energy norm $\left(\int_\Omega|\nabla u-\nabla u_h|^2\,d\Omega\right)^{1/2}$ of the discretization error associated to a finite element approximation of the model problem (3.13). This estimator replaces the exact gradient ∇u in the latter norm with a corresponding reconstruction obtained through a suitable post-processing of the discrete solution u_h. Over the years, several "recipes" have been proposed in the literature for the construction of the gradient ∇u (see, e.g., [ZZ92, Rod94, PWY90, LW94, NZ04, BMMP06]). The same procedure illustrated in Sect. 4.6.1 that leads to the reconstructed $\mathbf{g}_{h^*}$ defined in (4.85) can be used here for this purpose. Thus, having chosen a reconstruction, say $G_R(u_h)$, of ∇u, the Zienkiewicz and Zhu-type estimator is represented by the quantity $\eta=\left(\int_\Omega|G_R(u_h)-\nabla u_h|^2\,d\Omega\right)^{1/2}$. Clearly, to each new definition of $G_R(u_h)$ corresponds a new error estimator. For this reason, a posteriori error estimators with such structure are commonly called *recovery-based*. •

4.6.5 A posteriori estimates of a functional of the error

In the previous section, the adjoint problem (4.76) was used in a purely formal way, because the error e_h, that represents its forcing term, is unknown.

There exists another family of a posteriori estimators of the error, again based on the adjoint problem, which, instead, explicitly use the information provided by the latter (see, e.g., [Ran99]). In such case, an estimate is provided for a suitable functional J of the error e_h, instead of for a suitable norm of e_h. This prerogative turns out to be particularly useful whene one wants to provide significant estimates of the error for quantities of physical relevance, such as, for instance, resistance or drag in the case of bodies immersed in fluids, average values of concentration, strains, deformations, fluxes, etc. For this purpose, it will be sufficient to operate a suitable choice for the functional J. This type of adaptivity is called *goal-oriented*. To illustrate this new paradigm, let us still refer to the Poisson problem (3.13) and assume that we want to control the error of a given functional $J : H_0^1(\Omega) \to \mathbb{R}$ of the solution u. Let us consider the following weak formulation of the corresponding adjoint problem

$$\text{find } \phi \in V : \quad \int_{\Omega} \nabla\phi \cdot \nabla w \, d\Omega = J(w) \qquad \forall\, w \in V, \tag{4.108}$$

with $V = \mathrm{H}_0^1(\Omega)$. By using the Galerkin orthogonality and proceeding as done in the previous section, we find

$$\begin{aligned} J(e_h) = \int_{\Omega} \nabla e_h \cdot \nabla\phi \, d\Omega = \sum_{K \in \mathcal{T}_h} \Bigg[\int_K (f + \Delta u_h)\,(\phi - \phi_h)\, d\Omega \\ - \frac{1}{2} \int_{\partial K} \left[\frac{\partial u_h}{\partial n} \right] (\phi - \phi_h)\, d\gamma \Bigg], \end{aligned} \tag{4.109}$$

where $\phi_h \in V_h$ is typically a convenient interpolant of ϕ. By using the Cauchy-Schwarz inequality on each element K, we obtain

$$\begin{aligned} |J(e_h)| &= \left| \int_{\Omega} \nabla e_h \cdot \nabla\phi \, d\Omega \right| \\ &\leq \sum_{K \in \mathcal{T}_h} \left(\|f + \Delta u_h\|_{\mathrm{L}^2(K)} \, \|\phi - \phi_h\|_{\mathrm{L}^2(K)} + \frac{1}{2} \left\| \left[\frac{\partial u_h}{\partial n} \right] \right\|_{\mathrm{L}^2(\partial K)} \|\phi - \phi_h\|_{\mathrm{L}^2(\partial K)} \right) \\ &\leq \sum_{K \in \mathcal{T}_h} \left[\rho_K(u_h) \max\left(\frac{1}{h_K} \|\phi - \phi_h\|_{\mathrm{L}^2(K)}, \frac{1}{h_K^{1/2}} \|\phi - \phi_h\|_{\mathrm{L}^2(\partial K)} \right) \right], \end{aligned}$$

$\rho_K(u_h)$ being defined according to (4.98). We now introduce the so-called *local weights*

$$\omega_K(\phi) = \max\left(\frac{1}{h_K} \|\phi - \phi_h\|_{\mathrm{L}^2(K)}, \frac{1}{h_K^{1/2}} \|\phi - \phi_h\|_{\mathrm{L}^2(\partial K)} \right). \tag{4.110}$$

Thus,

$$|J(e_h)| \leq \sum_{K \in \mathcal{T}_h} \rho_K(u_h)\,\omega_K(\phi). \tag{4.111}$$

We can observe that, in contrast to the residue-type estimates introduced in Sects. 4.6.2 and 4.6.4, the estimate (4.111) depends not only on the discrete solution u_h

but also on the solution ϕ of the dual problem. In particular, having considered the local estimator $\rho_K(u_h)\omega_K(\phi)$, we can say that, while the residue $\rho_K(u_h)$ measures how the discrete solution approximates the differential problem under exam, the weight $\omega_K(\phi)$ takes into account how this information is propagated in the domain as an effect of the chosen functional. Hence, the grids obtained for different choices of the functional J, i.e. of the forcing term of the adjoint problem (4.108), will be different even if we start from the same differential problem (for more details, we refer to Example 12.12). Moreover, to make the estimate (4.111) efficient, we proceed by replacing the norms $\|\phi-\phi_h\|_{L^2(K)}$ and $\|\phi-\phi_h\|_{L^2(\partial K)}$ in (4.110) with suitable estimates of the interpolation error, having chosen ϕ_h as a suitable interpolant of the dual solution ϕ.

We point out two particular cases. Choosing $J(w) = \int_\Omega w e_h d\Omega$ in (4.108) we would find again the estimate (4.107) for the L^2-norm of the discretization error, provided of course that we can guarantee that the elliptic regularity result (4.74), stated in Lemma 4.6, is true. Instead, if we are interested in controlling e_h at a point $\mathbf{x}$ of Ω, it will be indeed sufficient to define J as $J(w) = {}_{W'}\langle\delta_{\mathbf{x}}, w\rangle_W$, with $W = H^1_0(\Omega)\cap C^0(\overline{\Omega})$ and $\delta_{\mathbf{x}}$ being Dirac's delta function at $\mathbf{x}$ (see Chapter 2).

Remark 4.10. The a posteriori analysis of this section, as well as that of the previous Sects. 4.6.2 and 4.6.4, can be extended to the case of more complex differential problems, like for instance transport and diffusion problems, and more general boundary conditions (see Example 12.12). The procedure remains basically the same. What changes is the definition of the local residue (4.98) and of the generalized jump (4.94). Indeed, while $\rho_K(u_h)$ directly depends on the differential formulation of the problem under exam, $[\partial u_h/\partial n]$ will need to take into account the conditions assigned on the boundary. •

For a thorough description of the adaptivity techniques provided up to now and for a presentation of other possible adaptive techniques, we refer the reader to [Ver96, Ran99, AO00].

4.7 Exercises

1. *Heat transfer in a thin rod.*
   Let us consider a thin rod of length L, having temperature t_0 at the endpoint $x = 0$ and insulated at the other endpoint $x = L$. Let us suppose that the cross-section of the rod has constant area equal to A and that the perimeter of A is p. The temperature t of the rod at a generic point $x \in (0,L)$ then satisfies the following mixed boundary-value problem

$$\begin{cases} -kAt'' + \sigma p t = 0, & x \in (0,L), \\ t(0) = t_0, & t'(L) = 0, \end{cases} \tag{4.112}$$

   having denoted by k the thermal conductivity coefficient and by σ the convective transfer coefficient.

Verify that the exact solution of this problem is

$$t(x) = t_0 \frac{\cosh[m(L-x)]}{\cosh(mL)},$$

with $m = \sqrt{\sigma p/kA}$. Write the weak formulation of (4.112), then its Galerkin-finite element approximation. Show how the approximation error in the $\mathrm{H}_0^1(0,L)$-norm depends on the parameters k, σ, p and t_0.
Finally, solve this problem using linear and quadratic finite elements on uniform grids, then evaluate the approximation error.

2. *Temperature of a fluid between two parallel plates.*
We consider a viscous fluid located between two horizontal parallel plates, at a distance of $2H$. Suppose that the upper plate, which has temperature t_{sup}, moves at a relative speed of U with respect to the lower one, having temperature t_{inf}. In such case the temperature $t : (0,2H) \to \mathbb{R}$ of the fluid satisfies the following Dirichlet problem

$$\begin{cases} -\dfrac{d^2t}{dy^2} = \alpha(H-y)^2, & y \in (0,2H), \\ t(0) = t_{inf}, & t(2H) = t_{sup}, \end{cases}$$

where $\alpha = \frac{4U^2\mu}{H^4k}$, k being the thermal conductivity coefficient and μ the viscosity of the fluid. Find the exact solution $t(y)$, then write the weak formulation and the Galerkin finite element formulation.
[*Solution*: the exact solution is

$$t(y) = -\frac{\alpha}{12}(H-y)^4 + \frac{t_{inf}-t_{sup}}{2H}(H-y) + \frac{t_{inf}+t_{sup}}{2} + \frac{\alpha H^4}{12}.]$$

3. *Deformation of a rope.*
Let us consider a rope with tension T and unit length, fixed at the endpoints. The function $u(x)$, measuring the vertical displacement of the rope when subject to a transversal charge of intensity w, satisfies the following Dirichlet problem

$$\begin{cases} -u'' + \dfrac{k}{T}u = \dfrac{w}{T} & \text{in } (0,1), \\ u(0) = 0, & u(1) = 0, \end{cases}$$

having indicated with k the elasticity coefficient of the rope. Write the weak formulation and the Galerkin-finite element formulation.

4. Prove Property 4.1.
[*Solution:* it suffices to observe that $a_{ij} = a(\varphi_j, \varphi_i)\ \forall i,j$.]

5. Prove (4.12).
[*Solution:* since the form is symmetric, the procedure contained in Remark 3.2 can be repeated, noting that the solution u_h satisfies $a(u_h,v_h) = a(u,v_h)$ for each $v_h \in V_h$. We deduce therefore that u_h minimizes $J(v_h) = a(v_h,v_h) - 2a(u,v_h)$ and therefore

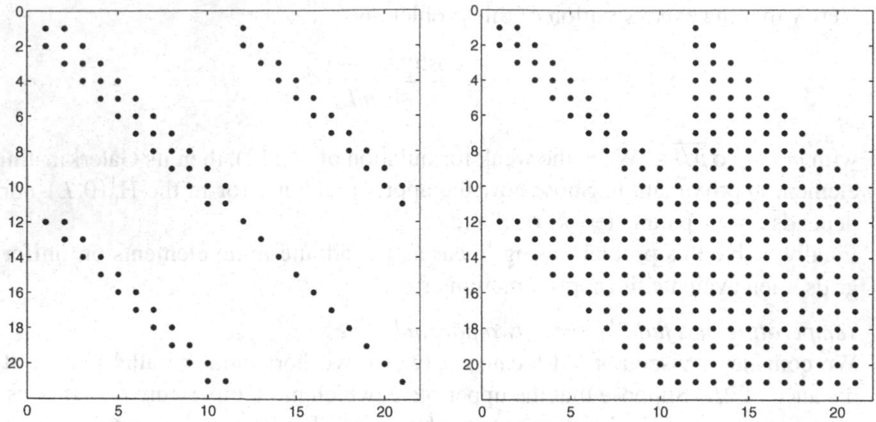

Fig. 4.25. Left: the sparsity pattern of the Galerkin finite element matrix associated to a discretization using 10 elements of the one-dimensional Poisson problem with quadratic finite elements. Unknowns are numbered as explained in Exercise 6. Right: the pattern of the L and U factors of A. Note that, because of the fill-in, the number of non-null finite elements has increased from 81 in the matrix to 141 in the factors

also $J^*(v_h) = J(v_h) + a(u,u) = a(u - v_h, u - v_h)$ (the last equality is made possible thanks to the symmetry of the bi-linear form). On the other hand,

$$\sqrt{\alpha}\|u - v_h\|_V \leq \sqrt{a(u - v_h, u - v_h)} \leq \sqrt{M}\|u - v_h\|_V,$$

hence the desired result.]

6. Given a partition of an interval (a,b) into $N+1$ sub-intervals, suppose to number first the endpoints of the single sub-intervals and then their midpoints. Is this labelling more or less convenient than the one introduced in Sect. 4.3 for the discretization of the Poisson problem with finite elements in X_h^2? Suppose to solve the linear system by a factorization method.
[*Solution:* the obtained matrix still has only five diagonals different from zero, as the one obtained using the numbering proposed in Sect. 4.3. However, it features a higher bandwidth. Consequently, in case it is factorized, it is subject to a larger fill-in, as shown in Fig. 4.25.]

7. Consider the following one-dimensional boundary-value problem

$$\begin{cases} -(\alpha u')' + \gamma u = f, & 0 < x < 1, \\ u = 0 & \text{at } x = 0, \\ \alpha u' + \delta u = 0 & \text{at } x = 1, \end{cases}$$

where $\alpha = \alpha(x)$, $\gamma = \gamma(x)$, $f = f(x)$ are assigned functions with $0 \leq \gamma(x) \leq \gamma_1$ and $0 < \alpha_0 \leq \alpha(x) \leq \alpha_1$ $\forall x \in [0,1]$, while $\delta \in \mathbb{R}$. Moreover, suppose that $f \in \mathrm{L}^2(0,1)$.

Write the problem's weak formulation specifying the appropriate functional spaces and hypotheses on the data to guarantee existence and uniqueness of the solution. Suppose to find an approximate solution u_h using the linear finite element method. What can be said about the existence, stability and accuracy of u_h?
[*Solution*: we seek $u \in V = \{v \in \mathrm{H}^1(0,1) : v(0) = 0\}$ such that $a(u,v) = F(v)$ $\forall v \in V$ where

$$a(u,v) = \int_0^1 \alpha u'v'\,dx + \int_0^1 \gamma uv\,dx + \delta u(1)v(1), \quad F(v) = \int_0^1 fv\,dx.$$

The existence and uniqueness of the solution of the weak problem are guaranteed if the hypotheses of the Lax-Milgram lemma hold. The form $a(\cdot,\cdot)$ is continuous as we have

$$|a(u,v)| \leq 2\max(\alpha_1,\gamma_1)\|u\|_V\|v\|_V + |\delta|\,|v(1)|\,|u(1)|,$$

from which, considering that $u(1) = \int_0^1 u'\,dx$, we obtain

$$|a(u,v)| \leq M\|u\|_V\|v\|_V \quad \text{with } M = 3\max(\alpha_1,\gamma_1,|\delta|).$$

We have coercivity if $\delta \geq 0$, for in such case we find

$$a(u,u) \geq \alpha_0\|u'\|^2_{\mathrm{L}^2(0,1)} + u^2(1)\delta \geq \alpha_0\|u'\|^2_{\mathrm{L}^2(0,1)}.$$

To find the inequality in $\|\cdot\|_V$ invoking the Poincaré inequality (2.13), it suffices to prove that

$$\frac{1}{1+C_\Omega^2}\|u\|_V^2 \leq \|u'\|^2_{\mathrm{L}^2(0,1)},$$

and then to conclude that

$$a(u,u) \geq \alpha^*\|u\|_V^2 \quad \text{with } \alpha^* = \frac{\alpha_0}{1+C_\Omega^2}.$$

The fact that F is a linear and continuous functional can be verified immediately. The finite element method is a Galerkin method with $V_h = \{v_h \in X_h^1 : v_h(0) = 0\}$. Consequently, thanks to Corollaries 4.1, 4.2 we deduce that the solution u_h exists and is unique. From the estimate (4.72) we furthermore deduce that, since $r = 1$, the error measured in the norm of V will tend to zero linearly with respect to h.]

8. Consider the following two-dimensional boundary-value problem

$$\begin{cases} -\mathrm{div}(\alpha\nabla u) + \gamma u = f & \text{in } \Omega \subset \mathbb{R}^2, \\ u = 0 & \text{on } \Gamma_D, \\ \alpha\nabla u \cdot \mathbf{n} = 0 & \text{on } \Gamma_N, \end{cases}$$

Ω being a bounded open domain having regular boundary $\partial\Omega = \Gamma_D \cup \Gamma_N$, with $\overset{\circ}{\Gamma}_D \cap \overset{\circ}{\Gamma}_N = \emptyset$ and unit outgoing normal $\mathbf{n}$; $\alpha \in \mathrm{L}^\infty(\Omega)$, $\gamma \in \mathrm{L}^\infty(\Omega)$, and $f \in \mathrm{L}^2(\Omega)$ are three assigned functions with $\gamma(\mathbf{x}) \geq 0$ and $0 < \alpha_0 \leq \alpha(\mathbf{x})$ a.e. in Ω.
Analyze the existence and uniqueness of the weak solution and the stability of the solution obtained using the Galerkin method. Suppose that $u \in \mathrm{H}^4(\Omega)$. Which polynomial degree would it be convenient to attain by using a finite element approximation?
[*Solution*: the weak problem consists in finding $u \in V = \mathrm{H}^1_{\Gamma_D}$ such that $a(u,v) = F(v)\ \forall v \in V$, where

$$a(u,v) = \int_\Omega \alpha \nabla u \nabla v \, d\Omega + \int_\Omega \gamma u v \, d\Omega, \quad F(v) = \int_\Omega f v \, d\Omega.$$

The bilinear form is continuous; indeed

$$\begin{aligned} |a(u,v)| &\leq \int_\Omega \alpha |\nabla u||\nabla v| \, d\Omega + \int_\Omega |\gamma||u|\,|v| \, d\Omega \\ &\leq \|\alpha\|_{\mathrm{L}^\infty(\Omega)} \|\nabla u\|_{\mathrm{L}^2(\Omega)} \|\nabla v\|_{\mathrm{L}^2(\Omega)} + \|\gamma\|_{\mathrm{L}^\infty(\Omega)} \|u\|_{\mathrm{L}^2(\Omega)} \|v\|_{\mathrm{L}^2(\Omega)} \\ &\leq M \|u\|_V \|v\|_V, \end{aligned}$$

having taken $M = 2\max\{\|\alpha\|_{\mathrm{L}^\infty(\Omega)}, \|\gamma\|_{\mathrm{L}^\infty(\Omega)}\}$. Moreover, it is coercive (see the solution to Exercise 7) with coercivity constant given by $\alpha^* = \frac{\alpha_0}{1+C_\Omega^2}$. Since F is a linear and bounded functional, owing to the Lax-Milgram lemma the weak solution exists and is unique. As far as the Galerkin method is concerned, we introduce a subspace V_h of V with finite dimension. Then there exists a unique solution u_h of the Galerkin problem: find $u_h \in V_h$ such that $a(u_h, v_h) = F(v_h)\ \forall v_h \in V_h$. Moreover, by Corollary 4.2 we have stability. As far as the choice of the optimal polynomial degree r is concerned, it is sufficient to note that the exponent s appearing in (4.26) is the minimum between r and $p = 3$. Hence, it will be convenient to use elements of degree 3.]
The fundamental steps of a finite element code can be summarized as follows:

(a) input the data;
(b) build the grid $\mathcal{T}_h = \{K\}$;
(c) build the local matrices A_K and the right-hand side elements f_K;
(d) assemble the global matrix A and the one of the source term $\mathbf{f}$;
(e) solve the linear system $A\mathbf{u} = \mathbf{f}$;
(f) post-process the results.

Suppose we use linear finite elements and consider the patch in Fig. 4.26.

a) Referring to steps (c) and (d), explicitly write the matrix T_K allowing to pass from the local matrix A_K to the global matrix A via a transformation of the kind $T_K^T A_K T_K$. What is the dimension of such matrix?
b) What sparsity pattern characterizes the matrix A associated to the patch in Fig. 4.26?

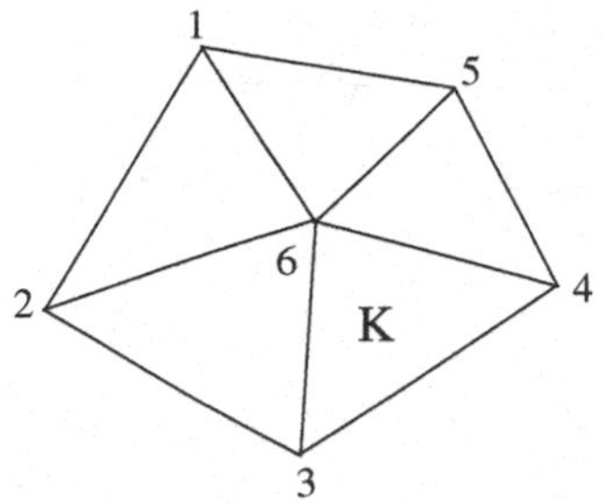

Fig. 4.26. Patch of elements for the assembly of the global matrix A

c) Write the elements of the matrix A explicitly as a function of the elements of the local matrices A_K.
d) In the case of a general grid $\mathcal{T}_h$ with $\mathtt{N}_\mathtt{V}$ vertices and $\mathtt{N}_\mathtt{T}$ triangles, what dimension does the global matrix A have in the case of linear and quadratic finite elements, respectively?

For a more exaustive treatment of this subject, we refer to Chapter 11.

9. Prove the results summarized in Table 3.1 by using the Lagrange identity (3.42).

5

Parabolic equations

In this chapter we consider parabolic equations of the form

$$\frac{\partial u}{\partial t} + Lu = f, \qquad \mathbf{x} \in \Omega,\ t > 0, \tag{5.1}$$

where Ω is a domain of $\mathbb{R}^d$, $d = 1,2,3$, $f = f(\mathbf{x},t)$ is a given function, $L = L(\mathbf{x})$ is a generic elliptic operator acting on the unknown $u = u(\mathbf{x},t)$. When solved only for a bounded temporal interval, say for $0 < t < T$, the region $Q_T = \Omega \times (0,T)$ is called *cylinder* in the space $\mathbb{R}^d \times \mathbb{R}^+$ (see Fig. 5.1). In the case where $T = +\infty$, $Q = \{(\mathbf{x},t) : \mathbf{x} \in \Omega, t > 0\}$ will be an infinite cylinder.
Equation (5.1) must be completed by assigning an initial condition

$$u(\mathbf{x},0) = u_0(\mathbf{x}), \qquad \mathbf{x} \in \Omega, \tag{5.2}$$

together with boundary conditions, which can take the following form

$$\begin{aligned} &u(\mathbf{x},t) = \varphi(\mathbf{x},t), && \mathbf{x} \in \Gamma_D \text{ and } t > 0, \\ &\frac{\partial u(\mathbf{x},t)}{\partial n} = \psi(\mathbf{x},t), && \mathbf{x} \in \Gamma_N \text{ and } t > 0, \end{aligned} \tag{5.3}$$

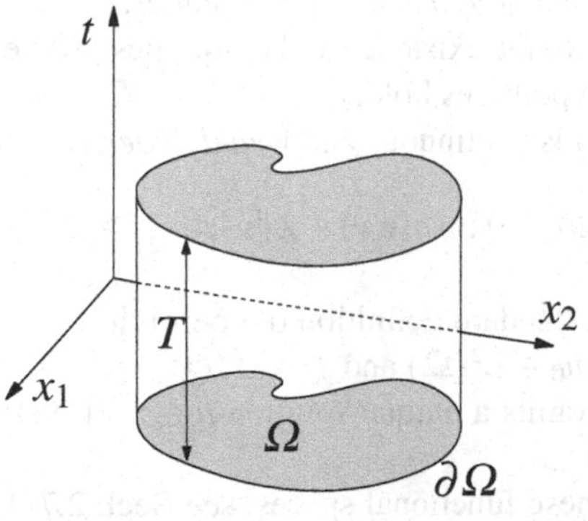

Fig. 5.1. The cylinder $Q_T = \Omega \times (0,T)$, $\Omega \subset \mathbb{R}^2$

A. Quarteroni: *Numerical Models for Differential Problems*, 2nd Ed.
MS&A – Modeling, Simulation & Applications 8
DOI 10.1007/978-88-470-5522-3_5, © Springer-Verlag Italia 2014

where u_0, φ and ψ are given functions and $\{\Gamma_D, \Gamma_N\}$ provides a boundary partition, that is $\Gamma_D \cup \Gamma_N = \partial\Omega$, $\overset{\circ}{\Gamma}_D \cap \overset{\circ}{\Gamma}_N = \emptyset$. For obvious reasons, Γ_D is called Dirichlet boundary and Γ_N Neumann boundary.
In the one-dimensional case, the problem

$$\begin{aligned} &\frac{\partial u}{\partial t} - \nu \frac{\partial^2 u}{\partial x^2} = f, \quad 0 < x < d, \quad t > 0, \\ &u(x,0) = u_0(x), \quad 0 < x < d, \\ &u(0,t) = u(d,t) = 0, \quad t > 0, \end{aligned} \tag{5.4}$$

describes the evolution of the temperature $u(x,t)$ at point x and time t of a metal bar of length d occupying the interval $[0,d]$, whose thermal conductivity is ν and whose endpoints are kept at a constant temperature of zero degrees. The function u_0 describes the initial temperature, while f represents the heat generated (per unit length) by the bar. For this reason, (5.4) is called *heat equation*. For a particular case, see Example 1.5 of Chapter 1.

5.1 Weak formulation and its approximation

In order to solve problem (5.1)–(5.3) numerically, we will introduce a weak formulation, as we did to handle elliptic problems.
We proceed formally, by multiplying for each $t > 0$ the differential equation by a test function $v = v(\mathbf{x})$ and integrating on Ω. We set $V = \mathrm{H}^1_{\Gamma_D}(\Omega)$ (see (3.26)) and for each $t > 0$ we seek $u(t) \in V$ such that

$$\int_\Omega \frac{\partial u(t)}{\partial t} v \, d\Omega + a(u(t), v) = \int_\Omega f(t) v d \, d\Omega \qquad \forall v \in V, \tag{5.5}$$

where $u(0) = u_0$, $a(\cdot,\cdot)$ is the bilinear form associated to the elliptic operator L, and where we have supposed for simplicity $\varphi = 0$ and $\psi = 0$. The modification of (5.5) in the case where $\varphi \neq 0$ and $\psi \neq 0$ is left to the reader.
A sufficient condition for the existence and uniqueness of the solution to problem (5.5) is that the following hypotheses hold:
the bilinear form $a(\cdot,\cdot)$ is continuous and *weakly coercive*, that is

$$\exists \lambda \geq 0, \ \exists \alpha > 0 : \quad a(v,v) + \lambda \|v\|^2_{\mathrm{L}^2(\Omega)} \geq \alpha \|v\|^2_V \quad \forall v \in V,$$

yielding for $\lambda = 0$ the standard definition of coercivity.
Moreover, we require $u_0 \in \mathrm{L}^2(\Omega)$ and $f \in \mathrm{L}^2(Q)$.
Then, problem (5.5) admits a unique solution $u \in \mathrm{L}^2(\mathbb{R}^+; V) \cap C^0(\mathbb{R}^+; \mathrm{L}^2(\Omega))$, with $V = \mathrm{H}^1_{\Gamma_D}(\Omega)$.
For the definition of these functional spaces, see Sect. 2.7. For the proof, see [QV94, Sect. 11.1.1].

Some a priori estimates of the solution u will be provided in the following section.

We now consider the Galerkin approximation of problem (5.5):
for each $t > 0$, find $u_h(t) \in V_h$ such that

$$\int_{\Omega} \frac{\partial u_h(t)}{\partial t} v_h \, d\Omega + a(u_h(t), v_h) = \int_{\Omega} f(t) v_h \, d\Omega \quad \forall v_h \in V_h \tag{5.6}$$

with $u_h(0) = u_{0h}$, where $V_h \subset V$ is a suitable space of finite dimension and u_{0h} is a convenient approximation of u_0 in the space V_h. Such problem is called *semi-discretization* of (5.5), as the temporal variable has not yet been discretized.
To provide an algebraic interpretation of (5.6) we introduce a basis $\{\varphi_j\}$ for V_h (as we did in the previous chapters), and we observe that it suffices that (5.6) is verified for the basis functions in order to be satisfied by all the functions of the subspace. Moreover, since for each $t > 0$ the solution to the Galerkin problem belongs to the subspace as well, we will have

$$u_h(\mathbf{x}, t) = \sum_{j=1}^{N_h} u_j(t) \varphi_j(\mathbf{x}),$$

where the coefficients $\{u_j(t)\}$ represent the unknowns of problem (5.6).

Denoting by $\dot{u}_j(t)$ the derivatives of the function $u_j(t)$ with respect to time, (5.6) becomes

$$\int_{\Omega} \sum_{j=1}^{N_h} \dot{u}_j(t) \varphi_j \varphi_i \, d\Omega + a\left(\sum_{j=1}^{N_h} u_j(t) \varphi_j, \varphi_i\right) = \int_{\Omega} f(t) \phi_i \, d\Omega, \qquad i = 1, 2, \ldots, N_h,$$

that is

$$\sum_{j=1}^{N_h} \dot{u}_j(t) \underbrace{\int_{\Omega} \varphi_j \varphi_i \, d\Omega}_{m_{ij}} + \sum_{j=1}^{N_h} u_j(t) \underbrace{a(\varphi_j, \varphi_i)}_{a_{ij}} = \underbrace{\int_{\Omega} f(t) \phi_i \, d\Omega}_{f_i(t)}, \qquad i = 1, 2, \ldots, N_h. \tag{5.7}$$

If we define the vector of unknowns $\mathbf{u} = (u_1(t),\, u_2(t),\, \ldots, u_{N_h}(t))^T$, the *mass matrix* $\mathrm{M} = [m_{ij}]$, the stiffness matrix $\mathrm{A} = [a_{ij}]$ and the right-hand side vector $\mathbf{f} = (f_1(t),\, f_2(t), \ldots, f_{N_h}(t))^T$, the system (5.7) can be rewritten in matrix form as

$$\mathrm{M}\dot{\mathbf{u}}(t) + \mathrm{A}\mathbf{u}(t) = \mathbf{f}(t).$$

For the numerical solution of this ODE system, many finite difference methods are available. See, e.g., [QSS07, Chap. 11]. Here we limit ourselves to considering the so-called θ-method. The latter discretizes the temporal derivative by a simple difference quotient and replaces the other terms with a linear combination of the value at time t^k

and of the value at time t^{k+1}, depending on the real parameter θ $(0 \leq \theta \leq 1)$,

$$\mathrm{M}\frac{\mathbf{u}^{k+1}-\mathbf{u}^k}{\Delta t}+\mathrm{A}[\theta\mathbf{u}^{k+1}+(1-\theta)\mathbf{u}^k] = \theta\mathbf{f}^{k+1}+(1-\theta)\mathbf{f}^k. \tag{5.8}$$

As usual, the real positive parameter $\Delta t = t^{k+1} - t^k$, $k = 0, 1, \ldots$, denotes the discretization step (here assumed to be constant), while the superscript k indicates that the quantity under consideration refers to the time t^k. Let us see some particular cases of (5.8):

- for $\theta = 0$ we obtain the *forward Euler* (or *explicit* Euler) method

$$\mathrm{M}\frac{\mathbf{u}^{k+1}-\mathbf{u}^k}{\Delta t}+\mathrm{A}\mathbf{u}^k = \mathbf{f}^k$$

  which is accurate to order one with respect to Δt;
- for $\theta = 1$ we have the *backward Euler* (or *implicit* Euler) method

$$\mathrm{M}\frac{\mathbf{u}^{k+1}-\mathbf{u}^k}{\Delta t}+\mathrm{A}\mathbf{u}^{k+1} = \mathbf{f}^{k+1},$$

  also of first order with respect to Δt;
- for $\theta = 1/2$ we have the *Crank-Nicolson* (or *trapezoidal*) method

$$\mathrm{M}\frac{\mathbf{u}^{k+1}-\mathbf{u}^k}{\Delta t}+\frac{1}{2}\mathrm{A}\left(\mathbf{u}^{k+1}+\mathbf{u}^k\right) = \frac{1}{2}\left(\mathbf{f}^{k+1}+\mathbf{f}^k\right)$$

  which is of second order in Δt. (More precisely, $\theta = 1/2$ is the only value for which we obtain a second-order method.)

Let us consider the two extremal cases, $\theta = 0$ and $\theta = 1$. For both, we obtain a system of linear equations: if $\theta = 0$, the system to solve has matrix $\frac{\mathrm{M}}{\Delta t}$, in the second case it has matrix $\frac{\mathrm{M}}{\Delta t} + A$. We observe that the M matrix is invertible, being positive definite (see Exercise 1).

In the $\theta = 0$ case, if we make M diagonal, we actually decouple the system. This operation is performed by the so-called *lumping* of the mass matrix (see Sect. 12.5). However, this scheme is not unconditionally stable (see Sect. 5.4) and in the case where V_h is a subspace of finite elements we have the following stability condition (see Sect.~5.4)

$$\exists c > 0 \ : \ \Delta t \leq ch^2 \qquad \forall h > 0,$$

so Δt icannot be chosen irrespective of h.

In case $\theta > 0$, the system will have the form $K\mathbf{u}^{k+1} = \mathbf{g}$, where $\mathbf{g}$ is the source term and $\mathrm{K} = \frac{\mathrm{M}}{\Delta t} + \theta\mathrm{A}$. Such matrix is however invariant in time (the operator L, and therefore the matrix A, being independent of time); if the space mesh does not change, it can then be factorized once and for all at the beginning of the process. Since M is symmetric, if A is symmetric too, the K matrix associated to the system will also be symmetric. Hence, we can use, for instance, the Cholesky factorization, K=H H^T, H being lower triangular. At each time step, we will therefore have to solve two triangular

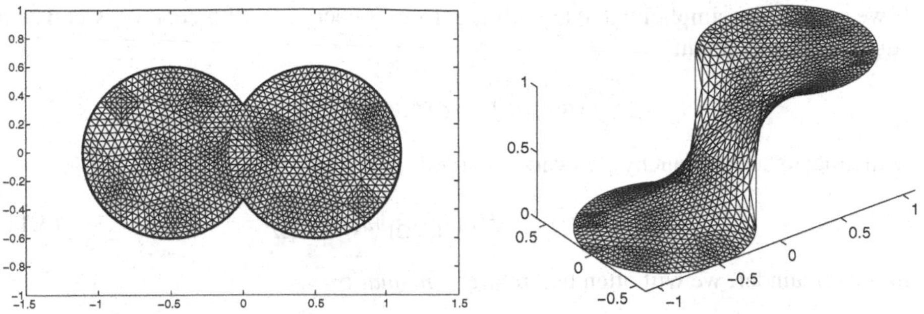

Fig. 5.2. Solution of the heat equation for the problem of Example 5.1

systems in N_h unknowns:

$$\mathrm{H}\mathbf{y} = \mathbf{g},$$
$$\mathrm{H}^T \mathbf{u}^{k+1} = \mathbf{y}$$

(see Chap. 7 and also [QSS07, Chap. 3]).

Example 5.1. Let us suppose to solve the heat equation $\frac{\partial u}{\partial t} - 0.1\Delta u = 0$ on the domain $\Omega \subset \mathbb{R}^2$ of Fig. 5.2 (left), which is the union of two circles of radius 0.5 and center $(-0.5, 0)$ resp. $(0.5, 0)$). We assign Dirichlet conditions on the whole boundary taking $u(\mathbf{x}, t) = 1$ for the points on $\partial\Omega$ for which $x_1 \geq 0$ and $u(\mathbf{x}, t) = 0$ if $x_1 < 0$. The initial condition is $u(\mathbf{x}, 0) = 1$ for $x_1 \geq 0$ and null elsewhere. In Fig. 5.2, we report the solution obtained at time $t = 1$. We have used linear finite elements in space and the implicit Euler method in time with $\Delta t = 0.01$. As it can be seen, the initial discontinuity has been regularized, in accordance with the boundary conditions. ■

5.2 A priori estimates

Let us consider problem (5.5); since the corresponding equations must hold for each $v \in V$, it will be legitimate to set $v = u(t)$ (t being given), solution of the problem itself, yielding

$$\int_\Omega \frac{\partial u(t)}{\partial t} u(t)\, d\Omega + a(u(t), u(t)) = \int_\Omega f(t) u(t)\, d\Omega \qquad \forall t > 0. \tag{5.9}$$

Considering the individual terms, we have

$$\int_\Omega \frac{\partial u(t)}{\partial t} u(t)\, d\Omega = \frac{1}{2}\frac{\partial}{\partial t} \int_\Omega |u(t)|^2 d\Omega = \frac{1}{2}\frac{\partial}{\partial t} \|u(t)\|^2_{L^2(\Omega)}. \tag{5.10}$$

If we assume for simplicity that the bilinear form is coercive (with coercivity constant equal to α), we obtain

$$a(u(t),u(t)) \geq \alpha \|u(t)\|_V^2,$$

while thanks to the Cauchy-Schwarz inequality, we find

$$(f(t),u(t)) \leq \|f(t)\|_{L^2(\Omega)} \|u(t)\|_{L^2(\Omega)}. \tag{5.11}$$

In the remainder, we will often use *Young's inequality*

$$\forall a,b \in \mathbb{R}, \qquad ab \leq \varepsilon a^2 + \frac{1}{4\varepsilon} b^2 \qquad \forall \varepsilon > 0, \tag{5.12}$$

which descends from the elementary inequality

$$\left(\sqrt{\varepsilon}\, a - \frac{1}{2\sqrt{\varepsilon}} b\right)^2 \geq 0.$$

Using first Poincaré' inequality (2.13) and Young's inequality, we obtain

$$\begin{aligned} \tfrac{1}{2}\tfrac{d}{dt}\|u(t)\|^2_{L^2(\Omega)} + \alpha \|\nabla u(t)\|^2_{L^2(\Omega)} &\leq \|f(t)\|_{L^2(\Omega)} \|u(t)\|_{L^2(\Omega)} \\ &\leq \tfrac{C_\Omega^2}{2\alpha} \|f(t)\|^2_{L^2(\Omega)} + \tfrac{\alpha}{2} \|\nabla u(t)\|^2_{L^2(\Omega)}. \end{aligned} \tag{5.13}$$

Then, by integrating in time we obtain, for all $t > 0$,

$$\|u(t)\|^2_{L^2(\Omega)} + \alpha \int_0^t \|\nabla u(s)\|^2_{L^2(\Omega)} ds \leq \|u_0\|^2_{L^2(\Omega)} + \frac{C_\Omega^2}{\alpha} \int_0^t \|f(s)\|^2_{L^2(\Omega)} ds. \tag{5.14}$$

This is an a priori energy estimate. Different kinds of a priori estimates can be obtained as follows. Note that

$$\frac{1}{2}\frac{d}{dt}\|u(t)\|^2_{L^2(\Omega)} = \|u(t)\|_{L^2(\Omega)} \frac{d}{dt} \|u(t)\|_{L^2(\Omega)}.$$

Then from (5.9), using (5.10) and (5.11) we obtain (still using the Poincaré inequality)

$$\begin{aligned} \|u(t)\|_{L^2(\Omega)} \frac{d}{dt}\|u(t)\|_{L^2(\Omega)} + \frac{\alpha}{C_\Omega} \|u(t)\|_{L^2(\Omega)} \|\nabla u(t)\|_{L^2(\Omega)} & \\ \leq \|f(t)\|_{L^2(\Omega)} \|u(t)\|_{L^2(\Omega)}, & \qquad t > 0. \end{aligned}$$

If $\|u(t)\|_{L^2(\Omega)} \neq 0$ (otherwise we should proceed differently, even though the final result is still true) we can divide by $\|u(t)\|_{L^2(\Omega)}$ and integrate in time to obtain

$$\|u(t)\|_{L^2(\Omega)} \leq \|u_0\|_{L^2(\Omega)} + \int_0^t \|f(s)\|_{L^2(\Omega)} ds, t > 0. \tag{5.15}$$

This is a further a priori estimate.

Let us now use the first inequality in (5.13), and integrate in time to yield

$$\|u(t)\|^2_{L^2(\Omega)} + 2\alpha \int_0^t \|\nabla u(s)\|^2 ds$$
$$\leq \|u_0\|^2_{L^2(\Omega)} + 2\int_0^t \|f(s)\|_{L^2(\Omega)} \|u(s)\|_{L^2(\Omega)} ds$$
$$\leq \|u_0\|^2_{L^2(\Omega)} + 2\int_0^t \|f(s)\|_{L^2(\Omega)} \cdot (\|u_0\|^2_{L^2(\Omega)} + \int_0^s \|f(\tau)\|_{L^2(\Omega)} d\tau) ds$$

(using (5.15))

$$= \|u_0\|^2_{L^2(\Omega)} + 2\int_0^t \|f(s)\|_{L^2(\Omega)} \|u_0\|_{L^2(\Omega)} + 2\int_0^t \|f(s)\|_{L^2(\Omega)} \int_0^s \|f(\tau)\|_{L^2(\Omega)} d\tau$$
$$= (\|u_0\|_{L^2(\Omega)} + \int_0^t \|f(s)\| ds)^2. \tag{5.16}$$

The latter equality follows upon noticing that

$$\|f(s)\|_{L^2(\Omega)} \int_0^s \|f(\tau)\|_{L^2(\Omega)} d\tau = \frac{d}{ds}(\int_0^s \|f(\tau)\|_{L^2(\Omega)} d\tau)^2.$$

We therefore conclude with the additional a priori estimate

$$\begin{aligned} &(\|u(t)\|^2_{L^2(\Omega)} + 2\alpha \textstyle\int_0^t \|\nabla u(s)\|^2_{L^2(\Omega)} ds)^{\frac{1}{2}} \\ &\quad \leq \|u_0\|_{L^2(\Omega)} + \textstyle\int_0^t \|f(s)\|_{L^2(\Omega)} ds, \qquad t > 0. \end{aligned} \tag{5.17}$$

We have seen that we can formulate the Galerkin problem (5.6) for problem (5.5) and that the latter, under suitable hypotheses, admits a unique solution. Similarly to what we did for problem (5.5) we can prove the following a priori (stability) estimates for the solution to problem (5.6):

$$\begin{aligned} &\|u_h(t)\|^2_{L^2(\Omega)} + \alpha \int_0^t \|\nabla u_h(s)\|^2_{L^2(\Omega)} ds \\ &\qquad \leq \|u_{0h}\|^2_{L^2(\Omega)} + \frac{C^2_\Omega}{\alpha} \int_0^t \|f(s)\|^2_{L^2(\Omega)} ds, \qquad t > 0. \end{aligned} \tag{5.18}$$

For its proof we can take, for every $t > 0$, $v_h = u_h(t)$ and proceed as we did to obtain (5.13). Then, by recalling that the initial data is $u_h(0) = u_{0h}$, we can deduce the following discrete counterparts of (5.15) and (5.17):

$$\|u_h(t)\|_{L^2(\Omega)} \leq \|u_{0h}(t)\|_{L^2(\Omega)} + \int_0^t \|f(s)\|_{L^2(\Omega)} ds, \qquad t > 0. \tag{5.19}$$

and

$$\begin{aligned} &(\|u_h(t)\|^2_{L^2(\Omega)} + 2\alpha \int_0^t \|\nabla u_h(s)\|^2_{L^2(\Omega)} ds)^{\frac{1}{2}} \\ &\qquad \leq \|u_{0h}\|_{L^2(\Omega)} + \int_0^t \|f(s)\|_{L^2(\Omega)} ds, \qquad t > 0. \end{aligned} \tag{5.20}$$

5.3 Convergence analysis of the semi-discrete problem

Let us consider problem (5.5) and its approximation (5.6). We want to prove the convergence of u_h to u in suitable norms.
By the coercivity hypotheses we can write

$$\begin{aligned}\alpha\|(u-u_h)(t)\|^2_{\mathrm{H}^1(\Omega)} &\leq a((u-u_h)(t),(u-u_h)(t)) \\ &= a((u-u_h)(t),(u-v_h)(t)) \\ &\quad + a((u-u_h)(t),(v_h-u_h)(t)) \qquad \forall v_h : v_h(t) \in V_h,\ \forall t>0.\end{aligned}$$

For the sake of clarity, we suppress the dependence from t. By subtracting equation (5.6) from equation (5.5) and setting $w_h = v_h - u_h$ we have

$$\left(\frac{\partial(u-u_h)}{\partial t}, w_h\right) + a(u-u_h, w_h) = 0,$$

where $(v,w) = \int_\Omega vw$ is the scalar product of $L^2(\Omega)$. Then

$$\alpha\|u-u_h\|^2_{\mathrm{H}^1(\Omega)} \leq a(u-u_h, u-v_h) - \left(\frac{\partial(u-u_h)}{\partial t}, w_h\right). \tag{5.21}$$

We analyze the two right-hand side terms separately:

- using the continuity of the form $a(\cdot,\cdot)$ and Young's inequality, we obtain

$$\begin{aligned}a(u-u_h,u-v_h) &\leq M\|u-u_h\|_{\mathrm{H}^1(\Omega)}\|u-v_h\|_{\mathrm{H}^1(\Omega)} \\ &\leq \frac{\alpha}{2}\|u-u_h\|^2_{\mathrm{H}^1(\Omega)} + \frac{M^2}{2\alpha}\|u-v_h\|^2_{\mathrm{H}^1(\Omega)};\end{aligned}$$

- writing w_h in the form $w_h = (v_h-u) + (u-u_h)$ we obtain

$$-\left(\frac{\partial(u-u_h)}{\partial t}, w_h\right) = \left(\frac{\partial(u-u_h)}{\partial t}, u-v_h\right) - \frac{1}{2}\frac{d}{dt}\|u-u_h\|^2_{\mathrm{L}^2(\Omega)}. \tag{5.22}$$

Replacing these two results in (5.21), we obtain

$$\frac{1}{2}\frac{d}{dt}\|u-u_h\|^2_{\mathrm{L}^2(\Omega)} + \frac{\alpha}{2}\|u-u_h\|^2_{\mathrm{H}^1(\Omega)} \leq \frac{M^2}{2\alpha}\|u-v_h\|^2_{\mathrm{H}^1(\Omega)} + (\frac{\partial(u-u_h)}{\partial t}, u-v_h).$$

Multiplying both sides by 2 and integrating in time between 0 and t we find

$$\begin{aligned}\|(u-u_h)(t)\|^2_{\mathrm{L}^2(\Omega)} + \alpha\int_0^t \|(u-u_h)(s)\|^2_{\mathrm{H}^1(\Omega)}\,ds &\leq \|(u-u_h)(0)\|^2_{\mathrm{L}^2(\Omega)} \\ + \frac{M^2}{\alpha}\int_0^t \|(u-v_h)(s)\|^2_{\mathrm{H}^1(\Omega)}\,ds &+ 2\int_0^t \left(\frac{\partial}{\partial t}(u-u_h)(s), (u-v_h)(s)\right)\,ds.\end{aligned} \tag{5.23}$$

Integrating by parts and using Young's inequality, we obtain

$$\int_0^t \left(\frac{\partial}{\partial t}(u-u_h)(s),(u-v_h)(s)\right) ds = -\int_0^t \left((u-u_h)(s),\frac{\partial}{\partial t}((u-v_h)(s))\right) ds$$
$$+((u-u_h)(t),(u-v_h)(t))-((u-u_h)(0),(u-v_h)(0))$$

$$\leq \int_0^t \|(u-u_h)(s)\|_{L^2(\Omega)}\, \|\frac{\partial((u-v_h)(s))}{\partial t}\|_{L^2(\Omega)}\, ds+\frac{1}{4}\|(u-u_h)(t)\|^2_{L^2(\Omega)}$$
$$+\|(u-v_h)(t)\|^2_{L^2(\Omega)}+\frac{1}{2}\|(u-u_h)(0)\|^2_{L^2(\Omega)}+\frac{1}{2}\|u(0)-v_h(0)\|^2_{L^2(\Omega)}.$$

From (5.23) we thus obtain

$$\begin{aligned}
&\frac{1}{2}\|(u-u_h)(t)\|^2_{L^2(\Omega)}+\alpha\int_0^t \|(u-u_h)(s)\|^2_{H^1(\Omega)}\, ds\\
&\leq 2\|(u-u_h)(0)\|^2_{L^2(\Omega)}+\frac{M^2}{\alpha}\int_0^t \|(u-v_h)(s)\|^2_{H^1(\Omega)}\, ds\\
&+2\int_0^t \|(u-u_h)(s)\|_{L^2(\Omega)}\|\frac{\partial((u-v_h)(s))}{\partial t}\|_{L^2(\Omega)}\, ds\\
&+2\|(u-v_h)(t)\|^2_{L^2(\Omega)}+\|u(0)-v_h(0)\|^2_{L^2(\Omega)}.
\end{aligned}\tag{5.24}$$

Let us now suppose that V_h is the space of finite elements of degree r, more precisely $V_h=\{v_h\in X_h^r : v_h|_{\Gamma_D}=0\}$, and let us choose, at each t, $v_h(t)=\Pi_h^r u(t)$, the interpolant of $u(t)$ in V_h (see (4.20)). Thanks to (4.69) we have, assuming that u is sufficiently regular,

$$h\|u(t)-\Pi_h^r u(t)\|_{H^1(\Omega)}+\|u(t)-\Pi_h^r u(t)\|_{L^2(\Omega)}\leq C_2 h^{r+1}|u(t)|_{H^{r+1}(\Omega)}.$$

Let us consider and bound from above some of the summands of the right-hand side of inequality (5.24):

$$E_1=2\|(u-u_h)(0)\|^2_{L^2(\Omega)}\leq C_1 h^{2r}|u_0|^2_{H^r(\Omega)}.$$

$$E_2=\frac{M^2}{\alpha}\int_0^t \|u(s)-v_h(s)\|^2_{H^1(\Omega)}\, ds\leq C_2 h^{2r}\int_0^t |u(s)|^2_{H^{r+1}(\Omega)}\, ds,$$

$$E_3=2\|u(t)-v_h(t)\|^2_{L^2(\Omega)}\leq C_3 h^{2r}|u(t)|^2_{H^r(\Omega)}.$$

Finally

$$E_4(s)=\|\frac{\partial(u(s)-v_h(s))}{\partial t}\|_{L^2(\Omega)}\leq C_4 h^r\left|\frac{\partial u(s)}{\partial t}\right|_{H^r(\Omega)}.$$

Consequently,

$$E_1+E_2+E_3+E_4\leq Ch^{2r}N(u),$$

where $N(u)$ is a suitable function depending on u and on $\frac{\partial u}{\partial t}$, and C is a suitable positive constant. In this way, from (5.24) we obtain the inequality

$$\|(u-u_h)(t)\|^2_{\mathrm{L}^2(\Omega)} + 2\alpha \int_0^t \|(u-u_h)(s)\|^2_{\mathrm{H}^1(\Omega)}\, ds$$
$$\leq Ch^{2r}N(u) + 2C_4h^r \int_0^t \left|\frac{\partial u(s)}{\partial t}\right|_{H^r(\Omega)} \|(u-u_h)(s)\|_{\mathrm{L}^2(\Omega)} ds.$$

Finally, applying the Gronwall lemma (Lemma 2.2 *ii*)), we obtain the a priori error estimate for all $t > 0$

$$\left\{\|(u-u_h)(t)\|^2_{\mathrm{L}^2(\Omega)} + 2\alpha \int_0^t \|u-u_h\|^2_{\mathrm{H}^1(\Omega)}\right\}^{1/2}$$
$$\leq \bar{C}h^r\left(\sqrt{N(u)} + \int_0^t \left|\frac{\partial u(s)}{\partial t}\right|_{H^r(\Omega)} ds\right) \forall t > 0 \quad (5.25)$$

for a suitable positive constant $\bar{C}$.

An alternative proof that does not make use of Gronwall' lemma goes as follows. If we subtract (5.6) from (5.5) and set $E_h = u - u_h$, we obtain that (the dependence of E_h on t is understood)

$$\left(\frac{\partial E_h}{\partial t}, v_h\right) + a(E_h, v_h) = 0\ \forall v_h \in V_h, \forall t > 0.$$

If, for the sake of simplicity, we suppose that $a(\cdot,\cdot)$ is symmetric, we can define the orthogonal projection operator

$$\Pi^r_{1,h} : V \to V_h \ :\ \forall w \in V,\ a(\Pi^r_{1,h}w - w, v_h) = 0 \qquad \forall v_h \in V_h. \quad (5.26)$$

Using the results seen in Chap. 3, we can prove (see [QV94, Sect. 3.5]) that there exists a constant $C > 0$ such that, $\forall w \in V \cap H^{r+1}(\Omega)$,

$$\|\Pi^r_{1,h}w - w\|_{\mathrm{H}^1(\Omega)} + h^{-1}\|\Pi^r_{1,h}w - w\|_{\mathrm{L}^2(\Omega)} \leq Ch^p|w|_{\mathrm{H}^{p+1}(\Omega)}, 0 \leq p \leq r. \quad (5.27)$$

Then we set

$$E_h = \sigma_h + e_h = (u - \Pi^r_{1,h}u) + (\Pi^r_{1,h}u - u_h). \quad (5.28)$$

Note that the orthogonal projection error σ_h can be bounded by inequality (5.27) and that e_h is an element of the subspace V_h. Then

$$(\frac{\partial e_h}{\partial t}, v_h) + a(e_h, v_h) = -(\frac{\partial \sigma_h}{\partial t}, v_h) - a(\sigma_h, v_h)\ \forall v_h \in V_h, \forall t > 0.$$

If we take at every $t > 0$, $v_h = e_h(t)$, and proceed as done in Sect. 5.2 to deduce the a priori estimates on the semi-discrete solution u_h, we obtain

$$\begin{aligned}\tfrac{1}{2}\tfrac{d}{dt}\|e_h(t)\|^2_{L^2(\Omega)} &+ \alpha\|\nabla e_h(t)\|^2_{L^2(\Omega)} \\ &\leq |a(\sigma_h(t), e_h(t))| + |(\tfrac{\partial}{\partial t}\sigma_h(t), e_h(t))|.\end{aligned} \quad (5.29)$$

Using the continuity of the bilinear form $a(\cdot,\cdot)$ (M being the continuity constant) and Young's inequality (5.12), we obtain

$$|a(\sigma_h(t), e_h(t))| \leq \frac{\alpha}{4}\|\nabla e_h(t)\|^2_{L^2(\Omega)} + \frac{M^2}{\alpha}\|\nabla\sigma_h(t)\|^2_{L^2(\Omega)}.$$

Moreover, using the Poincaré inequality and once more the Young's inequality it follows that

$$\begin{aligned}|(\frac{\partial}{\partial t}\sigma_h(t), e_h(t))| &\leq \|\frac{\partial}{\partial t}\sigma_h(t)\|_{L^2(\Omega)} C_\Omega \|\nabla e_h(t)\|_{L^2(\Omega)}\\ &\leq \frac{\alpha}{4}\|\nabla e_h(t)\|^2_{L^2(\Omega)} + \frac{C_\Omega^2}{\alpha}\|\frac{\partial}{\partial t}\sigma_h(t)\|^2_{L^2(\Omega)}.\end{aligned}$$

Using these bounds in (5.29) we obtain, after integrating with respect to t:

$$\begin{aligned}&\|e_h(t)\|^2_{L^2(\Omega)} + \alpha\int_0^t \|\nabla e_h(t)\|^2_{L^2(\Omega)} ds\\ &\leq \|e_h(0)\|^2_{L^2(\Omega)} + \frac{2M^2}{\alpha}\int_0^t \|\nabla\sigma_h(s)\|^2_{L^2(\Omega)} ds + \frac{2C_\Omega^2}{\alpha}\int_0^t \|\frac{\partial}{\partial t}\sigma_h(s)\|^2_{L^2(\Omega)} ds, \quad t > 0.\end{aligned}$$

At this point we can use (5.27) to bound the errors on the right-hand side:

$$\begin{aligned}&\|\nabla\sigma_h(t)\|_{L^2(\Omega)} \leq Ch^r |u(t)|_{H^{r+1}(\Omega)},\\ &\left\|\frac{\partial}{\partial t}\sigma_h(t)\right\|_{L^2(\Omega)} = \left\|\left(\frac{\partial u}{\partial t} - \Pi^r_{1,h}\frac{\partial u}{\partial t}\right)(t)\right\|_{L^2(\Omega)} \leq Ch^r \left|\frac{\partial u(t)}{\partial t}\right|_{H(\Omega)}.\end{aligned}$$

Finally, note that $\|e_h(0)\|_{L^2(\Omega)} \leq Ch^r |u_0|_{H^r(\Omega)}$, still using (5.27).
Since, for any norm $\|\cdot\|$,

$$\|u - u_h\| \leq \|\sigma_h\| + \|e_h\|$$

(owing to 5.28), using the previous estimates we can conclude that there exists a constant $C > 0$ independent of both t and h such that

$$\begin{aligned}&\{\|u(t) - u_h(t)\|^2_{L^2(\Omega)} + \alpha\int_0^t \|\nabla u(s) - \nabla u_h(s)\|^2_{L^2(\Omega)} ds\}^{1/2}\\ &\leq Ch^r\{|u_0|^2_{H^r(\Omega)} + \int_0^t |u(s)|^2_{H^{r+1}(\Omega)} ds + \int_0^t |\frac{\partial u(s)}{\partial t}|^2_{H^{r+1}(\Omega)} ds\}^{1/2}.\end{aligned}$$

Further error estimates are proven, e.g. in [QV94, Chap. 11].

5.4 Stability analysis of the θ-method

We now analyze the stability of the fully discretized problem.
Applying the θ-method to the Galerkin problem (5.6) we obtain

$$\left(\frac{u_h^{k+1}-u_h^k}{\Delta t}, v_h\right) + a\left(\theta u_h^{k+1} + (1-\theta)u_h^k, v_h\right)$$
$$= \theta F^{k+1}(v_h) + (1-\theta)F^k(v_h) \qquad \forall v_h \in V_h, \quad (5.30)$$

for each $k \geq 0$, with $u_h^0 = u_{0h}$; F^k indicates that the functional is evaluated at time t^k. We will limit ourselves to the case where $F = 0$ and start to consider the case of the implicit Euler method ($\theta = 1$) that is

$$\left(\frac{u_h^{k+1}-u_h^k}{\Delta t}, v_h\right) + a\left(u_h^{k+1}, v_h\right) = 0 \qquad \forall v_h \in V_h.$$

By choosing $v_h = u_h^{k+1}$, we obtain

$$(u_h^{k+1}, u_h^{k+1}) + \Delta t\, a\left(u_h^{k+1}, u_h^{k+1}\right) = (u_h^k, u_h^{k+1}).$$

By exploiting the following inequalities

$$a(u_h^{k+1}, u_h^{k+1}) \geq \alpha \|u_h^{k+1}\|_V^2, \qquad (u_h^k, u_h^{k+1}) \leq \frac{1}{2}\|u_h^k\|_{L^2(\Omega)}^2 + \frac{1}{2}\|u_h^{k+1}\|_{L^2(\Omega)}^2,$$

the former deriving from the coercivity of the bilinear form $a(\cdot,\cdot)$, and the latter from the Cauchy-Schwarz and Young inequalities, we obtain

$$\|u_h^{k+1}\|_{L^2(\Omega)}^2 + 2\alpha\Delta t \|u_h^{k+1}\|_V^2 \leq \|u_h^k\|_{L^2(\Omega)}^2. \quad (5.31)$$

By summing over k from 0 to $n-1$ we deduce that

$$\|u_h^n\|_{L^2(\Omega)}^2 + 2\alpha\Delta t \sum_{k=0}^{n-1} \|u_h^{k+1}\|_V^2 \leq \|u_{0h}\|_{L^2(\Omega)}^2.$$

When $f \neq 0$, using the discrete Gronwall lemma (see Sect. 2.7) it can be proved in a similar way that

$$\|u_h^n\|_{L^2(\Omega)}^2 + 2\alpha\Delta t \sum_{k=1}^{n} \|u_h^k\|_V^2 \leq C(t^n)\left(\|u_{0h}\|_{L^2(\Omega)}^2 + \sum_{k=1}^{n} \Delta t \|f^k\|_{L^2(\Omega)}^2\right). \quad (5.32)$$

Such relation is similar to (5.20), provided that the integrals $\int_0^t \cdot\, ds$ are approximated by a composite numerical integration formula with step Δt.

Finally, observing that $\|u_h^{k+1}\|_V \geq \|u_h^{k+1}\|_{L^2(\Omega)}$, we deduce from (5.31) that for each given $\Delta t > 0$,

$$\lim_{k\to\infty} \|u_h^k\|_{L^2(\Omega)} = 0,$$

that is the backward Euler method is absolutely stable without any restriction on the time step Δt.

Before analyzing the general case where θ is an arbitrary parameter ranging between 0 and 1, we introduce the following definition.
We say that the scalar λ is an *eigenvalue of the bilinear form* $a(\cdot,\cdot) : V \times V \mapsto \mathbb{R}$ and that $w \in V$ is its corresponding *eigenfunction* if it turns out that

$$a(w,v) = \lambda(w,v) \qquad \forall v \in V.$$

If the bilinear form $a(\cdot,\cdot)$ is symmetric and coercive, it has positive, real eigenvalues forming an infinite sequence; moreover, its eigenfunctions form a basis of the space V.

The eigenvalues and eigenfunctions of $a(\cdot,\cdot)$ can be approximated by finding the pairs $\lambda_h \in \mathbb{R}$ and $w_h \in V_h$ which satisfy

$$a(w_h, v_h) = \lambda_h(w_h, v_h) \quad \forall v_h \in V_h. \tag{5.33}$$

From an algebraic viewpoint, problem (5.33) can be formulated as follows

$$\mathbf{Aw} = \lambda_h \mathbf{Mw},$$

where A is the stiffness matrix and M the mass matrix. We are therefore dealing with a *generalized eigenvalue problem*.

Such eigenvalues are all positive and N_h in number (N_h being as usual the dimension of the subspace V_h); after ordering them in ascending order, $\lambda_h^1 \leq \lambda_h^2 \leq \ldots \leq \lambda_h^{N_h}$, we have

$$\lambda_h^{N_h} \longrightarrow \infty \qquad \text{for } N_h \longrightarrow \infty.$$

Moreover, the corresponding eigenfunctions form a basis for the subspace V_h and can be chosen to be *orthonormal* with respect to the scalar product of $\mathrm{L}^2(\Omega)$. This means that, denoting by w_h^i the eigenfunction corresponding to the eigenvalue λ_h^i, we have $(w_h^i, w_h^j) = \delta_{ij} \ \ \forall i,j = 1,\ldots,N_h$. Thus, each function $v_h \in V_h$ can be represented as follows

$$v_h(\mathbf{x}) = \sum_{j=1}^{N_h} v_j w_h^j(\mathbf{x})$$

and, thanks to the eigenfunction orthonormality,

$$\|v_h\|^2_{\mathrm{L}^2(\Omega)} = \sum_{j=1}^{N_h} v_j^2. \tag{5.34}$$

Let us consider an arbitrary $\theta \in [0,1]$ and let us limit ourselves to the case where the bilinear form $a(\cdot,\cdot)$ is symmetric (otherwise, although the final stability result holds in general, the following proof would not work, as the eigenfunctions would not necessarily form a basis). Let $\{w_h^i\}$ still denote the discrete (orthonormal) eigenfunctions of $a(\cdot,\cdot)$. Since $u_h^k \in V_h$, we can write

$$u_h^k(\mathbf{x}) = \sum_{j=1}^{N_h} u_j^k w_h^j(\mathbf{x}).$$

We observe that in this modal expansion, the u_j^k no longer represent the nodal values of u_h^k. If we now set $F = 0$ in (5.30) and take $v_h = w_h^i$, we find

$$\frac{1}{\Delta t}\sum_{j=1}^{N_h}[u_j^{k+1} - u_j^k]\left(w_h^j, w_h^i\right) + \sum_{j=1}^{N_h}[\theta u_j^{k+1} + (1-\theta)u_j^k]a(w_h^j, w_h^i) = 0,$$

for each $i = 1,\ldots,N_h$. For each pair $i, j = 1,\ldots,N_h$ we have

$$a(w_h^j, w_h^i) = \lambda_h^j(w_h^j, w_h^i) = \lambda_h^j\delta_{ij} = \lambda_h^i,$$

and thus, for each $i = 1,\ldots,N_h$,

$$\frac{u_i^{k+1} - u_i^k}{\Delta t} + [\theta u_i^{k+1} + (1-\theta)u_i^k]\lambda_h^i = 0.$$

Solving now for u_i^{k+1}, we find

$$u_i^{k+1} = u_i^k \frac{1-(1-\theta)\lambda_h^i\Delta t}{1+\theta\lambda_h^i\Delta t}.$$

Recalling (5.34), we can conclude that for the method to be absolutely stable, we must impose the inequality

$$\left|\frac{1-(1-\theta)\lambda_h^i\Delta t}{1+\theta\lambda_h^i\Delta t}\right| < 1,$$

that is

$$-1-\theta\lambda_h^i\Delta t < 1-(1-\theta)\lambda_h^i\Delta t < 1+\theta\lambda_h^i\Delta t.$$

Hence,

$$-\frac{2}{\lambda_h^i\Delta t} - \theta < \theta - 1 < \theta.$$

The second inequality is always verified, while the first one can be rewritten as

$$2\theta - 1 > -\frac{2}{\lambda_h^i\Delta t}.$$

If $\theta \geq 1/2$, the left-hand side is non-negative, while the right-hand side is negative, so the inequality holds for each Δt. Instead, if $\theta < 1/2$, the inequality is satisfied (hence the method is stable) only if

$$\Delta t < \frac{2}{(1-2\theta)\lambda_h^i}. \tag{5.35}$$

As such relation must hold for all the eigenvalues λ_h^i of the bilinear form, it will suffice to require that it holds for the largest among them, which we have supposed to be $\lambda_h^{N_h}$. To summarize, we have:

- *if* $\theta \geq 1/2$, *the* θ*-method is unconditionally stable, i.e. it is stable for each* Δt*;*
- *if* $\theta < 1/2$, *the* θ*-method is stable only for* $\Delta t \leq \frac{2}{(1-2\theta)\lambda_h^{N_h}}$.

Thanks to the definition of eigenvalue (5.33) and to the continuity property of $a(\cdot,\cdot)$, we deduce

$$\lambda_h^{N_h} = \frac{a(w_{N_h}, w_{N_h})}{\|w_{N_h}\|^2_{L^2(\Omega)}} \leq \frac{M\|w_{N_h}\|^2_V}{\|w_{N_h}\|^2_{L^2(\Omega)}} \leq M(1+C^2h^{-2}).$$

The constant $C > 0$ which appears in the latter step derives from the following *inverse inequality*

$$\exists C > 0 \; : \; \|\nabla v_h\|_{L^2(\Omega)} \leq Ch^{-1}\|v_h\|_{L^2(\Omega)} \qquad \forall v_h \in V_h,$$

for whose proof we refer to [QV94, Chap. 3].
Hence, for h small enough, $\lambda_h^{N_h} \leq Ch^{-2}$. In fact, we can prove that $\lambda_h^{N_h}$ is indeed of the order of h^{-2}, that is

$$\lambda_h^{N_h} = max_i \lambda_h^i \simeq ch^{-2}.$$

Keeping this into account, we obtain that for $\theta < 1/2$ the method is absolutely stable only if

$$\Delta t \leq C(\theta)h^2, \tag{5.36}$$

where $C(\theta)$ denotes a positive constant depending on θ. The latter relation implies that for $\theta < 1/2$, Δt cannot be chosen arbitrarily but is bound to the choice of h.

5.5 Convergence analysis of the θ-method

We can prove the following convergence theorem

Theorem 5.1. *Under the hypothesis that* u_0, f *and the exact solution are sufficiently regular, the following* a priori error estimate *holds:* $\forall n \geq 1$,

$$\|u(t^n) - u_h^n\|^2_{L^2(\Omega)} + 2\alpha\Delta t \sum_{k=1}^{n} \|u(t^k) - u_h^k\|^2_V \leq C(u_0, f, u)(\Delta t^{p(\theta)} + h^{2r}),$$

where $p(\theta) = 2$ *if* $\theta \neq 1/2$, $p(1/2) = 4$ *and* C *depends on its arguments but not on* h *and* Δt.

Proof. The proof is carried out by comparing the solution of the fully discretized problem (5.30) with that of the semi-discrete problem (5.6), using the stability result (5.32) as well as the decay rate of the truncation error of the time discretization. For simplicity, we will limit ourselves to considering the backward Euler method (corresponding to $\theta = 1$)

$$\frac{1}{\Delta t}(u_h^{k+1} - u_h^k, v_h) + a(u_h^{k+1}, v_h) = (f^{k+1}, v_h) \qquad \forall v_h \in V_h. \tag{5.37}$$

We refer the reader to [QV94], Sect. 11.3.1, for the proof in the general case.

Let $\Pi_{1,h}^r$ be the orthogonal projector operator introduced in (5.26). Then

$$\|u(t^k) - u_h^k\|_{\mathrm{L}^2(\Omega)} \le \|u(t^k) - \Pi_{1,h}^r u(t^k)\|_{\mathrm{L}^2(\Omega)} + \|\Pi_{1,h}^r u(t^k) - u_h^k\|_{\mathrm{L}^2(\Omega)}. \tag{5.38}$$

The first term can be estimated by referring to (5.27). To analyze the second term, where $\varepsilon_h^k = u_h^k - \Pi_{1,h}^r u(t^k)$, we obtain

$$\frac{1}{\Delta t}(\varepsilon_h^{k+1} - \varepsilon_h^k, v_h) + a(\varepsilon_h^{k+1}, v_h) = (\delta^{k+1}, v_h) \qquad \forall v_h \in V_h, \tag{5.39}$$

having set, $\forall v_h \in V_h$,

$$(\delta^{k+1}, v_h) = (f^{k+1}, v_h) - \frac{1}{\Delta t}(\Pi_{1,h}^r(u(t^{k+1}) - u(t^k)), v_h) - a(u(t^{k+1}), v_h) \tag{5.40}$$

and having exploited on the last summand the orthogonality (5.26) of the operator $\Pi_{1,h}^r$. The sequence $\{\varepsilon_h^k,\ k = 0, 1\ldots\}$ satisfies problem (5.39), which is similar to (5.37) (provided that we take δ^{k+1} instead of f^{k+1}). By adapting the stability estimate (5.32), we obtain, for each $n \ge 1$,

$$\|\varepsilon_h^n\|_{\mathrm{L}^2(\Omega)}^2 + 2\alpha\Delta t \sum_{k=1}^{n} \|\varepsilon_h^k\|_V^2 \le C(t^n)\left(\|\varepsilon_h^0\|_{\mathrm{L}^2(\Omega)}^2 + \sum_{k=1}^{n} \Delta t \|\delta^k\|_{\mathrm{L}^2(\Omega)}^2\right). \tag{5.41}$$

The norm associated to the initial time-level can easily be estimated: for instance, if $u_{0h} = \Pi_h^r u_0$ is the finite element interpolant of u_0, by suitably using the estimates (4.69) and (5.27) we obtain

$$\begin{aligned} \|\varepsilon_h^0\|_{\mathrm{L}^2(\Omega)} &= \|u_{0h} - \Pi_{1,h}^r u_0\|_{\mathrm{L}^2(\Omega)} \\ &\le \|\Pi_h^r u_0 - u_0\|_{\mathrm{L}^2(\Omega)} + \|u_0 - \Pi_{1,h}^r u_0\|_{\mathrm{L}^2(\Omega)} \le C h^r |u_0|_{\mathrm{H}^r(\Omega)}. \end{aligned} \tag{5.42}$$

Let us now focus on estimating the norm $\|\delta^k\|_{\mathrm{L}^2(\Omega)}$. We note that, thanks to (5.5),

$$(f^{k+1}, v_h) - a(u(t^{k+1}), v_h) = \left(\frac{\partial u(t^{k+1})}{\partial t}, v_h\right).$$

This allows us to rewrite (5.40) as

$$\begin{aligned} (\delta^{k+1}, v_h) &= \left(\frac{\partial u(t^{k+1})}{\partial t}, v_h\right) - \frac{1}{\Delta t}(\Pi_{1,h}^r(u(t^{k+1}) - u(t^k)), v_h) \\ &= \left(\frac{\partial u(t^{k+1})}{\partial t} - \frac{u(t^{k+1}) - u(t^k)}{\Delta t}, v_h\right) + \left(\left(I - \Pi_{1,h}^r\right)\left(\frac{u(t^{k+1}) - u(t^k)}{\Delta t}\right), v_h\right). \end{aligned} \tag{5.43}$$

Using the Taylor formula with the remainder in integral form, we have

$$\frac{\partial u(t^{k+1})}{\partial t} - \frac{u(t^{k+1}) - u(t^k)}{\Delta t} = \frac{1}{\Delta t} \int_{t^k}^{t^{k+1}} (s - t^k) \frac{\partial^2 u}{\partial t^2}(s)\, ds, \tag{5.44}$$

having made suitable regularity requirements on the function u with respect to the temporal variable. By now using the fundamental theorem of calculus and exploiting the commutativity between the projection operator $\Pi^r_{1,h}$ and the temporal derivative, we obtain

$$\left(I - \Pi^r_{1,h}\right)\left(u(t^{k+1}) - u(t^k)\right) = \int_{t^k}^{t^{k+1}} \left(I - \Pi^r_{1,h}\right)\left(\frac{\partial u}{\partial t}\right)(s)\, ds. \tag{5.45}$$

By choosing $v_h = \delta^{k+1}$ in (5.43), thanks to (5.44) and (5.45), we can deduce the following upper bound

$$\begin{aligned}
&\|\delta^{k+1}\|_{L^2(\Omega)} \\
&\le \left\| \frac{1}{\Delta t} \int_{t^k}^{t^{k+1}} (s - t^k) \frac{\partial^2 u}{\partial t^2}(s)\, ds \right\|_{L^2(\Omega)} + \left\| \frac{1}{\Delta t} \int_{t^k}^{t^{k+1}} \left(I - \Pi^r_{1,h}\right)\left(\frac{\partial u}{\partial t}\right)(s)\, ds \right\|_{L^2(\Omega)} \\
&\le \int_{t^k}^{t^{k+1}} \left\| \frac{\partial^2 u}{\partial t^2}(s) \right\|_{L^2(\Omega)} ds + \frac{1}{\Delta t} \int_{t^k}^{t^{k+1}} \left\| \left(I - \Pi^r_{1,h}\right)\left(\frac{\partial u}{\partial t}\right)(s) \right\|_{L^2(\Omega)} ds.
\end{aligned} \tag{5.46}$$

By reverting to the stability estimate (5.41) and exploiting (5.42) and the estimate (5.46) with suitably scaled indices, we have

$$\begin{aligned}
\|\varepsilon^n_h\|^2_{L^2(\Omega)} \le\ & C(t^n)\left(h^{2r} |u_0|^2_{H^r(\Omega)} + \sum_{k=1}^{n} \Delta t \left[\left(\int_{t^{k-1}}^{t^k} \left\| \frac{\partial^2 u}{\partial t^2}(s) \right\|_{L^2(\Omega)} ds \right)^2 \right.\right. \\
+\ & \left.\left. \frac{1}{\Delta t^2} \left(\int_{t^{k-1}}^{t^k} \left\| \left(I - \Pi^r_{1,h}\right)\left(\frac{\partial u}{\partial t}\right)(s) \right\|_{L^2(\Omega)} ds \right)^2 \right] \right),
\end{aligned}$$

Then, using the Cauchy-Schwarz inequality and estimate (5.27) for the projection operator $\Pi^r_{1,h}$, we obtain

$$\begin{aligned}
\|\varepsilon^n_h\|^2_{L^2(\Omega)} \le C(t^n)\left(h^{2r} |u_0|^2_{H^r(\Omega)} + \sum_{k=1}^{n} \Delta t \left[\Delta t \int_{t^{k-1}}^{t^k} \left\| \frac{\partial^2 u}{\partial t^2}(s) \right\|^2_{L^2(\Omega)} ds \right.\right. \\
\left.\left. + \frac{1}{\Delta t^2} \left(\int_{t^{k-1}}^{t^k} h^r \left| \frac{\partial u}{\partial t}(s) \right|_{H^r(\Omega)} ds \right)^2 \right] \right)
\end{aligned}$$

$$\leq C(t^n)\left(h^{2r}\,|u_0|^2_{\mathrm{H}^r(\Omega)} + \Delta t^2 \sum_{k=1}^{n} \int_{t^{k-1}}^{t^k} \left\| \frac{\partial^2 u}{\partial t^2}(s) \right\|^2_{\mathrm{L}^2(\Omega)} ds \right.$$

$$\left. + \frac{1}{\Delta t} h^{2r} \sum_{k=1}^{n} \Delta t \int_{t^{k-1}}^{t^k} \left| \frac{\partial u}{\partial t}(s) \right|^2_{\mathrm{H}^r(\Omega)} ds \right). \tag{5.47}$$

The result now follows using (5.38) and estimate (5.27). ◇

More stability and convergence estimates can be found in [Tho84].

5.6 Exercises

1. Verify that the mass matrix M introduced in (5.7) is positive definite.
2. Consider the problem:

$$\begin{cases} \dfrac{\partial u}{\partial t} - \dfrac{\partial}{\partial x}\left(\alpha \dfrac{\partial u}{\partial x}\right) - \beta u = 0 & \text{in } Q_T = (0,1) \times (0,\infty), \\ u = u_0 & \text{for } x \in (0,1),\ t = 0, \\ u = \eta & \text{for } x = 0,\ t > 0, \\ \alpha \dfrac{\partial u}{\partial x} + \gamma u = 0 & \text{for } x = 1,\ t > 0, \end{cases}$$

where $\alpha = \alpha(x)$, $u_0 = u_0(x)$ are given functions and $\beta, \gamma, \eta \in \mathbb{R}$ (with positive β).

   a) Prove existence and uniqueness of the weak solution for varying γ, providing suitable limitations on the coefficients and suitable regularity hypotheses on the functions α and u_0.
   b) Introduce the spatial semi-discretization of the problem using the Galerkin-finite element method, and carry out its stability and convergence analysis.
   c) In the case where $\gamma = 0$, approximate the same problem with the explicit Euler method in time and carry out its stability analysis.

3. Consider the following problem: find $u(x,t)$, $0 \leq x \leq 1$, $t \geq 0$, such that

$$\begin{cases} \dfrac{\partial u}{\partial t} + \dfrac{\partial v}{\partial x} = 0, & 0 < x < 1,\ t > 0, \\ v + \alpha(x)\dfrac{\partial u}{\partial x} - \gamma(x) u = 0, & 0 < x < 1,\ t > 0, \\ v(1,t) = \beta(t),\ u(0,t) = 0, & t > 0, \\ u(x,0) = u_0(x), & 0 < x < 1, \end{cases}$$

where α, γ, β, u_0 are given functions.

a) Introduce an approximation based on finite elements of degree two in x and the implicit Euler method in time and prove its stability.
b) How will the error behave as a function of the parameters h and Δt?
c) Suggest a way to provide an approximation for v starting from the one for u as well as its approximation error.

4. Consider the following (diffusion-transport-reaction) initial-boundary value problem: find $u : (0,1) \times (0,T) \to \mathbb{R}$ such that

$$\begin{cases} \dfrac{\partial u}{\partial t} - \dfrac{\partial}{\partial x}\left(\alpha \dfrac{\partial u}{\partial x}\right) + \dfrac{\partial}{\partial x}(\beta u) + \gamma u = 0, & 0 < x < 1,\ 0 < t < T, \\ u = 0 & \text{for } x = 0,\ 0 < t < T, \\ \alpha \dfrac{\partial u}{\partial x} + \delta u = 0 & \text{for } x = 1,\ 0 < t < T, \\ u(x,0) = u_0(x), & 0 < x < 1,\ t = 0, \end{cases}$$

where $\alpha = \alpha(x)$, $\beta = \beta(x)$, $\gamma = \gamma(x)$, $\delta = \delta(x)$, $u_0 = u_0(x)$, $x \in [0,1]$ are given functions.

a) Write its weak formulation.
b) In addition to the hypotheses:

$$a.\ \ \exists \beta_0,\ \alpha_0,\ \alpha_1 > 0 : \ \forall x \in (0,1)\ \alpha_1 \geq \alpha(x) \geq \alpha_0,\ \beta(x) \leq \beta_0,$$

$$b.\ \ \tfrac{1}{2}\beta'(x) + \gamma(x) \geq 0 \quad \forall x \in (0,1),$$

provide further possible hypotheses on the data so that the problem is well-posed. Moreover, give an a priori estimate of the solution. Treat the same problem with non-homogeneous Dirichlet data $u = g$ for $x = 0$ and $0 < t < T$.
c) Consider a semi-discretization based on the linear finite elements method and prove its stability.
d) Finally, provide a full discretization where the temporal derivative is approximated using the implicit Euler scheme and prove its stability.

5. Consider the following fourth-order initial-boundary value problem: find $u : \Omega \times (0,T) \to \mathbb{R}$ such that

$$\begin{cases} \dfrac{\partial u}{\partial t} - \operatorname{div}(\mu \nabla u) + \Delta^2 u + \sigma u = 0 & \text{in } \Omega \times (0,T), \\ u(\mathbf{x},0) = u_0 & \text{in } \Omega, \\ \dfrac{\partial u}{\partial n} = u = 0 & \text{on } \Sigma_T = \partial\Omega \times (0,T), \end{cases}$$

where $\Omega \subset \mathbb{R}^2$ is a bounded open domain with "regular" boundary $\partial\Omega$, $\Delta^2 = \Delta\Delta$ is the bi-harmonic operator, $\mu(\mathbf{x})$, $\sigma(\mathbf{x})$ and $u_0(\mathbf{x})$ are known functions defined in

Ω. It is known that

$$\sqrt{\int_{\Omega} |\Delta u|^2 \mathrm{d}\Omega} \simeq \|u\|_{\mathrm{H}^2(\Omega)} \qquad \forall\, u \in \mathrm{H}^2_0(\Omega)\,,$$

that is the two norms are equivalent, where

$$\mathrm{H}^2_0(\Omega) = \{u \in \mathrm{H}^2(\Omega) : \; u = \partial u / \partial n = 0 \text{ on } \partial\Omega\}. \tag{5.48}$$

a) Write its weak formulation and verify that the solution exists and is unique, formulating suitable regularity hypotheses on the data.
b) Consider a semi-discretization based on triangular finite elements and provide the minimum degree that such elements must have in order to solve the given problem adequately. (We note that, if $\mathcal{T}_h$ is a triangulation of Ω and $v_{h|K}$ is a polynomial for each $K \in \mathcal{T}_h$, then $v_h \in \mathrm{H}^2(\Omega)$ if and only if $v_h \in C^1(\overline{\Omega})$, that is v_h and its first derivatives are continuous across the interfaces of the elements of $\mathcal{T}_h$.)

6

Generation of 1D and 2D grids

As we have seen, the finite element method for the solution of partial differential equations requires a "triangulation" of the computational domain, i.e. a partition of the domain in simpler geometric entities (for instance, triangles or quadrangles in two dimensions, tetrahedra, prisms or hexahedra in three dimensions), called the elements, which verify a number of conditions. Similar partitions stand at the base of other approximation methods, such as the finite volume method (see Chapter 9) and the spectral element method (see Chapter 10). The set of all elements is the so-called computational grid (or, simply, grid, or mesh).

In this chapter, for simplicity, we focus on the main partitioning techniques for one- and two-dimensional domains, with no ambition of completeness. If necessary, we will refer the reader to the relevant specialized literature. We will deal only with the case of polygonal domains; for computational domains with curved boundaries, the interested reader can consult [Cia78], [BS94], [GB98]. The techniques exposed for the 2D case can be extended to three-dimensional domains.

6.1 Grid generation in 1D

Suppose that the computational domain Ω be an interval (a,b). The most elementary partition in sub-intervals is the one where the step h is constant. Having chosen the number of elements, say N, we set $h = \frac{b-a}{N}$ and introduce the points $x_i = x_0 + ih$, with $x_0 = a$ and $i = 0, \dots, N$. Such points $\{x_i\}$ are called "vertices" in analogy to the two-dimensional case, where they will actually be the vertices of the triangles whose union covers the domain Ω. The partition thus obtained is called grid. The latter is *uniform* as it is composed by elements of the same length.

In the more general case, we will use non-uniform grids, possibly generated according to a given law. Among the possible different procedures, we illustrate a fairly general one. Let a strictly positive function $\mathcal{H} : [a,b] \rightarrow \mathbb{R}^+$, called *spacing function*, be assigned and let us consider the problem of generating a partition of the interval $[a,b]$ having $N+1$ vertices x_i. The value $\mathcal{H}(x)$ represents the desired spacing in corre-

A. Quarteroni: *Numerical Models for Differential Problems*, 2nd Ed.
MS&A – Modeling, Simulation & Applications 8
DOI 10.1007/978-88-470-5522-3_6, © Springer-Verlag Italia 2014

spondence of the point x. For instance, if $\mathcal{H} = h$ (constant), with $h = (b-a)/M$ for a given integer M, we fall exactly in the preceding case of the uniform grid, with $N = M$. More generally, we compute $\mathcal{N} = \int_a^b \mathcal{H}^{-1}(x)\,dx$ and we set $N = \max(1, [\mathcal{N}])$, where $[\mathcal{N}]$ denotes the integer part of $\mathcal{N}$, i.e. the largest positive integer smaller than or equal to $\mathcal{N}$. Note that the resulting grid will have at least one element. Then we set $\kappa = \frac{N}{\mathcal{N}}$ and look for the points x_i such that

$$\kappa \int_a^{x_i} \mathcal{H}^{-1}(x)\,dx = i,$$

for $i = 0,\dots,N$. The constant κ is a positive correction factor, with a value as close as possible to 1, whose purpose is to guarantee that N is indeed an integer. In fact, for a given $\mathcal{H}$, the number N of elements is itself an unknown of the problem. Instead, the $\mathcal{H}^{-1}$ function defines a density function: to higher values of $\mathcal{H}^{-1}$ correspond denser nodes, and conversely, to smaller values of $\mathcal{H}^{-1}$ correspond sparser nodes.

Obviously, if we wish to construct a grid with a given number N of elements, as well as a given variation on $[a,b]$, is sufficient to renormalize the spacing function so that the integral on (a,b) of the corresponding density is exactly equal to N. In any case, to compute the points x_i, it is useful to introduce the following Cauchy problem

$$y'(x) = \kappa\mathcal{H}^{-1}(x),\, x \in (a,b), \text{ with } y(a) = 0.$$

The points x_i will then be defined by the relation $y(x_i) = i$, for $i = 1,\dots,N-1$. Then, it will automatically follow that $x_0 = a$ and $x_N = b$. We will then be able to use a numerical solution method to find the roots of the functions $f_j(x) = y(x) - j$, for each value of $j \in \{1,\dots,N-1\}$ (see e.g. [QSS07]).

Besides being quite general, this procedure can be easily extended to the generation of vertices on the curved boundary of a two-dimensional domain, as we will see in Sect. 6.4.2.

In the case where $\mathcal{H}$ does not exhibit excessive variations in the interval (a,b), we can also use a simplified procedure which consists in computing a set of preliminary points $\tilde{x}_i$, for $i = 0,\dots,N$, defined as follows:

1. Set $\tilde{x}_0 = a$ and define $\tilde{x}_i = \tilde{x}_{i-1} + \mathcal{H}(\tilde{x}_{i-1})$, $i = 1,2,\dots$, until finding the value M such that $\tilde{x}_M \geq b$ and $\tilde{x}_{M-1} < b$;
2. if $\tilde{x}_M - b \leq b - \tilde{x}_{M-1}$ set $N = M$, otherwise define $N = M-1$.

Then the final set of vertices are obtained by setting

$$x_i = x_{i-1} + k\mathcal{H}(\tilde{x}_{i-1}), \quad i = 1,\dots,N,$$

with $x_0 = a$ and $k = (b - x_{N-1})/(x_N - x_{N-1})$.

The MATLAB program **mesh_1d** allows to construct a grid on an interval with endpoints `a` and `b` with step specified in the macro `H`, using the previous simplified algorithm. For instance, with the following MATLAB commands:

```
a = 0; b = 1; H = '0.1';
coord = mesh_1d(a,b,H);
```

we create a uniform grid on $[0,1]$ with 10 sub-intervals with step $h = 0.1$.

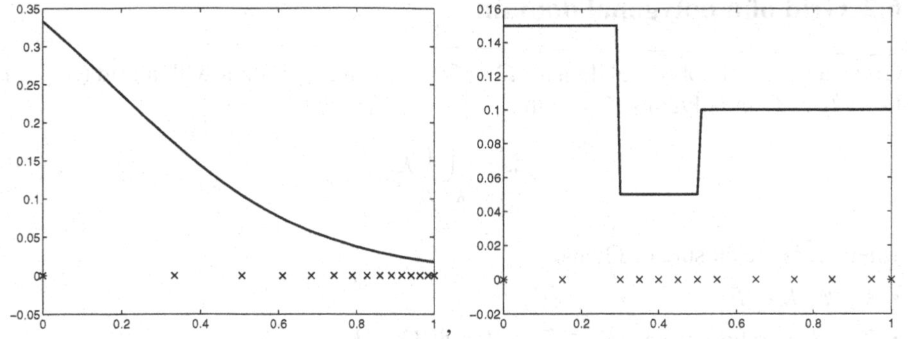

Fig. 6.1. On the left-hand side, the behaviour of the grid step (on the x-axis) associated to the function `H = '1/(exp(4*x)+2)'`, on the right-hand side the one relating to the function `H = '.1*(x<3) + .05*(x>5) + .05'`. The graph also reports the corresponding vertex distributions

Setting $\mathtt{H} =' \mathtt{1/(exp(4*x)+2)}'$ we obtain a grid that becomes finer when approaching the second endpoint of the interval, while for $\mathtt{H} =' .1*(\mathtt{x} < .3) + .05*(\mathtt{x} > .5) + .05'$ we obtain a grid with a discontinuously varying step (see Fig. 6.1).

Program 1 – mesh_1d: Constructs a one-dimensonal grid on an interval `[a,b]` following the spacing function `H`

```
function coord = mesh_1d(a,b,H)

coord = a;
while coord(end) < b
  x = coord(end);
  xnew = x + eval(H);
  coord = [coord, xnew];
end
if (coord(end) - b) > (b - coord(end-1))
  coord = coord(1:end-1);
end
coord_old = coord;
kappa = (b - coord(end-1))/(coord(end) - coord(end-1));
coord = a;
for i = 1:length(coord_old)-1
  x = coord_old(i);
  coord(i+1) = x + kappa*eval(H);
end
```

We point out that in case $\mathcal{H}$ is determined by an error estimate, Program 1 will allow to perform grid adaptivity.

We now tackle the problem of constructing the grid for two-dimensional domains.

6.2 Grid of a polygonal domain

Given a bounded polygonal domain Ω in $\mathbb{R}^2$, we can associate it with a grid (or partition) $\mathcal{T}_h$ of Ω in polygons K such that

$$\overline{\Omega} = \bigcup_{K \in \mathcal{T}_h} K,$$

where $\overline{\Omega}$ is the closure of Ω, and

- $\overset{\circ}{K} \neq \emptyset \; \forall K \in \mathcal{T}_h$;
- $\overset{\circ}{K}_1 \cap \overset{\circ}{K}_2 = \emptyset$ for each $K_1, K_2 \in \mathcal{T}_h$ such that $K_1 \neq K_2$;
- if $F = K_1 \cap K_2 \neq \emptyset$ with $K_1, K_2 \in \mathcal{T}_h$ and $K_1 \neq K_2$, then F is either a whole edge or a vertex of the grid;
- having denoted by h_K the diameter of K for each $K \in \mathcal{T}_h$, we define $h = \max_{K \in \mathcal{T}_h} h_K$.

We have denoted by $\overset{\circ}{K} = K \setminus \partial K$ the interior of K. The grid $\mathcal{T}_h$ is also called *mesh*, or sometimes *triangulation* (in a broad sense) of $\overline{\Omega}$.

The constraints imposed on the grid by the first two conditions are obvious: in particular, the second one requires that given two distinct elements, their interiors do not overlap. The third condition limits the admissible triangulations to the so-called *conforming* ones. To illustrate the concept, we represent in Fig. 6.2 a conforming (left) and nonconforming (right) triangulation. In the remainder, we will only consider conforming triangulations. However, there exist very specific finite element approximations, not considered in the present book, which use nonconforming grids, i.e. grids that do not satisfy the third condition. These methods are therefore more flexible, at least as far as the choice of the computational grid is concerned. They allow, among other things, the coupling of grids constructed from elements of different nature, for instance triangles and quadrilaterals. The fourth condition links the parameter h to the maximum diameter of the elements of $\mathcal{T}_h$.

For reasons linked to the interpolation error theory recalled in Chapter 4, we will only consider *regular* triangulations $\mathcal{T}_h$, i.e. the ones for which, for each element $K \in \mathcal{T}_h$, the ratio between the diameter h_K and the sphericity ρ_K (i.e. the diameter of the inscribed circle) is less that a given constant. More precisely, the grids satisfy Property

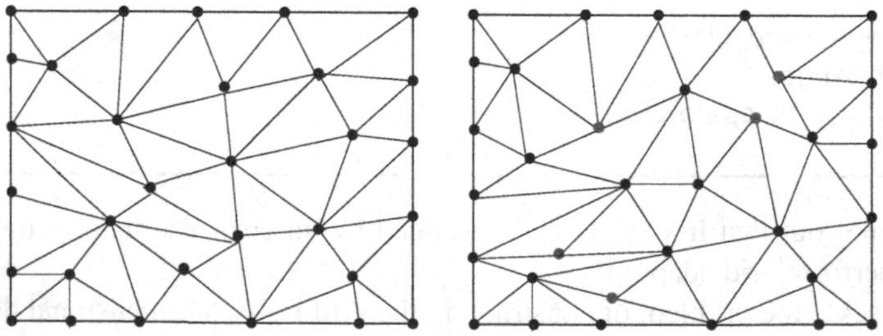

Fig. 6.2. Example of conforming (left) and nonconforming (right) grid

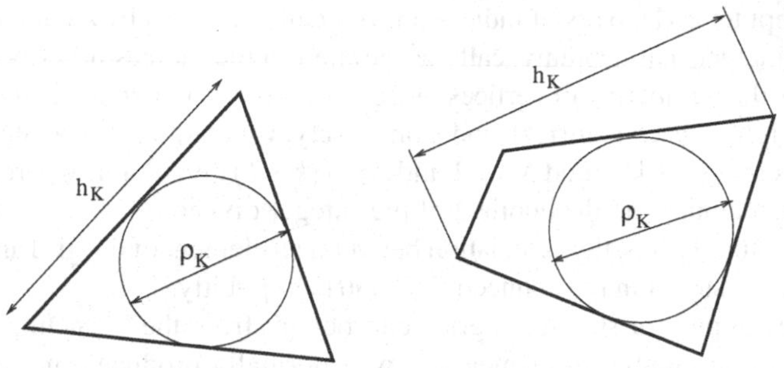

Fig. 6.3. Diameter and sphericity for a triangular (left) and quadrilateral element (right)

(4.37). Fig. 6.3 illustrates the meaning of diameter and sphericity for a triangular or quadrilateral element.

In actual applications, it is customary to distinguish between *structured* and *unstructured* grids. Structured grids basically use quadrangular elements and are characterized by the fact that access to the vertices adjacent to a given node (or to the elements adjacent to a given element) is immediate. Indeed, it is possible to establish a bijective relationship between the vertices of the grid and the pairs of integer numbers $(i, j), i = 1, \ldots, N_i, \quad j = 1, \ldots, N_j$ such that, given the node of indices (i, j), the four adjacent vertices are in correspondence with the indices $(i-1, j)$, $(i+1, j)$, $(i, j-1)$ and $(i, j+1)$. The total number of vertices is therefore $N_i N_j$. An analogous association can be established between the elements of the grid and the pairs (I, J), $I = 1, \ldots, N_i - 1, \quad J = 1, \ldots, N_j - 1$. Moreover, it is possible to identify directly the vertices corresponding to each element, without having to memorize the connectivity matrix explicitly (the latter is the matrix which, for each element, provides its vertex numbering). Fig. 6.4 (left) illustrates such situation.

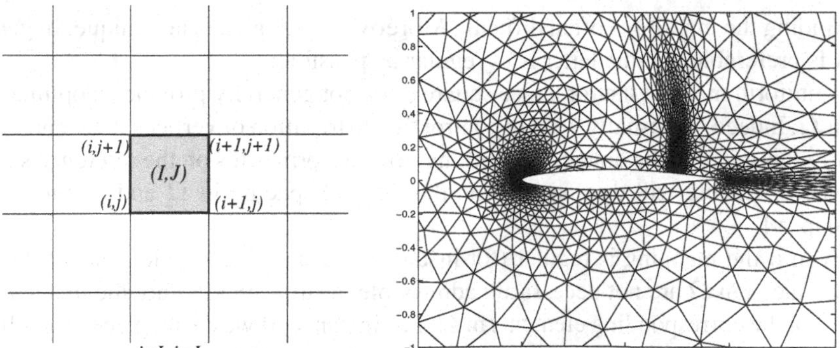

Fig. 6.4. (Left) (I,J)-Numbering of the vertices of an element in a structured grid. (Right) A non-structured triangular grid in an external region of an airfoil, adapted to improve the accuracy of the numerical solution for a given flow condition

In a computer code, pairs of indices are typically replaced by a numbering formed by a single integer that is biunivocally associated to the indices described above. For instance, for the numbering of vertices, we can choose to associate the integer number $k = i + (j-1)N_i$ to each pair (i, j), and, conversely, we uniquely associate to the vertex k the indices $i = ((k-1) \mod N_i) + 1$ and $j = ((k-1) \operatorname{div} N_i) + 1$, where mod and div denote the remainder and the quotient of the integer division.

In unstructured grids, the association between an element of the grid and its vertices must instead be stored in the connectivity matrix explicitly.

Code developed for structured grids can benefit from the "structure" of the grid, and, for an equal number of elements, it will normally produce a more efficient algorithm, both in terms of memory and in terms of computational time, with respect to a similar scheme on a non-structured grid. In contrast, non-structured grids offer a greater flexibility both from the viewpoint of a triangulation of domains of complex shape and for the possibility to locally refine/derefine the grid. Fig. 6.4 (right) shows an example of a non-structured grid whose spacing has been adapted to the specific problem under exam. Such localized refinements would be difficult, or impossible, to obtain using a structured type of grid (unless we use nonconforming grids).

Non-structured two-dimensional grids are generally formed by triangles, although it is possible to have quadrangular non-structured grids.

6.3 Generation of structured grids

The most elementary idea to generate a structured grid on a domain Ω of arbitrary shape consists in finding a regular and invertible map $\mathcal{M}$ between the square $\widehat{\Omega} = [0,1] \times [0,1]$ (which we will call reference square) and $\overline{\Omega}$. Note that the map must be regular up to the boundary (a requirement that can in some cases be relaxed). We proceed by generating a uniform - say - reticulation in the reference square, then we use the mapping $\mathcal{M}$ to transform the coordinates of the vertices in $\widehat{\Omega}$ into the corresponding ones in $\overline{\Omega}$.

There are different aspects of this procedure to be considered with due care.

1. Finding the map $\mathcal{M}$ is often difficult. Moreover, such map is not unique. In general it is preferable that the latter is as regular as possible.
2. A uniform mesh of the reference square does not generally provide an optimal grid in Ω. Indeed, we usually want to control the distribution of vertices in Ω, and generally this can only be done by generating non-uniform grids on the reference square, whose spacing will depend both on the desired spacing in Ω and on the chosen map $\mathcal{M}$.
3. Even if the mapping is regular (for instance of class C^1), the elements of the grid produced in Ω are not necessarily admissible, as the latter are not the image under $\mathcal{M}$ of the corresponding elements in $\widehat{\Omega}$. For instance, if we desire piecewise bilinear ($\mathbb{Q}_1$) finite elements in Ω, the edges of the latter will need to be parallel to the Cartesian axes, while the image of a mesh $\mathbb{Q}_1$ on the reference square produces curved edges in Ω if the mapping is nonlinear. In other words, the map is made effective only on the vertices, not on the edges, of the grid of $\widehat{\Omega}$.

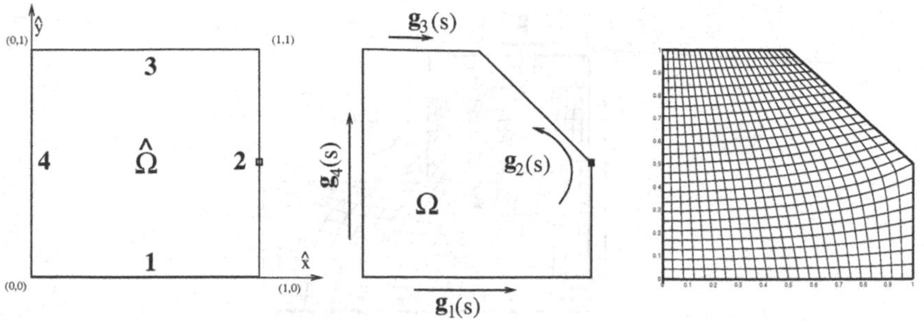

Fig. 6.5. Construction of a structured grid: on the left-hand side, identification of the map on the boundary; on the right-hand side, grid corresponding to a uniform partitioning of the reference square into 24×24 elements

An option to construct the map $\mathcal{M}$ consists in using the transfinite interpolation (10.3) that will be illustrated in Chap. 10. Such methodology is however not always easily applicable. We will therefore illustrate in the remainder a more general methodology, which we will apply to a specific example, and refer to the specific literature [TWM85, TSW99] for further examples and details.

Suppose we have a domain Ω whose boundary can be divided in four consecutive parts $\Gamma_1, \ldots, \Gamma_4$, as illustrated in Fig. 6.5 for a particularly simple domain. Moreover, suppose we can describe such portions of $\partial\Omega$ via four parametric curves $\mathbf{g}_1, \ldots, \mathbf{g}_4$ oriented as in the figure, where the parameter s varies between 0 and 1 on each curve. This construction allows us to create a bijective map between the sides of the reference square and the domain boundary. Indeed, we will associate each curve to the corresponding side of the square, as exemplified in Fig. 6.5. We now need to understand how to extend the mapping to the whole $\widehat{\Omega}$.

Remark 6.1. Note that the curves $\mathbf{g}_i$, $i = 1, \ldots, 4$, are generally not differentiable on all of $(0,1)$, but can exhibit a finite number of "corners" where $\frac{d\mathbf{g}_i}{ds}$ is undefined. In Fig. 6.5, for instance, the curve $\mathbf{g}_2$ is not differentiable at the "corner" marked by a small black square. •

An option to construct the map $\mathcal{M} : (\widehat{x}, \widehat{y}) \mapsto (x, y)$ consists in solving the following elliptic system in $\widehat{\Omega}$:

$$-\frac{\partial^2 \mathbf{x}}{\partial \widehat{x}^2} - \frac{\partial^2 \mathbf{x}}{\partial \widehat{y}^2} = 0 \quad \text{in } \widehat{\Omega} = (0,1)^2, \tag{6.1}$$

with boundary conditions

$$\mathbf{x}(\widehat{x}, 0) = \mathbf{g}_1(\widehat{x}),\ \mathbf{x}(\widehat{x}, 1) = \mathbf{g}_3(\widehat{x}), \quad \widehat{x} \in (0,1),$$
$$\mathbf{x}(1, \widehat{y}) = \mathbf{g}_2(\widehat{y}),\ \mathbf{x}(0, \widehat{y}) = \mathbf{g}_4(\widehat{y}), \quad \widehat{y} \in (0,1).$$

The vertices of a grid in the reference square can then be transformed into the vertices of a grid in Ω. Note that the solution of problem (6.1) will generally be found by using a numerical method, for instance via a finite difference (or finite element)

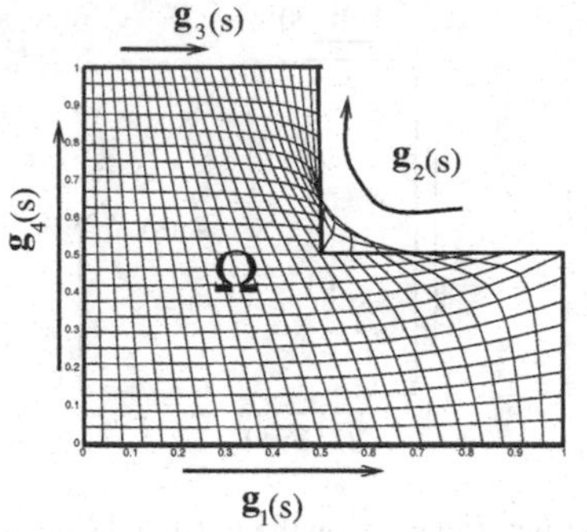

Fig. 6.6. Triangulation of a non-convex domain. Identification of the boundary map and mesh obtained by solving the elliptic problem (6.1)

scheme. Moreover, to abide by the geometry of the boundary of Ω suitably, it is necessary to ensure that a vertex is generated at each "edge". In Fig. 6.5 (right) we illustrate the result of this methodology to the domain in Fig. 6.5 (left). It can be noted that the grid corresponding to a regular partition of the reference square is not particularly satisfactory if, for instance, we want to have a higher distribution of vertices at the edge.

Moreover, the methodology described above is not applicable to non-convex domains. Indeed, let us consider Fig. 6.6 where we show an L-shaped domain, with the corresponding boundary partition, and the grid obtained by solving problem (6.1) starting from a regular partition of the reference domain. It is evident that such grid is unacceptable.

To solve such problems, we can proceed in several (not mutually exclusive) ways:

- we use in $\widehat{\Omega}$ a non-uniform grid, that accounts for the geometric features of Ω;
- we use a different map $\mathcal{M}$, obtained, for instance, by solving the following new differential problem instead of (6.1)

$$-\alpha\frac{\partial^2 \mathbf{x}}{\partial \widehat{x}^2} - \beta\frac{\partial^2 \mathbf{x}}{\partial \widehat{y}^2} + \gamma\mathbf{x} = \mathbf{f} \quad \text{in } \widehat{\Omega}, \tag{6.2}$$

  where $\alpha > 0$, $\beta > 0$, $\gamma \geq 0$ and $\mathbf{f}$ are suitable functions of $\widehat{x}$ and $\widehat{y}$. They are chosen depending on the geometry of Ω and in order to control the vertex distribution;
- we partition Ω in sub-domains that are triangulated separately. This technique is normally known as *blockwise structured grid generation*. If we wish the global grid to be conforming, we need to be very careful on how to distribute the number of vertices on the boundaries of the interfaces between the different sub-domains. The problem can become extremely complex when the number of sub-domains is very large.

Methods of the type illustrated above are called *elliptic schemes of grid generation*, as they are based on the solution of elliptic equations, such as (6.1) and (6.2).

The interested reader is referred to the above-cited specialized literature.

6.4 Generation of non-structured grids

We will here consider the generation of non-structured grids with triangular elements. The two main algorithms used for this purpose are the *Delaunay triangulation* and the *advancing front* technique.

6.4.1 Delaunay triangulation

A triangulation of a set of n points of $\mathbb{R}^2$ is a Delaunay triangulation if the disc circumscribed to each triangle contains no vertex (see Fig. 6.7).

A Delaunay triangulation features the following properties:

1. given a set of points, the Delaunay triangulation is unique, except for specific situations where M points (with $M > 3$) lie on a circle;
2. among all possible triangulations, the Delaunay triangulation is the one maximizing the minimum angle of the grid triangles (this is called the max-min regularity property);
3. the set composed by the union of triangles is the convex figure of minimum surface that encloses the given set of points (and is called convex hull).

The third property makes the Delaunay algorithm inapplicable to non-convex domains, at least in its original form.

However, there exists a variant, called *Constrained Delaunay Triangulation (CDT)*, that allows to fix *a priori* a set of the grid edges to generate: the resulting grid necessarily associates such edges to some triangle. In particular, we can a priori impose those edges which define the boundary of the grid.

In order to better specify the concept of CDT, we state beforehand the following definition: given two points P_1 and P_2, we will say that these are reciprocally *visible* if the segment P_1P_2 passes through none of the boundary sides (or, more generally, the edges we want to fix a priori). A constrained Delaunay triangulation satisfies the following property: the interior of the circle circumscribed to each triangle K contains no vertex visible from an internal point to K.

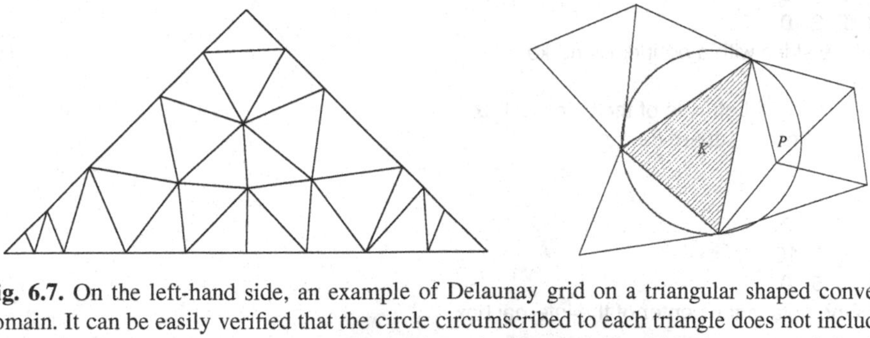

Fig. 6.7. On the left-hand side, an example of Delaunay grid on a triangular shaped convex domain. It can be easily verified that the circle circumscribed to each triangle does not include any vertex of the grid. On the right-hand side, a detail of a grid which does not satisfy the Delaunay condition: indeed, the vertex P falls inside the circle circumscribed to the triangle K

Once again, it can be proved that such triangulation is unique and satisfies the max-min regularity property. The CDT is therefore not a proper Delaunay triangulation, as some of its triangles could contain vertices belonging to the initial set. In any case, the vertices are only the original ones specified in the set, and no further vertices are added. However, two variants are possible: the *Conforming Delaunay Triangulation* and the *Conforming Constrained Delaunay Triangulation* (or CCDT). The former is a triangulation where each triangle is a Delaunay triangulation, but each edge to be fixed can be further subdivided in sub-segments; in this case, new vertices can be added to obtain shorter segments. The additional vertices are often necessary to guarantee the max-min Delaunay property and at the same time to ensure that each prescribed side is correctly represented. The second variant represents a triangulation where the triangles are of the constrained Delaunay type. Also in this case, we can add additional vertices, and the edges to be fixed cannot be divided in smaller segments. In the latter case, however, the aim is not to guarantee that the edges are preserved, but to improve the triangles' quality.

Among the available software for the generation of Delaunay grids, or their variants, `Triangle` [She] allows to generate Delaunay triangulations, with the option to modulate the regularity of the resulting grids in terms of maximal and minimal angles of the triangles. The geometry is given as input to `Triangle` in the form of a graph, called Planar Straight Line Graph (PSLG). Such codification is written in an input file with extension `.poly`: the latter basically contains a list of vertices and edges, but can also include information on cavities or concavities present in the geometry.

A sample `.poly` file is reported below.

```
# A box with eight vertices in 2D, no attribute, one boundary marker
8 2 0 1
# Vertices of the external box
1  0 0  0
2  0 3  0
3  3 0  0
4  3 3  0
# Vertices of the internal box
5  1 1  0
6  1 2  0
7  2 1  0
8  2 2  0
# Five sides with a boundary marker
5 1
1  1 2  5 # Left side of the external box
# Sides of the square cavity

2  5 7  0
3  7 8  0
4  8 6  10
5  6 5  0
# One hole in the center of the internal box
1
1  1.5 1.5
```

The example above illustrates a geometry representing a square with a square hole. The first part of the file lists the vertices, while the second one defines the sides to fix. The first line declares that eight vertices are going to follow, that the spatial dimension of the grid is two (we are in $\mathbb{R}^2$), that no other attribute is associated to the vertices and that a boundary marker is defined on each point. The attributes represent possible physical properties relating to the mesh nodes, such as conductibility and viscosity values, etc. The boundary markers are integer-valued flags which can be used within a computational code to assign suitable boundary conditions at different vertices. The following lines display the eight vertices, with their abscissae and ordinates, followed by the boundary marker value, zero in this case. The first line of the second part declares that there are five sides ensuing, and that on each of them a value will be specified for the boundary marker. Then, five boundary sides follow one another, specified by their respective endpoints, and by the value of the boundary marker. In the final section of the file, a hole is defined by specifying the center coordinates, in the last line, preceded by the progressive numbering (in this case, limited to 1) of the holes.

The constrained Delaunay grid associated to this geometry, say `box.poly`, is obtained via the command

```
triangle -pc box
```

The parameter `-p` declares that the input file is a `.poly`, while the option `-c` prevents that the concavities are removed, as would normally happen without it. De facto, this option forces the triangulation of the convex hull of the PSLG graph. The result will be the creation of three files, `box.1.poly`, `box.1.node` and `box.1.ele`. The first file contains the description of the sides of the produced triangulation, the second one contains the node description, and the latter defines the connectivity of the generated elements. For the sake of conciseness, we will not describe the format of these three files in detail. Finally, we point out that the numerical value, 1 in this example, that separates the name of these three files from their respective extensions, plays the role of an iteration counter: `Triangle` can indeed successively refine or modify the triangulations produced time after time. The resulting triangulation is depicted in Fig. 6.8. A software attached to `Triangle`, called `Show Me`, allows to visualize the outputs of `Triangle`. For instance, Fig 6.8 (left) is obtained via the command

```
showme box
```

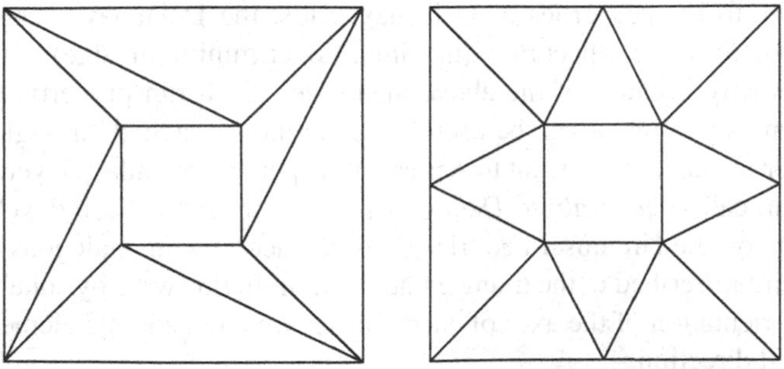

Fig. 6.8. Delaunay triangulation of a square with a square hole: CDT on the left-hand side, CCDT on the right-hand side

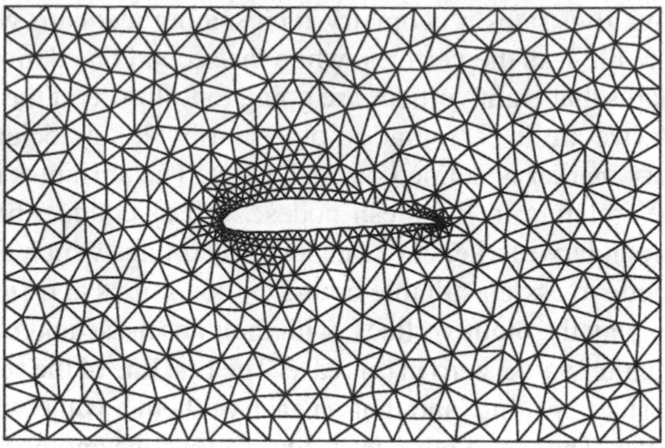

Fig. 6.9. Delaunay triangulation of a naca 4415 airfoil

To obtain a constrained conforming triangulation we must specify the command `triangle` with other parameters, such as `-q`, `-a` or `-u`. The first one imposes a constraint on the minimum angle, the second one fixes a maximum value for the surface of the triangles, while the third one forces the dimension of the triangles, typically through an external function which the user must provide. For example, via the command

```
triangle -pcq20 box
```

we obtain the constrained conforming Delaunay triangulation reported in Fig. 6.8 (right), characterized by a minimum angle of 20°. Finally, the conforming Delaunay triangulation is obtained by further specifying the option `-D`. A more complex example is represented in Fig. 6.9. The command used

```
triangle -pca0.01q30 naca
```

fixes the minimum angle to 30° and the maximum surface of the generated triangles to 0.01. The initial PSLG file `naca.poly` describes the geometry via 65 vertices, as many sides and one cavity. The final mesh consists of 711 vertices, 1283 elements and 137 edges on the boundary.
We refer to the wide on-line documentation and to the detailed help of `Triangle` for several further usage options of the software.

Returning to the properties of Delaunay grids, the Delaunay triangulation does not allow to control the aspect ratio (maximum over minimum edge) of the generated elements, exactly because of the above-mentioned max-min property. On the other hand, in some situations it can be useful to generate "stretched" triangles in a given direction, for instance if we want to represent properly a boundary layer. To this end, the algorithm called *generalized Delaunay triangulation* has been developed, where the condition on the circumscribed triangle is replaced by an analogous condition on the ellipse circumscribed to the triangle under exam. In this way, by suitably ruling the length and orientation of the axes of each ellipse, we can generate elements stretched in the desired direction.

The most currently used algorithms for the generation of Delaunay grids are incremental, i.e. they generate a sequence of Delaunay grids by adding a vertex at a time.

Hence, it is necessary to find procedures providing the new vertices in accordance with the desired grid spacing, and stopping such procedure as soon as the grid generated this way results to be unsatisfactory. For further details, [GB98] and [TSW99, Chap. 16] can be consulted, among others. A detailed description of the geometric properties of the constrained Delaunay triangulation, both for domains of $\mathbb{R}^2$ and of $\mathbb{R}^3$, can be found in [BE92].

6.4.2 Advancing front technique

We roughly described another widely used technique used for the generation of non-structured grids, the *advancing front* technique. A necessary ingredient is the knowledge of the desired spacing to be generated for the grid elements. Let us then suppose that a spacing function $\mathcal{H}$, defined on $\overline{\Omega}$, provides for each point P of $\overline{\Omega}$ the dimensions of the grid desired in that point, for instance, through the diameter h_K of the elements that must be generated in a neighborhood of P. If we want to control the shape aspect of the generated elements, $\mathcal{H}$ will have a more complex shape. In fact, it will be a positive definite symmetric tensor, i.e. $\mathcal{H} : \Omega \to \mathbb{R}^{2\times 2}$ such that, for each point P of the domain, the (perpendicular) eigenvectors of $\mathcal{H}$ denote the direction of maximum and minimum stretching of the triangles that will need to be generated in the neighborhood of P, while the eigenvalues (more precisely, the square roots of the eigenvalue inverses), characterize the two corresponding spacings (see [GB98]). In the remainder, we will only consider the case where $\mathcal{H}$ is a scalar function.

The first operation to perform is to generate the vertices along the domain boundary. Let us suppose that $\partial\Omega$ is described as the union of parametric curves $\mathbf{g}_i(s)$, $i = 1,\dots N$, for instance splines or polygonal splits. For simplicity, we assume that, for each curve, the parameter s varies between 0 and 1. If we wish to generate $N_i + 1$ vertices along the curve $\mathbf{g}_i$ it is sufficient to create a vertex for all the values of s for which the function

$$f_i(s) = \int_0^s \mathcal{H}^{-1}(\mathbf{g}_i(\tau)) \left|\frac{d\mathbf{g}_i}{ds}(\tau)\right| d\tau$$

takes integer values. More precisely, the curvilinear coordinates $s_i^{(j)}$ of the nodes to generate along the curve $\mathbf{g}_i$ satisfy the relations

$$f_i(s_i^{(j)}) = j, \quad j = 0,\cdots,N_i \text{ with the constraints } s_i^{(0)} = 0, s_i^{(N_i)} = 1.$$

The procedure is similar to the one described in Sect. 6.1. Note that the term $|\frac{d\mathbf{g}_i}{ds}|$ accounts for the intrinsic metric of the curve.

This being done, the advancing front process can start. The latter is described by a data structure that contains the list of the sides defining the boundary between the already triangulated portion of Ω and the one yet to be. At the beginning of the process, the front contains the boundary sides.

During the process of grid generation, each side of the front is available to create a new element, which is constructed by connecting the chosen side with a new or previously existing vertex of the grid. The choice whether to use an existing vertex or to create a new one depends on several factors, among which the compatibility

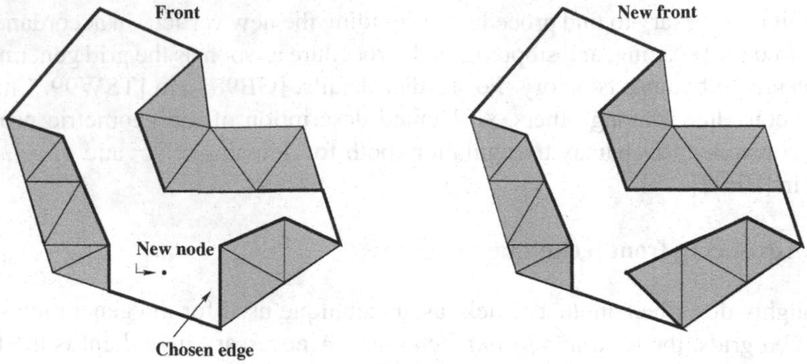

Fig. 6.10. Advancement of the front. The previously triangulated part of the domain has been shaded

between the dimension and the shape of the element that would be generated and the ones provided by the spacing function $\mathcal{H}$. Moreover, the new element must not intersect any side of the front.

Once the new element has been generated, its new sides will be "added" to the front so that the latter describes the new boundary between the triangulated and non-triangulated part, while the initial side is removed from the data list. In this way, during the generation process the front will progress from the already triangulated zones toward the zone yet to be triangulated (see Fig. 6.10).

The general advancing front algorithm hence consists of the following steps:

1. define the boundary of the domain to be triangulated;
2. initialize the front by a piecewise linear curve conforming to the boundary;
3. choose the side to be removed from the front using some criterion (typically the choice of the shortest side provides good quality meshes);
4. for the side, say AB, chosen this way:
   a) select the "candidate" vertex C, i.e. the point inside the domain whose distance from AB is prescribed by the desired spacing function $\mathcal{H}$;
   b) seek an already existing point C' on the front in a suitable neighbourhood of C. If the search is successful, C' becomes the new candidate point C. Continue the search;
   c) establish whether the triangle ABC intersects some other side of the front. If so, select a new candidate point from the front and start back from step 4.b);
5. add the new point C, the new edges and the new triangle ABC to the corresponding lists;
6. erase the edge AB from the front and add the new edges;
7. if the front is non-empty, continue from point 3.

It is obvious that if we wish the computational cost to be a linear function of the number of generated elements, it will be necessary to make the above-described operations as independent as possible from the number of dimensions of the grid we

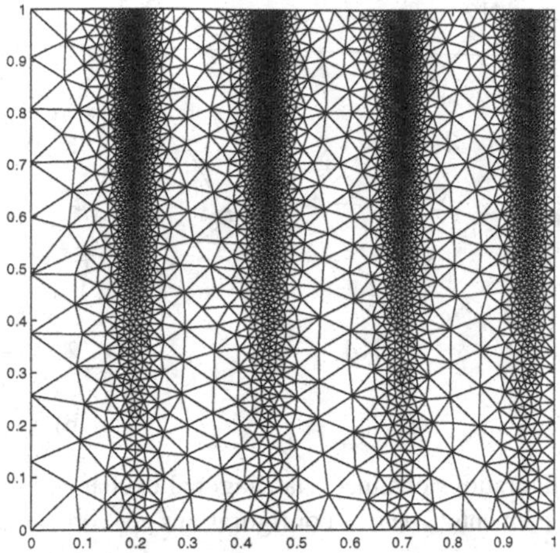

Fig. 6.11. Advancing front technique. Example of non-uniform spacing

are generating and, in particular, from the dimensions of the advancing front. Such an objective is not trivial, especially because operations such as the control of the intersection of a new triangle, or the search for the vertices of the front close to a generic point, span the whole front. We refer for this to the specialized literature, and in particular to Chaps. 14 and 17 of [TSW99].

As previously pointed out in the algorithm description, the quality of the generated grid depends on the procedure of choice of the front edge on which to generate the new triangle. In particular, a frequently adopted technique consists in choosing the side with the smallest length: intuitively, this also allows to satisfy non-uniform spacing requirements, without risking that the zones where a more dense node distributions is required are overwritten by triangles associated to a coarser spacing. An example of mesh obtained through such technique, in correspondence of the choice $\mathcal{H}(x,y) = e^{4\sin(8\pi x)}e^{-2y}$, is represented in Fig. 6.11.

By implementing the suitable tricks and data structures, the advancing algorithm provides a grid whose spacing is coherent with the requested one, with computational times almost proportional to the number of generated elements.

The advancing front technique can also be used for the generation of quadrangular grids.

6.5 Regularization techniques

Once the grid has been generated, a post-processing can be necessary in order to improve its regularity. Some methods allow to transform the grid via operations that improve the triangles' shape. In particular, we will examine regularization techniques that modify either the topological features (by diagonal exchange) or the geometrical features (by node displacement).

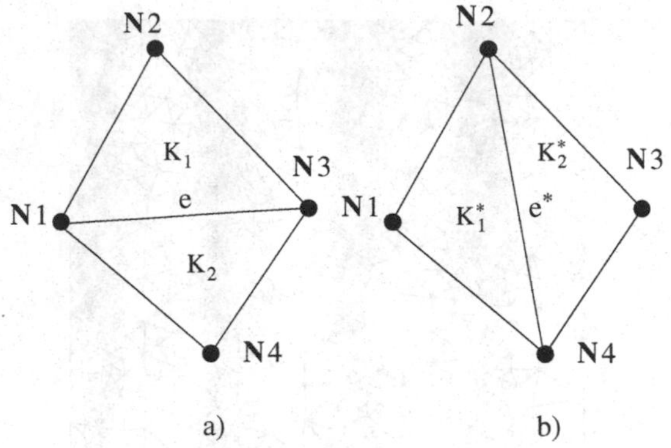

Fig. 6.12. The two configurations obtained via diagonal exchange in the convex quadrilateral formed by two adjacent elements. The two configurations are compared based on an optimality criterion

6.5.1 Diagonal swap

The exchange of diagonals is a technique allowing to modify the topology of the grid without changing the position and number of its vertices. Such technique is based on the fact that a quadrilateral can be subdivided into a couple of triangles sharing a common side in two different ways (see Fig. 6.12).
In general, diagonal exchange is used to improve the quality of non-structured grids by following a given optimality criterion. Suppose, for example, the goal is to avoid angles that are too big, as when the sum of two opposite angles is bigger than π. Exchanging the diagonals would in this case solve the problem.

A general scheme for a possible diagonal exchange algorithm is obtained by defining the optimality criterion at the element level, under the form of an appropriate non-negative function $S: K \rightarrow \mathbb{R}^+ \cup \{0\}$ that takes value 0 when K has the "optimal" shape and dimension. For instance, we can use

$$S(K) = \left| \frac{|K|}{\sum_{i=1}^{3} |e_i^K|^2} - \frac{\sqrt{3}}{12} \right|, \tag{6.3}$$

where $|K|$ denotes the size of K, e_i^K represents a generic side of K and $|e_i^K|$ is its length. Using this function, we privilege triangles that are close to being equilateral, for which $S(K) = 0$. Thus, we will generally obtain a grid as regular as possible, which does not take the spacing into account. With reference to Fig. 6.12, the algorithm will proceed as follows:

1. *Cycle 0:* set the exchanged side counter to zero: $swap = 0$;
2. span all internal sides e of the current mesh;
3. if the two triangles adjacent to e form a convex quadrilateral:
   a) compute $G = S^2(K_1) + S^2(K_2) - \left[S^2(K_1^*) + S^2(K_2^*)\right]$;

b) if $G \geq \tau$, with $\tau > 0$ a predetermined, then execute the diagonal exchange (hence modify the current grid) and set $swap = swap + 1$;

4. if $swap > 0$ start back from *Cycle 0*. Otherwise, the procedure terminates.

It can be easily verified that this algorithm necessarily terminates in a finite number of steps because, for each diagonal exchange, the positive quantity $\sum_K S^2(K)$, where the sum is extended to all the triangles of the current grid, is reduced by the finite quantity G (note that, although the grid is modified, at each diagonal exchange the number of elements and sides remains unchanged).

Remark 6.2. It is not always a good option to construct the optimality function S at the element level. For instance, based on the available data structures, S can also be associated to the nodes or to the sides of the grid. •

The diagonal exchange technique is also the basis for a widely used algorithm (the Lawson algorithm) for the Delaunay triangulation. It can indeed be proved that starting from *any* triangulation of a convex domain, the corresponding Delaunay triangulation (which, we recall, is unique) can be obtained through a finite number of diagonal exchanges. Moreover, the maximum number of necessary swaps for this purpose can be determined a priori and is a function of the number of grid vertices. The technique (and convergence results) can be extended to constrained Delaunay triangulations, through a suitable modification of the algorithm. We refer to the specialized literature, for instance [GB98], for the details.

6.5.2 Node displacement

Another method to improve the quality of the grid consists in moving its points without modifying its topology. Let us consider an internal vertex P and the polygon $\mathcal{K}_P$ constituted by the union of the grid elements containing it. The set $\mathcal{K}_P$ is often called "patch" associated to P and has been considered in Sect. 4.6. A regularization technique, called *Laplacian regularization*, or *barycentrization*, consists in moving P to the center of gravity of $\mathcal{K}_P$, that is in computing its new position $\mathbf{x}_P$ as follows:

$$\mathbf{x}_P = |\mathcal{K}_P|^{-1} \int_{\mathcal{K}_P} \mathbf{x}\, d\mathbf{x}$$

(see Fig. 6.13). This procedure will obviously be iterated on all the internal vertices of the mesh and repeated several times. In case of convergence, the final grid is the one minimizing the quantity

$$\sum_P \int_{\mathcal{K}_P} (\mathbf{x}_P - \mathbf{x})^2 d\mathbf{x}, \tag{6.4}$$

where the sum is extended to all the internal vertices of the grid. The name of such procedure derives from the known property of harmonic functions (those in the kernel of the Laplacian) which take in a point of the domain a value equal to that of the average on a closed curve containing the point.

The final grid will generally depend on the order with which the vertices are displaced, one after the other. Moreover, note that this procedure can provide an unac-

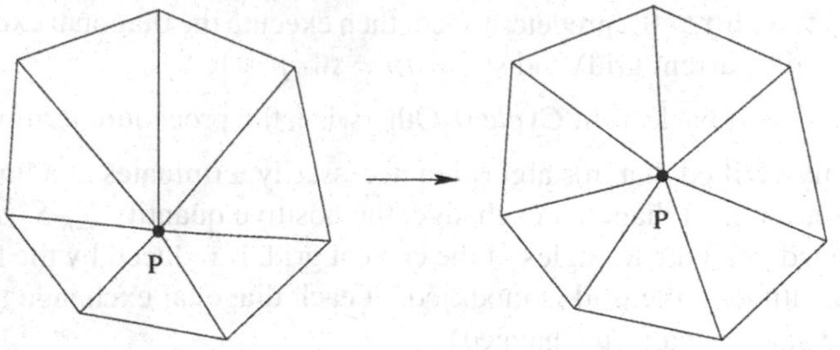

Fig. 6.13. Displacement of a point to the center of gravity of the convex polygon $\mathcal{K}_P$ formed by the union of the elements containing P

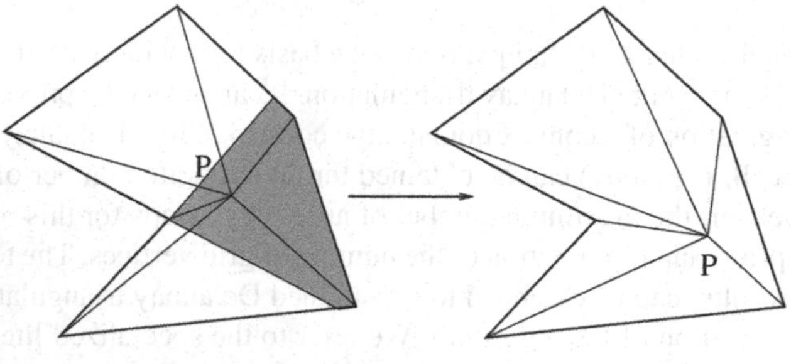

Fig. 6.14. Modification of the Laplacian regularization algorithm for concave patches. On the left-hand side, the initial patch; on the right-hand side, the modification due to regularization. We have shaded the concave polygon $\mathcal{C}_P$

ceptable grid if $\mathcal{K}_P$ is a concave polygon, as $\mathbf{x}_P$ can fall out of the polygon. We present an extension of the procedure that is suitable for generic patches of elements. Consider Fig. 6.14, which shows a concave patch $\mathcal{K}_P$. We define $\mathcal{C}_P$ as the locus of points of $\mathcal{K}_P$ "visible" to all boundary points of $\mathcal{K}_P$, that is $\mathcal{C}_P = \{A \in \mathcal{K}_P : AB \subset \mathcal{K}_P, \forall B \in \partial\mathcal{K}_P\}$; note that $\mathcal{C}_P$ is always convex. The modification of the regularization algorithm consists in placing P not in the center of gravity of $\mathcal{K}_P$, but in that of $\mathcal{C}_P$, as illustrated in Fig. 6.14. Clearly, in the case of convex patches, we have $\mathcal{C}_P = \mathcal{K}_P$. The set $\mathcal{C}_P$ can be constructed in a computationally efficient manner by using suitable algorithms, whose description is beyond the scope of this book.

Another option consists in displacing the vertex to the center of gravity of the boundary of $\mathcal{K}_P$ (or $\mathcal{C}_P$ in the case of concave patches), i.e. in setting

$$\mathbf{x}_P = |\partial\mathcal{K}_P|^{-1} \int_{\partial\mathcal{K}_P} \mathbf{x}\, d\mathbf{x}.$$

This is equivalent to minimizing the square of the distance between the vertex P and the sides forming the patch boundary.

A further technique, often found in the literature, consists in displacing each internal vertex to the center of gravity of the vertices belonging to the associated patch, i.e.

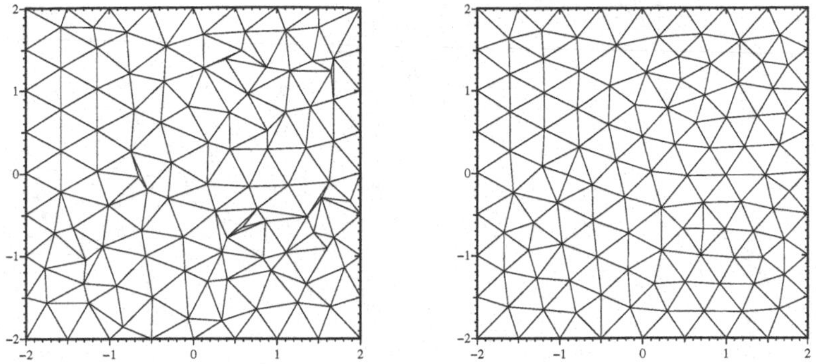

Fig. 6.15. Example of regularization through both diagonal exchange and node displacement

in computing the new position of each internal vertex P via

$$\mathbf{x}_P = \Big(\sum_{\substack{N\in\mathcal{K}_P\\ N\neq P}} \mathbf{x}_N\Big) \Big/ \Big(\sum_{\substack{N\in\mathcal{K}_P\\ N\neq P}} 1\Big),$$

where the sum is extended to all the vertices N belonging to the patch. Despite being the simplest methodology, the latter often yields bad results, in particular if the distribution of vertices inside the patch is very irregular. Moreover, it is more difficult to extend it to concave patches. Thus the two previous procedures are preferable. In Fig. 6.15 we present an example of successive application of both of the above-described regularization techniques. Note that the regularization algorithms presented here tend to uniform the grid, and therefore to destroy its thickenings or coarsenings due for instance to grid adaptivity procedures such as the ones described in Chap. 4. However, it is possible to modify them to account for a non-uniform spacing. For instance, a weighted barycentrization can be used, i.e. by setting

$$\mathbf{x}_P = \left(\int_{\mathcal{K}_P} \mu(\mathbf{x})\, d\mathbf{x}\right)^{-1} \int_{\mathcal{K}_P} \mu(\mathbf{x})\mathbf{x}\, d\mathbf{x},$$

where the strictly positive weight function μ depends on the grid spacing function. In the case of non-uniform spacing, μ will take larger values in the zones where the grid must be finer. When choosing for instance $\mu = \mathcal{H}^{-1}$ the resulting grid (approximately) minimizes

$$\sum_P \int_{\mathcal{K}_P} \left[\mathcal{H}^{-1}(\mathbf{x})(\mathbf{x}_P - \mathbf{x})\right]^2 d\mathbf{x},$$

where the sum is extended to the internal vertices.

Also concerning the diagonal exchange procedure we can take the spacing into account when evaluating the "optimal" configuration, for instance by suitably changing the definition of function $S(K)$ in (6.3).

7

Algorithms for the solution of linear systems

This chapter serves as a quick and elementary introduction of some of the basic algorithms that are used to solve a system of linear algebraic equations. For a more thorough presentation we advise the reader to refer to, e.g., [QSS07, Chaps. 3 and 4], [Saa96] and [vdV03].

A system of m linear equations in n unknowns is a set of algebraic relations of the form

$$\sum_{j=1}^{n} a_{ij}x_j = b_i, \quad i = 1,\ldots,m \tag{7.1}$$

x_j being the unknowns, a_{ij} the system's coefficients and b_i given numbers. System (7.1) will more commonly be written in matrix form

$$\mathbf{A}\mathbf{x} = \mathbf{b}, \tag{7.2}$$

having denoted by $\mathrm{A} = (a_{ij}) \in \mathbb{R}^{m\times n}$ the coefficient matrix, $\mathbf{b}=(b_i) \in \mathbb{R}^m$ being the right hand side vector and $\mathbf{x}=(x_i) \in \mathbb{R}^n$ the unknown vector. We call *solution* of (7.2) any n-tuple of values x_i verifying (7.1).

In the following sections we recall some numerical techniques for the solution of (7.2) in the case where $m = n$; we will obviously suppose that A is non-singular, i.e. that $\det(\mathrm{A}) \neq 0$. Numerical methods are called *direct* if they lead to the solution of the system in a finite number of operations, or *iterative* if they require a (theoretically) infinite number.

7.1 Direct methods

The solution of a linear system can be found through the Gauss elimination method (GEM), where the initial system $\mathrm{A}\mathbf{x}=\mathbf{b}$ is reduced in n steps to an equivalent system (i.e. having the same solution) of the form $\mathrm{A}^{(n)}\mathbf{x} = \mathbf{b}^{(n)}$, where $\mathrm{A}^{(n)} = \mathrm{U}$ is a nonsingular upper triangular matrix and $\mathbf{b}^{(n)}$ is a new source term. It will be possible to solve

A. Quarteroni: *Numerical Models for Differential Problems*, 2nd Ed.
MS&A – Modeling, Simulation & Applications 8
DOI 10.1007/978-88-470-5522-3_7, © Springer-Verlag Italia 2014

the latter system with a computational cost of the order of n^2 operations, through the following backward substitution algorithm:

$$
\begin{aligned}
x_n &= \frac{b_n^{(n)}}{u_{nn}}, \\
x_i &= \frac{1}{u_{ii}}\left(b_i^{(n)} - \sum_{j=i+1}^{n} u_{ij}x_j\right), \qquad i = n-1,\ldots,1.
\end{aligned}
\tag{7.3}
$$

Denoting by $\mathrm{A}^{(1)}\mathbf{x} = \mathbf{b}^{(1)}$ the original system, the kth step of GEM is achieved via the following formulae:

$$
\begin{aligned}
m_{ik} &= \frac{a_{ik}^{(k)}}{a_{kk}^{(k)}}, & i &= k+1,\ldots,n, \\
a_{ij}^{(k+1)} &= a_{ij}^{(k)} - m_{ik}a_{kj}^{(k)}, & i,j &= k+1,\ldots,n \\
b_i^{(k+1)} &= b_i^{(k)} - m_{ik}b_k^{(k)}, & i &= k+1,\ldots,n.
\end{aligned}
\tag{7.4}
$$

We note that in this way, the elements $a_{ij}^{(k+1)}$ with $j = k$ and $i = k+1,\ldots,n$ are null. The elements m_{ik} are called *multipliers*, while the denominators $a_{kk}^{(k)}$ are named *pivotal elements*. The GEM can obviously be achieved only if all the pivotal elements are non null. This happens, for instance, for symmetric positive definite matrices and for strict diagonal dominant ones. In general, it will be necessary to resort to the *pivoting* method, i.e. to the swapping of rows (and/or columns) of $\mathrm{A}^{(k)}$, in order to ensure that the element $a_{kk}^{(k)}$ be non-null.
To complete the Gauss eliminations, we need $2(n-1)n(n+1)/3 + n(n-1)$ flops, to which we must add n^2 flops to solve the upper triangular system $\mathrm{U}\mathbf{x} = \mathbf{b}^{(n)}$ via the backward substitution method. Hence, about $(2n^3/3 + 2n^2)$ flops are needed to solve the linear system via the GEM. More simply, by neglecting lower order terms in n, it can be said that the Gaussian elimination process requires $2n^3/3$ flops.

It can be verified that the GEM is equivalent to factorizing the matrix A, i.e. to rewriting A as the product LU of two matrices. The matrix U, upper triangular, coincides with the matrix $\mathrm{A}^{(n)}$ obtained at the end of the elimination process. The matrix L is lower triangular, its diagonal elements are equal to 1 while the ones located in the remaining lower triangular portion are equal to the multipliers.

Once the matrices L and U are known, the solution of the initial linear system simply involves the (successive) solution of the two triangular systems

$$\mathrm{L}\mathbf{y} = \mathbf{b}, \quad \mathrm{U}\mathbf{x} = \mathbf{y}.$$

Obviously, the computational cost of the factorization process is the same as the one required by the GEM. The advantages of such a reinterpretation are evident: as L and

U depend on A only, and not on the known term, the same factorization can be used to solve different linear systems having the same matrix A, but a variable known term **b** (think for instance of the discretization of a linear parabolic problem by an implicit method where at each time step it is necessary to solve a system with the same matrix all the time, but with a different constant term). Consequently, as the computational cost is concentrated in the elimination procedure (about $2n^3/3$ flops), we have in this way a considerable reduction in the number of operations when we want to solve several linear systems having the same matrix.

If A is a positive-definite, symmetric matrix, the LU factorization can be conveniently specialized. Indeed, there exists only one upper triangular matrix H with positive elements on the diagonal such that

$$\mathrm{A} = \mathrm{H}^{\mathrm{T}}\mathrm{H}. \tag{7.5}$$

Equation (7.5) is the so-called Cholesky factorization. The elements h_{ij} of H^T are given by the following formulae: $h_{11} = \sqrt{a_{11}}$ and, for $i = 2, \dots, n$,

$$h_{ij} = \left(a_{ij} - \sum_{k=1}^{j-1} h_{ik} h_{jk} \right) / h_{jj}, \quad j = 1, \dots, i-1,$$

$$h_{ii} = \left(a_{ii} - \sum_{k=1}^{i-1} h_{ik}^2 \right)^{1/2}.$$

This algorithm only requires about $n^3/3$ flops, i.e. it saves about twice the computing time of the LU factorization and about half the memory.

Let us now consider the particular case of a linear system with non-singular *tridiagonal* matrix A of the form

$$\mathrm{A} = \begin{bmatrix} a_1 & c_1 & & \mathbf{0} \\ b_2 & a_2 & \ddots & \\ & \ddots & \ddots & c_{n-1} \\ \mathbf{0} & & b_n & a_n \end{bmatrix}.$$

In this case, the matrices L and U of the LU factorization of A are bidiagonal matrices of the type

$$\mathrm{L} = \begin{bmatrix} 1 & & & \mathbf{0} \\ \beta_2 & 1 & & \\ & \ddots & \ddots & \\ \mathbf{0} & & \beta_n & 1 \end{bmatrix}, \qquad \mathrm{U} = \begin{bmatrix} \alpha_1 & c_1 & & \mathbf{0} \\ & \alpha_2 & \ddots & \\ & & \ddots & c_{n-1} \\ \mathbf{0} & & & \alpha_n \end{bmatrix}.$$

The unknown coefficients α_i and β_i can be easily computed by the following equations:

$$\alpha_1 = a_1, \quad \beta_i = \frac{b_i}{\alpha_{i-1}}, \quad \alpha_i = a_i - \beta_i c_{i-1}, \; i = 2, \ldots, n.$$

This algorithm is named *Thomas' algorithm* and can be seen as a particular kind of LU factorization without pivoting.

7.2 Iterative methods

Iterative methods aim at constructing the solution $\mathbf{x}$ of a linear system as the limit of a sequence $\{\mathbf{x}^{(n)}\}$ of vectors. To obtain the single elements of the sequence, computing the residue $\mathbf{r}^{(n)} = \mathbf{b} - A\mathbf{x}^{(n)}$ of the system is required. In the case where the matrix is full and of order n, the computational cost of an iterative method is therefore of the order of n^2 operations per iteration. Such cost must be compared with the approximately $2n^3/3$ operations required by a direct method. Consequently, iterative methods are competitive with direct methods only if the number of necessary iterations to reach convergence (within a given tolerance) is independent of n or depends on n in a sub-linear way.

Other considerations in the choice between an iterative method and a direct one intervene as soon as the matrix is sparse.

7.2.1 Classical iterative methods

A general strategy to construct iterative methods is based on an additive decomposition, called splitting, starting from a matrix A of the form A=P−N, where P and N are two suitable matrices and P is non-singular. For reasons which will become evident in the remainder, P is also called *preconditioning matrix* or *preconditioner.*

Precisely, given $\mathbf{x}^{(0)}$, we obtain $\mathbf{x}^{(k)}$ for $k \geq 1$ by solving the new systems

$$\mathrm{P}\mathbf{x}^{(k+1)} = \mathrm{N}\mathbf{x}^{(k)} + \mathbf{b}, \quad k \geq 0 \tag{7.6}$$

or, equivalently,

$$\mathbf{x}^{(k+1)} = \mathrm{B}\mathbf{x}^{(k)} + \mathrm{P}^{-1}\mathbf{b}, \quad k \geq 0 \tag{7.7}$$

having denoted by $\mathrm{B} = \mathrm{P}^{-1}\mathrm{N}$ the *iteration matrix.*

We are interested in *convergent* iterative methods, i.e. such that $\lim_{k\to\infty} \mathbf{e}^{(k)} = \mathbf{0}$ for each choice of the *initial vector* $\mathbf{x}^{(0)}$, having denoted by $\mathbf{e}^{(k)} = \mathbf{x}^{(k)} - \mathbf{x}$ the error. With a recursive argument we find

$$\mathbf{e}^{(k)} = \mathrm{B}^k \mathbf{e}^{(0)}, \qquad \forall k = 0, 1, \ldots \tag{7.8}$$

so can conclude that an iterative method of the form (7.6) is convergent if and only if $\rho(\mathrm{B}) < 1$, $\rho(\mathrm{B})$ being the spectral radius of the iteration matrix B, i.e. the maximum modulus of the eigenvalues of B.

Equation (7.6) can also be formulated in the form

$$\mathbf{x}^{(k+1)} = \mathbf{x}^{(k)} + \mathrm{P}^{-1}\mathbf{r}^{(k)}, \tag{7.9}$$

having denoted by

$$\mathbf{r}^{(k)} = \mathbf{b} - \mathrm{A}\mathbf{x}^{(k)} \tag{7.10}$$

the *residue* at step k. Equation (7.9) thus expresses the fact that to update the solution at step $k+1$, it is necessary to solve a linear system with matrix P. Hence, beside being non-singular, P must be invertible at a low computational cost if we want to prevent the overall cost of the scheme from increasing excessively (obviously, in the limit case where P is equal to A and N=0, method (7.9) converges in only one iteration, but at the cost of a direct method).

Let us now see how to accelerate the convergence of the iterative methods (7.6) by exploiting the latter form. We denote by

$$\mathrm{R_P} = \mathrm{I} - \mathrm{P}^{-1}\mathrm{A}$$

the iteration matrix associated to method (7.9). Equation (7.9) can be generalized by introducing a suitable relaxation (or acceleration) parameter α. In this way, we obtain the *stationary Richardson methods* (or, simply, *Richardson* methods), of the form

$$\mathbf{x}^{(k+1)} = \mathbf{x}^{(k)} + \alpha\mathrm{P}^{-1}\mathbf{r}^{(k)}, \qquad k \geq 0. \tag{7.11}$$

More generally, supposing α to be dependent on the iteration index, we obtain the *non-stationary Richardson methods* given by

$$\mathbf{x}^{(k+1)} = \mathbf{x}^{(k)} + \alpha_k\mathrm{P}^{-1}\mathbf{r}^{(k)}, \qquad k \geq 0. \tag{7.12}$$

If we set $\alpha = 1$, we can recover two classical iterative methods: the Jacobi method if $\mathrm{P} = \mathrm{D(A)}$ (the diagonal part of A), the Gauss-Seidel method if $\mathrm{P} = \mathrm{L(A)}$ (the lower triangular part of A).
The iteration matrix at step k for such methods is given by

$$\mathrm{R}(\alpha_k) = \mathrm{I} - \alpha_k\mathrm{P}^{-1}\mathrm{A},$$

(note that the latter depends on k). In the case where P=I, the methods under exam will be called *non preconditioned.*

We can rewrite (7.12) (and therefore also (7.11)) in a form of greater computational interest. Indeed, having set $\mathbf{z}^{(k)} = \mathrm{P}^{-1}\mathbf{r}^{(k)}$ (the so-called *preconditioned residue*), we have that $\mathbf{x}^{(k+1)} = \mathbf{x}^{(k)} + \alpha_k\mathbf{z}^{(k)}$ and $\mathbf{r}^{(k+1)} = \mathbf{b} - \mathrm{A}\mathbf{x}^{(k+1)} = \mathbf{r}^{(k)} - \alpha_k\mathrm{A}\mathbf{z}^{(k)}$. To summarize, a non-stationary Richardson method at step $k+1$ requires the following operations:

$$\begin{aligned} &\text{solving the linear system } \mathrm{P}\mathbf{z}^{(k)} = \mathbf{r}^{(k)},\\ &\text{computing the acceleration parameter } \alpha_k,\\ &\text{updating the solution } \mathbf{x}^{(k+1)} = \mathbf{x}^{(k)} + \alpha_k \mathbf{z}^{(k)},\\ &\text{updating the residue } \mathbf{r}^{(k+1)} = \mathbf{r}^{(k)} - \alpha_k \mathrm{A}\mathbf{z}^{(k)}. \end{aligned} \tag{7.13}$$

As far as the convergence of the stationary Richardson method (for which $\alpha_k = \alpha$, for each $k \geq 0$) is concerned, the following result holds:

Property 7.1. *If* P *is a non-singular matrix, the stationary Richardson method* (7.11) *is convergent if and only if*

$$\frac{2\mathrm{Re}\lambda_i}{\alpha|\lambda_i|^2} > 1 \quad \forall i = 1,\ldots,n, \tag{7.14}$$

λ_i being the eigenvalues of $\mathrm{P}^{-1}\mathrm{A}$.
Moreover, if we suppose that $\mathrm{P}^{-1}\mathrm{A}$ *has positive real eigenvalues, ordered in such a way that* $\lambda_1 \geq \lambda_2 \geq \ldots \geq \lambda_n > 0$*, then the stationary Richardson method* (7.11) *converges if and only if* $0 < \alpha < 2/\lambda_1$*. Having set*

$$\alpha_{opt} = \frac{2}{\lambda_1 + \lambda_n}, \tag{7.15}$$

the spectral radius of the iteration matrix R_α *is minimal if* $\alpha = \alpha_{opt}$*, with*

$$\rho_{opt} = \min_{\alpha}\left[\rho(\mathrm{R}_\alpha)\right] = \frac{\lambda_1 - \lambda_n}{\lambda_1 + \lambda_n}. \tag{7.16}$$

If P and A are both symmetric and positive definite, it can be proved that the Richardson method converges monotonically with respect to the vector norms $\|\cdot\|_2$ and $\|\cdot\|_{\mathrm{A}}$. We recall that $\|\mathbf{v}\|_2 = (\sum_{i=1}^n v_i^2)^{1/2}$ and $\|\mathbf{v}\|_{\mathrm{A}} = (\sum_{i,j=1}^n v_i a_{ij} v_j)^{1/2}$.
In this case, thanks to (7.16), we can relate ρ_{opt} with the condition number introduced in Sect. 4.5.2 in the following way:

$$\rho_{opt} = \frac{K_2(\mathrm{P}^{-1}\mathrm{A}) - 1}{K_2(\mathrm{P}^{-1}\mathrm{A}) + 1}, \quad \alpha_{opt} = \frac{2\|\mathrm{A}^{-1}\mathrm{P}\|_2}{K_2(\mathrm{P}^{-1}\mathrm{A}) + 1}. \tag{7.17}$$

The importance of the choice of the preconditioner P in a Richardson method is therefore clear. We refer to Chap. 4 of [QSS07] for some examples of preconditioners.

7.2.2 Gradient and conjugate gradient methods

The optimal expression of the acceleration parameter α, indicated in (7.15), turns out to be of little practical utility, as it requires knowing the maximum and minimum eigenvalues of the matrix $\mathrm{P}^{-1}\mathrm{A}$. In the particular case of positive definite symmetric matrices, it is however possible to evaluate the optimal acceleration parameter in a *dynamic* way, that is as a function of quantities computed by the method itself at step k, as we show below.

First of all, we observe that in the case where A is a symmetric positive definite matrix, solving system (7.2) is equivalent to finding the minimum $\mathbf{x} \in \mathbb{R}^n$ of the quadratic form

$$\Phi(\mathbf{y}) = \frac{1}{2}\mathbf{y}^T\mathrm{A}\mathbf{y} - \mathbf{y}^T\mathbf{b},$$

called *energy of system* (7.2).
The problem is thus reduced to determining the minimum point $\mathbf{x}$ of Φ starting from a point $\mathbf{x}^{(0)} \in \mathbb{R}^n$ and, consequently, choosing suitable directions along which to move to approach the solution $\mathbf{x}$ as quickly as possible. The optimal direction, joining $\mathbf{x}^{(0)}$ and $\mathbf{x}$, is obviously unknown a priori: we will therefore have to move from $\mathbf{x}^{(0)}$ along another direction $\mathbf{d}^{(0)}$ and fix a new point $\mathbf{x}^{(1)}$ on the latter, then repeat the procedure until convergence.

At the generic step k we will then determine $\mathbf{x}^{(k+1)}$ as

$$\mathbf{x}^{(k+1)} = \mathbf{x}^{(k)} + \alpha_k \mathbf{d}^{(k)}, \tag{7.18}$$

α_k being the value fixing the length of the step along $\mathbf{d}^{(k)}$. The most natural idea, consisting in taking as downhill direction that of the greatest increase of Φ, given by $\mathbf{r}^{(k)} = -\nabla\Phi(\mathbf{x}^{(k)})$, leads to the *gradient* or *steepest descent method*.
The latter leads to the following algorithm: given $\mathbf{x}^{(0)} \in \mathbb{R}^n$, and having set $\mathbf{r}^{(0)} = \mathbf{b} - \mathrm{A}\mathbf{x}^{(0)}$, for $k = 0, 1, \ldots$ until convergence, we compute

$$\alpha_k = \frac{\mathbf{r}^{(k)T}\mathbf{r}^{(k)}}{\mathbf{r}^{(k)T}\mathrm{A}\mathbf{r}^{(k)}}, \qquad \mathbf{x}^{(k+1)} = \mathbf{x}^{(k)} + \alpha_k \mathbf{r}^{(k)},$$
$$\mathbf{r}^{(k+1)} = \mathbf{r}^{(k)} - \alpha_k \mathrm{A}\mathbf{r}^{(k)}.$$

Its preconditioned version takes the following form: given $\mathbf{x}^{(0)} \in \mathbb{R}^n$, and having set $\mathbf{r}^{(0)} = b - \mathrm{A}\mathbf{x}^{(0)}$, $\mathbf{z}^{(0)} = P^{-1}\mathbf{r}^{(0)}$, for $k = 0, 1, \ldots$ until convergence, we compute

$$\alpha_k = \frac{\mathbf{z}^{(k)T}\mathbf{r}^{(k)}}{\mathbf{z}^{(k)T}\mathrm{A}\mathbf{z}^{(k)}}, \qquad \mathbf{x}^{(k+1)} = \mathbf{x}^{(k)} + \alpha_k \mathbf{z}^{(k)}.$$
$$\mathbf{r}^{(k+1)} = \mathbf{r}^{(k)} - \alpha_k \mathrm{A}\mathbf{z}^{(k)}, \quad P\mathbf{z}^{(k+1)} = \mathbf{r}^{(k+1)}.$$

As far as the convergence properties of the descent method are concerned, the following result holds

Theorem 7.1. *If* A *is symmetric and positive definite, the gradient method converges for each value of the initial datum* $\mathbf{x}^{(0)}$ *and*

$$\|\mathbf{e}^{(k+1)}\|_{\mathrm{A}} \leq \frac{K_2(\mathrm{A})-1}{K_2(\mathrm{A})+1}\|\mathbf{e}^{(k)}\|_{\mathrm{A}}, \qquad k=0,1,\ldots \tag{7.19}$$

where $\|\cdot\|_{\mathrm{A}}$ *is the previously defined energy norm.*

A similar result, with $K_2(\mathrm{A})$ replaced by $K_2(\mathrm{P}^{-1}\mathrm{A})$, holds also in the case of the preconditioned gradient method, as long as we assume that P is also symmetric and positive definite.

An even more effective alternative consists in using the *conjugate gradient method*, where the descent directions no longer coincide with that of the residue. In particular, having set $\mathbf{p}^{(0)} = \mathbf{r}^{(0)}$, we seek directions of the form

$$\mathbf{p}^{(k+1)} = \mathbf{r}^{(k+1)} - \beta_k \mathbf{p}^{(k)}, \quad k=0,1,\ldots \tag{7.20}$$

where the parameters $\beta_k \in \mathbb{R}$ are to be determined so that

$$(\mathrm{A}\mathbf{p}^{(j)})^T \mathbf{p}^{(k+1)} = 0, \quad j=0,1,\ldots,k. \tag{7.21}$$

Directions of this type are called A-orthogonal. The method in the preconditioned case then takes the form: given $\mathbf{x}^{(0)} \in \mathbb{R}^n$, having set $\mathbf{r}^{(0)} = \mathbf{b} - \mathrm{A}\mathbf{x}^{(0)}$, $\mathbf{z}^{(0)} = \mathrm{P}^{-1}\mathbf{r}^{(0)}$ and $\mathbf{p}^{(0)} = \mathbf{z}^{(0)}$, the k-th iteration, with $k = 0,1\ldots$, is

$$\begin{aligned}
\alpha_k &= \frac{\mathbf{p}^{(k)^T}\mathbf{r}^{(k)}}{(\mathrm{A}\mathbf{p}^{(k)})^T\mathbf{p}^{(k)}},\\
\mathbf{x}^{(k+1)} &= \mathbf{x}^{(k)} + \alpha_k \mathbf{p}^{(k)},\\
\mathbf{r}^{(k+1)} &= \mathbf{r}^{(k)} - \alpha_k \mathrm{A}\mathbf{p}^{(k)},\\
\mathrm{P}\mathbf{z}^{(k+1)} &= \mathbf{r}^{(k+1)},\\
\beta_k &= \frac{(\mathrm{A}\mathbf{p}^{(k)})^T\mathbf{z}^{(k+1)}}{\mathbf{p}^{(k)^T}\mathrm{A}\mathbf{p}^{(k)}},\\
\mathbf{p}^{(k+1)} &= \mathbf{z}^{(k+1)} - \beta_k \mathbf{p}^{(k)}.
\end{aligned}$$

The parameter α_k is chosen in order to guarantee that the error $\|\mathbf{e}^{(k+1)}\|_{\mathrm{A}}$ be minimized along the descent direction $\mathbf{p}^{(k)}$. The parameter β_k, instead, is chosen so that the new direction $\mathbf{p}^{(k+1)}$ is A-conjugate to $\mathbf{p}^{(k)}$, that is $(\mathrm{A}\mathbf{p}^{(k)})^T\mathbf{p}^{(k+1)} = 0$. Indeed, it can be proved (thanks to the induction principle) that if the latter relation is verified, then so are all the ones in (7.21) relative to $j = 0,...,k-1$. For a complete justification of the method, see e.g. [QSS07, Chap. 4] or [Saa96].

It can be proved that the conjugate gradient method converges in exact arithmetics in at most n steps, and that

$$\|\mathbf{e}^{(k)}\|_{\mathrm{A}} \leq \frac{2c^k}{1+c^{2k}}\|\mathbf{e}^{(0)}\|_{\mathrm{A}}, \tag{7.22}$$

with

$$c = \frac{\sqrt{K_2(\mathrm{P}^{-1}\mathrm{A})}-1}{\sqrt{K_2(\mathrm{P}^{-1}\mathrm{A})}+1}. \tag{7.23}$$

Consequently, in the absence of roundoff errors, the CG method can be seen as a direct method as it terminates after a finite number of operations.
On the other hand, for matrices of large dimension, it is usually applied as an iterative method and is arrested when an error estimator (as for instance the relative residue) is less than a given tolerance.
Thanks to (7.23), the dependence on the reduction factor of the error on the matrix' condition number is more favourable than the one of the gradient method (due to the presence of the square root of $K_2(P^{-1}\mathrm{A})$).

It can be noted that the number of iterations required for convergence (up to a prescribed tolerance) is proportional to $\frac{1}{2}\sqrt{K_2(P^{-1}A)}$ for the preconditioned conjugate gradient method, a clear improvement with respect to $\frac{1}{2}K_2(P^{-1}A)$ for the preconditioned gradient method. Of course, the PCG method is costlier per iteration, both in CPU time and storage.

7.2.3 Krylov subspace methods

Generalizations of the gradient method in the case where the matrix A is not symmetric lead to the so-called Krylov methods. Notable examples are the GMRES method and the conjugate bigradient method BiCG, as well as its stabilized version, the BiCGSTAB method. We send the interested reader to [QSS07, Chap. 4], [Saa96] and [vdV03].

Here we briefly review the GMRES (generalized minimal residual) method. We start by a revisitation of the Richardson method (7.13) with $\mathrm{P}=\mathrm{I}$; the residual at the k-th step can be related to the initial residual by

$$\mathbf{r}^{(k)} = \prod_{j=0}^{k-1}(\mathrm{I}-\alpha_j\mathrm{A})\mathbf{r}^{(0)} = p_k(\mathrm{A})\mathbf{r}^{(0)}, \tag{7.24}$$

where $p_k(\mathrm{A})$ is a polynomial in A of degree k. If we introduce the space

$$K_m(\mathrm{A};\mathbf{v}) = \mathrm{span}\{\mathbf{v},\mathrm{A}\mathbf{v},\ldots,\mathrm{A}^{m-1}\mathbf{v}\}, \tag{7.25}$$

it immediately appears from (7.24) that $\mathbf{r}^{(k)} \in K_{k+1}(\mathrm{A};\mathbf{r}^{(0)})$. The space defined in (7.25) is called the *Krylov subspace* of order m associated with the matrix A and the vector $\mathbf{v}$. It is a subspace of the span of all vectors $\mathbf{u} \in \mathbb{R}^n$ that can be written as $\mathbf{u} = p_{m-1}(\mathrm{A})\mathbf{v}$, where p_{m-1} is a polynomial in A of degree $\leq m-1$.

Similarly, the iterate $\mathbf{x}^{(k)}$ of the Richardson method can be represented as follows

$$\mathbf{x}^{(k)} = \mathbf{x}^{(0)} + \sum_{j=0}^{k-1} \alpha_j \mathbf{r}^{(j)},$$

whence $\mathbf{x}^{(k)}$ belongs to the space

$$W_k = \{\mathbf{v} = \mathbf{x}^{(0)} + \mathbf{y},\ \mathbf{y} \in K_k(\mathrm{A};\mathbf{r}^{(0)})\}. \tag{7.26}$$

Notice also that $\sum_{j=0}^{k-1} \alpha_j \mathbf{r}^{(j)}$ is a polynomial in A of degree less than $k-1$. In the non-preconditioned Richardson method we are thus looking for an approximate solution to $\mathbf{x}$ in the space W_k. More generally, one can devise methods that search for approximate solutions of the form

$$\mathbf{x}^{(k)} = \mathbf{x}^{(0)} + q_{k-1}(\mathrm{A})\mathbf{r}^{(0)}, \tag{7.27}$$

where q_{k-1} is a polynomial selected in such a way that $\mathbf{x}^{(k)}$ is, in a sense that must be made precise, the best approximation of $\mathbf{x}$ in W_k. A method that looks for a solution of the form (7.27) with W_k defined as in (7.26) is called a *Krylov method*.

A first question concerning Krylov subspace iterations is whether the dimension of $K_m(\mathrm{A};\mathbf{v})$ increases as the order m grows. A partial answer is provided by the following result.

Property 7.2. *Let* $\mathrm{A} \in \mathbb{R}^{n\times n}$ *and* $\mathbf{v} \in \mathbb{R}^n$. *The Krylov subspace* $K_m(\mathrm{A};\mathbf{v})$ *has dimension equal to* m *iff the degree of* $\mathbf{v}$ *with respect to* A, *denoted by* $\deg_{\mathrm{A}}(\mathbf{v})$, *is not less than* m; *the degree of* $\mathbf{v}$ *is defined as the minimum degree of a monic nonnull polynomial* p *in* A *for which* $p(\mathrm{A})\mathbf{v} = \mathbf{0}$.

The dimension of $K_m(\mathrm{A};\mathbf{v})$ is thus equal to the minimum between m and the degree of $\mathbf{v}$ with respect to A and, as a consequence, the dimension of the Krylov subspaces is certainly a nondecreasing function of m. The degree of $\mathbf{v}$ cannot be greater than n due to the Cayley-Hamilton theorem (see [QSS07, Sect. 1.7]).

Example 7.1. Consider the 4×4 matrix $\mathrm{A} = \mathrm{tridiag}_4(-1,2,-1)$. The vector $\mathbf{v} = [1,1,1,1]^T$ has degree 2 with respect to A since $p_2(\mathrm{A})\mathbf{v} = \mathbf{0}$ with $p_2(\mathrm{A}) = \mathrm{I}_4 - 3\mathrm{A} + \mathrm{A}^2$ (I_4 is the 4×4 identity matrix), while there is no monic polynomial p_1 of degree 1 for which $p_1(\mathrm{A})\mathbf{v} = \mathbf{0}$. As a consequence, all Krylov subspaces from $K_2(\mathrm{A};\mathbf{v})$ on have dimension equal to 2. The vector $\mathbf{w} = [1,1,-1,1]^T$ has, instead, degree 4 with respect to A. ■

For a fixed m, it is possible to compute an orthonormal basis for $K_m(\mathrm{A};\mathbf{v})$ using the so-called *Arnoldi algorithm*.

Setting $\mathbf{v}_1 = \mathbf{v}/\|\mathbf{v}\|_2$, this method generates an orthonormal basis $\{\mathbf{v}_i\}$ for $K_m(\mathrm{A};\mathbf{v}_1)$ using the Gram-Schmidt procedure (see [QSS07, Sect. 3.4.3]). For $k = 1,\ldots,m$, the

Arnoldi algorithm computes

$$\begin{aligned} h_{ik} &= \mathbf{v}_i^T \mathrm{A}\mathbf{v}_k, & i &= 1,2,\ldots,k, \\ \mathbf{w}_k &= \mathrm{A}\mathbf{v}_k - \sum_{i=1}^{k} h_{ik}\mathbf{v}_i, & h_{k+1,k} &= \|\mathbf{w}_k\|_2. \end{aligned} \tag{7.28}$$

If $\mathbf{w}_k = \mathbf{0}$ the process terminates and in such a case we say that a *breakdown* of the algorithm has occurred; otherwise, we set $\mathbf{v}_{k+1} = \mathbf{w}_k/\|\mathbf{w}_k\|_2$ and the algorithm restarts, incrementing k by 1.

It can be shown that if the method terminates at step m then the vectors $\mathbf{v}_1,\ldots,\mathbf{v}_m$ form a basis for $K_m(\mathrm{A};\mathbf{v})$. In such a case, if we denote by $\mathrm{V}_m \in \mathbb{R}^{n\times m}$ the matrix whose columns are the vectors $\mathbf{v}_i$, we have

$$\mathrm{V}_m^T \mathrm{A}\mathrm{V}_m = \mathrm{H}_m, \quad \mathrm{V}_{m+1}^T \mathrm{A}\mathrm{V}_m = \widehat{\mathrm{H}}_m, \tag{7.29}$$

where $\widehat{\mathrm{H}}_m \in \mathbb{R}^{(m+1)\times m}$ is the upper Hessenberg matrix whose entries h_{ij} are given by (7.28), and $\mathrm{H}_m \in \mathbb{R}^{m\times m}$ is the restriction of $\widehat{\mathrm{H}}_m$ to the first m rows and m columns.

The algorithm terminates at an intermediate step $k < m$ iff $\deg_{\mathrm{A}}(\mathbf{v}_1) = k$. As for the stability of the procedure, all the considerations valid for the Gram-Schmidt method hold. For more efficient and stable computational variants of (7.28), we refer to [Saa96].

We are now ready to solve the linear system (7.2) by a Krylov method. We look for the iterate $\mathbf{x}^{(k)}$ under the form (7.27); for a given $\mathbf{r}^{(0)}$, $\mathbf{x}^{(k)}$ is in the unique element in W_k which satisfies a criterion of minimal distance from $\mathbf{x}$. The criterion for selecting $\mathbf{x}^{(k)}$ is precisely the distinguishing feature of a Krylov method.

The most natural idea consists in searching for $\mathbf{x}^{(k)} \in W_k$ as the vector which minimizes the Euclidean norm of the error. This approach, however, does not work in practice since $\mathbf{x}^{(k)}$ would depend on the (unknown) solution $\mathbf{x}$. Two alternative strategies can be pursued:

1. compute $\mathbf{x}^{(k)} \in W_k$ by enforcing that the residual $\mathbf{r}^{(k)}$ is orthogonal to any vector in $K_k(\mathrm{A};\mathbf{r}^{(0)})$, i.e., we look for $\mathbf{x}^{(k)} \in W_k$ such that

$$\mathbf{v}^T(\mathbf{b} - \mathrm{A}\mathbf{x}^{(k)}) = 0 \qquad \forall \mathbf{v} \in K_k(\mathrm{A};\mathbf{r}^{(0)}); \tag{7.30}$$

2. compute $\mathbf{x}^{(k)} \in W_k$ by minimizing the Euclidean norm of the residual $\|\mathbf{r}^{(k)}\|_2$, i.e.

$$\|\mathbf{b} - \mathrm{A}\mathbf{x}^{(k)}\|_2 = \min_{\mathbf{v}\in W_k} \|\mathbf{b} - \mathrm{A}\mathbf{v}\|_2. \tag{7.31}$$

Satisfying (7.30) leads to the Arnoldi method for linear systems (more commonly known as FOM, *full orthogonalization method*), while satisfying (7.31) yields the GMRES (*generalized minimal residual*) method.

We shall assume that k steps of the Arnoldi algorithm have been carried out, so that an orthonormal basis for $K_k(\mathrm{A};\mathbf{r}^{(0)})$ has been generated and stored into the column vectors of the matrix V_k with $\mathbf{v}_1 = \mathbf{r}^{(0)}/\|\mathbf{r}^{(0)}\|_2$. In such a case the new iterate $\mathbf{x}^{(k)}$ can always be written as

$$\mathbf{x}^{(k)} = \mathbf{x}^{(0)} + \mathrm{V}_k \mathbf{z}^{(k)}, \tag{7.32}$$

where $\mathbf{z}^{(k)}$ must be selected according to a suitable criterion that we are going to specify. Consequently we have

$$\mathbf{r}^{(k)} = \mathbf{r}^{(0)} - \mathrm{AV}_k \mathbf{z}^{(k)}. \tag{7.33}$$

Since $\mathbf{r}^{(0)} = \mathbf{v}_1 \|\mathbf{r}^{(0)}\|_2$, and using (7.29), relation (7.33) becomes

$$\mathbf{r}^{(k)} = \mathrm{V}_{k+1}(\|\mathbf{r}^{(0)}\|_2 \mathbf{e}_1 - \widehat{\mathrm{H}}_k \mathbf{z}^{(k)}), \tag{7.34}$$

where $\mathbf{e}_1$ is the first vector of the canonical basis of $\mathbb{R}^{k+1}$. Therefore, in the GMRES method the solution at step k can be computed through (7.32), provided

$$\mathbf{z}^{(k)} \text{ minimizes } \| \, \|\mathbf{r}^{(0)}\|_2 \mathbf{e}_1 - \widehat{\mathrm{H}}_k \mathbf{z}^{(k)} \|_2 \tag{7.35}$$

(we note that the matrix V_{k+1} appearing in (7.34) does not alter the value of $\|\cdot\|_2$, since it is orthogonal). Having to solve at each step a least-squares problem of size k, the GMRES method will be the more effective, the smaller is the number of iterations.

Similarly to the CG method, the GMRES method has the finite termination property, that is it terminates at most after n iterations, yielding the exact solution (in exact arithmetic). Indeed, the kth iterate minimizes the residual in the Krylov subspace K_k. Since every subspace is contained in the next, the residual decreases monotonically. After n iterations, where n is the size of the matrix A, the Krylov space K_n is the whole of $\mathbb{R}^n$ and hence the GMRES method arrives at the exact solution. Premature stops are due to a breakdown in Arnoldi's orthonormalization algorithm. More precisely, we have the following result.

Property 7.3. *A breakdown occurs for the GMRES method at a step m (with $m < n$) if and only if the computed solution $\mathbf{x}^{(m)}$ coincides with the exact solution to the system.*

However, the idea is that after a small number of iterations (relatively to n), the vector $\mathbf{x}^{(k)}$ is already a good approximation of the exact solution. This is confirmed by the convergence results that we describe later in this section.

To improve the efficiency of the GMRES algorithm it is necessary to devise a stopping criterion which does not require the explicit evaluation of the residual at each step. This is possible, provided that the linear system with upper Hessenberg matrix $\widehat{\mathrm{H}}_k$ is appropriately solved.

In practice, the matrix $\widehat{\mathrm{H}}_k$ in (7.29) is transformed into an upper triangular matrix $\mathrm{R}_k \in \mathbb{R}^{(k+1)\times k}$ with $r_{k+1,k} = 0$ and such that $\mathrm{Q}_k^T \mathrm{R}_k = \widehat{\mathrm{H}}_k$, where Q_k is a matrix obtained as the product of k Givens rotations. Then, since Q_k is orthogonal, minimizing $\| \|\mathbf{r}^{(0)}\|_2 \mathbf{e}_1 - \widehat{\mathrm{H}}_k \mathbf{z}^{(k)} \|_2$ is equivalent to minimizing $\|\mathbf{f}_k - \mathrm{R}_k \mathbf{z}^{(k)}\|_2$, with $\mathbf{f}_k = \mathrm{Q}_k \|\mathbf{r}^{(0)}\|_2 \mathbf{e}_1$. It can also be shown that the $k+1$-th component of $\mathbf{f}_k$ is, in absolute value, the Euclidean norm of the residual at the k-th step.

As FOM, the GMRES method entails a high computational effort and a large amount of memory, unless convergence occurs after few iterations. For this reason, two variants of the algorithm are available, one named GMRES(m) and based on the *restart* after m steps, with $\mathbf{x}(m)$ as initial guess, the other named Quasi-GMRES or QGMRES and based on stopping the Arnoldi orthogonalization process. It is worth noting that these two methods do not enjoy Property 7.3.

The convergence analysis of GMRES is not trivial, and we report just some of the more elementary results here. If A is positive definite, i.e., its symmetric part A_S has positive eigenvalues, then the k-th residual decreases according to the following bound

$$\|\mathbf{r}^{(k)}\|_2 \leq \sin^k(\beta) \|\mathbf{r}^{(0)}\|_2 , \tag{7.36}$$

where $\cos(\beta) = \lambda_{\min}(L_S)/\|L\|$ with $\beta \in [0, \pi/2)$. Moreover, GMRES(m) converges for all $m \geq 1$. In order to obtain a bound on the residual at a step $k \geq 1$, let us assume that the matrix A is diagonalizable

$$A = T \Lambda T^{-1} ,$$

where Λ is the diagonal matrix of the eigenvalues $\{\lambda_j\}_{j=1,\ldots,n}$, and $T = [\omega^1, \ldots, \omega^n]$ is the matrix whose columns are the right eigenvectors of A. Under these assumptions, the residual norm after k steps of GMRES satisfies

$$\|\mathbf{r}^{(k)}\| \leq K_2(T) \delta \|\mathbf{r}^{(0)}\| ,$$

where $K_2(T) = \|T\|_2 \|T^{-1}\|_2$ is the condition number of T and

$$\delta = \min_{p \in \mathbb{P}_k, p(0)=1} \max_{1 \leq i \leq k} |p(\lambda_i)| .$$

Moreover, suppose that the initial residual is well represented by the first m eigenvectors, i.e., $\mathbf{r}^0 = \sum_{j=1}^m \alpha_j \omega^j + \mathbf{e}$, with $\|\mathbf{e}\|$ small in comparison to $\|\sum_{j=1}^m \alpha_j \omega^j\|$, and assume that if some complex ω^j appears in the previous sum, then its conjugate $\overline{\omega}^j$ appears as well. Then

$$\|\mathbf{r}^{(k)}\| \leq K_2(T) c_k \|\mathbf{e}\| ,$$

$$c_k = \max_{p > k} \prod_{j=1}^{k} \left| \frac{\lambda_p - \lambda_j}{\lambda_j} \right| .$$

Very often, c_k is of order one; hence, k steps of GMRES reduce the residual norm to the order of $\|\mathbf{e}\|$ provided that $\kappa_2(T)$ is not too large.

In general, as highlighted from the previous estimate, the eigenvalue information alone is not enough, and information on the eigensystem is also needed. If the eigensystem is orthogonal, as for normal matrices, then $K_2(T) = 1$, and the eigenvalues retain all the information about convergence. Otherwise, upper bounds for $\|\mathbf{r}^{(k)}\|$ can be provided in terms of both spectral and pseudospectral information, as well as the so-called *field of values* of A

$$\mathcal{F}(A) = \{\mathbf{v}^* A\mathbf{v} \mid \|\mathbf{v}\| = 1\}.$$

If $0 \notin \mathcal{F}(A)$, then estimate (7.36) can be improved by replacing $\lambda_{\min}(A_S)$ with $\mathrm{dist}(0, \mathcal{F}(A))$.

An extensive discussion on the convergence of GMRES and GMRES(m) can be found in [Saa96], [Emb99], [Emb03], [TE05], and [vdV03].

The GMRES method can of course be implemented for a preconditioned system. We provide here an implementation of the preconditioned GMRES method with a left preconditioner P.

Preconditioned GMRES (PGMRES) Method

Initialize

$\mathbf{x}^{(0)}, P\mathbf{r}^{(0)} = \mathbf{f} - A\mathbf{x}^{(0)}, \beta = \|\mathbf{r}^{(0)}\|_2, \mathbf{x}^{(1)} = \mathbf{r}^{(0)}/\beta.$

Iterate

$$\begin{array}{l}
\textit{For } j = 1, \dots, k \textit{ Do} \\
\quad \text{Compute } P\mathbf{w}^{(j)} = A\mathbf{x}^{(j)} \\
\quad \textit{For } i = 1, \dots, j \textit{ Do} \\
\quad\quad g_{ij} = (\mathbf{x}^{(i)})^T \mathbf{w}^{(j)} \\
\quad\quad \mathbf{w}^{(j)} = \mathbf{w}^{(j)} - g_{ij}\mathbf{x}_i \\
\quad \textit{End Do} \\
\quad g_{j+1,j} = \|\mathbf{w}^{(j)}\|_2 \\
\quad (\textit{if } g_{j+1,j} = 0 \text{ set } k = j \text{ and } \textit{Goto } (1)) \\
\quad \mathbf{x}^{(j+1)} = \mathbf{w}^{(j)}/g_{j+1,j} \\
\textit{End Do} \\
V_k = [\mathbf{x}^{(1)}, \dots, \mathbf{x}^{(k)}],\ \hat{H}_k = \{g_{ij}\},\ 1 \le j \le k,\ 1 \le i \le j+1; \\
(1) \text{ Compute } \mathbf{z}^{(k)} \text{, the minimizer of } \|\beta\mathbf{e}_1 - \hat{H}_k\mathbf{z}\| \\
\text{Set } \mathbf{x}^{(k)} = \mathbf{x}^{(0)} + V_k\mathbf{z}^{(k)}
\end{array} \tag{7.37}$$

More generally, as proposed by Saad (1996), a variable preconditioner P_k can be used at the k-th iteration, yielding the so-called *flexible GMRES* method. The use of a variable preconditioner is especially interesting in those situations where the preconditioner is not explicitly given, but implicitly defined, for instance, as an approximate Jacobian in a Newton iteration or by a few steps of an inner iteration process (see Chapter 16). Another meaningful case is the one of domain decomposition preconditioners (of either Schwarz or Schur type) where the preconditioning step involves one or several substeps of local solves in the subdomains (see Chapter 18).

Several considerations for the practical implementation of GMRES, its relation with FOM, how to restart GMRES, and the Householder version of GMRES can be found in [Saa96].

Remark 7.1 (Projection methods). Denoting by Y_k and L_k two generic m-dimensional subspaces of $\mathbb{R}^n$, we call *projection method* a process which generates an approximate solution $\mathbf{x}^{(k)}$ at step k, enforcing that $\mathbf{x}^{(k)} \in Y_k$ and that the residual $\mathbf{r}^{(k)} = \mathbf{b} - \mathrm{A}\mathbf{x}^{(k)}$ be orthogonal to L_k. If $Y_k = L_k$, the projection process is said to be *orthogonal*, and *oblique* otherwise (see [Saa96]).

The Krylov subspace iterations can be regarded as being projection methods. For instance, the Arnoldi method (see [Saa96]) is an orthogonal projection method where $L_k = Y_k = K_k(\mathrm{A};\mathbf{r}^{(0)})$, while the GMRES method is an oblique projection method with $Y_k = K_k(\mathrm{A};\mathbf{r}^{(0)})$ and $L_k = \mathrm{A}Y_k$. It is worth noticing that some classical methods introduced in previous sections fall into this category. For example, the Gauss-Seidel method is an orthogonal projection method where at the k-th step $K_k(\mathrm{A};\mathbf{r}^{(0)}) = \mathrm{span}\{\mathbf{e}_k\}$, with $k = 1,\ldots,n$. The projection steps are carried out cyclically from 1 to n until convergence. •

7.2.4 The Multigrid method

The Multigrid (MG) method is an iterative algorithm to solve the algebraic system associated to a certain grid by making use of one or several additional coarser grids. For the sake of simplicity, we describe only the case of a two-grid algorithm. Thus we suppose that (7.2) represents the algebraic system arising from, say, a finite element approximation of a boundary-value problem on a (fine) grid $\mathcal{T}_h$. For the sake of clarity we can rewrite (7.2) as

$$A_h\mathbf{u}_h = \mathbf{b}_h \tag{7.38}$$

where, as usual, $h = \max_{k\in\mathcal{T}_h} \mathrm{diam}(K), A_h$ is the stiffness FE matrix, $\mathbf{b}_h$ is the right hand side, $\mathbf{u}_h$ the vector of nodal values. Let $\mathcal{T}_H$ represent a coarse grid such that $\mathcal{T}_h$ can be regarded as a refinement of $\mathcal{T}_H$, for instance so that the vertices of $\mathcal{T}_h$ are obtained as the midpoints of edges from $\mathcal{T}_H$. In that case, $h = H/2$.
The generic iteration of the MG algorithm on these two grids consists of

1. *Pre-smoothing step:*
   perform m_1 (≥ 1) iterations on the fine grid using an iterative algorithm (e.g. Jacobi, or Gauss-Seidel, or Richardson),
   $$\mathbf{u}_h^{(l)} = S_h(\mathbf{u}_h^{(l-1)}, \mathbf{b}_h) \qquad l = 1,\ldots,m_1,$$
   for a suitable $\mathbf{u}_h^{(0)}$;
2. *Residual computation:*
   $$\mathbf{r}_h = \mathbf{b}_h - A_h\mathbf{u}_h^{(m_1)};$$
3. *Restriction to the coarse grid:*
   $$\mathbf{r}_H = I_h^H\mathbf{r}_h$$
   where $I_h^H : \mathbb{R}^{N_h} \longrightarrow \mathbb{R}^{N_H}$ is a *fine-to-coarse* operator, N_h is the number of unknowns of the fine-grid problem, N_H that of the coarse-grid one;

4. *Solution of the coarse-grid problem:*

$$A_H \mathbf{e}_H = \mathbf{r}_H$$

where A_H is the stiffness matrix associated with the FE discretization on the coarse grid $\mathcal{T}_H$;

5. *Coarse-grid correction:*

$$\mathbf{u}_h^{(m_1+1)} = \mathbf{u}_h^{(m_1)} + I_H^h \mathbf{e}_H,$$

where $I_H^h : \mathbb{R}^{N_H} \to \mathbb{R}^{N_h}$ is a *coarse-to-fine* operator;

6. *Post-smoothing step:*
Perform m_2 (≥ 1) iterations on the fine grid:

$$\mathbf{u}_h^{(l)} = S_h(\mathbf{u}_h^{(l-1)}, \mathbf{b}_h) \qquad l = m_1+1, \ldots, m_1+m_2+1.$$

The two inter-grid operators (i.e. matrices) are adjoint to each other.

Let V_h be the finite element space associated with $\mathcal{T}_h$, V_H that associated with $\mathcal{T}_H$. For every $\mathbf{w} \in \mathbb{R}^{N_H}$, let $w_H = \sum_{i=1}^{N_H} w_i \varphi_i^H$ be the corresponding FE function in V_H. Similarly, for every $\mathbf{v} \in \mathbb{R}^{N_h}$, let $v_h = \sum_{j=1}^{N_h} v_j \varphi_j^h$ be the corresponding FE function in V_h (see (4.7)). Here φ_i^h (resp. φ_j^H) indicate the Lagrangian basis functions in V_h (resp. V_H). Let $\mathcal{I}_H^h : V_H \to V_h$ be the operator corresponding to I_H^h, that is

$$\mathcal{I}_H^h w_H = v_h \quad \text{iff} \quad I_H^h \mathbf{w} = \mathbf{v}.$$

Typically $\mathcal{I}_H^h$ is the natural injection, in the sense that

$$\mathcal{I}_H^h w_H = w_H \qquad \forall w_H \in V_H.$$

This means that their nodal values $(\mathcal{I}_H^h w_H)(\mathbf{N}_j^h) = w_H(\mathbf{N}_j^h)$ are the same at all nodes $\mathbf{N}_j^h$ of $\mathcal{T}_h$. The entries of I_H^h are therefore $(I_H^h)_{ij} = \varphi_i^H(\mathbf{N}_j^h)$, $j = 1, \ldots, N_h$, $i = \cdots, N_H$. The operator I_h^H is the weighted transpose of I_H^h, that is

$$(I_h^H \mathbf{v}, \mathbf{w})_H = (\mathbf{v}, I_H^h \mathbf{w})_h \qquad \forall \mathbf{v} \in \mathbb{R}^{N_h}, \mathbf{w} \in \mathbb{R}^{N_H},$$

where we have introduced the weigthed (mesh-dependent) inner products:

$$(\mathbf{v}, \mathbf{w})_h = h^2 \sum_{j=1}^{N_h} v_j w_j \qquad \forall\, \mathbf{v}, \mathbf{w} \in \mathbb{R}^{N_h},$$

$$(\mathbf{x}, \mathbf{y})_H = H^2 \sum_{i=1}^{N_H} x_i y_i \qquad \forall\, \mathbf{x}, \mathbf{y} \in \mathbb{R}^{N_H}.$$

A simple algebraic calculation shows that

$$(I_h^H)_{ij} = \frac{h^2}{H^2} (I_H^h)_{ji}, \quad i = 1, \ldots, N_H, \; j = \ldots, N_h,$$

that is

$$I_h^H = \frac{h^2}{H^2}(I_H^h)^T.$$

What we have described is a two-grid V-cycle algorithm. As a matter of fact, a pictorial representation of one step of this algorithm, where the fine is a high level whereas the coarse is a low level, looks like a "V" workflow.
It is easy to see that this iterative algorithm can be associated with the following iteration matrix

$$MG = S_h^{m_2}(I_h - I_H^h(A_H)^{-1}I_h^H A_h)S_h^{m_1}, \qquad (7.39)$$

where $I_h \in \mathbb{R}^{N_h \times N_h}$ is the identity matrix and S_h is the smoothing iteration matrix (e.g., the Richardson matrix R_P or $R(\alpha_k)$ on the fine grid).

The convergence analysis of the two-grid V-cycle algorithm, as well as that of the more general multi-grid case with either V- or W-cycle, is carried out, e.g., in [Hac] and, for finite element discretization, in [BS94].

8

Elements of finite element programming

In this chapter we focus more deeply on a number of aspects relating to the translation of the finite-element method into computer code. This *implementation* process can hide some pitfalls. Beyond the syntactic requirements of a given programming language, the need for a high computational efficiency leads to an implementation that is generally not the immediate translation of what has been seen during the theoretical presentation. Efficiency depends on many factors, including the language used and the architecture on which one works[1]. Personal experience can play a role as fundamental as learning from a textbook. Moreover, although spending time searching for a bug in the code or for a more efficient data structure can sometimes appear to be a waste of time, it (almost) never is. For this reason, we wish to propose the present chapter as a sort of "guideline" for trials that the reader can perform on his own, rather than a chapter to be studied in the traditional sense.

A final note about the chapter style. The approach followed here is to provide *general* guidelines: obviously, each problem has specific features that can be exploited in a careful way for a yet more efficient implementation.

8.1 Working steps of a finite element code

The execution of a Finite-Element computation can be logically split into four working steps (Fig. 8.1).

1. *Pre-processing*. This step consists in setting up the problem and coding its computational domain, which, as seen in Chapter 4, requires the construction of the *mesh* (or grid). In general, setting aside the trivial cases (for instance in one dimension), the construction of an adequate mesh is a numerical problem of considerable inter-

[1] Currently, engineering applications involving scientific computing are running on parallel architectures with hundreds or thousands of Central Processor Units (CPUs) or Graphical Processor Units (GPUs), and this requires specific coding techniques. This topic is beyond the scope of the present book.

A. Quarteroni: *Numerical Models for Differential Problems*, 2nd Ed.
MS&A – Modeling, Simulation & Applications 8
DOI 10.1007/978-88-470-5522-3_8, © Springer-Verlag Italia 2014

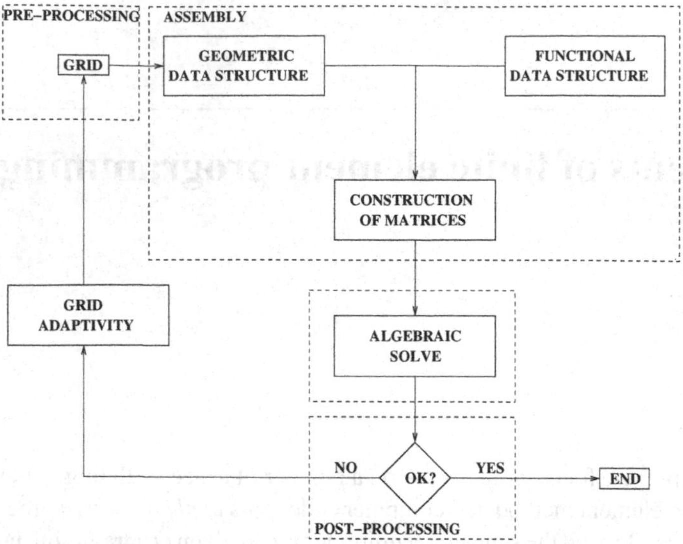

Fig. 8.1. Working steps of a finite element code

est, for which *ad hoc* techniques have been developed. Generally, this operation is performed by dedicated programs or modules within a solver, where great effort has been recently devolved to the aspect of interface and interfacing with CAD (*Computer Aided Design*) software. Chapter 6 is dedicated to the fundamental techniques for grid generation.

2. *Assembly.* In this phase, we construct the "functional" data structures, starting from the "geometric" ones obtained by the mesh and by the user's choices concerning the desired type of finite elements to be used. Moreover, based on the problem we want to solve and on its boundary conditions, we compute the stiffness matrix associated to the discretization (see Chapters 4 and 12). In an unsteady problem, this operation may need to be included in the time advancing loop, when the matrix depends on time (like for instance for the linearization of nonlinear problems, see Chapters 5 and 16). Strictly speaking, the term "assembly" refers to the construction of the matrix of the linear system, moving from the local computation performed on the reference element to the global one that concurs to determine the matrix associated to the discretized problem. Fig. 8.2 summarizes the different operations during the assembly phase for the preparation of the algebraic system.
3. *Solution of the algebraic system.* The core of the solution of any finite-element computation is represented by the solution of a linear system. As previously said, this will eventually be part of a temporal cycle (based on an implicit discretization method) or of an iterative cycle arising from the linearization of a nonlinear problem. The choice of the solution method is generally left to the user. For this reason, it is very important that the user understands the problem under exam, which, as we have seen in Chapter 4, has implications on the structure of the matrix (for in-

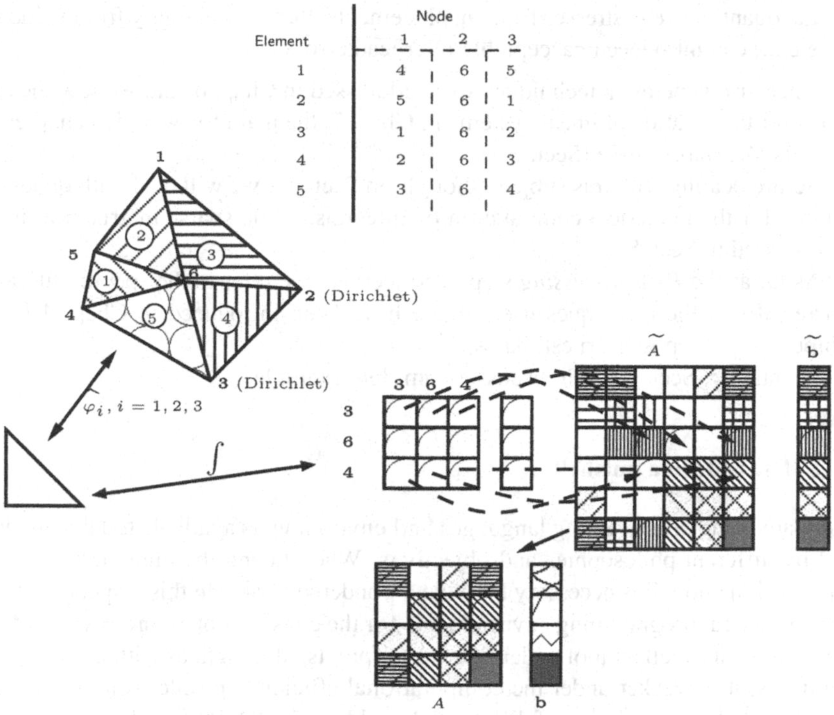

Fig. 8.2. Scheme of the assembly. The geometric and topological information (top table), suitably stored, describes the grid. Through the mapping on the reference element, we compute the discretization matrix $\widetilde{A}$ and of the term $\widetilde{\mathbf{b}}$, first by proceeding element by element (local computation) and then, by exploiting the additivity of integration operation, we update the global matrix. The symbols representing each element of the matrix are obtained through the overlap of the symbols used to define each element of the mesh. Finally, we implement the boundary conditions, which ideally remove the degrees of freedom with associated Dirichlet conditions, getting to the final structures A and $\mathbf{b}$. As we will see, the operation is often implemented in a different way

stance, symmetry and positivity); on the other hand she/he should be aware of the available methods to perform an optimal choice (which rarely is the default one). This is why in Chapter 7 we recalled the main properties of numerical methods for the solution of linear systems.
Nowadays, a host of very efficient computational libraries exist for the solution of various types of linear systems, hence the trend in the coding phase is generally to include such libraries rather than implementing from scratch. Among others, we remind PetSC (see [Pet]), UMFPACK [UMF], and TriLinos [Tri].

4. *Post-processing.* Since the amount of numerical data generated by a finite-element code might be huge, a post-processing is often necessary in order to present results that are concise and in a usable format. However, this may not be a trivial task. In particular, a reckless post-processing for the a posteriori computation of differen-

tial quantities (e.g. stresses from displacements, fluxes or vorticity from velocities, etc ...) can introduce unacceptable overhead errors.

Since grid generation techniques were addressed in Chap. 6, and we saw the algorithms for the solution of linear systems in Chap. 7, the main focus of this chapter will be on the *Assembly* phase (Sect. 8.4).

Before dealing with this subject, though, in Sect. 8.2 we will deal with quadrature formula for the numerical computation of integrals, while sparse matrix storing will be discussed in Sect. 8.3.

As far as the *Post-processing* step is concerned, we refer to the specific literature, and recall that the techniques used above have been introduced in Chap. 4 for the computation of a posteriori estimates.

Eventually, Sect. 8.6 will discuss a complete example.

8.1.1 The code in a nutshell

There are many programming languages and environments available today, characterized by different philosophies and objectives. When facing the implementation of a numerical method, it is necessary to make a pondered choice in this respect. Amongst the most useful programming environments for the construction of prototypes, Matlab is certainly an excellent tool under many viewpoints, although, as with all interpreted languages, it is weaker under the computational efficiency profile. Another environment targeted to the solution of differential problems in 2D through the finite-element method is FreeFem++ (see www.freefem.org). This environment comprises all four phases indicated above in a single package (free and usable under different operating systems). Its particularly captivating syntax reduces the gap between coding and theoretical formulation by bringing the former significantly closer to the latter. This operation has a clear "educational" merit, which is to quickly produce simulations also for non trivial problems. However, the computational costs and the difficulty of implementing new strategies that require an extension of the syntax can be penalizing in actual cases of interest. In [FSV05] several solved examples and problems are solved with FreeFem++.

Among compiled programming languages, Fortran (Fortran 77 in particular) is traditionally the one that has had the biggest success in the numerical domain, because it generates very efficient executable codes. Recently, the abstraction feature that is intrinsic to the object-oriented programming philosophy has proven to be very suitable for finite element programming. The level of abstraction made possible by far-reaching mathematical tools seems to find an excellent counterpart in the abstraction of object-oriented programming, based on the design of data types made by the user (more than on operations to perform, as in procedural programming) and on their polymorphism (see e.g. [LL00, Str00]). However, the computational cost of such an abstraction has sometimes reduced the interest for a theoretically attractive programming style. The latter is often operationally weak for science problems, where computational efficiency is (almost) always crucial. This has required the development of more sophisticated programming techniques (for instance, Expression Templates), that avoid the cost associated to the interpretation of abstract objects to become too heavy during code execution (see e.g. [Vel95, Fur97, Pru06, DV09]). Hence, besides Fortran, languages like C++ (born as an object-oriented improvement of the language C) are nowadays more and more frequent in the scientific domain; amongst others, we recall FEniCS OpenFOAM and LifeV..

In the code excerpts presented below we will refer to C++. An accurate examination of the code (which we will henceforth call "Programs" for simplicity) requires some basic knowledge of C++, for which we refer to [LL00]. However, as we want to use this chapter as a basis for autonomous experiments, it is not essential to master the C++ syntax to understand the text; a mild familiarity with its basic syntax will be enough for the reader who might prefer a different language.

8.2 Numerical computation of integrals

The effective numerical computation of the integrals in the finite element formulation is typically performed via *quadrature formulae*. For an introduction to the subject of numerical quadrature, we refer to basic numerical analysis textbooks (e.g. [QSS07]). Here, it will suffice to recall that a generic quadrature formula has the form

$$\int_K f(\mathbf{x})d\mathbf{x} \approx \sum_{iq=1}^{nqn} f(\mathbf{x}_{iq})w_{iq}$$

where K denotes the region over which we integrate (typically an element of the finite element grid), nqn is the number of quadrature nodes for the selected formula, $\mathbf{x}_{iq}$ are the coordinates of the *quadrature nodes* and w_{iq} are the *weights*. Typically, the accuracy of the formula and its computational cost grow with the number of quadrature nodes. As we will see in Chapter 10, Sects. 10.2.2 and 10.2.3, the formulae which guarantee the best accuracy for the same number of nodes are the *Gaussian* ones.

The computation of integrals is generally performed on the reference element (where the expression of basis functions is known) through a suitable change of variable (Sect. 4.3).

Let us denote with $\hat{x}_i$ and x_i (for $i = 1,\dots,d$) the coordinates on the reference element $\hat{K}$ and those on the generic element K, respectively. Integration in the reference space will then require the knowledge of the Jacobian matrices $J_K(\hat{\mathbf{x}})$ of the geometric transformation F_K that maps the reference element $\hat{K}$ on the element K (see Fig. 4.14),

$$J_K(\hat{\mathbf{x}}) = \left[\frac{\partial x_i}{\partial \hat{x}_j}(\hat{\mathbf{x}})\right]_{i,j=1}^{d}.$$

We then have

$$\int_K f(\mathbf{x})d\mathbf{x} = \int_{\hat{K}} \hat{f}(\hat{\mathbf{x}})|\mathrm{det}J_K(\hat{\mathbf{x}})|d\hat{\mathbf{x}} \approx \sum_q \hat{f}(\hat{\mathbf{x}}_q)|\mathrm{det}J_K(\hat{\mathbf{x}}_q)|\hat{w}_q, \tag{8.1}$$

where $\hat{f} = f \circ F_K$, and $\hat{w}_q$ are the weights on the reference element. In case of integrals involving derivatives, denoting with $\widehat{J}_K(\mathbf{x})$ the Jacobian matrix associated to F_K^{-1}, i.e.

$$\widehat{J}_K(\mathbf{x}) = \left[\frac{\partial \hat{x}_i}{\partial x_j}(\mathbf{x})\right]_{i,j=1}^{d},$$

we have, for $j = 1, \ldots, d$,

$$\frac{\partial f}{\partial x_i}(\mathbf{x}) = \sum_{j=1}^{d} \frac{\partial \hat{f}}{\partial \hat{x}_j}(\hat{\mathbf{x}}) \frac{\partial \hat{x}_j}{\partial x_i}(\mathbf{x}), \quad \nabla_x f(\mathbf{x}) = \left[\widehat{J}_K(\mathbf{x})\right]^T \nabla_{\hat{x}} \hat{f}(\hat{\mathbf{x}}).$$

We can prove that

$$\left[\widehat{J}_K(\mathbf{x})\right]^T = \frac{1}{\det J_K(\hat{\mathbf{x}})} J_K^{cof}(\hat{\mathbf{x}}),$$

$J_K^{cof}(\hat{\mathbf{x}})$ being the matrix of the cofactors of the elements of $J_K(\hat{\mathbf{x}})$, i.e. (in the two-dimensional case)

$$J_K^{cof}(\hat{\mathbf{x}}) = \begin{bmatrix} \dfrac{\partial x_2}{\partial \hat{x}_2}(\hat{\mathbf{x}}) & -\dfrac{\partial x_2}{\partial \hat{x}_1}(\hat{\mathbf{x}}) \\ -\dfrac{\partial x_1}{\partial \hat{x}_2}(\hat{\mathbf{x}}) & \dfrac{\partial x_1}{\partial \hat{x}_1}(\hat{\mathbf{x}}) \end{bmatrix}.$$

The gradient of the function f can thus be expressed in terms of the variables in the reference space as following

$$\nabla_x f(\mathbf{x}) = \frac{1}{\det J_K(\hat{\mathbf{x}})} J_K^{cof}(\hat{\mathbf{x}}) \nabla_{\hat{x}} \hat{f}(\hat{\mathbf{x}}).$$

Denoting with α and β the indices of two generic basis functions, the typical element of the stiffness matrix can therefore be computed as follows

$$\begin{aligned} &\int_K \nabla_x \varphi_\alpha(\mathbf{x}) \nabla_x \varphi_\beta(\mathbf{x}) d\mathbf{x} = \\ &\int_{\hat{K}} \left(J_K^{cof}(\hat{\mathbf{x}}) \nabla_{\hat{x}} \hat{\varphi}_\alpha(\hat{\mathbf{x}})\right) \left(J_K^{cof}(\hat{\mathbf{x}}) \nabla_{\hat{x}} \hat{\varphi}_\beta(\hat{\mathbf{x}})\right) \frac{1}{|\det J_K(\hat{\mathbf{x}})|} d\hat{\mathbf{x}} \simeq \\ &\sum_q \left[\frac{\hat{w}_q}{|\det J_K(\hat{\mathbf{x}}_q)|} \sum_{j=1}^{d} \left(\sum_{l=1}^{d} \left[J_K^{cof}(\hat{\mathbf{x}}_q)\right]_{jl} \frac{\partial \hat{\varphi}_\alpha}{\partial \hat{x}_l}(\hat{\mathbf{x}}_q)\right) \left(\sum_{m=1}^{d} \left[J_K^{cof}(\hat{\mathbf{x}}_q)\right]_{jm} \frac{\partial \hat{\varphi}_\beta}{\partial \hat{x}_m}(\hat{\mathbf{x}}_q)\right)\right]. \end{aligned} \tag{8.2}$$

Note that the matrices J_K, and consequently the matrices J_K^{cof}, are constant on the element K if K is a triangle or a rectangle in 2D (a tetrahedron or a parallelepiped in 3D) with no curved boundaries.

The class coding a quadrature formula stores quadrature nodes and their associated weights. In the effective integral computation we will then obtain the necessary mapping information for the actual computation, which depends on the geometry of K.

The choice of a quadrature formula responds to two (conflicting) needs:

1. On one hand, the higher the accuracy is, the smaller is the integration error eventually affecting the overall quality of the numerical computation; a proper choice of the quadrature rule, based on the concept of *degree of exactness*, may make the numerical integration error vanish. We control for problems whose differential operator has constant (or polynomial) coefficients.
2. On the other hand, a larger number of *nqn* nodes and in increase of the computational cost of assembly is necessary to obtain an increase of accuracy.

The appropriate synthesis of these two needs evidently depends on the requirements of the problem we want to solve, as well as on the accuracy and speed specifications to execute the computation.

8.2.1 Numerical integration using barycentric coordinates

The numerical evaluation of integrals on simplexes (intervals in 1D, triangles in 2D, tetrahedra in 3D) can profit from the use of the barycentric coordinates that were introduced in Sect. 4.4.3. To start with, we observe that the following exact integration formulas hold (see, e.g., [Aki94, Chap. 9] or [Hug00, Chap. 3]):

in 1D

$$\int_{\widehat{K}_1} \lambda_0^a \lambda_1^b d\omega = \frac{a!b!}{(a+b+1)!} \text{lenght}(\widehat{K}_1).$$

in 2D

$$\int_{\widehat{K}_2} \lambda_0^a \lambda_1^b \lambda_2^c d\omega = \frac{a!b!c!}{(a+b+c+2)!} 2\text{Area}(\widehat{K}_2),$$

in 3D

$$\int_{\widehat{K}_3} \lambda_0^a \lambda_1^b \lambda_2^c \lambda_3^d d\omega = \frac{a!b!c!d!}{(a+b+c+d+3)!} 6\text{Vol}(\widehat{K}_3),$$

More in general,

$$\int_{\hat{K}_d} \prod_{i=0}^{d} \lambda_i^{n_i} d\omega = \frac{\prod_{i=0}^{d} n_i!}{(\sum_{i=0}^{d} n_i + d)!} d!|\hat{K}_d| \tag{8.3}$$

where $\hat{K}_d$ is a d-dimensional standard simplex, $|\hat{K}_d|$ denotes its measure, $\{n_i, 0 \leq i \leq d\}$ is a set of non-negative integers.

These formulas are useful when dealing with finite-element approximations of the boundary-value problems for the exact computation of polynomial integrals in the characteristic Lagrangian basis functions.

For the sake of an example, Table 8.1 shows the weights and nodes for some popular quadrature formulas in 2D. Table 8.2 gives some formulas for a tetrahedron. These formulas are symmetric: we must consider all possible permutations of the barycentric coordinates to obtain the full list of nodes.

For the reader's convenience we have written, next to the total number *nqn* of nodes, the multiplicity *m* of each quadrature node, i.e. the number of nodes generated

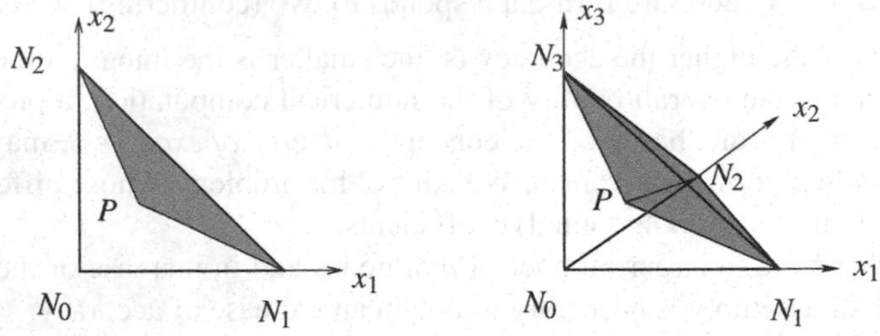

Fig. 8.3. The barycentric coordinate λ_i of the point P represents the ratio between the volume of the tetrahedron having as vertices P and the vertices of the face opposite to N_i (in the figure, right, we have shadowed the tetrahedron with vertices P, N_1, N_2, N_3 opposite N_0) and the total volume of the tetrahedron

Table 8.1. Nodes and weights for the quadrature formulae on triangles. The nodes are expressed through their barycentric coordinates. The weights do not take into account the measure of the reference element (which is equal to 1/2 in this case)

nqn	barycentric coordinates λ_j			*m*	w_j	*r*
1	1/3	1/3	1/3	1	1	1
3	1	0	0	3	1/3	1
3	2/3	1/3	1/3	3	1/3	1
4	1/3	1/3	1/3	1	-0.5625	2
	0.6	0.2	0.2	3	0.52083	
6	0.65902762237	0.23193336855	0.10903900907	6	1/6	2
6	0.81684757298	0.09157621351	0.09157621351	3	0.10995174366	3
	0.10810301817	0.44594849092	0.44594849092	3	0.22338158968	

Table 8.2. Nodes and weights for quadrature formulae on tetrahedra. The nodes are expressed using their barycentric coordinates. The weights do not take into account the measure of the reference element (equal to 1/6 in this case)

nqn	barycentric coordinates λ_j				*m*	w_j	*r*
1	1/4	1/4	1/4	1/4	1	1	1
4	0.58541020	0.13819660	0.13819660	0.13819660	4	1/4	2
5	1/4	1/4	1/4	1/4	1	-16/20	3
	1/2	1/6	1/6	1/6	4	9/20	

by the permutations. We have also provided the exactness degree r, that is the largest positive integer r for which all polynomials of degree $\leq r$ are integrated exactly by the quadrature formula at hand.

Let us see two simple examples. Suppose we want to compute:

$$I = \int_K f(\mathbf{x}) d\mathbf{x} = \int_{\widehat{K}} \widehat{f}(\widehat{\mathbf{x}}) |J|(\widehat{\mathbf{x}}) d\widehat{\mathbf{x}}.$$

Using the weights and nodes of the first row of the table, we obtain:

$$I \simeq \frac{1}{2}\widehat{f}(\frac{1}{3},\frac{1}{3})J(\frac{1}{3},\frac{1}{3}) = \text{Area}(K)f(\overline{\mathbf{x}}),$$

where the coefficient $1/2$ represents the area of the reference element, and $\overline{\mathbf{x}}$ is the node with barycentric coordinates $\lambda_1 = \lambda_2 = \lambda_3 = 1/3$ corresponding to the center of gravity of the triangle. The corresponding formula is the well-known *composite midpoint formula*.

To use the formula in the second row we note that $m = 3$, hence we have indeed 3 quadrature nodes whose barycentric coordinates are obtained via cyclic permutation:

$$(\lambda_0 = 1, \lambda_1 = 0, \lambda_2 = 0),\ (\lambda_0 = 0, \lambda_1 = 1, \lambda_2 = 0),\ (\lambda_0 = 0, \lambda_1 = 0, \lambda_2 = 1).$$

Hence for each triangle K we obtain

$$\begin{aligned}\int_K f(\mathbf{x})d\mathbf{x} &\simeq \frac{1}{2}\frac{1}{3}\left[\widehat{f}(0,0)|\det J(0,0)| + \widehat{f}(1,0)|\det J(1,0)| + \widehat{f}(0,1)|\det J(0,1)|\right] \\ &= \text{Area}(K)\sum_{i=0}^{2}\frac{1}{3}f(\mathbf{N}_i),\end{aligned}$$

$\mathbf{N}_0, \mathbf{N}_1, \mathbf{N}_2$ being the vertices of the triangle K, corresponding to the barycentric coordinates (0,0), (1,0) and (0,1) respectively. The corresponding formula therefore yields the *composite trapezoidal formula*. Both formulae have exactness degree equal to 1.

Other quadrature formulae for the computation of integrals for different finite elements can be found in [Hug00], [Str71].

Remark 8.1. When using quadrilateral or prismatic elements, nodes and weights of the quadrature formulae can be obtained as the tensor product of the Gauss quadrature formulae for the one-dimensional interval, see Sect. 10.2 (and also [CHQZ06]). •

8.3 Storage of sparse matrices

As seen in Chapter 4, finite element matrices are sparse. The distribution of non-null elements is retained by the so-called sparsity pattern (also called *graph*) of the matrix. The pattern depends on the computational grid, on the finite element type and on the numbering of the nodes adopted. The efficient storage of a matrix therefore consists in the storage of its non-null elements, according to the positioning given by the pattern. The discretization of different differential problems sharing the same computational grid and the same type of finite elements leads to matrices with the same graph. Hence in an object-oriented programming logic it can be useful to separate the storage of the graph (which can become a "data type" defined by the user, i.e. a class) from the storage of the values of each matrix. In this way, a matrix can be seen as a data structure for the storage of its values, together with a pointer to the graph associated to it. The pointer only stores the position in the memory where the pattern is stored, hence occupying a minimal amount of memory. Different matrices may therefore share the same graph, without useless storage duplications of the pattern.

In practice, there are several techniques to store sparse matrices efficiently, i.e. the position and value of non-null elements. At this juncture we should observe that, in this context, the adjective "efficient" does not only refer to the lower memory occupation that can be achieved, but also to the speed in accessing memory for each element. A storage format requiring the least possible memory waste is likely to be slower in accessing a desired value. Indeed, a higher storage compactness is typically obtained after finding the position in the memory by accessing the data structures that store the graph. The more intermediate passages are necessary, the longer the access time to the desired element will be. Precisely for the need of finding the right compromise, different storage techniques have been proposed in the literature, with different prerogatives. A review of these, with many comments and remarks, can be found e.g. in [FSV05], Appendix A. Here we just recall a widely used format for the storage of sparse square matrices, i.e. the CSR (*Compact Sparse Row*) format.

We denote by n the size of the matrix to be stored and by nz the number of its non-null elements. In the CSR format, a matrix is identified by three vectors. Two – denoted hereafter R and C – have integer values and form the so-called *pattern*, i.e. the graph of the sparsity of the matrix, which depends on the mesh and the finite element space; the third vector A contains the non-zero entries of the matrix. More precisely, A has size nz and it stores the entries in a row-wise order. The vector C has size nz, as well. The entry $C(k)$ contains the column of the entry $A(k)$. Finally, the vector R has size n and contains pointers to the vector C, indicating the beginning of each row. In other terms, $R(k)$ contains the position of the vector J (and A) where row k of the matrix begins. To be more concrete, we illustrate this on an example (see Fig. 8.4) where $n = 5$ and $nz = 17$.

$$
\begin{array}{cc}
 & \begin{array}{ccccc} 0 & 1 & 2 & 3 & 4 \end{array} \\
A = & \left[\begin{array}{ccccc}
a & 0 & f & 0 & g \\
0 & b & k & m & 0 \\
h & l & c & 0 & r \\
0 & n & 0 & d & p \\
i & 0 & s & q & e
\end{array}\right]
\begin{array}{c} 0 \\ 1 \\ 2 \\ 3 \\ 4 \end{array}
\end{array}
$$

We point out that the numbering of rows and columns in matrices and vectors starts from 0, following the C++ syntax.

The vectors representing this matrix in CSR format are reported in Fig. 8.4. Here, the different background colors in the vectors refer to the different rows of the matrix, and the lines connect the pointers of the vector R to the corresponding entries in the other vectors. In this way, if we want to retrieve a nonzero entry – for instance, in the second row – we start from position $R(1)$ (because of the C++ numbering) through the entry $R(2) - 1$, since $R(2)$ points to the beginning of the third row. The entries $A(R(1),\ldots,R(2)-1)$ are the values of the second row, whose corresponding columns are in $C(R(1),\ldots,R(2)-1)$. In this way it is relatively easy to access a matrix row-wise, whereas the column-wise access is more involved. The column-wise format CSC (Compressed Sparse Colum) is similarly formulated. In some libraries, an even more

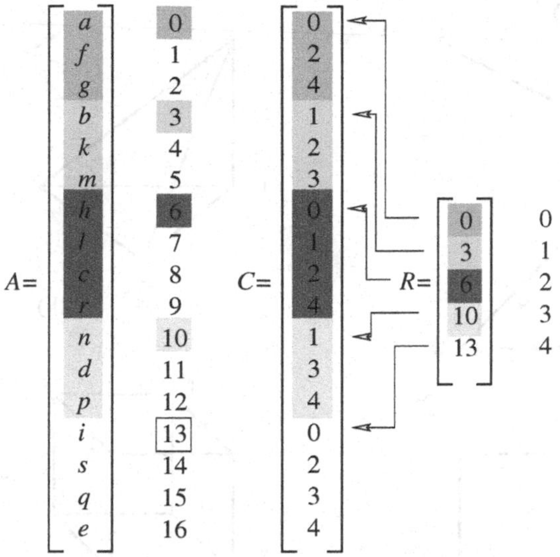

Fig. 8.4. Vectors representing the CSR format of Matrix A

compact format called Modified Sparse Row (MSR) is used for square matrices. In [FSV05] the latter format is extended to accommodate both row-wise and column-wise matrix access with the same computational cost.

8.4 Assembly

The assembly is the sequence of different operations leading to the construction of the matrix associated to the discretized problem. For doing this, we need two types of information:

1. *geometric*, typically contained in the mesh file;
2. *functional*, relative to the representation of the solution via finite elements.

In Fig. 8.5 we report possible reference geometries with their local vertex numbering. Tetrahedra represent the 3D extension of the triangular elements considered in Chapter 4. Prismatic elements extend in 3D the quadrilateral geometric elements in 2D which will be introduced in Chapter 10, see e.g. [Hug00, Chap. 3], for a complete description.

Geometric and functional information, suitably coded, is then used to construct the matrix of the discretized problem. As opposed to what would seem natural in the definition of Lagrangian finite elements, the matrix is constructed over cyclic permutations of the elements instead of nodes. The reason for this *element-oriented* approach, as opposed to the *node-oriented* one, is essentially linked to computational efficiency matters. The analytical expression of a base function associated to a node varies on each element sharing that node. Prior to the computation of the integrals, it would be

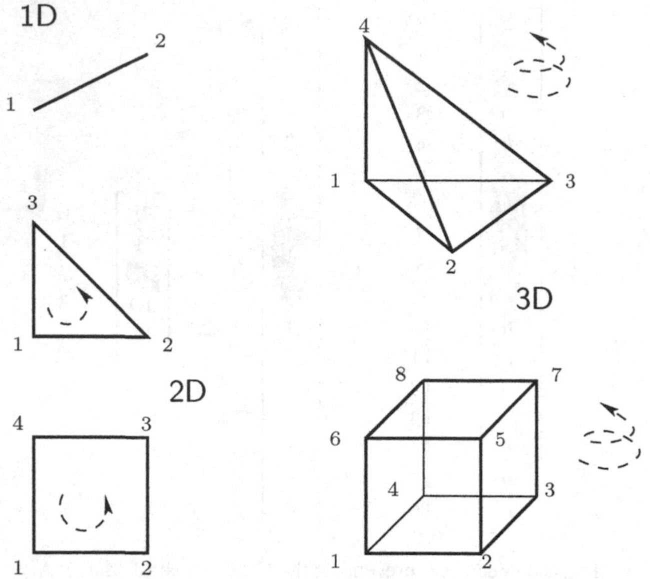

Fig. 8.5. Illustration of some reference elements available in LifeV with (conventional) local numbering of nodes

necessary to cycle over the nodes and detect the analytical expression of the appropriate basis functions to each element. Hence, we would have to cycle on the nodes and locate the analytic expression of the appropriate basis function for each different element, before carrying out the computation of the integrals. In terms of code, this means that the body of the cycle must be filled with *conditional branches*, that is instructions of the type if...then...elseif...then...else... within the assembly cycle. These are "costly" operations in computational terms, especially if they lie within a cycle (and thus are carried out several times), as is clear from the large number of micro-assembler instructions required in the compilation phase to expand a conditional branch, with respect to any other instruction, see e.g. [HVZ97]. As we will see, by exploiting the additivity of integration, the element-oriented approach allows to bypass this obstacle in a smart way.

In particular (see Chap. 4) the construction of the problem matrix can take place in two conceptual steps, within a cycle on the grid elements:

1. construction of the matrix and of the right-hand side that discretize the differential operator on the element at hand (*local* matrix and vector);
2. update of the *global* matrix and the right-hand side, by exploiting the additivity of the integration operation.

There also exist different approaches to the problem: in some cases, the matrix is not constructed, but its effects are directly computed when it multiplies a vector, as happens when the linear system is solved by iterative methods; for reasons of space, here we deal with the more standard approach.

As previously noted, the construction of the local matrix is carried out by integrating on the reference element $\widehat{K}$, using suitable quadrature formulae. Once the matrix and known terms have been constructed, the boundary conditions are imposed; in particular, imposing Dirichlet conditions does not necessarily require the technique seen in Sects. 3.2.2 and 4.5, which consisted in removing the degrees of freedom associated to such conditions after the lift is constructed.

This should explain why assembly is a complicated phase. In the following sections we will discuss the above aspects, though in little detail for reasons of space. First, we will treat the data structures for the coding of geometric (Sect. 8.4.1) and functional (Sect. 8.4.2) information. The computation of the geometric mapping between reference element and current element provides the opportunity of introducing *isoparametric elements* (Sect. 8.4.3). The effective computation of the local matrix and known term and their use in the construction of the global system are treated in Sect. 8.4.4. Finally, in Sect. 8.4.5 we will mention implementation techniques for the lifting of the boundary datum.

8.4.1 Coding geometrical information

In terms of data structures, the *mesh* can be seen as a collection of geometric elements and topological information. The former can be constructed by aggregating classes for the definition of points (i.e. zero-dimensional geometric elements), edges (one-dimensional geometric elements), faces (2D) and finally volumes (3D).

Starting from base classes coding these geometrical entities, a mesh will be a class collecting the elements. In fact the geometric structure should be supplemented by the following elements:

1. topological information allowing to characterize the elements in the grid, i.e. the connectivity among nodes, with respect to a conventional numbering of the latter. The convention for the possible elements in LifeV is illustrated in Fig. 8.5; to "visit" the elements of a grid efficiently, we can also add to each given element information on the adjacent elements;
2. specific information allowing to locate the degrees of freedom on the boundary; this simplifies handling the boundary condition prescription; note that we typically associate to each boundary geometric element an indicator that will subsequently be associated to a specific boundary condition.

Starting from the reference geometric class, we then retrieve the current geometric elements, according to the possible mappings treated in Sect. 8.4.3. For instance Program 2 gives a portion of a class for a linear (affine) tetrahedron built upon a tetrahedric basic reference shape with some functional information on the associated degrees of freedom.

Program 2 – LinearTetra: Class for the coding of tetrahedra obtained via affine geometric transformation of the reference element

```
class LinearTetra:
      public Tetra
{

public:
  typedef Tetra BasRefSha;
  typedef LinearTriangle GeoBShape;
  static const UInt numPoints = 4;
  static const UInt nbPtsPerVertex = 1;
  static const UInt nbPtsPerEdge = 0;
  static const UInt nbPtsPerFace = 0;
  static const UInt nbPtsPerVolume = 0;
}
```

Currently, no standards are available for the mesh file format. Each mesh generator has its own format. Typically, we expect such a file to contain the vertex coordinates, the connectivity associating the vertices to the geometric elements and the list of boundary elements, with corresponding indicator to be used for defining boundary conditions. The functions with the data serving as boundary conditions, instead, are generally assigned separately.

Remark 8.2. Multi-physics or multi-model problems are becoming a relevant component of scientific computation: think for instance of fluid-structure interaction problems, or the coupling of problems where the full (and computationally costlier) differential model is used only in a specific region of interest, and coupled it with simpler models in the remaining regions. These applications and, more generally, the need to develop parallel computation algorithms, have motivated the development of techniques for the solution of differential problems through *domain decomposition* (see Chap. 18 and the more comprehensive presentations [QV99, TW05]). In this case, the resulting mesh is the collection of subdomain meshes, together with topological information about subdomain interfaces. In this chapter, however, we will refer to single-domain problems only. •

8.4.2 Coding of functional information

As seen in Chapter 4, basis functions are defined on a reference element. For instance, for tetrahedra, this element coincides with the unit simplex (see Fig. 8.5). The coding of a reference element will basically include pointers to functions for determining basis functions and their derivatives.

In Program 3 we report the functions for the definition of linear finite elements on tetrahedra. For the sake of space, we provide the code for the first derivatives of the first basis functions only.

Program 3 – fctP13D: Basis functions for a linear tetrahedric element

```
Real fct1_P1_3D( cRRef x, cRRef y, cRRef z ){return 1 -x - y - z;}
Real fct2_P1_3D( cRRef x, cRRef, cRRef ){return x;}
Real fct3_P1_3D( cRRef, cRRef y, cRRef ){return y;}
Real fct4_P1_3D( cRRef, cRRef, cRRef z ){return z;}

Real derfct1_1_P1_3D( cRRef, cRRef, cRRef ){return -1;}
Real derfct1_2_P1_3D( cRRef, cRRef, cRRef ){return -1;}
Real derfct1_3_P1_3D( cRRef, cRRef, cRRef ){return -1;}
....
```

Once the reference element is instantiated, functional information will be available both for the representation of the solution and for the definition of the geometric mapping between reference element and current element, as we explain in the following section.

Having defined the geometric element and the type of finite elements we want to use, we are now able to construct the problem's degrees of freedom. This means assigning to each mesh element the numbering of the degrees of freedom lying on the element and the pattern of the local matrix; the latter is generally full, although it can always contain null elements.

8.4.3 Mapping between reference and physical element

In Chapter 4 we saw how convenient it is to write basis functions, quadrature formulae and, therefore, compute integrals with respect to a reference element. It can thus be interesting to examine some practical methods to construct and code such coordinate change. For further details, we refer to [Hug00]. Let us now limit ourselves to considering the case of triangular and tetrahedric elements.

A first type of coordinate transformation is the *affine* one. Basically, the mapping between $\hat{\mathbf{x}}$ and $\mathbf{x}$ can be expressed via a matrix B and a vector $\mathbf{c}$ (see Sect. 4.5.3 and Fig. 8.6)

$$\mathbf{x} = B\hat{\mathbf{x}} + \mathbf{c}. \tag{8.4}$$

In this way, we trivially have that $J = B$ (constant on each element). If the node distribution generated by the grid generator is correct, the determinant of J is always positive, which guarantees there are no degenerate cases (for instance, four vertices on the same plane in a tetrahedron) and that there are no incorrect permutations in the nodes corresponding to the mapping. The expressions of B and $\mathbf{c}$ can be obtained from those of the node coordinates. Indeed, let us suppose that the nodes, numbered *locally* 1,2,3,4, of the reference tetrahedron correspond to the nodes of the mesh numbered as i, k, l, m, respectively.

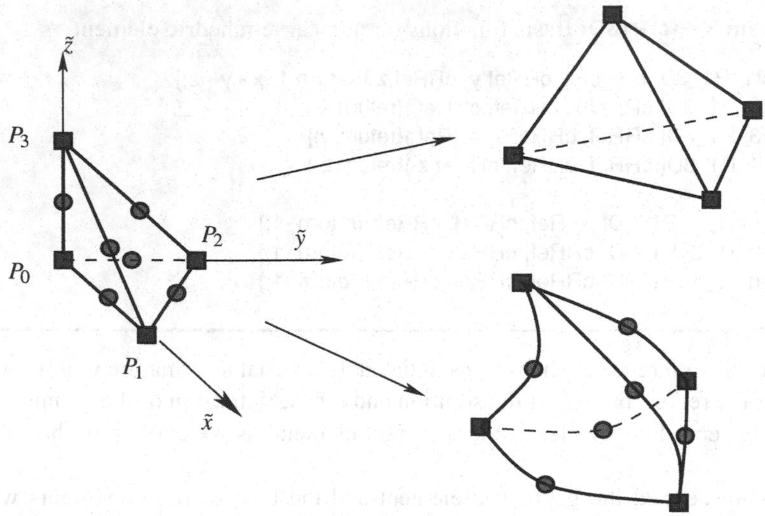

Fig. 8.6. Mapping between the reference tetrahedron and the current one. The top map is affine, the bottom one quadratic

We then have:

$$\begin{cases} x_i = c_1 & y_i = c_2 & y_i = c_3 \\ x_k = b_{11} + x_i & y_k = b_{12} + y_i & z_k = b_{13} + y_i \\ x_l = b_{21} + x_i & y_l = b_{22} + y_i & z_l = b_{23} + y_i \\ x_m = b_{31} + x_i & y_m = b_{32} + y_i & z_m = b_{33} + y_i \end{cases} \tag{8.5}$$

from which we obtain the expressions for B and $\mathbf{c}$.

However, there exists a more efficient way to represent the transformation: being element-wise linear, it can be represented via the basis functions of linear Lagrangian finite elements. Indeed, we can write:

$$x = \sum_{j=0}^{3} X_j \widehat{\varphi}_j(\widehat{x},\widehat{y},\widehat{z}),\ y = \sum_{j=0}^{3} Y_j \widehat{\varphi}_j(\widehat{x},\widehat{y},\widehat{z}),\ z = \sum_{j=0}^{3} Z_j \widehat{\varphi}_j(\widehat{x},\widehat{y},\widehat{z}). \tag{8.6}$$

The elements of the Jacobian matrix of the transformation are immediately computed:

$$J = \begin{bmatrix} \sum_{j=1}^{4} X_j \frac{\partial \widehat{\varphi}_j}{\partial \widehat{x}} & \sum_{j=1}^{4} X_j \frac{\partial \widehat{\varphi}_j}{\partial \widehat{y}} & \sum_{j=1}^{4} X_j \frac{\partial \widehat{\varphi}_j}{\partial \widehat{z}} \\ \sum_{j=1}^{4} Y_j \frac{\partial \widehat{\varphi}_j}{\partial \widehat{x}} & \sum_{j=1}^{4} Y_j \frac{\partial \widehat{\varphi}_j}{\partial \widehat{y}} & \sum_{j=1}^{4} Y_j \frac{\partial \widehat{\varphi}_j}{\partial \widehat{z}} \\ \sum_{j=1}^{4} Z_j \frac{\partial \widehat{\varphi}_j}{\partial \widehat{x}} & \sum_{j=1}^{4} Z_j \frac{\partial \widehat{\varphi}_j}{\partial \widehat{y}} & \sum_{j=1}^{4} Z_j \frac{\partial \widehat{\varphi}_j}{\partial \widehat{z}} \end{bmatrix}. \tag{8.7}$$

When in a Lagrangian finite element the same basis functions are used for the definition of the geometric mapping, we say that we are dealing with *iso-parametric* elements

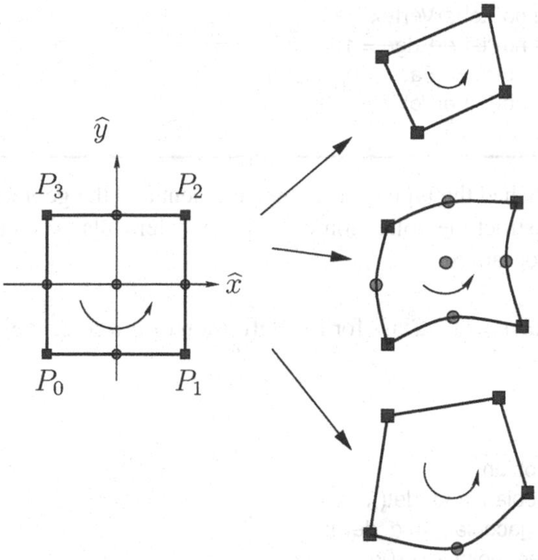

Fig. 8.7. Mapping between the reference quadrilateral and the current element: affine (top), isoparametric (middle), hybrid (bottom). The latter is constructed with 5 nodes, in order to have a biquadratic transformation for the nodes of a single side

(see Figs. 8.6 and 8.7). In the case at hand, this is a consequence of having chosen linear finite elements and affine geometric transformations. When we take finite elements of degree higher than 1, we can consider two kinds of mapping:

- affine finite elements: in this case, the geometric transformation is still described by the affine transformations (8.6), although the functional information relative to the solution is described by quadratic functions of higher degree; the boundary of the discretized domain Ω_h, in this case, is still polygonal (polyhedral);
- isoparametric finite elements: the geometric transformation is described by the same basis functions used to represent the solution; hence the elements in the physical space *Oxyz* will generally have curved sides;

The definition of a quadratic mapping starting from the tetrahedric reference element allows for instance to create tetrahedric quadratic geometric elements, coded in the class QuadraticTetra reported in Program 4.

Program 4 – QuadraticTetra: Class for the definition of quadratic tetrahedric elements

```
 class QuadraticTetra: public Tetra
{
public:
  typedef Tetra BasRefSha;
  typedef QuadraticTriangle GeoBShape;
  static const UInt numPoints = 10;
```

```
   static const UInt nbPtsPerVertex = 1;
   static const UInt nbPtsPerEdge = 1;
   static const UInt nbPtsPerFace = 0;
   static const UInt nbPtsPerVolume = 0;
};
```

Having established the type of reference element and the geometrical mappings, it is possible to construct the collection of "current" elements. The current element can be coded as in Program 5.

Program 5 – CurrentFE: Class for the definition of the current element

```
class CurrentFE
{
private:
   void _comp_jacobian();
   void _comp_jacobian_and_det();
   void _comp_inv_jacobian_and_det();
   void _comp_quad_point_coor();

   template <class GEOELE>
   void _update_point( const GEOELE& geoele );

   //! compute phiDer
   void _comp_phiDer();
   //! compute the second derivative phiDer2
   void _comp_phiDer2();
   //! compute phiDer and phiDer2
   void _comp_phiDerDer2();


   UInt _currentId;
public:
   CurrentFE( const RefFE& _refFE, const GeoMap& _geoMap, const QuadRule& _qr );
   const int nbGeoNode;
   const int nbNode;
   const int nbCoor;
   const int nbQuadPt;
   const int nbDiag;
   const int nbUpper;
   const int nbPattern;
   const RefFE& refFE;
   const GeoMap& geoMap;
   const QuadRule& qr;

};
```

As it can be seen, the class contains information relating to the reference element, to the geometric mapping that generates it and to the quadrature formula that will be used for the computation of the integrals.

In particular, (8.7) proves to be very efficient in the coding phase, which we report in Program 6. The computation of the Jacobian is carried out at the quadrature nodes required for the integral computation (Sec. 8.2).

Program 6 – comp-jacobian: Section of the class storing the current elements which computes the Jacobian of the transformation between current element and reference element

```
void CurrentFE::_comp_jacobian()
{
   Real fctDer;
   // GeoMap derivatives:
   for ( int ig = 0;ig < nbQuadPt;ig++ )
   {
      for ( int icoor = 0;icoor < nbCoor;icoor++ )
      {
         for ( int jcoor = 0;jcoor < nbCoor;jcoor++ )
         {
            fctDer = 0.;
            for ( int j = 0;j < nbGeoNode;j++ )
            {
               fctDer += point( j, icoor ) * dPhiGeo( j, jcoor, ig );
            }
            jacobian( icoor, jcoor, ig ) = fctDer;}}}
            }
```

In the case of quadrilateral and prismatic elements, several of the previous concepts can be extended, by referring e.g. to bilinear or biquadratic mappings. However, guaranteeing that the map is invertible is more difficult: for more details, see [FSV05].

There are cases where it can be convenient to use finite elements of different degree with respect to different coordinates. This is possible using quadrilateral structured grids, where we can construct an element having a biquadratic polynomial on one side, and bilinear polynomials on the remaining sides. In the case of an isoparametric coding of the geometrical mapping, this leads to having, say, quadrilateral elements with three straight sides and one curved side. To this end, we point out that [Hug00, Chap. 4], reports the "incremental" implementation of a quadrilateral element that, starting from a four-node bilinear setting, is enriched by other degrees of freedom up to the biquadratic 9-node element.

8.4.4 Construction of local and global systems

This phase is the core of the construction of the discretization of differential operators. As an example, let us take the code in Program 7, which constructs the discretization of the elliptic differential equation $\mu\triangle u+\sigma u=f$.

The overall operation is articulated in a cycle over all the elements of the mesh aMesh. After setting to zero the elementary matrix and vector, these structures are filled incrementally, first with the discretization of the stiffness (diffusion) operator and then with the mass operator (reaction). The source subroutine handles the local right-hand side vector. The assemb subroutines handle the update of the computation in the global matrix, as previously indicated in Fig. 8.2.

In this phase, to avoid checking whether a degree of freedom is on the boundary using conditional branches within the loop, we ignore boundary conditions.

Program 7 – assemble: Code for assembling the discretization of a diffusion-reaction problem $-\mu\triangle u+\sigma u=f$, where f is denoted by sourceFct

```
Real mu=1., sigma=0.5;
ElemMat elmat(fe.nbNode,1,1);
ElemVec elvec(fe.nbNode,1);
for(UInt i = 1; i<=aMesh.numVolumes(); i++){
  fe.updateFirstDerivQuadPt(aMesh.volumeList(i));
    //<- computes the necessary information for numerical integration

  elmat.zero();
  elvec.zero();
  stiff(mu,elmat,fe);
  mass(sigma,elmat,fe);
  source(sourceFct,elvec,fe,0);
  assemb_mat(A,elmat,fe,dof,0,0);
  assemb_vec(F,elvec,fe,dof,0);
}
```

Let us see in detail a possible implementation of the local computation and of the global update separately.

Computation of the local matrices

Program 8 reports the implementation of the computation of the local matrix of the diffusion operator and of the right-hand side of the linear system.

In particular, we first assemble the diagonal contributions and then the extra-diagonal ones of the local matrix, thus looping over the quadrature nodes. The "core" loop operation is:

```
s += fe.phiDer( iloc, icoor, ig ) * fe.phiDer( jloc, icoor, ig )
             * fe.weightDet( ig )*coef;
```

The instruction

mat(iloc, jloc) += s;

updates the term i, j of the local matrix incrementally: upon call of the ensuing subroutine mass(), the contribution of the reaction operator will be added to the one previously computed.

We proceed in a similar way in source for the computation of the local vector of known terms.

Program 8 – stiff-source: Subroutines for the computation of the second derivative and local-level computation of the right-hand side

```
void stiff( Real coef,
        ElemMat& elmat, const CurrentFE& fe,
        const Dof& dof,
        const ScalUnknown<Vector>& U,Real t)
{
 int iblock=0,jblock=0;
 ElemMat::matrix_view mat = elmat.block( 0,0 ); //initialize local matrix
 int iloc, jloc, i, icoor, ig, iu;
 double s, coef_s, x, y, z;
 ID eleId=fe.currentId();

 // Diagonal elements
 for ( i = 0;i < fe.nbDiag;i++ )
 {
  iloc = fe.patternFirst( i );s = 0;
  for ( ig = 0;ig < fe.nbQuadPt;ig++ ) // numerical integration
  {
   fe.coorQuadPt(x,y,z,ig);// definition of the quadrature formula
   for ( icoor = 0;icoor < fe.nbCoor;icoor++ ) // core of the assembly
    s += fe.phiDer( iloc, icoor, ig ) * fe.phiDer( iloc, icoor, ig )
          * fe.weightDet( ig )*coef(t,x,y,z,uPt);
  }
  mat( iloc, iloc ) += s;
  }
 //Extra-diagonal elements
 for ( i = fe.nbDiag;i < fe.nbDiag + fe.nbUpper;i++ )
  {
   iloc = fe.patternFirst( i );
   jloc = fe.patternSecond( i );s = 0;
   for ( ig = 0;ig < fe.nbQuadPt;ig++ )
    {
     fe.coorQuadPt(x,y,z,ig);
     for ( icoor = 0;icoor < fe.nbCoor;icoor++ )
          s += fe.phiDer( iloc, icoor, ig ) * fe.phiDer( jloc, icoor, ig ) *
             fe.weightDet( ig )*coef;
    }
```

```
  coef_s = s;
  mat( iloc, jloc ) += coef_s; //incremental
  mat( jloc, iloc ) += coef_s; //local matrix update
 // recall that the operator is SYMMETRIC!
  }}

void source( Real (*fct)(Real,Real,Real,Real,Real),
          ElemVec& elvec, const CurrentFE& fe,
          const Dof& dof,
          const ScalUnknown<Vector>& U,Real t)
{
 int iblock=0;
 int i, ig;
 ElemVec::vector_view vec = elvec.block( iblock );
 Real s;
 ID eleId=fe.currentId();
 int iu;

 for ( i = 0;i < fe.nbNode;i++ )
 {
  s = 0.0;
  for ( ig = 0;ig < fe.nbQuadPt;ig++ )
   {
    s += fe.phi( i, ig ) *
       fct(t, fe.quadPt( ig, 0 ),fe.quadPt( ig, 1 ),fe.quadPt( ig, 2 )) *
            fe.weightDet( ig );
     }
  vec( i ) += s; //right hand side computation  }}
```

Update of the global matrix

Program 9 contains the update of the global matrix starting from the local ones. The crucial point is the identification of the position of the nodes that compose the current element, on which we have just computed the local matrix within the global one. This operation is performed by looking up the dof.localToGlobal Tables, which contain this type of operation.

For the update of the right-hand side, we perform a similar operation. Obviously, the additivity of the integral requires the operation to be performed by adding the different contributions: this explains the += in the update of the vector (corresponding to V[ig]=V[ig]+ vec(i) and to the analogous term in M.setmatinc, which stands for *set matrix incrementally*.

Program 9 – assemb: Assembly of the global matrix and of the right-hand side

```
template <typename Matrix, typename DOF>
void
assemb_mat( Matrix& M, ElemMat& elmat, const CurrentFE& fe, const DOF& dof)
```

```
    ElemMat::matrix_view mat = elmat.block(0,0);
    UInt totdof = dof.numTotalDof();
    int i, j, k;
    UInt ig, jg;
    UInt eleId = fe.currentId();
    for ( k = 0 ; k < fe.nbPattern ; k++ )

        i = fe.patternFirst( k );
        j = fe.patternSecond( k );
        ig = dof.localToGlobal( eleId, i + 1 ) - 1;
        jg = dof.localToGlobal( eleId, j + 1 ) - 1;
        M.set_mat_inc( ig, jg, mat( i, j ) );

template <typename DOF, typename Vector, typename ElemVec>
void
assemb_vec( Vector& V, ElemVec& elvec, const CurrentFE& fe, const DOF& dof)

    UInt totdof = dof.numTotalDof();
    typename ElemVec::vector_view vec = elvec.block( iblock );
    int i;
    UInt ig;
    UInt eleId = fe.currentId();
    for ( i = 0 ; i < fe.nbNode ; i++ )

        ig = dof.localToGlobal( eleId, i + 1 ) - 1;
        V[ ig ] += vec( i );
```

8.4.5 Boundary conditions prescription

The need to store sparse matrices efficiently must be compensated by the need to access and manipulate the matrix itself, as we have previously noticed for the CSR format, for instance in the phase of setting the boundary conditions. In a finite element code, the matrix is typically assembled regardless of boundary conditions, so as not to introduce conditional branches within the assembly. Boundary conditions are then introduced by modifying the algebraic system. Imposing Neumann and Robin-type conditions basically translates into the computation of suitable boundary integrals (or, in one-dimensional cases, of values evaluated at the boundary). For instance, Program 10 implements the computation of integrals on the surface for Neumann-type conditions specified in function Bcb. The integral requires a suitable quadrature formula that allows to update the known term b. The structure bdLocalToGlobal allows to transfer the information for each boundary element having Neumann degrees of freedom at the global right-hand side.

Program 10 – BcNaturalManage: Subroutine for handling Neumann-type boundary conditions

```
template <typename VectorType, typename MeshType, typename DataType>

void bcNaturalManage( VectorType& b, const MeshType& mesh, const Dof& dof, \\
const BCBase& BCb, CurrentBdFE& bdfem, const DataType& t )
{
UInt nDofF = bdfem.nbNode;
UInt totalDof = dof.numTotalDof();
UInt nComp = BCb.numberOfComponents();

const IdentifierNatural* pId;
ID ibF, idDof, icDof, gDof;
Real sum;

DataType x, y, z;
// Loop on the type of boundary conditions
 for ( ID i = 1; i <= BCb.list_size(); ++i )
  {
  pId = static_cast< const IdentifierNatural* >( BCb( i ) );
  // Number of current boundary face
   ibF = pId->id();
  // definition of information on the face
   bdfem.updateMeas( mesh.boundaryFace( ibF ) );
  // Loop on degrees of freedom per face
   for ( ID idofF = 1; idofF <= nDofF; ++idofF )
   {
   // Loop on the involved unknown components
   for ( ID j = 1; j <= nComp; ++j )
    {
    //global Dof
    idDof = pId->bdLocalToGlobal( idofF ) + ( BCb.component( j ) - 1 ) * totalDof;
    // Loop on quadrature nodes
    for ( int l = 0; l < bdfem.nbQuadPt; ++l )
    {
     bdfem.coorQuadPt( x, y, z, l ); // quadrature point coordinates
    // Contribution in the known term
     b[ idDof - 1 ] += bdfem.phi( int( idofF - 1 ), l ) * BCb( t, x, y, z, BCb.component( j ) ) *
     bdfem.weightMeas( l );
     }}}}}
```

Handling Dirichlet (essential) boundary conditions is more complex (see Fig. 8.2). There are various strategies for this operation, some of which are treated in [FSV05]. The most coherent approach to what is prescribed by the theory consists in removing the rows and columns referring to the nodes associated to the Dirichlet boundary conditions from the system obtained during assembly, thus correcting the known term by using the values of the Dirichlet datum we want to impose.

In fact, this coincides with the operation of lifting the boundary datum through a piecewise polynomial function of the chosen degree for the finite elements, and whose support is limited to the only layer of elements of the triangulation that face the boundary (see Fig. 4.13 in Sect. 4.5.1)

This way of proceeding has the advantage of reducing the dimension of the problem to the effective number of degrees of freedom, however its practical implementation is problematic. Indeed, while for $1D$ problems, due to the natural ordering of degrees of freedom, the optional rows and columns to be removed are always the first and last one, for multi-dimensional problems the implementation involves eliminating rows and columns whose numbering can be arbitrary, a difficult operation to handle efficiently. It must also be noted that this operation substantially modifies the *pattern* of the matrix, and this can be inconvenient in case we want to share the latter among several matrices in order to save memory. For this reason, we prefer to consider the Dirichlet condition to be imposed at a given node k_D as an equation of the form $u_{k_D} = g_{k_D}$ replacing the k_D-th row of the original system. To avoid modifying the matrix pattern, this substitution must be inserted by annihilating the extra-diagonal row elements, except for the diagonal one, which is set to 1, while the corresponding entry in the right-hand side is set to g_{k_D}.

This operation only requires row-wise access to the matrix, for which the CSR format is particularly efficient.

8.5 Integration in time

Among the different methods to integrate in time, we analyzed the θ method in the previous chapters, and pointed out a number of other methods, in particular BDF (*Backward Difference Formulas*) methods implemented in LifeV. An introduction of these methods can be found in [QSS07]. We here recall some of their basic aspects.

Given the system of ordinary differential equations:

$$M\frac{d\mathbf{u}}{dt} = \mathbf{f} - A\mathbf{u}$$

and the associated initial datum $\mathbf{u}(t=0) = \mathbf{u}_0$, a BDF method is an implicit multi-step method of the form

$$\frac{\alpha_0}{\Delta t} M\mathbf{U}^{n+1} + A\mathbf{U}^{n+1} = \mathbf{f}^{n+1} + \sum_{j=1}^{p} \frac{\alpha_j}{\Delta t}\mathbf{U}^{n+1-j}, \tag{8.8}$$

for suitable $p \geq 1$, where the coefficients are determined so that:

$$\frac{\partial \mathbf{U}}{\partial t}\Big|_{t=t^{n+1}} = \frac{\alpha_0}{\Delta t}\mathbf{U}^{n+1} - \sum_{j=1}^{p} \frac{\alpha_j}{\Delta t}\mathbf{U}^{n+1-j} + \mathcal{O}(\Delta t^p).$$

Table 8.3. Coefficients α_i for the BDF methods ($p = 1,2,3$) and coefficients β_i for time-extrapolation

p	α_0	α_1	α_2	α_3	β_0	β_1	β_2
1	1	1	–	–	1	–	–
2	3/2	2	−1/2	–	2	−1	–
3	11/6	3	−3/2	1/3	3	−3	1

Here, $\Delta t > 0$ is the time-step, $t^n = n\Delta t$, and $\mathbf{U}^n$ stands for $\mathbf{U}$ at time t^n. In Table 8.3 (left) we report the coefficients for $p = 1$ (implicit Euler method), 2, 3.

If the matrix A is a function of $\mathbf{u}$, that is when problem (8.8) is nonlinear, BDF methods, being implicit, can be very costly, for they require at each time step the solution of the nonlinear algebraic system in $\mathbf{U}^{n+1}$

$$\frac{\alpha_0}{\Delta t} M\mathbf{U}^{n+1} + A(\mathbf{U}^{n+1})\mathbf{U}^{n+1} = \mathbf{f}^{n+1} + \sum_{j=1}^{p} \frac{\alpha_j}{\Delta t}\mathbf{U}^{n+1-j}.$$

A possible trade-off that significantly reduces computational costs, without switching to a completely explicit method (whose stability properties can in general be unsatisfactory), is to solve the linear system

$$\frac{\alpha_0}{\Delta t} M\mathbf{U}^{n+1} + A(\mathbf{U}^{*})\mathbf{U}^{n+1} = \mathbf{f}^{n+1} + \sum_{j=1}^{p} \frac{\alpha_j}{\Delta t}\mathbf{U}^{n+1-j}$$

where $\mathbf{U}^*$ approximates $\mathbf{U}^{n+1}$ using the solutions known from the previous steps. We basically set

$$\mathbf{U}^* = \sum_{j=0}^{p} \beta_j \mathbf{U}^{n-j} = \mathbf{U}^{n+1} + \mathcal{O}(\Delta t^p),$$

for suitable "extrapolation" coefficients β_j. The objective is to reduce the computational costs without dramatically reducing neither the region of absolute stability of the implicit scheme nor the overall accuracy of the time-advancing scheme. Table 8.3 reports the coefficients β_j.

The coding of a BDF time integrator can at this point be performed using a dedicated class, reported in Program 11, whose members are:

1. the indicator of the order p which also states the dimension of the vectors $\boldsymbol{\alpha}$ and $\boldsymbol{\beta}$;
2. the vectors $\boldsymbol{\alpha}$ and $\boldsymbol{\beta}$;
3. the unknowns matrix given by aligning the vectors $\mathbf{U}^n, \mathbf{U}^{n-1}, \ldots \mathbf{U}^{n+1-p}$. The size of each vector, i.e. the number of rows of such matrix (which has p columns) is stored in the size index.

Having assembled the matrices A and M, the time-advancing scheme will be performed by computing the matrix $\frac{\alpha_0}{\Delta t}M + A$, the right-hand side $\mathbf{f}^{n+1} + \sum_{j=1}^{p} \frac{\alpha_j}{\Delta t}\mathbf{U}^{n+1-j}$ and solving system (8.8). In particular, in the implementation presented in Program 11, the function time der computes the term $\sum_{j=1}^{p} \frac{\alpha_j}{\Delta t}\mathbf{U}^{n+1-j}$ by accessing the vector $\boldsymbol{\alpha}$ and the unknowns matrix. In case the problem is nonlinear, we can access to the vector $\boldsymbol{\beta}$ via the function extrap().

Having computed the solution at the new time step, the unknowns matrix has to "make room for it", by shifting all of its columns to the right, so that the first column is the solution just computed. This operation is performed by the function shift right, which basically copies the next-to-last column of unknowns into the last one, the third from the bottom into the second-last one and so on until the solution computed is fully stored.

Program 11 – Bdf: Base class for costructing Bdf time integration methods

```
class Bdf
{
public:
   Bdf( const UInt p );
   ~Bdf();
   void initialize_unk( Vector u0 );
   void shift_right( Vector const& u_curr );

   Vector time_der( Real dt ) const;
   Vector extrap() const;
   double coeff_der( UInt i ) const;
   double coeff_ext( UInt i ) const;
   const std::vector<Vector>& unk() const;
   void showMe() const;

private:
   UInt _M_order;
   UInt _M_size;
   Vector _M_alpha;
   Vector _M_beta;
   std::vector<Vector> _M_unknowns;
};


Bdf::Bdf( const UInt p )
   :
   _M_order( p ),
   _M_size( 0 ),
   _M_alpha( p + 1 ),
   _M_beta( p )
{
   if ( n <= 0 || n > BDF_MAX_ORDER )
   {
  // Error handling for requesting a wrong or non-implemented order
  ....
     }
   switch ( p )
   {
```

```
case 1:
   _M_alpha[ 0 ] = 1.; // implicit Euler
   _M_alpha[ 1 ] = 1.;
   _M_beta[ 0 ] = 1.; // u at time n+1 approximated by u at time n
   break;
case 2:
   _M_alpha[ 0 ] = 3. / 2.;
   _M_alpha[ 1 ] = 2.;
   _M_alpha[ 2 ] = -1. / 2.;
   _M_beta[ 0 ] = 2.;
   _M_beta[ 1 ] = -1.;
   break;
case 3:
   _M_alpha[ 0 ] = 11. / 6.;
   _M_alpha[ 1 ] = 3.;
   _M_alpha[ 2 ] = -3. / 2.;
   _M_alpha[ 3 ] = 1. / 3.;
   _M_beta[ 0 ] = 3.;
   _M_beta[ 1 ] = -3.;
   _M_beta[ 2 ] = 1.;
   break;
}
_M_unknowns.resize( p ); //number of columns of matrix _M_unknowns}
```

8.6 A complete example

We conclude this chapter with the listing of a program written for the solution of the parabolic diffusion-reaction problem:

$$\begin{cases} \dfrac{\partial u}{\partial t} - \mu(t)\Delta u + \sigma(t)u = f, & \mathbf{x} \in \Omega, \quad 0 < t \le 10, \\ u = g_1, & \mathbf{x} \in \Gamma_{10} \cup \Gamma_{11}, \quad 0 < t \le 10, \\ u = g_2, & \mathbf{x} \in \Gamma_{20} \cup \Gamma_{21}, \quad 0 < t \le 10, \\ \nabla u \cdot \mathbf{n} = 0, & \mathbf{x} \in \Gamma_{50}, \quad 0 < t \le 10, \\ u = u_0, & \mathbf{x} \in \Omega, \quad t = 0, \end{cases}$$

where Ω is a cubic domain and $\partial\Omega = \Gamma_{10} \cup \Gamma_{11} \cup \Gamma_{20} \cup \Gamma_{21} \cup \Gamma_{50}$. Precisely, the numerical codes on the various boundary portions are:

$$\begin{aligned} &\Gamma_{20} : x = 0,\ 0 < y < 1,\ 0 < z < 1; \\ &\Gamma_{21} : x = 0,\ (y = 0,\ 0 < z < 1) \cup (y = 1,\ 0 < z < 1) \\ &\qquad\qquad \cup (z = 0,\ 0 < y < 1) \cup (z = 0,\ 0 < y < 1); \\ &\Gamma_{10} : x = 1,\ 0 < y < 1,\ 0 < z < 1; \\ &\Gamma_{11} : x = 1,\ (y = 0,\ 0 < z < 1) \cup (y = 1,\ 0 < z < 1) \\ &\qquad\qquad \cup (z = 0,\ 0 < y < 1) \cup (z = 0,\ 0 < y < 1); \\ &\Gamma_{50} : \partial\Omega \backslash \{\Gamma_{20} \cup \Gamma_{21} \cup \Gamma_{10} \cup \Gamma_{11} \cup \Gamma_{50}\}. \end{aligned}$$

In particular, $\mu(t) = t^2$, $\sigma(t) = 2$, $g_1(x,y,z,t) = g_2(x,y,z,t) = t^2 + x^2$, $u_0(x,y,z) = 0$, $f = 2t + 2x^2$. The exact solution is precisely $t^2 + x^2$ and the test is made on a cubic grid of 6007 elements with quadratic affine tetrahedra, for a total of 9247 degrees of freedom. The time step is $\Delta t = 0.5$, the order of the BDF scheme is 3.

Program 12 contains the main program for this example (originally based on the library LifeV) and has been enriched by comments to help the reading, although obviously not everything will be immediately clear just by reading the preceding sections. Coherently with the spirit with which this chapter has been designed, we invite the reader to try to write her/his own code or run the tutorials of libraries such as LifeV, FEniCS or OpenFOAM.

Program 12 – main.cpp: Solution of a parabolic problem on a cubic domain

```
int main() {
  using namespace std;
  {

    // ==================================================
    // Definition  of the boundary conditions (associated with a file main.hpp)
    // ==================================================

    BCFunctionBase gv1(g1); // Function g1
    BCFunctionBase gv2(g2); // Function g2
    BCHandler BCh(2); // Two boundary conditions are imposed
    // To the two conditions, we associate the numerical codes 10 and 20
    // contained in the computational grid
    BCh.addBC("Dirichlet1",  10, Essential, Scalar, gv1);
    BCh.addBC("Dirichlet2",  20, Essential, Scalar, gv2);

    // ==================================================
    // Information on the geometric mapping and on the numerical integration
    // ==================================================

    const GeoMap& geoMap  = geoLinearTetra;
    const QuadRule& qr   = quadRuleTetra64pt;

    const GeoMap& geoMapBd = geoLinearTria;
    const QuadRule& qrBd  = quadRuleTria3pt;

    //P2 elements
    const RefFE& refFE   = feTetraP2;
    const RefFE& refBdFE  = feTriaP2;

    // ================================
    // Structure of the mesh
    // ================================

    RegionMesh3D<LinearTetra> aMesh;
```

```
GetPot datafile( "data" ); //information on the mesh file
// and other information is contained in a file named "data"
long int  m=1;
std::string mesh_type = datafile( "mesh_type", "INRIA" );
string mesh_dir = datafile( "mesh_dir", "." );
string fname=mesh_dir+datafile( "mesh_file", "cube_6007.mesh" );

 readMppFile(aMesh,fname,m); // grid reading

aMesh.updateElementEdges();
aMesh.updateElementFaces();
aMesh.showMe();

// ==========================================
// Definition of the current finite element, equipped with
// geometric mapping and quadrature rule
// ==========================================

CurrentFE fe(refFE,geoMap,qr);
CurrentBdFE feBd(refBdFE,geoMapBd,qrBd);

// ==========================================
// Definition of the degrees of freedom (DOF) of the problem
// and of the specific boundary conditions
// ==========================================

Dof dof(refFE);
dof.update(aMesh);
BCh.bdUpdate( aMesh,  feBd, dof );
UInt dim = dof.numTotalDof();
dof.showMe();

// ================================
// Initialization of the unknown vectors
// U and of known term F
// ================================

ScalUnknown<Vector> U(dim), F(dim);
U=ZeroVector( dim );
F=ZeroVector( dim );

// ==============================================
// Definition of the parameters for the integration in time
// always specified in "data" and read from there
// ==============================================

Real Tfin = datafile( "bdf/endtime", 10.0 );
Real delta_t = datafile( "bdf/timestep", 0.5 );
```

```
Real t0 = 0.;
UInt ord_bdf = datafile( "bdf/order", 3 );;
Bdf bdf(ord_bdf);
Real coeff=bdf.coeff_der(0)/delta_t;

bdf.showMe();

// ===========================================================
// Construction of the pattern and of the time-independent matrices
// ===========================================================
//  pattern for stiff operator

CSRPatt pattA(dof);

CSRMatr<double> A(pattA);
CSRMatr<double> M(pattA);
M.zeros();

cout << "*** Matrix computation          : "<<endl;
chrono.start();
//
SourceFct sourceFct;
ElemMat elmat(fe.nbNode,1,1);
ElemVec elvec(fe.nbNode,1);
for(UInt i = 1; i<=aMesh.numVolumes(); i++){
   fe.updateJacQuadPt(aMesh.volumeList(i));
   elmat.zero();
   mass(1.,elmat,fe);
   assemb_mat(M,elmat,fe,dof,0,0); // Mass matrix M
}


// ========================================
// TIME LOOP
// ========================================

int count=0;
bdf.initialize_unk(u0,aMesh,refFE,fe,dof,t0,delta_t,1);

for (Real t=t0+delta_t;t<=Tfin;t+=delta_t)
{
   A.zeros();
   F=ZeroVector( F.size() );
   // ========================================
   // Assembly and
   // Update of the known term with the solution of
   // the preceding steps
   // ========================================
```

```
Real visc=nu(t);// mu and sigma depend on time
Real s=sigma(t);
for(UInt i = 1; i<=aMesh.numVolumes(); i++){
   fe.updateFirstDerivQuadPt(aMesh.volumeList(i));
   elmat.zero();
   elvec.zero();
   mass(coeff+s,elmat,fe);
   stiff(visc,elmat,fe);
   source(sourceFct,elvec,fe,t,0);
   assemb_mat(A,elmat,fe,dof,0,0);
   assemb_vec(F,elvec,fe,dof,0);
}

// Handling of the right hand side
F += M*bdf.time_der(delta_t);

// ========================================
// Prescription of the boundary conditions
// ========================================

chrono.start();
bcManage(A,F,aMesh,dof,BCh,feBd,1.,t);

chrono.stop();
chrono.start();
Linear_solve(U.giveVec(), F.giveVec(), options, params, NULL,
      (int *)pattA.giveRaw_bindx(), NULL, NULL, NULL,
      A.giveRaw_value(), data_org,
      status, proc_config);

// ======================================
// Writing of the post-processing file
// ======================================

count++;
index << count;
wr_medit_ascii_scalar( "U" + index.str() + ".bb", U.giveVec(), dim );
wr_medit_ascii( "U" + index.str() + ".mesh", aMesh);

// ==================================================
// In this test case we know the analytic solution
// (specified in main.hpp)
// and we want to compute the errors in different norms
// ==================================================

AnalyticalSol analyticSol;

Real normL2=0., normL2diff=0., normL2sol=0.;
Real normH1=0., normH1diff=0., normH1sol=0.;
```

```
      for(UInt i=1; i<=aMesh.numVolumes(); ++i){
        fe.updateFirstDeriv(aMesh.volumeList(i));

      normL2    += elem_L2_2(U,fe,dof);
      normL2sol  += elem_L2_2(analyticSol,fe,t,( UInt )U.nbcomp());
      normL2diff += elem_L2_diff_2(U,analyticSol,fe, dof, t,( UInt )U.nbcomp());

      normH1    += elem_H1_2(U,fe,dof);
      normH1sol  += elem_H1_2(analyticSol,fe,t,U.nbcomp());
      normH1diff += elem_H1_diff_2(U,analyticSol,fe,dof,t,U.nbcomp());
      }

      normL2    = sqrt(normL2);
      normL2sol  = sqrt(normL2sol);
      normL2diff = sqrt(normL2diff);

      normH1    = sqrt(normH1);
      normH1sol  = sqrt(normH1sol);
      normH1diff = sqrt(normH1diff);

      bdf.shift_right(U);


    } // END OF TIME LOOP
  }
  return EXIT_SUCCESS;
}
```

This is what we have obtained after running the code

$||U||_{L^2} = 0.655108$
$||sol||_{L^2} = 0.655108$
$||U - sol||_{L^2} = 1.49398\text{e-}09$
$||U - sol||_{L^2}/||sol||_{L^2} = 2.28051\text{e-}09$
$||U||_{H^1} = 1.32759$
$||sol||_{H^1} = 1.32759$
$||U - sol||_{H^1} = 8.09782\text{e-}09$
$||U - sol||_{H^1}/||sol||_{H^1} = 6.09963\text{e-}09$

Note that the errors are to be attributed exclusively to the linear system's solution: as the exact solution is a parabolic function in time and space, the choice of finite elements of degree 2 and of the BDF scheme of order 3 guarantees that the discretization errors are non-null. Fig. 8.8 illustrates the results visualized by Medit.

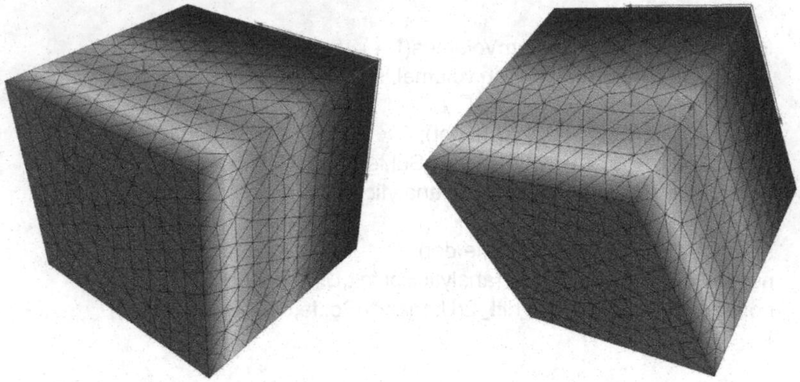

Fig. 8.8. Results of the simulation after 5 (left) and 20 (right) time steps

9

The finite volume method

The *finite volume* method is a very popular method for the space discretization of partial differential problems in conservation form. For an in-depth presentation of the method, we suggest the monographs [LeV02a] and [Wes01].
As a paradigm to describe the method and illustrate its main features, let us consider the following scalar equation

$$\frac{\partial u}{\partial t} + \mathrm{div}(\mathbf{F}(u)) = s(u), \quad \mathbf{x} \in \Omega,\ t > 0 \tag{9.1}$$

where $u : (\mathbf{x}, t) \to \mathbb{R}$ denotes the unknown, $\mathbf{x} \in \Omega \subset \mathbb{R}^d$ $(d = 1,2,3)$, $\mathbf{F}$ is a given vector function, linear or nonlinear, called flux, s is a given source function. If the flux $\mathbf{F}$ contains terms depending on the first derivatives of u, the differential problem is a second-order one. The differential equation (9.1) must be completed by the initial condition $u(\mathbf{x}, 0) = u_0(\mathbf{x})$, $\mathbf{x} \in \Omega$ for $t = 0$, as well as by suitable boundary conditions, on the whole boundary $\partial\Omega$ if problem (9.1) is a second-order one, or just on a subset $\partial\Omega^{in}$ of $\partial\Omega$ (the inflow boundary) in the case of first-order problems. As we will see in Chapter 15 (see Sect. 15.1 and Sect. 15.4), this type of differential equations are called *conservation laws*. Typically, the finite volume method operates on equations written in conservation form such as (9.1).

The diffusion-transport equations that will be addressed in Chapter 12, the pure transport equations of Chapters 13–15, and the parabolic ones examined in Chapter 5, can all be considered as special cases of (9.1). Indeed, all partial differential equations deriving from physical conservation laws can be expressed in conservation form.

With some additional effort, we can obviously consider the vector case, where the unknown $\mathbf{u}$ and the source $\mathbf{s}$ are vector functions with p components, while the flux $\mathbf{F}$ is now a tensor with dimension $p \times d$. In particular, also the Navier-Stokes equations and the Euler equations for compressible flows that will be considered in Sect. 15.4 can be rewritten in conservative form. A finite volume approximation of free-surface incompressible flows for real life applications will be discussed in Sect. 16.11.

A. Quarteroni: *Numerical Models for Differential Problems*, 2nd Ed.
MS&A – Modeling, Simulation & Applications 8
DOI 10.1007/978-88-470-5522-3_9, © Springer-Verlag Italia 2014

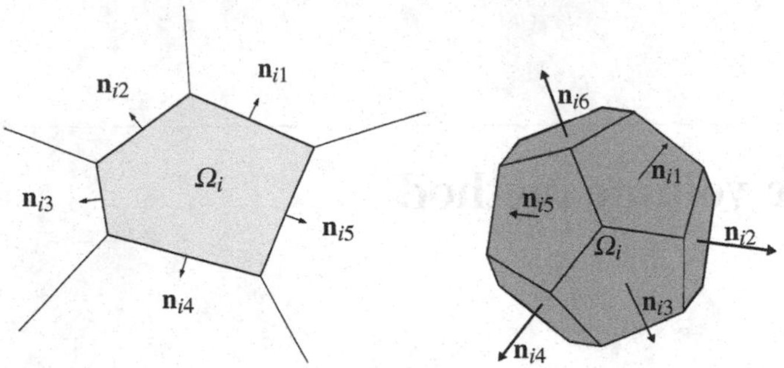

Fig. 9.1. A control volume in 2D (left) and 3D (right)

9.1 Some basic principles

The preliminary step towards a finite volume discretization of (9.1) consists in identifying a set of polyhedra $\Omega_i \subset \Omega$ with diameter less than h, called *control volumes* (or *control cells*), $i = 1,\ldots,M$, such that $\cup_i \overline{\Omega}_i = \overline{\Omega}$ (we will assume for simplicity that the domain Ω is polygonal, otherwise $\cup_i \overline{\Omega}_i$ will be its approximation). See Fig. 9.1 for an example of control volume. We will furthermore suppose the cells to be pairwise disjoint, this being the most commonly used case, although such restriction is not required, in principle, by the method.

Equation (9.1) is integrated on each Ω_i; using the divergence theorem we obtain the system of ordinary differential equations

$$\frac{\partial}{\partial t}\int\limits_{\Omega_i} u\, d\Omega + \int\limits_{\partial\Omega_i} \mathbf{F}(u)\cdot\mathbf{n}_i\, d\gamma = \int\limits_{\Omega_i} s(u)\, d\Omega, \quad i = 1,\ldots,M. \tag{9.2}$$

We have denoted by $\mathbf{n}_i$ the unit outward normal of $\partial\Omega_i$. In two dimensions, let us denote by L_i the number of straight sides of Ω_i (in Fig. 9.1 $L_i = 5$) and by $\mathbf{n}_{ij}$, $j = 1,\ldots,L_i$, the (constant) outward unit normal vector to the side l_{ij} of $\partial\Omega_i$. Then (9.2) can be rewritten as

$$\frac{\partial}{\partial t}\int\limits_{\Omega_i} u\, d\Omega + \sum_{j=1}^{L_i}\int\limits_{l_{ij}} \mathbf{F}(u)\cdot\mathbf{n}_{ij}\, d\gamma = \int\limits_{\Omega_i} s(u)\, d\Omega, \quad i = 1,\ldots,M. \tag{9.3}$$

Several issues have to be addressed:

- which geometrical shape should the control volumes have;
- how to represent the unknown u in each control volume, that is which are its degrees of freedom and where should they be placed;
- how to approximate the (volume and surface) integrals;
- how to represent the flux $\mathbf{F}(u)$ on each side, as a function of the values of the unknown u on the control volumes adjacent to the side.

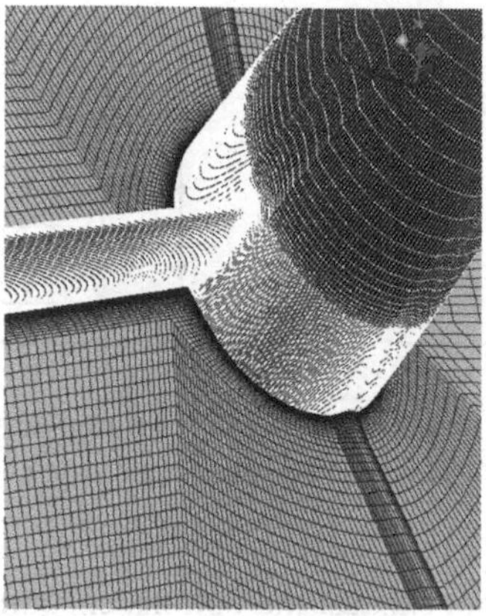

Fig. 9.2. An example of blockwise structured mesh

For the construction of the control volumes, we usually start from a triangulation $\mathcal{T}_h$ of the domain into polygons called elements, say $\{K_m\}$, of the same kind. Typically these are triangles or quadrilaterals in 2D, tetrahedra or hexahedra in 3D, as we saw in Chapter 4 when using finite elements. The grid can be structured, blockwise-structured (with either disjoint or overlapping blocks), or unstructured. Structured grids are bounded to domains of relatively simple shape, in order for the whole domain, or each block in which it is subdivided, to be mapped to a rectangle or a cube. In Fig. 9.2 we display a block structured grid on the surface of the appendages of a yacht.

Once the domain has been triangulated, we have two possibilities.
In the so-called *cell-centered* method, the elements $\{K_m\}$ of $\mathcal{T}_h$ directly serve as control volumes. Consequently, the unknowns are associated to an internal point on each element, typically the barycenter, called *node*. However, this apparently natural choice of control volumes has a disadvantage: as there are no nodes lying on the boundary of Ω, imposing the boundary conditions will require special actions, which we will examine later on. To account for such inconvenient, we can construct control volumes around the elements of $\mathcal{T}_h$, where we will place the unknowns. This yields to the so-called *vertex-centered* schemes.
Sometimes, in multifield problems with several unknowns, both techniques are used at the same time to place different unknowns at different nodes. In this case, we will say that *staggered grids* are used; we will present a remarkable example in Sect. 16.11, devoted to the discretization of Navier-Stokes equations.

A basic example on a structured quadrangular grid is reported in Fig. 9.3, where we also show the control volumes for *cell-centered* and *vertex-centered* schemes. The latter are defined by the squares

$$\Omega_i^V = \{\mathbf{x} \in \Omega \ : \ \|\mathbf{x} - \mathbf{x}_i\|_\infty < h/2\}, \quad \Omega_i = \Omega_i^V \cap \Omega,$$

where $\{\mathbf{x}_i\}$ are the vertices of the squares $\{K_m\}$ of the initial grid $\mathcal{T}_h$, which coincide in this case with the nodes of the control volumes, and h is the uniform length of the element edges.

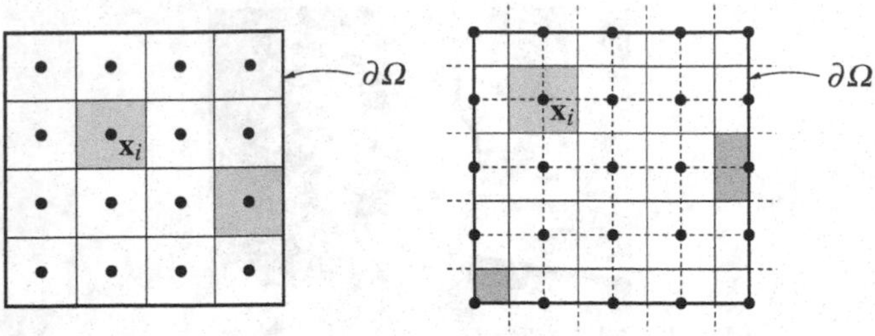

Fig. 9.3. Control volumes (in grey) generated by a partition of a square domain Ω with square finite elements of edge h. Left: *cell-centered* case. Right: *vertex-centered* case

These two choices do not exhaust the options encountered in the practice. Sometimes the variables are placed on each edge (or face, in 3D) of the grid $\mathcal{T}_h$, and the corresponding control volume is formed by the elements of $\mathcal{T}_h$ adjacent to the edge (or face).

In general terms, a finite volume approach is simple to implement: the discretization cells can be chosen in a very general form, the solution is typically assumed to be a constant function in each control volume, the Neumann boundary conditions are imposed in a natural way, and the very formulation of the problem expresses the local conservation of the amount $\int_{\Omega_i} u \, d\Omega$. The potential drawbacks are the objective difficulty in drawing high-order schemes, the need to treat essential (Dirichlet) boundary conditions, in particular for the cell-centered methods; finally, the mathematical analysis is less simple than in the case of Galerkin methods, as a direct application of variational techniques used for the former is not straightforward.

9.2 Construction of control volumes for vertex-centered schemes

In the case the original triangulation $\mathcal{T}_h$ is made of triangular unstructured elements in 2D or tetrahedric ones in 3D, the construction of control volumes around the vertices of $\mathcal{T}_h$ is not straightforward. In principle, we could choose as control volume Ω_i the set of all elements containing the vertex $\mathbf{x}_i$. However, this would generate control volumes with non-null intersection, a permitted-but not desirable-situation.
We can thus take advantage of some geometrical concepts. Let us consider for example a bounded polygonal domain $\Omega \subset \mathbb{R}^2$, and let $\{\mathbf{x}_i\}_{i \in \mathcal{P}}$ be a set of points, which we will call nodes, of $\overline{\Omega}$. Here $\mathcal{P}$ denotes a set of indexes. These points are typically the ones where we intend to provide an approximation of the solution u. We associate to each node the polygon

$$\Omega_i^V = \{\mathbf{x} \in \mathbb{R}^2 \,:\, |\mathbf{x} - \mathbf{x}_i| < |\mathbf{x} - \mathbf{x}_j| \;\; \forall j \in \mathcal{P}, \, j \neq i\}, \tag{9.4}$$

with $i \in \mathcal{P}$. The set $\{\Omega_i^V, i \in \mathcal{P}\}$ is called *Voronoi diagram*, or *Voronoi tessellation*, associated to the set of points $\{\mathbf{x}_i\}_{i \in \mathcal{P}}$; Ω_i^V is called i-th *Voronoi polygon*. For an example see Fig. 9.4. The polygons thus obtained are convex, but not necessarily bounded (con-

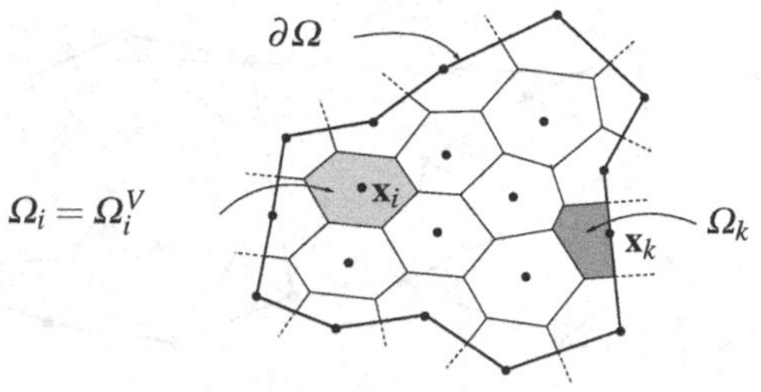

Fig. 9.4. A Voronoi diagram

sider for instance the ones adjacent to the boundary). Their vertices are called *Voronoi vertices*; a vertex is said *regular* when it is the meeting point of three Voronoi polygons, and *degenerate* when it is shared by at least four polygons. A Voronoi diagram with only regular vertices is in turn called *regular*.

At this point, we can define the control volumes Ω_i introduced in the previous section as

$$\Omega_i = \Omega_i^V \cap \Omega, \quad i \in \mathcal{P}. \tag{9.5}$$

For each $i \in \mathcal{P}$, we denote by $\mathcal{P}_i$ the set of indexes of the nodes adjacent to $\mathbf{x}_i$, i.e.

$$\mathcal{P}_i = \{j \in \mathcal{P} \backslash \{i\} \ : \ \partial\Omega_i \cap \partial\Omega_j \neq \emptyset\}.$$

Moreover, we denote by $l_{ij} = \partial\Omega_i \cap \partial\Omega_j$, $j \in \mathcal{P}_i$, a side of the boundary of Ω_i shared by an adjacent control volume, and by m_{ij} its length. If the Voronoi diagram is regular, we have $m_{ij} > 0$. In this case, if we connect each node $\mathbf{x}_i$ with the nodes of $\mathcal{P}_i$, we obtain a triangulation of Ω coinciding with the Delaunay triangulation (see Sect. 6.4.1) of the convex hull of the nodes. In case there are degenerate vertices in the Voronoi tessellation, from this procedure we can still obtain a Delaunay triangulation, provided we triangulate suitably the polygons Ω_i constructed around the degenerate vertices. Clearly, if Ω is convex, the above-described process directly provides a Delaunay triangulation of Ω, see e.g. Fig. 9.5. The inverse procedure is also possible, noting that the vertices of the Voronoi diagram correspond to the centres of the circles circumscribed to the triangles (the circumcenters) of the corresponding Delaunay triangulation. The

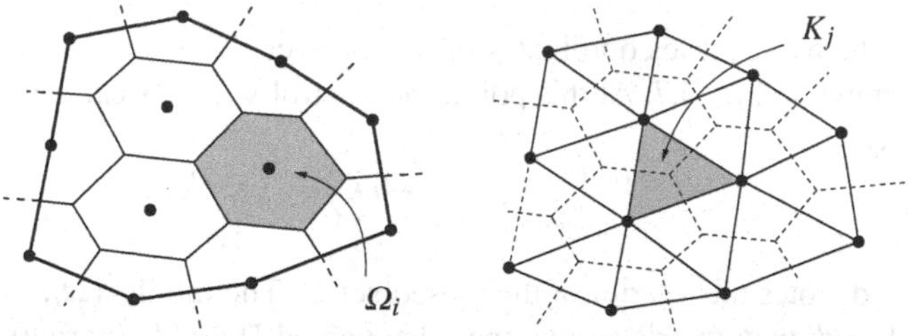

Fig. 9.5. Delaunay triangulation (right) obtained from a Voronoi diagram (left). The dots indicate the nodes $\{\mathbf{x}_i\}_{i\in\mathcal{P}}$

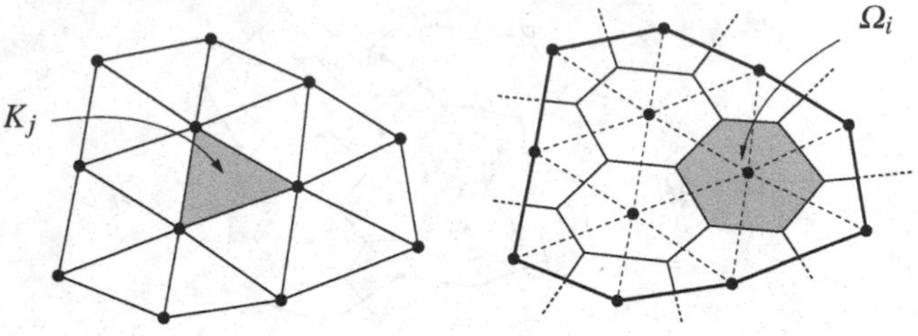

Fig. 9.6. Voronoi diagram (right) obtained starting from a Delaunay triangulation (left)

triangle axes thus form the sides of the tessellation. The latter therefore represents a possible set of control volumes associated to a given Delaunay triangulation (see e.g. Fig. 9.6).

The Voronoi diagram and the Delaunay triangulation are dual to one another: Voronoi vertices correspond one-to-one to elements (triangles) of the Delaunay triangulation, and, conversely, Delaunay vertices correspond to the polygons of the tessellation, hence to the nodes.

There are two interesting properties which are worth highlighting. The first one is that the center of the circumscribed circle to an acute triangle K lies within the closure of K. Hence if the Delaunay triangulation has no obtuse angles, the vertices of the corresponding Voronoi diagram are all contained in $\overline{\Omega}$. The second is that if we denote by $\mathbf{v}_i$, $i = 1,2,3$, the vertices of the non-obtuse triangle K, and by $\Omega_{i,K} = \Omega_i \cap K$ the portion of the control volume Ω_i included in K, then we have the following inequalities between the measures of K and $\Omega_{i,K}$

$$\frac{1}{4}|K| \leq |\Omega_{i,K}| \leq \frac{1}{2}|K|, \quad i = 1,2,3. \tag{9.6}$$

An alternative to the construction based on the Voronoi diagram, which does not require a Delaunay triangulation, consists in starting from a triangulation $\mathcal{T}_h$ of Ω formed by any kind of triangles including obtuse ones. If K is the generic triangle of $\mathcal{T}_h$ with vertices $\mathbf{v}_i$, $i = 1,2,3$, we now define

$$\Omega_{i,K} = \{\mathbf{x} \in K \,:\, \lambda_j(\mathbf{x}) < \lambda_i(\mathbf{x}),\ j \neq i\}$$

where λ_j are the barycentric coordinates of K (see Sect. 4.4.3 for their definition). An example is shown in Fig. 9.7. At this point, the control volumes can be defined in the following way

$$\Omega_i = \mathrm{int}\Big(\bigcup_{\{K \,:\, \mathbf{v}_i \in \partial K\}} \overline{\Omega}_{i,K} \Big), \quad i \in \mathcal{P},$$

where $\mathrm{int}(\mathcal{D})$ denotes the interior of the closed set $\mathcal{D}$. The family $\{\Omega_i,\ i \in \mathcal{P}\}$ defines the so-called *median dual grid* (sometimes also named *Donald diagram*). See Fig. 9.8 for an example. Consequently, we can define the quantities l_{ij}, m_{ij} and $\mathcal{P}_i$ as for the Voronoi diagram. Now the elements l_{ij} are not necessarily straight segments.

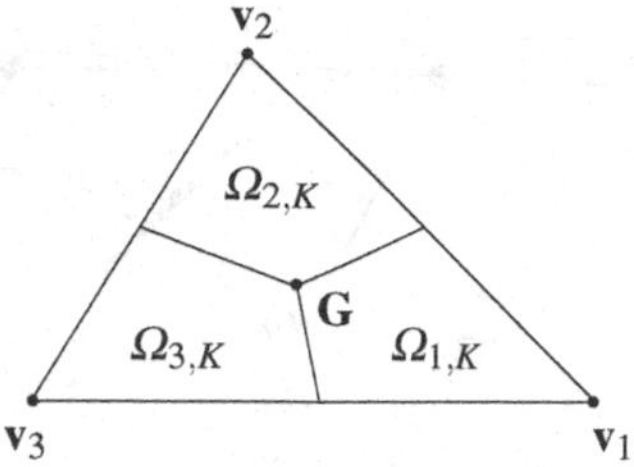

Fig. 9.7. A triangle K, its center of gravity $\mathbf{G} = \frac{1}{3}(\mathbf{v}_1 + \mathbf{v}_2 + \mathbf{v}_3)$, and the polygons $\Omega_{i,K}$

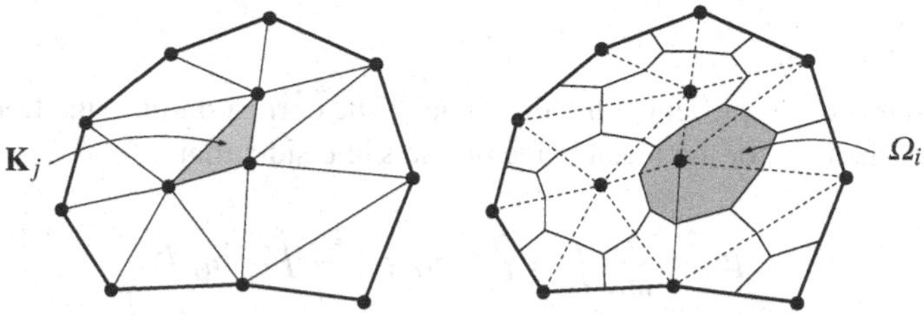

Fig. 9.8. Triangulation of the domain (left) and median dual grid, or Donald diagram (right)

9.3 Discretization of a diffusion-transport-reaction problem

Let us consider for the sake of an example equation (9.1) where

$$\mathbf{F}(u) = -\mu \nabla u + \mathbf{b}\, u, \quad s(u) = f - \sigma\, u. \tag{9.7}$$

This is a time-dependent diffusion-transport-reaction equation written in conservation form, similar to the one described at the beginning of Chapter 12. The functions f, μ, σ and $\mathbf{b}$ are given, and fulfill the hypotheses made at the beginning of Chapter 12. As in the case of problem (12.1), we will suppose for simplicity that u satisfies a homogeneous Dirichlet boundary condition, $u = 0$ on $\partial\Omega$. Let us suppose that Ω is partitioned by a Voronoi diagram and consider the corresponding Delaunay triangulation (an instance is provided in Fig. 9.5). What follows can in fact be extended to other types of finite volumes; for that, it will be sufficient to consider the set of the inner indexes only, $\mathcal{P}_{int} = \{i \in \mathcal{P} \,:\, \mathbf{x}_i \in \Omega\}$, because u vanishes on the boundary. Integrating the assigned equation on the control volume Ω_i as we did in (9.3) and using the divergence theorem, we find

$$\frac{\partial}{\partial t} \int_{\Omega_i} u\, d\Omega + \sum_{j=1}^{L_i} \int_{l_{ij}} \Big(-\mu \frac{\partial u}{\partial \mathbf{n}_{ij}} + \mathbf{b} \cdot \mathbf{n}_{ij}\, u \Big)\, d\gamma = \int_{\Omega_i} \Big(f - \sigma\, u \Big)\, d\Omega, \quad i \in \mathcal{P}_{int}. \tag{9.8}$$

In order to approximate the line integrals, a typical strategy consists in approximating the functions μ and $\mathbf{b} \cdot \mathbf{n}_{ij}$ using piecewise constants, and precisely

$$\mu\big|_{l_{ij}} \simeq \mu_{ij} = \text{const} > 0, \quad \mathbf{b} \cdot \mathbf{n}_{ij}\big|_{l_{ij}} \simeq b_{ij} = \text{const}. \tag{9.9}$$

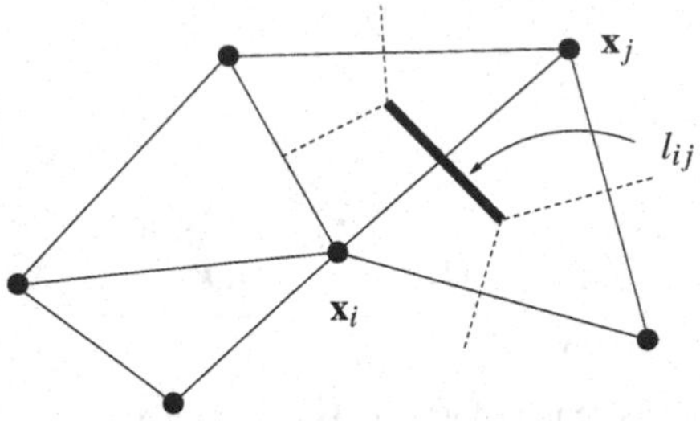

Fig. 9.9. The segment l_{ij}

Such constants can represent either the value of the corresponding function at the midpoint of segment l_{ij}, or the mean value on the same side, that is

$$\mu_{ij} = \frac{1}{m_{ij}} \int_{l_{ij}} \mu \, d\gamma, \quad b_{ij} = \frac{1}{m_{ij}} \int_{l_{ij}} \mathbf{b} \cdot \mathbf{n}_{ij} \, d\gamma.$$

As far as the normal derivatives are concerned, an option consists in approximating them using incremental ratios of the type

$$\frac{\partial u}{\partial \mathbf{n}_{ij}} \simeq \frac{u(\mathbf{x}_j) - u(\mathbf{x}_i)}{|\mathbf{x}_j - \mathbf{x}_i|}$$

(see e.g. Fig. 9.9). This formula is exact if u is linear on the segment connecting $\mathbf{x}_i$ and $\mathbf{x}_j$. Finally, regarding the approximation of the integral of u on l_{ij}, we replace $u\big|_{l_{ij}}$ by a constant obtained by a linear convex combination, that is

$$u\big|_{l_{ij}} \simeq \rho_{ij} \, u(\mathbf{x}_i) + (1 - \rho_{ij}) \, u(\mathbf{x}_j),$$

with $\rho_{ij} \in [0, 1]$ a parameter to be defined. Using the previous approximations, and denoting by u_i the approximation of the unknown value $u(\mathbf{x}_i)$, we can derive from (9.8) the following approximate equations

$$\begin{aligned} m_i \frac{du_i}{dt} & + \sum_{j=1}^{L_i} m_{ij} \{ -\mu_{ij} \frac{u_j - u_i}{\delta_{ij}} + b_{ij} [\rho_{ij} \, u_i + (1 - \rho_{ij}) \, u_j] \} \\ & + m_i \, \sigma_i \, u_i = m_i \, f_i, \quad i \in \mathcal{P}, \end{aligned} \tag{9.10}$$

having denoted by m_i the measure of Ω_i, by σ_i and f_i the values of σ and f at $\mathbf{x}_i$ and by δ_{ij} the distance between $\mathbf{x}_i$ and $\mathbf{x}_j$. Note that (9.10) can be written in the form

$$m_i \frac{du_i}{dt} + \sum_{j=1}^{L_i} m_{ij} H_{ij}(u_i, u_j) + m_i \, \sigma_i \, u_i = m_i \, f_i, \tag{9.11}$$

where H_{ij} is the so-called *numerical flux* representing the contribution of the approximation of the flux through the side l_{ij}. The concept of numerical flux is relevant also in the context of finite difference schemes for hyperbolic equations, as we will see in Chapters 13 and 15. Some of the features of the numerical flux also translate into scheme properties. For instance, to have a conservative scheme, it will be necessary that $H_{ij}(u_i,u_j) = -H_{ji}(u_j,u_i)$.

9.4 Analysis of the finite volume approximation

The system of equations (9.10) can be rewritten in the form of a discrete variational problem by proceeding in the following way. For each $i = 1,\dots,\overset{\circ}{M}$, the i-th equation is multiplied by a real number v_i then by summing over the index i we obtain

$$\begin{aligned}\sum_{i=1}^{\overset{\circ}{M}} m_i v_i \frac{du_i}{dt} &+ \sum_{i=1}^{\overset{\circ}{M}} v_i \sum_{j=1}^{L_i} m_{ij}\{-\mu_{ij}\frac{u_j-u_i}{\delta_{ij}} + b_{ij}[\rho_{ij}u_i + (1-\rho_{ij})u_j]\} \\ &+ \sum_{i=1}^{\overset{\circ}{M}} m_i\sigma_i v_i u_i = \sum_{i=1}^{\overset{\circ}{M}} m_i v_i f_i.\end{aligned} \tag{9.12}$$

Let us now denote by V_h the space of piecewise-linear continuous functions with respect to the Delaunay triangulation $\mathcal{T}_h$, which vanish at the boundary $\partial\Omega$ (see (4.17)). From a set of values v_i we can univocally reconstruct a function $v_h \in V_h$ that interpolates such values at the nodes $\mathbf{x}_i$, that is

$$v_h \in V_h \,:\, v_h(\mathbf{x}_i) = v_i, \quad i = 1,\dots,\overset{\circ}{M}\,.$$

In a similar way, let $u_h \in V_h$ be the function interpolating the values u_i at $\mathbf{x}_i$. Then, (9.12) is rewritten equivalently in the following discrete variational form: for each $t > 0$, find $u_h = u_h(t) \in V_h$ such that

$$(\frac{\partial}{\partial t}u_h, v_h)_h + a_h(u_h, v_h) = (f, v_h)_h \quad \forall v_h \in V_h, \tag{9.13}$$

having introduced the internal scalar product $(w_h, v_h)_h = \sum_{i=1}^{\overset{\circ}{M}} m_i v_i w_i$ and having denoted by $a_h(u_h, v_h)$ the bilinear form appearing in the left-hand side of (9.12). We have thus interpreted the finite volume approximation as a particular case of the generalized Galerkin method for the assigned problem. As far as the choice of the coefficients ρ_{ij} for the linear combination is concerned, an option is to use $\rho_{ij} = 1/2$, which corresponds to using a finite difference of the centered type for the convective term. As we will see in Chapter 13, this strategy is adequate when the so called *local Péclet number*

$$\mathbb{P}\mathrm{e}_{ij} = \frac{b_{ij}\delta_{ij}}{\mu_{ij}}$$

(see (12.23)) is less than 1 for every pair i, j. If this is not the case, a more careful choice of the coefficients ρ_{ij} for the convex combination is required. In general, $\rho_{ij} = \varphi(\mathbb{P}\mathrm{e}_{ij})$, where φ is a function of the local Péclet number with values in $[0,1]$, that can be chosen as follows: if $\varphi(z) = 1/2\,[\text{sign}(z)+1]$ gives a stabilization of *upwind* type, while choosing $\varphi(z) = 1-(1-z/(e^z-1))/z$ we will have a stabilization of *exponential-fitting* type. (A similar kind of stabilization will be used in Sect. 12.6 in the context of finite difference approximation of diffusion-transport equations.) By this choice, we can show that the bilinear form $a_h(\cdot,\cdot)$ is V_h-elliptic, uniformly with respect to h, under the usual hypothesis that the coefficients of the problem satisfy the positivity condition $1/2\,\text{div}(\mathbf{b})+\sigma \geq \beta_0 = \text{const} \geq 0$.

Precisely, in this case, supposing further that $\mu \geq \mu_0 = \text{const} > 0$,

$$a_h(v_h, v_h) \geq \mu_0\, |v_h|^2_{\mathrm{H}^1(\Omega)} + \beta_0\,(v_h, v_h)_h.$$

Moreover, as $(v_h, v_h)_h$ is uniformly equivalent to the exact scalar product (v_h, v_h) for functions of V_h, this ensures the stability of problem (9.13). Finally, the method is linearly convergent with respect to h. In particular

$$\|u-u_h\|_{\mathrm{H}^1(\Omega)} \leq Ch\left(\|u\|_{\mathrm{H}^2(\Omega)} + |\nabla f|_{\mathrm{L}^\infty(\Omega)}\right)$$

under the assumption that the norms on the right are bounded. For the proof, see e.g. [KA00]. We suggest the same reference for an analysis of other properties of the method, such as monotonicity and conservation.

9.5 Implementation of boundary conditions

As previously stated, the differential problem under exam must be completed by suitable boundary conditions. For a problem written in conservation form, natural boundary conditions would be to impose the fluxes, i.e.

$$\mathbf{F}(u)\cdot\mathbf{n} = h \quad \text{on } \Gamma_N \subset \partial\Omega.$$

For their implementation in the framework of finite volumes it is sufficient to act on the numerical flux relating to the boundary sides, imposing

$$H_{ik} = H(u_i, u_k) = h(\mathbf{x}_{ik}) \quad \text{if } l_{ik} \subset \Gamma_N,$$

where $\mathbf{x}_{ik}$ is a suitable point (typically the midpoint) of l_{ik}.

On the other hand, essential (Dirichlet) conditions of the form

$$u = g \quad \text{on } \Gamma_D \subset \partial\Omega,$$

are immediate to implement in the context of vertex-centered schemes, for it is sufficient to add the corresponding equation for the nodes lying on Γ_D. As previously noted, the matter is more complicated for cell-centered schemes, as in this case there are no nodes on the boundary. An option is to impose the conditions *weakly*, in a similar way to what we will illustrate, although in a different context, in Sect. 14.3.1. This is a matter of suitably modifying the numerical fluxes on the sides, imposing

$$H_{ik} = H(u_i, g(\mathbf{x}_{ik})) \quad \text{if } l_{ik} \subset \Gamma_D.$$

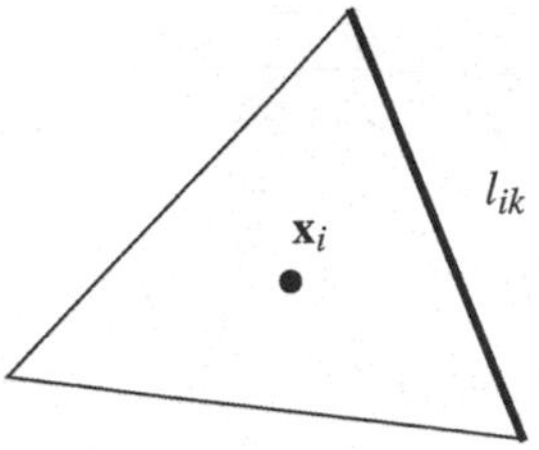

Fig. 9.10. The numerical flux on the side l_{ik} belonging to the Dirichlet boundary is computed in order to implement the boundary condition

Fig. 9.10 illustrates the situation for a cell-centered control volume adjacent to the boundary.

In practice, however, Dirichlet boundary conditions for cell-centered finite volumes are often implemented using the so-called *ghost nodes*. For each side l_{ik} on the boundary, we generate additional nodes, external to the domain, to which the corresponding boundary values are assigned. In this way, the computation of numerical fluxes is formally the same also for the boundary sides.

10

Spectral methods

As we have seen in Chapter 4, when we approximate boundary-value problems using the finite element method, the order of convergence is anyhow limited by the degree of the polynomials used, also in the case where solutions are very regular. In this chapter we will introduce *spectral methods*, for which the convergence rate is only limited by the regularity of the solution of the problem (and is exponential for analytical solutions). For a detailed analysis we refer to [CHQZ06, CHQZ07, Fun92, BM92].

10.1 The spectral Galerkin method for elliptic problems

The main feature that distinguishes finite elements from spectral methods in their classical "single-domain" version, is that the latter use global polynomials on the computational domain Ω, instead of piecewise polynomials. This is no longer true in the case of the spectral element method.

For each positive integer N, we denote by $\mathbb{Q}_N$ the space of polynomials with real coefficients and degree less than or equal to N with respect to *each* of the variables. Thus in one dimension we will denote by

$$\mathbb{Q}_N(I)=\left\{v(x)=\sum_{k=0}^{N}a_k x^k,\quad a_k\in\mathbb{R}\right\} \tag{10.1}$$

the space of polynomials of degree $\leq N$ on the interval $I\subset\mathbb{R}$, while in two dimensions,

$$\mathbb{Q}_N(\Omega)=\left\{v(\mathbf{x})=\sum_{k,m=0}^{N}a_{km}x_1^k x_2^m,\quad a_{km}\in\mathbb{R}\right\} \tag{10.2}$$

will denote the same space, but on the open set $\Omega\subset\mathbb{R}^2$. We note that while in one dimension $\mathbb{Q}_N=\mathbb{P}_N$, in several dimensions this does not happen. In particular, $\dim\mathbb{Q}_N=(N+1)^2$, while, as already seen in Sect. 4.4.1, $\dim\mathbb{P}_N=(N+1)(N+2)/2$.

Suppose we want to approximate the solution u of an elliptic problem which admits the variational formulation (4.1). Using a *spectral Galerkin method* (SM), the

A. Quarteroni: *Numerical Models for Differential Problems*, 2nd Ed.
MS&A – Modeling, Simulation & Applications 8
DOI 10.1007/978-88-470-5522-3_10, © Springer-Verlag Italia 2014

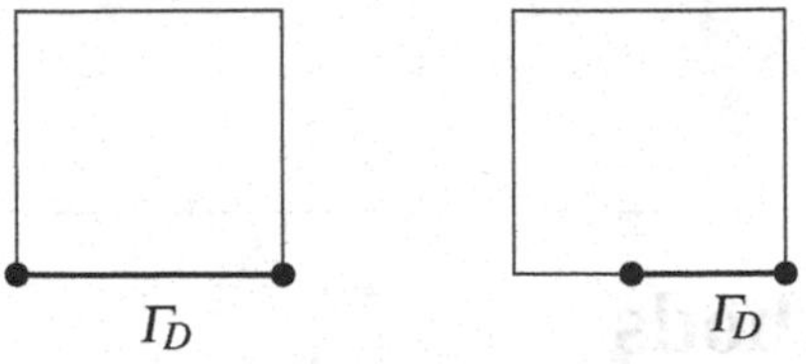

Fig. 10.1. Acceptable (left) and unacceptable (right) Dirichlet boundaries for the spectral method SM

space V will be approximated by a space $V_N \subset \mathbb{Q}_N$ and the approximate solution will consequently be indicated by u_N. In particular, if we suppose that V is $\mathrm{H}^1_{\Gamma_D}(\Omega)$ (the space defined in (3.27)), V_N will denote the set of polynomials of $\mathbb{Q}_N$ that vanish on the boundary portion Γ_D where a Dirichlet condition is prescribed, that is

$$V_N = \{v_N \in \mathbb{Q}_N : \quad v_N|_{\Gamma_D} = 0\}.$$

It is evident that $V_N \subset V$. The spectral Galerkin method SM will therefore be formulated on the subspace V_N. However, there is an issue in the definition of V_N: in the multi-dimensional case it is indeed not possible (in general) to require that a polynomial v_N vanishes only on an arbitrary part of the boundary of Ω. For instance, if Ω is the square $(-1,1)^2$, it is impossible to construct a polynomial that is null only on a portion of a boundary edge without it being null on the whole edge (see Fig. 10.1). This does not prevent a polynomial from vanishing on one whole side of the square or on all sides without necessarily being null in the whole of Ω (for instance, $v_2(\mathbf{x}) = (1-x_1^2)(1-x_2^2)$ is null only on the boundary of Ω).
For this reason, in the two-dimensional case we limit our attention to square domains (or, more generally, to domains that are reducible, through appropriate transformation, to the reference square $\widehat{\Omega} = (-1,1)^2$) and we suppose that the boundary's portion Γ_D is formed by the union of one or more sides of the domain.
However, the spectral method can be extended to the case of a domain Ω composed by the union of quadrilaterals Ω_k, each of which can be reduced to the reference square $\widehat{\Omega}$ via an invertible transformation $\varphi_k : \widehat{\Omega} \to \Omega_k$ (see Fig. 10.2), leading to the so-called *spectral element method (SEM)*, that was introduced by A.T. Patera [Pat84]. It is evident that in such a context it will be possible to require that the solution vanish on portions of the boundary given by the union of sides of the quadrilateral, but naturally not on portions of sides (see Fig. 10.2). In the SEM case, the discrete space has the following form

$$V_N^C = \{v_N \in C^0(\overline{\Omega}) : \; v_N|_{\Omega_k} \circ \varphi_k \in \mathbb{Q}_N(\widehat{\Omega})\}.$$

Example 10.1. A particularly important two-dimensional mapping is the transfinite interpolation (called *Gordon-Hall transformation* as well as *Coons patch*). The mapping φ_k is in this case expressed as a function of the invertible mappings $\pi_k^{(i)} : (-1,1) \to \Gamma_i$ (for $i = 1,\ldots,4$) that define the four sides of the computational domain Ω_k (see

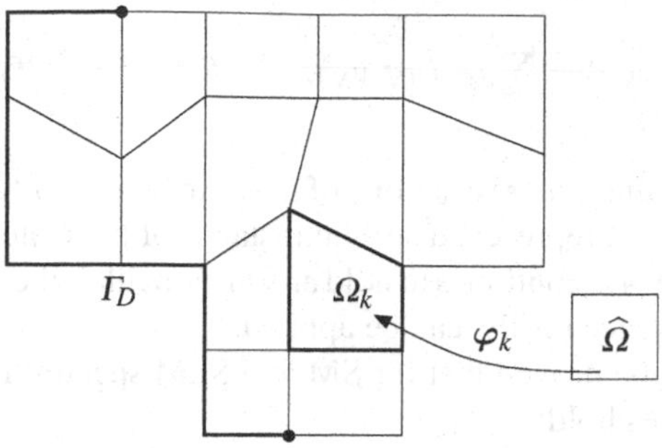

Fig. 10.2. Decomposition of the solution domain and acceptable boundary conditions for the SEM

Fig. 10.3). The transformation takes the following form

$$\begin{aligned}\varphi_k(\xi,\eta) &= \frac{1-\eta}{2}\pi_k^1(\xi)+\frac{1+\eta}{2}\pi_k^3(\xi)\\ &+\frac{1-\xi}{2}[\pi_k^4(\eta)-\frac{1+\eta}{2}\pi_k^3(1)-\frac{1-\eta}{2}\pi_k^1(-1)]\\ &+\frac{1+\xi}{2}[\pi_k^2(\eta)-\frac{1+\eta}{2}\pi_k^2(1)-\frac{1-\eta}{2}\pi_k^2(-1)].\end{aligned} \tag{10.3}$$

The transfinite interpolation therefore allows to consider computational domains Ω characterized by domains with non-straight edges. For more examples of transformations, see [CHQZ07]. ■

The approximation of problem (4.1) using the Galerkin spectral method (SM) is the following

$$\text{find } u_N \in V_N: \quad a(u_N, v_N) = F(v_N) \quad \forall v_N \in V_N,$$

while the spectral element one (SEM) will be

$$\text{find } u_N \in V_N^C: \quad a_C(u_N, v_N) = F_C(v_N) \quad \forall v_N \in V_N^C, \tag{10.4}$$

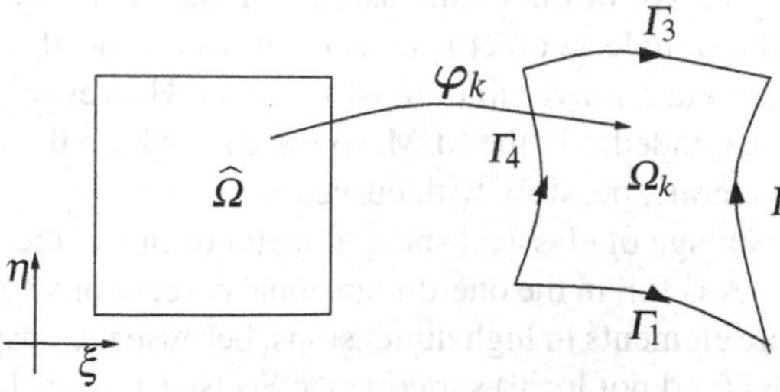

Fig. 10.3. The transformation φ_k in the case of the transfinite interpolation

where

$$a_C(u_N, v_N) = \sum_k a_{\Omega_k}(u_N, v_N), \qquad F_C(v_N) = \sum_k F_{\Omega_k}(v_N),$$

$a_{\Omega_k}(\cdot,\cdot)$ and $F_{\Omega_k}(\cdot)$ being the restrictions of $a(\cdot,\cdot)$ and $F(\cdot)$ to Ω_k.

Since these methods represent a special instance of the Galerkin method (4.2), the analysis made in Sect. 4.2 continues to hold and in particular, the existence, uniqueness, stability and convergence results can be applied.

Moreover, it can be proved that for SM and SEM spectral methods the following a priori error estimates hold:

Theorem 10.1. *Let $u \in V$ be the exact solution of the variational problem* (4.1) *and suppose that $u \in \mathrm{H}^{s+1}(\Omega)$, for some $s \geq 0$. If u_N is the corresponding approximate solution obtained via the SM, the following estimate holds*

$$\|u - u_N\|_{\mathrm{H}^1(\Omega)} \leq C_s N^{-s} \|u\|_{\mathrm{H}^{s+1}(\Omega)},$$

N being the degree of the approximating polynomials and C_s a constant that does not depend on N, but can depend on s. If u_N is, instead, the solution obtained via SEM , then we have

$$\|u - u_N\|_{\mathrm{H}^1(\Omega)} \leq C_s H^{\min(N,s)} N^{-s} \|u\|_{\mathrm{H}^{s+1}(\Omega)},$$

H being the maximum length of the sides of the macroelements Ω_k.

As opposed to what happens for the finite element method, a greater regularity of the solution leads to an increase in convergence rate, even supposing that the polynomial degree N is fixed. In particular, if u is analytical, the order of convergence of the spectral method becomes more than algebraic, i.e. exponential: more precisely,

$$\exists \gamma > 0: \quad \|u - u_N\|_{\mathrm{H}^1(\Omega)} \leq C \exp(-\gamma N).$$

Also in the case where u has finite regularity, it is still possible to obtain from the spectral method the maximal convergence rate allowed by the regularity of the exact solution: this is a clear advantage of spectral methods over finite elements.
The main limitation (in two or three dimensions) of classical spectral methods is that they can only handle simple geometries: rectangles or quadrilaterals which can be mapped into a square via an invertible transformation. However, as previously mentioned, they can be extended, via the SEM, to the case where the domain is given by the union of quadrilaterals, possibly with curved sides.

A further disadvantage of classical spectral methods lies in the fact that the associated stiffness matrix A is full in the one-dimensional case, or anyhow much less sparse than the one for finite elements in high dimensions, because the basis functions of such methods have global (and not local) support, see Sects. 10.2 and 10.3. The associated system of equations is generally more costly to solve.

Finally, the computational cost required to compute the elements of the stiffness matrix of the right-hand side must not be underestimated, as we are dealing with high degree polynomials. We will sort out this issue in the next section by using well-chosen Gaussian numerical integration.

Remark 10.1. In Sect. 10.5 at the end of this chapter, we will provide the algebraic formulation of the SEM for a one-dimensional problem. In particular, we will introduce the basis functions for the space V_N^C of composite polynomials. •

Remark 10.2. The SEM formulation is not so different from the p version of the finite element method. In both cases, the number of subdomains Ω_k is fixed while the local degree of polynomials (called N in the case of SEM, p in the finite element case) is increased locally, in order to improve the accuracy of the numerical approximation. For further details, we refer the interested reader to [CHQZ07, Sch98]. •

10.2 Orthogonal polynomials and Gaussian numerical integration

In this section we introduce the mathematical ingredients that allow to construct numerical integration formulae of Gaussian type. As previously anticipated, such formulae are the basis of pseudo-spectral methods, but also of spectral element methods that make use of numerical integration formulae.

10.2.1 Orthogonal Legendre polynomials

Let us consider a function $f:(-1,1)\rightarrow\mathbb{R}$. We recall that the space $\mathrm{L}^2(-1,1)$ is defined by (see Sect. 2.3.1)

$$\mathrm{L}^2(-1,1)=\left\{f:(-1,1)\rightarrow\mathbb{R}:\|f\|_{\mathrm{L}^2(-1,1)}=\left(\int_{-1}^{1}f^2(x)\,dx\right)^{1/2}<\infty\right\}.$$

Its scalar product is given by

$$(f,g)=\int_{-1}^{1}f(x)g(x)dx.$$

The *orthogonal Legendre polynomials* $L_k\in\mathbb{P}_k$, for $k=0,1,\ldots$, constitute a sequence for which the following orthogonality property is satisfied

$$(L_k,L_m)=\begin{cases}0 & \text{if } m\neq k,\\ (k+\frac{1}{2})^{-1} & \text{if } m=k.\end{cases}$$

They are linearly independent and form a basis for $\mathrm{L}^2(-1,1)$. Consequently, each function $f\in\mathrm{L}^2(-1,1)$ admits the series expansion

$$f(x)=\sum_{k=0}^{\infty}\widehat{f}_kL_k(x) \tag{10.5}$$

known as *Legendre series*. This is a *modal* representation of f. The Legendre coefficients $\widehat{f}_k$ can easily be computed by exploiting the orthogonality of Legendre polynomials. Indeed, we have

$$\begin{aligned}(f,L_k) &= \int_{-1}^{1} f(x)L_k(x)\,dx = \int_{-1}^{1}\left(\sum_{i=0}^{\infty}\widehat{f}_i L_i(x)L_k(x)\right)dx\\ &= \sum_{i=0}^{\infty}\left(\int_{-1}^{1} L_i(x)L_k(x)\,dx\right)\widehat{f}_i = \widehat{f}_k\,\|L_k\|^2_{\mathrm{L}^2(-1,1)}.\end{aligned}$$

Hence,

$$\widehat{f}_k = (f,L_k)/\|L_k\|^2_{\mathrm{L}^2(-1,1)} = (k+\frac{1}{2})\int_{-1}^{1} f(x)L_k(x)dx \tag{10.6}$$

from which the so-called *Parseval identity* immediately descends

$$\|f\|^2_{\mathrm{L}^2(-1,1)} = \sum_{k=0}^{\infty}(\widehat{f}_k)^2\,\|L_k\|^2_{\mathrm{L}^2(-1,1)}.$$

It is possible to compute the Legendre polynomials recursively via the following three-term relation:

$$\begin{aligned}L_0 &= 1, \qquad L_1 = x,\\ L_{k+1} &= \frac{2k+1}{k+1}xL_k - \frac{k}{k+1}L_{k-1}, \qquad k=1,2,\ldots\end{aligned}$$

(In Fig. 10.4, the graphs of the polynomials L_k, with $k=2,\ldots,5$, are drawn). The

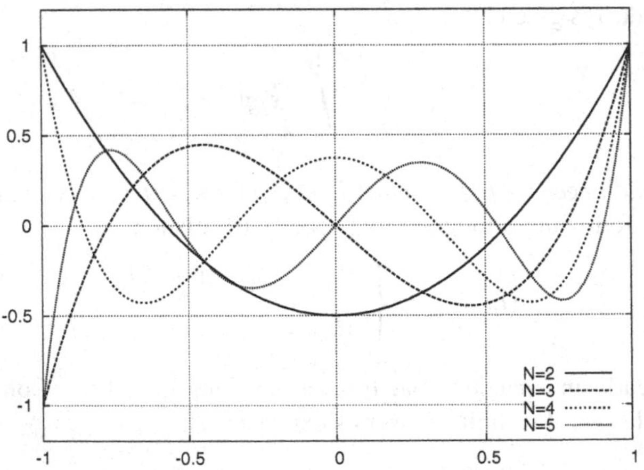

Fig. 10.4. The Legendre polynomials of degree $k=2,3,4,5$

Legendre series of any $f \in L^2(-1,1)$ converges to f in $L^2(-1,1)$ norm. Denoting by

$$f_N(x) = \sum_{k=0}^{N} \widehat{f}_k L_k(x)$$

the *N-th truncation of the Legendre series* of f, this means that

$$\lim_{N\to\infty} \|f - f_N\|_{L^2(-1,1)} = 0, \qquad (10.7)$$

that is

$$\lim_{N\to\infty} \Big\| \sum_{k=N+1}^{\infty} \widehat{f}_k L_k \Big\|_{L^2(-1,1)} = 0.$$

Thanks to the Parseval identity, we have that

$$\|f - f_N\|^2_{L^2(-1,1)} = \sum_{k=N+1}^{\infty} (\widehat{f}_k)^2 \|L_k\|^2_{L^2(-1,1)} = \sum_{k=N+1}^{\infty} \frac{(\widehat{f}_k)^2}{k+\frac{1}{2}},$$

hence condition (10.7) is equivalent to

$$\lim_{N\to\infty} \sum_{k=N+1}^{\infty} \frac{(\widehat{f}_k)^2}{k+\frac{1}{2}} = 0.$$

Moreover, it can be proved that if $f \in H^s(-1,1)$, for some $s \geq 1$, then it is possible to find a suitable constant $C_s > 0$, independent of N, such that

$$\|f - f_N\|_{L^2(-1,1)} \leq C_s \left(\frac{1}{N}\right)^s \|f^{(s)}\|_{L^2(-1,1)},$$

i.e. we have convergence of order s, with respect to $1/N$.

At this point, we can prove that f_N is the orthogonal projection of f on $\mathbb{Q}_N$ with respect to the scalar product of $L^2(-1,1)$, that is

$$(f - f_N, p) = 0 \quad \forall\, p \in \mathbb{Q}_N. \qquad (10.8)$$

First of all we note that

$$(f - f_N, L_m) = \left(\sum_{k=N+1}^{\infty} \widehat{f}_k L_k, L_m \right) = \sum_{k=N+1}^{\infty} \widehat{f}_k (L_k, L_m).$$

Since the polynomials L_k, with $0 \leq k \leq N$, form a basis for the space $\mathbb{Q}_N$, every polynomial $p \in \mathbb{Q}_N$ can be expanded with respect to this basis. Equation (10.8) follows noticing that for $m \leq N$, $(L_k, L_m) = 0 \quad \forall k \geq N+1$ because of orthogonality.
In particular, from (10.8) it follows that f_N is the function which minimizes the distance of f from $\mathbb{Q}_N$, that is

$$\|f - f_N\|_{L^2(-1,1)} \leq \|f - p\|_{L^2(-1,1)} \quad \forall\, p \in \mathbb{Q}_N. \qquad (10.9)$$

For this purpose, we start by observing that

$$\|f - f_N\|^2_{L^2(-1,1)} = (f - f_N, f - f_N) = (f - f_N, f - p) + (f - f_N, p - f_N)$$

for each $p \in \mathbb{Q}_N$ and that $(f - f_N, p - f_N) = 0$ by the orthogonality property (10.8). Consequently,

$$\|f - f_N\|^2_{L^2(-1,1)} = (f - f_N, f - p) \quad \forall\, p \in \mathbb{Q}_N,$$

from which, applying the Cauchy-Schwarz inequality, we obtain

$$\|f - f_N\|^2_{L^2(-1,1)} \leq \|f - f_N\|_{L^2(-1,1)} \, \|f - p\|_{L^2(-1,1)} \quad \forall\, p \in \mathbb{Q}_N,$$

i.e. (10.9).

10.2.2 Gaussian integration

Gaussian integration formulae are the ones which, having fixed the number of quadrature nodes, allow to obtain the highest *exactness degree* (see [QSS07]). The latter is the highest integer r such that all polynomials of degree less than or equal to r are integrated *exactly* by the formula at hand. We will start by introducing such formulae on the interval $(-1,1)$, and then extend them to the case of a generic interval.

We denote by N the number of nodes. We call Gauss-Legendre quadrature nodes the *zeroes* $\{\bar{x}_1, \ldots, \bar{x}_N\}$ *of the Legendre polynomial* L_N. In the presence of such a set of nodes, we will consider the following quadrature formula (called interpolatory of Gauss-Legendre)

$$I^{GL}_{N-1} f = \int_{-1}^{1} \Pi^{GL}_{N-1} f(x)\, dx, \tag{10.10}$$

$\Pi^{GL}_{N-1} f$ being the polynomial of degree $N-1$ interpolating f at the nodes $\bar{x}_1, \ldots, \bar{x}_N$. We denote by $\overline{\psi}_k \in \mathbb{Q}_{N-1}$, $k = 1,\ldots,N$, the characteristic Lagrange polynomials associated to the Gauss-Legendre nodes,

$$\overline{\psi}_k(\bar{x}_j) = \delta_{kj}, \quad j = 1,..,N.$$

The quadrature formula (10.10) then takes the following expression

$$\int_{-1}^{1} f(x)\, dx \simeq I^{GL}_{N-1} f = \sum_{k=1}^{N} \bar{\alpha}_k f(\bar{x}_k), \quad \text{with } \bar{\alpha}_k = \int_{-1}^{1} \overline{\psi}_k(x) dx,$$

and is called Gauss-Legendre quadrature formula (GL).

To find the nodes $\bar{t}_k$ and the weights $\bar{\delta}_k$ characterizing such formula on a generic interval $[a,b]$, it will be sufficient to refer to the relation

$$\bar{t}_k = \frac{b-a}{2}\bar{x}_k + \frac{a+b}{2}$$

for the former, while, for the latter, it can easily be verified that

$$\bar{\delta}_k = \frac{b-a}{2}\,\bar{\alpha}_k.$$

The *exactness degree* of these formulae is equal to $2N-1$ (and is the maximum possible for formulae with $N-1$ nodes). This means that

$$\int_a^b f(x)dx = \sum_{k=1}^{N} \bar{\delta}_k f(\bar{t}_k) \quad \forall f \in \mathbb{Q}_{2N-1}.$$

10.2.3 Gauss-Legendre-Lobatto formulae

A feature of the Gauss-Legendre integration formulae is to have all quadrature nodes internal to the integration interval. In the case of differential problems this makes the imposition of boundary conditions on the end points of the interval problematic.

To overcome such difficulty, the so-called Gauss-Lobatto formulae are introduced, particularly the Gauss-Legendre-Lobatto (GLL) formulae. There, nodes, relative to the interval $(-1,1)$, are represented by the end points of the interval themselves and by the maximum and minimum points of the Legendre polynomial of degree N, i.e. by the zeroes of the first derivative of the polynomial L_N.
We denote such nodes by $\{x_0 = -1, x_1, \dots, x_{N-1}, x_N = 1\}$. Therefore, we have

$$L'_N(x_i) = 0, \quad \text{for } i = 1, \dots, N-1. \tag{10.11}$$

(In this chapter the symbol "$'$" denotes a derivative with respect to x.) Let ψ_i be the corresponding characteristic polynomials, that is

$$\psi_i \in \mathbb{Q}_N : \quad \psi_i(x_j) = \delta_{ij}, \quad 0 \le i,j \le N, \tag{10.12}$$

whose analytical expression is given by

$$\psi_i(x) = \frac{-1}{N(N+1)}\,\frac{(1-x^2)L'_N(x)}{(x-x_i)L_N(x_i)}, \qquad i = 0,\dots,N \tag{10.13}$$

(see Fig. 10.5 for the graphs of the characteristic polynomials ψ_i, for $i = 0,\dots,4$ in the case where $N = 4$). The functions $\psi_i(x)$ are the counterpart of the Lagrangian basis functions $\{\varphi_i\}$ of the finite elements introduced in Sect. 4.3. Given a function $f \in C^0([-1,1])$, its interpolation polynomial $\Pi_N^{GLL} f \in \mathbb{Q}_N$ at the GLL nodes is identified by the relation

$$\Pi_N^{GLL} f(x_i) = f(x_i), \quad 0 \le i \le N\,. \tag{10.14}$$

It has the following expression

$$\Pi_N^{GLL} f(x) = \sum_{i=0}^{N} f(x_i)\psi_i(x). \tag{10.15}$$

It can be proved, thanks to the non-uniform distribution of the nodes $\{x_i\}$, that $\Pi_N^{GLL} f$ converges towards f when $N \to \infty$. Moreover, the following error estimate is satisfied:

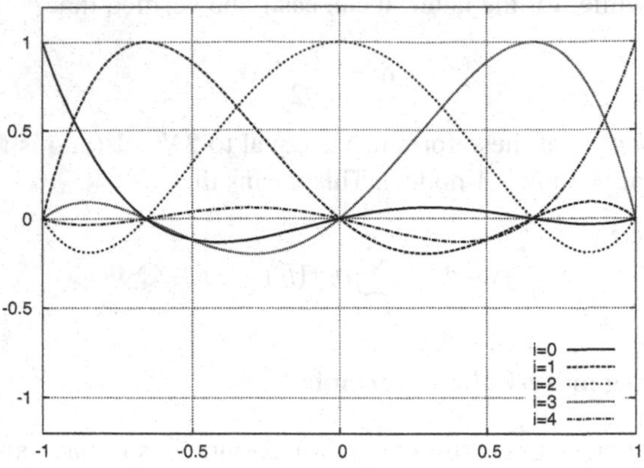

Fig. 10.5. The characteristic polynomials $\psi_i, i = 0, \ldots, 4$ of degree 4 corresponding to the Gauss-Legendre-Lobatto nodes

if $f \in \mathrm{H}^s(-1,1)$, for some $s \geq 1$,

$$\|f - \Pi_N^{GLL} f\|_{\mathrm{L}^2(-1,1)} \leq C_s \left(\frac{1}{N}\right)^s \|f^{(s)}\|_{\mathrm{L}^2(-1,1)}, \tag{10.16}$$

where C_s is a constant depending on s but not on N. More generally (see [CHQZ06]),

$$\|f - \Pi_N^{GLL} f\|_{\mathrm{H}^k(-1,1)} \leq C_s \left(\frac{1}{N}\right)^{s-k} \|f\|_{\mathrm{H}^s(-1,1)}, \quad s \geq 1, \; k = 0, 1. \tag{10.17}$$

In Fig. 10.6 (left), we show the convergence curves for the interpolation error of two different functions.

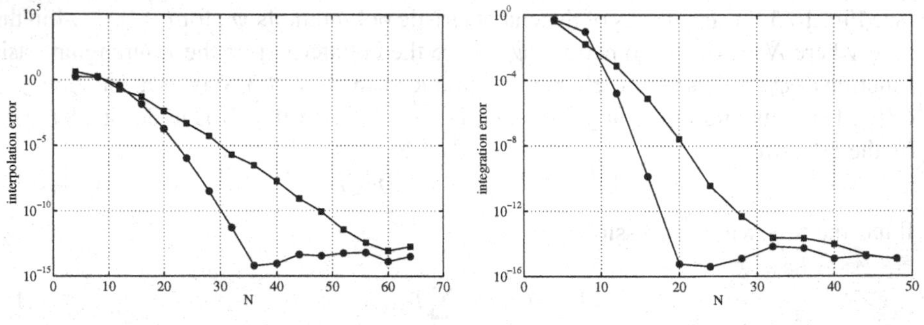

Fig. 10.6. Behaviour of the interpolation (left) and integration (right) error in the GLL nodes as a function of the degree N for the two functions $f_1(x) = \cos(4\pi x)$ (•) and $f_2(x) = 4\cos(4x)\exp(\sin(4x))$ (▪) on the interval $(-1, 1)$

By using $\Pi_N^{GLL} f$ instead of $\Pi_{N-1}^{GLL} f$ we can define the following Gauss-Legendre-Lobatto (GLL) integration formula, in alternative to (10.10)

$$I_N^{GLL} f = \int_{-1}^{1} \Pi_N^{GLL} f(x) dx = \sum_{k=0}^{N} \alpha_k f(x_x). \tag{10.18}$$

The new weights are $\alpha_i = \int_{-1}^{1} \psi_i(x)\, dx$ and take the following expression

$$\alpha_i = \frac{2}{N(N+1)} \frac{1}{L_N^2(x_i)}. \tag{10.19}$$

The GLL formula has *exactness degree* equal to $2N-1$, that is it integrates exactly all polynomials of degree $\leq 2N-1$,

$$\int_{-1}^{1} f(x)dx = I_N^{GLL} f \quad \forall f \in \mathbb{Q}_{2N-1}. \tag{10.20}$$

This is the maximum degree obtainable when $N+1$ nodes are used, 2 of which assigned a priori. Moreover, using the interpolation estimate (10.16), the following integration error estimate can be proved: if $f \in \mathrm{H}^s(-1,1)$, with $s \geq 1$,

$$\left| \int_{-1}^{1} f(x)\, dx - I_N^{GLL} f \right| \leq C_s \left(\frac{1}{N}\right)^s \|f^{(s)}\|_{\mathrm{L}^2(-1,1)},$$

where C_s is independent of N but can depend, in general, on s. This means that the more regular the function f is, the higher is the order of convergence of the integration formula. In Fig. 10.6 (right) we report the integration error for two different functions (the same ones considered for the left graph).

If we now consider a generic interval (a,b) instead of (-1,1), nodes and weights in (a,b) take the following expression

$$t_k = \frac{b-a}{2} x_k + \frac{a+b}{2}, \quad \delta_k = \frac{b-a}{2} \alpha_k.$$

Formula (10.20) generalizes as follows

$$\int_{a}^{b} f(x)dx \simeq \sum_{k=0}^{N} \delta_k f(t_k). \tag{10.21}$$

The properties of exactness and accuracy remain unchanged.

10.3 G-NI methods in one dimension

Let us consider the following one-dimensional elliptic problem with homogeneous Dirichlet data

$$\begin{cases} Lu = -(\mu u')' + \sigma u = f, & -1 < x < 1, \\ u(-1) = 0, & u(1) = 0, \end{cases} \tag{10.22}$$

$\mu(x) \geq \mu_0 > 0$ and $\sigma(x) \geq 0$, in order to have an associated bilinear form that is coercive in $\mathrm{H}_0^1(-1,1)$.
The *spectral Galerkin method* (SM) is written as

$$\text{find } u_N \in V_N : \int_{-1}^{1} \mu u_N' v_N' \, dx + \int_{-1}^{1} \sigma u_N v_N \, dx = \int_{-1}^{1} f v_N \, dx \quad \forall\, v_N \in V_N, \tag{10.23}$$

with

$$V_N = \{v_N \in \mathbb{Q}_N : v_N(-1) = v_N(1) = 0\}. \tag{10.24}$$

The G-NI (*Galerkin with Numerical Integration*) method is obtained by approximating the integrals in (10.23) via the GLL integration formulae. This amounts to substituting the scalar product (f,g) in $\mathrm{L}^2(-1,1)$ by the *discrete GLL scalar product* (for continuous functions)

$$(f,g)_N = \sum_{i=0}^{N} \alpha_i f(x_i) g(x_i), \tag{10.25}$$

where the x_i and the α_i are defined according to (10.11) and (10.19). Hence, the G-NI method is written as

$$\text{find } u_N^* \in V_N : \ (\mu u_N^{*\prime}, v_N')_N + (\sigma u_N^*, v_N)_N = (f, v_N)_N \quad \forall\, v_N \in V_N. \tag{10.26}$$

Due to the numerical integration, in general $u_N^* \neq u_N$, that is the solutions of the SM and G-NI methods do not coincide.

However, thanks to the exactness property (10.20), we will have

$$(f,g)_N = (f,g) \quad \text{provided that} \quad fg \in \mathbb{Q}_{2N-1}. \tag{10.27}$$

If we consider the particular case where in (10.22) μ is a constant and $\sigma = 0$, the G-NI problem becomes

$$\mu({u^*}'_N, v_N')_N = (f, v_N)_N. \tag{10.28}$$

In some very particular cases, the spectral and the G-NI methods coincide. This is for instance the case of (10.28), where f is a polynomial with degree equal at most to $N-1$. It is simple to verify that the two methods coincide thanks to the exactness relation (10.27).

Generalizing to the case of more complex differential formulations having different boundary conditions (Neumann, or mixed), the G-NI problem is written as

$$\text{find } u_N^* \in V_N : \quad a_N(u_N^*, v_N) = F_N(v_N) \quad \forall\, v_N \in V_N, \tag{10.29}$$

where $a_N(\cdot,\cdot)$ and $F_N(\cdot)$ are obtained starting from the bilinear form $a(\cdot,\cdot)$ and from the known term $F(\cdot)$ of the spectral Galerkin problem, by substituting the exact integrals with the GLL integration formulae. V_N is the space of polynomials of degree N that vanish on the boundary points (provided that there are any) on which Dirichlet conditions are imposed.

Observe that, due of the fact that the bilinear form $a_N(\cdot,\cdot)$ and the functional $F_N(\cdot)$ are no longer the ones associated to the initial problem, what we obtain is no longer a Galerkin approximation method, and the relative theoretical results cannot be applied (in particular, the Céa lemma, see Lemma 4.1).
In general, a method derived from a Galerkin method, either spectral or with finite elements, where numerical integrals replace exact ones, will be called *generalized Galerkin method (GG)*. For the corresponding analysis we will resort to the Strang lemma (see Sect. 10.4.1 and also [Cia78, QV94]) .

10.3.1 Algebraic interpretation of the G-NI method

The functions ψ_i, with $i = 1,2,\ldots,N-1$, introduced in Sect. 10.2.3 consititute a basis for the space V_N, as they are all null in $x_0 = -1$ and $x_N = 1$. We can therefore provide for the solution u_N^* of the G-NI problem (10.29) the *nodal* representation

$$u_N^*(x) = \sum_{i=1}^{N-1} u_N^*(x_i)\psi_i(x).$$

In analogy with the finite-element method, this means we identify the unknowns of our problem with the values taken by u_N^* at the nodes x_i (now coinciding with the Gauss-Legendre-Lobatto nodes). Moreover, for problem (10.29) to be verified for each $v_N \in V_N$, it will be sufficient that it be verified for each basis function ψ_i. We will therefore have

$$\sum_{j=1}^{N-1} u_N^*(x_j)\, a_N(\psi_j,\psi_i) = F_N(\psi_i), \qquad i = 1,2,\ldots,N-1,$$

which we can rewrite

$$\sum_{j=1}^{N-1} a_{ij} u_N^*(x_j) = f_i, \qquad i = 1,2,\ldots,N-1,$$

that is, in matrix form,

$$\mathbf{A}\mathbf{u}_N^* = \mathbf{f} \qquad (10.30)$$

where

$$\mathrm{A} = (a_{ij}) \quad \text{with} \quad a_{ij} = a_N(\psi_j,\psi_i), \quad \mathbf{f} = (f_i) \quad \text{with} \quad f_i = F_N(\psi_i),$$

and where $\mathbf{u}_N^*$ denotes the vector of unknown coefficients $u_N^*(x_j)$, for $j = 1,\ldots,N-1$. In the particular case of problem (10.26), we would obtain

$$a_{ij} = (\mu\psi_j', \psi_i')_N + \alpha_i \sigma(x_i)\delta_{ij}, \quad f_i = (f,\psi_i)_N = \alpha_i f(x_i),$$

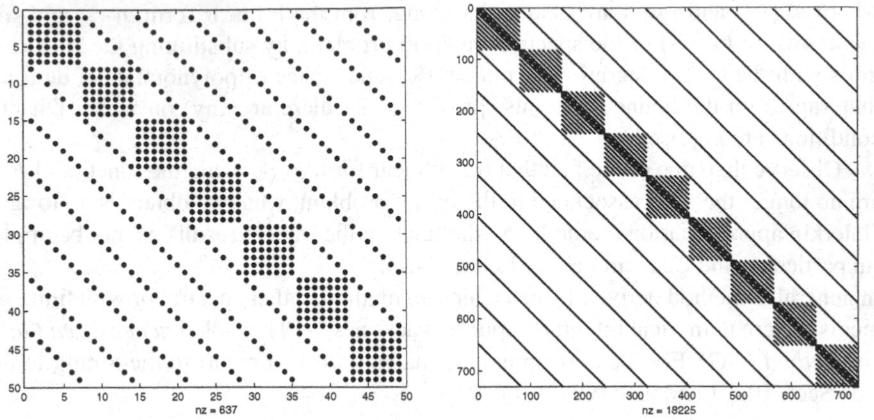

Fig. 10.7. Pattern of the matrix A for the G-NI method, the 2D (left) and 3D (right) case: nz denotes the number of non-null elements in the matrix

for each $i, j = 1, \ldots, N-1$. The matrix in 1D is full due to the presence of the diffusive term. Indeed, the reactive term only contributes to the diagonal. In more dimensions, the matrix A has a block structure, and the diagonal blocks are full. See Fig. 10.7, reporting the *pattern* relating to matrix A in 2D and 3D. Finally, we observe that the condition number we would get in the absence of numerical integration results is, in general, much larger, namely $O(N^4)$. Moreover, the matrix A turns out to be ill-conditioned, with condition number $O(N^3)$. For the solution of system (10.30) it is therefore convenient to resort, especially in 2D and 3D, to a suitably preconditioned iterative method. By choosing as a preconditioner the matrix of linear finite elements associated to the same bilinear form $a(\cdot,\cdot)$ and to the GLL nodes, we obtain a preconditioned matrix whose conditioning is independent of N ([CHQZ06]). At the top of Fig. 10.8 we report the condition number (as a function of N) of the matrix A and of the matrix obtained by preconditioning A with different preconditioning matrices: the diagonal matrix of A, the one obtained from A through the incomplete Cholesky factor-

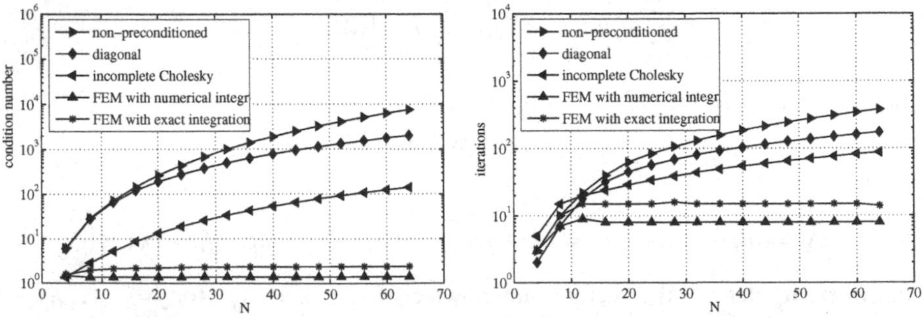

Fig. 10.8. Condition number (left) and iteration number (right), for different types of preconditioners

ization, the one obtained using linear finite elements by approximating integrals with the composite trapezoidal formula, and finally the exact one from finite elements. At the bottom of Fig. 10.8 we report the number of necessary iterations for the conjugate gradient method to converge in the different cases.

10.3.2 Conditioning of the stiffness matrix in the G-NI method

We seek estimates for the eigenvalues λ^N of the stiffness matrix A of the G-NI method

$$A\mathbf{u} = \lambda^N \mathbf{u}.$$

In the case of the simple second derivative operator, we have $A = (a_{ij})$, with $a_{ij} = (\psi_j', \psi_i')_N = (\psi_j', \psi_i')$, ψ_j being the j-th characteristic Lagrange function associated to the node x_j. Then,

$$\lambda^N = \frac{\mathbf{u}^T A\mathbf{u}}{\mathbf{u}^T\mathbf{u}} = \frac{\|u_x^N\|^2_{\mathrm{L}^2(-1,1)}}{\mathbf{u}^T\mathbf{u}}, \tag{10.31}$$

$u^N \in V_N$ being the only polynomial of the space V_N defined in (10.24) satisfying $u^N(x_j) = u_j$, for $j = 1, \ldots, N-1$, where $\mathbf{u} = (u_j)$. Thus, for each j, $u_j \int_i^i -1 x_j u_x^N(s)\, ds$; thanks to the Cauchy-Schwarz inequality we obtain the bound

$$|u_j| \le \left(\int_{-1}^{x_j} |(u^N)'(s)|^2\, ds \right)^{1/2} \left(\int_{-1}^{x_j} ds \right)^{1/2} \le \sqrt{2}\, \|(u^N)'\|_{\mathrm{L}^2(-1,1)}.$$

Hence

$$\mathbf{u}^T\mathbf{u} = \sum_{j=1}^{N-1} u_j^2 \le 2\,(N-1)\, \|(u^N)'\|^2_{\mathrm{L}^2(-1,1)},$$

which, thanks to (10.31), provides the lower bound

$$\lambda^N \ge \frac{1}{2(N-1)}. \tag{10.32}$$

An upper bound for λ^N can be obtained by recurring to the following *inverse inequality* for algebric polynomials (see [CHQZ06], Sect. 5.4.1)

$$\exists\, C > 0 \,:\, \forall\, p \in V_N, \quad \|p'\|_{\mathrm{L}^2(-1,1)} \le \sqrt{2} N \left(\int_{-1}^{1} \frac{p^2(x)}{1-x^2}\, dx \right)^{1/2}. \tag{10.33}$$

Then

$$\|(u^N)'\|^2_{\mathrm{L}^2(-1,1)} \le 2N^2 \int_{-1}^{1} \frac{[u^N(x)]^2}{1-x^2}\, dx = 2N^2 \sum_{j=1}^{N-1} \frac{[u^N(x_j)]^2}{1-x_j^2}\, \alpha_j, \tag{10.34}$$

where we use the exactness of the GLL integration formula (see (10.20)), as $[u^N]^2/(1-x^2) \in \mathbb{P}_{2N-2}$. Since for the coefficients α_j the following asymptotic estimate holds:

$\alpha_j/(1-x_j^2) \leq C$, for a suitable constant C independent of N, we can conclude, thanks to (10.31) and (10.34), that

$$\lambda^N \leq 2CN^2. \tag{10.35}$$

It can finally be proved that both estimates (10.32) and (10.35) are optimal as far as the asymptotic behaviour with respect to N is concerned.

10.3.3 Equivalence between G-NI and collocation methods

We want to prove that there exists a precise relationship between G-NI and *collocation methods*, i.e. those methods imposing the differential equation only at selected points of the computational interval. Let us consider once again the homogeneous Dirichlet problem (10.22), whose associated G-NI problem is written is the form (10.26).
We would like to counterintegrate by parts equation (10.26), but in order to do that, we must first rewrite the discrete scalar products as integrals. Let $\Pi_N^{GLL} : C^0([-1,1]) \mapsto \mathbb{Q}_N$ be the interpolation operator introduced in Sect. 10.2.3, which maps a continuous function to the corresponding interpolating polynomial at the Gauss-Legendre-Lobatto nodes.
Since the GLL integration formula uses the values of the function only at the integration nodes and since the function and its G-NI interpolant coincide there, we have

$$\sum_{i=0}^{N} \alpha_i f(x_i) = \sum_{i=0}^{N} \alpha_i \Pi_N^{GLL} f(x_i) = \int_{-1}^{1} \Pi_N^{GLL} f(x) dx,$$

where the latter equality descends from (10.20) as $\Pi_N^{GLL} f$ is integrated exactly, being a polynomial of degree N.
The discrete scalar product thus becomes a scalar product in $\mathrm{L}^2(-1,1)$, in the case where one of the two functions is a polynomial of degree strictly less than N, i.e.

$$(f,g)_N = (\Pi_N^{GLL} f, g)_N = (\Pi_N^{GLL} f, g) \quad \forall\, g \in \mathbb{Q}_{N-1}. \tag{10.36}$$

In this case, indeed, $\Pi_N^{GLL} f \in \mathbb{Q}_N$, $(\Pi_N^{GLL} f) g \in \mathbb{Q}_{2N-1}$ and therefore the integral is computed exactly. Integrating by parts the exact integrals, we obtain[1]

$$\begin{aligned}
(\mu u_N', v_N')_N &= (\Pi_N^{GLL}(\mu u_N'), v_N')_N = (\Pi_N^{GLL}(\mu u_N'), v_N') \\
&= -([\Pi_N^{GLL}(\mu u_N')]', v_N) + [\Pi_N^{GLL}(\mu u_N')\, v_N]_{-1}^{1} \\
&= -([\Pi_N^{GLL}(\mu u_N')]', v_N)_N,
\end{aligned}$$

where the last equality holds because v_N vanishes at the boundary and the terms which appear in the scalar product yield a polynomial whose total degree is equal to $2N-1$.
At this point, we can rewrite the G-NI problem as follows

$$\text{find } u_N \in V_N : \ (L_N u_N, v_N)_N = (f, v_N)_N \quad \forall\, v_N \in V_N, \tag{10.37}$$

[1] From now on, for simplicity of notation, we will denote the G-NI solution by u_N (instead of u_N^*), since there is no longer the risk to confuse it with the spectral solution.

where we have defined

$$L_N u_N = -[\Pi_N^{GLL}(\mu u_N')]' + \sigma u_N. \tag{10.38}$$

By imposing that (10.37) is valid for each basis function ψ_i, we obtain

$$(L_N u_N, \psi_i)_N = (f, \psi_i)_N, \quad i = 1, 2, \ldots, N-1.$$

Now we examine the i-th equation. The first term is

$$\begin{aligned} -([\Pi_N^{GLL}(\mu u_N')]', \psi_i)_N &= -\sum_{j=0}^{N} \alpha_j [\Pi_N^{GLL}(\mu u_N')]'(x_j)\, \psi_i(x_j) \\ &= -\alpha_i [\Pi_N^{GLL}(\mu u_N')]'(x_i), \end{aligned}$$

since $\psi_i(x_j) = \delta_{ij}$. Analogously, for the second term we have

$$(\sigma u_N, \psi_i)_N = \sum_{j=0}^{N} \alpha_j \sigma(x_j) u_N(x_j) \psi_i(x_j) = \alpha_i \sigma(x_i) u_N(x_i).$$

Finally, the right-hand side becomes

$$(f, \psi_i)_N = \sum_{j=0}^{N} \alpha_j f(x_j) \psi_i(x_j) = \alpha_i f(x_i).$$

Dividing by α_i the equation thus found, we obtain the following equivalent formulation of the G-NI problem

$$\begin{cases} L_N u_N(x_i) = f(x_i), & i = 1, 2, \ldots, N-1, \\ u_N(x_0) = 0, & u_N(x_N) = 0. \end{cases} \tag{10.39}$$

This is called a *collocation* problem as it is equivalent to placing at the internal nodes x_i the assigned differential equation (after approximating the operator L by L_N), and satisfying the boundary conditions at the boundary nodes.

We now introduce the *interpolation derivative*, $D_N(\Phi)$, of a continuous function Φ, as being the derivative of the interpolating polynomial $\Pi_N^{GLL}\Phi$ defined according to (10.14), i.e.

$$D_N(\Phi) = D[\Pi_N^{GLL}\Phi], \tag{10.40}$$

D being the symbol of exact differentiation. If we consider the differential operator L and replace all derivatives with the corresponding interpolation derivatives, we obtain a new operator, called *pseudo-spectral operator* L_N, that coincides with the one defined in (10.38). It follows that the G-NI method, introduced here as a generalized Galerkin method, can also be interpreted as a collocation method that operates directly on the differential part of the problem, analogously to what happens, for instance, in the case of finite differences. In this sense, finite differences can be considered as a less accurate version of the G-NI method, as the derivatives are approximated using formulae that use a small number of nodal values.

If the initial operator had been

$$Lu = (-\mu u')' + (bu)' + \sigma u,$$

then the corresponding pseudo-spectral operator would have been

$$L_N u_N = -D_N(\mu u_N') + D_N(b u_N) + \sigma u_N. \tag{10.41}$$

Had the boundary conditions for problem (10.22) been of Neumann type,

$$(\mu u')(-1) = g_-, \quad (\mu u')(1) = g_+,$$

the spectral Galerkin method would be formulated as follows

$$\text{find } u_N \in \mathbb{Q}_N \quad : \quad \int_{-1}^{1} \mu u_N' v_N' \, dx + \int_{-1}^{1} \sigma u_N v_N \, dx =$$

$$\int_{-1}^{1} f v_N \, dx \quad + \quad \mu(1) g_+ v_N(1) - \mu(-1) g_- v_N(-1) \quad \forall\, v_N \in \mathbb{Q}_N,$$

while the G-NI method would become

$$\text{find } u_N \in \mathbb{Q}_N \quad : \quad (\mu u_N', v_N')_N + (\sigma u_N, v_N)_N =$$

$$(f, v_N)_N \quad + \quad \mu(1) g_+ v_N(1) - \mu(-1) g_- v_N(-1) \quad \forall\, v_N \in \mathbb{Q}_N.$$

Its interpretation as a collocation method becomes: find $u_N \in \mathbb{Q}_N$ such that

$$L_N u_N(x_i) = f(x_i), \quad i = 1, \ldots, N-1,$$

$$(L_N u_N(x_0) - f(x_0)) - \frac{1}{\alpha_0}((\mu\, u_N')(-1) - g_-) = 0,$$

$$(L_N u_N(x_N) - f(x_N)) + \frac{1}{\alpha_N}((\mu\, u_N')(1) - g_+) = 0,$$

where L_N is defined in (10.38). Note that at the boundary nodes the Neumann condition is satisfied up to the equation residual $L_N u_N - f$ multiplied by the coefficient of the GLL formula, which is an infinitesimal of order 2 with respect to $1/N$.
In Fig. 10.9 (taken from [CHQZ06]) we report the error in the $\mathrm{H}^1(-1,1)$-norm (left) and the absolute value of the difference $(\mu\, u_N')(\pm 1) - g_\pm$ (right), that can be regarded as the error made on the fulfillment of the Neumann boundary condition, for different values of N. Both errors decay exponentially when N increases. Moreover, we report the errors obtained by using the Galerkin finite element approximations of degree $r = 1, 2, 3$.

Finally, it can be useful to observe that the interpolation derivative (10.40) can be represented through a matrix $\mathrm{D} \in \mathbb{R}^{(N+1)\times(N+1)}$, called *matrix of the interpolation derivative*, associating to any vector $\mathbf{v} \in \mathbb{R}^{N+1}$ of nodal values $v_i = \Phi(x_i)$, $i = 0, \ldots, N$, the vector $\mathbf{w} = \mathrm{D}\mathbf{v}$ whose components are the nodal values of the polynomial $D_N(\Phi)$,

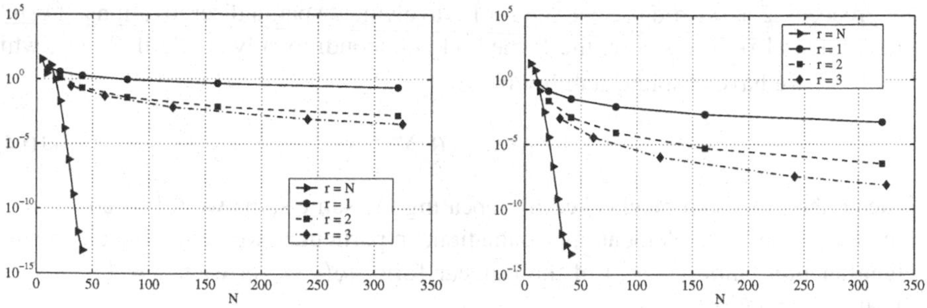

Fig. 10.9. Error in $H^1(-1,1)$ (left) and error on the Neumann datum (right) for varying N

i.e. $w_i = (D_N(\Phi))(x_i)$, $i = 0,\ldots,N$. The elements of D are (see [CHQZ06])

$$D_{ij} = \psi_j'(x_i) = \begin{cases} \dfrac{L_N(x_i)}{L_N(x_j)}\dfrac{1}{x_i - x_j}, & i,j = 0,\ldots,N,\, i \neq j, \\ -\dfrac{(N+1)N}{4}, & i = j = 0, \\ \dfrac{(N+1)N}{4}, & i = j = N, \\ 0 & \text{otherwise,} \end{cases}$$

where $d_0 = d_N = 2$ and $d_j = 1$ for $j = 1,\ldots,N-1$.

10.3.4 G-NI for parabolic equations

When we consider time-dependent problems, the spectral G-NI method can be used for the spatial approximation. For the discretization of the time derivative we can then apply a finite difference scheme. In this section, we consider one specific instance, the θ-method that was introduced in Sect. 5.1.

The θ-method applied to the G-NI spatial discretization of the homogeneous Dirichlet problem (5.4), defined on the space interval $-1 < x < 1$, is formulated as follows:
for each $k \geq 0$, find $u_N^k \in V_N = \{v_N \in \mathbb{Q}_N \,:\, v_N(-1) = v_N(1) = 0\}$ such that

$$\left(\frac{u_N^{k+1} - u_N^k}{\Delta t}, v_N\right)_N + a_N(\theta u_N^{k+1} + (1-\theta)u_N^k, v_N)$$
$$= \theta\,(f^{k+1}, v_N)_N + (1-\theta)\,(f^k, v_N)_N \qquad \forall v_N \in V_N,$$

with $u_N^0 = u_{0,N} \in V_N$ being a convenient approximation of u_0 (for instance, the interpolant $\Pi_N^{GLL} u_0$ introduced in (10.14)). As usual, $(\cdot,\cdot)_N$ denotes the discrete scalar product obtained using the Gauss-Legendre-Lobatto (GLL) numerical integration formula, while $a_N(\cdot,\cdot)$ is the approximation of the bilinear form $a(\cdot,\cdot)$ obtained by replacing the exact integrals with the above-mentioned numerical integration formula.

By proceeding as we did in Sect. 5.4 for finite element spatial discretizations, it can be proved that also in this case, the θ-method is unconditionally stable if $\theta \geq \frac{1}{2}$, while for $\theta < \frac{1}{2}$ we have absolute stability if

$$\Delta t \leq C(\theta) N^{-4}. \tag{10.42}$$

Indeed, the proof can be checked by repeating the same steps we followed earlier in the case of the finite element approximation. In particular, we define the eigenvalue-eigenfunction pairs (λ_j, w_N^j) of the bilinear form $a_N(\cdot,\cdot)$, for each $j = 1,\ldots,N-1$, through the relation

$$w_N^j \in V_N \, : \, a_N(w_N^j, v_N) = \lambda_j \, (w_N^j, v_N) \qquad \forall v_N \in V_N.$$

Hence

$$\lambda_j = \frac{a_N(w_N^j, w_N^j)}{\|w_N^j\|_N^2}.$$

Using the continuity of the bilinear form $a_N(\cdot,\cdot)$, we find

$$\lambda_j \leq \frac{M \, \|w_N^j\|^2_{\mathrm{H}^1(-1,1)}}{\|w_N^j\|_N^2}.$$

We now recall the following inverse inequality for algebraic polynomials ([CHQZ06])

$$\exists \, C_I > 0 \, : \, \|v_N'\|^2_{\mathrm{L}^2(-1,1)} \leq C_I \, \|v_N\|^2_{\mathrm{L}^2(-1,1)} \qquad \forall v_N \in \mathbb{Q}_N.$$

Then

$$\lambda_j \leq \frac{C_I^2 M N^4 \, \|w_N^j\|^2_{\mathrm{L}^2(-1,1)}}{\|w_N^j\|_N^2}.$$

Recalling the equivalence property (10.54), we conclude that

$$\lambda_j \leq 3 C_I^2 M N^4 \qquad \forall j = 1,\ldots,N-1.$$

Inequality (10.42) is now obtained using the stability condition (5.35) (with the finite element eigenvalues $\{\lambda_h^i\}$ replaced by the λ_j's). Moreover, we have the following convergence estimate, for $n \geq 1$ and $\Omega = (-1,1)$,

$$\|u(t^n) - u_N^n\|_{\mathrm{L}^2(\Omega)} \leq \widetilde{C}(t^n) \Big[N^{-r} \Big(|u_0|_{\mathrm{H}^r(\Omega)} + \int_0^{t^n} \Big| \frac{\partial u}{\partial t}(s) \Big|_{\mathrm{H}^r(\Omega)} ds$$
$$+ |u(t^n)|_{\mathrm{H}^r(\Omega)} \Big) + \Delta t \int_0^{t^n} \Big\| \frac{\partial^2 u}{\partial t^2}(s) \Big\|_{\mathrm{L}^2(\Omega)} ds \Big].$$

For the proof, refer to [CHQZ06, Chap. 7].

10.4 Generalization to the two-dimensional case

Let us consider as a domain the unit square $\Omega = (-1,1)^2$. Since Ω is the tensor product of the one-dimensional interval $(-1,1)$, it is natural to choose as nodes the points $\mathbf{x}_{ij}$ whose coordinates both coincide with the one-dimensional GLL nodes x_i,

$$\mathbf{x}_{ij} = (x_i, x_j), \quad i,j = 0,\dots,N,$$

while we take as weights the product of the corresponding one-dimensional weights

$$\alpha_{ij} = \alpha_i \alpha_j, \quad i,j = 0,\dots,N.$$

The Gauss-Legendre-Lobatto (GLL) integration formula in two dimensions is therefore defined by

$$\int_{\Omega} f(\mathbf{x})\, d\Omega \simeq \sum_{i,j=0}^{N} \alpha_{ij} f(\mathbf{x}_{ij}),$$

while the discrete scalar product is given by

$$(f,g)_N = \sum_{i,j=0}^{N} \alpha_{ij} f(\mathbf{x}_{ij}) g(\mathbf{x}_{ij}). \tag{10.43}$$

Analogously to the one-dimensional case it can be proved that the integration formula (10.43) is exact whenever the integrand function is a polynomial of degree at most $2N-1$. In particular, this implies that

$$(f,g)_N = (f,g) \qquad \forall\, f,g \text{ such that } fg \in \mathbb{Q}_{2N-1}.$$

In this section, for each N, $\mathbb{Q}_N$ denotes the space of polynomials of degree less than or equal to N with respect to each of the variables, introduced in (10.2).
We now consider as an example the problem

$$\begin{cases} Lu = -\mathrm{div}(\mu \nabla u) + \sigma u = f & \text{in } \Omega = (-1,1)^2, \\ u = 0 & \text{on } \partial\Omega. \end{cases}$$

By assuming that $\mu(\mathbf{x}) \geq \mu_0 > 0$ and $\sigma(\mathbf{x}) \geq 0$, the corresponding bilinear form is coercive in $\mathrm{H}^1_0(\Omega)$. Its G-NI approximation is given by

$$\text{find } u_N \in V_N: \qquad a_N(u_N, v_N) = F_N(v_N) \qquad \forall\, v_N \in V_N,$$

where

$$V_N = \{v \in \mathbb{Q}_N : v|_{\partial\Omega} = 0\},$$

$$a_N(u,v) = (\mu \nabla u, \nabla v)_N + (\sigma u, v)_N$$

and

$$F_N(v_N) = (f, v_N)_N.$$

As shown in the one-dimensional case, also in higher dimensions the G-NI formulation is equivalent to a collocation method where the operator L is replaced by L_N, the pseudo-spectral operator obtained by approximating each derivative by the corresponding interpolation derivative (10.40).
In the case of spectral element methods, we will need to generalize the GLL numerical integration formula on each element Ω_k. This can be done thanks to the transformation $\varphi_k : \widehat{\Omega} \to \Omega_k$ (see Fig. 10.2). Indeed, we can first of all generate the GLL nodes on the generic element Ω_k, by setting

$$\mathbf{x}_{ij}^{(k)} = \varphi_k(\mathbf{x}_{ij}), \quad i,j = 0,\ldots,N,$$

then defining the corresponding weights

$$\alpha_{ij}^{(k)} = \alpha_{ij}|\det J_k| = \alpha_{ij}\frac{|\Omega_k|}{4}, \quad i,j = 0,\ldots,N,$$

having denoted by J_k the Jacobian of the transformation φ_k and by $|\Omega_k|$ the measure of Ω_k. The GLL integration formula on Ω_k hence becomes

$$\int_{\Omega_k} f(x)\, d\mathbf{x} \simeq I_{N,k}^{GLL}(f) = \sum_{i,j=0}^{N} \alpha_{ij}^{(k)} f(\mathbf{x}_{ij}^{(k)}). \tag{10.44}$$

The spectral element formulation with Gaussian numerical integration, which we will denote by SEM-NI, becomes

$$\text{find } u_N \in V_N^C : \qquad a_{C,N}(u_N, v_N) = F_{C,N}(v_N) \quad \forall v_N \in V_N^C. \tag{10.45}$$

We have set

$$a_{C,N}(u_N, v_N) = \sum_k a_{\Omega_k,N}(u_N, v_N)$$

where $a_{\Omega_k,N}(u_N, v_N)$ is the approximation of $a_{\Omega_k}(u_N, v_N)$ obtained by approximating each integral on Ω_k that appears in its bilinear form via the GLL numerical integration formula in Ω_k (10.44). The term $F_{C,N}$ is defined in a similar way, and precisely $F_{C,N}(v_N) = \sum_k F_{\Omega_k,N}(v_N)$, where $F_{\Omega_k,N}$ is obtained, in turn, by replacing $\int_{\Omega_k} f v_N\, d\mathbf{x}$ with the formula $I_{N,k}^{GLL}(f v_N)$ for each k.

Remark 10.3. Fig. 10.10 summarizes rather schematically the origin of the different approximation schemes evoked up to now. In the case of finite differences, we have denoted by L_Δ the discretization of the operator through finite difference schemes applied to the various derivatives appearing in the definition of L. •

10.4.1 Convergence of the G-NI method

As observed in the one-dimensional case, the G-NI method can be considered as a generalized Galerkin method. For the latter, the analysis of convergence is based on the following general result:

Lemma 10.1 (Strang). *Consider the problem*

$$\text{find } u \in V: \quad a(u,v) = F(v) \quad \forall v \in V, \tag{10.46}$$

where V is a Hilbert space with norm $\|\cdot\|_V$, $F \in V'$ a linear and bounded functional on V and $a(\cdot,\cdot) : V \times V \to \mathbb{R}$ a bilinear, continuous and coercive form on V.
Consider an approximation of (10.46) *that can be formulated through the following generalized Galerkin problem*

$$\text{find } u_h \in V_h: \quad a_h(u_h, v_h) = F_h(v_h) \quad \forall v_h \in V_h, \tag{10.47}$$

$\{V_h,\ h > 0\}$ being a family of finite-dimensional subspaces of V.
Let us suppose that the discrete bilinear form $a_h(\cdot,\cdot)$ is continuous on $V_h \times V_h$, and uniformly coercive on V_h, that is

$$\exists \alpha^* > 0 \text{ independent of } h \text{ such that } a_h(v_h, v_h) \geq \alpha^* \|v_h\|_V^2 \qquad \forall\, v_h \in V_h.$$

Furthermore, let us suppose that F_h is a linear and bounded functional on V_h. Then:

1. *there exists a unique solution u_h to problem* (10.47)*;*
2. *such solution depends continuously on the data, i.e. we have*
$$\|u_h\|_V \leq \frac{1}{\alpha^*} \sup_{v_h \in V_h \setminus \{0\}} \frac{F_h(v_h)}{\|v_h\|_V};$$
3. *finally, the following a priori error estimate holds*

$$\begin{aligned} \|u - u_h\|_V \leq \inf_{w_h \in V_h} & \left\{ \left(1 + \frac{M}{\alpha^*}\right) \|u - w_h\|_V \right. \\ & \left. + \frac{1}{\alpha^*} \sup_{v_h \in V_h \setminus \{0\}} \frac{|a(w_h, v_h) - a_h(w_h, v_h)|}{\|v_h\|_V} \right\} \\ & + \frac{1}{\alpha^*} \sup_{v_h \in V_h \setminus \{0\}} \frac{|F(v_h) - F_h(v_h)|}{\|v_h\|_V}, \end{aligned} \tag{10.48}$$

M being the continuity constant of the bilinear form $a(\cdot,\cdot)$.

Proof. The assumptions of the Lax-Milgram lemma for problem (10.47) are satisfied, so the solution of such problem exists and is unique. Moreover,

$$\|u_h\|_V \leq \frac{1}{\alpha^*} \|F_h\|_{V_h'},$$

$\|F_h\|_{V_h'} = \sup_{v_h \in V_h \setminus \{0\}} \frac{F_h(v_h)}{\|v_h\|_V}$ being the norm of the dual space V_h' of V_h.

Let us now prove the error inequality (10.48). Let $u_h \in V_h$ be the solution of problem (10.47) and let w_h be any function of the subspace V_h. Setting $\sigma_h = u_h - w_h \in V_h$,

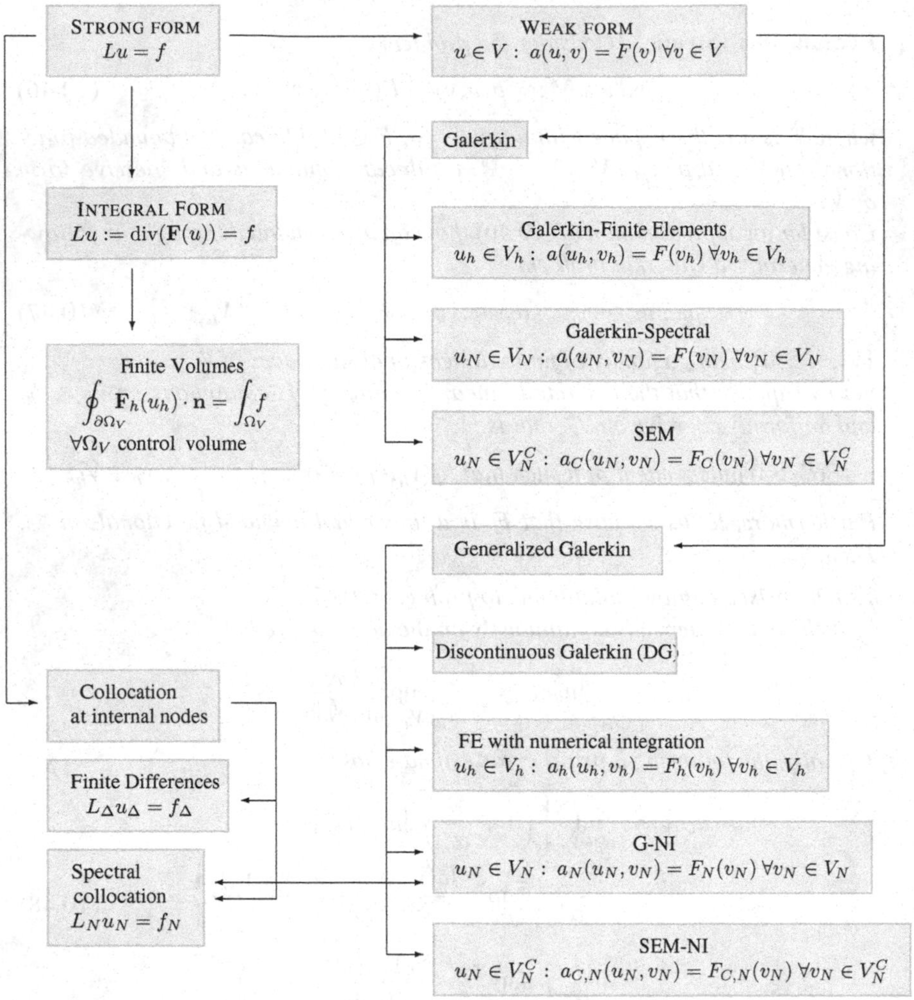

Fig. 10.10. Reference frame for the main numerical methods addressed in this book

we have:

$$\begin{aligned}
\alpha^* \|\sigma_h\|_V^2 &\leq a_h(\sigma_h, \sigma_h) \quad \text{[by the coercivity of } a_h\text{]} \\
&= a_h(u_h, \sigma_h) - a_h(w_h, \sigma_h) \\
&= F_h(\sigma_h) - a_h(w_h, \sigma_h) \quad \text{[thanks to (10.47)]} \\
&= F_h(\sigma_h) - F(\sigma_h) + F(\sigma_h) - a_h(w_h, \sigma_h) \\
&= [F_h(\sigma_h) - F(\sigma_h)] + a(u, \sigma_h) - a_h(w_h, \sigma_h) \quad \text{[thanks to (10.46)]} \\
&= [F_h(\sigma_h) - F(\sigma_h)] + a(u - w_h, \sigma_h) + [a(w_h, \sigma_h) - a_h(w_h, \sigma_h)]. \qquad (10.49)
\end{aligned}$$

If $\sigma_h \neq 0$, (10.49) can be divided by $\alpha^*\|\sigma_h\|_V$, to give

$$\|\sigma_h\|_V \leq \frac{1}{\alpha^*}\left\{\frac{|a(u-w_h,\sigma_h)|}{\|\sigma_h\|_V} + \frac{|a(w_h,\sigma_h)-a_h(w_h,\sigma_h)|}{\|\sigma_h\|_V} + \frac{|F_h(\sigma_h)-F(\sigma_h)|}{\|\sigma_h\|_V}\right\}$$

$$\leq \frac{1}{\alpha^*}\left\{M\|u-w_h\|_V + \sup_{v_h\in V_h\setminus\{0\}}\frac{|a(w_h,v_h)-a_h(w_h,v_h)|}{\|v_h\|_V} + \sup_{v_h\in V_h\setminus\{0\}}\frac{|F_h(v_h)-F(v_h)|}{\|v_h\|_V}\right\} \quad \text{[by the continuity of } a\text{]}.$$

If $\sigma_h = 0$ such inequality is still valid (it states that 0 is smaller than a sum of positive terms), although the proof breaks down.

We can now estimate the error between the solution u of (10.46) and the solution u_h of (10.47). Since

$$u - u_h = (u - w_h) - \sigma_h,$$

we obtain

$$\|u-u_h\|_V \leq \|u-w_h\|_V + \|\sigma_h\|_V \leq \|u-w_h\|_V + \frac{1}{\alpha^*}\left\{M\|u-w_h\|_V + \sup_{v_h\in V_h\setminus\{0\}}\frac{|a(w_h,v_h)-a_h(w_h,v_h)|}{\|v_h\|_V} + \sup_{v_h\in V_h\setminus\{0\}}\frac{|F_h(v_h)-F(v_h)|}{\|v_h\|_V}\right\}$$

$$= \left(1+\frac{M}{\alpha^*}\right)\|u-w_h\|_V + \frac{1}{\alpha^*}\sup_{v_h\in V_h\setminus\{0\}}\frac{|a(w_h,v_h)-a_h(w_h,v_h)|}{\|v_h\|_V} + \frac{1}{\alpha^*}\sup_{v_h\in V_h\setminus\{0\}}\frac{|F_h(v_h)-F(v_h)|}{\|v_h\|_V}.$$

If the previous inequality holds $\forall w_h \in V_h$, it also holds when taking the infimum when w_h varies in V_h. Hence, we obtain (10.48). ◇

By observing the right-hand side of inequality (10.48), we can recognize three different contributions to the approximation error $u - u_h$: the first is the best approximation error, the second is the error deriving from the approximation of the bilinear form $a(\cdot,\cdot)$ using the discrete bilinear form $a_h(\cdot,\cdot)$, and the third is the error arising from the approximation of the linear functional $F(\cdot)$ by the discrete linear functional $F_h(\cdot)$.

Remark 10.4. If in the preceding proof we choose $w_h = u_h^*$, u_h^* being the solution to the Galerkin problem

$$u_h^* \in V_h : \; a(u_h^*, v_h) = F(v_h) \quad \forall\, v_h \in V_h,$$

then the term $a(u - w_h, \sigma_h)$ is null thanks to (10.46), (4.1). It is therefore possible to obtain the following estimate, alternative to (10.48)

$$\begin{aligned} \|u - u_h\|_V \leq \|u - u_h^*\|_V & \\ + \frac{1}{\alpha^*} \sup_{v_h \in V_h \setminus \{0\}} & \frac{|a(u_h^*, v_h) - a_h(u_h^*, v_h)|}{\|v_h\|_V} \\ + \frac{1}{\alpha^*} \sup_{v_h \in V_h \setminus \{0\}} & \frac{|F(v_h) - F_h(v_h)|}{\|v_h\|_V}. \end{aligned}$$

The latter highlights the fact that the error due to the generalized Galerkin method can be bounded by the error of the Galerkin method plus the errors induced by the use of numerical integration for the computation of both $a(\cdot,\cdot)$ and $F(\cdot)$. •

We now want to apply Strang's lemma to the G-NI method, to verify its convergence. For simplicity we will only consider the one-dimensional case. Obviously, V_h will be replaced by V_N, u_h by u_N, v_h by v_N and w_h by w_N.
First of all, we begin by computing the error of the GLL numerical integration formula

$$E(g, v_N) = (g, v_N) - (g, v_N)_N,$$

g and v_N being a generic continuous function and a generic polynomial of $\mathbb{Q}_N$, respectively. By introducing the interpolation polynomial $\Pi_N^{GLL} g$ defined according to (10.14), we obtain

$$\begin{aligned} E(g, v_N) &= (g, v_N) - (\Pi_N^{GLL} g, v_N)_N \\ &= (g, v_N) - (\Pi_{N-1}^{GLL} g, v_N) + \underbrace{(\overbrace{\Pi_{N-1}^{GLL} g}^{\in \mathbb{Q}_{N-1}}, \overbrace{v_N}^{\in \mathbb{Q}_N})}_{\in \mathbb{Q}_{2N-1}} - (\Pi_N^{GLL} g, v_N)_N \\ &= (g, v_N) - (\Pi_{N-1}^{GLL} g, v_N) \\ &\quad + (\Pi_{N-1}^{GLL} g, v_N)_N - (\Pi_N^{GLL} g, v_N)_N \;\; [\text{by } (10.27)] \\ &= (g - \Pi_{N-1}^{GLL} g, v_N) + (\Pi_{N-1}^{GLL} g - \Pi_N^{GLL} g, v_N)_N. \end{aligned} \tag{10.50}$$

The first summand of the right-hand side can be bounded from above using the Cauchy-Schwarz inequality as follows

$$|(g - \Pi_{N-1}^{GLL} g, v_N)| \leq \|g - \Pi_{N-1}^{GLL} g\|_{\mathrm{L}^2(-1,1)} \|v_N\|_{\mathrm{L}^2(-1,1)}. \tag{10.51}$$

To find an upper bound for the second summand, we must first introduce the two following lemmas, for the proof of which we refer to [CHQZ06]:

Lemma 10.2. *The discrete scalar product* $(\cdot,\cdot)_N$ *defined in* (10.25) *is a scalar product on* $\mathbb{Q}_N$ *and, as such, it satisfies the Cauchy-Schwarz inequality*

$$|(\varphi,\psi)_N| \le \|\varphi\|_N \|\psi\|_N, \tag{10.52}$$

where the discrete norm $\|\cdot\|_N$ *is given by*

$$\|\varphi\|_N = \sqrt{(\varphi,\varphi)_N} \quad \forall\, \varphi \in \mathbb{Q}_N. \tag{10.53}$$

Lemma 10.3. *The "continuous" norm of* $\mathrm{L}^2(-1,1)$ *and the "discrete" norm* $\|\cdot\|_N$ *defined in* (10.53) *verify the inequalities*

$$\|v_N\|_{\mathrm{L}^2(-1,1)} \le \|v_N\|_N \le \sqrt{3}\|v_N\|_{\mathrm{L}^2(-1,1)} \quad \forall\, v_N \in \mathbb{Q}_N, \tag{10.54}$$

hence they are uniformly equivalent *on* $\mathbb{Q}_N$.

By using first (10.53) and then (10.54) we obtain

$$\begin{aligned}|(\Pi_{N-1}^{GLL}g - \Pi_N^{GLL}g, v_N)_N| &\le \|\Pi_{N-1}^{GLL}g - \Pi_N^{GLL}g\|_N\, \|v_N\|_N \\ &\le 3\left[\|\Pi_{N-1}^{GLL}g - g\|_{\mathrm{L}^2(-1,1)} + \|\Pi_N^{GLL}g - g\|_{\mathrm{L}^2(-1,1)}\right]\|v_N\|_{\mathrm{L}^2(-1,1)}.\end{aligned}$$

Using such inequality and (10.51), from (10.50) we can obtain the following upper bound

$$|E(g,v_N)| \le \left[4\|\Pi_{N-1}^{GLL}g - g\|_{\mathrm{L}^2(-1,1)} + 3\|\Pi_N^{GLL}g - g\|_{\mathrm{L}^2(-1,1)}\right]\|v_N\|_{\mathrm{L}^2(-1,1)}.$$

Using the interpolation estimate (10.17), we have that

$$|E(g,v_N)| \le C\left[\left(\frac{1}{N-1}\right)^s + \left(\frac{1}{N}\right)^s\right]\|g\|_{\mathrm{H}^s(-1,1)}\, \|v_N\|_{\mathrm{L}^2(-1,1)},$$

provided that $g \in \mathrm{H}^s(-1,1)$, for some $s \ge 1$. Finally, as for each $N \ge 2$, $1/(N-1) \le 2/N$, the Gauss-Legendre-Lobatto integration error results to be bound as

$$|E(g,v_N)| \le C\left(\frac{1}{N}\right)^s \|g\|_{\mathrm{H}^s(-1,1)}\|v_N\|_{\mathrm{L}^2(-1,1)}, \tag{10.55}$$

for each $g \in \mathrm{H}^s(-1,1)$ and for each polynomial $v_N \in \mathbb{Q}_N$.

At this point we are ready to evaluate the various contributions that intervene in (10.48). We anticipate that this analysis will be carried out in the case where suitable simplifying hypotheses are introduced on the differential problem under exam. We begin with the simplest term, i.e. the one associated with the functional F, supposing to consider a problem with homogeneous Dirichlet boundary conditions, in order to obtain $F(v_N) = (f, v_N)$ and $F_N(v_N) = (f, v_N)_N$. Provided that $f \in \mathrm{H}^s(-1,1)$ for some $s \geq 1$, then,

$$
\begin{aligned}
&\sup_{v_N \in V_N \setminus \{0\}} \frac{|F(v_N) - F_N(v_N)|}{\|v_N\|_V} = \sup_{v_N \in V_N \setminus \{0\}} \frac{|(f, v_N) - (f, v_N)_N|}{\|v_N\|_V} \\
&= \sup_{v_N \in V_N \setminus \{0\}} \frac{|E(f, v_N)|}{\|v_N\|_V} \leq \sup_{v_N \in V_N \setminus \{0\}} \frac{C\left(\frac{1}{N}\right)^s \|f\|_{\mathrm{H}^s(-1,1)} \|v_N\|_{\mathrm{L}^2(-1,1)}}{\|v_N\|_V} \qquad (10.56) \\
&\leq C\left(\frac{1}{N}\right)^s \|f\|_{\mathrm{H}^s(-1,1)},
\end{aligned}
$$

having exploited relation (10.55) and having bounded the norm in $\mathrm{L}^2(-1,1)$ by that in $\mathrm{H}^s(-1,1)$.

As for the contribution arising from the approximation of the bilinear form,

$$
\sup_{v_N \in V_N \setminus \{0\}} \frac{|a(w_N, v_N) - a_N(w_N, v_N)|}{\|v_N\|_V},
$$

we cannot explicitly evaluate it without referring to a particular differential problem. We then choose, as an example, the one-dimensional diffusion-reaction problem (10.22), supposing moreover that μ and σ are constant. Incidentally, such problem satisfies homogeneous Dirichlet boundary conditions, in accordance with what was requested for deriving estimate (10.56). In such case, the associated bilinear form is

$$
a(u, v) = (\mu u', v') + (\sigma u, v),
$$

while its G-NI approximation is given by

$$
a_N(u, v) = (\mu u', v')_N + (\sigma u, v)_N.
$$

We must then evaluate

$$
a(w_N, v_N) - a_N(w_N, v_N) = (\mu w_N', v_N') - (\mu w_N', v_N')_N + (\sigma w_N, v_N) - (\sigma w_N, v_N)_N.
$$

Since $w_N' v_N' \in \mathbb{Q}_{2N-2}$, if we suppose that μ is constant, the product $\mu w_N' v_N'$ is integrated exactly by the GLL integration formula, that is $(\mu w_N', v_N') - (\mu w_N', v_N')_N = 0$.
We now observe that

$$
(\sigma w_N, v_N) - (\sigma w_N, v_N)_N = E(\sigma w_N, v_N) = E(\sigma(w_N - u), v_N) + E(\sigma u, v_N),
$$

and therefore, using (10.55), we obtain

$$
|E(\sigma(w_N - u), v_N)| \leq C\left(\frac{1}{N}\right) \|\sigma(w_N - u)\|_{\mathrm{H}^1(-1,1)} \|v_N\|_{\mathrm{L}^2(-1,1)},
$$

$$
|E(\sigma u, v_N)| \leq C\left(\frac{1}{N}\right)^s \|\sigma u\|_{\mathrm{H}^s(-1,1)} \|v_N\|_{\mathrm{L}^2(-1,1)}.
$$

On the other hand, since σ is also constant, setting $w_N = \Pi_N^{GLL} u$ and using (10.17), we obtain

$$\|\sigma(w_N - u)\|_{\mathrm{H}^1(-1,1)} \leq C\|u - \Pi_N^{GLL} u\|_{\mathrm{H}^1(-1,1)} \leq C\left(\frac{1}{N}\right)^{s-1} \|u\|_{\mathrm{H}^s(-1,1)}.$$

Hence,

$$\sup_{v_N \in V_N \setminus \{0\}} \frac{|a(w_N, v_N) - a_N(w_N, v_N)|}{\|v_N\|_V} \leq C^* \left(\frac{1}{N}\right)^s \|u\|_{\mathrm{H}^s(-1,1)}. \tag{10.57}$$

We still need to estimate the first summand of (10.48). Having chosen $w_N = \Pi_N^{GLL} u$ and exploiting (10.17) again, we obtain that

$$\|u - w_N\|_V = \|u - \Pi_N^{GLL} u\|_{\mathrm{H}^1(-1,1)} \leq C\left(\frac{1}{N}\right)^s \|u\|_{\mathrm{H}^{s+1}(-1,1)} \tag{10.58}$$

provided that $u \in \mathrm{H}^{s+1}(-1,1)$, for a suitable $s \geq 1$. To conclude, thanks to (10.56), (10.57) and (10.58), from (10.48) applied to the G-NI approximation of problem (10.22), and under the previous hypotheses, we find the following error estimate

$$\|u - u_N\|_{\mathrm{H}^1(-1,1)} \leq C\left(\frac{1}{N}\right)^s \left(\|f\|_{\mathrm{H}^s(-1,1)} + \|u\|_{\mathrm{H}^{s+1}(-1,1)}\right).$$

The convergence analysis just carried out for the model problem (10.22) can be generalized (with a few technical difficulties) to the case of more complex differential problems and different boundary conditions.

Example 10.2 (Problem with regularity depending on a parameter). Let us consider the following (trivial but instructive) problem

$$\begin{cases} -u'' = 0, & x \in (0,1], \\ -u'' = -\alpha(\alpha-1)(x-1)^{\alpha-2}, & x \in (1,2), \\ u(0) = 0, & u(2) = 1, \end{cases}$$

with $\alpha \in \mathbb{N}$. The exact solution is null in $(0,1)$ and equals $(x-1)^\alpha$ for $x \in (1,2)$. Thus it belongs to $\mathrm{H}^\alpha(0,2)$, but not to $\mathrm{H}^{\alpha+1}(0,2)$. We report in Table 10.1 the behaviour of the error in $\mathrm{H}^1(0,2)$ norm with respect to N using a G-NI method for three different values of α. As it can be seen, when the regularity increases, so does the order of convergence of the spectral method with respect to N, as stated by the theory. In the same table we report the results obtained using linear finite elements (this time N denotes the number of elements). The order of convergence of the finite element method remains linear in either case. ■

Example 10.3. Let us take the second example illustrated in Sect. 4.6.3, this time using the spectral element method. Let us consider a partition of the domain into four spectral elements of degree 8 as shown in the left of Fig. 10.11. The solution obtained

Table 10.1. Behaviour of the error of the G-NI spectral method for varying polynomial degree N and solution regularity index (left). Behaviour of the error of the linear finite element method for varying number of intervals N and solution regularity index (right)

N	$\alpha=2$	$\alpha=3$	$\alpha=4$
4	0.5931	0.2502	0.2041
8	0.3064	0.0609	0.0090
16	0.1566	0.0154	$7.5529 \cdot 10^{-4}$
32	0.0792	0.0039	$6.7934 \cdot 10^{-5}$

N	$\alpha=2$	$\alpha=3$
4	0.4673	0.5768
8	0.2456	0.3023
16	0.1312	0.1467
32	0.0745	0.0801

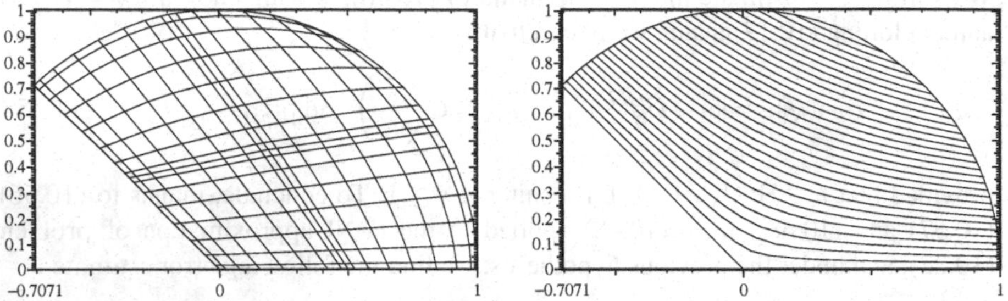

Fig. 10.11. The grid (left) and the solution isolines obtained using the spectral finite element method (right) for the problem in Example 10.3

(Fig. 10.11, left) does not exhibit any inaccuracy in proximity of the origin, as opposed to the solution obtained using finite elements in the absence of grid adaptivity (compare with Fig. 4.24, left). ■

10.5 G-NI and SEM-NI methods for a one-dimensional model problem

Let us consider the one-dimensional diffusion-reaction problem

$$-[(1+x^2)\,u'(x)]' + \cos(x^2)\,u(x) = f(x), \quad x \in (-1,1), \tag{10.59}$$

together with mixed-type boundary conditions

$$u(-1) = 0, \quad u'(1) = 1.$$

The goal of this section is to discuss in detail how to formulate the G-NI and SEM-NI approximations. For the former, we will also provide the corresponding matrix formulation as well as a stability analysis.

10.5.1 The G-NI method

The weak formulation of problem (10.59) is:

$$\text{find} \quad u \in V : a(u,v) = F(v) \quad \forall v \in V,$$

$V = \{v \in \mathrm{H}^1(-1,1) : v(-1) = 0\}$, $a : V \times V \longrightarrow \mathbb{R}$ and $F : V \longrightarrow \mathbb{R}$ being the bilinear form and the linear functional, respectively, defined by

$$a(u,v) = \int_{-1}^{1} (1+x^2)\, u'(x)\, v'(x)\, dx + \int_{-1}^{1} \cos(x^2)\, u(x)\, v(x)\, dx,$$

$$F(v) = \int_{-1}^{1} f(x)\, v(x)\, dx + 2\, v(1).$$

The spectral-Galerkin formulation (SM) takes the following form

$$\text{find} \quad u_N \in V_N \quad \text{such that} \quad a(u_N, v_N) = F(v_N) \quad \forall v_N \in V_N, \tag{10.60}$$

with

$$V_N = \{v_N \in \mathbb{Q}_N \; : \; v_N(-1) = 0\} \subset V. \tag{10.61}$$

In order to obtain the corresponding G-NI formulation, it is sufficient to approximate in (10.60) all scalar products on $\mathrm{L}^2(-1,1)$ with the GLL discrete scalar product defined in (10.25). We then have

$$\text{find} \quad u_N^* \in V_N : a_N(u_N^*, v_N) = F_N(v_N) \quad \forall v_N \in V_N, \tag{10.62}$$

having set

$$\begin{aligned} a_N(u,v) &= \left((1+x^2)\, u', v'\right)_N + \left(\cos(x^2)\, u, v\right)_N \\ &= \sum_{i=0}^{N} (1+x_i^2)\, u'(x_i)\, v'(x_i)\, \alpha_i + \sum_{i=1}^{N} \cos(x_i^2)\, u(x_i)\, v(x_i)\, \alpha_i \end{aligned} \tag{10.63}$$

and

$$F_N(v) = (f, v)_N + 2\, v(1) = \sum_{i=1}^{N} f(x_i)\, v(x_i)\, \alpha_i + 2\, v(1). \tag{10.64}$$

Note that this requires f to be continuous. We observe that the index i of the last sum in (10.63) and of the sum in (10.64) starts from 1, instead of 0, since $v(x_0) = v(-1) = 0$. Moreover, the formulations SM (10.60) and G-NI (10.62) never coincide. Consider, for instance, the diffusive term $(1+x^2)\,(u_N^*)'\, v_N'$: this is a polynomial of degree $2N$. Since the GLL integration formula has exactness degree 2N-1, the discrete scalar product (10.25) will not return the exact value of the corresponding continuous scalar product $\left((1+x^2)\,(u_N^*)', v_N'\right)$.

To obtain the matrix formulation of the G-NI approximation, we denote by ψ_i, for $i = 1, \ldots, N$, the characteristic polynomials associated to all GLL nodes except to the one where a Dirichlet boundary condition is assigned, $x_0 = -1$. Such polynomials constitute a basis for the space V_N introduced in (10.61). This allows us, in the first place, to write the solution u_N^* of the G-NI formulation as

$$u_N^*(x) = \sum_{j=1}^{N} u_N^*(x_j)\,\psi_j(x).$$

Secondly, we can choose in (10.62) $v_N = \psi_i$, $i = 1,\ldots,N$, obtaining

$$a_N(u_N^*, \psi_i) = F_N(\psi_i), \quad i = 1,\ldots,N,$$

i.e.

$$\sum_{j=1}^{N} u_N^*(x_j)\,a_N(\psi_j, \psi_i) = F_N(\psi_i), \quad i = 1,\ldots,N.$$

In matrix form,

$$A\,\mathbf{u}_N^* = \mathbf{f},$$

having $\mathbf{u}_N^* = (u_N^*(x_i))$, $A = (a_{ij})$, with

$$\begin{aligned} a_{ij} = a_N(\psi_j, \psi_i) &= \sum_{k=0}^{N} (1+x_k^2)\,\psi_j'(x_k)\,\psi_i'(x_k)\,\alpha_k + \sum_{k=1}^{N} \cos(x_k^2)\,\psi_j(x_k)\,\psi_i(x_k)\,\alpha_k \\ &= \sum_{k=0}^{N} (1+x_k^2)\,\psi_j'(x_k)\,\psi_i'(x_k)\,\alpha_k + \cos(x_i^2)\,\alpha_i\,\delta_{ij}, \end{aligned}$$

and

$$\begin{aligned} \mathbf{f} = (f_i), \ \text{con} \ f_i = F_N(\psi_i) &= (f, \psi_i)_N + 2\,\psi_i(1) \\ &= \sum_{k=1}^{N} f(x_k)\,\psi_i(x_k)\,\alpha_k + 2\,\psi_i(1) \\ &= \begin{cases} \alpha_i\,f(x_i) & \text{for} \quad i = 1,\ldots,N-1, \\ \alpha_N\,f(1)+2 & \text{for} \quad i = N. \end{cases} \end{aligned}$$

We recall that the matrix A, besides being ill-conditioned, is full due to the presence of the diffusive term.

Finally, we can verify that the G-NI method (10.62) can be reformulated as a suitable collocation method. To this end, we wish to rewrite the discrete formulation (10.62) in continuous form in order to counterintegrate by parts, i.e. to return to the initial differential operator. In order to do this, we will resort to the interpolation operator Π_N^{GLL} defined in (10.15), recalling in addition that the discrete scalar product (10.25) coincides with the continuous one on $\mathrm{L}^2(-1,1)$ if the product of the two integrand functions is a polynomial of degree $\leq 2N-1$ (see (10.36)).

We then accurately rewrite the first summand of $a_N(u_N^*, v_N)$, ignoring the $*$ to simplify the notation. Thanks to (10.36) and integrating by parts, we have

$$
\begin{aligned}
&\big((1+x^2)\,u_N',v_N'\big)_N \\
&\quad= \big(\Pi_N^{GLL}((1+x^2)\,u_N'),v_N'\big)_N = \big(\Pi_N^{GLL}((1+x^2)\,u_N'),v_N'\big) \\
&\quad= -\big([\Pi_N^{GLL}((1+x^2)\,u_N')]',v_N\big) + \Pi_N^{GLL}((1+x^2)\,u_N')(1)\,v_N(1) \\
&\quad= -\big([\Pi_N^{GLL}((1+x^2)\,u_N')]',v_N\big)_N + \Pi_N^{GLL}((1+x^2)\,u_N')(1)\,v_N(1).
\end{aligned}
$$

Hence, we can reformulate (10.62) as

$$
\begin{aligned}
\text{find}\quad u_N \in V_N\ :\ \big(L_N u_N, v_N\big)_N &= (f,v_N)_N \\
&+\ \big(2-\Pi_N^{GLL}((1+x^2)\,u_N')(1)\big)\,v_N(1)\quad \forall v_N \in V_N,
\end{aligned}
\tag{10.65}
$$

with

$$
L_N u_N = -\big[\Pi_N^{GLL}((1+x^2)\,u_N')\big]' + \cos(x^2)\,u_N = -D_N\big((1+x^2)\,u_N'\big) + \cos(x^2)\,u_N,
$$

D_N being the interpolation derivative introduced in (10.40). We now choose (10.65) $v_N = \psi_i$. For $i = 1,\ldots,N-1$, we have

$$
\begin{aligned}
\big(L_N u_N,\psi_i\big)_N &= \big(-\big[\Pi_N^{GLL}((1+x^2)\,u_N')\big]',\psi_i\big)_N + \big(\cos(x^2)\,u_N,\psi_i\big)_N \\
&= -\sum_{j=1}^{N-1}\alpha_j\,\big[\Pi_N^{GLL}((1+x^2)\,u_N')\big]'(x_j)\,\psi_i(x_j) + \sum_{j=1}^{N-1}\alpha_j\cos(x_j^2)\,u_N(x_j)\,\psi_i(x_j) \\
&= -\alpha_i\,\big[\Pi_N^{GLL}((1+x^2)\,u_N')\big]'(x_i) + \alpha_i\cos(x_i^2)\,u_N(x_i) = (f,\psi_i)_N \\
&= \sum_{j=1}^{N-1}\alpha_j\,f(x_j)\,\psi_i(x_j) = \alpha_i\,f(x_i),
\end{aligned}
$$

that is, exploiting the definition of the L_N operator and dividing everything by α_i,

$$
L_N u_N(x_i) = f(x_i), \qquad i = 1,\ldots,N-1. \tag{10.66}
$$

Having set $v_N = \psi_N$ in (10.65), we obtain instead

$$
\begin{aligned}
\big(L_N u_N,\psi_N\big)_N &= -\alpha_N\,\big[\Pi_N^{GLL}((1+x^2)\,u_N')\big]'(x_N) + \alpha_N\cos(x_N^2)\,u_N(x_N) \\
&= (f,\psi_N)_N + 2 - \Pi_N^{GLL}((1+x^2)\,u_N')(1) \\
&= \alpha_N\,f(x_N) + 2 - \Pi_N^{GLL}((1+x^2)\,u_N')(1),
\end{aligned}
$$

or, dividing all by α_N,

$$
L_N u_N(x_N) = f(x_N) + \frac{1}{\alpha_N}\big(2-\Pi_N^{GLL}((1+x^2)\,u_N')(1)\big). \tag{10.67}
$$

Equations (10.66) and (10.67) therefore provide the collocation in all the nodes (except the potential boundary ones where Dirichlet conditions are assigned) of the given differential problem, after approximating of the differential operator L using operator L_N.

Finally, we analyze the stability of formulation (10.62). Since we are dealing with a generalized Galerkin-type of approach, we will have to resort to the Strang

Lemma 10.1. This guarantees that, for the solution u_N^* of (10.62), the estimate

$$\|u_N^*\|_V \leq \frac{1}{\alpha^*} \sup_{v_N \in V_N \setminus \{0\}} \frac{|F_N(v_N)|}{\|v_N\|_V} \tag{10.68}$$

holds, α^* being the (uniform) coercivity constant associated to the discrete bilinear form $a_N(\cdot,\cdot)$. We apply this result to problem (10.59), by computing first of all α^*. By exploiting the definition (10.53) of the discrete norm $\|\cdot\|_N$ and the equivalence relation (10.54), we have

$$\begin{aligned} a_N(u_N,u_N) &= \big((1+x^2)\,u_N', u_N'\big)_N + \big(\cos(x^2)\,u_N, u_N\big)_N \\ &\geq \big(u_N', u_N'\big)_N + \cos(1)\,\big(u_N, u_N\big)_N = \|u_N'\|_N^2 + \cos(1)\,\|u_N\|_N^2 \\ &\geq \|u_N'\|^2_{\mathrm{L}^2(-1,1)} + \cos(1)\,\|u_N\|^2_{\mathrm{L}^2(-1,1)} \geq \cos(1)\,\|u_N\|_V^2, \end{aligned}$$

having moreover exploited the relations

$$\begin{aligned} \min_j(1+x_j^2) &\geq \min_{x\in[-1,1]}(1+x^2) = 1, \\ \min_j \cos(x_j^2) &\geq \min_{x\in[-1,1]} \cos(x^2) = \cos(1). \end{aligned}$$

This allows us to identify α^* using the value $\cos(1)$. At this point, we can evaluate the quotient $|F_N(v_N)|/\|v_N\|_V$ in (10.68). Indeed, we have

$$\begin{aligned} |F_N(v_N)| &= |(f,v_N)_N + 2v_N(1)| \leq \|f\|_N\,\|v_N\|_N + 2\,|v_N(1)| \\ &\leq \sqrt{3}\,\|f\|_N\,\|v_N\|_V + 2\left|\int_{-1}^{1} v_N'(x)\,dx\right| \leq \sqrt{3}\,\|f\|_N\,\|v_N\|_V + 2\sqrt{2}\,\|v_N\|_V, \end{aligned}$$

having once more used the equivalence property (10.54) together with the Cauchy-Schwarz inequality in its discrete (10.52) and continuous (3.7) versions. We can thus conclude that

$$\frac{|F_N(v_N)|}{\|v_N\|_V} \leq \sqrt{3}\,\|f\|_N + 2\sqrt{2},$$

that is, returning to the stability estimate (10.68),

$$\|u_N^*\|_V \leq \frac{1}{\cos(1)}\left[\sqrt{3}\,\|f\|_N + 2\sqrt{2}\right].$$

Finally, we note that $\|f\|_N \leq 2\,\|f\|_{C^0([-1,1])}\ \forall f \in C^0([-1,1])$.

10.5.2 The SEM-NI method

Starting from problem (10.59), we now want to consider its SEM-NI formulation, i.e. a spectral element formulation that uses the integration formulae of type GLL in each element. Moreover, we propose to provide a basis for the space where such formulation will be implemented.

We first introduce a partition of the interval $(-1,1)$ in M (≥ 2) disjoint sub-intervals $\Omega_m = (\overline{x}_{m-1}, \overline{x}_m)$, with $m = 1, \ldots, M$, denoting by $h_m = \overline{x}_m - \overline{x}_{m-1}$ the width of the m-th interval, and setting $h = \max_m h_m$. The SEM formulation of problem (10.59) takes the form

$$\text{find} \quad u_N \in V_N^C : a(u_N, v_N) = F(v_N) \quad \forall v_N \in V_N^C, \tag{10.69}$$

with

$$V_N^C = \{v_N \in C^0([-1,1]) \,:\, v_N\big|_{\Omega_m} \in \mathbb{Q}_N,\ \forall m = 1,\ldots,M,\ v_N(-1) = 0\}.$$

We note that the functional space V_N^C of the SEM approach loses the "global" nature that is instead typical of a SM formulation. Similarly to what happens in the case of finite element approximations, we now have piecewise polynomial functions. By exploiting the partition $\{\Omega_m\}$, we can rewrite formulation (10.69) in the following way

$$\text{find} \quad u_N \in V_N^C : \sum_{m=1}^{M} a_{\Omega_m}(u_N, v_N) = \sum_{m=1}^{M} F_{\Omega_m}(v_N) \quad \forall v_N \in V_N^C, \tag{10.70}$$

where

$$\begin{aligned} a_{\Omega_m}(u_N, v_N) &= a(u_N, v_N)\big|_{\Omega_m} \\ &= \int_{\overline{x}_{m-1}}^{\overline{x}_m} (1+x^2)\, u_N'(x)\, v_N'(x)\, dx + \int_{\overline{x}_{m-1}}^{\overline{x}_m} \cos(x^2)\, u_N(x)\, v_N(x)\, dx, \end{aligned}$$

while

$$F_{\Omega_m}(v_N) = F(v_N)\big|_{\Omega_m} = \int_{\overline{x}_{m-1}}^{\overline{x}_m} f(x) v_N(x)\, dx + 2 v_N(1)\delta_{mM}.$$

The SEM-NI formulation can be obtained at this point by replacing in (10.70) the continuous scalar products by the discrete GLL scalar product (10.25):

$$\text{find} \quad u_N^* \in V_N^C : \sum_{m=1}^{M} a_{N,\Omega_m}(u_N^*, v_N) = \sum_{m=1}^{M} F_{N,\Omega_m}(v_N) \quad \forall v_N \in V_N^C,$$

where

$$\begin{aligned} a_{N,\Omega_m}(u,v) &= \left((1+x^2)\,u', v'\right)_{N,\Omega_m} + \left(\cos(x^2)\,u, v\right)_{N,\Omega_m}, \\ F_{N,\Omega_m}(v) &= (f, v)_{N,\Omega_m} + 2v(1)\delta_{mM}, \\ (u,v)_{N,\Omega_m} &= \sum_{i=0}^{N} u(x_i^{(m)})\, v(x_i^{(m)})\, \alpha_i^{(m)}, \end{aligned}$$

$x_i^{(m)}$ being the i-th GLL node of the sub-interval Ω_m and $\alpha_i^{(m)}$ the corresponding integration weight.

Starting from the reference element $\widehat{\Omega} = (-1,1)$ (which, in the case under exam, coincides with the domain Ω of problem (10.59)) and calling

$$\varphi_m(\xi) = \frac{h_m}{2}\xi + \frac{\overline{x}_m + \overline{x}_{m-1}}{2}, \qquad \xi \in [-1,1],$$

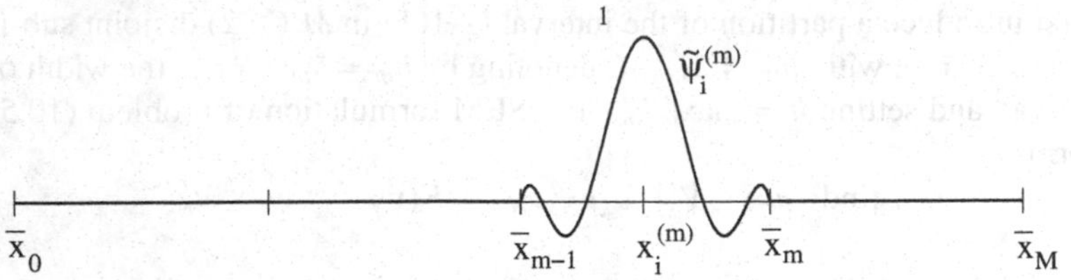

Fig. 10.12. basis function $\widetilde{\psi}_i^{(m)}$ associated to the internal node $x_i^{(m)}$

the affine map from $\widehat{\Omega}$ into Ω_m, for $m = 1,\ldots,M$, we will have

$$x_i^{(m)} = \varphi_m(x_i) = \alpha_i^{(m)} = \frac{h_m}{2}\alpha_i, \quad i = 0,...N \tag{10.71}$$

that is $x_i^{(m)}$ is the image, through the mapping φ_m, of the i-th GLL node of $\widehat{\Omega}$.

We introduce, on each Ω_m, the set $\{\psi_i^{(m)}\}_{i=0}^N$ of basis functions, such that

$$\psi_i^{(m)}(x) = \psi_i(\varphi_m^{-1}(x)) \quad \forall x \in \Omega_m,$$

ψ_i being the characteristic polynomial introduced in (10.12) and (10.13) associated to node x_i of GLL in $\widehat{\Omega}$. Having now a basis for each sub-interval Ω_m, we can write the solution u_N of the SEM on each Ω_m as

$$u_N(x) = \sum_{i=0}^{N} u_i^{(m)} \psi_i^{(m)}(x) \quad \forall x \in \Omega_m, \tag{10.72}$$

where $u_i^{(m)} = u_N(x_i^{(m)})$.
Since we want to define a global basis for the space V_N^C, we start by defining the basis functions associated to the internal nodes of Ω_m, for $m = 1,\ldots,M$. For this purpose, it will be sufficient to extend trivially, outside Ω_m, each basis function $\psi_i^{(m)}$, yielding

$$\widetilde{\psi}_i^{(m)}(x) = \begin{cases} \psi_i^{(m)}(x), & x \in \Omega_m \\ 0, & \text{otherwise.} \end{cases}$$

These are, overall, $(N-1)M$ functions that behave as shown in Fig. 10.12. For each end node $\bar{x}_m$ of the Ω_m sub-domains, with $m = 1,\ldots,M-1$, we define instead the basis function

$$\psi_m^*(x) = \begin{cases} \psi_N^{(m)}(x), & x \in \Omega_m \\ \psi_0^{(m+1)}(x), & x \in \Omega_{m+1} \\ 0, & \text{otherwise,} \end{cases}$$

obtained by "pasting" $\psi_N^{(m)}$ and $\psi_0^{(m+1)}$ together (see Fig. 10.13). In particular, we observe that ψ_0^* is not needed, since a homogeneous Dirichlet condition is assigned

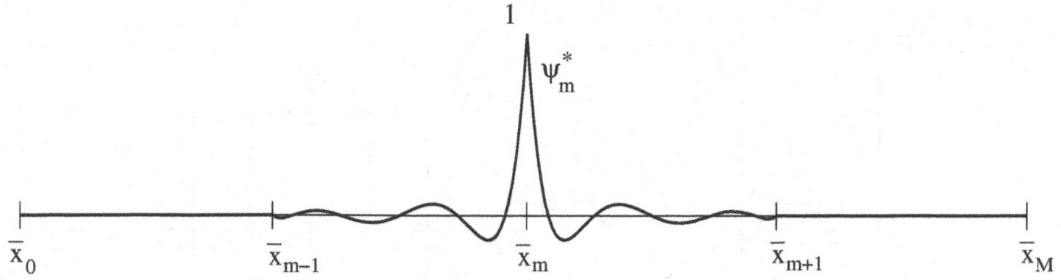

Fig. 10.13. basis function ψ_m^* associated to the internal node $\bar{x}_m$

at $\bar{x}_0 = -1$, whereas we need ψ_M^* that we indicate with $\psi_N^{(M)}$. Thus, by the choice of boundary conditions made, there exist M basis functions associated to the endpoints of the sub-intervals Ω_m. (Had Dirichlet conditions been applied at both endpoints of Ω, we would have had the $(M-1)$ functions ψ_m^*, $m = 1,...,M-1$.)
Hence, we have $n = (N-1)M + M$ basis functions for the space V_N^C altogether. Each function $u_N \in V_N^C$ can then be expressed in the following way

$$u_N(x) = \sum_{m=1}^{M} u_m^\Gamma \, \psi_m^*(x) + \sum_{m=1}^{M} \sum_{i=1}^{N-1} u_i^{(m)} \, \widetilde{\psi}_i^{(m)}(x),$$

with $u_m^\Gamma = u_N(\bar{x}_m)$ and $u_i^{(m)}$ defined as in (10.72). This way, the Dirichlet boundary condition is respected.

10.6 Spectral methods on triangles and tetrahedra

As we have seen, the use of spectral methods on quadrilaterals in two dimensions (or parallelepipeds in three dimensions) is made possible by the use of tensor products of one-dimensional functions (on the reference interval $[-1,1]$) and of the one-dimensional Gaussian numerical integration formulae. Since a few years, however, we are witnessing a growth of interest in the use of spectral-type methods also on geometries that do not have tensor product structure, such as, for instance, triangles in 2D and tetrahedra, prisms or pyramids in 3D.

We briefly describe Dubiner's pioneering idea [Dub91] to introduce polynomial bases of high degree on triangles, later extended in [KS05] to the three-dimensional case. We consider the reference triangle

$$\widehat{T} = \{(x_1,x_2) \in \mathbb{R}^2 \; : \; -1 < x_1,x_2 \; ; \; x_1 + x_2 < 0\}$$

and the reference square

$$\widehat{Q} = \{(\xi_1,\xi_2) \in \mathbb{R}^2 \; : \; -1 < \xi_1,\xi_2 < 1\}.$$

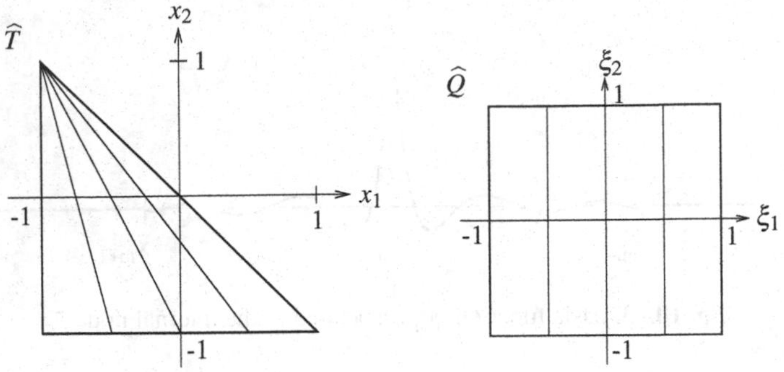

Fig. 10.14. Transformation of the reference triangle $\widehat{T}$ into the reference square $\widehat{Q}$. The slanting segments are transformed into vertical segments

The transformation

$$(x_1,x_2) \to (\xi_1,\xi_2), \quad \xi_1 = 2\frac{1+x_1}{1-x_2} - 1, \quad \xi_2 = x_2 \tag{10.73}$$

is a bijection between $\widehat{T}$ and $\widehat{Q}$. Its inverse is given by

$$(\xi_1,\xi_2) \to (x_1,x_2), \quad x_1 = \frac{1}{2}(1+\xi_1)(1-\xi_2) - 1, \quad x_2 = \xi_2.$$

As highlighted in Fig. 10.14, the mapping $(x_1,x_2) \to (\xi_1,\xi_2)$ sends the ray in $\widehat{T}$ issuing from the vertex $(-1,1)$ and passing through the point $(x_1,-1)$ to the vertical segment of $\widehat{Q}$ of equation $\xi_1 = x_1$. The latter therefore becomes singular in $(-1,1)$. For this reason we call (ξ_1,ξ_2) the *collapsed Cartesian coordinates* of the point of the triangles having coordinates (x_1,x_2).
We denote by $\{J_k^{(\alpha,\beta)}(\xi),\ k \geq 0\}$ the family of Jacobi polynomials that are orthogonal with respect to the weight $w(\xi) = (1-\xi)^\alpha(1+\xi)^\beta$, for $\alpha,\beta \geq 0$. Hence,

$$\forall k \geq 0, \quad J_k^{(\alpha,\beta)} \in \mathbb{P}_k \quad \text{and} \quad \int_{-1}^{1} J_k^{(\alpha,\beta)}(\xi)\, J_m^{(\alpha,\beta)}(\xi)\, w(\xi)\, d\xi = 0 \quad \forall\, m \neq k. \tag{10.74}$$

We observe that, for $\alpha = \beta = 0$, $J_k^{(0,0)}$ coincides with the k-th Legendre polynomial L_k.
For each pair of integers $\mathbf{k} = (k_1,k_2)$ we define the so-called *warped tensor product* basis on $\widehat{Q}$

$$\Phi_{\mathbf{k}}(\xi_1,\xi_2) = \Psi_{k_1}(\xi_1)\,\Psi_{k_1,k_2}(\xi_2), \tag{10.75}$$

with $\Psi_{k_1}(\xi_1) = J_{k_1}^{(0,0)}(\xi_1)$ and $\Psi_{k_1,k_2}(\xi_2) = (1-\xi_2)^{k_1} J_{k_2}^{(2k_1+1,0)}(\xi_2)$. Note that $\Phi_{\mathbf{k}}$ is a polynomial of degree k_1 in ξ_1 and k_1+k_2 in ξ_2.

By now applying mapping (10.73), we find the following function defined on $\widehat{T}$

$$\varphi_{\mathbf{k}}(x_1,x_2) = \Phi_{\mathbf{k}}(\xi_1,\xi_2) = J_{k_1}^{(0,0)}\Big(2\frac{1+x_1}{1-x_2}-1\Big)(1-x_2)^{k_1} J_{k_2}^{(2k_1+1,0)}(x_2). \tag{10.76}$$

This is a polynomial of total degree k_1+k_2 in the variables x_1, x_2, i.e. $\varphi_{\mathbf{k}} \in \mathbb{P}_{k_1+k_2}(\widehat{T})$. The orthogonality of the Jacobi polynomials (10.74), for each $m \neq k$, allows to prove that

$$\begin{aligned}\int_{\widehat{T}} \varphi_{\mathbf{k}}(x_1,x_2)\varphi_{\mathbf{m}}(x_1,x_2)\,dx_1\,dx_2 = \frac{1}{2}\Big(\int_{-1}^{1} J_{k_1}^{(0,0)}(\xi_1)\,J_{m_1}^{(0,0)}(\xi_1)\,d\xi_1\Big)\cdot \\ \Big(\int_{-1}^{1} J_{k_2}^{(2k_1+1,0)}(\xi_2)\,J_{m_2}^{(2m_1+1,0)}(\xi_2)\,(1-\xi_2)^{k_1+m_1+1}\,d\xi_2\Big) = 0.\end{aligned} \tag{10.77}$$

Hence, $\{\varphi_{\mathbf{k}}\,:\, 0 \le k_1,k_2,\ k_1+k_2 \le N\}$ constitutes an *orthogonal (modal) basis* for the space of polynomials $\mathbb{P}_N(\widehat{T})$, with dimension $\frac{1}{2}(N+1)(N+2)$.
The orthogonality property is undoubtedly convenient as it allows to diagonalize the mass matrix (see Chap. 5). However, with the modal basis described above, imposing the boundary conditions (in case the computational domain is a triangle $\widehat{T}$), as well as satisfying the continuity conditions on the interelements (in case spectral element methods with triangular elements are used) results to be uncomfortable. A possible remedy consists in *adapting* such basis by generating a new one, which we will denote by $\{\varphi_{\mathbf{k}}^{ba}\}$; *ba* stands for *boundary adapted*. In order to obtain it, we will start by replacing the one-dimensional Jacobi basis $J_k^{(\alpha,0)}(\xi)$ (with $\alpha = 0$ or $2k+1$) with the adapted basis constituted by:

- two boundary functions: $\frac{1+\xi}{2}$ and $\frac{1-\xi}{2}$;
- $(N-1)$ bubble functions: $\big(\frac{1+\xi}{2}\big)\big(\frac{1-\xi}{2}\big)J_{k-2}^{(\alpha,\beta)}(\xi)$, $k = 2,\dots,N$, for suitable fixed $\alpha,\beta \ge 1$.

These one-dimensional bases are then used as in (10.75) instead of the non-adapted Jacobi polynomials. This way, we find vertex-type, edge-type and bubble functions. Precisely:

- vertex-type functions:

$$\Phi^{V_1}(\xi_1,\xi_2) = \Big(\frac{1-\xi_1}{2}\Big)\Big(\frac{1-\xi_2}{2}\Big) \qquad \text{(vertex } V_1 = (-1,-1)\text{)},$$

$$\Phi^{V_2}(\xi_1,\xi_2) = \Big(\frac{1+\xi_1}{2}\Big)\Big(\frac{1-\xi_2}{2}\Big) \qquad \text{(vertex } V_2 = (1,-1)\text{)},$$

$$\Phi^{V_3}(\xi_1,\xi_2) = \frac{1+\xi_2}{2} \qquad \text{(vertex } V_3 = (-1,1)\text{)};$$

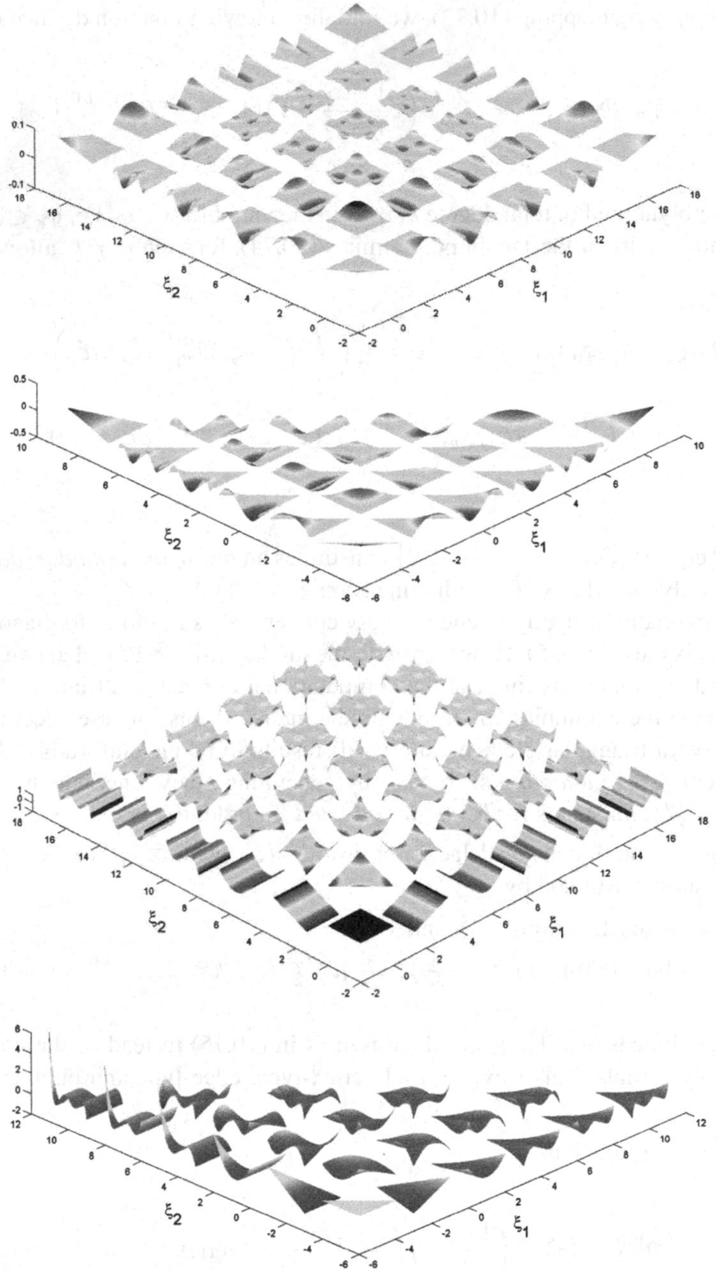

Fig. 10.15. Basis functions of degree $N = 5$: *boundary-adapted* bases on the square (first from the top) and on the triangle (second from top) associated to the values $\beta = 1$ and $\delta = 0$; Jacobi basis $J_k^{(\alpha,\beta)}$ on the square (second from the bottom) corresponding to the values $\alpha = \beta = 0$ (Legendre case); Dubiner basis functions $\{\Phi_{\mathbf{k}}\}$ on the triangle (bottom)

- edge-type functions:

$$\Phi_{K_1}^{V_1V_2}(\xi_1,\xi_2) = \left(\frac{1-\xi_1}{2}\right)\left(\frac{1+\xi_1}{2}\right)J_{k_1-2}^{(\beta,\beta)}(\xi_1)\left(\frac{1-\xi_2}{2}\right)^{k_1}, \qquad 2 \le k_1 \le N,$$

$$\Phi_{K_2}^{V_1V_3}(\xi_1,\xi_2) = \left(\frac{1-\xi_1}{2}\right)\left(\frac{1-\xi_2}{2}\right)\left(\frac{1+\xi_2}{2}\right)J_{k_2-2}^{(\beta,\beta)}(\xi_2), \qquad 2 \le k_2 \le N,$$

$$\Phi_{K_2}^{V_2V_3}(\xi_1,\xi_2) = \left(\frac{1+\xi_1}{2}\right)\left(\frac{1-\xi_2}{2}\right)\left(\frac{1+\xi_2}{2}\right)J_{k_2-2}^{(\beta,\beta)}(\xi_2), \qquad 2 \le k_2 \le N;$$

- bubble-type functions:

$$\Phi_{k_1,k_2}^{\beta}(\xi_1,\xi_2) = \left(\frac{1-\xi_1}{2}\right)\left(\frac{1+\xi_1}{2}\right)J_{k_1-2}^{(\beta,\beta)}(\xi_1)\cdot$$

$$\left(\frac{1-\xi_2}{2}\right)^{k_1}\left(\frac{1+\xi_2}{2}\right)J_{k_2-2}^{(2k_1-1+\delta,\beta)}(\xi_2),$$

$2 \le k_1, k_2,\ k_1+k_2 \le N.$

Although the choice $\beta = \delta = 2$ ensures the orthogonality of the bubble functions, generally we prefer the choice $\beta = 1$, $\delta = 0$, which guarantees a good sparsity of the mass and stiffness matrices and an acceptable condition number for the stiffness matrix for second-order differential operators.
In Fig. 10.15 we report some examples of bases on triangles corresponding to different choices of β and δ and different values of the degree N.
Using these modal bases, we can now set up a spectral Galerkin approximation for a boundary-value problem set on the triangle $\widehat{T}$, or a SEM-type method on a domain Ω partitioned in triangular elements. We refer the interested reader to [CHQZ06, CHQZ07, KS05].

10.7 Exercises

1. Prove inequality (10.52).
2. Prove property (10.54).
3. Write the weak formulation of problem

$$\begin{cases} -((1+x)u'(x))' + u(x) = f(x), & 0 < x < 1, \\ u(0) = \alpha, \qquad u(1) = \beta, \end{cases}$$

   and the linear system resulting from its discretization using the G-NI method.
4. Approximate the problem

$$\begin{cases} -u''(x) + u'(x) = x^2, & -1 < x < 1, \\ u(-1) = 1, \qquad u'(1) = 0, \end{cases}$$

   using the G-NI method and analyze its stability and convergence.

5. Write the G-NI approximation of the problem

$$\begin{cases} Lu(x) = -(\mu(x)u'(x))' + (b(x)u(x))' + \sigma(x)u(x) = f(x), & -1 < x < 1, \\ \mu(\pm 1)u'(\pm 1) = 0. \end{cases}$$

Find the conditions on the data under which the pseudo-spectral approximation is stable. Moreover, verify that the following relations hold:

$$L_N u_N(x_j) = f(x_j), \quad j = 1, \dots, N-1,$$

$$\mu(1)\, u_N'(1) = \alpha_N (f - L_N u_N)(1),$$

$$\mu(-1)\, u_N'(-1) = -\alpha_0 (f - L_N u_N)(-1),$$

L_N being the pseudo-spectral operator defined in (10.41).

6. Consider the problem

$$\begin{cases} -\mu \Delta u + \mathbf{b} \cdot \nabla u - \sigma u = f & \text{in } \Omega = (-1,1)^2, \\ u(\mathbf{x}) = u_0 & \text{for } x_1 = -1, \\ u(\mathbf{x}) = u_1 & \text{for } x_1 = 1, \\ \nabla u(\mathbf{x}) \cdot \mathbf{n}(\mathbf{x}) = 0 & \text{for } x_2 = -1 \text{ and } x_2 = 1, \end{cases}$$

where $\mathbf{x} = (x_1, x_2)^T$, $\mathbf{n}$ is the outgoing normal of Ω, $\mu = \mu(\mathbf{x})$, $\mathbf{b} = \mathbf{b}(\mathbf{x})$, $\sigma = \sigma(\mathbf{x})$, $f = f(\mathbf{x})$ are assigned functions, and u_0, u_1 are given constants.
Provide sufficient conditions on the data to guarantee the existence and uniqueness of the weak solution, and give an a priori estimate. Then approximate the weak problem using the G-NI method, providing an analysis of its stability and convergence.

7. Prove the stability condition (10.42) in the case of the pseudo-spectral approximation of the equation (5.4) (replacing the interval $(0,1)$ with $(-1,1)$).
[*Solution*: follow a similar procedure to that explained in Sect. 5.4 for the finite element solution and invoke the properties reported in Lemmas 10.2 and 10.3.]

8. Consider the parabolic heat equation

$$\begin{cases} \dfrac{\partial u}{\partial t} - \dfrac{\partial^2 u}{\partial x^2} = 0, & -1 < x < 1, t > 0, \\ u(x,0) = u_0(x), & -1 < x < 1, \\ u(-1,t) = u(1,t) = 0, & t > 0. \end{cases}$$

Approximate it using the G-NI method in space and the implicit Euler method in time and its stability study.

11

Discontinuous element methods (DG and mortar)

Up to now we have considered Galerkin methods with subspaces of continuous polynomial functions, either within the finite element method (Chapter 3) or the spectral element method (Chapter 10). This chapter deals with approximation techniques based on subspaces of polynomials that are discontinuous between elements. We will, in particular, introduce the so-called *Discontinuous Galerkin* method (DG) and the *mortar* method. We will carry out this for the Poisson problem first, and then generalize to the case of diffusion and transport problems (see Chapter 12). To maintain the presentation general we will consider a partition of the computational domain into disjoint subdomains that may be either finite or spectral elements.

11.1 The discontinuous Galerkin method (DG) for the Poisson problem

Let us consider the Poisson problem together with homogeneous Dirichlet boundary conditions (3.13) in a domain $\Omega \subset \mathbb{R}^2$ divided in the union of M disjoint elements Ω_m, $m = 1, \ldots, M$. We wish to attain an alternative weak formulation to the usual one, that will serve as starting point for the DG method. To simplify the discussion we assume the exact solution to be sufficiently regular, for instance $u \in H_0^1(\Omega) \cap H^2(\Omega)$, so that all operations below make sense. Define the space

$$W^0 = \{v \in W : v|_{\partial\Omega} = 0\},$$

where

$$W = \{v \in L^2(\Omega) : v|_{\Omega_m} \in H^1(\Omega_m), m = 1, \ldots, M\}.$$

By Green's formula we have, for every $w \in W^0$,

$$\sum_{m=1}^{M} (-\Delta u, v)_{\Omega_m} = \sum_{m=1}^{M} \left((\nabla u, \nabla v)_{\Omega_m} - \int_{\partial\Omega_m} v \nabla u \cdot \mathbf{n}_m \right), \tag{11.1}$$

A. Quarteroni: *Numerical Models for Differential Problems*, 2nd Ed.
MS&A – Modeling, Simulation & Applications 8
DOI 10.1007/978-88-470-5522-3_11, © Springer-Verlag Italia 2014

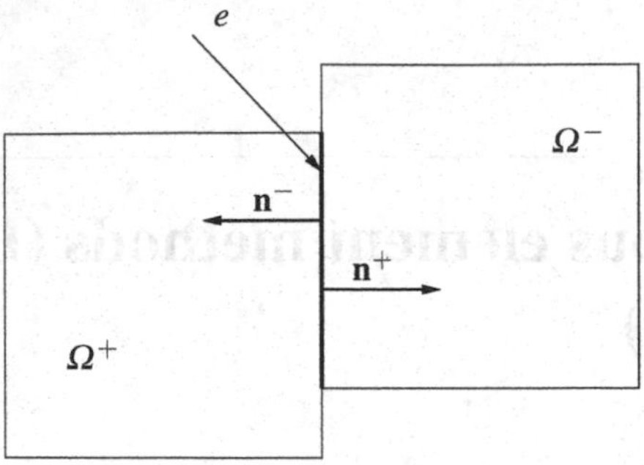

Fig. 11.1. An "edge" e separating two neighbouring subdomains (or elements)

where $\mathbf{n}_m$ is the outward unit normal to $\partial\Omega_m$ and $(\cdot,\cdot)_{\Omega_m}$ denotes the scalar product of $L^2(\Omega_m)$. Calling $\mathcal{E}_\delta$ the union of all internal edges, i.e. the interfaces separating the subdomains (outermost edges may be neglected as v vanishes on them), we can rearrange terms to obtain

$$-\sum_{m=1}^{M}\int_{\partial\Omega_m} v\nabla u\cdot\mathbf{n}_m = -\sum_{e\in\mathcal{E}_\delta}\int_e (v^+\nabla u^+\cdot\mathbf{n}^+ + v^-\nabla u^-\cdot\mathbf{n}^-)|_e\,, \tag{11.2}$$

in which the signs "+" and "−" label the information according to the two possible normal orientations (see, e.g., Fig. 11.1).

We will use the following notation to denote mean values and jumps on elements' edges:

$$\{v\} = \frac{v^+ + v^-}{2}, \qquad [v] = v^+\mathbf{n}^+ + v^-\mathbf{n}^-,$$
$$\{\!\{\nabla w\}\!\} = \frac{(\nabla w)^+ + (\nabla w)^-}{2}, \qquad [\![\nabla w]\!] = (\nabla w)^+\cdot\mathbf{n}^+ + (\nabla w)^-\cdot\mathbf{n}^-.$$

Notice how the above convention guarantees that the definition of the jump operator will not depend on how subdomains (elements) are numbered. A little algebraic manipulation eventually gives

$$\begin{aligned} v^+\nabla u^+\cdot\mathbf{n}^+ + v^-\nabla u^-\cdot\mathbf{n}^- &= 2[v]\cdot\{\!\{\nabla u\}\!\} - (v^+\nabla u^-\cdot\mathbf{n}^+ + v^-\nabla u^+\cdot\mathbf{n}^-)\\ &= 2[v]\cdot\{\!\{\nabla u\}\!\} + 2[\![\nabla u]\!]\{v\}\\ &\quad - (v^+\nabla u^+\cdot\mathbf{n}^+ + v^-\nabla u^-\cdot\mathbf{n}^-) \end{aligned}$$

and so

$$v^+\nabla u^+\cdot\mathbf{n}^+ + v^-\nabla u^-\cdot\mathbf{n}^- = [v]\cdot\{\!\{\nabla u\}\!\} + [\![\nabla u]\!]\{v\}. \tag{11.3}$$

Using (11.2) and (11.3), from (11.1) we obtain that the solution to the Poisson problem (3.13) satisfies

$$\sum_{m=1}^{M}(\nabla u,\nabla v)_{\Omega_m} - \sum_{e\in\mathcal{E}_\delta}\int_e ([v]\cdot\{\!\{\nabla u\}\!\} + [\![\nabla u]\!]\{v\}) = \sum_{m=1}^{M}(f,v)_{\Omega_m} \quad \forall v\in W^0.$$

Now we introduce the discrete space

$$W_\delta = \{v_\delta \in W : v_\delta|_{\Omega_m} \in \mathbf{P}_r(\Omega_m), m = 1, ..., M\},$$

$\mathbf{P}_r(\Omega_m)$ being a space of "polynomials" on Ω_m. More precisely, $\mathbf{P}_r(\Omega_m) = \mathbb{P}_r$ if Ω_m is a simplex (2D triangle or 3D tetrahedron), while $\mathbf{P}_r(\Omega_m) = \mathbb{Q}_r \circ F_m(\Omega_m)$ if Ω_m is a spectral element (a quadrilateral in 2D, a parallelepiped in 3D, cf. Ch. 10). At last, let W_δ^0 be the following subspace of W_δ

$$W_\delta^0 = \{v_\delta \in W_\delta : v_\delta|_{\partial\Omega} = 0\}.$$

Note that the term $[\![\nabla u]\!]\{v\}$ in (11.1) is null because if $u \in H_0^1(\Omega) \cap H^2(\Omega)$ then $[\![\nabla u]\!] = 0$ on every edge $e \in \mathcal{E}_\delta$. This fact together with expression (11.1) motivates the following DG approximation for problem (3.13): find $u_\delta \in W_\delta^0$ satisfying

$$\sum_{m=1}^{M} (\nabla u_\delta, \nabla v_\delta)_{\Omega_m} - \sum_{e \in \mathcal{E}_\delta} \int_e [v_\delta] \cdot \{\!\{\nabla u_\delta\}\!\} - \tau \sum_{e \in \mathcal{E}_\delta} \int_e [u_\delta] \cdot \{\!\{\nabla v_\delta\}\!\}$$
$$+ \sum_{e \in \mathcal{E}_\delta} \gamma |e|^{-1} \int_e [u_\delta] \cdot [v_\delta] = \sum_{m=1}^{M} (f, v_\delta)_{\Omega_m} \quad \forall v_\delta \in W_\delta^0, \quad (11.4)$$

where $\gamma = \gamma(r)$ is a suitable positive constant (depending on the local polynomial degree), $|e|$ is the length of $e \in \mathcal{E}_\delta$ and τ is a suitable fixed number. The additional new terms $\tau[u_\delta] \cdot \{\!\{\nabla v_\delta\}\!\}$ and $\gamma|e|^{-1}[u_\delta] \cdot [v_\delta]$ do not undermine strong consistency (since $[u] = 0$ if u is the exact Poisson solution), beside warranting greater generality and improved stability features.

Formulation (11.4), introduced at the end of the 70s, is called *Interior Penalty* (IP) ([Whe78, Arn82]). In case $\tau = 1$, the method preserves the symmetry and the resulting formulation is known as SIPG method (*Symmetric Interior Penalty Galerkin*) [Whe78, Arn82]. For $\tau \neq 1$ the bilinear form is no longer symmetric, and the special values $\tau = -1$ and $\tau = 0$ respectively lead to the NIPG method (*Non-symmetric Interior Penalty Galerkin*) [RWG99] and the IIPG method (*Incomplete Interior Penalty Galerkin*) [DSW04]. Whereas the former is stable for any given $\gamma > 0$, SIPG and IIPG require, in order to reach a stable formulation, a sufficiently large penalty parameter γ.

Several variants of formulation (11.4) have been proposed within the context of approximations by finite elements. Here we will only briefly describe the most classical situations, and refer to the article [ABCM02] of Arnold, Brezzi, Cockburn, Marini both for a general overview and a detailed study of stability and convergence.

A first version consists in replacing the last term on the left side of (11.4) with the following stabilization term

$$\sum_{e \in \mathcal{E}_\delta} \gamma \int_e r_e([u_\delta]) \cdot r_e([v_\delta]). \quad (11.5)$$

Above $r_e(\cdot)$ is a suitable extension operator that, from the jump of a function $[v_\delta]$ across $e \in \mathcal{E}_\delta$, generates a function $r_e([v_\delta])$ with non-zero support on the elements having e as edge. See [BRM+97] and [ABCM02] for full details.

A second variant (cf. [Ste98]) replaces the averages $\{\!\{\nabla w\}\!\}$ in (11.4) by the averages with relaxation

$$\{\!\{\nabla w\}\!\}_\theta = \theta \nabla w^+ + (1-\theta)\nabla w^-, \qquad 0 \le \theta \le 1.$$

Up to this point we have imposed the homogeneous Dirichlet condition "strongly". In order to add the boundary constraints, say $u = g$ on $\partial\Omega$, in weak form ("à la Nitsche" [Nit71]), as is more natural for DG-like approximations, we write the discrete formulation (11.4) in W_δ rather than in W_δ^0, and add on the left side the following contributions to the boundary edges $e \subseteq \partial\Omega$

$$-\sum_{e\subseteq\partial\Omega}\int_e v_\delta \nabla u_\delta \cdot \mathbf{n} - \tau \sum_{e\subseteq\partial\Omega}\int_e (u_\delta - g_\delta)\nabla v_\delta \cdot \mathbf{n} + \sum_{e\subseteq\partial\Omega} \gamma |e|^{-1} \int_e (u_\delta - g_\delta) v_\delta, \qquad u_\delta, v_\delta \in W_\delta.$$

The positive constant $\gamma = \gamma(r)$ is the same of (11.4), and g_δ is a convenient approximation of g. The first term, arising naturally from integration by parts, ensures the method is strongly consistent, while the second term makes the formulation symmetric if $\tau = 1$ and non-symmetric if $\tau = -1, 0$. The last term penalizes the trace of the discrete solution u_δ and makes it "approach" the Dirichlet datum. Observe how incorporating these terms does not affect the method's strong consistency.

The DG formulation with boundary conditions imposed weakly thus becomes: find $u_\delta \in W_\delta$ such that

$$\begin{aligned}
&\sum_{m=1}^{M} (\nabla u_\delta, \nabla v_\delta)_{\Omega_m} - \sum_{e\in\mathcal{E}_\delta}\int_e [v_\delta]\cdot\{\!\{\nabla u_\delta\}\!\} - \tau \sum_{e\in\mathcal{E}_\delta}\int_e [u_\delta]\cdot\{\!\{\nabla v_\delta\}\!\} \\
&\quad - \sum_{e\subseteq\partial\Omega}\int_e v_\delta \nabla u_\delta \cdot \mathbf{n} - \tau \sum_{e\subseteq\partial\Omega}\int_e u_\delta \nabla v_\delta \cdot \mathbf{n} + \sum_{e\subseteq\partial\Omega} \gamma |e|^{-1}\int_e u_\delta v_\delta \\
&= \sum_{m=1}^{M} (f, v_\delta)_{\Omega_m} - \tau \sum_{e\subseteq\partial\Omega}\int_e g_\delta \nabla v_\delta \cdot n + \sum_{e\subseteq\partial\Omega} \gamma |e|^{-1}\int_e g_\delta v_\delta \qquad \forall v_\delta \in W_\delta.
\end{aligned} \tag{11.6}$$

We shall refer to the latter formulation as the DG-N method (N for Nitsche). Clearly, if the Dirichlet datum g is zero the last two terms on the right will not show up.

Concerning the accuracy of method (11.6) for discretizing the Poisson problem (3.13) with homogeneous Dirichlet boundary conditions, let us introduce the so-called energy norm

$$|\!|\!| u_\delta |\!|\!| = \left(\sum_{m=1}^{M} \int_{\Omega_m} |\nabla u_\delta|^2 + \sum_{e\in\mathcal{E}_\delta} \gamma |e|^{-1}\int_e [u_\delta]^2 + \sum_{e\subseteq\partial\Omega} \gamma |e|^{-1}\int_e |u_\delta|^2 \right)^{1/2}. \tag{11.7}$$

For formulation (11.4), where boundary conditions are imposed strongly, the last term is missing. It can be proved that if the exact solution is sufficiently regular, the SIPG method ($\tau = 1$) converges with optimal convergence rate both for the $L^2(\Omega)$ norm and for (11.7), as long as the penalty parameter γ is large enough. Better said, for finite elements of degree r one has

$$h|||u-u_\delta||| + \|u-u_\delta\|_{L^2(\Omega)} \leq Ch^{r+1}|u|_{H^{r+1}(\Omega)}, \tag{11.8}$$

where C is an appropriate positive number that depends on r (for a proof see [ABCM02], for example). As always, r is the polynomial degree employed on each element Ω_m. For the non-symmetric methods NIPG and IIPG, as these schemes are not strongly consistent on the adjoint problem, one cannot get optimal L^2 estimates. In many cases, nevertheless, both methods exhibit optimal rates of convergence when the degree of the approximation is odd and grids are sufficiently regular (see e.g. [OBB98]).

For all variants of the DG-N method we have seen one can prove that if $u \in H^{s+1}(\Omega)$, $s \geq 1$, and if the polynomial degree r satisfies $r \geq s$, the error can be estimated in energy norm (11.7) as follows

$$|||u-u_\delta||| \leq C\left(\frac{h}{r}\right)^s r^{1/2}|u|_{H^{s+1}(\Omega)}, \tag{11.9}$$

where C is a suitable positive constant that does not depend on r. For the SIPG method ($\tau = 1$) and the IIPG method ($\tau = 0$) estimate (11.9) holds as long as the penalty parameter γ is taken large enough. In particular, convergence in r is exponential when the exact solution u is analytic. Let us also remark, by comparison with the known results for spectral elements, that (11.9) is sub-optimal with respect to the approximation degree r, due to the presence of the factor $r^{1/2}$. More details can be found in [RWG99, RWG01, HSS02, PS03], for instance.

In certain special situations one can attain optimal estimates in r. For the two-dimensional case with quadrilateral grids, for example, [GS05] provides optimal estimates in energy norm under the assumption that the solutions belongs locally to an enriched Sobolev space. Different estimates, still in two dimensions and with quadrilateral grids, were proved in [SW10] without extra regularity hypotheses, but under homogeneous Dirichlet boundary conditions.

We close the section by observing that sometimes formulation (11.6) is stable even without the penalty term for jumps, i.e. choosing γ=0 for the internal edges and for the external ones as well. Rivière, Wheeler and Girault [RWG99] proved that the non-symmetric version ($\tau = -1$), known in the literature as the Baumann-Oden method [OBB98], is stable and provides optimal estimates of the error (in energy norm) if the approximation degree r satisfies $r \geq 2$. In that case one uses a special interpolation operator (Morley operator) for which $\{\!\{\nabla_h(u-u^I)\}\!\} = 0$ on each edge. In the article of Brezzi and Marini [BM06] (see also [BS08]) it was proved that the Baumann-Oden method (in its non-symmetric incarnation, with $\tau = -1$) in two dimensions with triangular grids is stable, provided we add to the space of linear polynomials a bubble function for each element. The Baumann-Oden method was shown in [ABM09] to be stable (always in 2 dimensions) when adding to linear polynomials $n-2$ bubbles for each element, for decompositions involving n-gons (polygons with n edges). At last, we mention that Burman *et al.* [BEMS07] proved the 1-dimensional symmetric variant ($\tau = 1$) need not be stabilized if $r \geq 2$.

To learn more on DG-type methods the reader should consult, for example, [Riv08], [HW08], [ABCM02], [Woh01].

11.1.1 Numerical results for the DG approximation of Poisson problem

We present in this section numerical results coming from applying the discontinuous Galerkin method to the homogeneous Dirichlet problem (3.13) on $\Omega = (0,1)^2$, where the forcing term f is such that the exact solution reads $u(x,y) = (x-x^2)\exp(3x)\sin(2\pi y)$. We discussed method (11.4) with $\tau = 1$ and penalty constant $\gamma = 10r^2$. This choice makes sure the SIPG method is well posed. Then the $\{\Omega_m\}$ are nothing but the finite elements (triangles) and r is the polynomial degree on each element. In the ensuing numerical experiments $\mathcal{E}_\delta$ is thus the union of all inner edges in the grid. Errors were computed in $L^2(\Omega)$ norm and in the energy norm introduced in (11.7). Fig. 11.2 (left) shows the (normalized) errors computed on a sequence of triangular grids made by linear elements ($r = 1$). As predicted by (11.8), the error tends to zero linearly in energy norm and quadratically in $L^2(\Omega)$. In Fig. 11.2 (right) we can read the (normalized) errors computed on a sequence of Cartesian grids with bi-quadratic elements ($r = 2$) and bi-cubic ones ($r = 3$). The approximation error in norm (11.7) tends to zero when $h \to 0$, and the convergence order equals r.

In the framework of spectral elements we can attain a DG-SEM formulation starting from a partition of Ω in quadrilaterals, using formulation (11.6), and replacing the volume integrals $(\cdot,\cdot)_{\Omega_m}$ with local LGL quadrature formulae; similarly for the integrals extended over the edges of spectral elements.

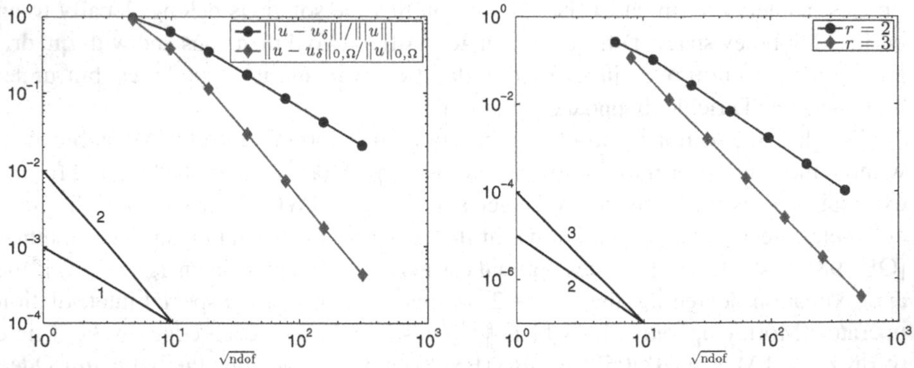

Fig. 11.2. Study of the convergence of method (11.4) ($\tau = 1$, $\gamma = 10r^2$). Left: errors in energy norm (11.7) and $L^2(\Omega)$ norm ($r = 1$, structured triangular grids). Right: errors in energy norm from biquadratic and bi-cubic elements on a sequence of Cartesian grids

11.2 The mortar method

An alternative to the DG technique is based on the so-called mortar method, originating in the framework of spectral element methods (SEM).

Let us consider again the Poisson problem (3.13) in a domain $\Omega \subset \mathbb{R}^2$ with homogeneous Dirichlet boundary conditions.
Define on Ω a partition into pairwise-disjoint non-empty open subregions $\Omega_i \subset \Omega$, $i = 1,\ldots,M$ such that $\overline{\Omega} = \cup_{i=1}^M \overline{\Omega}_i$. Then let $\Gamma_{ij} = \Gamma_{ji} = \partial\Omega_i \cap \partial\Omega_j$ be the interface between Ω_i and Ω_j, $1 \le i \ne j \le M$, and define $\Gamma = \cup_{ij}\Gamma_{ij}$ to be their union (Fig. 11.3).

Solving (3.13) by a mortar method means finding a discrete solution u_δ that, inside every subregion Ω_i, is continuous and polynomial (globally or locally), and that fulfills a continuity condition on the interface Γ, called *weak* or *integral*: namely, that for every i, j with $1 \le i \ne j \le M$

$$\int_{\Gamma_{ij}} (u_\delta|_{\Omega_i} - u_\delta|_{\Omega_j})\psi = 0 \qquad \forall \psi \in \tilde{\Lambda}, \tag{11.10}$$

where $\tilde{\Lambda}$ is a suitable finite-dimensional space that depends on the discretization chosen on the Ω_i. To (11.10) are then added strong continuity constraints at certain points lying on the interface Γ.

Note that equations (11.10) do not force the solution's jump to vanish on the interface, but they prescribe that its L^2 projection on $\tilde{\Lambda}$ be zero. Consequently, in contrast to what happens for an approximation of Galerkin type (see Chaps. 4 and 10), $u_\delta \notin H^1_0(\Omega)$ in general, but rather $u_\delta \in L^2(\Omega)$ with $u_\delta|_{\Omega_i} \in H^1(\Omega_i)$.

To simplify the discussion, let us consider a partition of Ω in $M = 2$ subregions Ω_1, Ω_2 and call Γ the interface, so that $\Gamma = \Gamma_{12} = \Gamma_{21}$. In Ω_i, $i = 1,2$, we define a further partition $\mathcal{T}_i = \cup_k T_{i,k}$ in triangles or quadrilaterals $T_{i,k}$ as explained in Sect. 6.2.

If the elements $T_{i,k}$ are quadrilaterals we also require each $T_{i,k}$ to be the image of the reference element $\widehat{T} = (-1,1)^2$ under a smooth bijection $\varphi_{i,k}$, see Sect. 10.1. Given polynomial interpolation degrees $N_i \ge 1$ in each Ω_i, define

$$V_{N_i}(T_{i,k}) = \{v \in C^0(\overline{T_{i,k}}) : v \circ \varphi_{i,k} \in \mathbb{Q}_{N_i}(\widehat{T})\}$$

where $\mathbb{Q}_N(\widehat{T})$ is the space of degree N polynomials in every variable in the reference element $\widehat{T}$ (cf. definition (10.2)).

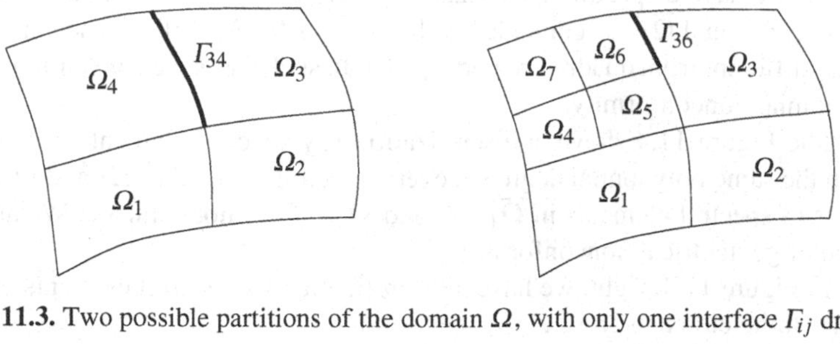

Fig. 11.3. Two possible partitions of the domain Ω, with only one interface Γ_{ij} drawn

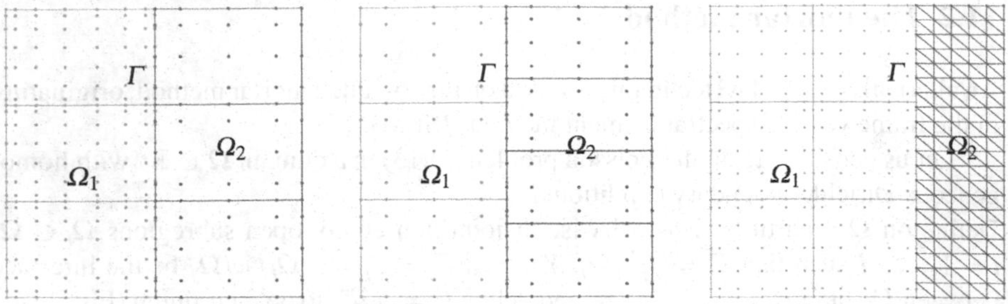

Fig. 11.4. Left: discretization by quadrilateral spectral elements with nonconformity of polynomial type; middle: geometric nonconformity by spectral elements; right: discretization by spectral elements and finite elements

For a triangular partition, instead, we consider the finite-element spaces of Sect. 4.5, and for $i = 1,2$ and any $T_{i,k} \in \mathcal{T}_i$ set

$$X_{\delta_i}(T_{i,k}) = \begin{cases} \mathbb{P}_{r_i}(T_{i,k}) \text{ for finite elements of degree } r_i \text{ on triangles,} \\ V_{N_i}(T_{i,k}) \text{ for spectral elements of degree } N_i \text{ on quadrilaterals.} \end{cases} \tag{11.11}$$

The parameter δ_i implicitly depends on the degree (r_i or N_i) and on the maximum diameter h of the elements of $\mathcal{T}_i$.
The finite-dimensional spaces induced by the local discretizations in Ω_i, $i = 1,2$, are then

$$V_{i,\delta_i} = \{v_\delta \in C^0(\overline{\Omega}_i) : v_\delta|_{T_{i,k}} \in X_{\delta_i}(T_{i,k}), \forall T_{i,k} \in \mathcal{T}_i\}, \tag{11.12}$$

and the discrete solution u_δ must be seeked in

$$Y_\delta = \{v_\delta \in L^2(\Omega) : v_\delta^{(i)} = v_{\delta|\Omega_i} \in V_{i,\delta_i}, \text{ for } i = 1,2\}.$$

As the space Y_δ does not retain the information on how to match the functions $v_\delta^{(i)}$ on the interface, we must introduce a subspace $V_\delta \subset Y_\delta$ of functions satisfying (11.10) and search for the mortar solution u_δ inside V_δ.

Let us observe first that the choice of mesh and polynomial degree in one subdomain is completely independent of the choice in the other subdomain, as Figure 11.4 explains.
On the left we have a spectral discretization in either Ω_i, where the edges of the elements of $\overline{\Omega}_1 \cap \Gamma$ and $\overline{\Omega}_2 \cap \Gamma$ coincide but the degrees N_1, N_2 differ (hence interpolating nodes are different from quadrature nodes, too). In such cases we conventionally speak of polynomial nonconformity.
The middle Figure 11.4 shows a discretization (by spectral elements in both spaces Ω_i) with the same polynomial degree in every spectral element of Ω_1 and Ω_2, but now the edges of spectral elements in $\overline{\Omega}_1 \cap \Gamma$ and $\overline{\Omega}_2 \cap \Gamma$ do not coincide. We are then in presence of geometrical nonconformity.
At last, in Figure 11.4, right, we have a discretization by spectral elements on Ω_1 and triangular finite elements in Ω_2.

Indicate with $\mathcal{M}_i$ the set of nodes induced by the discretization chosen in Ω_i. In the spectral case these are images on every $T_{i,k}$ of the Legendre-Gauss-Lobatto points defined on $\hat{T}$ (see Sect. 10.2.3). The same discretizations induce two distinct sets of nodes (not necessarily disjoint) on Γ which we indicate by $\mathcal{M}_i^\Gamma = \mathcal{M}_i \cap \Gamma$, and two sets $\mathcal{U}_i^\Gamma$ of degrees of freedom on Γ, whose elements are the values of the functions $u_\delta^{(i)}$ at the nodes of $\mathcal{M}_i^\Gamma$.

Of the two sets of degrees of freedom on Γ, one, called *mortar* or *master* set, is picked to play an active role in the problem's formulation, meaning that its degrees of freedom are primal unknowns for the problem. The other set, called *non-mortar* or *slave*, characterizes the space $\tilde{\Lambda}$ onto which the continuity condition projects. The degrees of freedom of the sets $\mathcal{U}_1^\Gamma$ and $\mathcal{U}_2^\Gamma$ will depend on each other via a linear relationship dictated by the integral conditions (11.10).

Label with $m, s \in \{1,2\}$ the master set $\mathcal{U}_m^\Gamma$ and the slave set $\mathcal{U}_s^\Gamma$ respectively, and let $\mathcal{N}_{master}$, $\mathcal{N}_{slave}$ be their cardinalities. The subscripts m and s pass on to subdomains, polynomial degrees and all other quantities in the picture, so for instance we will have Ω_m, N_m and so on for the master domain, Ω_s, N_s for the slave domain. In our study $\mathcal{U}_1^\Gamma$ plays the master role, while $\mathcal{U}_2^\Gamma$ that of the slave.

The choice of finite or spectral elements for the discretization of the problem requires separate arguments from now on. So let us suppose to only have either spectral elements or finite elements in both Ω_1, Ω_2, for the time being; we will see in Section 11.8 how to treat the mixed case.

11.2.1 Characterization of the space of constraints by spectral elements

Denote by $\mathcal{E}_s^\Gamma$ the collection of edges of spectral elements in the slave domain Ω_s that lie on Γ, and set

$$\widetilde{\Lambda}_\delta = \text{span}\{\psi \in L^2(\Gamma) : \psi|_e \in \mathbb{P}_{N_s-2} \quad \forall e \in \mathcal{E}_s^\Gamma\} \tag{11.13}$$

and

$$P_\delta = \{\mathbf{p} \in \mathcal{M}_s^\Gamma : \mathbf{p} \text{ is the endpoint of an edge } e \in \mathcal{E}_s^\Gamma\}. \tag{11.14}$$

For the definition of $\widetilde{\Lambda}_\delta$ to make any sense we have to take $N_s \geq 2$. (The case $N_s = 1$ can be reduced to the finite element formulation of type $\mathbb{Q}_1$.)

We want to characterize the space $\widetilde{\Lambda}_\delta$ in terms of a basis whose $L^2(\Gamma)$-functions have support on one edge only $e \in \mathcal{E}_s^\Gamma$, and that on this edge coincide with Lagrange's characteristic polynomials of degree $N_s - 2$ associated to the $N_s - 1$ Legendre-Gauss quadrature nodes on e (see [CHQZ06, formula (2.3.10)]). Figure 11.5, left, shows a function ψ_l of the basis of $\widetilde{\Lambda}_\delta$ supported on the second edge of $\mathcal{E}_s^\Gamma$ and associated to the first Legendre-Gauss node in e.

It is straightforward to see that the dimension of $\widetilde{\Lambda}_\delta$ equals $(N_s - 1)$ times the cardinality of $\mathcal{E}_s^\Gamma$, and that $\dim(\widetilde{\Lambda}_\delta) + \dim(P_\delta) = \mathcal{N}_{slave}$. In the example of Fig. 11.6 we fixed $N_s = 4$, so $\dim(P_\delta) = 3$, $\dim(\widetilde{\Lambda}_\delta) = 6$ and $\mathcal{N}_{slave} = 9$.

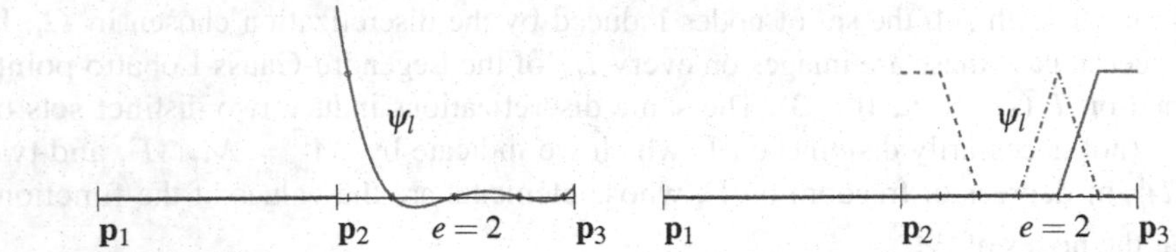

Fig. 11.5. Left: a function in $\widetilde{\Lambda}_\delta$ for a spectral element discretization; right: functions in $\widetilde{\Lambda}_\delta$ for a finite element discretization

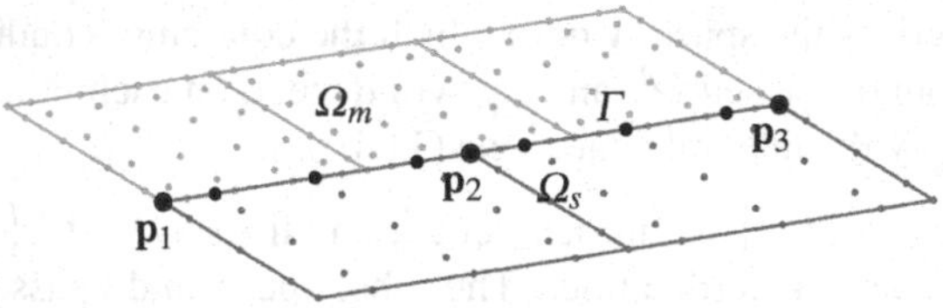

Fig. 11.6. The black dots on Γ represent the nodes of the slave set $\mathcal{M}_s^\Gamma$

11.2.2 Characterization of the space of constraints by finite elements

Now we denote by $\mathcal{E}_s^\Gamma$ the set of edges of Ω_s that lie on Γ, as in Fig. 11.7, left, and by $\mathcal{E}_{s,\delta}^\Gamma$ the edges of the triangles $T_{s,k}$ of the slave set on Γ (Figure 11.7, right).
The set P_δ is defined as in (11.14), whilst the projection space of the solution's jump is

$$\begin{aligned}\widetilde{\Lambda}_\delta = \text{span}\{\psi \in L^2(\Gamma) : \ \psi|_e \in \mathbb{P}_{r_s}(e) \quad \forall e \in \mathcal{E}_{s,\delta}^\Gamma \ \text{ such that } \ \overline{e} \cap P_\delta = \emptyset, \\ \psi|_e \in \mathbb{P}_{r_s-1}(e) \quad \forall e \in \mathcal{E}_{s,\delta}^\Gamma \ \text{ such that } \ \overline{e} \cap P_\delta \neq \emptyset\},\end{aligned} \qquad (11.15)$$

where $\mathbb{P}_r(e)$ are degree r polynomials in one variable on the interval e.

Figure 11.7, right, depicts a generic function of the space $\widetilde{\Lambda}_\delta$ over a rectification of the interface Γ.

To characterize a basis for (11.15) in presence of finite elements $\mathbb{P}_1$, we indicate by $\mathbf{x}_j^e$ $(j = 1, \ldots, N_s + 1)$ the nodes belonging to $\mathcal{M}_s^\Gamma \cap \overline{e}$, where e is an edge of $\mathcal{E}_s^\Gamma$ (see Figure 11.7, left).

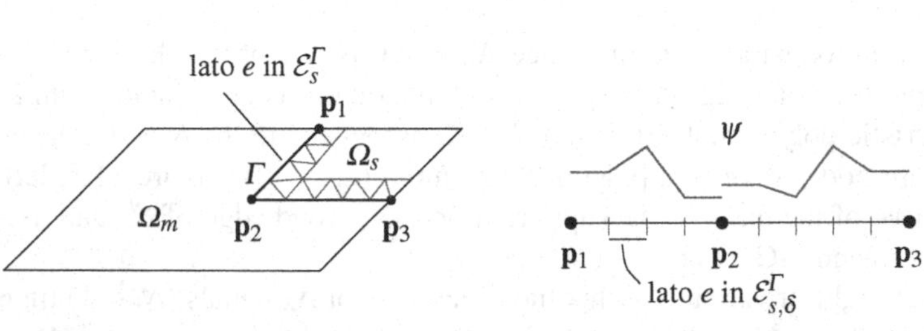

Fig. 11.7. Left: a decomposition of Ω with finite element discretization; right: a function of the space (11.15) on a rectification of the interface Γ

The basis functions of $\widetilde{\Lambda}_\delta$ are functions in $L^2(\Gamma)$ that have one edge $e \in \mathcal{E}_s^\Gamma$ as support, and are associated to the inner nodes of e; for $j = 3, \ldots, N_s - 1$ they coincide with the piecewise-linear functions of the Lagrangian basis associated to nodes $\mathbf{x}_j^e$, while for $j = 2$ (resp. $j = N_s$) the function ψ_j equals 1 on the segment $[\mathbf{x}_1^e, \mathbf{x}_2^e]$ (resp. $[\mathbf{x}_{N_s}^e, \mathbf{x}_{N_s+1}^e]$) and on the rest of e it coincides with the piecewise-linear characteristic Lagrange function for node $\mathbf{x}_2^e$ (resp. $\mathbf{x}_{N_s}^e$).
Figure 11.5, right, shows basis functions ψ_l of the space (11.15), with support on the second edge of $\mathcal{E}_s^\Gamma$.

11.3 Mortar formulation for the Poisson problem

At this juncture we can characterize the space V_δ in which to look for the mortar solution to problem (3.13). We will say a function $v_\delta \in Y_\delta$ satisfies the mortar conditions if

$$\begin{cases} \displaystyle\int_\Gamma (v_\delta^{(m)} - v_\delta^{(s)})\psi = 0 & \forall \psi \in \widetilde{\Lambda}_\delta, \\ v_\delta^{(m)}(\mathbf{p}) = v_\delta^{(s)}(\mathbf{p}) & \forall \mathbf{p} \in P_\delta \end{cases} \tag{11.16}$$

and we also set

$$V_\delta = \{v_\delta \in Y_\delta : v_\delta \text{ satisfies conditions (11.16) and } v_\delta = 0 \text{ on } \partial\Omega\}. \tag{11.17}$$

The mortar formulation of (3.13) is thus

$$\text{find } u_\delta \in V_\delta : \qquad \sum_{i=1}^{2} a_i(u_\delta^{(i)}, v_\delta^{(i)}) = \sum_{i=1}^{2} \int_{\Omega_i} f v_\delta^{(i)} \qquad \forall v_\delta \in V_\delta, \tag{11.18}$$

where $a_i(u_\delta^{(i)}, v_\delta^{(i)})$ is the restriction to Ω_i of the bilinear form $a(u_\delta, v_\delta)$, or possibly of a discretization of it by quadrature formulae of Gaussian type whenever a Galerkin-type formulation with numerical integration is used (G-NI, see Chapter 10).

The mortar solution therefore satisfies a weak continuity condition on each *slave* edge ($e \in \mathcal{E}_s^\Gamma$) of the interface Γ, and a pointwise matching condition at the endpoints ($\mathbf{p} \in P_\delta$) of *slave* edges. In [Bel99, BM94] an alternative mortar formulation was proposed where conditions $(11.16)_2$ are absent. This formulation is not very favourable, computationally-speaking, for domains in $\mathbb{R}^3$.

When dealing with a spectral approximation, if the partitions in Ω_m and Ω_s are geometrically conforming on Γ and the spectral interpolation degree N_m on the master domain is not larger than the slave degree N_s, then conditions (11.16) force strong continuity on all of Γ. Only if $N_m > N_s$ is the solution u_δ discontinuous on the interface. The literature labels as mortar or nonconforming only this latter approximation, in which the discrete solution u_δ does not belong to the same space as the continuous solution.

For finite element approximations, even when the interpolation degrees in Ω_m and Ω_s are equal, we have nonconformity each time the sets $\mathcal{M}_m^\Gamma$ and $\mathcal{M}_s^\Gamma$ differ.

Whenever we adopt the spectral element discretization it can be proved ([BMP94]) that if the solution u of the continuous problem (3.13) and the function f are regular enough on each subdomain Ω_i, i.e. $u_{|\Omega_i} \in H^{\sigma_i}(\Omega_i)$ with $\sigma_i > \frac{3}{2}$ and $f_{|\Omega_i} \in H^{\rho_i}(\Omega_i)$ with $\rho_i > 1$, $i = 1, \ldots, M$, then

$$|||u - u_\delta||| \leq C \left(\sum_{i=1}^{M} N_i^{1-\sigma_i} \|u_{|\Omega_i}\|_{H^{\sigma_i}(\Omega_i)} + N_i^{-\rho_i} \|f_{|\Omega_i}\|_{H^{\rho_i}(\Omega_i)} \right), \tag{11.19}$$

where $|||v|||$ represents the so-called H^1 *broken norm*, meaning

$$|||v||| = \left(\sum_{i=1}^{M} \|v_{|\Omega_i}\|^2_{H^1(\Omega_i)} \right)^{1/2}.$$

For finite elements, calling h_i the maximum diameter of the triangles $T_{i,k}$, one can prove that if $u_{|\Omega_i} \in H^{\sigma_i}(\Omega_i)$ with $\sigma_i > \frac{3}{2}$, then

$$|||u - u_\delta||| \leq C \sum_{i=1}^{M} h_i^{\min\{\sigma_i, r_i+1\}-1} \|u_{|\Omega_i}\|_{H^{\sigma_i}(\Omega_i)}. \tag{11.20}$$

11.4 Choosing basis functions

To solve problem (11.18), let us discuss how one can define the basis functions $v_\delta \in V_\delta$. Denote with

- $\varphi_{k'}^{(m)}$, $k' = 1, \ldots, \mathcal{N}_{\Omega_m}$, Lagrange's characteristic functions in Ω_m associated to the nodes of $\mathcal{M}_m \setminus \mathcal{M}_m^\Gamma$; these belong to the space V_{m,δ_m} and vanish identically on Γ;
- $\varphi_{k''}^{(s)}$, $k'' = 1, \ldots, \mathcal{N}_{\Omega_s}$, Lagrange's characteristic functions in Ω_s associated to the nodes of $\mathcal{M}_s \setminus \mathcal{M}_s^\Gamma$; they belong to V_{s,δ_s} and are null on Γ;
- $\mu_k^{(m)}$, $k = 1, \ldots, \mathcal{N}_{master}$, Lagrange's characteristic functions in Ω_m associated to the nodes of $\mathcal{M}_m^\Gamma$; they belong to V_{m,δ_m} and vanish at the nodes of $\mathcal{M}_m \setminus \mathcal{M}_m^\Gamma$;
- $\mu_j^{(s)}$, $j = 1, \ldots, \mathcal{N}_{slave}$, Lagrange's characteristic functions in Ω_s associated to the nodes of $\mathcal{M}_s^\Gamma$; these belong to V_{s,δ_s} and are zero on the nodes of $\mathcal{M}_s \setminus \mathcal{M}_s^\Gamma$;
- μ_k, $k = 1, \ldots, \mathcal{N}_{master}$, the basis functions associated to active (or master) nodes of Γ and thus defined

$$\mu_k = \begin{cases} \mu_k^{(m)} \in V_{m,\delta_m} & \text{such that } \mu_k^{(m)}(\mathbf{x}) = 0 \quad \forall \mathbf{x} \in \mathcal{M}_m \setminus \mathcal{M}_m^\Gamma \\ \tilde{\mu}_k^{(s)} \in V_{s,\delta_s} & \text{such that } \tilde{\mu}_k^{(s)}(\mathbf{x}) = 0 \quad \forall \mathbf{x} \in \mathcal{M}_s \setminus \mathcal{M}_s^\Gamma, \end{cases} \tag{11.21}$$

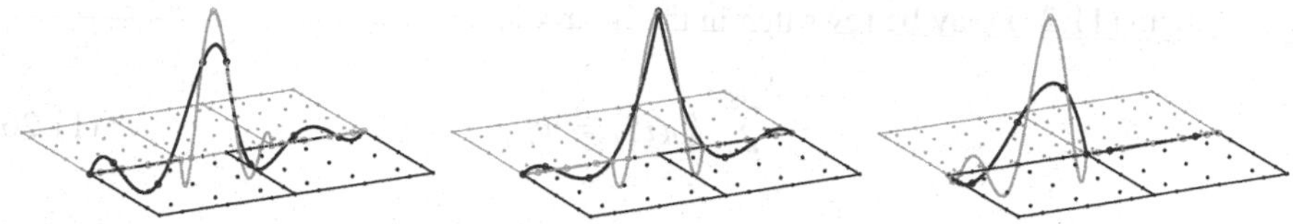

Fig. 11.8. Restriction to the interface of three functions μ_k for a spectral element approximation. In grey the trace associated to *master* degrees of freedom, in black the *slave* trace

where $\tilde{\mu}_k^{(s)}$ are $\mathcal{N}_{master}$ functions in V_{s,δ_s} that we can write as linear combinations of the $\mu_j^{(s)}$ via a rectangular matrix $\Xi = [\xi_{jk}]$

$$\tilde{\mu}_k^{(s)} = \sum_{j=1}^{\mathcal{N}_{slave}} \xi_{jk}\mu_j^{(s)}, \qquad \text{for } k = 1,\dots,\mathcal{N}_{master}. \tag{11.22}$$

It is easy to check that

$$V_\delta = \text{span}\{\varphi_{k'}^{(m)},\ \varphi_{k''}^{(s)},\ \mu_k\}. \tag{11.23}$$

Figure 11.8 shows the restrictions to Γ of three different functions μ_k for a spectral element discretization. On the left we have the function μ_k associated to a node in $\mathcal{M}_m^\Gamma$ but not in $\mathcal{M}_s^\Gamma$ for a geometrically nonconforming partition; on the right the same but for a conforming partition. In the middle the function μ_k associated to a node in $\mathcal{M}_m^\Gamma \cap \mathcal{M}_s^\Gamma$.

While we can eliminate the functions $\varphi_{k'}^{(m)}$ and $\varphi_{k''}^{(s)}$ associated to the nodes of $\partial\Omega$ by using the Dirichlet conditions, we keep all $\mu_k^{(m)}$, $\mu_j^{(s)}$ and μ_k, because the functions $\tilde{\mu}_k^{(s)}$ also depend on the $\mu_j^{(s)}$ associated to the nodes of $\partial\Omega \cap \partial\Gamma$ if the Dirichlet conditions are non-homogeneous.

As the functions $\varphi_{k'}^{(m)}$ and $\varphi_{k''}^{(s)}$ vanish on Γ, imposing $v_\delta \in V_\delta$, i.e. the mortar conditions (11.16), is the same as asking that for $k = 1,\dots,\mathcal{N}_{master}$

$$\begin{cases} \displaystyle\int_\Gamma (\mu_k^{(m)} - \tilde{\mu}_k^{(s)})\psi_\ell \quad \forall \psi_\ell \in \widetilde{\Lambda}_\delta \\ \mu_k^{(m)}(\mathbf{p}) = \tilde{\mu}_k^{(s)}(\mathbf{p}) \quad \forall \mathbf{p} \in P_\delta. \end{cases} \tag{11.24}$$

Equivalently, by (11.22),

$$\begin{cases} \displaystyle\sum_{j=1}^{\mathcal{N}_{slave}} \xi_{jk} \int_\Gamma \mu_j^{(s)}\psi_\ell = \int_\Gamma \mu_k^{(m)}\psi_\ell \quad \forall \psi_\ell \in \widetilde{\Lambda}_\delta \\ \displaystyle\sum_{j=1}^{\mathcal{N}_{slave}} \xi_{jk}\mu_j^{(s)}(\mathbf{p}) = \mu_k^{(m)}(\mathbf{p}) \qquad \forall \mathbf{p} \in P_\delta, \end{cases} \tag{11.25}$$

still for $k = 1,\dots,\mathcal{N}_{master}$.

System (11.25) may be rewritten in the matrix form

$$\sum_{j=1}^{\mathcal{N}_{slave}} \xi_{jk} P_{\ell j} = \Phi_{\ell k} \tag{11.26}$$

where P is a square matrix of dimension $\mathcal{N}_{slave}$, and Φ a rectangular matrix with $\mathcal{N}_{slave}$ rows and $\mathcal{N}_{master}$ columns whose entries arise from the relations in (11.25).

It can be proved that P is non-singular, because the pair $\mathbb{P}_{N-2} - \mathbb{P}_N$ satisfies an inf-sup condition in case of spectral element approximation (see [BM92]); and similarly for finite elements ([Bel99]), where we have an inf-sup condition on $\Lambda_\delta - \mathbb{P}_1$.

Therefore the matrix Ξ can be found by solving the linear system

$$P\Xi = \Phi. \tag{11.27}$$

The computation of the entries $P_{\ell j} = \int_\Gamma \mu_j^{(s)} \psi_\ell$ and $\Phi_{\ell k} = \int_\Gamma \mu_k^{(m)} \psi_\ell$ using suitably accurate quadrature formulae is crucial in order to ensure optimal error estimates (see (11.19) and (11.20)).

11.5 Choosing quadrature formulae for spectral elements

The entries $P_{\ell j}$ depend only on the discretization in the slave domain, so we may rewrite

$$P_{\ell j} = \int_\Gamma \mu_j^{(s)} \psi_\ell = \sum_{e \in \mathcal{E}_s^\Gamma} \int_e \mu_j^{(s)} \psi_\ell$$

and use the Gauss-Legendre-Lobatto quadrature formulae with $N_s + 1$ nodes on each edge $e \in \mathcal{E}_s^\Gamma$. These formulae are exact to degree $2N_s - 1$, and since $\mu_j^{(s)}|_e \in \mathbb{P}_{N_s}$ and $\psi_\ell|_e \in \mathbb{P}_{N_s-2}$, they compute the terms $P_{\ell j}$ exactly.

To compute the elements $\Phi_{\ell k}$ we need to specify whether there is geometric conformity on Γ or not. If yes, the set $\mathcal{E}_m^\Gamma$ (the edges of the elements in Ω_m lying on Γ) coincides with $\mathcal{E}_s^\Gamma$ and we can write

$$\Phi_{\ell k} = \int_\Gamma \mu_k^{(m)} \psi_\ell = \sum_{e \in \mathcal{E}_s^\Gamma} \int_e \mu_k^{(m)} \psi_\ell.$$

Since $\mu_k^{(m)}|_e \in \mathbb{P}_{N_m}$ and $\psi_\ell|_e \in \mathbb{P}_{N_s-2}$ on each edge $e \in \mathcal{E}_s^\Gamma$, to compute exactly the integrals on $e \in \mathcal{E}_s^\Gamma$ we can use the Legendre-Gauss quadrature formulae on $N_q + 1$ nodes with $N_q = \max\{N_s, N_m\}$, because these formulae are exact to degree $2N_q + 1$.

In a geometrically nonconforming setting, on the other hand, $\mathcal{E}_s^\Gamma$ and $\mathcal{E}_m^\Gamma$ do not coincide and composite integration over the edges of $\mathcal{E}_s^\Gamma$ (or $\mathcal{E}_m^\Gamma$) always induces a big quadrature error. Suppose, in fact, we choose a partition associated to $\mathcal{E}_m^\Gamma$ to compute the integrals $\Phi_{\ell k}$. For each $e \in \mathcal{E}_m^\Gamma$ we have $\mu_k^{(m)}|_e \in \mathbb{P}_{N_m}$, but not necessarily $\psi_\ell|_e \in \mathbb{P}_{N_s-2}$; quite the opposite, actually, for ψ_ℓ might be discontinuous on $e \in \mathcal{E}_m^\Gamma$

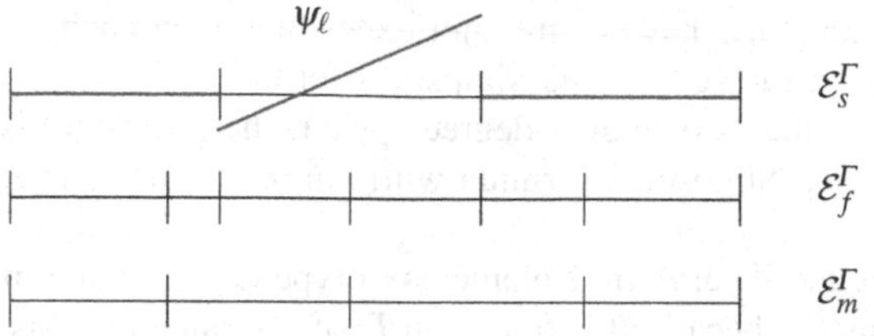

Fig. 11.9. The partitions $\mathcal{E}_s^\Gamma$, $\mathcal{E}_m^\Gamma$, $\mathcal{E}_f^\Gamma$ on Γ and a function $\psi_\ell|_e$ in case of spectral element approximation with $N_s = 3$

(Figure 11.9). A similar thing happens if we take the partition of $\mathcal{E}_s^\Gamma$ instead of that of $\mathcal{E}_m^\Gamma$.

So let us build a new partition $\mathcal{E}_f^\Gamma$, that will be finer than either $\mathcal{E}_m^\Gamma$ and $\mathcal{E}_s^\Gamma$ and whose every edge $\tilde{e} \in \mathcal{E}_f^\Gamma$ is contained in one edge only of $\mathcal{E}_m^\Gamma$ and one only of $\mathcal{E}_s^\Gamma$ (Figure 11.9). Then we can write

$$\Phi_{\ell k} = \int_\Gamma \mu_k^{(m)} \psi_\ell = \sum_{\tilde{e} \in \mathcal{E}_f^\Gamma} \int_{\tilde{e}} \mu_k^{(m)} \psi_\ell,$$

and since $\mu_k^{(m)}|_{\tilde{e}} \in \mathbb{P}_{N_m}$ and $\psi_\ell|_{\tilde{e}} \in \mathbb{P}_{N_s-2}$ on each $\tilde{e} \in \mathcal{E}_f^\Gamma$, we can use on the edges $\tilde{e}$ a Legendre-Gauss quadrature formula over $N_q + 1$ nodes with $N_q = \max\{N_s, N_m\}$, and so compute the $\Phi_{\ell k}$ exactly.

11.6 Choosing quadrature formulae for finite elements

Let us consider the case of finite elements $\mathbb{P}_1$, and recall that $\mathcal{E}_s^\Gamma$ denotes the set of edges of Γ, while $\mathcal{E}_{s,\delta}^\Gamma$ is the set of edges of the triangles $T_{s,k} \in \mathcal{T}_s$ lying on Γ (Figure 11.7).

The elements $P_{\ell j}$, depending only on the discretization of the slave domain, are now

$$P_{\ell j} = \int_\Gamma \mu_j^{(s)} \psi_\ell = \sum_{e \in \mathcal{E}_{s,\delta}^\Gamma} \int_e \mu_j^{(s)} \psi_\ell,$$

and as on each edge $e \in \mathcal{E}_{s,\delta}^\Gamma$ the product $\mu_j^{(s)} \psi_\ell$ is polynomial of degree ≤ 2, we can integrate exactly on edges $e \in \mathcal{E}_{s,\delta}^\Gamma$ using Simpson's formula (see, e.g., [QSS07]).

To compute the elements $\Phi_{\ell k}$ we proceed in analogy to what we did in the spectral case without geometric conformity on Γ.

Build a partition $\mathcal{E}_f^\Gamma$ finer than $\mathcal{E}_m^\Gamma$ and $\mathcal{E}_{s,\delta}^\Gamma$, such that each side $\tilde{e} \in \mathcal{E}_f^\Gamma$ lies only on one edge of $\mathcal{E}_m^\Gamma$ and one edge of $\mathcal{E}_{s,\delta}^\Gamma$, with the result that

$$\Phi_{\ell k} = \int_\Gamma \mu_k^{(m)} \psi_\ell = \sum_{\tilde{e} \in \mathcal{E}_f^\Gamma} \int_{\tilde{e}} \mu_k^{(m)} \psi_\ell.$$

Both $\mu_k^{(m)}|_{\tilde{e}}$ and $\psi_\ell|_{\tilde{e}}$ have degree not exceeding 1 on each $\tilde{e} \in \mathcal{E}_f^\Gamma$, and we can integrate exactly on each edge $\tilde{e}$ via Simpson's formula.

When finite elements of higher degree appear, the procedure is similar, with the proviso of replacing Simpson's formula with a more accurate one, like a Legendre-Gauss formula.

The case of quadrilateral finite elements of type $\mathbb{Q}_1$ is treated in the same manner of finite elements $\mathbb{P}_1$, because the traces on Γ of Lagrangian basis functions $\mathbb{Q}_1$ and $\mathbb{P}_1$ coincide, and the space $\widetilde{\Lambda}_\delta$ is defined alike for both elements.

11.7 Solving the linear system of the mortar method

The coefficients ξ_{ij} found by solving system (11.27) ensure that the functions μ_k of (11.21)–(11.22) satisfy the constraints of the space V_δ, and once the *master* degrees of freedom $\lambda_k^{(m)} \in \mathcal{U}_m$ are known it is possible to compute the *slave* degrees of freedom $\lambda_j^{(s)} \in \mathcal{U}_s$ using

$$\boldsymbol{\lambda}^{(s)} = \Xi \boldsymbol{\lambda}^{(m)} \tag{11.28}$$

where $\boldsymbol{\lambda}^{(s)} = [\lambda_j^{(s)}]_{j=1}^{\mathcal{N}_{slave}}$ and $\boldsymbol{\lambda}^{(m)} = [\lambda_k^{(m)}]_{k=1}^{\mathcal{N}_{master}}$.

When the discretization is conforming on Γ, the matrix Ξ coincides with the identity matrix of dimension $\mathcal{N}_{master} = \mathcal{N}_{slave}$.

By (11.23) every function in V_δ can be written

$$v_\delta(\mathbf{x}) = \sum_{k'=1}^{\mathcal{N}_1} u_{k'}^{(m)} \varphi_{k'}^{(m)}(\mathbf{x}) + \sum_{k''=1}^{\mathcal{N}_2} u_{k''}^{(s)} \varphi_{k''}^{(s)}(\mathbf{x}) + \sum_{k=1}^{\mathcal{N}_{master}} \lambda_k^{(m)} \mu_k(\mathbf{x}).$$

Now varying $v_\delta \in \text{span}\{\varphi_{k'}^{(m)}, \varphi_{k''}^{(s)}, \mu_k\}$ and defining vectors $\mathbf{u}^{(m)} = [u_{k'}^{(m)}]^T$, $\mathbf{u}^{(s)} = [u_{k''}^{(s)}]^T$, the mortar system (11.18) reads

$$\begin{bmatrix} A_{mm} & 0 & A_{m,\Gamma_m} \\ 0 & A_{ss} & A_{s,\Gamma_s}\Xi \\ A_{\Gamma_m,m} & \Xi^T A_{\Gamma_s,s} & A_{\Gamma_m,\Gamma_m} + \Xi^T A_{\Gamma_s,\Gamma_s}\Xi \end{bmatrix} \begin{bmatrix} \mathbf{u}^{(m)} \\ \mathbf{u}^{(s)} \\ \boldsymbol{\lambda}^{(m)} \end{bmatrix} = \begin{bmatrix} \mathbf{f}_m \\ \mathbf{f}_s \\ \mathbf{f}_{\Gamma_m} + \Xi^T \mathbf{f}_{\Gamma_s} \end{bmatrix}. \tag{11.29}$$

Above, for $i \in \{m,s\}$, we introduced the matrices $(A_{ii})_{jk} = a_i(\varphi_k^{(i)}, \varphi_j^{(i)})$, $(A_{i,\Gamma_i})_{jk} = a_i(\mu_k^{(i)}, \varphi_j^{(i)})$, $(A_{\Gamma_i,i})_{kj} = a_i(\varphi_j^{(i)}, \mu_k^{(i)})$, $(A_{\Gamma_i,\Gamma_i})_{k\ell} = a_i(\mu_\ell^{(i)}, \mu_k^{(i)})$ and the vectors $(\mathbf{f}_i)_j = \int_{\Omega_i} f\varphi_j^{(i)}$, $(\mathbf{f}_{\Gamma_i})_\ell = \int_{\Omega_i} f\mu_\ell^{(i)}$ for the basis functions associated to the nodes of Ω_i not on the boundary $\partial\Omega$.

The matrix Ξ depends solely on the chosen discretization and is built once the discretization's parameters have been fixed.

System (11.29) can be solved by one of the direct of iterative methods seen in Chapter 7. Instead of solving the overall system (11.29), one can solve its Schur complement for the vector $\boldsymbol{\lambda}^{(m)}$, which consists in eliminating the unknowns $\mathbf{u}^{(m)}$, $\mathbf{u}^{(s)}$ from the system (see Section 18.3.1 for a thorough description). Let us define the following matrices (called local Schur complements)

$$\Sigma_i = A_{\Gamma_i,\Gamma_i} - A_{\Gamma_i,i}A_{ii}^{-1}A_{i,\Gamma_i}, \qquad \text{for } i = m,s, \tag{11.30}$$

and set

$$\Sigma = \Sigma_m + \Xi^T\Sigma_s\Xi, \qquad \chi = (\mathbf{f}_{\Gamma_m} - A_{mm}^{-1}\mathbf{f}_m) + \Xi^T(\mathbf{f}_{\Gamma_s} - A_{ss}^{-1}\mathbf{f}_s),$$

Now we follow the recipe:

- compute *master* degrees of freedom on Γ by solving

$$\Sigma\boldsymbol{\lambda}^{(m)} = \chi; \tag{11.31}$$

- determine *slave* degrees of freedom on Γ using the linear relationship (11.28);
- solve problems $A_{ii}\mathbf{u}^{(i)} = \mathbf{f}_i - A_{i,\Gamma_i}\boldsymbol{\lambda}^{(i)}$, $i = 1,2$, independently. This is equivalent to solving two Dirichlet problems with prescribed trace on the interface Γ.

Equation (11.31) is the discrete counterpart to the Steklov-Poincaré equation (18.26) (see Chap. 18) that expresses the continuity of fluxes through the interface, rather than strong continuity if the discretization on Γ is conforming, or weak continuity if the formulation on Γ is of mortar type.

System (11.31) is typically solved by iterative methods (such as the Conjugate Gradient, Bi-CGStab or GMRES), since local Schur complements Σ_i are not assembled explicitly due to the presence of the matrices A_{ii}^{-1}.

Various preconditioners have been suggested in the literature for the algebraic system resulting from the mortar formulation; for example [AMW99] studies a preconditioner for (11.31) based on the decomposition of the space of mortar traces in the direct sum of subspaces associated to the traces of the interfaces (in case of many subdomains) and on a coarse space that allows to reduce the lower frequences of the error.

At present, as there are only two subdomains and one interface, we have preconditioned system (11.31) with the matrix Σ_m defined in (11.30). For a spectral element discretization this preconditioner turns out to be optimal, in the sense that the number of iterations required by the iterative method to solve (11.31) up to a given tolerance is independent of the degrees N_i ($i = m,s$) on the *master* and *slave* domains (Figure 11.13). For finite element approximations the preconditioner Σ_m lowers the number of iterations needed for the method to converge, but now this number does depend on the discretization parameter h, as one can see from the results of Figure 11.14.

11.8 The mortar method for combined finite and spectral elements

Until this point we have looked at situations where the spaces X_{i,δ_i} ($i = m,s$) of (11.11) are of the same kind on both domains Ω_m, Ω_s, that is to say both of spectral type or of finite element type.

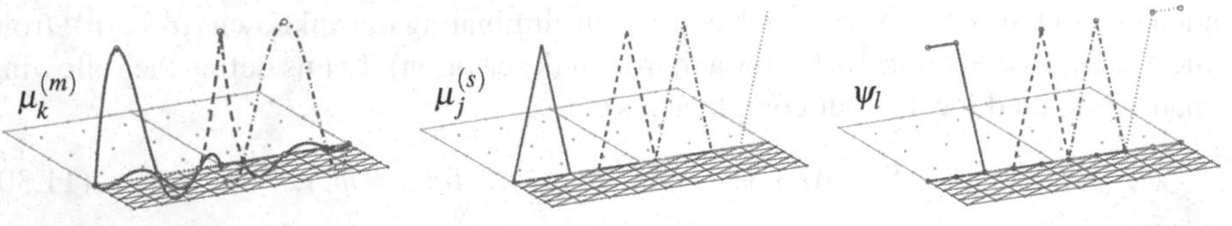

Fig. 11.10. The restrictions to Γ of $\mu_k^{(m)}$, $\mu_j^{(s)}$ and the functions ψ_l in the spectral master / slave finite elements case

Now we consider how the picture changes when we choose X_{m,δ_m} of finite element type while X_{s,δ_s} of spectral element type, or the other way around.

First of all, notice that the spaces V_{i,δ_i}, $i = m, s$, are naturally defined by the X_{i,δ_i}.

The definition of the space of constraints $\widetilde{\Lambda}_\delta$ is strictly related to the discretization adopted on the *slave* space, so $\widetilde{\Lambda}_\delta$ will be defined as in (11.13) if the discretization in Ω_s is spectral, or as in (11.15) with finite elements. The corresponding basis functions ψ_l will abide by the definitions of Sections 11.2.1 or 11.2.2 respectvely.

With these choices now the point is to compute accurately the integrals appearing in (11.25) that define the entries of the matrices P and Φ.

Computing the $P_{\ell j}$ is only a matter of the discretization chosen on the *slave* domain, so it is carried out as explained in Section 11.5 (if we have spectral elements in Ω_s) or Section 11.6 (with finite elements).

Computing the $\Phi_{\ell k}$ requires more care, instead, for it involves both discretizations in Ω_s (via the functions ψ_l) and in Ω_m (via the $\mu_k^{(m)}$). We will keep the two situations separate and discuss only finite elements of type $\mathbb{P}_1$.

Case 1: *spectral master / slave finite elements.*
The restrictions of the functions $\mu_k^{(m)}$ to the edges e of $\mathcal{E}_m^\Gamma$ are polynomials of degree N_m, while the restrictions of the ψ_l to the edges of $\mathcal{E}_{s,\delta}^\Gamma$ are polynomials of degree 1 at most (Figure 11.10). Let us produce a finer partition $\mathcal{E}_f^\Gamma$ than either $\mathcal{E}_m^\Gamma$ and $\mathcal{E}_{s,\delta}^\Gamma$ so that on every $\tilde{e} \in \mathcal{E}_f^\Gamma$ the restrictions of $\mu_k^{(m)}$, ψ_l are polynomials. The degree of the product $\mu_k^{(m)} \psi_l$ is at most $N_m + 1$, and to compute each integral $\int_{\tilde{e}} \mu_k^{(m)} \psi_l$ exactly we can use Legendre-Gauss quadrature formulae on $N_q + 1$ quadrature nodes in $\tilde{e}$, with $N_q = N_m/2$ if N_m is even and $N_q = (N_m + 1)/2$ if N_m is odd.

Case 2: *master finite elements / spectral slave.*
The restrictions of the $\mu_k^{(m)}$ to the edges e of $\mathcal{E}_{m,\delta}^\Gamma$ (the set of all edges of the triangles $T_{m,k}$ on Γ) are polynomials of degree one at most, while the restrictions of the ψ_l to the edges of $\mathcal{E}_s^\Gamma$ are polynomials of degree $N_s - 2$ (Figure 11.11). We generate a partition $\mathcal{E}_f^\Gamma$ that is finer than $\mathcal{E}_{m,\delta}^\Gamma$ and than $\mathcal{E}_s^\Gamma$, so that on every $\tilde{e} \in \mathcal{E}_f^\Gamma$ the restrictions of $\mu_k^{(m)}$ and ψ_l are polynomial. The degree of the product $\mu_k^{(m)} \psi_l$ is at most $N_s + 1$, and to compute each $\int_{\tilde{e}} \mu_k^{(m)} \psi_l$ exactly we may employ Legendre-Gauss quadrature formulae with $N_q + 1$ quadrature nodes in $\tilde{e}$, where $N_q = N_s/2$ if N_s is even and $N_q = (N_s + 1)/2$ if odd.

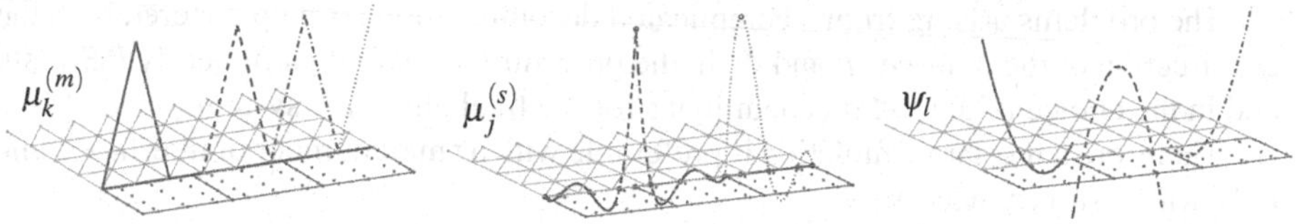

Fig. 11.11. The restrictions to Γ of $\mu_k^{(m)}$, $\mu_j^{(s)}$ and the functions ψ_l in the master finite elements / spectral slave case

Once the matrices P and Φ have been found using (11.27), we compute the matrix Ξ and solve the linear system (11.29) (or equivalently (11.31) as described in Section 11.7).

Concerning the analysis of the approximation error we have optimal convergence ([BMP94]), a result that generalizes estimates (11.19) and (11.20). Among the domains Ω_i ($i = 1,\ldots,M$) we distinguish those with spectral discretization Ω_i^{es}, $i = 1,\ldots,M^{es}$, from those with finite element discretization Ω_i^{ef}, $i = 1,\ldots,M^{ef}$.

If the solution u to the continuous problem (3.13) and the function f are regular enough on each subdomain Ω_i, i.e. $u_{|\Omega_i} \in H^{\sigma_i}(\Omega_i)$ with $\sigma_i > \frac{3}{2}$ and $f_{|\Omega_i^{es}} \in H^{\rho_i}(\Omega_i^{es})$ with $\rho_i > 1$ for all $i = 1,\ldots,M^{es}$, then

$$\begin{aligned} |||u - u_\delta||| \leq C\Big(& \sum_{i=1}^{M^{es}} N_i^{1-\sigma_i} \|u_{|\Omega_i}\|_{H^{\sigma_i}(\Omega_i^{es})} + N_i^{-\rho_i} \|f_{|\Omega_i}\|_{H^{\rho_i}(\Omega_i^{es})} \\ & + \sum_{i=1}^{M^{ef}} h_i^{\min\{\sigma_i, r_i+1\}-1} \|u_{|\Omega_i}\|_{H^{\sigma_i}(\Omega_i^{ef})}\Big). \end{aligned} \tag{11.32}$$

11.9 Generalization of the mortar method to multi-domain decompositions

Suppose we decompose the domain Ω in more than two subdomains. The previous sections' study of one interface must be repeated for every single interface of the decomposition. Hence for every interface Γ_{ij} we have to choose which between Ω_i and Ω_j is the *master* and which the *slave*, and then we must impose system (11.24) on Γ_{ij}.

So for every interface Γ_{ij} there is a "local" constraint space $\widetilde{\Lambda}_\delta$ that depends on the slave domain chosen on Γ_{ij}, and the overall, global space of constraints will be the Cartesian product of the local ones.

At the vertices of the subdomains Ω_i lying on the closure of Γ one imposes a continuity condition, in analogy to $(11.24)_2$.

Observe that a domain might be *master* for one interface and *slave* for another one, as in Figure 11.3, right: Ω_3 could for example be *master* for Γ_{36} and slave for Γ_{35} and Γ_{32}.

The problems arising from a complicated decomposition crop up concretely in the construction of the matrices P and Φ, in the procedure for solving efficiently $P\Xi = \Phi$, and in the choice of a good preconditioner for the final algebraic system.

Therefore a mortar formulation is preferable with a small number of interfaces, and only where strictly necessary.

11.10 Numerical results for the mortar method

Consider the Poisson problem

$$\begin{cases} -\Delta u = f & \text{in } \Omega = (0,2)^2 \\ u = g & \text{on } \partial\Omega, \end{cases} \tag{11.33}$$

where f and g are such that the exact solution reads $u(x,y) = \sin(\pi xy) + 1$. Subdivide Ω in two subdomains $\Omega_1 = (0,1) \times (0,2)$ and $\Omega_2 = (1,2) \times (0,2)$, on both of which we introduce a further uniform partition into rectangles, and then discretize by spectral elements.

Figure 11.12 displays the errors in *broken norm* between the mortar solution and the exact one, once Ω_1 has been appointed master. On the left the slave's degree $N_s = N_2 = 14$ is fixed, and the master's degree $N_m = N_1$ varies, whilst on the right $N_1 = 14$ is fixed in the master domain and the slave degree varies. The two curves refer to different partitions on the subdomains: the first one is geometrically conforming with 2×2 spectral elements in each Ω_i, the second has 2×3 spectral elements in Ω_m and 2×2 in Ω_s. In both cases the error converges exponentially until the error in the domain with fixed spectral degree prevails.

Figure 11.13 gives the number of iterations the preconditioned Bi-CGStab method needs in order to solve system (11.31), using Σ_m as preconditioner and with fixed tolerance $\varepsilon = 10^{-12}$ in the stopping test. Note that for a conforming discretization, convergence is reached after one iteration, while in the nonconforming setting more iter-

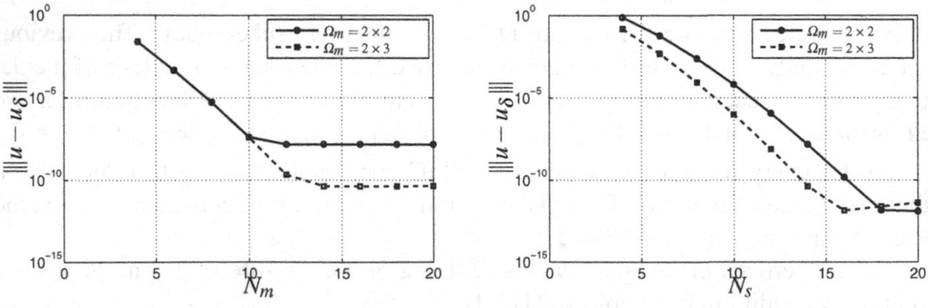

Fig. 11.12. Errors in *broken norm* for the solution to problem (11.33). $\Omega_s = 2 \times 2$ spectral elements. On the left the degree $N_s = 14$ on the slave domain Ω_s is fixed, on the right the degree $N_m = 14$ on the master domain Ω_m is fixed

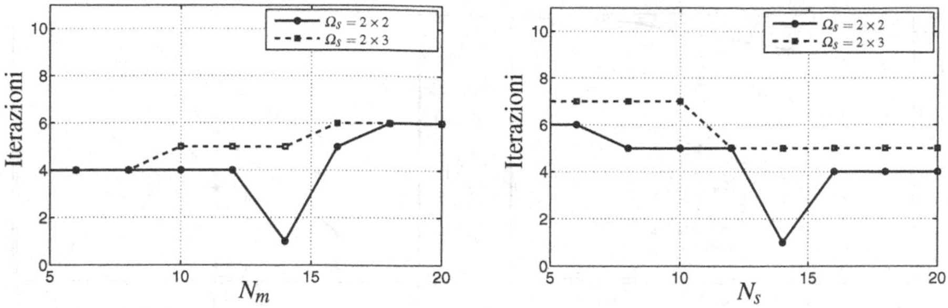

Fig. 11.13. Iterations of preconditioned Bi-CGstab to solve problem (11.33). $\Omega_m = 2\times 2$ spectral elements. The degree is fixed on the slave domain Ω_s on the left, on the master domain Ω_m on the right

ations are necessary, although their number is independent of the polynomial degrees N_m and N_s.

Figure 11.14 shows the numerical results for an approximation of problem (11.33) by finite elements $\mathbb{P}_1$, both in the master domain Ω_1 and in the slave domain Ω_2. The functions f, g and the subdomains are as in the previous case. In both Ω_i we assumed uniform triangulations made of $2n_i \times 2n_i$ triangles, with $n_m \neq n_s$: precisely, $n_m = 2k$ and $n_s = 3(k+2)$ for $k = 5, 10, 20, 40$.

In Figure 11.14 one can read the absolute errors in norm $H^1(\Omega_i)$ and the *broken-norm* error between the exact solution and the mortar solution, as the *mesh size* h_i varies: they decrease linearly with respect to h_i, in agreement with estimate (11.20). The number of iterations required by the preconditioned Bi-CGStab method, with preconditioner Σ_m and with given tolerance $\varepsilon = 10^{-12}$ in the stopping test, is independent of h_i, and turns out to be ≤ 6.

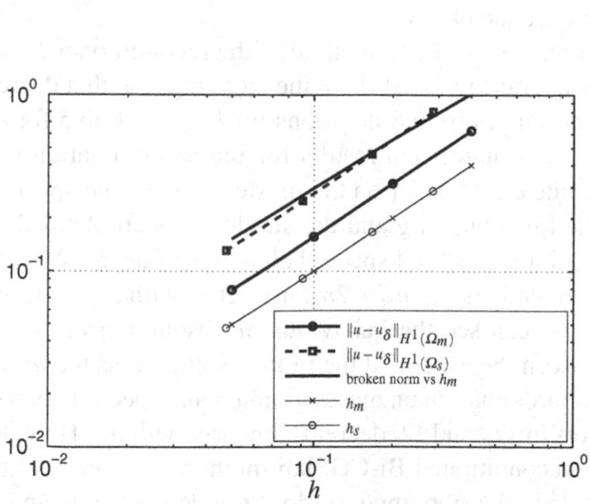

Fig. 11.14. Absolute errors in norm H^1 and in *broken norm* for the mortar approximation spectral master /slave finite elements.

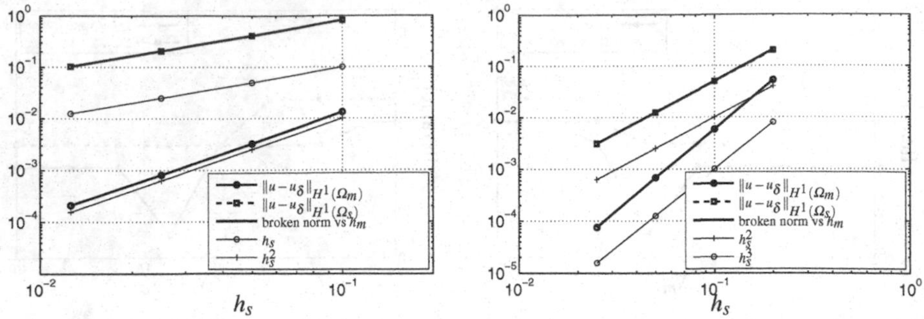

Fig. 11.15. Absolute errors in norm H^1 and *broken norm* for the mortar approximation spectral master /slave finite elements. On the left $\mathbb{Q}_6 - \mathbb{P}_1$, on the right $\mathbb{Q}_8 - \mathbb{Q}_2$. In both cases the error line in *broken norm* overlaps and practically hides the curve in norm $H^1(\Omega_s)$.

In Figure 11.15 we have the absolute errors in norm $H^1(\Omega_i)$ and the error in *broken norm* between exact and mortar solutions, as the *mesh size* h_s varies in Ω_s, relative to the approximation of problem (11.33) with spectral elements on the master domain and finite elements ($\mathbb{P}_1$ or $\mathbb{Q}_2$) on the slave domain. The functions f, g and the subdomains are defined as in the previous cases.

The errors on the left refer to a partition of Ω_m in 3×3 spectral elements of degree $N_m = 6$, of Ω_s in $2n_s \times 2n_s$ equal triangles, with $n_s = 20, 40, 80, 160$ ($h_s = 2/n_s$). The errors on the right refer to a partition of Ω_m in 3×3 spectral elements of degree $N_m = 8$, and of Ω_s in $n_s \times n_s$ equal quadrilaterals, with $n_s = 10, 20, 40, 80$ ($h_s = 2/n_s$).

We remark that the error in the slave domain of finite elements decreases like $h_s^{r_s}$ (r_s is the polynomial degree of the finite elements), whereas the error in the master domain does not reach the spectral case's accuracy because it is sensitive of the worse accuracy on the slave domain. However, it decreases as $h_s^{r_s+1}$ and the error in *broken norm* agrees with estimate (11.20).

The preconditioned Bi-CGStab method with preconditioner Σ_m needs a number of iterations, given a tolerance $\varepsilon = 10^{-12}$ in the stopping test, that decreases slightly with h_s in both tests, and ranges from 8 iterations for $h_s = 1/10$ to 5 for $h_s = 1/40$.

In Figure 11.16 the numerical results for the approximation of problem (11.33) are shown, with finite elements $\mathbb{P}_1$ on the master domain and spectral elements on the slave domain. The functions f, g and the subdomains are defined as in above cases. The domain Ω_s is divided in 4×4 spectral elements of degree $N_s = 6$, while in Ω_m we have uniform triangulations of $2n_m \times 2n_m$ triangles, with $n_m = 20, 40, 80, 160$.

In particular one can see the behaviour of absolute errors in norm $H^1(\Omega_i)$ and *broken-norm*, between the exact and the mortar solution, as the *mesh-size* h_s varies in Ω_s. With mortar approximation on master domain and spectral approximation on slave domain, both errors in Ω_m and Ω_s decrease linearly with h_m. Here, too, the number of iterations of the preconditioned Bi-CGStab method with preconditioner Σ_m, given a tolerance $\varepsilon = 10^{-12}$ in the stopping test, does not depend on h_i and is ≤ 8 in all tests.

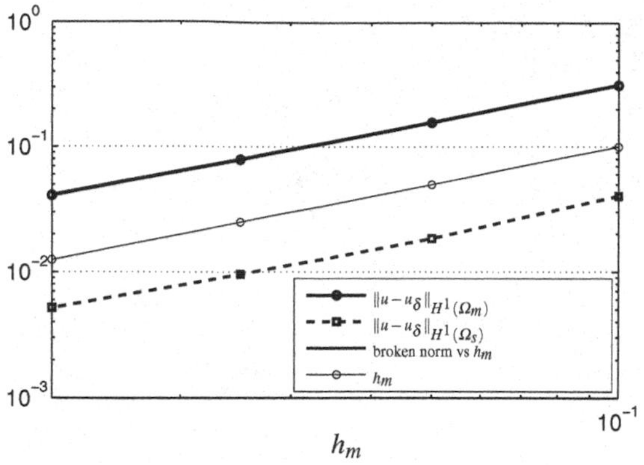

Fig. 11.16. Absolute errors in norm H^1 and *broken norm* for the mortar approximation master finite elements/spectral slave.

12

Diffusion-transport-reaction equations

In this chapter we consider problems of the following form

$$\begin{cases} Lu = -\mathrm{div}(\mu\nabla u) + \mathbf{b}\cdot\nabla u + \sigma u = f & \text{in } \Omega, \\ u = 0 & \text{on } \partial\Omega, \end{cases} \tag{12.1}$$

where μ, σ, f and $\mathbf{b}$ are given functions (or constants). In the most general case, we will suppose that $\mu \in \mathrm{L}^\infty(\Omega)$ with $\mu(\mathbf{x}) \geq \mu_0 > 0$, $\sigma \in \mathrm{L}^2(\Omega)$ with $\sigma(\mathbf{x}) \geq 0$ a.e. in Ω, $\mathbf{b} \in [\mathrm{L}^\infty(\Omega)]^2$ with $\mathrm{div}(\mathbf{b}) \in \mathrm{L}^2(\Omega)$, and $f \in \mathrm{L}^2(\Omega)$.
In many practical applications, the *diffusion* term $-\mathrm{div}(\mu\nabla u)$ is dominated by the *convection* term $\mathbf{b}\cdot\nabla u$ (also called *transport* term) or by the *reaction* term σu (also called the *absorption* term when σ is non-negative). In such cases, as we will see, the solution can give rise to *boundary layers*, that is regions, generally close to the boundary of Ω, where the solution is characterized by strong gradients.
To derive such models, and to capture the analogy with random walk processes, see e.g. [Sal08, Chap. 2.]

In this chapter we analyze the conditions ensuring the existence and uniqueness of the solution to problem (12.1). We also consider the Galerkin method, illustrate its difficulties in providing stable solutions in the presence of boundary layers, and finally propose alternative discretization methods for the approximation of (12.1).

12.1 Weak problem formulation

Let $V = \mathrm{H}_0^1(\Omega)$. By introducing the bilinear form $a : V \times V \mapsto \mathbb{R}$,

$$a(u,v) = \int_\Omega \mu\nabla u\cdot\nabla v\, d\Omega + \int_\Omega v\mathbf{b}\cdot\nabla u\, d\Omega + \int_\Omega \sigma uv\, d\Omega \qquad \forall u,v \in V, \tag{12.2}$$

the weak formulation of problem (12.1) becomes

$$\text{find } u \in V: \qquad a(u,v) = (f,v) \qquad \forall v \in V. \tag{12.3}$$

A. Quarteroni: *Numerical Models for Differential Problems*, 2nd Ed.
MS&A – Modeling, Simulation & Applications 8
DOI 10.1007/978-88-470-5522-3_12, © Springer-Verlag Italia 2014

In order to prove the existence and uniqueness of the solution of (12.3) we will put ourselves in the condition to apply the Lax-Milgram lemma.

To verify the coercivity of the bilinear form $a(\cdot,\cdot)$, we proceed separately on the single terms composing (12.2).
For the first term we have

$$\int_{\Omega} \mu \nabla v \cdot \nabla v \, d\Omega \geq \mu_0 \|\nabla v\|^2_{\mathrm{L}^2(\Omega)}. \tag{12.4}$$

As $v \in \mathrm{H}^1_0(\Omega)$, the Poincaré inequality holds (see (2.13))

$$\|v\|_{\mathrm{L}^2(\Omega)} \leq C_\Omega \|\nabla v\|_{\mathrm{L}^2(\Omega)}, \tag{12.5}$$

for a suitable positive constant C_Ω independent of v. Thus

$$\|v\|^2_{\mathrm{H}^1(\Omega)} = \|v\|^2_{\mathrm{L}^2(\Omega)} + \|\nabla v\|^2_{\mathrm{L}^2(\Omega)} \leq (1 + C^2_\Omega) \|\nabla v\|^2_{\mathrm{L}^2(\Omega)}$$

and therefore it follows from (12.4) that

$$\int_{\Omega} \mu \nabla v \cdot \nabla v \, d\Omega \geq \frac{\mu_0}{1 + C^2_\Omega} \|v\|^2_{\mathrm{H}^1(\Omega)}.$$

We now move to the convective term. Using Green's formula (3.16) yields

$$\int_{\Omega} v \mathbf{b} \cdot \nabla v \, d\Omega = \frac{1}{2} \int_{\Omega} \mathbf{b} \cdot \nabla(v^2) \, d\Omega = -\frac{1}{2} \int_{\Omega} v^2 \mathrm{div}(\mathbf{b}) \, d\Omega + \frac{1}{2} \int_{\partial\Omega} \mathbf{b} \cdot \mathbf{n} v^2 \, d\gamma$$

$$= -\frac{1}{2} \int_{\Omega} v^2 \mathrm{div}(\mathbf{b}) \, d\Omega,$$

as $v = 0$ on $\partial\Omega$, whence

$$\int_{\Omega} v \mathbf{b} \cdot \nabla v \, d\Omega + \int_{\Omega} \sigma v^2 \, d\Omega = \int_{\Omega} v^2 (-\frac{1}{2} \mathrm{div}(\mathbf{b}) + \sigma) \, d\Omega,$$

The last integral is certainly positive if we suppose that

$$-\frac{1}{2} \mathrm{div}(\mathbf{b}) + \sigma \geq 0 \quad \text{a.e. in } \Omega. \tag{12.6}$$

Consequently, the bilinear form $a(\cdot,\cdot)$ is coercive, as

$$a(v,v) \geq \alpha \|v\|^2_{\mathrm{H}^1(\Omega)} \quad \forall v \in V, \quad \text{with} \quad \alpha = \frac{\mu_0}{1 + C^2_\Omega}. \tag{12.7}$$

Let us now prove that the bilinear form $a(\cdot,\cdot)$ is continuous, that is there exists a positive constant M such that

$$|a(u,v)| \leq M \|u\|_{\mathrm{H}^1(\Omega)} \|v\|_{\mathrm{H}^1(\Omega)} \qquad \forall u, v \in V. \tag{12.8}$$

The first term on the right-hand side of (12.2) can be bounded as follows

$$\left|\int_\Omega \mu \nabla u \cdot \nabla v \, d\Omega\right| \leq \|\mu\|_{L^\infty(\Omega)} \|\nabla u\|_{L^2(\Omega)} \|\nabla v\|_{L^2(\Omega)} \\ \leq \|\mu\|_{L^\infty(\Omega)} \|u\|_{H^1(\Omega)} \|v\|_{H^1(\Omega)}, \tag{12.9}$$

having used the Hölder and Cauchy-Schwarz inequalities (see Sect. 2.5), as well as the inequality $\|\nabla w\|_{L^2(\Omega)} \leq \|w\|_{H^1(\Omega)}$ $\forall w \in H^1(\Omega)$. For the right-hand side, proceeding in a similar way we find

$$\left|\int_\Omega v\mathbf{b} \cdot \nabla u \, d\Omega\right| \leq \|\mathbf{b}\|_{L^\infty(\Omega)} \|v\|_{L^2(\Omega)} \|\nabla u\|_{L^2(\Omega)} \\ \leq \|\mathbf{b}\|_{L^\infty(\Omega)} \|v\|_{H^1(\Omega)} \|u\|_{H^1(\Omega)}. \tag{12.10}$$

Finally, for the third term we have, thanks to the Cauchy-Schwarz inequality,

$$\left|\int_\Omega \sigma uv \, d\Omega\right| \leq \|\sigma\|_{L^2(\Omega)} \|uv\|_{L^2(\Omega)} \leq \|\sigma\|_{L^2(\Omega)} \|u\|_{H^1(\Omega)} \|v\|_{H^1(\Omega)}. \tag{12.11}$$

Indeed, $\|uv\|_{L^2(\Omega)} \leq \|u\|_{L^4(\Omega)} \|v\|_{L^4(\Omega)} \leq \|u\|_{H^1(\Omega)} \|v\|_{H^1(\Omega)}$, having applied inequality (2.18) and exploited inclusions (2.19).
Summing (12.9), (12.10) and (12.11) term by term, property (12.8) follows by taking, e.g.,

$$M = \|\mu\|_{L^\infty(\Omega)} + \|\mathbf{b}\|_{L^\infty(\Omega)} + \|\sigma\|_{L^2(\Omega)}. \tag{12.12}$$

On the other hand, the right-hand side of (12.3) defines a bounded and linear functional thanks to the Cauchy-Schwarz inequality and to (12.5).
As the Lax-Milgram lemma hypotheses are verified (see Lemma 3.1), it follows that the solution of the weak problem (12.3) exists and is unique. Moreover, the following a priori estimates hold

$$\|u\|_{H^1(\Omega)} \leq \frac{1}{\alpha} \|f\|_{L^2(\Omega)}, \quad \|\nabla u\|_{L^2(\Omega)} \leq \frac{C_\Omega}{\mu_0} \|f\|_{L^2(\Omega)},$$

as consequences of (12.4), (12.7) and (12.5). The first is an immediate consequence of Corollary 3.1, the second one can easily be proven starting from equation $a(u,u) = (f,u)$ and using the Cauchy-Schwarz and Poincaré inequalities as well as (12.4) and (12.6).

The Galerkin approximation of problem (12.3) is

$$\text{find } u_h \in V_h : \qquad a(u_h, v_h) = (f, v_h) \qquad \forall v_h \in V_h, \tag{12.13}$$

where $\{V_h, h > 0\}$ is a suitable family of subspaces of $H^1_0(\Omega)$. By replicating the proof carried out above for the exact problem (12.3), the following estimates can be proved:

$$\|u_h\|_{H^1(\Omega)} \leq \frac{1}{\alpha} \|f\|_{L^2(\Omega)}, \qquad \|\nabla u_h\|_{L^2(\Omega)} \leq \frac{C_\Omega}{\mu_0} \|f\|_{L^2(\Omega)}.$$

These prove, in particular, that the gradient of the discrete solution (as well as that of the weak solution u) could be as large as μ_0 is small.
Moreover, the Galerkin error inequality (4.10) gives

$$\|u-u_h\|_V \leq \frac{M}{\alpha} \inf_{v_h \in V_h} \|u-v_h\|_V. \tag{12.14}$$

By the definitions of α and M (see (12.7) and (12.12)), the upper-bounding constant M/α becomes as large (and, correspondingly, the estimate (12.14) meaningless) as the ratio $\|\mathbf{b}\|_{L^\infty(\Omega)}/\|\mu\|_{L^\infty(\Omega)}$ (resp. the ratio $\|\sigma\|_{L^2(\Omega)}/\|\mu\|_{L^\infty(\Omega)}$) grows, which happens when the convective (resp. reactive) term dominates over diffusive one.

In such cases the Galerkin method can give inaccurate solutions, unless – as we will see – an extremely small discretization step h is used.

Remark 12.1. Problem (12.1) is known as the *non-conservative form* of the diffusion-transport(-reaction) problem, the conservative form being

$$\begin{cases} Lu = \mathrm{div}(-\mu\nabla u + \mathbf{b}u) + \sigma u = f & \text{in } \Omega \\ u = 0 & \text{on } \partial\Omega \end{cases} \tag{12.15}$$

If $\mathbf{b}$ is constant, the two formulations (12.1) and (12.15) are equivalent. The bilinear form associated to (12.15) is

$$a(u,v) = \int_\Omega (\mu\nabla u - \mathbf{b}u)\cdot\nabla v \, d\Omega + \int_\Omega \sigma uv \, d\Omega \qquad \forall u,v \in V. \tag{12.16}$$

It can be easily verified that the condition which ensures the coercivity of this bilinear form is

$$\frac{1}{2}\mathrm{div}(\mathbf{b}) + \sigma \geq 0 \qquad \text{a.e. in } \Omega. \tag{12.17}$$

Under these assumptions, the conclusions drawn for problem (12.1) (and for its approximations) also hold for problem (12.15). •

In order to evaluate more precisely the behaviour of the numerical solution provided by the Galerkin method, we analyze a one-dimensional problem.

12.2 Analysis of a one-dimensional diffusion-transport problem

Let us consider the following one-dimensional diffusion-transport problem

$$\begin{cases} -\mu u'' + bu' = 0, & 0 < x < 1, \\ u(0) = 0, & u(1) = 1, \end{cases} \tag{12.18}$$

μ and b being two positive constants.

Its weak formulation is

$$\text{find } u \in \mathrm{H}^1(0,1): \quad a(u,v) = 0 \quad \forall v \in \mathrm{H}^1_0(0,1), \tag{12.19}$$

with $u(0) = 0$, $u(1) = 1$, and $a(u,v) = \int_0^1 (\mu u'v' + bu'v)dx$. Following what indicated in Sect. 3.2.2, we can reformulate (12.19) by introducing a suitable lifting (or extension) of the boundary data. In this particular case, we can choose $R_g = x$. Having then set $\mathring{u} = u - R_g = u - x$, we can reformulate (12.19) in the following way

$$\text{find } \mathring{u} \in \mathrm{H}_0^1(0,1): \quad a(\mathring{u}, v) = F(v) \quad \forall v \in \mathrm{H}_0^1(0,1), \tag{12.20}$$

where $F(v) = -a(x,v) = -\int_0^1 bv\,dx$ represents the contribution due to the data lifting. We define the *global Péclet number* as the ratio

$$\mathbb{P}\mathrm{e}_g = \frac{bL}{2\mu},$$

L being the linear dimension of the domain (1 in our case). This ratio provides a measure of how the convective term dominates the diffusive one. As such it plays the same rôle as the Reynolds number in the Navier-Stokes equations, which we will see in Chapter 16. For a negative b, its absolute value should be used in the previous definition.

We start by computing the exact solution of such problem. Its associated characteristic equation

$$-\mu\lambda^2 + b\lambda = 0$$

has two roots, $\lambda_1 = 0$ and $\lambda_2 = b/\mu$. The general solution is therefore

$$u(x) = C_1 e^{\lambda_1 x} + C_2 e^{\lambda_2 x} = C_1 + C_2 e^{\frac{b}{\mu}x}.$$

By imposing the boundary conditions we find the constants C_1 and C_2, and therefore the solution

$$u(x) = \frac{\exp(\frac{b}{\mu}x) - 1}{\exp(\frac{b}{\mu}) - 1}.$$

Using the Taylor expansion for the exponentials, if $b/\mu \ll 1$ we obtain

$$u(x) = \frac{1 + \frac{b}{\mu}x + \cdots - 1}{1 + \frac{b}{\mu} + \cdots - 1} \simeq \frac{\frac{b}{\mu}x}{\frac{b}{\mu}} = x.$$

Thus, the solution lies near the straight line interpolating the boundary data (which is the solution corresponding to the case $b = 0$).
Conversely, if $b/\mu \gg 1$ the exponentials are very large, hence

$$u(x) \simeq \frac{\exp(\frac{b}{\mu}x)}{\exp(\frac{b}{\mu})} = \exp\left(-\frac{b}{\mu}(1-x)\right),$$

and the solution is close to zero on almost all of the interval, except for a neighborhood of the point $x = 1$, where it tends to 1 exponentially. Such neighborhood has a width

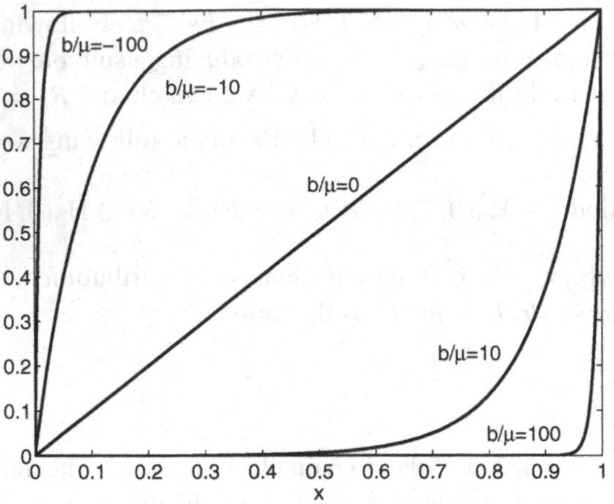

Fig. 12.1. Behaviour of the solution of problem (12.18) when varying the ratio b/μ. For completeness, we also highlight the solutions relating to the case where b is negative

of the order of μ/b and is therefore very small: the solution exhibits a *boundary layer* of width $\mathcal{O}(\frac{\mu}{b})$ in proximity of $x = 1$ (see Fig. 12.1), where the derivative behaves like b/μ, and is therefore unbounded if $\mu \to 0$.

Let us now suppose to use the Galerkin finite element method with piecewise-linear polynomials to approximate (12.19)

$$\text{find } u_h \in X_h^1 : \begin{cases} a(u_h, v_h) = 0 & \forall v_h \in \overset{\circ}{X}{}_h^1, \\ u_h(0) = 0,\ u_h(1) = 1, \end{cases} \tag{12.21}$$

where, denoting by x_i, for $i = 0, \ldots M$, the vertices of the partition introduced on $(0,1)$, we have set, coherently with (4.14),

$$X_h^r = \{v_h \in C^0([0,1]) : v_h\Big|_{[x_{i-1},x_i]} \in \mathbb{P}_r,\ i = 1,\ldots,M\},$$

$$\overset{\circ}{X}{}_h^r = \{v_h \in X_h^r : v_h(0) = v_h(1) = 0\},$$

for $r \geq 1$. Having chosen, for each $i = 1,\ldots,M-1$, $v_h = \varphi_i$ (the i-th basis function of X_h^1), we have

$$\int_0^1 \mu u_h' \varphi_i' \, dx + \int_0^1 b u_h' \varphi_i \, dx = 0.$$

Put differently, if we suppose the support of φ_i to be equal to $[x_{i-1}, x_{i+1}]$ and writing $u_h = \sum j = 1M - 1 u_j \varphi_j(x)$, we have:

$$\mu \left[u_{i-1} \int\limits_{x_{i-1}}^{x_i} \varphi_{i-1}' \varphi_i' \, dx + u_i \int\limits_{x_{i-1}}^{x_{i+1}} (\varphi_i')^2 \, dx + u_{i+1} \int\limits_{x_i}^{x_{i+1}} \varphi_{i+1}' \varphi_i' \, dx \right]$$

$$+b\left[u_{i-1}\int_{x_{i-1}}^{x_i}\varphi_{i-1}'\varphi_i\,dx+u_i\int_{x_{i-1}}^{x_{i+1}}\varphi_i'\varphi_i\,dx+u_{i+1}\int_{x_i}^{x_{i+1}}\varphi_{i+1}'\varphi_i\,dx\right]=0,$$

$\forall i=1,\ldots,M-1$. If the partition is uniform, that is $x_0=0$ and $x_i=x_{i-1}+h$, with $i=1,\ldots,M$, observing that $\varphi_i'(x)=\frac{1}{h}$ if $x_{i-1}<x<x_i$, $\varphi_i'(x)=-\frac{1}{h}$ if $x_i<x<x_{i+1}$, for $i=1,\ldots,M-1$, we obtain

$$\mu\left(-u_{i-1}\frac{1}{h}+u_i\frac{2}{h}-u_{i+1}\frac{1}{h}\right)+b\left(-u_{i-1}\frac{1}{h}\frac{h}{2}+u_{i+1}\frac{1}{h}\frac{h}{2}\right)=0,$$

that is

$$\frac{\mu}{h}(-u_{i-1}+2u_i-u_{i+1})+\frac{1}{2}b(u_{i+1}-u_{i-1})=0,\quad i=1,\ldots,M-1. \tag{12.22}$$

Reordering the terms we find

$$\left(\frac{b}{2}-\frac{\mu}{h}\right)u_{i+1}+\frac{2\mu}{h}u_i-\left(\frac{b}{2}+\frac{\mu}{h}\right)u_{i-1}=0,\quad i=1,\ldots,M-1.$$

Dividing by μ/h and defining the *local* (or "*grid*") *Péclet number*

$$\mathbb{P}\mathrm{e}=\frac{|b|h}{2\mu}, \tag{12.23}$$

we finally have

$$(\mathbb{P}\mathrm{e}-1)u_{i+1}+2u_i-(\mathbb{P}\mathrm{e}+1)u_{i-1}=0,\qquad i=1,\ldots,M-1. \tag{12.24}$$

This is a linear difference equation that admits exponential solutions of the form $u_i=\rho^i$ (see [QSS07]). Replacing such expression into (12.24), we obtain

$$(\mathbb{P}\mathrm{e}-1)\rho^2+2\rho-(\mathbb{P}\mathrm{e}+1)=0,$$

from which we get the two roots

$$\rho_{1,2}=\frac{-1\pm\sqrt{1+\mathbb{P}\mathrm{e}^2-1}}{\mathbb{P}\mathrm{e}-1}=\begin{cases}(1+\mathbb{P}\mathrm{e})/(1-\mathbb{P}\mathrm{e}),\\ 1.\end{cases}$$

Thanks to the linearity of (12.24), the general solution of such equation takes the form

$$u_i=A_1\rho_1^i+A_2\rho_2^i,$$

with A_1 and A_2 two arbitrary constants. By imposing the boundary conditions $u_0=0$ and $u_M=1$, we find

$$A_1=-A_2\ \text{and}\ A_2=\left(1-\left(\frac{1+\mathbb{P}\mathrm{e}}{1-\mathbb{P}\mathrm{e}}\right)^M\right)^{-1}.$$

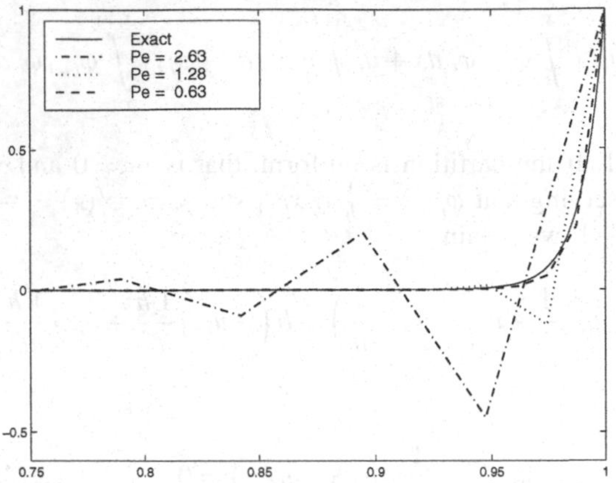

Fig. 12.2. Finite element solution of the diffusion-transport problem (12.18) with $\mathbb{P}\mathrm{e}_g = 50$ for different values of the local Péclet number

To conclude, the solution of problem (12.21) has the following nodal values

$$u_i = \frac{1 - \left(\dfrac{1+\mathbb{P}\mathrm{e}}{1-\mathbb{P}\mathrm{e}}\right)^i}{1 - \left(\dfrac{1+\mathbb{P}\mathrm{e}}{1-\mathbb{P}\mathrm{e}}\right)^M}, \qquad i = 0, \ldots, M.$$

We observe that, if $\mathbb{P}\mathrm{e} > 1$, the term within brackets is negative and the approximate solution becomes oscillatory, as opposed to the exact solution that is monotone! This phenomenon is displayed in Fig. 12.2 where the solution of (12.24), for different values of the local Péclet number, is compared to the exact solution for a case where the global Péclet number is equal to 50. As it can be observed, the higher the Péclet number gets, the more the behaviour of the approximate solution differs from that of the exact solution, with oscillations that become more and more noticeable in proximity of the boundary layer.

The most obvious remedy to this misbehaviour would be to choose a sufficiently small grid-size h, in order to ensure $\mathbb{P}\mathrm{e} < 1$. However, this strategy is not always convenient: for instance, if $b = 1$ and $\mu = 1/5000$, we should take $h < 1/2500$, that is introduce at least 2500 intervals on $(0,1)$! In particular, such strategy would require an unreasonably high number of nodal points for boundary-value problems in several dimensions. A more suitable remedy consists in using an a-priori adaptive procedure that refines the grid only in proximity of the boundary layer. Several strategies are availabel for this purpose. Among the better known, we mention the so-called type B (for Bakhvâlov) or type S (for Shishkin) grids. See e.g. [GRS07].

Alternative grid adaptive strategies, both a-priori and a-posteriori, especially useful for multidimensional problems, are those described in Sect. 4.6.

12.3 Analysis of a one-dimensional diffusion-reaction problem

Let us now consider a one-dimensional diffusion-reaction problem

$$\begin{cases} -\mu u'' + \sigma u = 0, & 0 < x < 1, \\ u(0) = 0, & u(1) = 1, \end{cases} \tag{12.25}$$

with μ and σ positive constants, whose solution is

$$u(x) = \frac{\sinh(\alpha x)}{\sinh(\alpha)} = \frac{e^{\alpha x} - e^{-\alpha x}}{e^{\alpha} - e^{-\alpha}}, \text{ with } \alpha = \sqrt{\sigma/\mu}.$$

Also in this case, if $\sigma/\mu \gg 1$ there is a boundary layer for $x \to 1$, with thickness of order $\sqrt{\mu/\sigma}$, where the first derivative becomes unbounded for $\mu \to 0$ (note, for instance, the exact solution for the case displayed in Fig. 12.3). Also in this case, it is interesting to define the global Péclet number, which takes the form

$$\mathbb{P}\mathrm{e}_g = \frac{\sigma L^2}{6\mu},$$

L still being the linear dimension of the domain (1 in our case).
The Galerkin finite element approximation of (12.25) reads

$$\text{find } u_h \in X_h^r \text{ such that } a(u_h, v_h) = 0 \quad \forall v_h \in \overset{\circ}{X_h^r}, \tag{12.26}$$

for $r \geq 1$, with $u_h(0) = 0$ and $u_h(1) = 1$ and $a(u_h, v_h) = \int_0^1 (\mu u_h' v_h' + \sigma u_h v_h)dx$. Equivalently, by setting $\overset{\circ}{u}_h = u_h - x$, and $F(v_h) = -a(x, v_h) = -\int_0^1 x v_h dx$, we have

$$\text{find } \overset{\circ}{u}_h \in V_h \text{ such that } a(\overset{\circ}{u}_h, v_h) = F(v_h) \quad \forall v_h \in V_h, \tag{12.27}$$

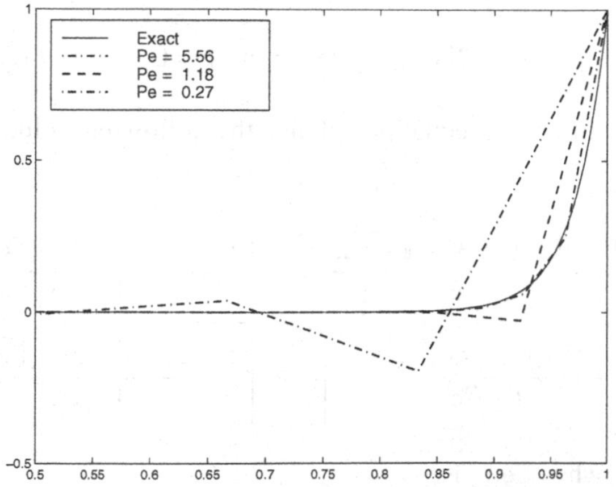

Fig. 12.3. Comparison between the numerical solution and the exact solution of the diffusion-reaction problem (12.25) with $\mathbb{P}\mathrm{e}_g = 200$. The numerical solution has been obtained using the Galerkin-linear finite elements method on uniform grids

with $V_h = \overset{\circ}{X}{}_h^r$. For the sake of simplicity, let us consider problem (12.26) with piecewise linear elements (that is $r = 1$) on a uniform partition. The equation associated to the generic basis function $v_h = \varphi_i$, $i = 1, \ldots, M-1$, is

$$\int_0^1 \mu u_h' \varphi_i' \, dx + \int_0^1 \sigma u_h \varphi_i \, dx = 0.$$

By carrying out our computation in a similar way to what we did in the previous section, and observing that

$$\int_{x_{i-1}}^{x_i} \varphi_{i-1}\, \varphi_i \, dx = \frac{h}{6}, \qquad \int_{x_{i-1}}^{x_{i+1}} \varphi_i^2 \, dx = \frac{2}{3}h, \qquad \int_{x_i}^{x_{i+1}} \varphi_i\, \varphi_{i+1} \, dx = \frac{h}{6},$$

we obtain

$$\mu \left(-u_{i-1}\frac{1}{h} + u_i \frac{2}{h} - u_{i+1}\frac{1}{h} \right) + \sigma \left(u_{i-1}\frac{h}{6} + u_i \frac{2}{3}h + u_{i+1}\frac{h}{6} \right) = 0, \qquad (12.28)$$

that is

$$\left(\frac{h}{6}\sigma - \frac{\mu}{h} \right) u_{i+1} + \left(\frac{2}{3}\sigma h + \frac{2\mu}{h} \right) u_i + \left(\frac{h}{6}\sigma - \frac{\mu}{h} \right) u_{i-1} = 0.$$

Dividing by μ/h and defining the following local Péclet number

$$\mathbb{P}\mathrm{e} = \frac{\sigma h^2}{6\mu}, \qquad (12.29)$$

we finally have

$$(\mathbb{P}\mathrm{e} - 1)u_{i+1} + 2(1 + 2\mathbb{P}\mathrm{e})u_i + (\mathbb{P}\mathrm{e} - 1)u_{i-1} = 0, \qquad i = 1, \ldots, M-1.$$

This three-term difference equation admits the following solutions for each $i = 0, \ldots, M$,

$$u_i = \frac{\left[\dfrac{1 + 2\mathbb{P}\mathrm{e} + \sqrt{3\mathbb{P}\mathrm{e}(\mathbb{P}\mathrm{e} + 2)}}{1 - \mathbb{P}\mathrm{e}} \right]^i - \left[\dfrac{1 + 2\mathbb{P}\mathrm{e} - \sqrt{3\mathbb{P}\mathrm{e}(\mathbb{P}\mathrm{e} + 2)}}{1 - \mathbb{P}\mathrm{e}} \right]^i}{\left[\dfrac{1 + 2\mathbb{P}\mathrm{e} + \sqrt{3\mathbb{P}\mathrm{e}(\mathbb{P}\mathrm{e} + 2)}}{1 - \mathbb{P}\mathrm{e}} \right]^M - \left[\dfrac{1 + 2\mathbb{P}\mathrm{e} - \sqrt{3\mathbb{P}\mathrm{e}(\mathbb{P}\mathrm{e} + 2)}}{1 - \mathbb{P}\mathrm{e}} \right]^M},$$

again oscillatory when $\mathbb{P}\mathrm{e} > 1$.
The problem is therefore critical when $\frac{\sigma}{\mu} \gg 1$, that is when the diffusion coefficient is very small with respect to the reaction one (see the example reported in Fig. 12.3).

12.4 Finite elements and finite differences (FD)

We want to analyze the behaviour of the finite difference method (FD, in short) applied to the solution of diffusion-transport and diffusion-reaction problems, and highlight analogies and differences with the finite element method (FE, in short). We will limit ourselves to the *one-dimensional* case and we will consider a *uniform mesh*.

Let us consider problem (12.18) once more and let us approximate it via finite differences. In order to generate a local discretization error of the same magnitude for both terms, we will approximate the derivatives by using the following centred incremental ratios:

$$u'(x_i) = \frac{u(x_{i+1}) - u(x_{i-1})}{2h} + \mathcal{O}(h^2), \qquad i = 1,\dots,M-1, \quad (12.30)$$

$$u''(x_i) = \frac{u(x_{i+1}) - 2u(x_i) + u(x_{i-1})}{h^2} + \mathcal{O}(h^2), \quad i = 1,\dots,M-1. \quad (12.31)$$

In both cases, as highlighted, the remainder is an infinitesimal with respect to the step size h, as it can be easily proven by invoking the truncated Taylor series (see, e.g., [QSS07]). By replacing in (12.18) the exact derivatives with these incremental ratios (thus ignoring the infinitesimal error), we find the following scheme

$$\begin{cases} -\mu \dfrac{u_{i+1} - 2u_i + u_{i-1}}{h^2} + b\dfrac{u_{i+1} - u_{i-1}}{2h} = 0, \quad i = 1,\dots,M-1, \\ u_0 = 0, \quad u_M = 1. \end{cases} \qquad (12.32)$$

For each i, the unknown u_i provides an approximation for the nodal value $u(x_i)$. Multiplying by h, we obtain the same equation (12.22) obtained using linear finite elements on the same uniform grid.

Let us now consider the diffusion-reaction problem (12.25). Proceeding in an analogous way, its approximation using finite differences yields

$$\begin{cases} -\mu \dfrac{u_{i+1} - 2u_i + u_{i-1}}{h^2} + \sigma u_i = 0, \qquad i = 1,\dots,M-1, \\ u_0 = 0, \qquad u_M = 1. \end{cases} \qquad (12.33)$$

The above equation is different from (12.28), which was obtained using linear finite elements: instead the reaction term, appearing in (12.33) with the diagonal contribution σu_i, yields in (12.28) the sum of three different contributions

$$\sigma\left(u_{i-1}\frac{h}{6} + u_i\frac{2}{3}h + u_{i+1}\frac{h}{6}\right).$$

Hence the two methods FE and FD are *not* equivalent in this case. We observe that the solution obtained using the FD scheme (12.33) does not display oscillations, whichever value is chosen for the discretization step h. As a matter of fact, the solution of (12.33) is

$$u_i = (\rho_1^M - \rho_2^M)^{-1}(\rho_1^i - \rho_2^i),$$

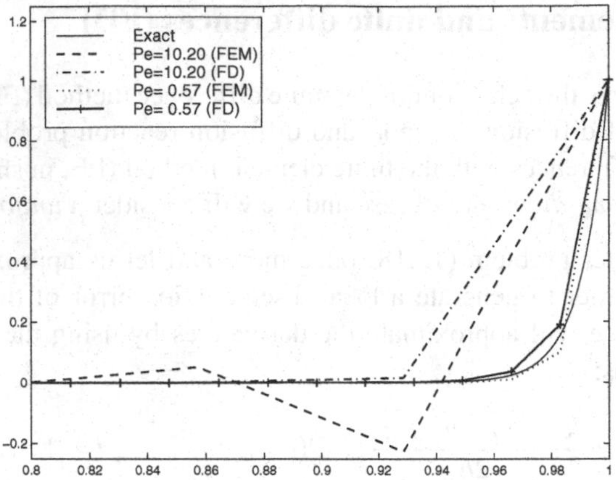

Fig. 12.4. Comparison between the numerical solutions of the one-dimensional diffusion-transport equation (12.25) with $\mathbb{P}e_g = 2000$ obtained using the Galerkin-linear finite element method (FEM) and the finite difference method (FD), for different values of the local Péclet number

with

$$\rho_{1,2} = \frac{\gamma}{2} \pm \left(\frac{\gamma^2}{4} - 1\right)^{\frac{1}{2}} \text{ and } \gamma = 2 + \frac{\sigma h^2}{\mu}.$$

The i-th powers now have a positive basis, guaranteeing a monotone behaviour of the sequence $\{u_i\}$. This differs from what we have seen in Sect. 12.3 for the FE, for which it is necessary to choose $h \leq \sqrt{\frac{6\mu}{\sigma}}$ to guarantee that the local Péclet number (12.29) is less than 1. See the example reported in Fig. 12.4 for a comparison between a finite element approximation and a finite difference approximation.

12.5 The mass-lumping technique

In the case of the reaction-diffusion problem, we can obtain the same result as with finite differences by using linear finite elements, provided that we resort to the so-called *mass-lumping* technique, thanks to which the *mass matrix*

$$\mathrm{M} = (m_{ij}), \qquad m_{ij} = \int_0^1 \varphi_j \varphi_i \, dx,$$

which is tridiagonal, is approximated using a diagonal matrix M_L, called *condensed* or *lumped matrix*. To this end we use the following trapezoidal quadrature formula on each interval (x_i, x_{i+1}), for each $i = 0, \ldots, M-1$

$$\int_{x_i}^{x_{i+1}} f(x)\, dx \simeq \frac{h}{2}(f(x_i) + f(x_{i+1})).$$

Thanks to the properties of finite element basis functions, we then find:

$$\begin{aligned}
\int_{x_{i-1}}^{x_i} \varphi_{i-1}\varphi_i\,dx &\simeq \tfrac{h}{2}\left[\varphi_{i-1}(x_{i-1})\varphi_i(x_{i-1})+\varphi_{i-1}(x_i)\varphi_i(x_i)\right]=0,\\
\int_{x_{i-1}}^{x_{i+1}} \varphi_i^2\,dx &= 2\textstyle\int_{x_{i-1}}^{x_i}\varphi_i^2\,dx \simeq 2\frac{h}{2}\left[\varphi^2{}_i(x_{i-1})+\varphi^2{}_i(x_i)\right]=h,\\
\int_{x_i}^{x_{i+1}} \varphi_i\varphi_{i+1}\,dx &\simeq \tfrac{h}{2}\left[\varphi_i(x_i)\varphi_{i+1}(x_i)+\varphi_i(x_{i+1})\varphi_{i+1}(x_{i+1})\right]=0.
\end{aligned}$$

Using the previous formulae to approximate the mass matrix coefficients, we get to the following diagonal matrix M_L whose elements are the sums of the elements of each row of M, i.e.

$$\mathrm{M}_L = \mathrm{diag}(\widetilde{m}_{ii}), \quad \text{with} \quad \widetilde{m}_{ii} = \sum_{j=i-1}^{i+1} m_{ij}. \tag{12.34}$$

Note that, thanks to the following *partition of unity* property of the basis functions

$$\sum_{j=0}^{M} \varphi_j(x) = 1 \quad \forall x \in [0,1], \tag{12.35}$$

the elements of M_L take the following expression on the interval $[0,1]$

$$\widetilde{m}_{ii} = \int_0^1 \varphi_i\,dx, \qquad i=0,\ldots,M.$$

Their values are reported in Exercise 3 for finite elements of degree $1,2,3$.
If the terms of order zero are replaced in the following way

$$\int_0^1 \sigma\, u_h\, \varphi_i\,dx = \sigma \sum_{j=1}^{M-1} u_j \int_0^1 \varphi_j\,\varphi_i\,dx = \sigma \sum_{j=1}^{M-1} m_{ij}\,u_j \simeq \sigma\,\widetilde{m}_{ii}\,u_i,$$

the finite element problem produces solutions coinciding with those of finite differences, hence monotone solutions for each value of h. Moreover, replacing M with M_L does not reduce the order of accuracy of the method.

The process of mass lumping (12.34) can be generalized to the two-dimensional case when linear elements are used. For quadratic finite elements, instead, the above-mentioned procedure consisting in summing by rows would generate a singular mass matrix M_L (see Example 12.1). An alternative diagonalization strategy consists in using the matrix $\widehat{\mathrm{M}} = \mathrm{diag}(\widehat{m}_{ii})$ with elements

$$\widehat{m}_{ii} = \frac{m_{ii}}{\sum_j m_{jj}}.$$

In the one-dimensional case, for linear and quadratic finite elements, the matrices $\widehat{\mathrm{M}}$ and M_L coincide, while they differ for cubic elements (see Exercise 3). The matrix $\widehat{\mathrm{M}}$ is non-singular also for Lagrangian finite elements of high order, while it can turn out to be singular when using non-Lagrangian finite elements, for instance when using hierarchical bases. In the latter case, we resort to more sophisticated mass-lumping procedures. Indeed, a number of diagonalization techniques able to generate non-singular matrices have been elaborated also for finite elements of high degree. See for example [CJRT01].

Example 12.1. The mass matrix for the $\mathbb{P}_2$ finite elements on the reference triangle with vertices $(0,0)$, $(1,0)$ and $(0,1)$ is given by

$$\mathrm{M} = \frac{1}{180}\begin{bmatrix} 6 & -1 & -1 & 0 & -4 & 0 \\ -1 & 6 & -1 & 0 & 0 & -4 \\ -1 & -1 & 6 & -4 & 0 & 0 \\ 0 & 0 & -4 & 32 & 16 & 16 \\ -4 & 0 & 0 & 16 & 32 & 16 \\ 0 & -4 & 0 & 16 & 16 & 32 \end{bmatrix},$$

while the lumped mass matrices are given by

$$\mathrm{M}_L = \frac{1}{180}\mathrm{diag}(0\ 0\ 0\ 60\ 60\ 60),$$

$$\widehat{\mathrm{M}} = \frac{1}{114}\mathrm{diag}(6\ 6\ 6\ 32\ 32\ 32).$$

As it can be noticed the matrix M_L is singular. ■

The mass-lumping technique is also used in other contexts, for instance in the solution of parabolic problems (see Chap. 5) when finite-element spatial discretizations and finite-difference explicit temporal discretizations (e.g., the forward-Euler method) are used. In such case, lumping the mass matrix that arises from the discretization of the temporal derivative can conduct to the solution of a diagonal system, with corresponding reduction of the computational cost.

12.6 Decentred FD schemes and artificial diffusion

The comparative analysis with finite differences allowed us to find a remedy to the oscillatory behaviour of finite element solutions in the case of a diffusion-reaction problem. We now wish to find a remedy for the case of the diffusion-transport problem (12.18) as well.

Let us consider finite differences. The oscillations in the numerical solution arise from the fact that we use a centred finite difference (CFD) scheme for the discretization of the transport term. Since the latter is non-symmetric, this suggests to discretize the first derivative at a point x_i with a decentred incremental ratio where the value at x_{i-1} intervenes if the field is positive, and at x_{i+1} in the opposite case.

This technique is called *upwinding* and the resulting scheme, called *upwind scheme* (FDUP, in short) in the case $b > 0$ is written as

$$-\mu\frac{u_{i+1} - 2u_i + u_{i-1}}{h^2} + b\frac{u_i - u_{i-1}}{h} = 0, \qquad i = 1, \ldots, M-1. \qquad (12.36)$$

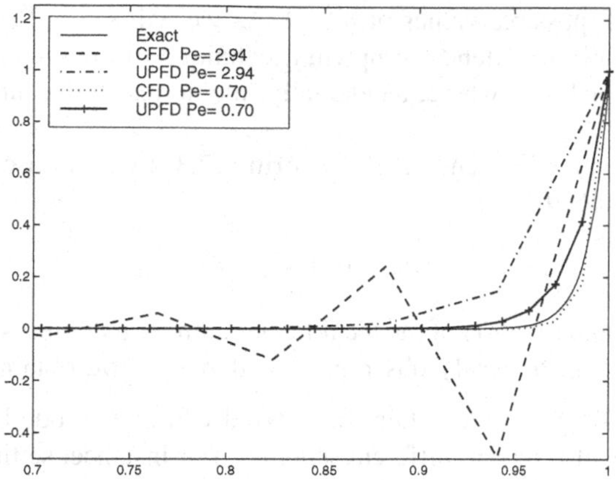

Fig. 12.5. Solution obtained using the centred (CFD) and upwind (UPFD) finite difference scheme for the one-dimensional diffusion-transport equation (12.18) with $\mathbb{P}\mathrm{e}_g = 50$ and two different local Péclet numbers. Also in the presence of high local Péclet numbers, one can notice the stabilizing effect of the artificial diffusion introduced by the upwind scheme, inevitably accompanied by a loss of accuracy

(See Fig. 12.5 for an example of application of the upwind scheme). The price to pay is a reduction of the order of convergence, because the decentred incremental ratio introduces a local discretization error $\mathcal{O}(h)$ (see (12.31)), as opposed to $\mathcal{O}(h^2)$ in the CFD case.

We now observe that

$$\frac{u_i - u_{i-1}}{h} = \frac{u_{i+1} - u_{i-1}}{2h} - \frac{h}{2}\frac{u_{i+1} - 2u_i + u_{i-1}}{h^2},$$

that is, the decentred incremental ratio to approximate the first derivative can be written as the sum of a centred incremental ratio plus a term proportional to the discretization of the second derivative, still with a centred incremental ratio. Thus, the upwind scheme can be reinterpreted as a centred finite difference scheme where an *artificial diffusion* term proportional to h has been introduced. As a matter of fact, (12.36) is equivalent to

$$-\mu_h \frac{u_{i+1} - 2u_i + u_{i-1}}{h^2} + b\frac{u_{i+1} - u_{i-1}}{2h} = 0, \qquad i = 1, \ldots, M-1, \tag{12.37}$$

where $\mu_h = \mu(1 + \mathbb{P}\mathrm{e})$, $\mathbb{P}\mathrm{e}$ being the local Péclet number introduced in (12.23). Scheme (12.37) corresponds to the discretization using a CFD scheme of the *perturbed problem*

$$-\mu_h u'' + bu' = 0. \tag{12.38}$$

The viscosity "correction" $\mu_h - \mu = \mu\mathbb{P}\mathrm{e} = \dfrac{bh}{2}$ is called *numerical viscosity* or *artificial viscosity*. The new local Péclet number associated to the scheme (12.37) is

$$\mathbb{P}\mathrm{e}^* = \frac{bh}{2\mu_h} = \frac{\mathbb{P}\mathrm{e}}{(1 + \mathbb{P}\mathrm{e})},$$

so $\mathbb{P}e^* < 1$ for all possible values of $h > 0$. As we will see in the next section, this interpretation allows to extend the upwind technique to finite elements, and also to the two-dimensional case where, incidentally, the notion of decentred differentiation is not obvious.

More generally, in a CFD scheme of the form (12.37) we can use the following numerical viscosity coefficient

$$\mu_h = \mu(1 + \phi(\mathbb{P}e)), \tag{12.39}$$

where ϕ is a suitable function of the local Péclet number that must satisfy the property $\lim_{t \to 0+} \phi(t) = 0$. It can be easily observed that if $\phi = 0$, we obtain the CFD method (12.32), while if $\phi(t) = t$, we obtain the upwind UPFD method (12.36) (or (12.37)). Different choices of ϕ lead to different schemes. For instance, setting

$$\phi(t) = t - 1 + B(2t), \tag{12.40}$$

where B is the so-called *Bernoulli function* defined as

$$B(t) = \frac{t}{e^t - 1} \quad \text{if } t > 0, \quad \text{and} \quad B(0) = 1,$$

we obtain the *exponential fitting* scheme, generally attributed to Scharfetter and Gummel or to Iljin (in fact, it was originally introduced by Allen and Southwell [AS55]). See also Sect. 12.8.7 for more on this method.

Having denoted by ϕ^U, resp. ϕ^{SG}, the two functions determined by the choices $\phi(t) = t$ and $\phi(t) = t - 1 - B(2t)$, we observe that $\phi^{SG} \simeq \phi^U$ if $\mathbb{P}e \to +\infty$, while $\phi^{SG} = \mathcal{O}(\mathbb{P}e^2)$ and $\phi^U = \mathcal{O}(\mathbb{P}e)$ if $\mathbb{P}e \to 0^+$ (see Fig. 12.6).

It can be verified that for each given μ and b the Scharfetter-Gummel scheme is a second order scheme (with respect to h) and, because of this, it is sometimes called upwind scheme with optimal viscosity. In fact, it can also be verified, in the case where f is constant – more generally, it is sufficient that f is constant on each interval $[x_i, x_{i+1}]$) – that the numerical solution produced by this scheme is nodally exact. This means that

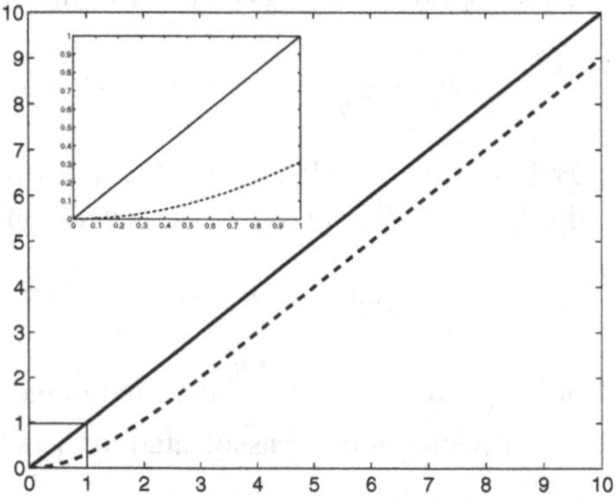

Fig. 12.6. The functions ϕ^U (solid line) and ϕ^{SG} (etched line) versus the local Péclet number

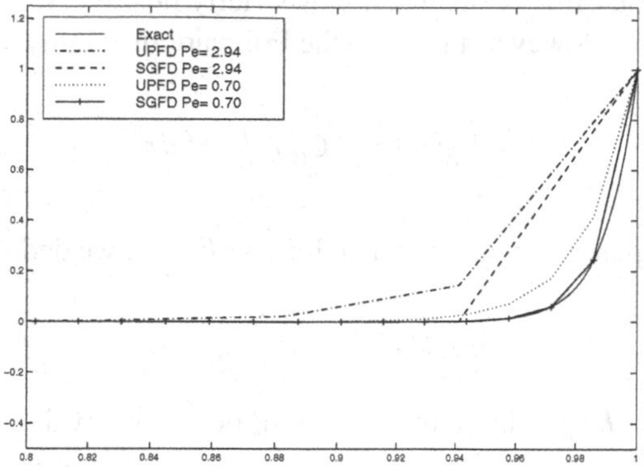

Fig. 12.7. Comparison between the exact solution and those obtained by the upwind scheme (UPFD) and the Scharfetter and Gummel one (SGFD) in the case where $\mathbb{P}e_g = 50$

it coincides exactly with the solution u at each discretization node inside the interval $(0,1)$, that is we have

$$u_i = u(x_i) \quad \text{for } i = 1,\dots,M-1,$$

independently of the choice of h (see Fig. 12.7).
We observe that the local Péclet number associated with the coefficient (12.39) is

$$\mathbb{P}e^* = \frac{bh}{2\mu_h} = \frac{\mathbb{P}e}{(1+\phi(\mathbb{P}e))},$$

and is therefore always less than 1, for each value of h.

Remark 12.2. The matrix associated with the upwind and the exponential fitting scheme is an M-matrix *regardless* of the value of h; hence, the numerical solution has a monotone behaviour (see [QSS07, Chap. 1]). •

12.7 Eigenvalues of the diffusion-transport equation

Let us consider the operator $Lu = -\mu u'' + bu'$ associated to problem (12.18) on a generic interval (α,β). Its eigenvalues λ solve $Lu = \lambda u$, $\alpha < x < \beta$, $u(\alpha) = u(\beta) = 0$, u being an eigenfunction. Such eigenvalues, in general, will be complex because of the presence of the first-order term bu'. Supposing $\mu > 0$ constant (and b variable, a priori), we have

$$\mathrm{Re}(\lambda) = \frac{\int_\alpha^\beta Lu\bar{u}\,dx}{\int_\alpha^\beta |u|^2\,dx} = \frac{\mu\int_\alpha^\beta |u'|^2\,dx - \frac{1}{2}\int_\alpha^\beta b'|u|^2\,dx}{\int_\alpha^\beta |u|^2\,dx}. \tag{12.41}$$

It can be inferred that if μ is small and b' is strictly positive, the real part of λ is not necessarily positive. However, thanks to the Poincaré inequality

$$\int_{\alpha}^{\beta} |u'|^2\,dx \geq C_{\alpha,\beta} \int_{\alpha}^{\beta} |u|^2\,dx, \tag{12.42}$$

with $C_{\alpha,\beta}$ being a positive constant depending on $\beta-\alpha$, we deduce from (12.41) that

$$\mathrm{Re}(\lambda) \geq C_{\alpha,\beta}\,\mu - \frac{1}{2}\,b'_{max},$$

where $b'_{max} = \max_{\alpha\leq s\leq\beta} b'(s)$. Thus only a finite number of eigenvalues can have a negative real part. In particular, let us observe that

$$\mathrm{Re}(\lambda) > 0 \quad \text{if } b \text{ is constant or if} \quad b'(x) \leq 0 \quad \forall x \in [\alpha,\beta].$$

The same kind of lower bound can be obtained for the eigenvalues associated to the Galerkin-finite element approximation of the problem at hand. The latter are the solution of the problem

$$\text{find } \lambda_h \in \mathbb{C},\ u_h \in V_h: \int_{\alpha}^{\beta} \mu u'_h v'_h\,dx + \int_{\alpha}^{\beta} b u'_h v_h\,dx = \lambda_h \int_{\alpha}^{\beta} u_h v_h\,dx \ \ \forall v_h \in V_h, \tag{12.43}$$

where $V_h = \{v_h \in X_h^r \ :\ v_h(\alpha) = v_h(\beta) = 0\}$. To prove this, it suffices to take again $v_h = \overline{u}_h$ in (12.43) and proceed as previously.
We can instead obtain an upper bound by choosing again $v_h = \overline{u}_h$ in (12.43) and taking the modulus in both members:

$$|\lambda_h| \leq \frac{\mu \|u'_h\|^2_{\mathrm{L}^2(\alpha,\beta)} + \|b\|_{\mathrm{L}^\infty(\alpha,\beta)}\,\|u'_h\|_{\mathrm{L}^2(\alpha,\beta)}\,\|u_h\|_{\mathrm{L}^2(\alpha,\beta)}}{\|u_h\|^2_{\mathrm{L}^2(\alpha,\beta)}}.$$

By using the *inverse inequality* (4.52) in the one-dimensional case

$$\exists\, C_I = C_I(r) > 0 \ :\ \forall v_h \in X_h^r, \quad \|v'_h\|_{\mathrm{L}^2(\alpha,\beta)} \leq C_I h^{-1}\,\|v_h\|_{\mathrm{L}^2(\alpha,\beta)}, \tag{12.44}$$

we easily find that

$$|\lambda_h| \leq \mu\, C_I^2 h^{-2} + \|b\|_{\mathrm{L}^\infty(\alpha,\beta)}\, C_I h^{-1}.$$

If, instead, we use a Legendre G-NI spectral approximation of the same problem on the usual reference interval $(-1,1)$ (see Sect. 10.3), the eigenvalue problem takes the following form

$$\begin{aligned} &\text{find } \lambda^N \in \mathbb{C},\ u_N \in \mathbb{P}_N^0: \\ &\big(\mu u'_N, v'_N\big)_N + \big(b u'_N, v_N\big)_N = \lambda^N \big(u_N, v_N\big)_N \ \ \forall v_N \in \mathbb{P}_N^0, \end{aligned} \tag{12.45}$$

with $\mathbb{P}_N^0$ now being the space of algebraic polynomials of degree N vanishing at $x = \pm 1$, and $(\cdot,\cdot)_N$ the discrete GLL scalar product defined in (10.25). We will suppose, for simplicity, that b is also constant. Taking $v_N = \overline{u}_N$, we obtain

$$\mathrm{Re}(\lambda^N) = \frac{\mu \|u_N'\|^2_{\mathrm{L}^2(-1,1)}}{\|u_N\|^2_N},$$

and so $\mathrm{Re}(\lambda^N) > 0$. Thanks to the Poincaré inequality (12.42) (which holds in the interval $(-1,1)$ with constant $C_{\alpha,\beta} = \pi^2/4$), we obtain the lower bound

$$\mathrm{Re}(\lambda^N) > \mu \,\frac{\pi^2}{4}\, \frac{\|u_N\|^2_{\mathrm{L}^2(-1,1)}}{\|u_N\|^2_N}.$$

As u_N is a polynomial of degree at most N, thanks to (10.54) we obtain

$$\mathrm{Re}(\lambda^N) > \mu \,\frac{\pi^2}{12}.$$

Instead, using the following inverse inequality for algebraic polynomials

$$\exists\, C > 0 \,:\, \forall v_N \in \mathbb{P}_N, \quad \|v_N'\|_{\mathrm{L}^2(-1,1)} \leq C N^2 \|v_N\|_{\mathrm{L}^2(-1,1)} \tag{12.46}$$

(see [CHQZ06]) and once again (10.54), we find

$$\mathrm{Re}(\lambda^N) < C \mu N^4.$$

In fact, if $N > 1/\mu$, we can prove that the moduli of the eigenvalues of the diffusion-transport problem (12.45) behave like those of the pure diffusion problem, that is

$$C_1 N^{-1} \leq |\lambda^N| \leq C_2 N^{-2}.$$

For proofs and more details, see [CHQZ06, Sect. 4.3.3].

12.8 Stabilization methods

The Galerkin method introduced in the previous sections provides a centred approximation of the transport term. A possible way to use a decentred approximation consists in choosing test functions v_h in a different space from the one u_h belong to: by doing so, we obtain a method called *Petrov-Galerkin*, for which the analysis based on the Céa lemma no longer holds. We will analyze this approach more in detail in Sect. 12.8.2. In this section we will deal instead with the methods of *stabilized finite elements*. More precisely, instead of using the Galerkin finite element method (12.27) for the approximation of (12.14), we consider the *generalized Galerkin method*

$$\text{find } \overset{\circ}{u}_h \in V_h : \; a_h(\overset{\circ}{u}_h, v_h) = F_h(v_h) \quad \forall v_h \in V_h, \tag{12.47}$$

where

$$a_h(\overset{\circ}{u}_h, v_h) = a(\overset{\circ}{u}_h, v_h) + b_h(\overset{\circ}{u}_h, v_h) \text{ and } F_h(v_h) = F(v_h) + G_h(v_h). \tag{12.48}$$

The additional terms $b_h(\overset{\circ}{u}_h, v_h)$ and $G_h(v_h)$ have the purpose of eliminating (or at least reducing) the numerical oscillations produced by the Galerkin method (when the grid is not fine enough) and are therefore named *stabilization terms*. The latter depend parametrically on h.

Remark 12.3. We want to point out that the term "*stabilization*" is in fact inexact. The Galerkin method is indeed already stable, in the sense of the continuity of the solution with respect to the data of problem (see what has been proved, e.g. in Sect. 12.1 for problem (12.1)). In this case, stabilization must be understood as the aim of reducing (ideally, eliminating) the oscillations in the numerical solution when $\mathbb{P}e > 1$. •

Let us now see several possible ways to choose the stabilization terms.

12.8.1 Artificial diffusion and decentred finite element schemes

Based on what we have seen for finite differences, we can apply the Galerkin method to problem (12.18) (whose weak formulation is (12.20)) by replacing the viscosity coefficient μ with a new coefficient $\mu_h = \mu(1+\phi(\mathbb{P}e))$. This way, we end up adding to the original viscosity term μ an *artificial* (or *numerical*) *viscosity* equal to $\mu\phi(\mathbb{P}e)$, which depends on the discretization step h through the local Péclet number $\mathbb{P}e$.
This corresponds to choosing in (12.48) $G_h(v_h) = 0$ and

$$b_h(\overset{\circ}{u}_h, v_h) = \mu\phi(\mathbb{P}e)\int_0^1 \overset{\circ}{u}{}'_h\, v'_h\, dx. \tag{12.49}$$

Since

$$a_h(\overset{\circ}{u}_h, \overset{\circ}{u}_h) \geq \mu_h |\, \overset{\circ}{u}_h\, |^2_{\mathrm{H}^1(\Omega)}$$

and $\mu_h \geq \mu$, we can say that this formulation is "more coercive" (i.e. has a larger coercivity constant) than the standard Galerkin formulation which corresponds to taking $a_h = a$ and $F_h = F$ in (12.47).

The following result provides an a priori estimate of the error made by approximating the solution of problem (12.20) with that of (12.47), (12.48), (12.49).

Theorem 12.1. *Under the assumption that $u \in \mathrm{H}^{r+1}(\Omega)$, the error between the solution of problem* (12.20) *and that of the approximate problem* (12.47) *with artificial diffusion is bounded from above:*

$$\| \overset{\circ}{u} - \overset{\circ}{u}_h \|_{\mathrm{H}^1(\Omega)} \leq$$
$$C\frac{h^r}{\mu(1+\phi(\mathbb{P}e))}\| \overset{\circ}{u} \|_{\mathrm{H}^{r+1}(\Omega)} + \frac{\phi(\mathbb{P}e)}{1+\phi(\mathbb{P}e)}\| \overset{\circ}{u} \|_{\mathrm{H}^1(\Omega)}, \tag{12.50}$$

with C a suitable positive constant independent of h and μ.

Proof. We can take advantage of *Strang's lemma*, previously introduced in Sect. 10.4.1, thanks to which we obtain

$$\begin{aligned} \| \mathring{u} - \mathring{u}_h \|_{\mathrm{H}^1(\Omega)} \leq & \inf_{w_h \in V_h} \left\{ \left(1 + \frac{M}{\mu_h}\right) \| \mathring{u} - w_h \|_{\mathrm{H}^1(\Omega)} \right. \\ & \left. + \frac{1}{\mu_h} \sup_{v_h \in V_h, v_h \neq 0} \frac{|a(w_h, v_h) - a_h(w_h, v_h)|}{\|v_h\|_{\mathrm{H}^1(\Omega)}} \right\}. \end{aligned} \tag{12.51}$$

We choose $w_h = P_h^r \mathring{u}$, the orthogonal projection of $\mathring{u}$ on V_h with respect to the scalar product $\int_0^1 u'v'\,dx$ of $\mathrm{H}_0^1(\Omega)$, that is

$$P_h^r \mathring{u} \in V_h : \quad \int_0^1 (P_h^r \mathring{u} - \mathring{u})' v_h' dx = 0 \quad \forall v_h \in V_h.$$

It can be proved that (see [QV94, Chap. 3])

$$\|(P_h^r \mathring{u})'\|_{\mathrm{L}^2(\Omega)} \leq \|(\mathring{u})'\|_{\mathrm{L}^2(\Omega)} \quad \text{and} \quad \|P_h^r \mathring{u} - \mathring{u}\|_{\mathrm{H}^1(\Omega)} \leq C h^r \|\mathring{u}\|_{\mathrm{H}^{r+1}(\Omega)},$$

C being a constant independent of h. Thus, we can bound the first addendum of the right-hand side in (12.51) by $(C/\mu_h) h^r \|\mathring{u}\|_{\mathrm{H}^{r+1}(\Omega)}$.
Now, thanks to (12.49), we obtain

$$\frac{1}{\mu_h} \frac{|a(w_h, v_h) - a_h(w_h, v_h)|}{\|v_h\|_{\mathrm{H}^1(\Omega)}} \leq \frac{\mu}{\mu_h} \phi(\mathbb{P}\mathrm{e}) \frac{1}{\|v_h\|_{\mathrm{H}^1(\Omega)}} \left| \int_0^1 w_h' v_h'\, dx \right|.$$

Using the Cauchy-Schwarz inequality, and observing that

$$\|v_h'\|_{\mathrm{L}^2(\Omega)} \leq \|v_h\|_{\mathrm{H}^1(\Omega)} \quad \text{and that} \quad \|(P_h^r \mathring{u})'\|_{\mathrm{L}^2(\Omega)} \leq \|P_h^r \mathring{u}\|_{\mathrm{H}^1(\Omega)} \leq \|\mathring{u}\|_{\mathrm{H}^1(\Omega)},$$

we obtain

$$\frac{1}{\mu_h} \sup_{v_h \in V_h, v_h \neq 0} \frac{\left| a(P_h^r \mathring{u}, v_h) - a_h(P_h^r \mathring{u}, v_h) \right|}{\|v_h\|_{\mathrm{H}^1(\Omega)}} \leq \frac{\phi(\mathbb{P}\mathrm{e})}{1 + \phi(\mathbb{P}\mathrm{e})} \|\mathring{u}\|_{\mathrm{H}^1(\Omega)}.$$

Inequality (12.50) is therefore proved. ◇

Corollary 12.1. *For a given μ and for h tending to 0 we have*

$$\| \mathring{u} - \mathring{u}_h \|_{\mathrm{H}^1(\Omega)} \leq C_1 \left[h^r \| \mathring{u} \|_{\mathrm{H}^{r+1}(\Omega)} + \phi(\mathbb{P}\mathrm{e}) \| \mathring{u} \|_{\mathrm{H}^1(\Omega)} \right], \qquad (12.52)$$

where C_1 is a positive constant independent of h, while for a given h and μ tending to 0 we have

$$\| \mathring{u} - \mathring{u}_h \|_{\mathrm{H}^1(\Omega)} \leq C_2 \left[h^{r-1} \| \mathring{u} \|_{\mathrm{H}^{r+1}(\Omega)} + \| \mathring{u} \|_{\mathrm{H}^1(\Omega)} \right], \qquad (12.53)$$

where C_2 is a positive constant independent of h and μ.

Proof. We obtain (12.52) from (12.50) remembering that $\phi(\mathbb{P}\mathrm{e}) \to 0$ for any given μ when $h \to 0$. To obtain (12.53) it is sufficient to observe that, in the *upwind* case, $\phi^U(\mathbb{P}\mathrm{e}) = \mathbb{P}\mathrm{e}$, so

$$\mu(1 + \phi(\mathbb{P}\mathrm{e})) = \mu + \frac{b}{2}h \qquad \text{and} \qquad \frac{\phi(\mathbb{P}\mathrm{e})}{1 + \phi(\mathbb{P}\mathrm{e})} = \frac{h}{h + 2\mu/b}.$$

For the Scharfetter and Gummel method, $\phi^{SG}(\mathbb{P}\mathrm{e}) \simeq \phi^U(\mathbb{P}\mathrm{e})$ for a given h and μ tending to 0. ◇

In particular, for a given μ, the stabilized method generates an error that decays linearly in h (irrespectively of the degree r) when using the *upwind* viscosity, while with an artificial viscosity of Scharfetter and Gummel type, the convergence rate becomes quadratic if $r \geq 2$. This result follows from estimate (12.52), recalling that $\phi^U(\mathbb{P}\mathrm{e}) = \mathcal{O}(h)$ while $\phi^{SG}(\mathbb{P}\mathrm{e}) = \mathcal{O}(h^2)$ for a fixed μ and for $h \to 0$.

12.8.2 The Petrov-Galerkin method

An equivalent way to write the generalized Galerkin problem (12.47) with numerical viscosity is to reformulate it as a Petrov-Galerkin method, that is a method where the space of test functions is different from the space where the solution is sought. Precisely, the approximation takes the following form

$$\text{find } \mathring{u}_h \in V_h : \quad a(\mathring{u}_h, v_h) = F(v_h) \quad \forall v_h \in W_h, \qquad (12.54)$$

where $W_h \neq V_h$, while the bilinear form $a(\cdot,\cdot)$ is the same as in the initial problem. It can be verified that in the case of linear finite elements, that is for $r = 1$, problem (12.47)–(12.49) can be rewritten as (12.54), where W_h is the space generated by the functions $\psi_i(x) = \varphi_i(x) + B_i^\alpha$ (see Fig. 12.8, right). Here the $B_i^\alpha = \alpha B_i(x)$ are the so-called *bubble functions*, with

$$B_i(x) = \begin{cases} g\left(1 - \frac{x - x_{i-1}}{h}\right), & x_{i-1} \leq x \leq x_i, \\ -g\left(\frac{x - x_i}{h}\right), & x_i \leq x \leq x_{i+1}, \\ 0 & \text{otherwise,} \end{cases}$$

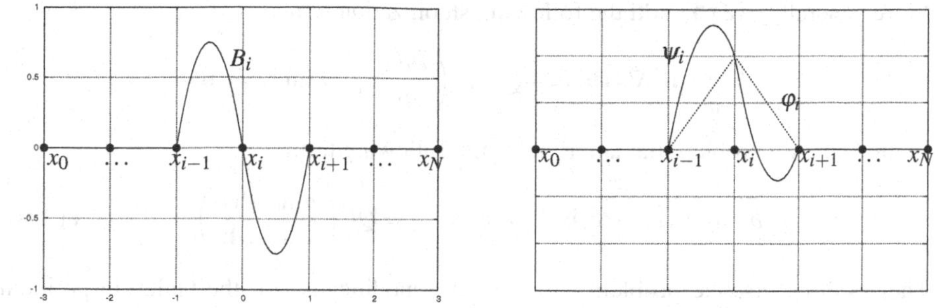

Fig. 12.8. Example of a bubble function B_i and of a basis function ψ_i of the space W_h

and $g(\xi) = 3\xi(1-\xi)$, with $0 \le \xi \le 1$ (see Fig. 12.8, left) [ZT00]. In the case of *upwind* finite differences we have $\alpha = 1$, while in the case of the Scharfetter-Gummel scheme we have $\alpha = \coth(\mathbb{P}\mathrm{e}) - 1/\mathbb{P}\mathrm{e}$. Note that the test functions lose their symmetry feature (with respect to the usual piecewise linear basis functions) under the effect of the convective field.

12.8.3 The artificial diffusion and streamline-diffusion methods in the two-dimensional case

The upwind artificial-viscosity method can be generalized to the case where we consider a two- or a three-dimensional problem of the type (12.1). In such case, it will suffice to modify the Galerkin approximation (12.13) by adding to the bilinear form (12.2) a term like

$$Qh \int_{\Omega} \nabla u_h \cdot \nabla v_h \, d\Omega \quad \text{for a chosen } Q > 0, \tag{12.55}$$

which corresponds to adding the artificial diffusion term $-Qh\Delta u$ to the initial problem (12.1). The corresponding method is called *upwind artificial diffusion.* This way an additional diffusion is introduced, not only in the direction of the field $\mathbf{b}$, as one should rightly do in order to stabilize the oscillations generated by the Galerkin method, but also in the orthogonal direction, which is not at all necessary. For instance, if we consider the two-dimensional problem

$$-\mu\Delta u + \frac{\partial u}{\partial x} = f \quad \text{in } \Omega, \quad u = 0 \quad \text{on } \partial\Omega,$$

where the transport field is given by the vector $\mathbf{b} = [1,0]^T$, the artificial diffusion term we would add is

$$-Qh\frac{\partial^2 u}{\partial x^2} \qquad \text{and not} \qquad -Qh\Delta u = -Qh\left(\frac{\partial^2 u}{\partial x^2} + \frac{\partial^2 u}{\partial y^2}\right).$$

More generally, we can add the following stabilization term

$$-Qh\mathrm{div}\left[(\mathbf{b}\cdot\nabla u)\,\mathbf{b}\right] = -Qh\mathrm{div}\left(\frac{\partial u}{\partial \mathbf{b}}\,\mathbf{b}\right), \text{ with } Q = |\mathbf{b}|^{-1}.$$

In the Galerkin problem the latter yields the following term

$$b_h(u_h, v_h) = Qh(\mathbf{b}\cdot\nabla u_h, \mathbf{b}\cdot\nabla v_h) = Qh\left(\frac{\partial u_h}{\partial \mathbf{b}}, \frac{\partial v_h}{\partial \mathbf{b}}\right). \tag{12.56}$$

The resulting discrete problem is therefore a modification of the Galerkin problem (12.13), called *streamline-diffusion* problem, and reads

$$\text{find } u_h \in V_h : \;\; a_h(u_h, v_h) = (f, v_h) \;\;\; \forall v_h \in V_h,$$

where

$$a_h(u_h, v_h) = a(u_h, v_h) + b_h(u_h, v_h).$$

Basically, we are adding a term proportional to the second derivative in the direction of the field $\mathbf{b}$ (also called *streamline*). Note that, in this case, the artificial viscosity coefficient is actually a *tensor*. As a matter of fact, the stabilization term $b_h(\cdot,\cdot)$ can be seen as the bilinear form associated to the operator $-\mathrm{div}(\mu_a\nabla u)$ with $[\mu_a]_{ij} = Qhb_ib_j$, b_i being the i-th component of $\mathbf{b}$.

Although the term (12.56) is less diffusive than (12.55), the accuracy is only $\mathcal{O}(h)$ also for the streamline-diffusion method. More accurate stabilization methods are described in Sects. 12.8.6, 12.8.8 and 12.8.9. To introduce them we will need some definitions that we will anticipate in Sects. 12.8.4 and 12.8.5.

12.8.4 Consistency and truncation error for the Galerkin and generalized Galerkin methods

Let us consider a generalized Galerkin problem of the form (12.48), and replace $\overset{\circ}{u}_h$ by u_h to recover more familiar notations. Note that this formulation can refer to a problem in any spatial dimension. We define a functional of the variable v_h

$$\tau_h(u; v_h) = a_h(u, v_h) - F_h(v_h), \tag{12.57}$$

whose norm

$$\tau_h(u) = \sup_{v_h\in V_h, v_h\neq 0} \frac{|\tau_h(u; v_h)|}{\|v_h\|_V} \tag{12.58}$$

is called the *truncation error* associated to the generalized Galerkin method (12.47). In accordance with the definitions given in Sect. 1.2, the generalized Galerkin method is said to be *consistent* if $\lim_{h\to 0} \tau_h(u) = 0$.

Moreover, we will say that it is *strongly* (or *fully*) *consistent* if the truncation error (12.58) is non-zero for each value of h.

The standard Galerkin method is strongly consistent, as seen in Chap. 4, since $\forall v_h \in V_h$

we have

$$\tau_h(u;v_h) = a(u,v_h) - F(v_h) = 0.$$

Instead, the generalized Galerkin method is only consistent (in general) as long as $a_h - a$ and $F_h - F$ "tend to zero" when h tends to zero, as guaranteed by Strang's lemma.
Concerning the upwind and streamline-diffusion methods, we have

$$\begin{aligned}\tau_h(u;v_h) &= a_h(u,v_h) - F(v_h)\\ &= a_h(u,v_h) - a(u,v_h) = \begin{cases} Qh(\nabla u, \nabla v_h) & \textit{(Upwind)},\\ Qh(\frac{\partial u}{\partial \mathbf{b}}, \frac{\partial v_h}{\partial \mathbf{b}}) & \textit{(Streamline-Diffusion)},\end{cases}\end{aligned}$$

hence both are consistent but *not* strongly consistent. Remarkable instances of strongly consistent methods will be introduced and analyzed in Sect. 12.8.6

12.8.5 Symmetric and skew-symmetric part of an operator

Let V be a Hilbert space and V' its dual. We will say that an operator $L: V \to V'$ is *symmetric* if

$$_{V'}\langle Lu, v\rangle_V = {}_V\langle u, Lv\rangle_{V'} \qquad \forall u, v \in V,$$

skew-symmetric when

$$_{V'}\langle Lu, v\rangle_V = -{}_V\langle u, Lv\rangle_{V'} \qquad \forall u, v \in V.$$

An operator can be split into the sum of a symmetric part L_S and a skew-symmetric part L_{SS},

$$Lu = L_S u + L_{SS} u.$$

Let us consider, for instance, the following diffusiont-transport-reaction operator

$$Lu = -\mu \Delta u + \mathrm{div}(\mathbf{b}u) + \sigma u, \qquad \mathbf{x} \in \Omega \subset \mathbb{R}^d, d \geq 2, \tag{12.59}$$

operating on the space $V = H_0^1(\Omega)$. Since

$$\begin{aligned}\mathrm{div}(\mathbf{b}u) &= \tfrac{1}{2}\mathrm{div}(\mathbf{b}u) + \tfrac{1}{2}\mathrm{div}(\mathbf{b}u)\\ &= \tfrac{1}{2}\mathrm{div}(\mathbf{b}u) + \tfrac{1}{2}u\mathrm{div}(\mathbf{b}) + \tfrac{1}{2}\mathbf{b}\cdot\nabla u,\end{aligned}$$

we can split L the following way

$$Lu = \underbrace{-\mu\Delta u + \left[\sigma + \frac{1}{2}\mathrm{div}(\mathbf{b})\right] u}_{L_S u} + \underbrace{\frac{1}{2}\left[\mathrm{div}(\mathbf{b}u) + \mathbf{b}\cdot\nabla u\right]}_{L_{SS} u}.$$

Note that the reaction coefficient has become $\sigma^* = \sigma + \frac{1}{2}\mathrm{div}(\mathbf{b})$. We can verify that the two parts are symmetric resp. skew-symmetric. Indeed, integrating twice by parts, we obtain, $\forall u, v \in V$,

$$\begin{aligned} {}_{V'}\langle L_S u, v\rangle_V &= \mu(\nabla u, \nabla v) + (\sigma^* u, v) \\ &= -\mu\, {}_V\langle u, \Delta v\rangle_{V'} + (u, \sigma^* v) \\ &= {}_V\langle u, L_S v\rangle_{V'}, \end{aligned}$$

$$\begin{aligned} {}_{V'}\langle L_{SS} u, v\rangle_V &= \frac{1}{2}(\mathrm{div}(\mathbf{b}u), v) + \frac{1}{2}(\mathbf{b}\cdot\nabla u, v) \\ &= -\frac{1}{2}(\mathbf{b}u, \nabla v) + \frac{1}{2}(\nabla u, \mathbf{b}v) \\ &= -\frac{1}{2}(u, \mathbf{b}\cdot\nabla v) - \frac{1}{2}(u, \mathrm{div}(\mathbf{b}v)) \\ &= -{}_V\langle u, L_{SS} v\rangle_{V'}, \end{aligned}$$

indicating by $(\cdot,\cdot)$ the scalar product of $L^2(\Omega)$.

Remark 12.4. We recall that any matrix A can be decomposed into the sum

$$\mathrm{A} = \mathrm{A}_S + \mathrm{A}_{SS},$$

where

$$\mathrm{A}_S = \frac{1}{2}(\mathrm{A} + \mathrm{A}^T)$$

is a symmetric matrix called the *symmetric part* of A and

$$\mathrm{A}_{SS} = \frac{1}{2}(\mathrm{A} - \mathrm{A}^T)$$

is a skew-symmetric matrix called the *skew-symmetric part* of A. •

12.8.6 Strongly consistent methods (GLS, SUPG)

We consider a diffusion-transport-reaction problem that we write in the abstract form $Lu = f$ in Ω, with $u = 0$ on $\partial\Omega$. Let us consider the corresponding weak formulation (12.3) with $a(\cdot,\cdot)$ being the bilinear form associated to L. A stabilized and strongly consistent method can be obtained by adding a further term to the Galerkin approximation (12.13), that is by considering the problem

$$\text{find } u_h \in V_h: \quad a(u_h, v_h) + \mathcal{L}_h(u_h, f; v_h) = (f, v_h) \quad \forall v_h \in V_h, \tag{12.60}$$

for a suitable form $\mathcal{L}_h$ satisfying

$$\mathcal{L}_h(u, f; v_h) = 0 \quad \forall v_h \in V_h. \tag{12.61}$$

(This is the case of the generalized Galerkin method (12.47), (12.48), provided we require $b_h(u, v_h) = G_h(v_h) \quad \forall v_h \in V_h$.) We observe that in (12.60) the form $\mathcal{L}_h$ depends both on the approximate solution u_h and on the forcing term f. A possible choice that verifies (12.61) is

$$\mathcal{L}_h(u_h, f; v_h) = \mathcal{L}_h^{(\rho)}(u_h, f; v_h) = \sum_{K \in \mathcal{T}_h} (Lu_h - f, \tau_K \mathcal{S}^{(\rho)}(v_h))_{L^2(K)},$$

where $(u, v)_{L^2(K)} = \int_K uv\, dK$, ρ and τ_K are parameters to be determined, and we have set

$$\mathcal{S}^{(\rho)}(v_h) = L_{SS}v_h + \rho L_S v_h,$$

where L_S and L_{SS} are the symmetric resp. skew-symmetric part of the operator L under exam. A possible choice for τ_K is

$$\tau_K = \delta \frac{h_K}{|\mathbf{b}(\mathbf{x})|} \qquad \forall \mathbf{x} \in K, \forall K \in \mathcal{T}_h \tag{12.62}$$

where $\mathbf{b}$ is the convective (or transport) field, h_K the diameter of the generic element K, and δ a dimensionless coefficient to be prescribed.
To verify that (12.60) is fully consistent, we note that

$$\tau_h(u; v_h) \quad = a(u, v_h) + \mathcal{L}_h^{(\rho)}(u, f; v_h) - (f, v_h)$$

is zero for all $v_h \in V_h$, thanks to (12.3) and property (12.61). Thus the truncation error (12.58) is null. Let us now see some particular cases associated to three different choices of the parameter ρ:

- if $\rho = 1$ we obtain the method called *Galerkin Least-Squares* (GLS), where

$$\mathcal{S}^{(1)}(v_h) = Lv_h.$$

  If we take $v_h = u_h$ we see that a term proportional to $\int_K (Lu_h)^2\, dK$ has been added on each triangle to the original bilinear form;

- if $\rho = 0$ we obtain the method named *Streamline Upwind Petrov-Galerkin* (SUPG) where

$$\mathcal{S}^{(0)}(v_h) = L_{SS}v_h;$$

- if $\rho = -1$ we obtain the so-called Douglas-Wang (DW) method where

$$\mathcal{S}^{(-1)}(v_h) = (L_{SS} - L_S)v_h.$$

If $\sigma = 0$ and we use $\mathbb{P}_1$ finite elements, the three previous methods coincide, as $-\Delta u_h|_K = 0 \quad \forall K \in \mathcal{T}_h$.

Let us now limit ourselves to the two most classical procedures, GLS ($\rho = 1$) and SUPG ($\rho = 0$) and to the problem written in conservative form (12.59). We define the "ρ norm"

$$\|v\|_{(\rho)} = \{\mu\|\nabla v\|^2_{L^2(\Omega)} + \|\sqrt{\gamma}v\|^2_{L^2(\Omega)} + \sum_{K\in\mathcal{T}_h}\Big((L_{SS}+\rho L_S)v, \tau_K \mathcal{S}^{(\rho)}(v)\Big)_{L^2(K)}\}^{\frac{1}{2}},$$

where γ is a function given by $\frac{1}{2}\mathrm{div}\mathbf{b}+\sigma$ when we use the conservative form (12.15) of the operator L, otherwise $\gamma = -\frac{1}{2}\mathrm{div}\mathbf{b}+\sigma$ when using the non-conservative form (12.1). In either case we assume that γ is a non-negative function. The following (stability) inequality holds: there exists an α^* depending on γ and on the coercivity constant α of $a(\cdot,\cdot)$ such that

$$\|u_h\|_{(\rho)} \leq \frac{C}{\alpha^*}\|f\|_{L^2(\Omega)}, \tag{12.63}$$

where C is a suitable constant (see for instance (12.79)). Moreover, the following error estimate holds

$$\|u-u_h\|_{(\rho)} \leq Ch^{r+1/2}|u|_{H^{r+1}(\Omega)}, \tag{12.64}$$

hence the order of accuracy of the method increases when the degree r of the polynomials we employ increases, as in the standard Galerkin method. The proofs of (12.63) and (12.64) in the case $\rho = 1$ will be provided in Sect. 12.8.8.

The choice of the stabilization parameter δ, measuring the amount of artificial viscosity, is extremely important. To this end, we report in Table 12.1 the range admitted for such parameter as a function of the chosen stabilized scheme. In the table, C_0 is the constant of the following *inverse inequality*

$$\sum_{K\in\mathcal{T}_h} h_K^2 \int_K |\Delta v_h|^2 dK \leq C_0\|\nabla v_h\|^2_{L^2(\Omega)} \quad \forall v_h \in X_h^r. \tag{12.65}$$

Obviously, $C_0 = C_0(r)$. Let us note that for linear finite elements $C_0 = 0$. In such a case, the constant δ in Table 12.1 is *not* subject to any upper bound. On the contrary, if we are interested in polynomials of higher degree, $r \geq 2$, then

$$C_0(r) = \bar{C}_0 r^{-4}. \tag{12.66}$$

For a more extensive analysis, we refer to [QV94, Chap. 8], and to [RST96]. We also suggest [Fun97] for the case of an approximation with spectral elements.

Table 12.1. Admissible values for the stabilization parameter δ

SUPG	$0 < \delta < 1/C_0$
GLS	$0 < \delta$
DW	$0 < \delta < 1/(2C_0)$

12.8.7 On the choice of the stabilization parameter τ_K

For linear finite elements ($r = 1$) another choice of the stabilization function τ_K, alternative to that in (12.62), is

$$\tau_K(\mathbf{x}) = \frac{h_K}{2|\mathbf{b}(\mathbf{x})|}\xi\,(\mathbb{P}\mathrm{e}_K) \quad \forall \mathbf{x} \in K, \quad \forall K \in \mathcal{T}_h, \tag{12.67}$$

where

$$\mathbb{P}\mathrm{e}_K(\mathbf{x}) = \frac{|\mathbf{b}(\mathbf{x})|\,h_K}{2\,\mu(\mathbf{x})} \quad \forall \mathbf{x} \in K, \quad \forall K \in \mathcal{T}_h, \tag{12.68}$$

is the local Péclet number (in analogy to definition (12.23) for dimension one), and the *upwind* function $\xi(\cdot)$ can be

$$\xi(\theta) = \coth(\theta) - 1/\theta, \quad \theta > 0 \tag{12.69}$$

for example. As $\lim_{\theta \to +\infty} \xi(\theta) = 1$ (cf. Fig. 12.9), if $\mathbb{P}\mathrm{e}_K(\mathbf{x}) \gg 1$ then (12.67) reduces, in the limit, to (12.62) with $\delta = 1/2$. Moreover, since $\theta \to 0$ implies $\xi(\theta) = \theta/3 + o(\theta)$, we have $\tau_K(\mathbf{x}) \to 0$ when $\mathbb{P}\mathrm{e}_K(\mathbf{x}) \ll 1$ (in fact no stabilization is necessary if the problem is diffusion-dominated). Other possibilities for the function τ_K are found in the literature. For instance, h_K can be replaced in (12.67)–(12.68) by the diameter of the element K along $\mathbf{b}$, or one can choose the *upwind* function $\xi(\cdot)$ to be $\xi(\theta) = \max\{0, 1 - 1/\theta\}$, or $\xi(\theta) = \min\{1, \theta/3\}$ (see [JK07] for more details).

Let us now give a heuristic explanation for the choice (12.67) of the stabilizing function τ_K. To this end, take the variational formulation (12.20) of the one-dimensional diffusion-transport problem (12.18). Given a uniform partition $\mathcal{T}_h$ of $\Omega = (0, 1)$ in N intervals of width $h = 1/N$, consider the SUPG method for the discretization of (12.20):

$$\text{find } \overset{\circ}{u}_h \in V_h \text{ such that} \quad a(\overset{\circ}{u}_h, v_h) = F(v_h) \quad \forall v_h \in V_h,$$

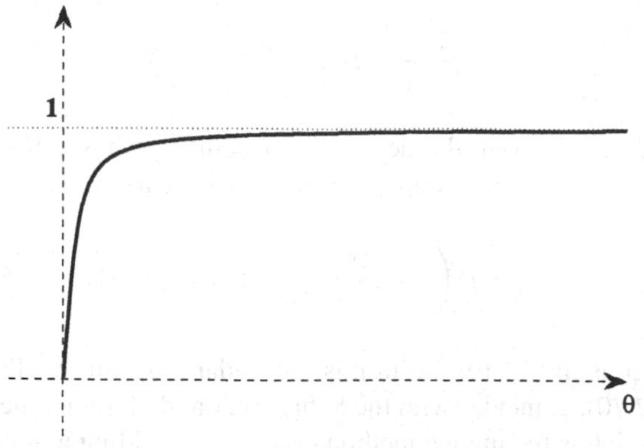

Fig. 12.9. *Upwind* function ξ defined in (12.69).

where $V_h \subset H_0^1(0,1)$ is the space of piecewise-linear continuous polynomials on $\mathcal{T}_h$, and

$$a(u,v) = \int_0^1 (\mu u'v' + bu'v)\,dx + \tau \int_0^1 |b|^2 u'v'\,dx \qquad \forall u,v \in V_h,$$
$$F(v) = -\int_0^1 bv\,dx \qquad \forall v \in V_h.$$

By defining $\mu_h = \mu(1+\tau|b|^2/\mu)$ the bilinear form $a(\cdot,\cdot)$ may be equivalently written as

$$a(u,v) = \int_0^1 \mu_h u'v'\,dx + \int_0^1 bu'v\,dx \qquad \forall u,v \in V_h.$$

Choose the parameter τ as

$$\tau = \frac{h}{2|b|}\left[\coth(\mathbb{P}\mathrm{e}) - \frac{1}{\mathbb{P}\mathrm{e}}\right], \tag{12.70}$$

where $\mathbb{P}\mathrm{e} = \frac{|b|h}{2\mu}$ is the local Péclet number. By virtue of these definitions we obtain

$$\begin{aligned}
\tau\frac{|b|^2}{\mu} &= \mathbb{P}\mathrm{e}\left[\coth(\mathbb{P}\mathrm{e}) - \frac{1}{\mathbb{P}\mathrm{e}}\right] \\
&= \mathbb{P}\mathrm{e}\coth(\mathbb{P}\mathrm{e}) - 1 \\
&= \mathbb{P}\mathrm{e} - 1 + \mathbb{P}\mathrm{e}(\coth(\mathbb{P}\mathrm{e}) - 1) \\
&= \mathbb{P}\mathrm{e} - 1 + B(2\mathbb{P}\mathrm{e}).
\end{aligned}$$

The final equality involves the identity:

$$\begin{aligned}
t(\coth(t) - 1) &= t\left[\frac{e^t + e^{-t}}{e^t - e^{-t}} - 1\right] = 2t\left[\frac{e^{-t}}{e^t - e^{-t}}\right] \\
&= \frac{2t}{e^{2t} - 1} = B(2t), \qquad t > 0,
\end{aligned}$$

which in turn descends from the definition of $\coth(\cdot)$. Above, $B(\cdot)$ is the Bernoulli function (cf. Sect. 12.6). To sum up, μ_h may be written as

$$\mu_h = \mu\left(1 + \tau\frac{|b|^2}{\mu}\right) = \mu\,(1 + \phi(\mathbb{P}\mathrm{e})),$$

having chosen ϕ as in (12.40). So in this particular case, the SUPG method, with τ chosen as in (12.70), coincides with the Scharfetter and Gummel method encountered in Sect. 12.6, which is the unique method capable of yielding a numerical solution to a constant-coefficient problem (with constant source) that is nodally exact.

Remark 12.5. If one employs polynomials of degree $r \geq 2$ (as in the hp formulation, or with spectral elements), a more coherent definition of the local Péclet number is this

$$\mathbb{P}\mathrm{e}_K^r = \frac{|\mathbf{b}(\mathbf{x})| h_K}{2\mu(\mathbf{x}) r},$$

while the corresponding stabilizing function (12.67) becomes

$$\tau_K(\mathbf{x}) = \frac{h_K}{2|\mathbf{b}| r} \xi(\mathbb{P}\mathrm{e}_K^r),$$

(see [GaAML04]). •

12.8.8 Analysis of the GLS method

In this section we want to prove the stability property (12.63) and the convergence property (12.64) in the case of the GLS method (hence for $\rho = 1$).

We suppose that the differential operator L has the form (12.59), with $\mu > 0$ and $\sigma \geq 0$ constant, and $\mathbf{b}$ being a vector function whose components are continuous (e.g. constant), with homogeneous Dirichlet boundary conditions being assigned. The bilinear form $a(\cdot,\cdot) : V \times V \to \mathbb{R}$ associated to the operator L is therefore

$$a(u,v) = \mu \int_\Omega \nabla u \cdot \nabla v \, d\Omega + \int_\Omega \mathrm{div}(\mathbf{b}u) v \, d\Omega + \int_\Omega \sigma u v \, d\Omega,$$

with $V = \mathrm{H}_0^1(\Omega)$. For simplicity, we suppose that there exist two constants γ_0 and γ_1 such that

$$0 < \gamma_0 \leq \gamma(\mathbf{x}) = \frac{1}{2}\mathrm{div}(\mathbf{b}(\mathbf{x})) + \sigma \leq \gamma_1 \quad \forall \mathbf{x} \in \Omega. \tag{12.71}$$

In this case the form $a(\cdot,\cdot)$ is coercive, as $a(v,v) \geq \mu \|\nabla v\|^2_{L^2(\Omega)} + \gamma_0 \|v\|^2_{L^2(\Omega)}$. Following the procedure developed in Sect. 12.8.5, we can write the symmetric and skew-symmetric parts associated to L as

$$L_S u = -\mu \Delta u + \gamma u, \quad L_{SS} u = \frac{1}{2}\Big(\mathrm{div}(\mathbf{b}u) + \mathbf{b} \cdot \nabla u\Big).$$

Moreover, we rewrite the stabilized formulation (12.60) by splitting $\mathcal{L}_h(u_h, f; v_h)$ in two terms, one containing u_h, the other f:

$$\text{find } u_h \in V_h : \quad a_h^{(1)}(u_h, v_h) = f_h^{(1)}(v_h) \quad \forall v_h \in V_h, \tag{12.72}$$

with

$$a_h^{(1)}(u_h, v_h) = a(u_h, v_h) + \sum_{K \in \mathcal{T}_h} \delta \left(L u_h, \frac{h_K}{|\mathbf{b}|} L v_h \right)_K \tag{12.73}$$

and

$$f_h^{(1)}(v_h) = (f, v_h) + \sum_{K \in \mathcal{T}_h} \delta \left(f, \frac{h_K}{|\mathbf{b}|} L v_h \right)_K. \tag{12.74}$$

We observe that, using these notations, the strong consistency property (12.61) is expressed via the equality

$$a_h^{(1)}(u, v_h) = f_h^{(1)}(v_h) \quad \forall v_h \in V_h. \tag{12.75}$$

We can now prove the following preliminary result.

Lemma 12.1. *The bilinear form $a_h^{(1)}(\cdot,\cdot)$ defined in* (12.73) *satisfies the following relation*

$$\begin{aligned} a_h^{(1)}(v_h, v_h) &= \mu\|\nabla v_h\|^2_{\mathrm{L}^2(\Omega)} + \|\sqrt{\gamma} v_h\|^2_{\mathrm{L}^2(\Omega)} \\ &+ \sum_{K\in\mathcal{T}_h} \delta \left(\frac{h_K}{|\mathbf{b}|} L v_h, L v_h\right)_K \quad \forall\, v_h \in V_h. \end{aligned} \tag{12.76}$$

This identity follows from definition (12.73) (having chosen $v_h = u_h$) and from (12.71). In the case under exam, the norm $\|\cdot\|_{(1)}$, which we here denote by the symbol $\|\cdot\|_{GLS}$ for convenience, becomes

$$\|v_h\|^2_{GLS} = \mu\|\nabla v_h\|^2_{\mathrm{L}^2(\Omega)} + \|\sqrt{\gamma} v_h\|^2_{\mathrm{L}^2(\Omega)} + \sum_{K\in\mathcal{T}_h} \delta \left(\frac{h_K}{|\mathbf{b}|} L v_h, L v_h\right)_K. \tag{12.77}$$

We can prove the following stability result.

Lemma 12.2. *Let u_h be the solution of the GLS method. Then for each $\delta > 0$ there exists a constant $C > 0$, independent of h, such that*

$$\|u_h\|_{GLS} \leq C\|f\|_{\mathrm{L}^2(\Omega)}.$$

Proof. We choose $v_h = u_h$ in (12.72). By exploiting Lemma 12.1 and definition (12.77), we can first write that

$$\|u_h\|^2_{GLS} = a_h^{(1)}(u_h, u_h) = f_h^{(1)}(u_h) = (f, u_h) + \sum_{K\in\mathcal{T}_h} \delta \left(f, \frac{h_K}{|\mathbf{b}|} L u_h\right)_K. \tag{12.78}$$

We look for an upper bound for the two right-hand side terms of (12.78) separately, by applying suitably the Cauchy-Schwarz and Young inequalities. We thus obtain

$$\begin{aligned} (f, u_h) &= \left(\frac{1}{\sqrt{\gamma}} f, \sqrt{\gamma} u_h\right) \leq \left\|\frac{1}{\sqrt{\gamma}} f\right\|_{\mathrm{L}^2(\Omega)} \|\sqrt{\gamma} u_h\|_{\mathrm{L}^2(\Omega)} \\ &\leq \frac{1}{4}\|\sqrt{\gamma} u_h\|^2_{\mathrm{L}^2(\Omega)} + \left\|\frac{1}{\sqrt{\gamma}} f\right\|^2_{\mathrm{L}^2(\Omega)}, \end{aligned}$$

$$\sum_{K\in\mathcal{T}_h} \delta \left(f, \frac{h_K}{|\mathbf{b}|} L u_h\right)_K = \sum_{K\in\mathcal{T}_h} \left(\sqrt{\delta \frac{h_K}{|\mathbf{b}|}}\, f, \sqrt{\delta \frac{h_K}{|\mathbf{b}|}}\, L u_h\right)_K$$

$$\leq \sum_{K\in\mathcal{T}_h} \left\| \sqrt{\delta \frac{h_K}{|\mathbf{b}|}} f \right\|_{L^2(K)} \left\| \sqrt{\delta \frac{h_K}{|\mathbf{b}|}} Lu_h \right\|_{L^2(K)}$$

$$\leq \sum_{K\in\mathcal{T}_h} \delta \left(\frac{h_K}{|\mathbf{b}|} f, f \right)_K + \frac{1}{4} \sum_{K\in\mathcal{T}_h} \delta \left(\frac{h_K}{|\mathbf{b}|} Lu_h, Lu_h \right)_K .$$

By summing the two previous upper bounds and by exploiting again definition (12.77), we have

$$\|u_h\|^2_{GLS} \leq \left\| \frac{1}{\sqrt{\gamma}} f \right\|^2_{L^2(\Omega)} + \sum_{K\in\mathcal{T}_h} \delta \left(\frac{h_K}{|\mathbf{b}|} f, f \right)_K + \frac{1}{4} \|u_h\|^2_{GLS},$$

that is, recalling that $h_K \leq h$,

$$\|u_h\|^2_{GLS} \leq \frac{4}{3} \left[\left\| \frac{1}{\sqrt{\gamma}} f \right\|^2_{L^2(\Omega)} + \sum_{K\in\mathcal{T}_h} \delta \left(\frac{h_K}{|\mathbf{b}|} f, f \right)_K \right] \leq C^2 \|f\|^2_{L^2(\Omega)},$$

having set

$$C = \left(\frac{4}{3} \max_{\mathbf{x}\in\Omega} \left(\frac{1}{\gamma} + \delta \frac{h}{|\mathbf{b}|} \right) \right)^{1/2}. \tag{12.79}$$

◇

We observe that the previous result is valid with the only constraint that the stabilization parameter δ be positive. In fact, such parameter might also vary for each element K. In this case, we would have δ_K instead of δ in (12.73) and (12.74), while the constant δ in (12.79) would have the meaning of $\max_{K\in\mathcal{T}_h} \delta_K$.

We now study the convergence of the GLS method.

Theorem 12.2. *We assume that the space V_h satisfies the following local approximation property: for each $v \in V \cap H^{r+1}(\Omega)$, there exists a function $\hat{v}_h \in V_h$ such that*

$$\|v - \hat{v}_h\|_{L^2(K)} + h_K |v - \hat{v}_h|_{H^1(K)} + h_K^2 |v - \hat{v}_h|_{H^2(K)} \leq C h_K^{r+1} |v|_{H^{r+1}(K)} \tag{12.80}$$

for each $K \in \mathcal{T}_h$. Moreover, we suppose that for each $K \in \mathcal{T}_h$ the local Péclet number of K is larger than 1,

$$\mathbb{P}e_K(\mathbf{x}) = \frac{|\mathbf{b}(\mathbf{x})|\, h_K}{2\mu} > 1 \quad \forall \mathbf{x} \in K. \tag{12.81}$$

Finally, we suppose that the inverse inequality (12.65) holds and that the stabilization parameter satisfies the relation $0 < \delta \leq 2C_0^{-1}$.
Then the following estimate holds for the error associated to the GLS method

$$\|u_h - u\|_{GLS} \leq C h^{r+1/2} |u|_{H^{r+1}(\Omega)}, \tag{12.82}$$

as long as $u \in H^{r+1}(\Omega)$.

Proof. First of all, we rewrite the error as follows

$$e_h = u_h - u = \sigma_h - \eta, \tag{12.83}$$

with $\sigma_h = u_h - \hat{u}_h$, $\eta = u - \hat{u}_h$, where $\hat{u}_h \in V_h$ is a function that depends on u and that satisfies property (12.80). If, for instance, $V_h = X_h^r \cap H_0^1(\Omega)$, we can choose $\hat{u}_h = \Pi_h^r u$, that is the finite element interpolant of u.

We start by estimating the norm $\|\sigma_h\|_{GLS}$. By exploiting the strong consistency of the GLS scheme given by (12.75), we obtain

$$||\sigma_h||_{GLS}^2 = a_h^{(1)}(\sigma_h,\sigma_h) = a_h^{(1)}(u_h - u + \eta, \sigma_h) = a_h^{(1)}(\eta,\sigma_h)$$

thanks to (12.72). Now, by definition (12.73) and thanks to the homogeneous Dirichlet boundary conditions it follows that

$$\begin{aligned}
a_h^{(1)}(\eta,\sigma_h) &= \mu \int_\Omega \nabla\eta\cdot\nabla\sigma_h\, d\Omega - \int_\Omega \eta\,\mathbf{b}\cdot\nabla\sigma_h\, d\Omega + \int_\Omega \sigma\,\eta\,\sigma_h\, d\Omega \\
&+ \sum_{K\in\mathcal{T}_h} \delta\Big(L\eta, \frac{h_K}{|\mathbf{b}|} L\sigma_h\Big)_K = \underbrace{\mu(\nabla\eta,\nabla\sigma_h)}_{\text{(I)}} \underbrace{- \sum_{K\in\mathcal{T}_h}(\eta, L\sigma_h)_K}_{\text{(II)}} + \underbrace{2(\gamma\eta,\sigma_h)}_{\text{(III)}} \\
&+ \underbrace{\sum_{K\in\mathcal{T}_h}(\eta,-\mu\Delta\sigma_h)_K}_{\text{(IV)}} + \underbrace{\sum_{K\in\mathcal{T}_h}\delta\left(L\eta, \frac{h_K}{|\mathbf{b}|}L\sigma_h\right)_K}_{\text{(V)}}.
\end{aligned}$$

We now bound the terms (I)-(V) separately. By carefully using the Cauchy-Schwarz and Young inequalities we obtain

$$\begin{aligned}
\text{(I)} &= \mu(\nabla\eta,\nabla\sigma_h) \le \frac{\mu}{4}||\nabla\sigma_h||^2_{\mathrm{L}^2(\Omega)} + \mu\|\nabla\eta\|^2_{\mathrm{L}^2(\Omega)}, \\
\text{(II)} &= -\sum_{K\in\mathcal{T}_h}(\eta, L\sigma_h)_K = -\sum_{K\in\mathcal{T}_h}\left(\sqrt{\frac{|\mathbf{b}|}{\delta\, h_K}}\,\eta, \sqrt{\frac{\delta\, h_K}{|\mathbf{b}|}}L\sigma_h\right)_K \\
&\le \frac{1}{4}\sum_{K\in\mathcal{T}_h}\delta\Big(\frac{h_K}{|\mathbf{b}|}L\sigma_h, L\sigma_h\Big)_K + \sum_{K\in\mathcal{T}_h}\Big(\frac{|\mathbf{b}|}{\delta\, h_K}\eta,\eta\Big)_K, \\
\text{(III)} &= 2(\gamma\eta,\sigma_h) = 2(\sqrt{\gamma}\eta,\sqrt{\gamma}\sigma_h) \le \frac{1}{2}\|\sqrt{\gamma}\sigma_h\|^2_{\mathrm{L}^2(\Omega)} + 2\|\sqrt{\gamma}\eta\|^2_{\mathrm{L}^2(\Omega)}.
\end{aligned}$$

For the term (IV), thanks again to the Cauchy-Schwarz and Young inequalities, hypothesis (12.81) and the inverse inequality (12.65), we obtain

$$\begin{aligned}
\text{(IV)} &= \sum_{K\in\mathcal{T}_h}(\eta,-\mu\Delta\sigma_h)_K \\
&\le \frac{1}{4}\sum_{K\in\mathcal{T}_h}\delta\,\mu^2\Big(\frac{h_K}{|\mathbf{b}|}\Delta\sigma_h,\Delta\sigma_h\Big)_K + \sum_{K\in\mathcal{T}_h}\Big(\frac{|\mathbf{b}|}{\delta\, h_K}\eta,\eta\Big)_K
\end{aligned}$$

$$\leq \frac{1}{8}\delta\mu \sum_{K\in\mathcal{T}_h} h_K^2 (\Delta\sigma_h,\Delta\sigma_h)_K + \sum_{K\in\mathcal{T}_h}\Big(\frac{|\mathbf{b}|}{\delta h_K}\eta,\eta\Big)_K$$
$$\leq \frac{\delta C_0\mu}{8}||\nabla\sigma_h||^2_{L^2(\Omega)} + \sum_{K\in\mathcal{T}_h}\Big(\frac{|\mathbf{b}|}{\delta h_K}\eta,\eta\Big)_K.$$

Term (V) can finally be bounded once again thanks to the Cauchy-Schwarz and Young inequalities as follows

$$\begin{aligned}(V) &= \sum_{K\in\mathcal{T}_h}\delta\Big(L\eta,\frac{h_K}{|\mathbf{b}|}L\sigma_h\Big)_K\\ &\leq \frac{1}{4}\sum_{K\in\mathcal{T}_h}\delta\Big(\frac{h_K}{|\mathbf{b}|}L\sigma_h,L\sigma_h\Big)_K + \sum_{K\in\mathcal{T}_h}\delta\Big(\frac{h_K}{|\mathbf{b}|}L\eta,L\eta\Big)_K.\end{aligned}$$

Thanks to these upper bounds and using once more the definition (12.77) , we obtain the following estimate

$$\begin{aligned}&||\sigma_h||^2_{GLS} = a_h^{(1)}(\eta,\sigma_h) \leq \frac{1}{4}||\sigma_h||^2_{GLS}\\ &+\frac{1}{4}\Big(||\sqrt{\gamma}\sigma_h||^2_{L^2(\Omega)} + \sum_{K\in\mathcal{T}_h}\delta\Big(\frac{h_K}{|\mathbf{b}|}L\sigma_h,L\sigma_h\Big)_K\Big) + \frac{\delta C_0\mu}{8}||\nabla\sigma_h||^2_{L^2(\Omega)}\\ &+\underbrace{\mu||\nabla\eta||^2_{L^2(\Omega)} + 2\sum_{K\in\mathcal{T}_h}\Big(\frac{|\mathbf{b}|}{\delta h_K}\eta,\eta\Big)_K + 2||\sqrt{\gamma}\eta||^2_{L^2(\Omega)} + \sum_{K\in\mathcal{T}_h}\delta\Big(\frac{h_K}{|\mathbf{b}|}L\eta,L\eta\Big)_K}_{\mathcal{E}(\eta)}\\ &\leq \frac{1}{2}||\sigma_h||^2_{GLS} + \mathcal{E}(\eta),\end{aligned}$$

having exploited, in the last passage, the assumption that $\delta \leq 2C_0^{-1}$. Then, we can state that

$$||\sigma_h||^2_{GLS} \leq 2\mathcal{E}(\eta).$$

We now estimate the term $\mathcal{E}(\eta)$, by bounding each of its summands separately. To this end, we will basically use the local approximation property (12.80) and the requirement formulated in (12.81) on the local Péclet number $\mathbb{P}\mathrm{e}_K$. Moreover, we observe that the constants C, introduced in the remainder, depend neither on h nor on $\mathbb{P}\mathrm{e}_K$, but can depend on other quantities such as the constant γ_1 in (12.71), the reaction constant σ, the norm $||\mathbf{b}||_{L^\infty(\Omega)}$, the stabilization parameter δ. We then have

$$\begin{aligned}\mu||\nabla\eta||^2_{L^2(\Omega)} &\leq C\mu h^{2r}|u|^2_{H^{r+1}(\Omega)}\\ &\leq C\frac{||\mathbf{b}||_{L^\infty(\Omega)}h}{2}h^{2r}|u|^2_{H^{r+1}(\Omega)} \leq Ch^{2r+1}|u|^2_{H^{r+1}(\Omega)},\end{aligned} \tag{12.84}$$

$$2\sum_{K\in\mathcal{T}_h}\left(\frac{|\mathbf{b}|}{\delta h_K}\eta,\eta\right)_K \leq C\frac{\|\mathbf{b}\|_{L^\infty(\Omega)}}{\delta}\sum_{K\in\mathcal{T}_h}\frac{1}{h_K}h_K^{2(r+1)}|u|^2_{\mathrm{H}^{r+1}(K)}$$
$$\leq Ch^{2r+1}|u|^2_{\mathrm{H}^{r+1}(\Omega)},$$

$$2\|\sqrt{\gamma}\eta\|^2_{\mathrm{L}^2(\Omega)} \leq 2\gamma_1\|\eta\|^2_{\mathrm{L}^2(\Omega)} \leq Ch^{2(r+1)}|u|^2_{\mathrm{H}^{r+1}(\Omega)}, \tag{12.85}$$

having exploited, for controlling the third summand, the assumption (12.71).

Finding an upper bound for the fourth summand of $\mathcal{E}(\eta)$ is slightly more difficult: first, by elaborating on the term $L\eta$, we have

$$\sum_{K\in\mathcal{T}_h}\delta\left(\frac{h_K}{|\mathbf{b}|}L\eta,L\eta\right)_K = \sum_{K\in\mathcal{T}_h}\delta\left\|\sqrt{\frac{h_K}{|\mathbf{b}|}}L\eta\right\|^2_{\mathrm{L}^2(K)}$$
$$= \sum_{K\in\mathcal{T}_h}\delta\left\|-\mu\sqrt{\frac{h_K}{|\mathbf{b}|}}\Delta\eta+\sqrt{\frac{h_K}{|\mathbf{b}|}}\mathrm{div}(\mathbf{b}\eta)+\sigma\sqrt{\frac{h_K}{|\mathbf{b}|}}\eta\right\|^2_{\mathrm{L}^2(K)}$$
$$\leq C\sum_{K\in\mathcal{T}_h}\delta\left(\left\|\mu\sqrt{\frac{h_K}{|\mathbf{b}|}}\Delta\eta\right\|^2_{\mathrm{L}^2(K)}+\left\|\sqrt{\frac{h_K}{|\mathbf{b}|}}\mathrm{div}(\mathbf{b}\eta)\right\|^2_{\mathrm{L}^2(K)}+\left\|\sigma\sqrt{\frac{h_K}{|\mathbf{b}|}}\eta\right\|^2_{\mathrm{L}^2(K)}\right). \tag{12.86}$$

Now, with a similar computation to the one performed to obtain estimates (12.84) and (12.85), it is easy to prove that the second and third summands of the left-hand side of (12.86) can be bounded using a term of the form $Ch^{2r+1}|u|^2_{\mathrm{H}^{r+1}(\Omega)}$, for a suitable choice of the constant C. For the first summand, we have

$$\sum_{K\in\mathcal{T}_h}\delta\left\|\mu\sqrt{\frac{h_K}{|\mathbf{b}|}}\Delta\eta\right\|^2_{\mathrm{L}^2(K)} \leq \sum_{K\in\mathcal{T}_h}\delta\frac{h_K^2\mu}{2}\|\Delta\eta\|^2_{\mathrm{L}^2(K)}$$
$$\leq C\delta\|\mathbf{b}\|_{L^\infty(\Omega)}\sum_{K\in\mathcal{T}_h}h_K^3\|\Delta\eta\|^2_{\mathrm{L}^2(K)} \leq Ch^{2r+1}|u|^2_{\mathrm{H}^{r+1}(\Omega)},$$

having again used conditions (12.80) and (12.81). The latter bound allows us to conclude that

$$\mathcal{E}(\eta) \leq Ch^{2r+1}|u|^2_{\mathrm{H}^{r+1}(\Omega)},$$

that is

$$\|\sigma_h\|_{GLS} \leq Ch^{r+1/2}|u|_{\mathrm{H}^{r+1}(\Omega)}. \tag{12.87}$$

Reverting to (12.83), to obtain the desired estimate for the norm $\|u_h-u\|_{GLS}$ we still have to estimate $\|\eta\|_{GLS}$. This evidently leads to estimating three contributions as in (12.84), (12.85) and (12.86), and eventually produces

$$\|\eta\|_{GLS} \leq Ch^{r+1/2}|u|_{\mathrm{H}^{r+1}(\Omega)}.$$

The desired estimate (12.82) follows by combining this result with (12.87). ◇

12.8.9 Stabilization through bubble functions

The generalized Galerkin method considered in the previous sections yields a stable numerical solution owing to the *enrichment* of the bilinear form $a(\cdot,\cdot)$. An alternative strategy consists of adopting a *richer subspace* than the standard one V_h. The idea is then to choose both the approximate solution and the test functions in the enriched space, therefore remaining within a classical Galerkin framework.
Referring to the usual diffusion-transport-reaction problem of the form (12.1), we introduce the finite dimensional space

$$V_h^b = V_h \oplus B,$$

where $V_h = X_h^r \cap H_0^1(\Omega)$ is the usual space and B is a finite-dimensional *space* of *bubble functions*, or

$$B = \{v_B \in H_0^1(\Omega) \, : \, v_B|_K = c_K \, b_K, \; b_K|_{\partial K} = 0 \text{ and } c_K \in \mathbb{R}\}.$$

On each element K we then add the correction term b_K, for which several different choices are possible. As we only wish to work on the initial grid $\mathcal{T}_h$ associated to the space V_h, a standard choice leads to defining $b_K = \lambda_1 \lambda_2 \lambda_3$ where the λ_i, for $i = 0, \ldots, 2$, are the barycentric coordinates on K, i.e. linear polynomials defined on K, vanishing on one of the sides of the triangle and taking the value 1 at the vertex opposed to such side. (See Sect. 4.4.3 for their definition). The function b_K coincides in this case with the so-called *cubic bubble* that takes value 0 on the boundary of K and positive values inside it (see Fig. 12.10 (left)).Hence c is the only degree of freedom associated to the triangle K (it will coincide, for instance, with the largest value taken by b_K on K, or with the value it takes in the center of gravity). (see Sect. 4.4.3).

Remark 12.6. In order to introduce a *computational subgrid* on the domain Ω (obtained as a suitable refinement of the mesh $\mathcal{T}_h$), we can adopt more complex definitions for the bubble function b_K. For instance, b_K could be a piecewise linear function, again defined on the element K and assuming the value 0 on the boundary of the triangle (like

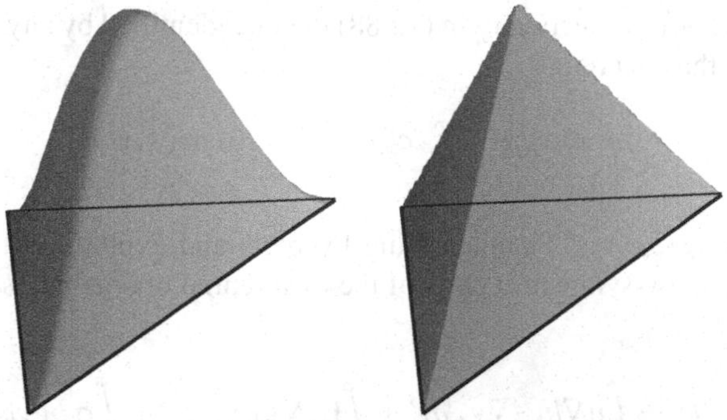

Fig. 12.10. Example of a cubic (left) and linear (right) bubble

the basis function of linear finite elements associated to some point inside K) (see Fig. 12.10 (right)) [EG04]. •

At this point, we can introduce the Galerkin approximation on the space V_h^b of the problem under exam, which will take the form

$$\text{find } u_h^b \in V_h^b: \quad a(u_h + u_b, v_h^b) = (f, v_h^b) \quad \forall v_h^b \in V_h^b, \tag{12.88}$$

with $a(\cdot,\cdot)$ being the bilinear form associated to the differential operator L.
We propose to rewrite (12.88) as a stabilized Galerkin scheme in V_h, by eliminating function u_b. So far we can only say that, in each element K, $u_b\big|_K = c_{b,K} b_K$ for a suitable (unknown) constant $c_{b,K}$, with $u_b \in B$.
We decompose both u_h^b and v_h^b sums of a function of V_h and a function of B, that is

$$u_h^b = u_h + u_b, \qquad v_h^b = v_h + v_b.$$

We first select as test function v_h^b in (12.88) the one identified by $v_h = 0$ and $v_b \in B$ such that

$$v_b = \begin{cases} b_K & \text{in } K, \\ 0 & \text{elsewhere.} \end{cases}$$

We then have

$$a(u_h + u_b, v_b) = a_K(u_h + c_{b,K} b_K, b_K),$$

having denoted by $a_K(\cdot,\cdot)$ the restriction of the bilinear form $a(\cdot,\cdot)$ to the element K. We can therefore rewrite (12.88) as

$$a_K(u_h, b_K) + c_{b,K} a_K(b_K, b_K) = (f, b_K)_K. \tag{12.89}$$

Exploiting the fact that b_K vanishes on the boundary of K, we can integrate by parts the first term of (12.89), obtaining $a_K(u_h, b_K) = (Lu_h, b_K)_K$, then the unknown value of the constant $c_{b,K}$, is

$$c_{b,K} = \frac{(f - Lu_h, b_K)_K}{a_K(b_K, b_K)}.$$

We now choose as test function v_h^b in (12.88) the one identified by any function $v_h \in V_h$ and by $v_b = 0$, thus obtaining

$$a(u_h, v_h) + \sum_{K \in \mathcal{T}_h} c_{b,K} a_K(b_K, v_h) = (f, v_h). \tag{12.90}$$

Let us rewrite $a_K(b_K, v_h)$. By integrating by parts and exploiting the definitions of symmetric and skew-symmetric parts of the differential operator L (see Sect. 12.8.5), we have

$$a_K(b_K, v_h) = \int_K \mu \nabla b_K \cdot \nabla v_h \, dK + \int_K \mathbf{b} \cdot \nabla b_K \, v_h \, dK + \int_K \sigma \, b_K v_h \, dK$$

$$= -\int_K \mu\, b_K \Delta v_h \, dK + \int_{\partial K} \mu\, b_K \nabla v_h \cdot \mathbf{n}\, d\gamma - \int_K b_K \nabla v_h \cdot \mathbf{b}\, dK$$

$$+ \int_{\partial K} \mathbf{b} \cdot \mathbf{n}\, v_h b_K \, d\gamma + \int_K \sigma\, b_K v_h \, dK = (b_K, (L_S - L_{SS}) v_h)_K.$$

We have exploited the property that the bubble function b_K vanishes on the boundary of the element K, and moreover that $\mathrm{div}(\mathbf{b}) = 0$. In a very similar way we can rewrite the denominator of the constant $c_{b,K}$ in the following way

$$a_K(b_K, b_K) = (L_S b_K, b_K)_K.$$

Reverting to (12.90), we thus have

$$a(u_h, v_h) + a_B(u_h, f; v_h) = (f, v_h) \quad \forall\, v_h \in V_h,$$

where

$$a_B(u_h, f; v_h) = \sum_{K \in \mathcal{T}_h} \frac{(Lu_h - f, b_K)_K (L_{SS} v_h - L_S v_h, b_K)_K}{(L_S b_K, b_K)_K}.$$

We have therefore found a stabilized Galerkin scheme, which can be formulated in the strongly consistent form (12.60). In the case where $\mathbf{b}$ is constant, we can identify it using a sort of generalized Douglas-Wang method.
By choosing a convenient bubble b_K and following an analogous procedure to the one illustrated above, it is also possible to define generalized SUPG and GLS methods (see [BFHR97]). Similar strategies based on the so-called *subgrid viscosity* can be successfully used. See [EG04] for an extensive analysis.

12.9 DG methods for diffusion-transport equations

The Discontinuous Galerkin method introduced in Chap. 11 for the Poisson problem can be extended to the diffusion-transport-reaction problem (12.15) (in conservation form) as follows: find $u_\delta \in W_\delta^0$ s.t.

$$\begin{aligned}
&\sum_{m=1}^{M} (\mu \nabla u_\delta, \nabla v_\delta)_{\Omega_m} - \sum_{e \in \mathcal{E}_\delta} \int_e [v_\delta] \cdot \{\!\{\mu \nabla u_\delta\}\!\} - \tau \sum_{e \in \mathcal{E}_\delta} \int_e [u_\delta] \{\!\{\mu \nabla v_\delta\}\!\} \\
&+ \sum_{e \in \mathcal{E}_\delta} \int_e \gamma |e|^{-1} [u_\delta] \cdot [v_\delta] - \sum_{m=1}^{M} (\mathbf{b} u_\delta, \nabla v_\delta)_{\Omega_m} + \sum_{e \in \mathcal{E}_\delta} \int_e \{\!\{\mathbf{b} u_\delta\}\!\}_{\mathbf{b}} \cdot [v_\delta] \qquad (12.91) \\
&+ \sum_{m=1}^{M} (\sigma u_\delta, v_\delta)_{\Omega_m} = \sum_{m=1}^{M} (f, v_\delta)_{\Omega_m},
\end{aligned}$$

where we have set

$$\{\!\{\mathbf{b}u_\delta\}\!\}_\mathbf{b} = \begin{cases} \mathbf{b}u_\delta^+ & \text{if } \mathbf{b}\cdot\mathbf{n}^+ > 0 \\ \mathbf{b}u_\delta^- & \text{if } \mathbf{b}\cdot\mathbf{n}^+ < 0 \\ \mathbf{b}\{u_\delta\} & \text{if } \mathbf{b}\cdot\mathbf{n}^+ = 0\,. \end{cases} \tag{12.92}$$

Observe that $\{\!\{\mathbf{b}u_\delta\}\!\}_\mathbf{b}\cdot[v_\delta] = 0$ if $\mathbf{b}\cdot\mathbf{n}^+ = 0$.

If the diffusion-transport-reaction equation is written in non-conservative form as in (12.1), by $\mathbf{b}\cdot\nabla u = \mathrm{div}(\mathbf{b}u) - \mathrm{div}(\mathbf{b})u$ it is sufficient to modify (12.91) by substituting the term

$$\sum_{m=1}^{M}(\sigma u_\delta, v_\delta)_{\Omega_m} \quad \text{with} \quad \sum_{m=1}^{M}(\eta u_\delta, v_\delta)_{\Omega_m},$$

where $\eta(\mathbf{x}) = \sigma(\mathbf{x}) - \mathrm{div}(\mathbf{b}(\mathbf{x}))$; this time we suppose that there exists a positive constant $\eta_0 > 0$ so that $\eta(\mathbf{x}) \geq \eta_0$ for almost every $\mathbf{x}\in\Omega$.

The DG method can easily be localized to every subdomain Ω_m (either a finite or a spectral element). Indeed, since test functions do not have to be continuous, for every m ($m = 1,\ldots,M$) we can choose test functions to vanish outside the element Ω_m. In this way the sum in (12.91) (or in (11.6)) is reduced to the single index m, while the one on the edges is reduced to the edges on the boundary of Ω_m. We spot here another peculiarity of the DG method: it fits well with local refinements, element by element, either grid-wise ("*h*-refinement"), or polynomial-wise ("*p*-refinement", p being the local polynomial degree, the same indicated with r for finite elements and with N for spectral elements methods).

In Chaps. 13,14,15 we will see that in hyperbolic problems the solutions can be discontinuous. This is the case when either the initial and/or boundary data are discontinuous and, more in general, for non-linear problems. In all these cases, approximations based on discontinuous polynomial functions are very appropriate. For further particulars on DG methods for diffusion-transport problems, see, e.g., [Coc99, HSS02, BMS04, BHS06, AM09].

Finally, let us point out that the previous description can be extended to a problem where the spatial operator is written as the divergence of a flux depending on ∇u, $\mathrm{div}(\phi(\nabla u))$. For this we only need to replace in every formula ∇u with the flux expression $\Phi(\nabla u)$.

We refer to Sect. 12.11 for an analysis of the numerical results obtained by the DG method (12.91).

12.10 Mortar methods for the diffusion-transport equations

Mortar methods, described in Chapter 11 for the Poisson problem, can be applied to the discretization of equations (12.1) and of those written in conservation form (12.15). The domain discretization and the choice of *master* and *slave* spaces can be carried out as described in Chapter 11 for the Poisson problem.

The bilinear forms $a_i(u_\delta^{(i)}, v_\delta^{(i)})$ showing up in equation (11.18) and in system (11.29) now read

$$a_i(u_\delta^{(i)}, v_\delta^{(i)}) = \int_{\Omega_i} \mu \nabla u_\delta^{(i)} \cdot \nabla v_\delta^{(i)} + \int_{\Omega_i} v_\delta^{(i)} \mathbf{b} \cdot \nabla u_\delta^{(i)} + \int_{\Omega_i} \sigma u_\delta^{(i)} v_\delta^{(i)}$$

in case of problem (12.1), and

$$a_i(u_\delta^{(i)}, v_\delta^{(i)}) = \int_{\Omega_i} \mu \nabla u_\delta^{(i)} \cdot \nabla v_\delta^{(i)} - \int_{\Omega_i} u_\delta^{(i)} \mathbf{b} \cdot \nabla v_\delta^{(i)} + \int_{\Omega_i} \sigma u_\delta^{(i)} v_\delta^{(i)} + \int_{\partial\Omega_i} \mathbf{b} \cdot \mathbf{n} u_\delta^{(i)} v_\delta^{(i)} \tag{12.93}$$

in case of the problem in conservation form (12.15), where the transport term has been integrated by parts.

However, we must point out that when $|\mathbf{b}(\mathbf{x})| \gg \mu$, the mortar solution, as seen in the previous chapter, does not always allow to attain optimal error estimates like (11.19) and (11.20), that is a global error that can be expressed as the sum of local errors, without requiring any compatibility constraints between the discretization of the subdomains sharing the same interface. More precisely, instabilities can arise due to the non-conformity at the interfaces, even if stabilization techniques, like GLS or SUPG, are called into play. To understand why that may occur, we first consider the diffusion-transport-reaction problem in conservation form (12.15) and its weak mortar formulation (as in (11.18)) with $a_i(u_\delta^{(i)}, v_\delta^{(i)})$ defined in (12.93). Add now the GLS stabilization by substituting the bilinear forms with the stabilized ones

$$a_{i,\delta}^{(1)}(u_\delta^{(i)}, v_\delta^{(i)}) = a_i(u_\delta^{(i)}, v_\delta^{(i)}) + \sum_{T_{i,k} \in \mathcal{T}_i} \left(L u_\delta^{(i)}, \tau_k L v_\delta^{(i)} \right) \tag{12.94}$$

(the index (1) in $a_{i,\delta}^{(1)}$ indicates the choice $\rho = 1$ in the stabilization, according to the formalism used in the previous sections) and then write

$$a_\delta^{(1)}(u_\delta, v_\delta) = \sum_{i=1}^{2} a_{i,\delta}^{(1)}(u_\delta^{(i)}, v_\delta^{(i)}).$$

(Source terms are stabilized in a similar way.)
Using the classical rules of integration, we have

$$\begin{aligned}\int_{\Omega_i} \mathrm{div}(\mathbf{b} u_\delta^{(i)}) u_\delta^{(i)} &= \frac{1}{2} \int_{\Omega_i} \mathrm{div}(\mathbf{b} u_\delta^{(i)}) u_\delta^{(i)} - \frac{1}{2} \int_{\Omega_i} u_\delta^{(i)} \mathbf{b} \cdot \nabla u_\delta^{(i)} + \frac{1}{2} \int_{\partial\Omega_i} \mathbf{b} \cdot \mathbf{n}_i (u_\delta^{(i)})^2 \\ &= \frac{1}{2} \int_{\Omega_i} (\mathrm{div}\mathbf{b})(u_\delta^{(i)})^2 + \frac{1}{2} \int_{\Gamma} \mathbf{b} \cdot \mathbf{n}_i (u_\delta^{(i)})^2\end{aligned}$$

where $\mathbf{n}_i$ indicates the outward unit vector on $\partial\Omega_i$ and the last boundary integral is now confined on the internal interface thanks to the homogeneous Dirichlet conditions.

Then

$$a_{i,\delta}^{(1)}(u_\delta^{(i)},u_\delta^{(i)}) = \mu\|\nabla u_\delta^{(i)}\|^2_{L^2(\Omega_i)} + \|\sqrt{\sigma+\frac{1}{2}\mathrm{div}\mathbf{b}}\, u_\delta^{(i)}\|^2_{L^2(\Omega_i)}$$
$$+\sum_{T_{i,k}\in\mathcal{T}_i}\left(Lu_\delta^{(i)},\tau_k Lu_\delta^{(i)}\right)+\frac{1}{2}\int_\Gamma \mathbf{b}\cdot\mathbf{n}_i(u_\delta^{(i)})^2$$

and

$$a_\delta^{(1)}(u_\delta,u_\delta) = \mu\|\nabla u_\delta\|^2_{L^2(\Omega)} + \|\sqrt{\sigma+\frac{1}{2}\mathrm{div}\mathbf{b}}\, u_\delta\|^2_{L^2(\Omega)}$$
$$+\sum_{\substack{i=1,2\\ T_{i,k}\in\mathcal{T}_i}}\left(Lu_\delta^{(i)},\tau_k Lu_\delta^{(i)}\right)+\frac{1}{2}\int_\Gamma \mathbf{b}\cdot\mathbf{n}_\Gamma((u_\delta^{(1)})^2-(u_\delta^{(2)})^2)$$

, since $\mathbf{n}_\Gamma=\mathbf{n}_1=-\mathbf{n}_2$ on the interface.

Because of the presence of the integral $\frac{1}{2}\int_\Gamma \mathbf{b}\cdot\mathbf{n}_\Gamma((u_\delta^{(1)})^2-(u_\delta^{(2)})^2)$, $a_\delta^{(1)}$ can fail to be coercive with respect to the norm $\|\cdot\|_{GLS}$ or any other discrete norm. To go around this problem, in [AAH+98] it is suggested to add to the bilinear form $a_\delta^{(1)}$ one more DG-like stabilization term on Γ,

$$\tilde{a}_\Gamma(u_\delta,v_\delta)=\int_\Gamma(\mathbf{b}\cdot\mathbf{n}_1)^-(u_\delta^{(2)}-u_\delta^{(1)})v_\delta^{(1)}+\int_\Gamma(\mathbf{b}\cdot\mathbf{n}_2)^-(u_\delta^{(1)}-u_\delta^{(2)})v_\delta^{(2)}.$$

Here x^- indicates the negative part of the number x (i.e. $x^-=(|x|-x)/2$). By splitting each integral on Γ into two terms according to the sign of $(\mathbf{b}\cdot\mathbf{n}_i)^-$, with the aid of simple algebraic operations we obtain

$$\tilde{a}_\Gamma(u_\delta,u_\delta)=\frac{1}{2}\int_\Gamma|\mathbf{b}\cdot\mathbf{n}_\Gamma|(u_\delta^{(1)}-u_\delta^{(2)})^2.$$

The form $a_\delta^{(1)}(u_\delta,u_\delta)+\tilde{a}_\Gamma(u_\delta,u_\delta)$ is therefore coercive with respect to the *GLS* norm (12.78).

12.11 Some numerical tests

We now present some numerical solutions obtained using the finite elements methods for the following two-dimensional diffusion-transport probem

$$\begin{cases} -\mu\Delta u+\mathbf{b}\cdot\nabla u=f & \text{in } \Omega=(0,1)\times(0,1),\\ u=g & \text{on } \partial\Omega, \end{cases} \tag{12.95}$$

where $\mathbf{b}=(1,1)^T$. To start with let us consider the following constant data: $f\equiv 1$ and

$g \equiv 0$. In this case the solution is characterized by a boundary layer near the edges $x = 1$ and $y = 1$. We have considered two different viscosity values: $\mu = 10^{-3}$ and $\mu = 10^{-5}$. For both problems we compare the solutions obtained using the standard and GLS Galerkin methods, respectively, by making two different choices for the uniform discretization step h: $1/20$ and $1/80$, respectively. In the GLS case we used the value (12.62) of $\mathcal{T}_K$. The combinations of the two values for μ and h yield four different values for the local Péclet number $\mathbb{P}\mathrm{e}$. As it can be observed by analyzing Figs. 12.11–12.14 (bearing in mind the different vertical scales) for increasing Péclet numbers, the solution provided by the standard Galerkin method denotes stronger and stronger fluctuations. The latter eventually overcome completely the numerical solution (see Fig. 12.14). On the other hand, the GLS method is able to provide an acceptable numerical solution even for extremely high values of $\mathbb{P}\mathrm{e}$ (even though it develops an *over-shoot* at the point $(1,1)$).

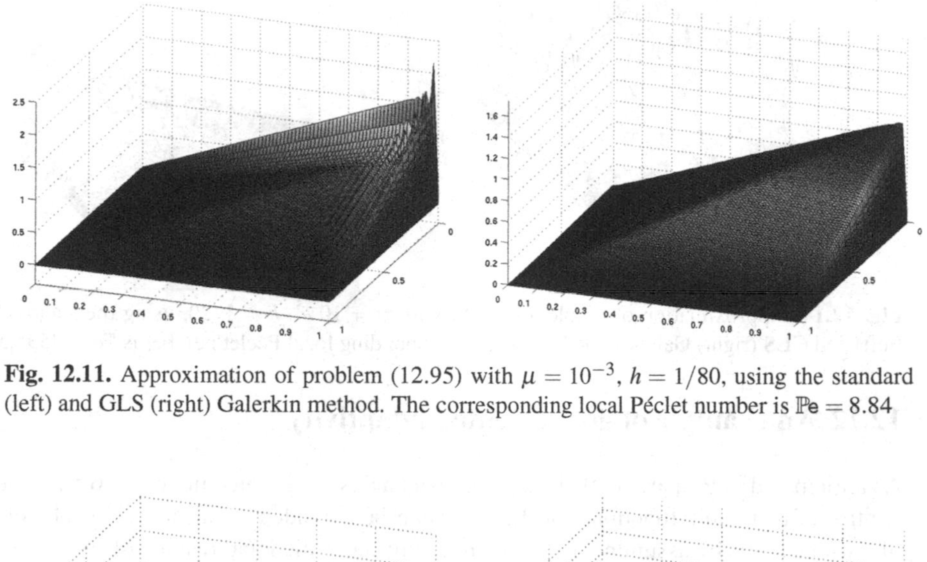

Fig. 12.11. Approximation of problem (12.95) with $\mu = 10^{-3}$, $h = 1/80$, using the standard (left) and GLS (right) Galerkin method. The corresponding local Péclet number is $\mathbb{P}\mathrm{e} = 8.84$

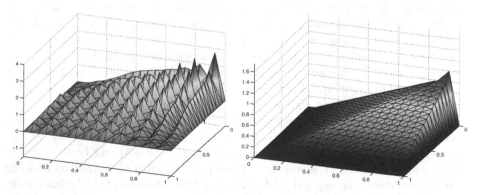

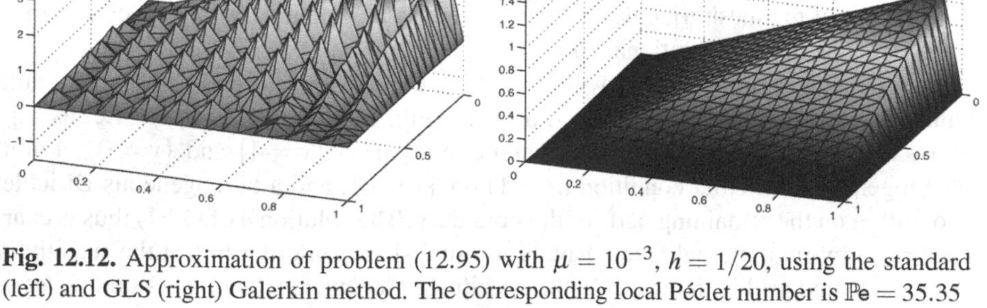

Fig. 12.12. Approximation of problem (12.95) with $\mu = 10^{-3}$, $h = 1/20$, using the standard (left) and GLS (right) Galerkin method. The corresponding local Péclet number is $\mathbb{P}\mathrm{e} = 35.35$

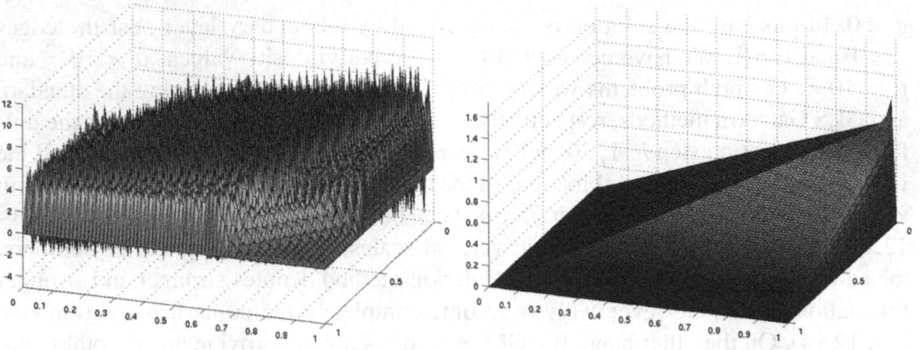

Fig. 12.13. Approximation of problem (12.95) with $\mu = 10^{-5}$, $h = 1/80$, using the standard (left) and GLS (right) Galerkin method. The corresponding local Péclet number is $\mathbb{P}\mathrm{e} = 883.88$

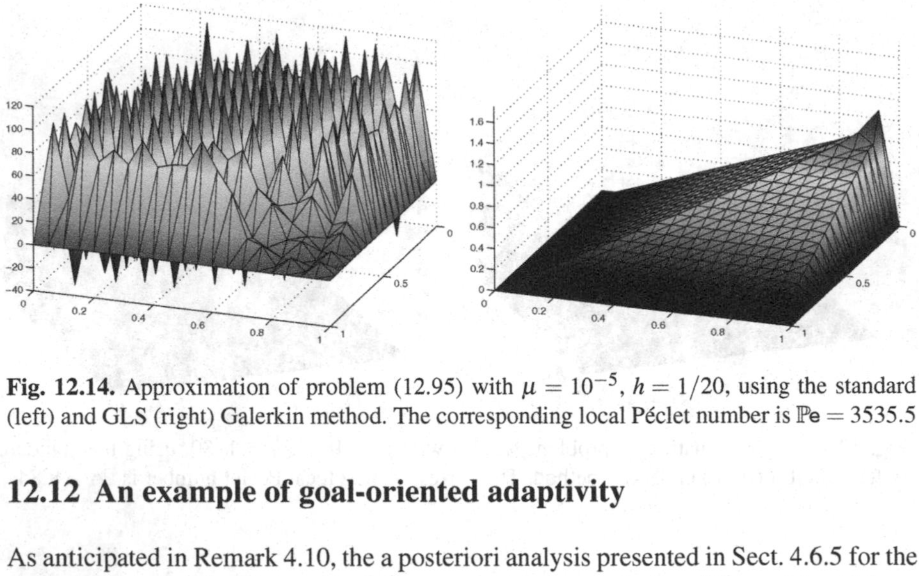

Fig. 12.14. Approximation of problem (12.95) with $\mu = 10^{-5}$, $h = 1/20$, using the standard (left) and GLS (right) Galerkin method. The corresponding local Péclet number is $\mathbb{P}\mathrm{e} = 3535.5$

12.12 An example of goal-oriented adaptivity

As anticipated in Remark 4.10, the a posteriori analysis presented in Sect. 4.6.5 for the control of a suitable functional of the error can be extended to differential problems of various kinds by assuming a suitable redefinition of the local residue (4.98) and of the generalized jump (4.94). A grid adaptation turns out to be particularly useful when dealing with diffusion-transport problems with dominant transport. Here, an accurate placement of the mesh triangles, e.g. at the (internal or boundary) layers, can dramatically reduce the computational cost.

Let us consider problem (12.1) with $\mu = 10^{-3}$, $\mathbf{b} = (y, -x)^T$, σ and f identically null, and Ω coinciding with the L-shaped domain (Fig. 12.15) $(0,4)^2 \backslash (0,2)^2$. Let us suppose to assign a homogeneous Neumann condition on $\{x = 4\}$ and $\{y = 0\}$, a non-homogeneous Dirichlet condition ($u = 1$) on $\{x = 0\}$, and a homogeneous Dirichlet condition on the remaining parts of the boundary. The solution u of (12.1) thus is characterized by two internal layers having a round shape. In order to test the sensitivity of the adapted grid with respect to the specific choice made for the functional J, let us

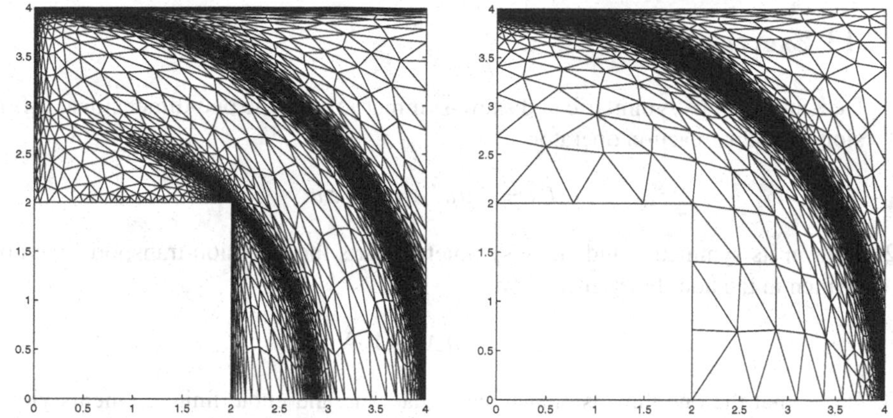

Fig. 12.15. Fourth adapted grid for the functional J_1 (left); second adapted grid for the functional J_2 (right)

consider the two following options:

$$J(v) = J_1(v) = \int_{\Gamma_1} \mathbf{b} \cdot \mathbf{n} v\, ds, \qquad \text{with} \qquad \Gamma_1 = \{x = 4\} \cup \{y = 0\},$$

for the control of the outgoing normal flow through $\{x = 4\}$ and $\{y = 0\}$, and

$$J(v) = J_2(v) = \int_{\Gamma_2} \mathbf{b} \cdot \mathbf{n} v\, ds, \qquad \text{with} \qquad \Gamma_2 = \{x = 4\},$$

if we are still interested in controlling the flow, but only through $\{x = 4\}$. Starting from a quasi-uniform initial grid of 1024 elements, we show in Fig. 12.15 the (anisotropic) grids obtained for the choice $J = J_1$ (left) resp. $J = J_2$ (right), at the fourth and second iteration of the adaptive process. Both boundary layers are responsible for the flow through Γ_1, and in fact the grid is refined in correspondence of the two layers. However, only the upper layer is "recognized" as carrying information to the flow along Γ_2. Finally, note the strongly anisotropic nature of the mesh in the figure. In order to follow not only the refinement but also the correct orientation of the grid, triangles are implemented in such a way to follow the directional properties (the boundary layers) of the solution. For further details, refer to [FMP04].

12.13 Exercises

1. Decompose in its symmetric and skew-symmetric parts the one-dimensional diffusion-transport-reaction operator

$$Lu = -\mu u'' + bu' + \sigma u.$$

2. Split in its symmetric and skew-symmetric parts the diffusion-transport operator written in the non-divergence form

$$Lu = -\mu \Delta u + \mathbf{b} \cdot \nabla u.$$

3. Prove that the one-dimensional linear, quadratic and cubic finite elements yield, in the reference interval $[0,1]$, the following condensed matrices, obtained via the *mass-lumping* technique:

$$r=1 \quad \mathrm{M}_L = \widehat{\mathrm{M}} = \frac{1}{2}\mathrm{diag}(1\ 1),$$

$$r=2 \quad \mathrm{M}_L = \widehat{\mathrm{M}} = \frac{1}{6}\mathrm{diag}(1\ 4\ 1),$$

$$r=3 \quad \begin{cases} \mathrm{M}_L = \dfrac{1}{8}\mathrm{diag}(1\ 3\ 3\ 1), \\ \widehat{\mathrm{M}} = \dfrac{1}{1552}\mathrm{diag}(128\ 648\ 648\ 128) = \mathrm{diag}\left(\dfrac{8}{97}, \dfrac{81}{194}, \dfrac{81}{194}, \dfrac{8}{97}\right). \end{cases}$$

4. Consider the problem

$$\begin{cases} -\varepsilon u''(x) + bu'(x) = 1, & 0 < x < 1, \\ u(0) = \alpha, & u(1) = \beta, \end{cases}$$

where $\varepsilon > 0$ and $\alpha, \beta, b \in \mathbb{R}$ are given. Find its finite element formulation with upwind artificial viscosity. Discuss its stability and convergence properties and compare them with that of the Galerkin-linear finite elements formulation.

5. Consider the problem

$$\begin{cases} -\varepsilon u''(x) + u'(x) = 1, \quad 0 < x < 1, \\ u(0) = 0, \qquad u'(1) = 1, \end{cases}$$

with $\varepsilon > 0$ given. Write its weak formulation and its approximation of Galerkin-finite element type. Verify that the scheme is stable and explain why.

6. Consider the problem

$$\begin{cases} -\mathrm{div}(\mu \nabla u) + \mathrm{div}(\boldsymbol{\beta} u) + \sigma u = f & \text{in } \Omega, \\ -\boldsymbol{\gamma} \cdot \mathbf{n} + \mu \nabla u \cdot \mathbf{n} = 0 & \text{on } \Gamma_N, \\ u = 0 & \text{on } \Gamma_D, \end{cases}$$

where Ω is an open subset of $\mathbb{R}^2$ with boundary $\Gamma = \Gamma_D \cup \Gamma_N$, $\Gamma_D \cap \Gamma_N = \emptyset$, $\mathbf{n}$ is the outgoing normal to Γ, $\mu = \mu(\mathbf{x}) > \mu_0 > 0$, $\sigma = \sigma(\mathbf{x}) > 0$, $f = f(\mathbf{x})$ are given scalar functions, $\boldsymbol{\beta} = \boldsymbol{\beta}(\mathbf{x})$, $\boldsymbol{\gamma} = \boldsymbol{\gamma}(\mathbf{x})$ are given vector functions.
Approximate it using the Galerkin-linear finite element method. State under which hypotheses on the coefficients μ, σ and $\boldsymbol{\beta}$ the method is inaccurate, and suggest the relevant remedies in the different cases.

7. Consider the one-dimensional diffusion-transport problem

$$\begin{cases} -(\mu u' - \psi' u)' = 1, \quad 0 < x < 1, \\ u(0) = u(1) = 0, \end{cases} \tag{12.96}$$

where μ is a positive constant and ψ a given function.

a) Study the existence and uniqueness of problem (12.96) by introducing suitable hypotheses on the function ψ, and propose a stable numerical approximation with finite elements.
b) Consider the variable change $u = \rho e^{\psi/\mu}$, ρ being an auxiliary unknown function. Study the existence and uniqueness of the weak solution of problem (12.96) in the new unknown ρ and provide its numerical approximation using the finite elements method.
c) Compare the two approaches followed in (a) and (b), both from the abstract viewpoint and from the numerical one.

8. Consider the diffusion-transport-reaction problem

$$\begin{cases} -\Delta u + \mathrm{div}(\mathbf{b}u) + u = 0 & \text{in } \Omega \subset \mathbb{R}^2, \\ u = \varphi & \text{on } \Gamma_D, \\ \dfrac{\partial u}{\partial n} = 0 & \text{on } \Gamma_N, \end{cases}$$

where Ω is an open bounded domain, $\partial\Omega = \Gamma_D \cup \Gamma_N$, $\Gamma_D \neq \emptyset$.
Prove the existence and uniqueness of the solution by making suitable regularity assumptions on the data $\mathbf{b} = (b_1(\mathbf{x}), b_2(\mathbf{x}))^T$ ($\mathbf{x} \in \Omega$) and $\varphi = \varphi(\mathbf{x})$ ($\mathbf{x} \in \Gamma_D$).
In the case where $|\mathbf{b}| \gg 1$, approximate the problem with the artificial diffusion-finite elements and SUPG-finite elements methods, discussing advantages and disadvantages with respect to the Galerkin finite element method.

9. Consider the problem

$$\begin{cases} -\displaystyle\sum_{i,j=1}^{2} \frac{\partial^2 u}{\partial x_i \partial x_j} + \beta \frac{\partial^2 u}{\partial x_1^2} + \gamma \frac{\partial^2 u}{\partial x_1 \partial x_2} + \delta \frac{\partial^2 u}{\partial x_2^2} + \eta \frac{\partial u}{\partial x_1} = f & \text{in } \Omega, \\ u = 0 & \text{on } \partial\Omega, \end{cases}$$

where β, γ, δ, η are given coefficients and f is a given function of $\mathbf{x} = (x_1, x_2) \in \Omega$.

a) Find the conditions on the data that ensure the existence and uniqueness of a weak solution.
b) Provide an approximation using the Galerkin finite element method and analyze its convergence.
c) Under which conditions on the data is the Galerkin problem symmetric?
In such case, provide suitable methods for the solution of the associated algebraic problem.

13

Finite differences for hyperbolic equations

In this chapter we deal with time-dependent problems of hyperbolic type. For their origin and an in-depth analysis see e.g. [Sal08, Chap. 4]. We will limit ourselves to considering the numerical approximation using the finite difference method, which was historically the first one to be applied to this type of equations. To introduce in a simple way the basic concepts of the theory, most of our presentation will concern problems depending on a single space variable. Finite element approximations will be addressed in Chapter 14, the extension to nonlinear problems in Chapter 15.

13.1 A scalar transport problem

Let us consider the following scalar hyperbolic problem

$$\begin{cases} \dfrac{\partial u}{\partial t} + a\dfrac{\partial u}{\partial x} = 0, \quad x \in \mathbb{R},\, t > 0, \\ u(x,0) = u_0(x), \quad x \in \mathbb{R}, \end{cases} \tag{13.1}$$

where $a \in \mathbb{R} \setminus \{0\}$. The solution of such problem is a wave travelling at velocity a, given by

$$u(x,t) = u_0(x - at), \quad t \geq 0.$$

We consider the curves $x(t)$ in the plane (x,t), solutions of the following ordinary differential equation

$$\begin{cases} \dfrac{dx}{dt} = a, \qquad t > 0, \\ x(0) = x_0, \end{cases}$$

for varying values of $x_0 \in \mathbb{R}$.

A. Quarteroni: *Numerical Models for Differential Problems,* 2nd Ed.
MS&A – Modeling, Simulation & Applications 8
DOI 10.1007/978-88-470-5522-3_13,

Such curves are called *characteristic lines* (often simply characteristics) and the solution along these lines remains constant, for

$$\frac{du}{dt} = \frac{\partial u}{\partial t} + \frac{\partial u}{\partial x}\frac{dx}{dt} = 0.$$

In the case of the more general problem

$$\begin{cases} \dfrac{\partial u}{\partial t} + a\dfrac{\partial u}{\partial x} + a_0 u = f, & x \in \mathbb{R}, t > 0, \\ u(x,0) = u_0(x), & x \in \mathbb{R}, \end{cases} \tag{13.2}$$

where a, a_0, and f are given functions of (x,t), the characteristic lines $x(t)$ are the solutions of the Cauchy problem

$$\begin{cases} \dfrac{dx}{dt} = a(x,t), & t > 0, \\ x(0) = x_0. \end{cases}$$

In such case, the solutions of (13.2) satisfy the following relation

$$\frac{d}{dt}u(x(t),t) = f(x(t),t) - a_0(x(t),t)u(x(t),t).$$

Therefore it is possible to extract the solution u by solving an ordinary differential equation on each characteristic curve (this approach leads to the so-called *characteristic method*).

Let us now consider problem (13.1) in a bounded interval. For instance, let us suppose $x \in [0,1]$ and $a > 0$. As u is constant on the characteristics, from Fig. 13.1 we deduce that the value of the solution at a point P coincides with the value of u_0 at the foot P_0 of the characteristic outgoing from P. Instead, the characteristic outgoing from Q intersects the straight line $x = 0$ for $t > 0$. The point $x = 0$ is therefore an inflow point at which we must necessarily assign the value of u. Note that if $a < 0$, the inflow point would be $x = 1$.

By referring to problem (13.1) it is useful to observe that if u_0 were a discontinuous function at x_0, then such discontinuity would propagate along the characteristic outgo-

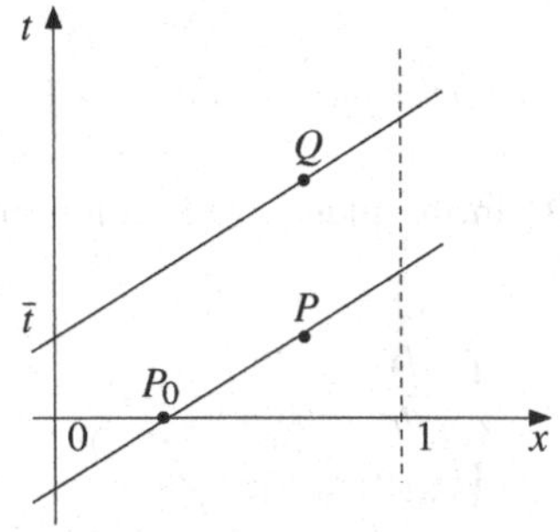

Fig. 13.1. Examples of characteristic lines (straight lines in this case) issuing from P and Q

ing from x_0 (this process can be rigorously formalized from a mathematical viewpoint by introducing the concept of *weak solution* for hyperbolic problems). In order to regularize the discontinuity, one could approximate the initial datum u_0 with a sequence of regular functions $u_0^\varepsilon(x), \varepsilon > 0$. However, this procedure is only effective if the hyperbolic problem is linear. The solutions of nonlinear hyperbolic problems can indeed develop discontinuities also for regular initial data (as we will see in Chapter 15). In this case the strategy (which also inspires numerical methods) is to regularize the differential equation itself, rather than the initial datum. We can consider the following diffusion-transport equation

$$\frac{\partial u^\varepsilon}{\partial t} + a\frac{\partial u^\varepsilon}{\partial x} = \varepsilon\frac{\partial^2 u^\varepsilon}{\partial x^2}, \quad x \in \mathbb{R},\ t > 0,$$

for small values of $\varepsilon > 0$, which can be regarded as a parabolic regularization of equation (13.1). If we set $u^\varepsilon(x,0) = u_0(x)$, we can prove that

$$\lim_{\varepsilon \to 0^+} u^\varepsilon(x,t) = u_0(x - at),\ t > 0,\ x \in \mathbb{R}.$$

13.1.1 An a priori estimate

Let us now return to the transport-reaction problem (13.2) on a bounded interval

$$\begin{cases} \dfrac{\partial u}{\partial t} + a\dfrac{\partial u}{\partial x} + a_0 u = f, & x \in (\alpha,\beta),\ t > 0, \\ u(x,0) = u_0(x), & x \in [\alpha,\beta], \\ u(\alpha,t) = \varphi(t), & t > 0, \end{cases} \tag{13.3}$$

where $a(x)$, $f(x,t)$ and $\varphi(t)$ are assigned functions; we have made the assumption that $a(x) > 0$, so that $x = \alpha$ is the inflow point (where to impose the boundary condition), while $x = \beta$ is the outflow point.

By multiplying the first equation of (13.3) by u, integrating in x and using the formula of integration by parts, we obtain for each $t > 0$

$$\frac{1}{2}\frac{d}{dt}\int_\alpha^\beta u^2\, dx + \int_\alpha^\beta (a_0 - \frac{1}{2}a_x)u^2\, dx + \frac{1}{2}(au^2)(\beta) - \frac{1}{2}(au^2)(\alpha) = \int_\alpha^\beta fu\, dx.$$

By supposing that there exists a $\mu_0 \geq 0$ such that

$$a_0 - \tfrac{1}{2}a_x \geq \mu_0 \quad \forall x \in [\alpha,\beta],$$

we find

$$\frac{1}{2}\frac{d}{dt}\|u(t)\|^2_{\mathrm{L}^2(\alpha,\beta)} + \mu_0\|u(t)\|^2_{\mathrm{L}^2(\alpha,\beta)} + \frac{1}{2}(au^2)(\beta) \leq \int_\alpha^\beta fu\, dx + \frac{1}{2}a(\alpha)\varphi^2(t).$$

If f and φ are identically zero, then

$$\|u(t)\|_{L^2(\alpha,\beta)} \leq \|u_0\|_{L^2(\alpha,\beta)} \quad \forall t>0.$$

In the case of the more general problem (13.2), if we suppose that $\mu_0 > 0$, thanks to the Cauchy-Schwarz and Young inequalities we have

$$\int_\alpha^\beta fu\, dx \leq \|f\|_{L^2(\alpha,\beta)}\|u\|_{L^2(\alpha,\beta)} \leq \frac{\mu_0}{2}\|u\|^2_{L^2(\alpha,\beta)} + \frac{1}{2\mu_0}\|f\|^2_{L^2(\alpha,\beta)}.$$

Integrating over time we get the following a priori estimate

$$\|u(t)\|^2_{L^2(\alpha,\beta)} + \mu_0 \int_0^t \|u(s)\|^2_{L^2(\alpha,\beta)}\, ds + a(\beta)\int_0^t u^2(\beta,s)\, ds$$
$$\leq \|u_0\|^2_{L^2(\alpha,\beta)} + a(\alpha)\int_0^t \varphi^2(s)\, ds + \frac{1}{\mu_0}\int_0^t \|f\|^2_{L^2(\alpha,\beta)}\, ds.$$

An alternative estimate that does not require the differentiability of $a(x)$, but uses, instead, the hypothesis that $a_0 \leq a(x) \leq a_1$ for two suitable positive constants a_0 and a_1, can be obtained by multiplying the equation by a^{-1},

$$a^{-1}\frac{\partial u}{\partial t} + \frac{\partial u}{\partial x} = a^{-1} f.$$

By multiplying by u and integrating between α and β we obtain, after a few simple steps,

$$\frac{1}{2}\frac{d}{dt}\int_\alpha^\beta a^{-1}(x)u^2(x,t)\, dx + \frac{1}{2}u^2(\beta,t) = \int_\alpha^\beta a^{-1}(x)f(x,t)u(x,t)dx + \frac{1}{2}\varphi^2(t).$$

If $f=0$ we immediately obtain

$$\|u(t)\|^2_a + \int_0^t u^2(\beta,s)ds = \|u_0\|^2_a + \int_0^t \varphi^2(s)ds,\ t>0.$$

We have defined

$$\|v\|_a = \left(\int_\alpha^\beta a^{-1}(x)v^2(x)dx\right)^{\frac{1}{2}}.$$

Thanks to the lower and upper bounds of a^{-1}, the latter is equivalent to the norm of $L^2(\alpha,\beta)$. On the other hand, if $f \neq 0$, we can proceed as follows

$$\|u(t)\|^2_a + \int_0^t u^2(\beta,s)\, ds \leq \|u_0\|^2_a + \int_0^t \varphi^2(s)\, ds + \int_0^t \|f\|^2_a\, ds + \int_0^t \|u(s)\|^2_a ds,$$

having used the Cauchy-Schwarz inequality.

By now applying Gronwall's lemma (see Lemma 2.2) we obtain, for each $t > 0$,

$$\|u(t)\|_a^2 + \int_0^t u^2(\beta,s)\, ds \le e^t \left(\|u_0\|_a^2 + \int_0^t \varphi^2(s) ds + \int_0^t \|f\|_a^2\, ds \right). \tag{13.4}$$

13.2 Systems of linear hyperbolic equations

Let us consider a linear system of the form

$$\begin{cases} \dfrac{\partial \mathbf{u}}{\partial t} + \mathrm{A}\dfrac{\partial \mathbf{u}}{\partial x} = \mathbf{0}, & x \in \mathbb{R},\ t > 0, \\ \mathbf{u}(0,x) = \mathbf{u}_0(x), & x \in \mathbb{R}, \end{cases} \tag{13.5}$$

where $\mathbf{u} : [0,\infty) \times \mathbb{R} \to \mathbb{R}^p$, $\mathrm{A} : \mathbb{R} \to \mathbb{R}^{p\times p}$ is a given matrix, and $\mathbf{u}_0 : \mathbb{R} \to \mathbb{R}^p$ is the initial datum.

Let us first consider the case where the coefficients of A are constant (i.e. independent of both x and t). System (13.5) is called *hyperbolic* if A can be diagonalized and has real eigenvalues. In such case, there exists a non-singular matrix $\mathrm{T} : \mathbb{R} \to \mathbb{R}^{p\times p}$ such that

$$\mathrm{A} = \mathrm{T}\Lambda\mathrm{T}^{-1},$$

where $\Lambda = \mathrm{diag}(\lambda_1, ..., \lambda_p)$, with $\lambda_i \in \mathbb{R}$ for $i = 1, \dots, p$, is the diagonal matrix of the eigenvalues of A while $\mathrm{T} = [\omega^1, \omega^2, \dots, \omega^p]$ is the matrix whose column vectors are the right eigenvectors of A, that is

$$\mathrm{A}\omega^k = \lambda_k \omega^k, \quad k = 1, \dots, p.$$

Through this similarity transformation it is possible to rewrite system (13.5) in the form

$$\frac{\partial \mathbf{w}}{\partial t} + \Lambda \frac{\partial \mathbf{w}}{\partial x} = \mathbf{0}, \tag{13.6}$$

where $\mathbf{w} = \mathrm{T}^{-1}\mathbf{u}$ are called *characteristic variables*. In this way we obtain p independent equations of the form

$$\frac{\partial w_k}{\partial t} + \lambda_k \frac{\partial w_k}{\partial x} = 0, \quad k = 1, \dots, p,$$

analogous in all to the equation of problem (13.1) (provided that we suppose a_0 and f null). The solution w_k is therefore constant along each *characteristic curve* $x = x(t)$, solution of the Cauchy problem

$$\begin{cases} \dfrac{dx}{dt} = \lambda_k, & t > 0, \\ x(0) = x_0. \end{cases} \tag{13.7}$$

Since the λ_k are constant, the characteristic curves are in fact the lines $x(t) = x_0 + \lambda_k t$ and the solutions read $w_k(x,t) = \psi_k(x - \lambda_k t)$, where ψ_k is a function of a single variable determined by the initial conditions. In the case of problem (13.5), we have that $\psi_k(x) = w_k(x,0)$, thus the solution $\mathbf{u} = \mathrm{T}\mathbf{w}$ will be of the form

$$\mathbf{u}(x,t) = \sum_{k=1}^{p} w_k(x - \lambda_k t, 0)\omega^k.$$

The latter is composed by p travelling, non-interacting waves.
Since a strictly hyperbolic system admits p different characteristic lines each point $(\bar{x},\bar{t})$ of the plane (x,t), $u(\bar{x},\bar{t})$ will only depend on the initial datum at the points $\bar{x} - \lambda_k\bar{t}$, for $k = 1,\ldots,p$. For this reason, the set of the p points that form the feet of the characteristics outgoing from $(\bar{x},\bar{t})$, that is

$$D(\bar{x},\bar{t}) = \{x \in \mathbb{R} \,|\, x = \bar{x} - \lambda_k\bar{t}\ ,\ k = 1,...,p\},\tag{13.8}$$

is called *domain of dependence* of the solution $\mathbf{u}$ at the point $(\bar{x},\bar{t})$.

In case we consider a bounded interval (α,β) instead of the whole real line, the sign of λ_k, $k = 1,\ldots,p$, denotes the inflow point for each of the characteristic variables. The function ψ_k in the case of a problem set on a bounded interval will be determined not only by the initial conditions, but also by the boundary conditions at the inflow of each characteristic variable. Having considered a point $(\bar{x},\bar{t})$ with $\bar{x} \in (\alpha,\beta)$ and $\bar{t} > 0$, if $\bar{x} - \lambda_k\bar{t} \in (\alpha,\beta)$ then $w_k(\bar{x},\bar{t})$ is determined by the initial condition, in particular we have $w_k(\bar{x},\bar{t}) = w_k(\bar{x} - \lambda_k\bar{t}, 0)$. Conversely, if $\bar{x} - \lambda_k\bar{t} \notin (\alpha,\beta)$ then the value of $w_k(\bar{x},\bar{t})$ will depend on the boundary condition (see Fig. 13.2):

$$\text{if } \lambda_k > 0\ ,\ w_k(\bar{x},\bar{t}) = w_k(\alpha, \frac{\bar{x} - \alpha}{\lambda_k}),$$

$$\text{if } \lambda_k < 0\ ,\ w_k(\bar{x},\bar{t}) = w_k(\beta, \frac{\bar{x} - \beta}{\lambda_k}).$$

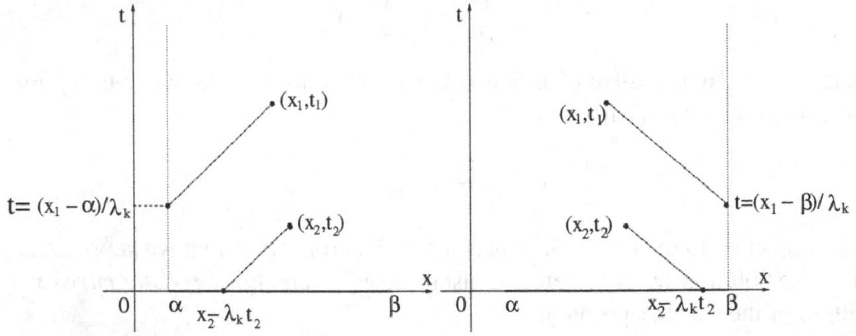

Fig. 13.2. The value of w_k at a point in the plane (x,t) depends either on the boundary condition or on the initial condition, depending on the value of $x - \lambda_k t$. Both signs of λ_k, the positive (right) and negative (left), are shown

As a consequence, the number of positive eigenvalues determines the number of boundary conditions to be assigned at $x = \alpha$, while at $x = \beta$ we will need to assign as many conditions as the number of negative eigenvalues.

In the case where the coefficients of the matrix A in (13.5) are functions of x and t, we denote respectively by

$$\mathrm{L} = \begin{bmatrix} \mathbf{l}_1^T \\ \vdots \\ \mathbf{l}_p^T \end{bmatrix} \quad \text{and} \quad \mathrm{R} = [\mathbf{r}_1 \ldots \mathbf{r}_p]$$

the matrices containing the left resp. right eigenvectors of A, whose elements satisfy the relations

$$\mathrm{A}\mathbf{r}_k = \lambda_k \mathbf{r}_k, \qquad \mathbf{l}_k^T \mathrm{A} = \lambda_k \mathbf{l}_k^T,$$

that is

$$\mathrm{AR} = \mathrm{R}\Lambda, \qquad \mathrm{LA} = \Lambda \mathrm{L}.$$

Without loss of generality, we can suppose that LR = I. Let us now suppose that there exists a vector function $\mathbf{w}$ satisfying the relations

$$\frac{\partial \mathbf{w}}{\partial \mathbf{u}} = \mathrm{R}^{-1}, \qquad \text{that is} \qquad \frac{\partial \mathbf{u}_k}{\partial \mathbf{w}} = \mathbf{r}_k, \qquad k = 1, \ldots, p.$$

Proceeding as we did initially, we obtain

$$\mathrm{R}^{-1}\frac{\partial \mathbf{u}}{\partial t} + \Lambda \mathrm{R}^{-1}\frac{\partial \mathbf{u}}{\partial x} = 0,$$

hence the new diagonal system (13.6). By reintroducing the characteristic curves (13.7) (the latter will no longer be straight lines as the eigenvalues λ_k vary for different values of x and t), $\mathbf{w}$ is constant along them. The components of $\mathbf{w}$ are therefore still called characteristic variables. As $\mathrm{R}^{-1} = \mathrm{L}$ (thanks to the normalization relation) we obtain

$$\frac{\partial w_k}{\partial \mathbf{u}} \cdot \mathbf{r}_m = \mathbf{l}_k \cdot \mathbf{r}_m = \delta_{km}, \qquad k, m = 1, \ldots, p.$$

The functions w_k, $k = 1, \ldots, p$ are called *Riemann invariants* of the hyperbolic system.

13.2.1 The wave equation

Let us consider the following second order hyperbolic equation

$$\frac{\partial^2 u}{\partial t^2} - \gamma^2 \frac{\partial^2 u}{\partial x^2} = f, \quad x \in (\alpha, \beta), \quad t > 0. \tag{13.9}$$

Let

$$u(x,0) = u_0(x) \quad \text{and} \quad \frac{\partial u}{\partial t}(x,0) = v_0(x), \quad x \in (\alpha, \beta),$$

be the initial data and let us suppose, moreover, that u is identically null at the boundary

$$u(\alpha,t)=0 \quad \text{and} \quad u(\beta,t)=0, \quad t>0. \tag{13.10}$$

In this case, u can represent the vertical displacement of a vibrating elastic chord with lenght $\beta-\alpha$, fixed at the endpoints, and γ is a coefficient that depends on the specific mass of the chord and on its tension. The chord is subject to a vertical force whose density is f. The functions $u_0(x)$ and $v_0(x)$ describe the initial displacement and the velocity of the chord.

For simplicity of notation, we denote by u_t the derivative $\frac{\partial u}{\partial t}$, by u_x the derivative $\frac{\partial u}{\partial x}$ and we use similar notations for the second derivatives.

Let us now suppose that f is null. From equation (13.9) we deduce that the kinetic energy of the system is preserved, that is (see Exercise 1)

$$\|u_t(t)\|^2_{\mathrm{L}^2(\alpha,\beta)}+\gamma^2\,\|u_x(t)\|^2_{\mathrm{L}^2(\alpha,\beta)}=\|v_0\|^2_{\mathrm{L}^2(\alpha,\beta)}+\gamma^2\,\|u_{0x}\|^2_{\mathrm{L}^2(\alpha,\beta)}. \tag{13.11}$$

With the change of variables

$$\omega_1=u_x, \quad \omega_2=u_t,$$

the wave equation (13.9) becomes the following first-order system

$$\frac{\partial \omega}{\partial t}+\mathrm{A}\frac{\partial \omega}{\partial x}=\mathbf{f}, \quad x\in(\alpha,\beta), \quad t>0, \tag{13.12}$$

where

$$\omega=\begin{bmatrix}\omega_1\\ \omega_2\end{bmatrix}, \quad \mathrm{A}=\begin{bmatrix}0 & -1\\ -\gamma^2 & 0\end{bmatrix}, \quad \mathbf{f}=\begin{bmatrix}0\\ f\end{bmatrix},$$

whose initial conditions are $\omega_1(x,0)=u_0'(x)$ and $\omega_2(x,0)=v_0(x)$.
Since the eigenvalues of A are distinct real numbers $\pm\gamma$ (representing the wave propagation rates), system (13.12) is hyperbolic.
Note that, also in this case, to regular initial data correspond regular solutions, while discontinuities in the initial data will propagate along the characteristic lines $\frac{dx}{dt}=\pm\gamma$.

13.3 The finite difference method

Out of simplicity we will now consider problem (13.1). To solve the latter numerically, we can use spatio-temporal discretizations based on the finite difference method. In this case, the half-plane $\{t>0\}$ is discretized choosing a temporal step Δt, a spatial discretization step h and defining the gridpoints (x_j,t^n) in the following way

$$x_j=jh, \qquad j\in\mathbb{Z}, \qquad t^n=n\Delta t, \qquad n\in\mathbb{N}.$$

Set

$$\lambda=\Delta t/h,$$

and let us define

$$x_{j+1/2} = x_j + h/2.$$

We seek discrete solutions u_j^n which approximate $u(x_j, t^n)$ for each j and n.

The hyperbolic initial value problems are often discretized in time using explicit methods. Of course, this imposes restrictions on the values of λ that implicit methods generally do not have. For instance, let us consider problem (13.1). Any explicit finite difference method can be written in the form

$$u_j^{n+1} = u_j^n - \lambda(H_{j+1/2}^n - H_{j-1/2}^n), \tag{13.13}$$

where $H_{j+1/2}^n = H(u_j^n, u_{j+1}^n)$ for a suitable function $H(\cdot,\cdot)$ called *numerical flux*.

The numerical scheme (13.13) is basically the outcome of the following consideration. Suppose that a is constant and let us write equation (13.1) in conservation form

$$\frac{\partial u}{\partial t} + \frac{\partial (au)}{\partial x} = 0,$$

au being the *flux* associated to the equation. By integrating in space, we obtain

$$\int_{x_{j-1/2}}^{x_{j+1/2}} \frac{\partial u}{\partial t}\, dx + [au]_{x_{j-1/2}}^{x_{j+1/2}} = 0, \quad j \in \mathbb{Z},$$

that is

$$\frac{\partial}{\partial t} U_j + \frac{(au)(x_{j+\frac{1}{2}}) - (au)(x_{j-\frac{1}{2}})}{h} = 0, \quad \text{where} \quad U_j = h^{-1} \int_{x_{j-\frac{1}{2}}}^{x_{j+\frac{1}{2}}} u(x)\, dx.$$

Equation (13.13) can now be interpreted as an approximation where the temporal derivative is discretized using the forward Euler finite difference scheme, U_j is replaced by u_j and $H_{j+1/2}$ is a suitable approximation of $(au)(x_{j+\frac{1}{2}})$.

13.3.1 Discretization of the scalar equation

In the context of explicit methods, numerical methods are distinguished by how the numerical flux H is chosen. In particular, we cite the following methods:

- **forward/centered Euler (FE/C)**

$$u_j^{n+1} = u_j^n - \frac{\lambda}{2} a(u_{j+1}^n - u_{j-1}^n), \tag{13.14}$$

that takes the form (13.13) provided we define

$$H_{j+1/2} = \frac{1}{2} a(u_{j+1} + u_j). \tag{13.15}$$

- **Lax-Friedrichs (LF)**

$$u_j^{n+1} = \frac{1}{2}(u_{j+1}^n + u_{j-1}^n) - \frac{\lambda}{2}a(u_{j+1}^n - u_{j-1}^n), \tag{13.16}$$

also of the form (13.13) with

$$H_{j+1/2} = \frac{1}{2}[a(u_{j+1} + u_j) - \lambda^{-1}(u_{j+1} - u_j)]. \tag{13.17}$$

- **Lax-Wendroff (LW)**

$$u_j^{n+1} = u_j^n - \frac{\lambda}{2}a(u_{j+1}^n - u_{j-1}^n) + \frac{\lambda^2}{2}a^2(u_{j+1}^n - 2u_j^n + u_{j-1}^n), \tag{13.18}$$

that can be rewritten in the form (13.13) provided that we take

$$H_{j+1/2} = \frac{1}{2}[a(u_{j+1} + u_j) - \lambda a^2(u_{j+1} - u_j)]. \tag{13.19}$$

- **Upwind (or forward/decentered Euler) (U)**

$$u_j^{n+1} = u_j^n - \frac{\lambda}{2}a(u_{j+1}^n - u_{j-1}^n) + \frac{\lambda}{2}|a|(u_{j+1}^n - 2u_j^n + u_{j-1}^n), \tag{13.20}$$

corresponding to the form (13.13) provided that we choose

$$H_{j+1/2} = \frac{1}{2}[a(u_{j+1} + u_j) - |a|(u_{j+1} - u_j)]. \tag{13.21}$$

The LF method represents a modification of the FE/C method consisting in replacing the nodal value u_j^n in (13.14) with the average of the previous nodal value u_{j-1}^n and of the following one, u_{j+1}^n.
The LW method derives from the Taylor expansion in time

$$u^{n+1} = u^n + (\partial_t u)^n \Delta t + (\partial_{tt} u)^n \frac{\Delta t^2}{2} + \mathcal{O}(\Delta t^3),$$

where $(\partial_t u)^n$ denotes the partial derivative of u at time t^n. Then, using equation (13.1), we replace $\partial_t u$ by $-a\partial_x u$, and $\partial_{tt} u$ by $a^2 \partial_{xx} u$. Neglecting the remainder $\mathcal{O}(\Delta t^3)$ and approximating the spatial derivatives with centered finite differences, we get to (13.18). Finally, the U method is obtained by discretizing the convective term $a\partial_x u$ of the equation with the upwind finite difference, as seen in Chapter 12, Sect. 12.6.

All of the previously introduced schemes are explicit. An example of implicit method is the following:

- **Backward/centered Euler (BE/C)**

$$u_j^{n+1} + \frac{\lambda}{2}a(u_{j+1}^{n+1} - u_{j-1}^{n+1}) = u_j^n. \tag{13.22}$$

Naturally, the implicit schemes can also be rewritten in a general form that is similar to (13.13) where H^n is replaced by H^{n+1}. In the specific case, the numerical flux will again be defined by (13.15).

The advantage of formulation (13.13) is that it can be extended easily to the case of more general hyperbolic problems.
In particular, we will examine the case of linear systems in Sect. 13.3.2. The extension to the case of nonlinear hyperbolic equations will instead be considered in Sect. 15.2. Finally, we point out the following schemes for approximating the wave equation (13.9), again in the case $f = 0$:

- **Leap-Frog**

$$u_j^{n+1} - 2u_j^n + u_j^{n-1} = (\gamma\lambda)^2(u_{j+1}^n - 2u_j^n + u_{j-1}^n). \tag{13.23}$$

- **Newmark**

$$u_j^{n+1} - 2u_j^n + u_j^{n-1} = \frac{(\gamma\lambda)^2}{4}\left(w_j^{n-1} + 2w_j^n + w_j^{n+1}\right), \tag{13.24}$$

where $w_j^n = u_{j+1}^n - 2u_j^n + u_{j-1}^n$.

13.3.2 Discretization of linear hyperbolic systems

Let us consider the linear system (13.5). Generalizing (13.13), a numerical scheme for a finite difference approximation can be written in the form

$$\mathbf{u}_j^{n+1} = \mathbf{u}_j^n - \lambda(\mathbf{H}_{j+1/2}^n - \mathbf{H}_{j-1/2}^n),$$

where $\mathbf{u}_j^n$ is the vector approximating $\mathbf{u}(x_j, t^n)$. Now, $\mathbf{H}_{j+1/2}$ is a *vector numerical flux*. Its formal expression can be easily derived by generalizing the scalar case and replacing a, a^2, and $|a|$ in (13.15), (13.17), (13.19), (13.21) respectively with A, A^2, and $|\mathrm{A}|$, where

$$|\mathrm{A}| = \mathrm{T}|\Lambda|\mathrm{T}^{-1},$$

$|\Lambda| = \mathrm{diag}(|\lambda_1|, ..., |\lambda_p|)$ and T is the matrix of eigenvectors of A.

For instance, transforming system (13.5) in p independent transport equations and approximating each of these with an upwind scheme for scalar equations, we obtain the following upwind numerical scheme for the initial system

$$\mathbf{u}_j^{n+1} = \mathbf{u}_j^n - \frac{\lambda}{2}\mathrm{A}(\mathbf{u}_{j+1}^n - \mathbf{u}_{j-1}^n) + \frac{\lambda}{2}|\mathrm{A}|(\mathbf{u}_{j+1}^n - 2\mathbf{u}_j^n + \mathbf{u}_{j-1}^n).$$

The numerical flux of such scheme is

$$\mathbf{H}_{j+\frac{1}{2}} = \frac{1}{2}[\mathrm{A}(\mathbf{u}_{j+1} + \mathbf{u}_j) - |\mathrm{A}|(\mathbf{u}_{j+1} - \mathbf{u}_j)].$$

The Lax-Wendroff method becomes

$$\mathbf{u}_j^{n+1} = \mathbf{u}_j^n - \frac{1}{2}\lambda\mathrm{A}(\mathbf{u}_{j+1}^n - \mathbf{u}_{j-1}^n) + \frac{1}{2}\lambda^2\mathrm{A}^2(\mathbf{u}_{j+1}^n - 2\mathbf{u}_j^n + \mathbf{u}_{j-1}^n)$$

and its numerical flux is

$$\mathbf{H}_{j+\frac{1}{2}} = \frac{1}{2}[\mathrm{A}(\mathbf{u}_{j+1} + \mathbf{u}_j) - \lambda\mathrm{A}^2(\mathbf{u}_{j+1} - \mathbf{u}_j)].$$

13.3.3 Boundary treatment

In case we want to discretize the hyperbolic equation (13.3) on a bounded interval, we will obviously need to use the inflow node $x = \alpha$ to impose the boundary condition, say $u_0^{n+1} = \varphi(t^{n+1})$, while at all other nodes x_j, $1 \leq j \leq m$ (including the outflow node $x_m = \beta$) we will write the finite difference scheme.

However, schemes using a centered discretization of the space derivative require a particular treatment at x_m. Indeed, they would require the value u_{m+1}, which is unavailable as it relates to the point with coordinates $\beta + h$, outside the integration interval. The problem can be solved in various ways. An option is to use only the upwind decentered discretization on the last node, as such discretization does not require knowing the datum in x_{m+1}; this approach however is only a first-order one. Alternatively, the value u_m^{n+1} can be obtained through extrapolation from the values available at the internal nodes. An example could be an extrapolation along the characteristic lines applied to a scheme for which $\lambda a \leq 1$; this provides $u_m^{n+1} = u_{m-1}^n \lambda a + u_m^n(1 - \lambda a)$.

A further option consists in applying the centered finite difference scheme to the outflow node x_m as well, and use, in place of u_{m+1}^n, an approximation based on a constant extrapolation ($u_{m+1}^n = u_m^n$), or on a linear one ($u_{m+1}^n = 2u_m^n - u_{m-1}^n$).

This matter becomes more problematic in the case of hyperbolic systems, where we must resort to compatibility equations. To shed more light on these aspects and to analyze their possible instabilities deriving from the numerical boundary treatment, the reader can refer to Strickwerda [Str89], [QV94, Chap. 14] and [LeV07].

13.4 Analysis of the finite difference methods

We analyze the consistency, stability and convergence properties of the finite difference methods we introduced previously.

13.4.1 Consistency and convergence

For a given numerical scheme, the local truncation error is the error generated by expecting the exact solution to verify the numerical scheme itself.

For instance, in the case of scheme (13.14), having denoted by u the solution of the exact problem (13.1), we can define the truncation error at the point (x_j, t^n) as follows

$$\tau_j^n = \frac{u(x_j, t^{n+1}) - u(x_j, t^n)}{\Delta t} + a \frac{u(x_{j+1}, t^n) - u(x_{j-1}, t^n)}{2h}.$$

If the *truncation error*

$$\tau(\Delta t, h) = \max_{j,n} |\tau_j^n|$$

tends to zero when Δt and h tend to zero, independently, then the numerical scheme will be said to be *consistent*.

Moreover, we will say that a numerical scheme is *accurate to order p in time* and *to order q in space* (for suitable integers p and q), if for a sufficiently regular solution of the exact problem we have

$$\tau(\Delta t, h) = \mathcal{O}(\Delta t^p + h^q).$$

Using Taylor expansions suitably, we can then see that the truncation error of the previously introduced methods is:

- **Euler (forward or backward) / centered:** $\mathcal{O}(\Delta t + h^2)$;
- **Upwind:** $\mathcal{O}(\Delta t + h)$;
- **Lax-Friedrichs:** $\mathcal{O}(\frac{h^2}{\Delta t} + \Delta t + h^2)$;
- **Lax-Wendroff:** $\mathcal{O}(\Delta t^2 + h^2 + h^2 \Delta t)$.

Finally, we will say that a scheme is *convergent* (in the maximum norm) if

$$\lim_{\Delta t, h \to 0} (\max_{j,n} |u(x_j, t^n) - u_j^n|) = 0.$$

Obviously, we can also consider weaker norms, such as $\|\cdot\|_{\Delta,1}$ and $\|\cdot\|_{\Delta,2}$, which we will introduce in (13.26).

13.4.2 Stability

We will say that a numerical method for a linear hyperbolic problem is *stable* if for each instant T there exists a constant $C_T > 0$ (possibly depending on T) such that for each $h > 0$, there exists $\delta_0 > 0$ (possibly dependent on h) such that for each $0 < \Delta t < \delta_0$ we have

$$\|\mathbf{u}^n\|_\Delta \leq C_T \|\mathbf{u}^0\|_\Delta, \tag{13.25}$$

for each n such that $n\Delta t \leq T$, and for each initial datum $\mathbf{u}_0$. Note that C_T should not depend on Δt and h. Often (always, in the case of explicit methods) we will have stability only if the temporal step is sufficiently small with respect to the spatial one, that is for $\delta_0 = \delta_0(h)$.

The notation $\|\cdot\|_\Delta$ denotes a suitable discrete norm, for instance

$$\|\mathbf{v}\|_{\Delta,p} = \left(h \sum_{j=-\infty}^{\infty} |v_j|^p \right)^{\frac{1}{p}} \quad \text{for } p = 1,2, \quad \|\mathbf{v}\|_{\Delta,\infty} = \sup_j |v_j|. \tag{13.26}$$

Note how $\|\mathbf{v}\|_{\Delta,p}$ represents an approximation of the $\mathrm{L}^p(\mathbb{R})$ norm, for $p = 1,2$ or $+\infty$.

The implicit backward/centered Euler scheme (13.22) is stable in the norm $\|\cdot\|_{\Delta,2}$ for any choice of the parameters Δt and h (see Exercise 2).

A scheme is called *strongly stable* with respect to the norm $\|\cdot\|_\Delta$ if

$$\|\mathbf{u}^n\|_\Delta \leq \|\mathbf{u}^{n-1}\|_\Delta, \tag{13.27}$$

for each n such that $n\Delta t \leq T$, and for each initial datum $\mathbf{u}_0$, which implies that (13.25) is verified with $C_T = 1$.

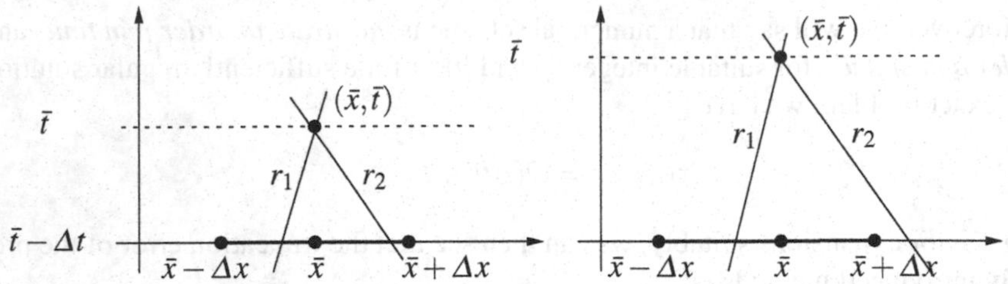

Fig. 13.3. Geometric interpretation of the CFL condition for a system with $p = 2$, where $r_i = \bar{x} - \lambda_i(t - \bar{t})$ $i = 1,2$. The CFL condition is satisfied on the left, and violated on the right

Remark 13.1. In the context of hyperbolic problems, one often wants long-time solutions (solutions with $T \gg 1$). Such cases usually require a strongly stable scheme, as this guarantees that the numerical solution is bounded for each value of T. •

As we will see, a necessary condition for the stability of an explicit numerical scheme of the form (13.13) is that the temporal and spatial discretization steps satisfy

$$|a\lambda| \leq 1, \text{ or } \Delta t \leq \frac{h}{|a|}, \tag{13.28}$$

called *CFL condition* (from Courant, Friedrichs and Lewy). The number $a\lambda$ is commonly called *CFL number* and is a physically dimensionless quantity (a being a velocity).

The geometrical interpretation of the CFL stability condition is the following. In a finite difference scheme, the value of u_j^{n+1} generally depends on the values u_{j+i}^n of u^n at the three points x_{j+i}, $i = -1,0,1$. Proceeding backwards, we deduce that the solution u_j^{n+1} will only depend on the initial data at the points x_{j+i}, for $i = -(n+1),...,(n+1)$ (see Fig. 13.3).

Calling *numerical domain of dependence* $D_{\Delta t}(x_j,t^n)$ the domain of dependence of u_j^n, which will therefore be called numerical dependence domain of u_j^n, the former will verify

$$D_{\Delta t}(x_j,t^n) \subset \{x \in \mathbb{R}: \ |x - x_j| \leq nh = \frac{t^n}{\lambda}\}.$$

Consequently, for each given point $(\bar{x},\bar{t})$ we have

$$D_{\Delta t}(\bar{x},\bar{t}) \subset \{x \in \mathbb{R}: \ |x - \bar{x}| \leq \frac{\bar{t}}{\lambda}\}.$$

In particular, taking the limit for $\Delta t \to 0$, and fixing λ, the numerical dependency domain becomes

$$D_0(\bar{x},\bar{t}) = \{x \in \mathbb{R}: \ |x - \bar{x}| \leq \frac{\bar{t}}{\lambda}\}.$$

The condition (13.28) is then equivalent to the inclusion

$$D(\bar{x},\bar{t}) \subset D_0(\bar{x},\bar{t}), \tag{13.29}$$

where $D(\overline{x},\overline{t})$ is the dependency domain of the exact solution defined in (13.8). Note that in the scalar case, $p = 1$ and $\lambda_1 = a$.

Remark 13.2. The CFL condition establishes, in particular, that there is no explicit finite different scheme for hyperbolic initial value problems that is unconditionally stable and consistent. Indeed, suppose the CFL condition is violated. Then there exists at least a point x^* in the dependency domain that does not belong to the numerical dependency domain. Then changing the initial datum to x^* will only modify the exact solution and not the numerical one. This implies non-convergence of the method and therefore also instability. Indeed, for a consistent method, the Lax-Richtmyer equivalence theorem states that stability is a necessary and sufficient condition for its convergence. •

Remark 13.3. In the case where $a = a(x,t)$ is no longer constant in (13.1), the CFL condition becomes

$$\Delta t \leq \frac{h}{\sup\limits_{x\in\mathbb{R},\, t>0} |a(x,t)|}.$$

If the spatial discretization step varies, we have

$$\Delta t \leq \min_k \frac{h_k}{\sup\limits_{x\in(x_k,x_{k+1}),\, t>0} |a(x,t)|},$$

with $h_k = x_{k+1} - x_k$. •

Referring to the hyperbolic system (13.5), the CFL stability condition, in analogy to (13.28), will be

$$\left|\lambda_k \frac{\Delta t}{h}\right| \leq 1, \quad k = 1,\ldots,p, \qquad \text{or, equivalently,} \qquad \Delta t \leq \frac{h}{\max_k |\lambda_k|},$$

where $\{\lambda_k,\ k = 1\ldots,p\}$ are the eigenvalues of A.

This condition, as well, can be written in the form (13.29). The latter expresses the requirement that each line of the form $x = \overline{x} - \lambda_k(\overline{t} - t)$, $k = 1,\ldots,p$, must intersect the horizontal line $t = \overline{t} - \Delta t$ at points $x^{(k)}$ which lie within the numerical dependency domain.

Theorem 13.1. *If the CFL condition (13.28) is satisfied, the upwind, Lax-Friedrichs and Lax-Wendroff schemes are strongly stable in the norm* $\|\cdot\|_{\Delta,1}$.

Proof. To prove the stability of the upwind scheme (13.20) we rewrite it in the following form (having supposed $a > 0$)

$$u_j^{n+1} = u_j^n - \lambda a(u_j^n - u_{j-1}^n).$$

Then

$$\|\mathbf{u}^{n+1}\|_{\Delta,1} \leq h\sum_j |(1-\lambda a)u_j^n| + h\sum_j |\lambda a u_{j-1}^n|.$$

Under the hypothesis (13.28) both values λa and $1-\lambda a$ are non-negative. Hence,

$$\|\mathbf{u}^{n+1}\|_{\Delta,1} \leq h(1-\lambda a)\sum_j |u_j^n| + h\lambda a\sum_j |u_{j-1}^n| = \|\mathbf{u}^n\|_{\Delta,1},$$

that is, inequality (13.25) holds with $C_T = 1$. The scheme is therefore strongly stable with respect to the norm $\|\cdot\|_\Delta = \|\cdot\|_{\Delta,1}$.

For the Lax-Friedrichs scheme, always under the CFL condition (13.28), we derive from (13.16) that

$$u_j^{n+1} = \frac{1}{2}(1-\lambda a)u_{j+1}^n + \frac{1}{2}(1+\lambda a)u_{j-1}^n,$$

so

$$\begin{aligned}\|\mathbf{u}^{n+1}\|_{\Delta,1} &\leq \frac{1}{2}h\left[\sum_j \left|(1-\lambda a)u_{j+1}^n\right| + \sum_j \left|(1+\lambda a)u_{j-1}^n\right|\right] \\ &\leq \frac{1}{2}(1-\lambda a)\|\mathbf{u}^n\|_{\Delta,1} + \frac{1}{2}(1+\lambda a)\|\mathbf{u}^n\|_{\Delta,1} = \|\mathbf{u}^n\|_{\Delta,1}.\end{aligned}$$

For the Lax-Wendroff scheme, the proof is analogous (see e.g. [QV94, Chap. 14] or [Str89]). ◇

Finally, we can prove that if the CFL condition is verified, the upwind scheme satisfies

$$\|\mathbf{u}^n\|_{\Delta,\infty} \leq \|\mathbf{u}^0\|_{\Delta,\infty} \quad \forall n \geq 0, \tag{13.30}$$

i.e. it is strongly stable in the norm $\|\cdot\|_{\Delta,\infty}$. Relation (13.30) is called *discrete maximum principle* (see Exercise 4).

Theorem 13.2. *The backward Euler scheme BE/C is strongly stable in the norm $\|\cdot\|_{\Delta,2}$, with no restriction on Δt. The forward Euler scheme FE/C, instead, is never strongly stable. However, it is stable with constant $C_T = e^{T/2}$ provided that we assume that Δt satisfies the following condition (more restrictive than the CFL condition)*

$$\Delta t \leq \left(\frac{h}{a}\right)^2. \tag{13.31}$$

Proof. We observe that

$$(B-A)B = \frac{1}{2}(B^2 - A^2 + (B-A)^2) \quad \forall A,B \in \mathbb{R}. \tag{13.32}$$

As a matter of fact

$$(B-A)B = (B-A)^2 + (B-A)A = \frac{1}{2}((B-A)^2 + (B-A)(B+A)).$$

Multiplying (13.22) by u_j^{n+1} we find

$$(u_j^{n+1})^2 + (u_j^{n+1} - u_j^n)^2 = (u_j^n)^2 - \lambda a(u_{j+1}^{n+1} - u_{j-1}^{n+1})u_j^{n+1}.$$

Observing that

$$\sum_{j\in\mathbb{Z}} (u_{j+1}^{n+1} - u_{j-1}^{n+1})u_j^{n+1} = 0 \tag{13.33}$$

(telescopic sum), we immediately obtain that $||\mathbf{u}^{n+1}||_{\Delta,2}^2 \le ||\mathbf{u}^n||_{\Delta,2}^2$, which is the result sought for the BE/C scheme.

Let us now move to the FE/C scheme and multiply (13.14) by u_j^n. Observing that

$$(B-A)A = \frac{1}{2}(B^2 - A^2 - (B-A)^2) \quad \forall A, B \in \mathbb{R}, \tag{13.34}$$

we find

$$(u_j^{n+1})^2 = (u_j^n)^2 + (u_j^{n+1} - u_j^n)^2 - \lambda a(u_{j+1}^n - u_{j-1}^n)u_j^n.$$

On the other hand, we obtain once again from (13.14) that

$$u_j^{n+1} - u_j^n = -\frac{\lambda a}{2}(u_{j+1}^n - u_{j-1}^n)$$

and therefore

$$(u_j^{n+1})^2 = (u_j^n)^2 + \left(\frac{\lambda a}{2}\right)^2 (u_{j+1}^n - u_{j-1}^n)^2 - \lambda a(u_{j+1}^n - u_{j-1}^n)u_j^n.$$

Now summing on j and observing that the last addendum yields a telescopic sum (hence it does not provide any contribution) we obtain, after multiplying by h,

$$||\mathbf{u}^{n+1}||_{\Delta,2}^2 = ||\mathbf{u}^n||_{\Delta,2}^2 + \left(\frac{\lambda a}{2}\right)^2 h \sum_{j\in\mathbb{Z}} (u_{j+1}^n - u_{j-1}^n)^2,$$

from which we infer that there is no value of Δt for which the method is strongly stable. However, as

$$(u_{j+1}^n - u_{j-1}^n)^2 \le 2\left[(u_{j+1}^n)^2 + (u_{j-1}^n)^2\right],$$

we find that, under the hypothesis (13.31),

$$||\mathbf{u}^{n+1}||_{\Delta,2}^2 \le (1 + \lambda^2 a^2)||\mathbf{u}^n||_{\Delta,2}^2 \le (1 + \Delta t)||\mathbf{u}^n||_{\Delta,2}^2.$$

By recursion, we find

$$\|\mathbf{u}^n\|^2_{\Delta,2} \leq (1+\Delta t)^n \|\mathbf{u}^0\|^2_{\Delta,2} \leq e^T \|\mathbf{u}^0\|^2_{\Delta,2},$$

where we have used the inequality

$$(1+\Delta t)^n \leq e^{n\Delta t} \leq e^T \quad \forall n \text{ such that } t^n \leq T.$$

We conclude that

$$\|\mathbf{u}^n\|_{\Delta,2} \leq e^{T/2} \|\mathbf{u}^0\|_{\Delta,2},$$

which is the stability result sought for the FE/C scheme. ◇

13.4.3 Von Neumann analysis and amplification coefficients

Von Neumann's analysis is useful to investigate the stability of a scheme in the norm $\|\cdot\|_{\Delta,2}$. To this purpose, we assume that the function $u_0(x)$ is 2π-periodic and thus it can be written as a Fourier series as follows

$$u_0(x) = \sum_{k=-\infty}^{\infty} \alpha_k e^{ikx}, \tag{13.35}$$

where

$$\alpha_k = \frac{1}{2\pi} \int_0^{2\pi} u_0(x)\, e^{-ikx}\, dx$$

is the k-th Fourier coefficient. Hence,

$$u_j^0 = u_0(x_j) = \sum_{k=-\infty}^{\infty} \alpha_k e^{ikjh}, \quad j = 0, \pm 1, \pm 2, \cdots$$

It can be verified that applying any of the difference schemes seen in Sect. 13.3.1 we get the following relation

$$u_j^n = \sum_{k=-\infty}^{\infty} \alpha_k e^{ikjh} \gamma_k^n, \quad j = 0, \pm 1, \pm 2, \ldots, \quad n \geq 1. \tag{13.36}$$

The number $\gamma_k \in \mathrm{C}$ is called *amplification coefficient* of the k-th frequency (or harmonic), and characterizes the scheme under exam. For instance, in the case of the forward centered Euler method (FE/C) we find

$$\begin{aligned} u_j^1 &= \sum_{k=-\infty}^{\infty} \alpha_k e^{ikjh} \left(1 - \frac{a\Delta t}{2h}(e^{ikh} - e^{-ikh})\right) \\ &= \sum_{k=-\infty}^{\infty} \alpha_k e^{ikjh} \left(1 - \frac{a\Delta t}{h} i \sin(kh)\right). \end{aligned}$$

Table 13.1. Amplification coefficient for the different numerical schemes in Sect. 13.3.1. We recall that $\lambda = \Delta t/h$

Scheme	γ_k
Forward/centered Euler	$1 - ia\lambda\sin(kh)$
Backward/centered Euler	$(1 + ia\lambda\sin(kh))^{-1}$
Upwind	$1 - \lvert a\rvert\lambda(1 - e^{-ikh})$
Lax-Friedrichs	$\cos kh - ia\lambda\sin(kh)$
Lax-Wendroff	$1 - ia\lambda\sin(kh) - a^2\lambda^2(1 - cos(kh))$

Hence,

$$\gamma_k = 1 - \frac{a\Delta t}{h} i\sin(kh) \qquad \text{and thus} \qquad |\gamma_k| = \left\{1 + \left(\frac{a\Delta t}{h}\sin(kh)\right)^2\right\}^{\frac{1}{2}}.$$

As there exist values of k for which $|\gamma_k| > 1$, there is no value of Δt and h for which the scheme is strongly stable.

Proceeding in a similar way for the other schemes, we find the coefficients reported in Table 13.1.

We will now see how the von Neumann analysis can be applied to study the stability of a numerical scheme with respect to the $\|\cdot\|_{\Delta,2}$ norm and to ascertain its dissipation and dispersion properties.

To this purpose, we prove the following result:

Theorem 13.3. *If there exist a number $\beta \geq 0$ and a positive integer m such that, for suitable choices of Δt and h, we have $|\gamma_k| \leq (1 + \beta\Delta t)^{\frac{1}{m}}$ for each k, then the scheme is stable with respect to the norm $\|\cdot\|_{\Delta,2}$ with stability constant $C_T = e^{\beta T/m}$. In particular, if we can take $\beta = 0$ (and therefore $|\gamma_k| \leq 1\ \forall k$) then the scheme is strongly stable with respect to the same norm.*

Proof. We will suppose that problem (13.1) is formulated on the interval $[0, 2\pi]$. In such interval, let us consider $N + 1$ equidistant nodes,

$$x_j = jh, \quad j = 0, \ldots, N, \quad \text{with} \quad h = \frac{2\pi}{N},$$

(N being an even positive integer) where to satisfy the numerical scheme (13.13). Moreover, we will suppose for simplicity that the initial datum u_0 is periodic. As the numerical scheme only depends on the values of u_0 at the nodes x_j, we can replace u_0 by the Fourier polynomial of order $N/2$,

$$\tilde{u}_0(x) = \sum_{k=-\frac{N}{2}}^{\frac{N}{2}-1} \alpha_k e^{ikx} \tag{13.37}$$

which interpolates it at the nodes. Note that $\tilde{u}_0$ is a periodic function with period 2π. We will have, thanks to (13.36),

$$u_j^0 = u_0(x_j) = \sum_{k=-\frac{N}{2}}^{\frac{N}{2}-1} \alpha_k e^{ikjh}, \quad u_j^n = \sum_{k=-\frac{N}{2}}^{\frac{N}{2}-1} \alpha_k \gamma_k^n e^{ikjh}.$$

We note that

$$\|\mathbf{u}^n\|_{\Delta,2}^2 = h \sum_{j=0}^{N-1} \sum_{k,m=-\frac{N}{2}}^{\frac{N}{2}-1} \alpha_k \overline{\alpha}_m (\gamma_k \overline{\gamma}_m)^n e^{i(k-m)jh}.$$

As

$$h \sum_{j=0}^{N-1} e^{i(k-m)jh} = 2\pi \delta_{km}, \quad -\frac{N}{2} \le k, m \le \frac{N}{2} - 1,$$

(see e.g. [QSS07, Lemma 10.2]) we find

$$\|\mathbf{u}^n\|_{\Delta,2}^2 = 2\pi \sum_{k=-\frac{N}{2}}^{\frac{N}{2}-1} |\alpha_k|^2 |\gamma_k|^{2n}.$$

Thanks to the assumption made on $|\gamma_k|$ we have

$$\|\mathbf{u}^n\|_{\Delta,2}^2 \le (1+\beta\Delta t)^{\frac{2n}{m}} 2\pi \sum_{k=-\frac{N}{2}}^{\frac{N}{2}-1} |\alpha_k|^2 = (1+\beta\Delta t)^{\frac{2n}{m}} \|\mathbf{u}^0\|_{\Delta,2}^2 \ \forall n \ge 0.$$

As $1+\beta\Delta t \le e^{\beta\Delta t}$, we deduce that

$$\|\mathbf{u}^n\|_{\Delta,2} \le e^{\frac{\beta\Delta t n}{m}} \|\mathbf{u}^0\|_{\Delta,2} = e^{\frac{\beta T}{m}} \|\mathbf{u}^0\|_{\Delta,2} \quad \forall n \quad \text{such that} \quad n\Delta t \le T.$$

This proves the theorem. ◇

Remark 13.4. Should strong stability be required, the condition $|\gamma_k| \le 1$ indicated in Theorem 13.3 is also necessary. •

In the case of the upwind scheme (13.20), as

$$|\gamma_k|^2 = [1 - |a|\lambda(1-\cos kh)]^2 + a^2\lambda^2 \sin^2 kh, \quad k \in \mathbb{Z},$$

we obtain

$$|\gamma_k| \le 1 \ \text{if} \ \Delta t \le \frac{h}{|a|}, \quad k \in \mathbb{Z}, \tag{13.38}$$

that is, the CFL condition guarantees strong stability in the $\|\cdot\|_{\Delta,2}$ norm.

Proceeding in a similar way, we can verify that (13.38) also holds for the Lax-Friedrichs scheme.

The centered backward Euler scheme BE/C instead is unconditionally strongly stable in the norm $\|\cdot\|_{\Delta,2}$, as $|\gamma_k| \leq 1$ for each k and for each possible choice of Δt and h, as we previously obtained in Theorem 13.2 by following a different procedure.

In the case of the centered forward Euler method FE/C we have

$$|\gamma_k|^2 = 1 + \frac{a^2 \Delta t^2}{h^2} \sin^2(kh) \leq 1 + \frac{a^2 \Delta t^2}{h^2}, \quad k \in \mathbb{Z}.$$

If $\beta > 0$ is a constant such that

$$\Delta t \leq \beta \frac{h^2}{a^2} \tag{13.39}$$

then $|\gamma_k| \leq (1 + \beta \Delta t)^{1/2}$. Hence, applying Theorem 13.3 (with $m = 2$) we deduce that the FE/C scheme is stable, albeit with a more restrictive CFL condition, as previously obtained following a different path in Theorem 13.2.

We can find a strong stability condition for the centered forward Euler method applied to the transport-reaction equation

$$\frac{\partial u}{\partial t} + a \frac{\partial u}{\partial x} + a_0 u = 0, \tag{13.40}$$

with $a_0 > 0$. In this case we have for each $k \in \mathbb{Z}$

$$|\gamma_k|^2 = 1 - 2a_0 \Delta t + \Delta t^2 a_0^2 + \lambda^2 \sin^2(kh) \leq 1 - 2a_0 \Delta t + \Delta t^2 a_0^2 + \left(\frac{a \Delta t}{h}\right)^2,$$

and thus the scheme is strongly stable in the $||.||_{\Delta,2}$ norm under the condition

$$\Delta t < \frac{2a_0}{a_0^2 + h^{-2} a^2}. \tag{13.41}$$

Example 13.1. In order to verify numerically the stability condition (13.41), we have considered equation (13.40) in the interval $(0,1)$ with periodic boundary conditions. We have chosen $a = a_0 = 1$ and the initial datum u_0 equal to 2 in the interval $(1/3, 2/3)$ and 0 elsewhere. As the initial datum is a square wave, its Fourier expansion has all its α_k coefficients not null. On the right of Fig. 13.4, we report $\|\mathbf{u}^n\|_{\Delta,2}$ in the time interval $(0, 2.5)$ for two values of Δt, one larger and one smaller than the critical value $\Delta t^* = 2/(1 + h^{-2})$, provided by (13.41). Note that for $\Delta t < \Delta t^*$ the norm is decreasing, while, in the opposite case, after an initial decrease it grows exponentially. Fig. 13.5 shows the result for $a_0 = 0$ obtained with FE/C using the same initial datum. In the figure on the left, we display the behaviour of $\|\mathbf{u}^n\|_{\Delta,2}$ for different values of h and using $\Delta t = 10h^2$, that is varying the time step based on the restriction provided by inequality (13.39) and taking $\beta = 10$. Note how the norm of the solution remains bounded for decreasing values of h. On the right-hand side of the same figure, we illustrate the result obtained for the same values of h taking as condition $\Delta t = 0.1h$, which corresponds to a constant CFL number equal to 0.1. In this case, the discrete norm of the numerical solution blows up as h decreases, as expected. ■

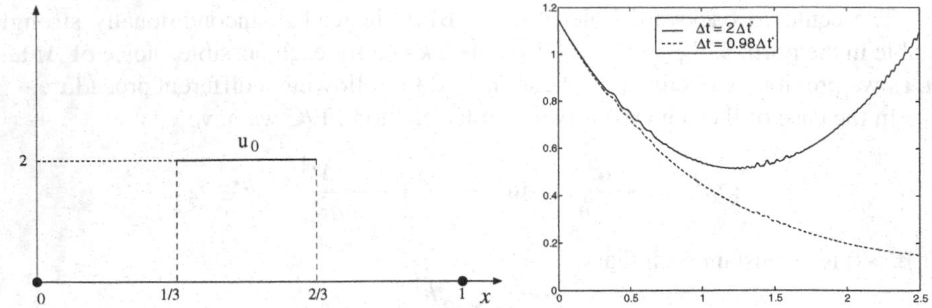

Fig. 13.4. The figure on the right displays the behaviour of $\|\mathbf{u}^n\|_{\Delta,2}$, where $\mathbf{u}^n$ is the solution of equation (13.40) (with $a = a_0 = 1$) obtained using the FE/C method, for two values of Δt, one smaller and one greater than the critical value Δt^*. On the left, the initial datum used

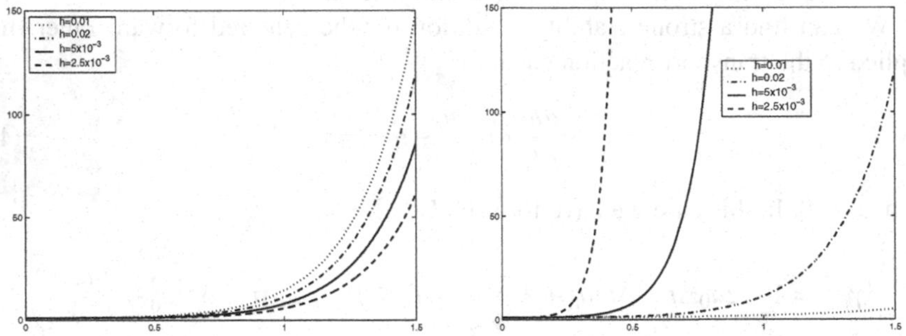

Fig. 13.5. Behaviour of $\|\mathbf{u}^n\|_{\Delta,2}$ where $\mathbf{u}^n$ is the solution obtained using the FE/C method for $a_0 = 0$ and for different values of h. On the left, the case where Δt satisfies the stability condition (13.39). On the right, the results obtained by maintaining the CFL number constant and equal to 0.1, violating condition (13.39)

13.4.4 Dissipation and dispersion

Besides allowing to enquire about the stability of a numerical scheme, the analysis of the amplification coefficients is also useful to study its dissipation and dispersion properties.
To clarify the matter better, let us consider the exact solution of problem (13.1); for such solution, we have the following relation

$$u(x,t^n) = u_0(x - an\Delta t), \quad n \geq 0, \quad x \in \mathbb{R},$$

with $t^n = n\Delta t$. In particular, using (13.35) we obtain

$$u(x_j,t^n) = \sum_{k=-\infty}^{\infty} \alpha_k e^{ikjh} (g_k)^n \quad \text{with} \quad g_k = e^{-iak\Delta t}. \tag{13.42}$$

Comparing (13.42) with (13.36) we can note that the amplification coefficient γ_k (generated by the specific numerical scheme) is the correspondent of g_k.

We observe that $|g_k| = 1$ for each $k \in \mathbb{Z}$, while $|\gamma_k| \leq 1$ in order to guarantee the strong stability of the scheme. Thus, γ_k is a *dissipative* coefficient. The smaller $|\gamma_k|$ is, the larger will be the reduction of the amplitude α_k and, consequently, the larger will be the dissipation of the numerical scheme.
The ratio $\varepsilon_a(k) = \frac{|\gamma_k|}{|g_k|}$ is called *amplification error* (or *dissipation error*) of the k-th harmonic associated to the numerical scheme (and in our case it coincides with the amplification coefficient).
Having set

$$\phi_k = kh,$$

as $k\Delta t = \lambda\phi_k$ we obtain

$$g_k = e^{-ia\lambda\phi_k}. \tag{13.43}$$

The real number ϕ_k, here expressed in radians, is called *phase angle* of the k-th harmonic. We rewrite γ_k in the following way

$$\gamma_k = |\gamma_k| e^{-i\omega\Delta t} = |\gamma_k| e^{-i\frac{\omega}{k}\lambda\phi_k},$$

and comparing such relation to (13.43), we can deduce that the ratio $\frac{\omega}{k}$ represents the *propagation rate* of the numerical scheme, relatively to the k-th harmonic. The ratio

$$\varepsilon_d(k) = \frac{\omega}{ka} = \frac{\omega h}{\phi_k a}$$

between the numerical propagation and the propagation a of the exact solution is called *dispersion error* ε_d relative to the k-th harmonic.

The amplification (or dissipation) error and the dispersion error for the numerical schemes analyzed up to now are function of the phase angle ϕ_k and of the CFL number $a\lambda$, as reported in Fig. 13.6. For symmetry reasons we have considered the interval $0 \leq \phi_k \leq \pi$ and we have used degrees instead of radians on the x-axis to indicate ϕ_k. Note how the forward/centered Euler scheme gives a curve of the amplification factor with values above one for all the CFL schemes we have considered, in accordance with the fact that such scheme is never strongly stable.

Example 13.2. In Fig. 13.7 we compare the numerical results obtained by solving equation (13.1) with $a = 1$ and initial datum u_0. The solutions are composed by a packet of two sinusoidal waves of equal length l centered at the origin ($x = 0$). In the figures on the left $l = 20h$, while in the right ones we have $l = 8h$. As $k = \frac{2\pi}{l}$, we have $\phi_k = \frac{2\pi}{l}h$ and therefore the values of the phase angle of the wave packet are $\phi_k = \pi/20$ on the left and $\phi_k = \pi/8$ on the right. The numerical solution has been computed for the value 0.75 of the CFL number, using the different (stable) schemes illustrated previously. We can note how the dissipative effect is very strong at high frequencies ($\phi_k = \pi/4$) and in particular for the first-order upwind, backward/centered Euler and Lax-Friedrichs methods.

In order to appreciate the dispersion effects, the solution for $\phi_k = \pi/4$ after 8 time steps is reported in Fig. 13.8. We can note how the Lax-Wendroff method is the least

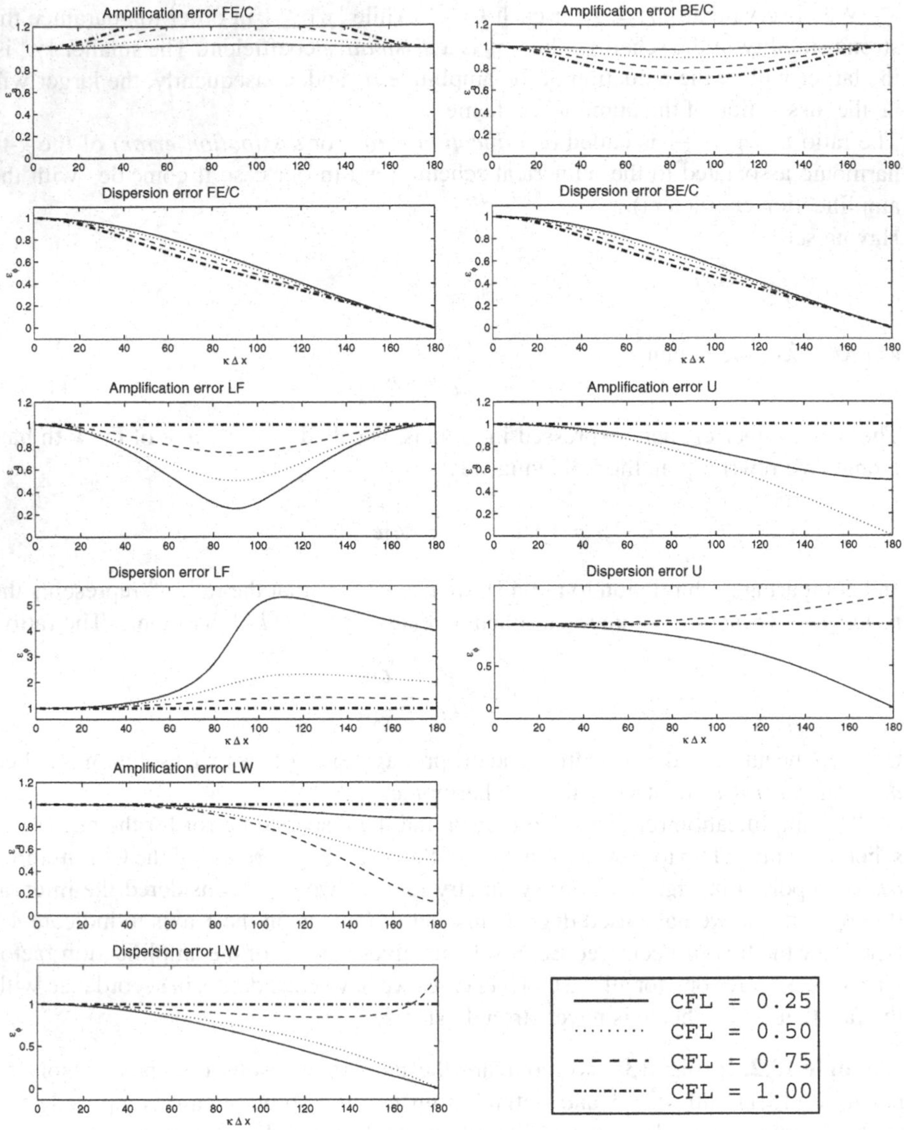

Fig. 13.6. Amplification and dispersion errors for different numerical schemes in terms of the phase angle $\phi_k = kh$ and for different values of the CFL number

dissipative. Moreover, by observing attentively the position of the numerical wave crests with respect to those of the numerical solution, we can verify that the Lax-Friedrichs method features a positive dispersion error. Indeed, the numerical wave anticipates the exact one. The upwind method is also weakly dispersive for a CFL

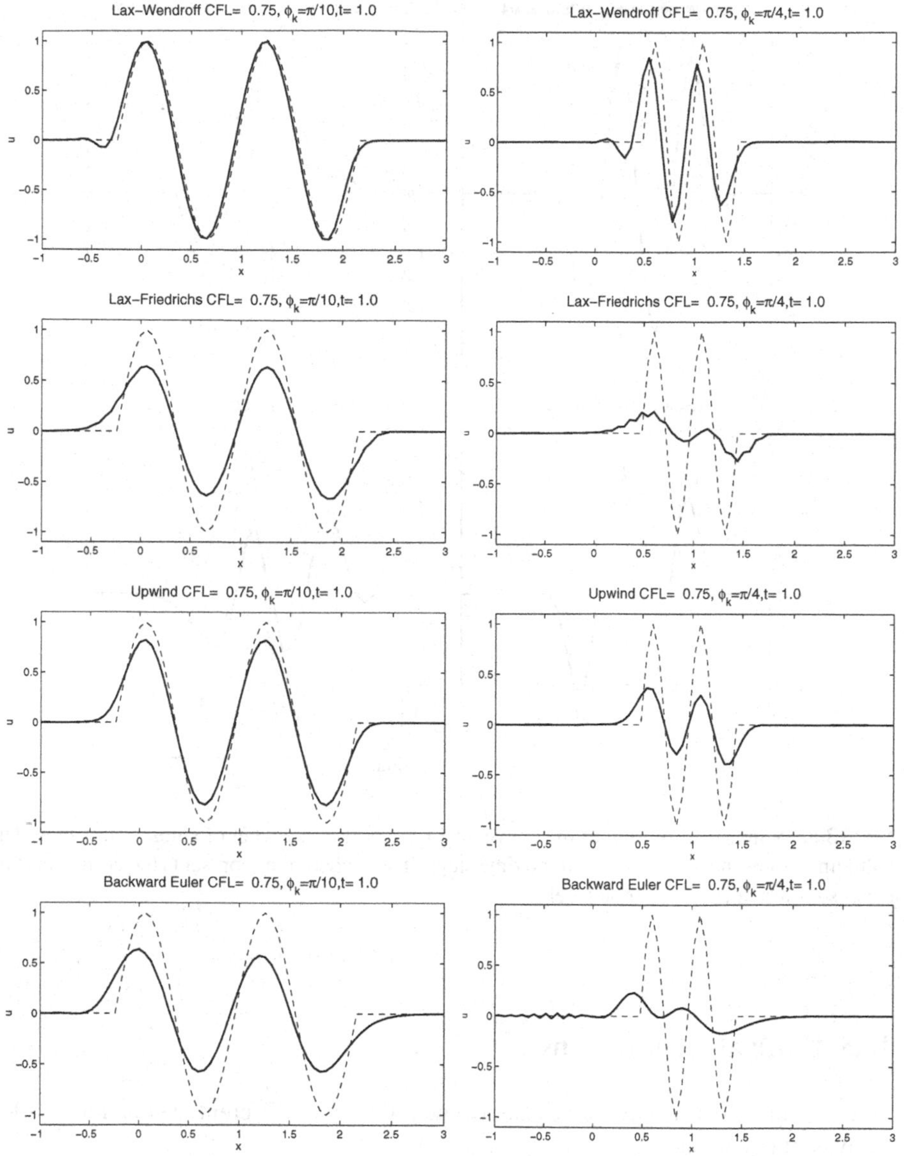

Fig. 13.7. Numerical solution of the convective transport equation of a sine wave packet with different wavelengths ($l = 20h$ left, $l = 8h$ right) obtained with different numerical schemes. The numerical solution for $t = 1$ is displayed by the solid line, while the exact solution at the same time instant is displayed by the dashed line

number equal to 0.75, while the dispersion of the Lax-Friedrichs and backward Euler methods is evident (even after only 8 time steps!). ■

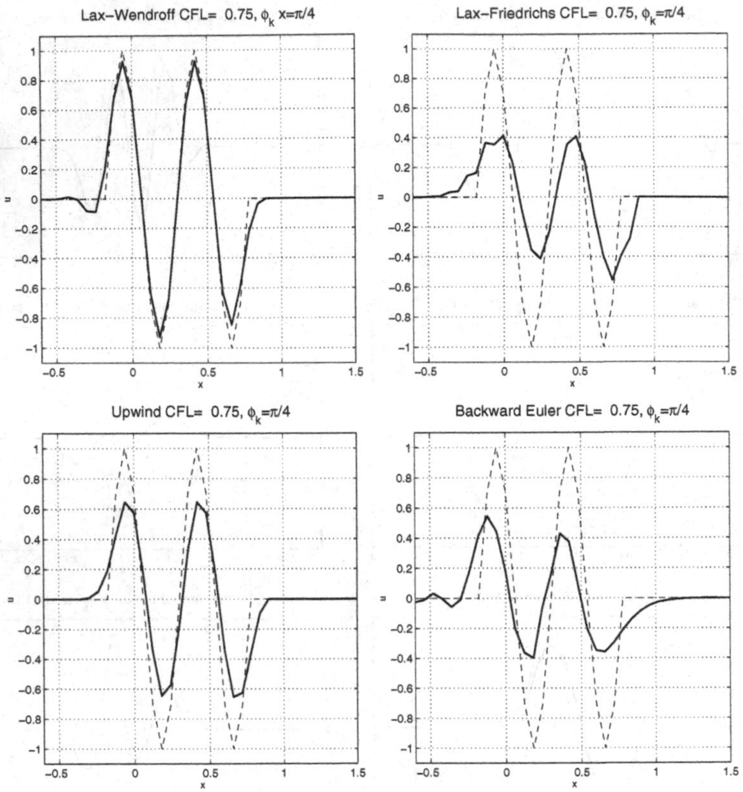

Fig. 13.8. Numerical solution of the convective transport of a packet of sinusoidal waves. The solid line represents the solution after 8 time steps. The etched line represents the corresponding exact solution at the same time level

13.5 Equivalent equations

To each numerical scheme, we can associate a family of differential equations, called equivalent equations.

13.5.1 The upwind scheme case

Let us first focus on the upwind scheme. Suppose there exists a regular function $v(x,t)$ satisfying the difference equation (13.20) at each point $(x,t) \in \mathbb{R} \times \mathbb{R}^+$ (and not only at the grid nodes (x_j,t^n)!). We can then write (in the case where $a > 0$)

$$\frac{v(x,t+\Delta t) - v(x,t)}{\Delta t} + a\frac{v(x,t) - v(x-h,t)}{h} = 0. \tag{13.44}$$

Using the Taylor expansions with respect to x and t relative to the point (x,t) and supposing that v is of class C^4 with respect to x and t, we can write

$$\frac{v(x,t+\Delta t)-v(x,t)}{\Delta t} = v_t + \frac{\Delta t}{2}v_{tt} + \frac{\Delta t^2}{6}v_{ttt} + \mathcal{O}(\Delta t^3),$$

$$a\frac{v(x,t)-v(x-h,t)}{h} = av_x - \frac{ah}{2}v_{xx} + \frac{ah^2}{6}v_{xxx} + \mathcal{O}(h^3),$$

where the right-hand side derivatives are all evaluated at (x,t). Thanks to (13.44) we deduce that, at each point (x,t), the function v satisfies the relation

$$v_t + av_x = R^U + \mathcal{O}(\Delta t^3 + h^3), \tag{13.45}$$

with

$$R^U = \frac{1}{2}(ah\,v_{xx} - \Delta t\,v_{tt}) - \frac{1}{6}(ah^2 v_{xxx} + \Delta t^2 v_{ttt}).$$

Formally differentiating such equation in t, we find

$$v_{tt} + av_{xt} = R^U_t + \mathcal{O}(\Delta t^3 + h^3).$$

Instead, differentiating it in x, we have

$$v_{xt} + av_{xx} = R^U_x + \mathcal{O}(\Delta t^3 + h^3). \tag{13.46}$$

Hence

$$v_{tt} = a^2 v_{xx} + R^U_t - aR^U_x + \mathcal{O}(\Delta t^3 + h^3), \tag{13.47}$$

which allows to obtain from (13.45)

$$v_t + av_x = \mu v_{xx} - \frac{1}{6}(ah^2 v_{xxx} + \Delta t^2 v_{ttt}) - \frac{\Delta t}{2}(R^U_t - aR^U_x) + \mathcal{O}(\Delta t^3 + h^3), \tag{13.48}$$

having set

$$\mu = \frac{1}{2}ah(1-(a\lambda)) \tag{13.49}$$

and, as usual, $\lambda = \Delta t/h$. Now, differentiating (13.47) with respect to t formally, and (13.46) with respect to x, we find

$$\begin{aligned} v_{ttt} &= a^2 v_{xxt} + R^U_{tt} - aR^U_{xt} + \mathcal{O}(\Delta t^3 + h^3) \\ &= -a^3 v_{xxx} + a^2 R^U_{xx} + R^U_{tt} - aR^U_{xt} + \mathcal{O}(\Delta t^3 + h^3). \end{aligned} \tag{13.50}$$

Moreover, we have that

$$\begin{aligned} R^U_t &= \frac{1}{2}ah\,v_{xxt} - \frac{\Delta t}{2}v_{ttt} - \frac{ah^2}{6}v_{xxxt} - \frac{\Delta t^2}{6}v_{tttt}, \\ R^U_x &= \frac{1}{2}ah\,v_{xxx} - \frac{\Delta t}{2}v_{ttx} - \frac{ah^2}{6}v_{xxxx} - \frac{\Delta t^2}{6}v_{tttx}. \end{aligned} \tag{13.51}$$

Using relations (13.50) and (13.51) in (13.48) we obtain

$$\begin{aligned} v_t + av_x = {} & \mu v_{xx} - \frac{ah^2}{6}\left(1 - \frac{a^2\Delta t^2}{h^2} - \frac{3a\Delta t}{2h}\right)v_{xxx} \\ & \underbrace{+\frac{\Delta t}{4}\left(\Delta t\, v_{ttt} - ah\, v_{xxt} - a\Delta t\, v_{ttx}\right)}_{(A)} \\ & + \frac{\Delta t}{12}(\Delta t^2 v_{tttt} - a\Delta t^2 v_{tttx} + ah^2 v_{xxxt} - a^2h^2 v_{xxxx}) \\ & - \frac{a^2\Delta t^2}{6}R^U_{xx} - \frac{\Delta t^2}{6}R^U_{tt} + a\frac{\Delta t^2}{6}R^U_{xt} + \mathcal{O}(\Delta t^3 + h^3). \end{aligned} \tag{13.52}$$

Let us now focus on the third derivatives of v contained in the term (A). Thanks to (13.50), (13.46) and (13.47), respectively, we find:

$$\begin{aligned} v_{ttt} &= -a^3 v_{xxx} + r_1, \\ v_{xxt} &= -a\, v_{xxx} + r_2, \\ v_{ttx} &= a^2 v_{xxx} + r_3, \end{aligned}$$

where r_1, r_2 and r_3 are terms containing derivatives of v of order no less than four, as well as terms of order $\mathcal{O}(\Delta t^3 + h^3)$. (Note that it follows from the definition of R^U that its derivatives of order two are expressed through derivatives of v of order no less than four.) Regrouping the coefficients that multiply v_{xxx}, we therefore deduce from (13.52) that

$$v_t + av_x = \mu v_{xx} + \nu v_{xxx} + R_4(v) + \mathcal{O}(\Delta t^3 + h^3), \tag{13.53}$$

having set

$$\nu = -\frac{ah^2}{6}(1 - 3a\lambda + 2(a\lambda)^2), \tag{13.54}$$

and having indicated with $R_4(v)$ the set of terms containing the derivatives of v of order at least four.

We can conclude that the function v satisfies, respectively, the equations:

$$v_t + av_x = 0 \tag{13.55}$$

if we neglect the terms containing derivatives of order above the first;

$$v_t + av_x = \mu v_{xx} \tag{13.56}$$

if we neglect the terms containing derivatives of order above the second;

$$v_t + av_x = \mu v_{xx} + \nu v_{xxx} \tag{13.57}$$

if we neglect the derivatives of order above the third. The coefficients μ and ν are in (13.49) and (13.54). Equations (13.55), (13.56) and (13.57) are called *equivalent equations* (at the first, second resp. third order) relative to the upwind scheme.

13.5.2 The Lax-Friedrichs and Lax-Wendroff case

Proceeding in a similar way, we can derive the equivalent equations of any numerical scheme. For instance, in the case of the Lax-Friedrichs scheme, having denoted by v a hypothetic function that verifies equation (13.16) at each point (x,t), having observed that

$$\frac{1}{2}\big(v(x+h,t)+v(x-h,t)\big) = v+\frac{h^2}{2}v_{xx}+\mathcal{O}(h^4),$$

$$\frac{1}{2}\big(v(x+h,t)-v(x-h,t)\big) = h\,v_x+\frac{h^3}{6}v_{xxx}+\mathcal{O}(h^4),$$

we obtain

$$v_t + av_x = R^{LF} + \mathcal{O}\Big(\frac{h^4}{\Delta t}+\Delta t^3\Big), \tag{13.58}$$

having set

$$R^{LF} = \frac{h^2}{2\Delta t}(v_{xx}-\lambda^2 v_{tt}) - \frac{ah^2}{6}(v_{xxx}+\frac{\lambda^2}{a}v_{ttt}).$$

Proceeding as we did previously, tedious computation allows us to deduce from (13.58) the equivalent equations (13.55)–(13.57), in this case having

$$\mu = \frac{h^2}{2\Delta t}(1-(a\lambda)^2), \quad \nu = \frac{ah^2}{3}(1-(a\lambda)^2).$$

In the case of the *Lax-Wendroff* scheme, the equivalent equations are characterized by the following parameters

$$\mu = 0, \qquad \nu = \frac{ah^2}{6}((a\lambda)^2-1).$$

13.5.3 On the meaning of coefficients in equivalent equations

In general, in the equivalent equations the term μv_{xx} represents a dissipation, while νv_{xxx} represents a dispersion. We can provide a heuristic proof of this by examining the solution to the problem

$$\begin{cases} v_t + av_x = \mu v_{xx} + \nu v_{xxx}, & x \in \mathbb{R},\ t>0, \\ v(x,0) = e^{ikx}, & k \in \mathbb{Z}. \end{cases} \tag{13.59}$$

By applying the Fourier transform we find, if $\mu = \nu = 0$,

$$v(x,t) = e^{ik(x-at)},$$

while for μ and ν arbitrary real numbers (with $\mu > 0$) we have

$$v(x,t) = e^{-\mu k^2 t} e^{ik[x-(a+\nu k^2)t]}.$$

Comparing these two relations, we see that for growing μ the modulus of the solution gets smaller. Such effect becomes more remarkable as the frequency k increases (a phenomenon we have already registered in the previous section, albeit with partly different arguments).

The term μv_{xx} in (13.59) therefore has a dissipative effect on the solution. In turn, ν modifies the propagation rate of the solution, increasing it in the $\nu > 0$ case, and decreasing it if $\nu < 0$. Also in this case, the effect is more notable at high frequencies. Hence, the third derivative term νv_{xxx} introduces a dispersive effect.

In general, in the equivalent equation, even-order spatial derivatives represent diffusive terms, while odd-order derivatives represent dispersive terms. For first-order schemes (such as the upwind scheme) the dispersive effect is often barely visible, as it is disguised by the dissipative one. Taking Δt and h of the same order, from (13.56) and (13.57) we evince that $\nu \ll \mu$ for $h \to 0$, as $\nu = O(h^2)$ and $\mu = O(h)$. In particular, if the CFL number is $\frac{1}{2}$, the third-order equivalent equation of the upwind method features a null dispersion, in accordance with the numerical results seen in the previous section.

Conversely, the dispersive effect is evident for the Lax-Friedrichs scheme, as well as for the Lax-Wendroff scheme which, being of second order, does not feature a dissipative term of type μv_{xx}. However, being stable, the latter cannot avoid being dissipative. Indeed, the fourth-order equivalent equation for the Lax-Wendroff scheme is

$$v_t + a v_x = \frac{ah^2}{6}[(a\lambda)^2 - 1]v_{xxx} - \frac{ah^3}{6} a\lambda[1-(a\lambda)^2]v_{xxxx},$$

where the last term is dissipative if $|a\lambda| < 1$, as one can easily verify by applying the Fourier transform. We then recover, also for the Lax-Wendroff scheme, the CFL condition.

13.5.4 Equivalent equations and error analysis

The technique applied to obtain the equivalent equation denotes a strong analogy with the so-called *backward analysis* that we encounter during the numerical solution of linear systems, where the computed (not exact) solution is interpreted as the exact solution of a perturbed linear system (see [QSS07, Chap. 3]). As a matter of fact, the perturbed system plays a similar role to that of the equivalent equation.

Moreover, we observe that an error analysis of a numerical scheme can be carried out by using the equivalent equation associated to it. Indeed, by generically denoting by $r = \mu v_{xx} + \nu v_{xxx}$ the right-hand side of the equivalent equation, by comparison with (13.1) we obtain the error equation

$$e_t + a e_x = r,$$

where $e = v - u$. Multiplying such equation by e and integrating in space and time (between 0 and t) we obtain

$$\|e(t)\|_{L^2(\mathbb{R})} \leq C(t)\left(\|e(0)\|_{L^2(\mathbb{R})} + \sqrt{\int_0^t \|r(s)\|^2_{L^2(\mathbb{R})} ds}\right), \quad t > 0$$

having used the a priori estimate (13.4). We can assume $e(0) = 0$ and therefore observe that $\|e(t)\|_{L^2(\mathbb{R})}$ tends to zero (for h and Δt tending to zero) with order 1 for the upwind or Lax-Friedrichs schemes, and with order 2 for the Lax-Wendroff scheme (having supposed v to be sufficiently regular).

13.6 Exercises

1. Verify that the solution to the problem (13.9)–(13.10) (with $f = 0$) satisfies identity (13.11).
   [*Solution:* Multiplying (13.9) by u_t and integrating in space we obtain

$$0 = \int_\alpha^\beta u_{tt} u_t \, dx - \int_\alpha^\beta \gamma^2 u_{xx} u_t \, dx = \frac{1}{2} \int_\alpha^\beta [(u_t)^2]_t \, dx + \int_\alpha^\beta \gamma^2 u_x u_{xt} \, dx - [\gamma^2 u_x u_t]_\alpha^\beta. \tag{13.60}$$

   As

$$\int_\alpha^\beta u_{tt} u_t \, dx = \frac{1}{2} \int_\alpha^\beta [(u_t)^2]_t \, dx \quad \text{and} \quad \int_\alpha^\beta \gamma^2 u_x u_{xt} \, dx = \frac{1}{2} \int_\alpha^\beta \gamma^2 [(u_x)^2]_t \, dx,$$

   integrating (13.60) in time we have

$$\int_\alpha^\beta u_t^2(t) \, dx + \int_\alpha^\beta \gamma^2 u_x^2(t) \, dx - \int_\alpha^\beta v_0^2 \, dx - \int_\alpha^\beta u_{0x}^2 \, dx = 0. \tag{13.61}$$

   Hence (13.11) immediately follows from the latter relation.]
2. Verify that the solution provided by the backward/centered Euler scheme (13.22) is unconditionally stable; more precisely,

$$\|\mathbf{u}\|_{\Delta,2} \leq \|\mathbf{u}^0\|_{\Delta,2} \quad \forall \Delta t, h > 0.$$

   [*Solution:* Note that, thanks to (13.32),

$$(u_j^{n+1} - u_j^n) u_j^{n+1} \geq \frac{1}{2} \left(|u_j^{n+1}|^2 - |u_j^n|^2 \right) \quad \forall j, n.$$

   Then, multiplying (13.22) by u_j^{n+1}, summing over the index j and using (13.33) we find

$$\sum_j |u_j^{n+1}|^2 \leq \sum_j |u_j^n|^2 \quad \forall n \geq 0,$$

   from which the result follows.]

3. Prove (13.30)
[*Solution:* We note that, in the case where $a > 0$, the upwind scheme can be rewritten in the form

$$u_j^{n+1} = (1 - a\lambda)u_j^n + a\lambda u_{j-1}^n.$$

Under hypothesis (13.28) both coefficients $a\lambda$ and $1 - a\lambda$ are non-negative, hence

$$\min(u_j^n, u_{j-1}^n) \leq u_j^{n+1} \leq \max(u_j^n, u_{j-1}^n).$$

Then

$$\inf_{l\in\mathbb{Z}}\{u_l^0\} \leq u_j^n \leq \sup_{l\in\mathbb{Z}}\{u_l^0\} \quad \forall j \in \mathbb{Z},\ \forall n \geq 0,$$

from which (13.30) follows.]
4. Study the accuracy of the Lax-Friedrichs scheme (13.16) for the solution of problem (13.1).
5. Study the accuracy of the Lax-Wendroff scheme (13.18) for the solution of problem (13.1).

14

Finite elements and spectral methods for hyperbolic equations

In this chapter we will illustrate how to apply Galerkin methods, and in particular the finite element method and the spectral one, to the spatial and/or temporal discretization of scalar hyperbolic equations. We will treat both continuous as well as discontinuous finite elements.

Let us consider the transport problem (13.3) and let us set for simplicity $(\alpha,\beta) = (0,1)$, $\varphi = 0$. Moreover, let us suppose that a is a positive constant and a_0 a non-negative constant.

To start with, we proceed with a spatial discretization based on continuous finite elements. We therefore attempt a semidiscretization of the following form:

$\forall t > 0$, find $u_h = u_h(t) \in V_h$ such that

$$\left(\frac{\partial u_h}{\partial t}, v_h\right) + a\left(\frac{\partial u_h}{\partial x}, v_h\right) + a_0\,(u_h, v_h) = (f, v_h)\ \forall v_h \in V_h, \qquad (14.1)$$

u_h^0 being the approximation of the initial datum. We have set

$$V_h = \{v_h \in X_h^r :\ v_h(0) = 0\}, \quad r \geq 1.$$

The space X_h^r is defined as in (4.14), provided that we replace (a,b) with $(0,1)$.

14.1 Temporal discretization

For the temporal discretization of problem (14.1) we will use finite difference schemes such as those introduced in Chapter 13.

As usual, we will denote by u_h^n, $n \geq 0$, the approximation of u_h at time $t^n = n\Delta t$.

14.1.1 The forward and backward Euler schemes

In case we use the forward Euler scheme, the discrete problem becomes:

A. Quarteroni: *Numerical Models for Differential Problems*, 2nd Ed.
MS&A – Modeling, Simulation & Applications 8
DOI 10.1007/978-88-470-5522-3_14, © Springer-Verlag Italia 2014

$\forall n \geq 0$, find $u_h^{n+1} \in V_h$ such that

$$\frac{1}{\Delta t}\left(u_h^{n+1}-u_h^n, v_h\right)+a\left(\frac{\partial u_h^n}{\partial x}, v_h\right)+a_0\left(u_h^n, v_h\right)=\left(f^n, v_h\right) \quad \forall v_h \in V_h, \tag{14.2}$$

where $(u,v)=\int_0^1 u(x)v(x)dx$ denotes as usual the scalar product of $\mathrm{L}^2(0,1)$.

In the case of the backward Euler method, instead of (14.2) we will have

$$\frac{1}{\Delta t}\left(u_h^{n+1}-u_h^n, v_h\right)+a\left(\frac{\partial u_h^{n+1}}{\partial x}, v_h\right)+a_0\left(u_h^{n+1}, v_h\right)=\left(f^{n+1}, v_h\right) \forall v_h \in V_h. \tag{14.3}$$

Theorem 14.1. *The backward Euler scheme is strongly stable with no restriction on Δt. Instead, the forward Euler method is strongly stable only for $a_0>0$, provided we suppose that*

$$\Delta t \leq \frac{2a_0}{(aCh^{-1}+a_0)^2} \tag{14.4}$$

for a given constant $C=C(r)$.

Proof. Choosing $v_h=u_h^n$ in (14.2), we obtain (in the case $f=0$)

$$\left(u_h^{n+1}-u_h^n, u_h^n\right)+\Delta t a\left(\frac{\partial u_h^n}{\partial x}, u_h^n\right)+\Delta t a_0\|u_h^n\|^2_{\mathrm{L}^2(0,1)}=0.$$

For the first term, we use the identity

$$(v-w,w)=\frac{1}{2}\left(\|v\|^2_{\mathrm{L}^2(0,1)}-\|w\|^2_{\mathrm{L}^2(0,1)}-\|v-w\|^2_{\mathrm{L}^2(0,1)}\right) \forall v,w \in \mathrm{L}^2(0,1) \tag{14.5}$$

which generalizes (13.34). For the second summand, integrating by parts and using the boundary conditions, we find

$$\left(\frac{\partial u_h^n}{\partial x}, u_h^n\right)=\frac{1}{2}(u_h^n(1))^2.$$

Thus, we obtain

$$\|u_h^{n+1}\|^2_{\mathrm{L}^2(0,1)}+a\Delta t(u_h^n(1))^2+2a_0\Delta t\|u_h^n\|^2_{\mathrm{L}^2(0,1)}=\|u_h^n\|^2_{\mathrm{L}^2(0,1)}+\|u_h^{n+1}-u_h^n\|^2_{\mathrm{L}^2(0,1)}. \tag{14.6}$$

We now seek an estimate for the term $\|u_h^{n+1}-u_h^n\|^2_{\mathrm{L}^2(0,1)}$. To this end, setting in (14.2)

$v_h = u_h^{n+1} - u_h^n$, we obtain

$$\|u_h^{n+1} - u_h^n\|^2_{L^2(0,1)} \leq \Delta t a \left| \left(\frac{\partial u_h^n}{\partial x}, u_h^{n+1} - u_h^n \right) \right| + \Delta t a_0 \left| (u_h^n, u_h^{n+1} - u_h^n) \right|$$

$$\leq \Delta t \left[a \|\frac{\partial u_h^n}{\partial x}\|_{L^2(0,1)} + a_0 \|u_h^n\|_{L^2(0,1)} \right] \|u_h^{n+1} - u_h^n\|_{L^2(0,1)}.$$

By now using the inverse inequality (12.44) (referred to the interval $(0,1)$), we obtain

$$\|u_h^{n+1} - u_h^n\|_{L^2(0,1)} \leq \Delta t \left(a C_I h^{-1} + a_0 \right) \|u_h^n\|_{L^2(0,1)}.$$

Finally, (14.6) becomes

$$\begin{aligned} &\|u_h^{n+1}\|^2_{L^2(0,1)} + a\Delta t (u_h^n(1))^2 \\ &\quad + \Delta t \left[2a_0 - \Delta t (a C_I h^{-1} + a_0)^2 \right] \|u_h^n\|^2_{L^2(0,1)} \leq \|u_h^n\|^2_{L^2(0,1)}. \end{aligned} \tag{14.7}$$

If (14.4) is satisfied, then $\|u_h^{n+1}\|_{L^2(0,1)} \leq \|u_h^n\|_{L^2(0,1)}$ and we therefore have strong stability in $L^2(0,1)$ norm.

In the case where $a_0 = 0$ the obtained stability condition is meaningless. However, if we suppose that

$$\Delta t \leq \frac{K h^2}{a^2 C_I^2},$$

for a given constant $K > 0$, then we can apply the discrete Gronwall lemma (see Lemma 2.3) to (14.7) and we find that the method is stable, with a stability constant which in this case depends on the final time T. Precisely,

$$\|u_h^n\|_{L^2(0,1)} \leq \exp(K t^n) \|u_h^0\|_{L^2(0,1)} \leq \exp(KT) \|u_h^0\|_{L^2(0,1)}.$$

In the case of the backward Euler method (14.3), we choose instead $v_h = u_h^{n+1}$. By using the relation

$$(v - w, v) = \frac{1}{2} \left(\|v\|^2_{L^2(0,1)} - \|w\|^2_{L^2(0,1)} + \|v - w\|^2_{L^2(0,1)} \right) \quad \forall v, w \in L^2(0,1) \tag{14.8}$$

which generalizes (13.32), we find

$$(1 + 2a_0 \Delta t) \|u_h^{n+1}\|^2_{L^2(0,1)} + a \Delta t (u_h^{n+1}(1))^2 \leq \|u_h^n\|^2_{L^2(0,1)}. \tag{14.9}$$

Hence we have strong stability in $L^2(0,1)$, unconditioned (that is for each Δt) and for each $a_0 \geq 0$. ◇

14.1.2 The upwind, Lax-Friedrichs and Lax-Wendroff schemes

The generalization to the finite elements case of the Lax-Friedrichs (LF), Lax-Wendroff (LW) and upwind (U) finite difference schemes can be attained in different ways.

We start by observing that (13.16), (13.18), and (13.20) can be rewritten in the following comprehensive form

$$\frac{u_j^{n+1}-u_j^n}{\Delta t}+a\frac{u_{j+1}^n-u_{j-1}^n}{2h}-\mu\frac{u_{j+1}^n-2u_j^n+u_{j-1}^n}{h^2}+a_0u_j^n=0. \tag{14.10}$$

(Note however that $a_0 = 0$ in (13.16), (13.18) and (13.20).) The second term is the discretization via centered finite differences of the convective term $au_x(t^n)$, while the third one is a numerical diffusion term and corresponds to the discretization via finite differences of $-\mu u_{xx}(t^n)$. The numerical viscosity coefficient μ is given by

$$\mu=\begin{cases} h^2/2\Delta t & \text{(LF)},\\ a^2\Delta t/2 & \text{(LW)},\\ ah/2 & \text{(U)}. \end{cases} \tag{14.11}$$

Equation (14.10) suggests the following finite element version for the approximation of problem (13.3): $\forall n\geq 0$, find $u_h^{n+1}\in V_h$ such that

$$\begin{aligned} &\frac{1}{\Delta t}\left(u_h^{n+1}-u_h^n,v_h\right)+a\left(\frac{\partial u_h^n}{\partial x},v_h\right)+a_0\left(u_h^n,v_h\right)\\ &+\mu\left(\frac{\partial u_h^n}{\partial x},\frac{\partial v_h}{\partial x}\right)-\mu\gamma\frac{\partial u_h^n}{\partial x}(1)v_h(1)=(f^n,v_h)\quad\forall v_h\in V_h, \end{aligned} \tag{14.12}$$

where $\gamma = 1,0$ depending on whether or not we want to take the boundary contribution into account when integrating by parts the numerical viscosity term.

For the stability analysis, in the case $\gamma=0$, $a_0=0$, $a>0$, let us set $v_h=u_h^{n+1}-u_h^n$, in order to obtain

$$\|u_h^{n+1}-u_h^n\|_{\mathrm{L}^2(0,1)}\leq \Delta t(a+\mu C_I h^{-1})\|\frac{\partial u_h^n}{\partial x}\|_{\mathrm{L}^2(0,1)},$$

thanks to inequality (4.52). Having now set $v_h=u_h^n$, thanks to (14.5) we obtain

$$\begin{aligned} \|u_h^{n+1}\|^2_{\mathrm{L}^2(0,1)}-\|u_h^n\|^2_{\mathrm{L}^2(0,1)}&+a\Delta t(u_h^n(1))^2+2\Delta t\mu\|\frac{\partial u_h^n}{\partial x}\|^2_{\mathrm{L}^2(0,1)}\\ &=\|u_h^{n+1}-u_h^n\|^2_{\mathrm{L}^2(0,1)}\leq \Delta t^2(a+\mu C_I h^{-1})^2\|\frac{\partial u_h^n}{\partial x}\|^2_{\mathrm{L}^2(0,1)}. \end{aligned}$$

A sufficient condition for strong stability (i.e. to obtain an estimate such as (13.27), with respect to $\|\cdot\|_{L^2(0,1)}$) is therefore

$$\Delta t\leq\frac{2\mu}{(a+\mu C_I h^{-1})^2}.$$

Thanks to (14.11), in the case of the upwind method this is equivalent to

$$\Delta t\leq\frac{h}{a}\left(\frac{1}{1+C_I/2}\right)^2.$$

In the case of linear finite elements, $C_I \simeq 2\sqrt{3}$, so we deduce that

$$\frac{a\Delta t}{h} \lesssim \left(\frac{1}{1+\sqrt{3}}\right)^2.$$

The stability analysis we have just developed is based on the *energy method* and, in this case, leads to sub-optimal results. A better indicator can be obtained by resorting to the von Neumann analysis, as we saw in Sect. 13.4.3. To this end we observe that, in the case of linear finite elements with constant spacing h, (14.12) with $f = 0$ can be rewritten in the following way for each internal node x_j:

$$\begin{aligned}\frac{1}{6}(u_{j+1}^{n+1} + 4u_j^{n+1} + u_{j-1}^{n+1}) + \frac{\lambda a}{2}(u_{j+1}^n - u_{j-1}^n) + \frac{a_0}{6}\Delta t(u_{j+1}^n + 4u_j^n + u_{j-1}^n)\\ -\mu\Delta t\frac{u_{j+1}^n - 2u_j^n + u_{j-1}^n}{h^2} = \frac{1}{6}(u_{j+1}^n + 4u_j^n + u_{j-1}^n).\end{aligned} \tag{14.13}$$

By comparing such relation to (14.10), we can note that the difference only resides in the term arising from the temporal derivative and from the term of order zero, and has to be attributed to the presence of the mass matrix in the case of finite elements. On the other hand, we saw in Sect. 12.5 that we can apply the mass-lumping technique to approximate the mass matrix using a diagonal matrix. By proceeding in this way, scheme (14.13) can effectively be reduced to (14.10) (see Exercise 1).

Remark 14.1. Note that relations (14.13) refer to the internal nodes. The approach used to handle boundary conditions with the finite element method generally yields different relations than those obtained via the finite difference method. •

These observations allow us to extend all the schemes seen in Sect. 13.3.1 to analogous schemes, generated by discretizations in space with continuous linear finite elements. To this end, it will be sufficient to replace the term $u_j^{n+1} - u_j^n$ with

$$\frac{1}{6}[(u_{j-1}^{n+1} - u_{j-1}^n) + 4(u_j^{n+1} - u_j^n) + (u_{j+1}^{n+1} - u_{j+1}^n)].$$

Thus, the general scheme (13.13) is replaced by

$$\frac{1}{6}(u_{j-1}^{n+1} + 4u_j^{n+1} + u_{j-1}^{n+1}) = \frac{1}{6}(u_{j-1}^n + 4u_j^n + u_{j-1}^n) - \lambda(H_{j+1/2}^{n*} - H_{j-1/2}^{n*}), \tag{14.14}$$

where

$$H_{j+1/2}^{n*} = \begin{cases} H_{j+1/2}^n & \text{for explicit time-advancing schemes,} \\ H_{j+1/2}^{n+1} & \text{for implicit time-advancing schemes.} \end{cases}$$

Note that, even if we had adopted a numerical flux corresponding to an explicit time-advancing scheme, the resulting scheme would no longer lead to a diagonal system (indeed, it becomes a tridiagonal one) because of the mass matrix terms. The use of an explicit time-advancing scheme for finite elements might seem inconvenient with respect to a similar full finite difference scheme. However, such a scheme has interesting features. In particular, let us consider its amplification and dispersion coefficients, using the von Neumann analysis illustrated in Sect. 13.4.3. To this end, let us suppose that the differential equation is defined on all of $\mathbb{R}$, or, alternatively, let us consider a bounded interval and assume periodic boundary conditions at its endpoints. In either case, we can assume that relation (14.14) holds for all values of the index j. A simple computation leads us to writing the following relation between the amplification coefficient γ_k of a finite difference scheme (see Table 13.1) and the amplification coefficient γ_k^{FEM} of the corresponding finite element scheme

$$\gamma_k^{\mathrm{FEM}} = \frac{3\gamma_k - 1 + \cos(\phi_k)}{2 + \cos(\phi_k)}, \tag{14.15}$$

where we denote again with ϕ_k the phase angle relative to the k-th harmonic (see Sect. 13.4.3).

We can thus compute the amplification and dispersion errors, which are reported in Fig. 14.1. Comparing them with the analogous errors relating to the corresponding finite difference scheme (reported in Fig. 13.6) we can make the following remarks. The forward Euler scheme is still unconditionally unstable (in the sense of strong stability). The upwind scheme (FEM) is strongly stable if the CFL number is less than $\frac{1}{3}$ (hence, a less restrictive result than the one found using the energy method), while the Lax-Friedrichs (FEM) method *never satisfies* the condition $\gamma_k^{FEM} \leq 1$ (in accordance with the result that we would find using the energy method in this case).
More generally, we can say that in the case of schemes with an explicit temporal treatment, the "finite element" version requires more restrictive stability conditions than the corresponding finite difference one. In particular, for the Lax-Wendroff finite element scheme, that we will denote with LW (FEM), the CFL number must now be less than $\frac{1}{\sqrt{3}}$, instead of 1 as in the finite differences case. However, the LW (FEM) scheme (for the CFL values for which it is stable), is slightly less diffusive and dispersive than the equivalent finite difference scheme, for a wide range of values of the phase angle $\phi_k = kh$. The implicit Euler scheme remains unconditionally stable also in the FEM version (coherently with what we obtained using the energy method in Sect. 14.1.1).

Example 14.1. The previous conclusions have been experimentally verified as follows. We have repeated the case of Fig. 13.7 (right), where we have now considered a CFL value of 0.5. The numerical solutions obtained via the classical Lax-Wendroff method (LW) and via LW (FEM) for $t = 2$ are reported in Fig. 14.2. We can note how the LW (FEM) scheme provides a solution that is more accurate and, especially, featuring a smaller phase error. This result is confirmed by the value of the $\|\cdot\|_{\Delta,2}$ norm of the error in the two cases. Indeed, by calling u the exact solution and u_{LW} resp. $u_{LW(FEM)}$ the one obtained using the two numerical schemes, $\|u_{LW} - u\|_{\Delta,2} = 0.78$, $\|u_{LW(FEM)} - u\|_{\Delta,2} = 0.49$.
Further tests conducted with non-periodic boundary conditions confirm the stability properties previously derived. ■

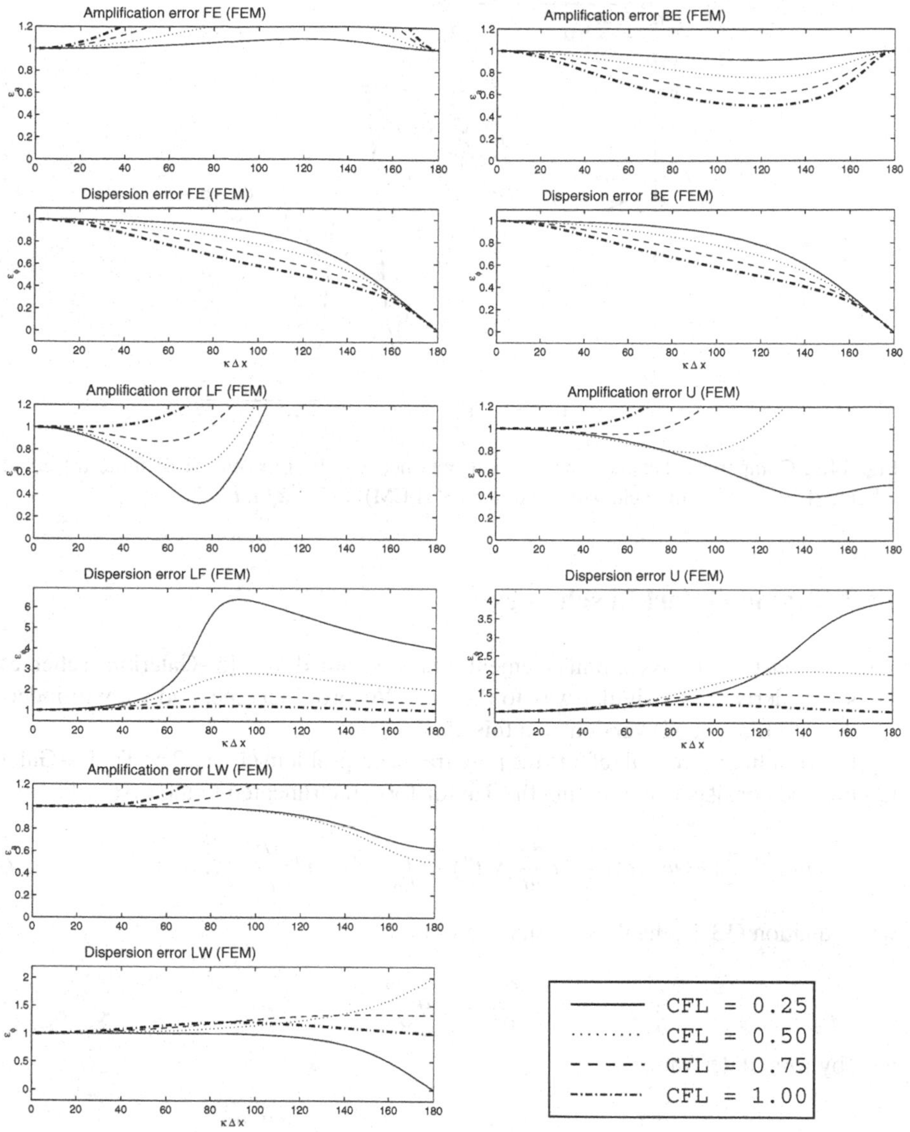

Fig. 14.1. Amplification and dispersion errors for several finite element schemes obtained from the general scheme (14.14)

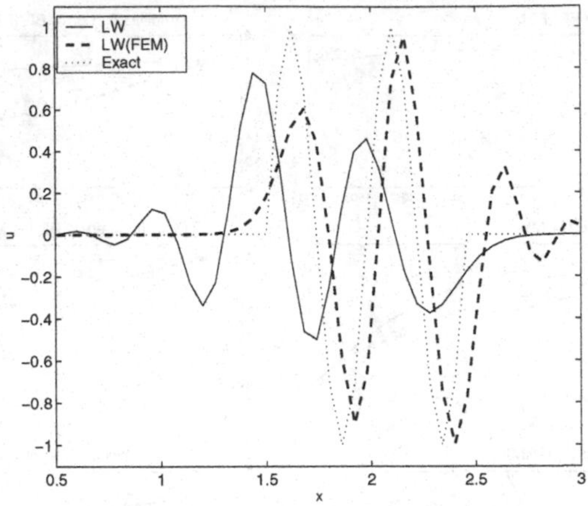

Fig. 14.2. Comparison between the solution obtained via the Lax-Wendroff finite difference scheme (LW) and its finite element version (LW (FEM)) ($\phi_k = \pi/4, t = 2$)

14.2 Taylor-Galerkin schemes

We now illustrate a class of finite element schemes named "Taylor-Galerkin" schemes. These are derived in a similar way to the Lax-Wendroff scheme, and we will indeed see that the LW (FEM) version is in this class.

For simplicity, we will refer to the pure transport problem (13.1). The Taylor-Galerkin method consists in combining the Taylor formula truncated to the first order

$$u(x,t^{n+1}) = u(x,t^n) + \Delta t \frac{\partial u}{\partial t}(x,t^n) + \int_{t^n}^{t^{n+1}} (s-t^n) \frac{\partial^2 u}{\partial t^2}(x,s)\,ds \tag{14.16}$$

with equation (13.1), thanks to which we obtain

$$\frac{\partial u}{\partial t} = -a \frac{\partial u}{\partial x}$$

and, by formal derivation,

$$\frac{\partial^2 u}{\partial t^2} = \frac{\partial}{\partial t}(-a \frac{\partial u}{\partial x}) = -a \frac{\partial}{\partial x}\frac{\partial u}{\partial t} = a^2 \frac{\partial^2 u}{\partial x^2}.$$

From (14.16) we then obtain

$$u(x,t^{n+1}) = u(x,t^n) - a\Delta t \frac{\partial u}{\partial x}(x,t^n) + a^2 \int_{t^n}^{t^{n+1}} (s-t^n) \frac{\partial^2 u}{\partial x^2}(x,s)\,ds. \tag{14.17}$$

We approximate the integral in the following way

$$\int_{t^n}^{t^{n+1}} (s-t^n) \frac{\partial^2 u}{\partial x^2}(x,s)\,ds \approx \frac{\Delta t^2}{2}\left[\theta \frac{\partial^2 u}{\partial x^2}(x,t^n) + (1-\theta)\frac{\partial^2 u}{\partial x^2}(x,t^{n+1})\right], \tag{14.18}$$

obtained by evaluating the first factor at $s = t^n + \frac{\Delta t}{2}$ and the second one through a linear combination using $\theta \in [0,1]$ as a parameter of its values in $s = t^n$ and $s = t^{n+1}$. We denote by $u^n(x)$ the approximating function $u(x,t^n)$.

Let us consider two remarkable situations. If $\theta = 1$, the resulting semi-discretized scheme is explicit in time and is written as

$$u^{n+1} = u^n - a\Delta t \frac{\partial u^n}{\partial x} + \frac{a^2\Delta t^2}{2}\frac{\partial^2 u^n}{\partial x^2}.$$

If we now discretize in space by finite differences or finite elements, we re-encounter the previously examined LW and LW (FEM) schemes.
Instead, if we take $\theta = \frac{2}{3}$, the approximation error in (14.18) becomes $O(\Delta t^4)$ (supposing that u has the required regularity). De facto, such choice corresponds to approximating $\frac{\partial^2 u}{\partial x^2}$ between t^n and t^{n+1} with its linear interpolant. The resulting semi-discretized scheme is written

$$\left[1 - \frac{a^2\Delta t^2}{6}\frac{\partial^2}{\partial x^2}\right] u^{n+1} = u^n - a\Delta t \frac{\partial u^n}{\partial x} + \frac{a^2\Delta t^2}{3}\frac{\partial^2 u^n}{\partial x^2} \tag{14.19}$$

and the truncation error of the semi-discretized scheme in time (14.19) is $\mathcal{O}(\Delta t^3)$.

At this point, a discretization in space using the finite element method leads to the following scheme, called Taylor-Galerkin (TG):

for $n = 0, 1, \ldots$ find $u_h^{n+1} \in V_h$ such that

$$\begin{aligned} A(u_h^{n+1}, v_h) = (u_h^n, v_h) - a\Delta t \left(\frac{\partial u_h^n}{\partial x}, v_h\right) - \frac{a^2\Delta t^2}{3}\left(\frac{\partial u_h^n}{\partial x}, \frac{\partial v_h}{\partial x}\right) \\ + \gamma \frac{a^2\Delta t^2}{3}\frac{\partial u_h^n}{\partial x}(1) v_h(1) \quad \forall v_h \in V_h, \end{aligned} \tag{14.20}$$

where

$$A(u_h^{n+1}, v_h) = (u_h^{n+1}, v_h) + \frac{a^2\Delta t^2}{6}\left(\frac{\partial u_h^{n+1}}{\partial x}, \frac{\partial v_h}{\partial x}\right) - \gamma \frac{a^2\Delta t^2}{6}\frac{\partial u_h^{n+1}}{\partial x}(1) v_h(1),$$

and $\gamma = 1, 0$ depending on whether or not we want to take into account the boundary contribution when integrating by parts the second derivative.

The latter yields a linear system whose matrix is

$$\mathrm{A} = \mathrm{M} + \frac{a^2(\Delta t)^2}{6}\mathrm{K},$$

M being the mass matrix and K being the stiffness matrix, possibly taking the boundary contribution as well into account (if $\gamma = 1$).

In the case of linear finite elements, the von Neumann analysis leads to the following amplification factor for scheme (14.20)

$$\gamma_k = \frac{2 + \cos(kh) - 2a^2\lambda^2(1 - \cos(kh)) + 3ia\lambda \sin(kh)}{2 + \cos(kh) + a^2\lambda^2(1 - \cos(kh))}. \tag{14.21}$$

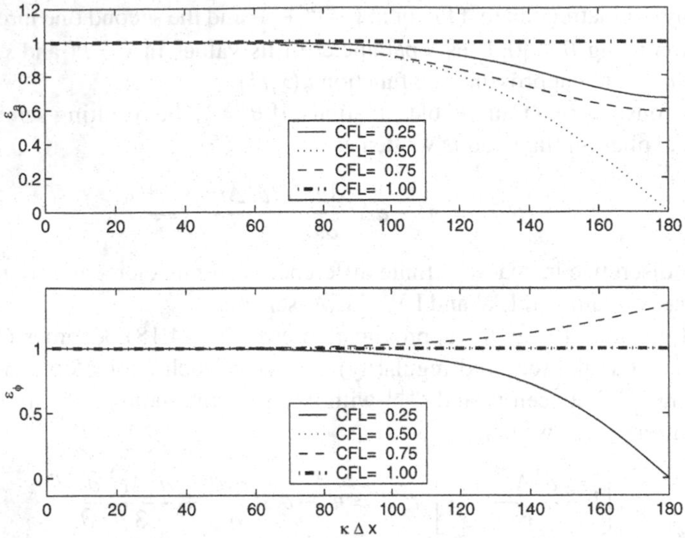

Fig. 14.3. Amplification (top) and dispersion (bottom) error of the Taylor-Galerkin scheme (14.20), as a function of the phase angle $\phi_k = kh$ and for different values of the CFL number

It can be proven that the scheme is strongly stable in $\|\cdot\|_{\Delta,2}$ under the CFL condition $\frac{a\Delta t}{h} \leq 1$. Thus, it has a *less restrictive* stability condition than the Lax-Wendroff (FEM) scheme.
Fig. 14.3 shows the behaviour of the amplification and dispersion error for the scheme (14.20), as a function of the phase angle, in analogy to what we have seen for other schemes in Sect. 13.4.4.

In the case of linear finite elements the truncation error of the TG scheme is $\mathcal{O}(\Delta t^3) + \mathcal{O}(h^2) + \mathcal{O}(h^2 \Delta t)$.

Example 14.2. To compare the accuracy of the schemes presented in the last two sections, we have considered the problem

$$\begin{cases} \dfrac{\partial u}{\partial t} + \dfrac{\partial u}{\partial x} = 0, & x \in (0,0.5), t > 0, \\ u(x,0) = u_0(x), & x \in (0,0.5), \end{cases}$$

with periodic boundary conditions, $u(0,t) = u(0.5,t)$, for $t > 0$. The initial datum is $u_0(x) = 2\cos(4\pi x) + \sin(20\pi x)$, and is illustrated in Fig. 14.4 (left). The latter superposes two harmonics, one with low frequency one and one with high frequency.

We have considered the Taylor-Galerkin, Lax-Wendroff (FEM), (finite difference) Lax-Wendroff and upwind schemes. In Fig. 14.4 (right) we show the error in discrete norm $\|u - u_h\|_{\Delta,2}$ obtained at time $t = 1$ for different values of Δt and with a fixed CFL number of 0.55. We can note a better convergence of the Taylor-Galerkin scheme, while the two versions of the Lax-Wendroff scheme show the same order of conver-

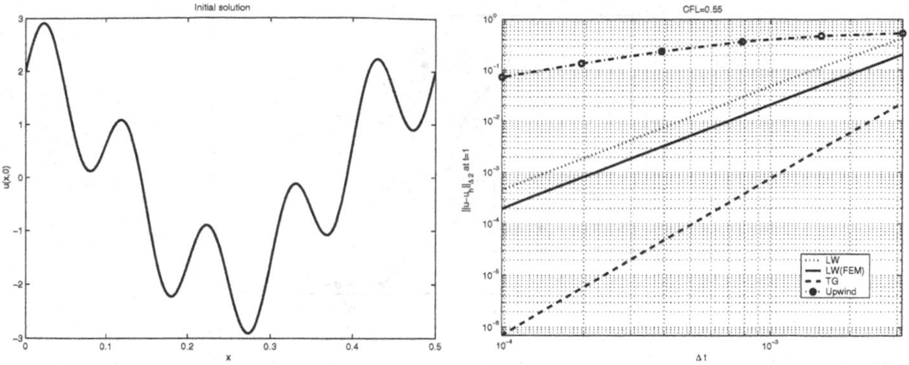

Fig. 14.4. Initial condition u_0 for the simulation of example 14.2 (left) and error $\|u - u_h\|_{\Delta,2}$ at $t = 1$ for varying Δt and fixed CFL for different numerical schemes (right)

gence, but with a smaller error for the finite element version. The upwind scheme is less accurate: it features a larger absolute error and a lower convergence rate. Moreover, it can be verified that for a fixed CFL, the error of the upwind scheme is $\mathcal{O}(\Delta t)$, that of both variants of the Lax-Wendroff scheme is $\mathcal{O}(\Delta t^2)$, while the error of the Taylor-Galerkin scheme is $\mathcal{O}(\Delta t^3)$. ∎

We report in Figs. 14.5 and 14.6 the numerical approximations and corresponding errors in maximum norm for the transport problem

$$\begin{cases} \dfrac{\partial u}{\partial t} - \dfrac{\partial u}{\partial x} = 0, & x \in (0, 2\pi), t > 0 \\ u(x,0) = \sin\big(\pi \cos(x)\big), & x \in (0, 2\pi) \end{cases}$$

having periodic boundary conditions. Such approximations are obtained using finite differences of order 2 and 4 (ufd2, ufd4), compact finite differences of order 4 and 6

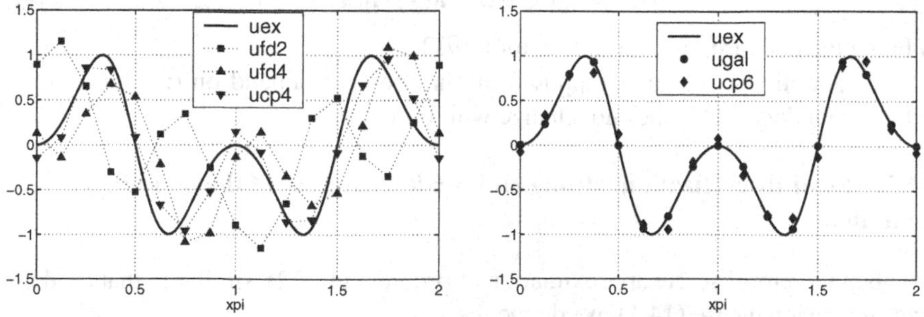

Fig. 14.5. Approximation of the solution of a wave propagation problem using finite difference methods (of order 2, 4), compact finite difference methods (of order 4 and 6) and with the Fourier Galerkin spectral method (from [CHQZ06])

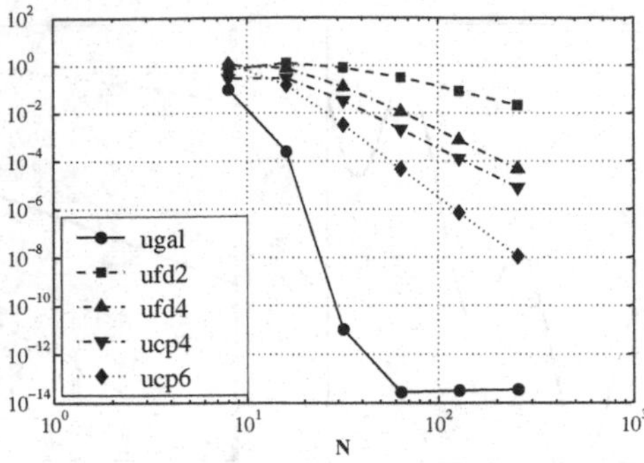

Fig. 14.6. Behaviour of the error in maximum norm for the different numerical methods reported in Fig. 14.5 (from [CHQZ06])

(ucp4, ucp6), and by the Galerkin spectral method with Fourier basis (ugal). For the sake of comparison, we also report the exact solution $u(x,t) = \sin\left(\pi \cos(x+t)\right)$ (uex).

14.3 The multi-dimensional case

Let us now move to the multi-dimensional case and consider the following first-order, linear and scalar hyperbolic transport-reaction problem in the domain $\Omega \subset \mathbb{R}^d$, with $d = 2,3$

$$\begin{cases} \dfrac{\partial u}{\partial t} + \mathbf{a} \cdot \nabla u + a_0 u = f, & \mathbf{x} \in \Omega,\ t > 0 \\ u = \varphi, & \mathbf{x} \in \partial\Omega^{in},\ t > 0, \\ u_{|t=0} = u_0, & \mathbf{x} \in \Omega, \end{cases} \tag{14.22}$$

where $\mathbf{a} = \mathbf{a}(\mathbf{x})$, $a_0 = a_0(\mathbf{x},t)$ (possibly zero), $f = f(\mathbf{x},t)$, $\varphi = \varphi(\mathbf{x},t)$ and $u_0 = u_0(\mathbf{x})$ are given functions. The inflow boundary $\partial\Omega^{in}$ is defined by

$$\partial\Omega^{in} = \{\mathbf{x} \in \partial\Omega :\ \mathbf{a}(\mathbf{x}) \cdot \mathbf{n}(\mathbf{x}) < 0\}, \tag{14.23}$$

$\mathbf{n}$ being the outward unit normal vector to $\partial\Omega$.

For simplicity, we have supposed that $\mathbf{a}$ does not depend on t; in this way, the inflow boundary $\partial\Omega^{in}$ does not change with time.

14.3.1 Semi-discretization: strong and weak treatment of the boundary conditions

To obtain a semi-discrete approximation of problem (14.22), similar to that used in the one-dimensional case (14.1), we define the spaces

$$V_h = X_h^r, \qquad V_h^{in} = \{v_h \in V_h :\ v_{h|\partial\Omega^{in}} = 0\},$$

where r is an integer ≥ 1 and X_h^r was introduced in (4.38).

We denote by $u_{0,h}$ and φ_h two suitable finite element approximations of u_0 and φ, respectively, and we consider the problem: for each $t > 0$ find $u_h(t) \in V_h$ such that

$$\begin{cases} \int\limits_{\Omega} \frac{\partial u_h(t)}{\partial t} v_h \, d\Omega + \int\limits_{\Omega} \mathbf{a} \cdot \nabla u_h(t) v_h \, d\Omega + \int\limits_{\Omega} a_0(t) u_h(t) v_h \, d\Omega \\ \qquad = \int\limits_{\Omega} f(t) v_h \, d\Omega \qquad \forall\, v_h \in V_h^{in}, \\ u_h(t) = \varphi_h(t) \quad \text{on } \partial\Omega^{in}, \end{cases} \tag{14.24}$$

with $u_h(0) = u_{0,h} \in V_h$.

To obtain a stability estimate, we assume for simplicity that φ, and therefore φ_h, is identically null. In this case $u_h(t) \in V_h^{in}$, and taking, for every t, $v_h = u_h(t)$, we have the following inequality

$$\begin{aligned} \|u_h(t)\|^2_{L^2(\Omega)} &+ \int_0^t \mu_0 \|u_h(\tau)\|^2_{L^2(\Omega)} \, d\tau + \int_0^t \int\limits_{\partial\Omega \setminus \partial\Omega^{in}} \mathbf{a} \cdot \mathbf{n} \, u_h^2(\tau) \, d\gamma \, d\tau \\ &\leq \|u_{0,h}\|^2_{L^2(\Omega)} + \int_0^t \frac{1}{\mu_0} \|f(\tau)\|^2_{L^2(\Omega)} d\tau \, . \end{aligned} \tag{14.25}$$

We have assumed that there exists a positive constant μ_0 such that, for all $t > 0$ and for each $\mathbf{x}$ in Ω

$$0 < \mu_0 \leq \mu(\mathbf{x}, t) = a_0(\mathbf{x}, t) - \frac{1}{2} \mathrm{div}\mathbf{a}(\mathbf{x}). \tag{14.26}$$

In the case where such hypothesis is not verified (for instance if $\mathbf{a}$ is a constant field and $a_0 = 0$), then by using the Gronwall Lemma 2.2 we obtain

$$\begin{aligned} \|u_h(t)\|^2_{L^2(\Omega)} &+ \int_0^t \int_{\partial\Omega \setminus \partial\Omega^{in}} \mathbf{a} \cdot \mathbf{n} u_h^2(\tau) d\gamma d\tau \\ &\leq \left(\|u_{0,h}\|^2_{L^2(\Omega)} + \int_0^t \|f(\tau)\|^2_{L^2(\Omega)} \, d\tau \right) \exp \int_0^t [1 + 2\mu^*(\tau)] \, d\tau, \end{aligned} \tag{14.27}$$

where we have set $\mu^*(t) = \max_{\mathbf{x} \in \bar{\Omega}} |\mu(\mathbf{x}, t)|$.

Supposing for simplicity that $f = 0$, if $u_0 \in H^{r+1}(\Omega)$ we have the following convergence result

$$\begin{aligned} \max_{t \in [0,T]} \|u(t) - u_h(t)\|_{L^2(\Omega)} &+ \left(\int_0^T \int_{\partial\Omega} |\mathbf{a} \cdot \mathbf{n}| \, |u(t) - u_h(t)|^2 \, d\gamma \, dt \right)^{1/2} \\ &\leq C h^r \|u_0\|_{H^{r+1}(\Omega)}. \end{aligned}$$

For the proofs, we refer to [QV94, Chap. 14], [Joh87] and to the references cited therein. In problem (14.24) the boundary condition has been imposed in a *strong* (or essential) way. An alternative option is the *weak* (or natural) treatment that derives from the integration by parts of the transport term in the first equation in (14.24), where we now consider $v_h \in V_h$ (i.e. we no longer require that the test function is null on the inflow boundary). We obtain

$$\int_{\Omega} \frac{\partial u_h(t)}{\partial t} v_h \, d\Omega - \int_{\Omega} \mathrm{div}(\mathbf{a} v_h)) u_h(t) \, d\Omega + \int_{\Omega} a_0 u_h(t) v_h \, d\Omega + \int_{\partial\Omega} \mathbf{a}\cdot\mathbf{n} u_h(t) v_h \, d\gamma = \int_{\Omega} f(t) v_h \, d\Omega.$$

The boundary condition is imposed by replacing u_h with φ_h on the inflow boundary part, obtaining

$$\int_{\Omega} \frac{\partial u_h(t)}{\partial t} v_h \, d\Omega - \int_{\Omega} \mathrm{div}(\mathbf{a} v_h) u_h(t) \, d\Omega + \int_{\Omega} a_0 u_h(t) v_h \, d\Omega + \int_{\partial\Omega\setminus\partial\Omega^{in}} \mathbf{a}\cdot\mathbf{n} u_h(t) v_h \, d\gamma$$
$$= \int_{\Omega} f(t) v_h \, d\Omega - \int_{\partial\Omega^{in}} \mathbf{a}\cdot\mathbf{n}\, \varphi_h(t) v_h \, d\gamma \;\; \forall v_h \in V_h. \tag{14.28}$$

Clearly, the solution u_h found in this way only satisfies the boundary condition in an approximate way.

A further option consists in counter-integrating (14.28) by parts, thus producing the following formulation: for each $t > 0$, find $u_h(t) \in V_h$ such that

$$\int_{\Omega} \frac{\partial u_h(t)}{\partial t} v_h \, d\Omega + \int_{\Omega} \mathbf{a}\cdot\nabla u_h(t) v_h \, d\Omega + \int_{\Omega} a_0 u_h(t) v_h \, d\Omega + \int_{\partial\Omega^{in}} v_h(\varphi_h(t) - u_h(t)) \mathbf{a}\cdot\mathbf{n} \, d\gamma$$
$$= \int_{\Omega} f(t) v_h \, d\Omega \;\; \forall v_h \in V_h. \tag{14.29}$$

We note that the formulations (14.28) and (14.29) are equivalent: the only difference is the way boundary terms are highlighted. In particular, the boundary integral in formulation (14.29) can be interpreted as a penalization term with which we evaluate how different u_h is from the data φ_h on the inflow boundary. Assuming that hypothesis (14.26) is still true, having chosen $v_h = u_h(t)$ in (14.29), integrating the convective term by parts and using the Cauchy-Schwarz and Young inequalities, we get the following stability estimate

$$\|u_h(t)\|^2_{\mathrm{L}^2(\Omega)} + \int_0^t \mu_0 \|u_h(\tau)\|^2_{\mathrm{L}^2(\Omega)} d\tau + \int_0^t \int_{\partial\Omega\setminus\partial\Omega^{in}} \mathbf{a}\cdot\mathbf{n} u_h^2(\tau) \, d\gamma \, d\tau$$
$$\leq \|u_{0,h}\|^2_{\mathrm{L}^2(\Omega)} + \int_0^t \int_{\partial\Omega^{in}} |\mathbf{a}\cdot\mathbf{n}| \varphi_h^2(\tau) d\gamma d\tau + \int_0^t \frac{1}{\mu_0} \|f(\tau)\|^2_{\mathrm{L}^2(\Omega)} d\tau. \tag{14.30}$$

In absence of hypothesis (14.26), inequality (14.30) would change in an analogous way to what we have previously seen, provided we use the Gronwall Lemma 2.2 as we did to derive (14.27).

Remark 14.2. In the case where the boundary condition for problem (14.22) takes the form $\mathbf{a}\cdot\mathbf{n}u = \psi$, we could again impose it weakly by adding a penalization term, that in such case would take the form

$$\int_{\partial\Omega^{in}} (\psi_h(t) - \mathbf{a}\cdot\mathbf{n}u_h(t))v_h\, d\gamma,$$

ψ_h being a suitable finite element approximation of the datum ψ. •

Alternatively to the strong and weak imposition of the boundary conditions, i.e. to formulations (14.24) and (14.29), we could adopt a Petrov-Galerkin approach by imposing in a strong way the condition $u_h(t) = \varphi_h(t)$ on the inflow boundary $\partial\Omega^{in}$, and requiring $v_h = 0$ on the outflow boundary $\partial\Omega^{out}$, yielding the following discrete formulation. Set $V_h^{out} = \{v_h \in V_h : v_{h|\partial\Omega^{out}} = 0\}$. Then for each $t > 0$ find $u_h(t) \in V_h = X_h^r$ such that

$$\begin{cases} \displaystyle\int_{\Omega} \frac{\partial u_h(t)}{\partial t} v_h\, d\Omega + \int_{\Omega} (\mathbf{a}\cdot\nabla u_h(t))\, v_h\, d\Omega + \int_{\Omega} a_0(t)\, u_h(t)\, v_h\, d\Omega \\ \qquad\qquad\qquad\qquad\qquad\qquad \displaystyle = \int_{\Omega} f(t)\, v_h\, d\Omega \quad \forall v_h \in V_h^{out}, \\ u_h(t) = \varphi_h(t) \quad \text{on } \partial\Omega^{in}. \end{cases}$$

We recall that for a Petrov-Galerkin formulation, the well-posedness analysis cannot be based on the Lax-Milgram lemma any longer.

Instead, if the inflow condition were inposed in a weak way, we would have the following formulation:
for each $t > 0$, find $u_h(t) \in V_h = X_h^r$ such that, for each $v_h \in V_h^{out}$,

$$\int_{\Omega} \frac{\partial u_h(t)}{\partial t} v_h\, d\Omega - \int_{\Omega} \mathrm{div}(\mathbf{a}v_h)\, u_h(t)\, d\Omega + \int_{\Omega} a_0(t)\, u_h(t)\, v_h\, d\Omega$$
$$= \int_{\Omega} f(t)\, v_h\, d\Omega - \int_{\partial\Omega^{in}} \mathbf{a}\cdot\mathbf{n}\,\varphi_h(t)\, v_h\, d\gamma.$$

For further details, the reader can refer to [QV94, Chap. 14].

14.3.2 Temporal discretization

For an illustrative purpose, let us limit ourselves to considering the Galerkin semi-discrete problem (14.24). Using the backward Euler scheme for the temporal discretization, we obtain the following fully discrete problem:

$\forall n \geq 0$ find $u_h^n \in V_h$ such that

$$\begin{cases} \frac{1}{\Delta t} \int\limits_{\Omega} (u_h^{n+1} - u_h^n) v_h \, d\Omega + \int\limits_{\Omega} \mathbf{a} \cdot \nabla u_h^{n+1} v_h \, d\Omega + \int\limits_{\Omega} a_0^{n+1} u_h^{n+1} v_h \, d\Omega \\ \qquad = \int\limits_{\Omega} f^{n+1} v_h \, d\Omega \quad \forall v_h \in V_h^{in}, \\ u_h^{n+1} = \varphi_h^{n+1} \text{ on } \partial\Omega^{in}, \end{cases}$$

with $u_h^0 = u_{0,h} \in V_h$ being a suitable approximation in V_h of the initial datum u_0.

Let us limit ourselves to the homogeneous case, where $f = 0$ and $\varphi_h = 0$ (in this case $u_h^n \in V_h^{in}$ for every $n \geq 0$). Having set $v_h = u_h^{n+1}$ and using identities (14.8) and (14.26), we obtain, for each $n \geq 0$

$$\frac{1}{2\Delta t} \left(\|u_h^{n+1}\|^2_{L^2(\Omega)} - \|u_h^n\|^2_{L^2(\Omega)} \right) + \frac{1}{2} \int_{\partial\Omega \setminus \partial\Omega^{in}} \mathbf{a} \cdot \mathbf{n} (u_h^{n+1})^2 \, d\gamma + \mu_0 \|u_h^{n+1}\|^2_{L^2(\Omega)} \leq 0.$$

For each $m \geq 1$, summing over n from 0 to $m-1$ we obtain

$$\|u_h^m\|^2_{L^2(\Omega)} + 2\Delta t \left(\mu_0 \sum_{n=0}^{m} \|u_h^n\|^2_{L^2(\Omega)} + \frac{1}{2} \sum_{n=0}^{m} \int_{\partial\Omega \setminus \partial\Omega^{in}} \mathbf{a} \cdot \mathbf{n} (u_h^n)^2 \, d\gamma \right) \leq \|u_{0,h}\|^2_{L^2(\Omega)}.$$

In particular, as $\mathbf{a} \cdot \mathbf{n} \geq 0$ on $\partial\Omega \setminus \partial\Omega^{in}$, we conclude that

$$\|u_h^m\|_{L^2(\Omega)} \leq \|u_{0,h}\|_{L^2(\Omega)} \quad \forall m \geq 0.$$

As expected, this method is strongly stable, with no condition on Δt and h.

We now consider the discretization in time using the forward Euler method

$$\begin{cases} \frac{1}{\Delta t} \int\limits_{\Omega} (u_h^{n+1} - u_h^n) v_h \, d\Omega + \int\limits_{\Omega} \mathbf{a} \cdot \nabla u_h^n v_h \, d\Omega + \int\limits_{\Omega} a_0^n u_h^n v_h \, d\Omega \\ \qquad = \int\limits_{\Omega} f^n v_h \, d\Omega \qquad \forall v_h \in V_h, \\ u_h^{n+1} = \varphi_h^{n+1} \quad \text{on } \partial\Omega^{in}. \end{cases} \tag{14.31}$$

We suppose again that $f = 0$, $\varphi = 0$ and that the condition (14.26) is verified. Moreover, we suppose that $\|\mathbf{a}\|_{L^\infty(\Omega)} < \infty$ and that, for each $t > 0$, $\|a_0\|_{L^\infty(\Omega)} < \infty$.

Setting $v_h = u_h^n$, exploiting identity (14.5) and integrating the convective term by parts, we obtain

$$\frac{1}{2\Delta t}\left(\|u_h^{n+1}\|^2_{L^2(\Omega)} - \|u_h^n\|^2_{L^2(\Omega)} - \|u_h^{n+1} - u_h^n\|^2_{L^2(\Omega)}\right)$$
$$+\int_{\partial\Omega\setminus\partial\Omega^{in}} \mathbf{a}\cdot\mathbf{n}(u_h^n)^2\, d\gamma + (-\frac{1}{2}\,\mathrm{div}(\mathbf{a}) + a_0^n, (u_h^n)^2) = 0,$$

and then, after a few steps,

$$\|u_h^{n+1}\|^2_{L^2(\Omega)} + 2\Delta t\int_{\partial\Omega\setminus\partial\Omega^{in}} \mathbf{a}\cdot\mathbf{n}(u_h^n)^2\, d\gamma + 2\Delta t\mu_0\|u_h^n\|^2_{L^2(\Omega)}$$
$$\leq \|u_h^n\|^2_{L^2(\Omega)} + \|u_h^{n+1} - u_h^n\|^2_{L^2(\Omega)}. \qquad (14.32)$$

It is now necessary to control the term $\|u_h^{n+1} - u_h^n\|^2_{L^2(\Omega)}$. To this end, we set $v_h = u_h^{n+1} - u_h^n$ in (14.31) and obtain

$$\|u_h^{n+1} - u_h^n\|^2_{L^2(\Omega)} = -\Delta t(\mathbf{a}\nabla u_h^n, u_h^{n+1} - u_h^n) - \Delta t(a_0^n u_h^n, u_h^{n+1} - u_h^n)$$
$$\leq \Delta t\|\mathbf{a}\|_{L^\infty(\Omega)}|(\nabla u_h^n, u_h^{n+1} - u_h^n)| + \Delta t\|a_0^n\|_{L^\infty(\Omega)}|(u_h^n, u_h^{n+1} - u_h^n)|$$
$$\leq \Delta t\|\mathbf{a}\|_{L^\infty(\Omega)}\|\nabla u_h^n\|_{L^2(\Omega)}\|u_h^{n+1} - u_h^n\|_{L^2(\Omega)} +$$
$$\Delta t\|a_0^n\|_{L^\infty(\Omega)}\|u_h^n\|_{L^2(\Omega)}\|u_h^{n+1} - u_h^n\|_{L^2(\Omega)}.$$

Using the inverse inequality (4.52), we obtain

$$\|u_h^{n+1} - u_h^n\|^2_{L^2(\Omega)} \leq \Delta t(C_I h^{-1}\|\mathbf{a}\|_{L^\infty(\Omega)} +$$
$$\|a_0^n\|_{L^\infty(\Omega)})\|u_h^n\|_{L^2(\Omega)}\|u_h^{n+1} - u_h^n\|_{L^2(\Omega)},$$

and then

$$\|u_h^{n+1} - u_h^n\|_{L^2(\Omega)} \leq \Delta t\left(C_I h^{-1}\|\mathbf{a}\|_{L^\infty(\Omega)} + \|a_0^n\|_{L^\infty(\Omega)}\right)\|u_h^n\|_{L^2(\Omega)}.$$

Using such results to find an upper bound for the term in (14.32), we have

$$\|u_h^{n+1}\|^2_{L^2(\Omega)} + 2\Delta t\int_{\partial\Omega\setminus\partial\Omega^{in}} \mathbf{a}\cdot\mathbf{n}(u_h^n)^2\, d\Omega +$$
$$\Delta t\left[2\mu_0 - \Delta t\ \left(C_I h^{-1}\|\mathbf{a}\|_{L^\infty(\Omega)} + \|a_0^n\|_{L^\infty(\Omega)}\right)^2\right]\|u_h^n\|^2_{L^2(\Omega)}$$
$$\leq \|u_h^n\|^2_{L^2(\Omega)}.$$

The integral on $\partial\Omega\setminus\partial\Omega^{in}$ is positive because of the hypotheses on the boundary conditions; hence, if

$$\Delta t \leq \frac{2\mu_0}{\left(C_I h^{-1}\|\mathbf{a}\|_{L^\infty(\Omega)} + \|a_0^n\|_{L^\infty(\Omega)}\right)^2} \qquad (14.33)$$

we have $\|u_h^{n+1}\|_{L^2(\Omega)} \leq \|u_h^n\|_{L^2(\Omega)}$, that is the scheme is strongly stable. Note that the stability condition (14.33) is of parabolic type, similar to the one found in (13.31) for the case of finite difference discretizations.

Remark 14.3. In the case where **a** is constant and $a_0 = 0$ we have that $\mu_0 = 0$, and the stability condition (14.33) can never be satisfied by a positive Δt. Thus, the result in (14.33) does not contradict the one we have previously found for the forward Euler scheme. •

14.4 Discontinuous finite elements

An alternative approach to the one adopted so far is based on the use of *discontinuous* finite elements. The resulting method is called the discontinuous Galerkin method (DG in short). This choice is motivated by the fact that, as we previously observed, the solutions of (even linear) hyperbolic problems can be discontinuous.

For a given mesh $\mathcal{T}_h$ of Ω, the space of discontinuous finite elements is

$$W_h = Y_h^r = \{v_h \in L^2(\Omega) \mid v_{h|K} \in \mathbb{P}_r, \ \forall K \in \mathcal{T}_h\}, \tag{14.34}$$

that is the space of piecewise polynomial functions of degree less than or equal to r, with $r \geq 0$, which are not necessarily continuous across the finite element interfaces.

14.4.1 The one-dimensional upwind DG method

In the case of the one-dimensional problem (13.3) in its simplest *upwind* version, the DG finite element method takes the following form: $\forall t > 0$, find a function $u_h = u_h(t) \in W_h$ such that

$$\begin{aligned}
&\int_\alpha^\beta \frac{\partial u_h(t)}{\partial t} v_h \, dx \\
&\quad + \sum_{i=0}^{m-1} \left[\int_{x_i}^{x_{i+1}} \left(a \frac{\partial u_h(t)}{\partial x} + a_0 u_h(t) \right) v_h \, dx + a(x_i)(u_h^+(t) - U_h^-(t))(x_i) v_h^+(x_i) \right] \\
&\qquad = \int_\alpha^\beta f(t) v_h \, dx \quad \forall v_h \in W_h,
\end{aligned} \tag{14.35}$$

where we have supposed that $a(x)$ is a continuous function. We have set, for each $t > 0$,

$$U_h^-(t)(x_i) = \begin{cases} u_h^-(t)(x_i), & i = 1, \ldots, m-1, \\ \varphi_h(t)(x_0), \end{cases} \tag{14.36}$$

where $\{x_i, \, i = 0, \cdots, m\}$ are the nodes, $x_0 = \alpha$, $x_m = \beta$, h is the maximal distance between two consecutive nodes, $v_h^+(x_i)$ denotes the right limit of v_h at x_i, $v_h^-(x_i)$ the left one. For simplicity of notation, the dependence of u_h and f on t will often be understood when this does not yield to ambiguities.

We now derive a stability estimate for the solution u_h of (14.35), supposing, for simplicity, that the forcing term f is identically null. Having then chosen $v_h = u_h$ in (14.35), we have (setting $\Omega = (\alpha, \beta)$)

$$\frac{1}{2}\frac{d}{dt}\|u_h\|^2_{L^2(\Omega)} + \sum_{i=0}^{m-1}\left[\int_{x_i}^{x_{i+1}} \left(\frac{a}{2}\frac{\partial}{\partial x}(u_h)^2 + a_0 u_h^2\right) dx + a(x_i)(u_h^+ - U_h^-)(x_i)u_h^+(x_i)\right] = 0.$$

Now, integrating the convective term by parts, we have

$$\frac{1}{2}\frac{d}{dt}\|u_h\|^2_{L^2(\Omega)} + \sum_{i=0}^{m-1}\int_{x_i}^{x_{i+1}} \left(a_0 - \frac{\partial}{\partial x}\left(\frac{a}{2}\right)\right) u_h^2 \, dx +$$
$$\sum_{i=0}^{m-1}\left[\frac{a}{2}(x_{i+1})\,(u_h^-(x_{i+1}))^2 + \frac{a}{2}(x_i)\,(u_h^+(x_i))^2 - a(x_i)\,U_h^-(x_i)\,u_h^+(x_i)\right] = 0. \quad (14.37)$$

Isolating the contribution associated to node x_0 and exploiting definition (14.36), we can rewrite the second sum in the previous equation as

$$\sum_{i=0}^{m-1}\left[\frac{a}{2}(x_{i+1})\,(u_h^-(x_{i+1}))^2 + \frac{a}{2}(x_i)\,(u_h^+(x_i))^2 - a(x_i)\,U_h^-(x_i)\,u_h^+(x_i)\right]$$
$$= \frac{a}{2}(x_0)\,(u_h^+(x_0))^2 - a(x_0)\,\varphi_h(x_0)\,u_h^+(x_0) + \frac{a}{2}(x_m)\,(u_h^-(x_m))^2 +$$
$$\sum_{i=1}^{m-1}\left[\frac{a}{2}(x_i)\left((u_h^-(x_i))^2 + (u_h^+(x_i))^2\right) - a(x_i)\,u_h^-(x_i)\,u_h^+(x_i)\right] \quad (14.38)$$
$$= \frac{a}{2}(x_0)\,(u_h^+(x_0))^2 - a(x_0)\,\varphi_h(x_0)\,u_h^+(\alpha) +$$
$$\frac{a}{2}(x_m)\,(u_h^-(x_m))^2 + \sum_{i=1}^{m-1}\frac{a}{2}(x_i)\,\left[u_h(x_i)\right]^2,$$

having denoted by $\left[u_h(x_i)\right] = u_h^+(x_i) - u_h^-(x_i)$ the jump of function u_h at node x_i. We now suppose, analogously to the multi-dimensional case (see (14.26)), that

$$\exists \gamma \geq 0 \text{ suchthat } a_0 - \frac{\partial}{\partial x}\left(\frac{a}{2}\right) \geq \gamma. \quad (14.39)$$

Returning to (14.37) and using the relation (14.38) and the Cauchy-Schwarz and Young inequalities, we have

$$\frac{1}{2}\frac{d}{dt}\|u_h\|^2_{L^2(\Omega)} + \gamma\|u_h\|^2_{L^2(\Omega)} + \sum_{i=1}^{m-1}\frac{a}{2}(x_i)\left[u_h(x_i)\right]^2 + \frac{a}{2}(x_0)\left(u_h^+(x_0)\right)^2 +$$
$$\frac{a}{2}(x_m)\left(u_h^-(x_m)\right)^2 = a(x_0)\,\varphi_h(x_0)\,u_h^+(x_0) \le \frac{a}{2}(x_0)\,\varphi_h^2(x_0) + \frac{a}{2}(x_0)\left(u_h^+(x_0)\right)^2,$$

that is, integrating with respect to time as well, $\forall t > 0$,

$$\begin{aligned}\|u_h(t)\|^2_{L^2(\Omega)} + 2\gamma\int_0^t \|u_h(t)\|^2_{L^2(\Omega)}\,dt + \sum_{i=1}^{m-1} a(x_i)\int_0^t \left[u_h(x_i,t)\right]^2 dt\, + \\ a(x_m)\left(u_h^-(x_m)\right)^2 \le \|u_{0,h}\|^2_{L^2(\Omega)} + a(x_0)\int_0^t \varphi_h^2(x_0,t)\,dt.\end{aligned} \qquad (14.40)$$

Such estimate represents the desired stability result.
Note that, in case the forcing term is no longer null, we can replicate the previous analysis by suitably using the Gronwall Lemma 2.2 to handle the contribution of f. This would lead to an estimate similar to (14.40), however this time the right-hand side of the inequality would become

$$e^t\left(\|u_{0,h}\|^2_{L^2(\Omega)} + a(x_0)\int_0^t \varphi_h^2(x_0,t)\,dt + \int_0^t (f(\tau))^2\,d\tau\right). \qquad (14.41)$$

In the case where the constant γ in inequality (14.39) is strictly positive, we could avoid using the Gronwall lemma, and attain an estimate such as (14.40) where in the first term 2γ is replaced by γ, while the second term takes the form (14.41) without the exponential e^t.

Because of the discontinuity of test functions, (14.35) can be rewritten in an equivalent way as follows, $\forall i = 0,\dots,m-1$,

$$\begin{aligned}\int_{x_i}^{x_{i+1}}\left(\frac{\partial u_h}{\partial t} + a\frac{\partial u_h}{\partial x} + a_0 u_h\right) v_h dx + a(u_h^+ - U_h^-)(x_i)v_h^+(x_i) \\ = \int_{x_i}^{x_{i+1}} f v_h dx \quad \forall v_h \in \mathbb{P}_r(I_i),\end{aligned} \qquad (14.42)$$

with $I_i = [x_i, x_{i+1}]$. In other terms, the approximation via discontinuous finite elements yields to element-wise "independent" relations; the only term connecting an element and its neighbours is the jump term $(u_h^+ - U_h^-)$ that can also be interpreted as the attribution of the boundary datum on the inflow boundary of the element under exam.

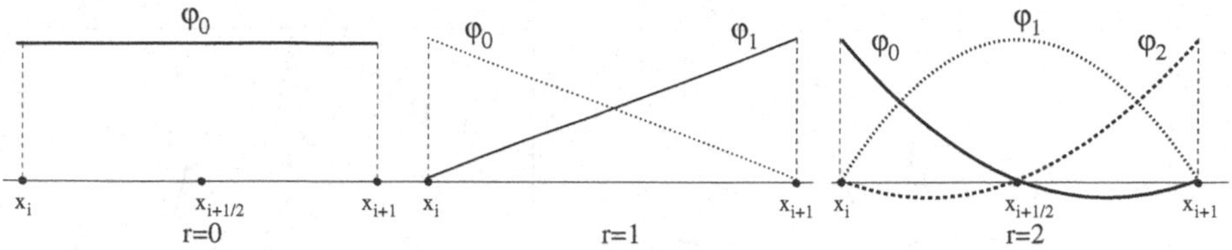

Fig. 14.7. The Lagrange bases for $r=0$, $r=1$ and $r=2$

We then have a set of small problems to be solved in each element, precisely $r+1$ equations for each interval $[x_i, x_{i+1}]$. Let us write them in compact form as

$$M_h \dot{\mathbf{u}}_h(t) + L_h \mathbf{u}_h(t) = \mathbf{f}_h(t) \qquad \forall t > 0, \quad \mathbf{u}_h(0) = \mathbf{u}_{0,h}, \tag{14.43}$$

M_h being the mass matrix, L_h the matrix associated to the bilinear form and to the jump relation, $\mathbf{f}_h$ the source term:

$$(M_h)_{pq} = \int_{x_i}^{x_{i+1}} \varphi_p \varphi_q \, dx, \quad (L_h)_{pq} = \int_{x_i}^{x_{i+1}} (a\varphi_{q,x} + a_0 \varphi_q)\, \varphi_p \, dx + (a\varphi_q \varphi_p)(x_i),$$

$$(\mathbf{f}_h)_p = \int_{x_i}^{x_{i+1}} f \varphi_p \, dx + aU_h^-(x_i)\varphi_p(x_i), \quad p,q = 0,\ldots,r.$$

We have denoted by $\{\varphi_q,\ q=0,\ldots,r\}$ a basis for $\mathbb{P}_r([x_i,x_{i+1}])$ and by $\mathbf{u}_h(t)$ the coefficients of $u_h(x,t)|_{[x_i,x_{i+1}]}$ in the basis $\{\varphi_q\}$. If we take the Lagrange basis we will have, for instance, the functions reported in Fig. 14.7 (for $r=0$, $r=1$ and $r=2$) and the values of $\{\mathbf{u}_h(t)\}$ are the ones taken by $u_h(t)$ at nodes ($x_{i+1/2}$ for $r=0$, x_i and x_{i+1} for $r=1$, x_i, $x_{i+1/2}$ and x_{i+1} for $r=2$). Note that all previous functions are identically null outside the interval $[x_i,x_{i+1}]$. Moreover, in the case of discontinuous finite elements it is perfectly acceptable to use polynomials of degree $r=0$, in which case the transport term $a\frac{\partial u_h}{\partial x}$ will provide a null contribution on each element.

With the aim of diagonalizing the mass matrix, it can be interesting to use as a basis for $\mathbb{P}_r([x_i,x_{i+1}])$ the Legendre polynomials $\varphi_q(x) = L_q(2(x-x_i)/h_i)$, $h_i = x_{i+1}-x_i$. The family $\{L_q, q=0,1,\ldots\}$ are the orthogonal Legendre polynomials defined over the interval $[-1,1]$, that we have introduced in Sect. 10.2.2. Indeed, in such a way we obtain $(M_h)_{pq} = \frac{h_i}{2p+1}\delta_{pq}$, $p,q=0,\ldots r$. Obviously, in this case the unknown values $\{\mathbf{u}_h(t)\}$ cannot be interpreted as nodal values of $u_h(t)$, but rather as the Legendre coefficients of the expansion of $u_h(t)$ in the new basis.

The diagonalization of the mass matrix turns out to be particularly interesting when we use explicit time advancing schemes (such as, e.g., second- and third-order Runge-Kutta schemes, introduced in Chapter 15). In this case, indeed, we will have a fully explicit problem to solve on each small interval.

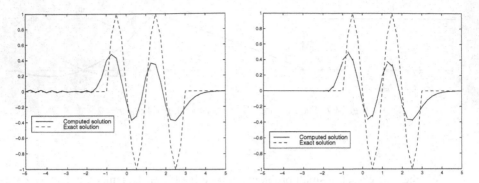

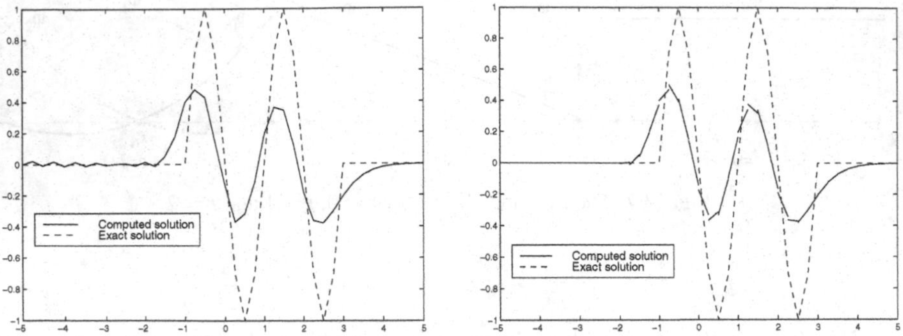

Fig. 14.8. Solution at time $t = 1$ of problem (14.44) with $\phi_k = \pi/2$, $h = 0.25$, obtained using continuous (left) and discontinuous (right) linear finite elements and backward Euler time discretization

For illustrative purposes, we present below some numerical results obtained for problem

$$\begin{cases} \dfrac{\partial u}{\partial t} + \dfrac{\partial u}{\partial x} = 0, & x \in (-5,5),\ t > 0, \\ u(-5,t) = 0, & t > 0, \end{cases} \tag{14.44}$$

using the initial condition

$$u(x,0) = \begin{cases} \sin(\pi x), & x \in (-2,2), \\ 0 & \text{otherwise}. \end{cases} \tag{14.45}$$

The problem has been discretized using linear finite elements in space, both continuous and discontinuous. For the temporal discretization, we have used the backward Euler scheme in both cases. We have chosen $h = 0.25$ and a time step $\Delta t = h$; for such value of h the phase number associated to the sinusoidal wave is $\phi_k = \pi/2$.

In Fig. 14.8 we report the numerical solution at time $t = 1$ together with the corresponding exact solution. The scheme has strong numerical diffusion, but also small oscillations towards the end in the case of continuous elements. Furthermore, we can observe that the numerical solution obtained using discontinuous finite elements, although being discontinuous, no longer features an oscillatory behaviour towards the end.

Let us now consider the following problem

$$\begin{cases} \dfrac{\partial u}{\partial t} + \dfrac{\partial u}{\partial x} = 0, & x \in (0,1),\ t > 0, \\ u(0,t) = 1, & t > 0, \\ u(x,0) = 0, & x \in [0,1], \end{cases} \tag{14.46}$$

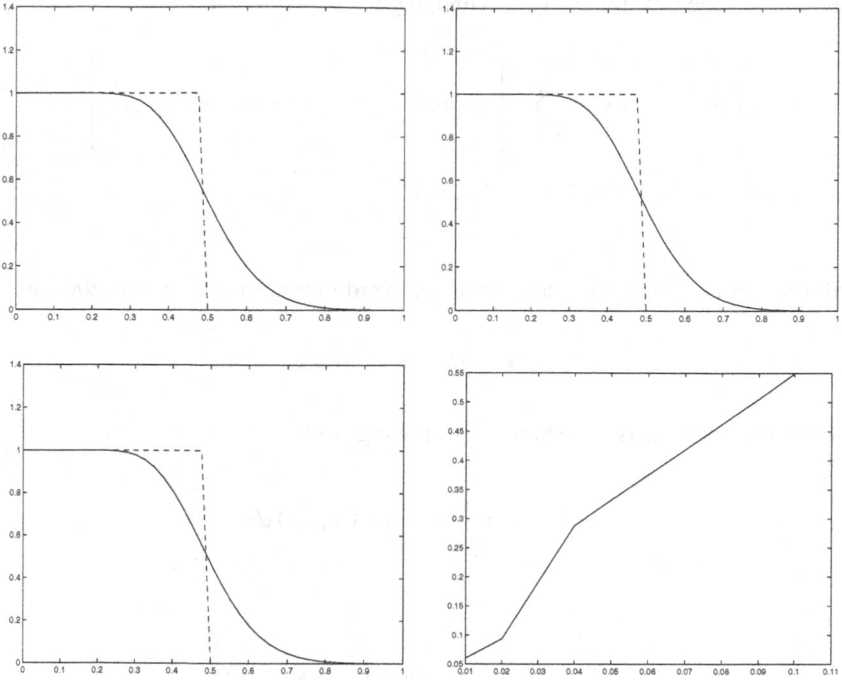

Fig. 14.9. Solution to problem (14.46) for $t = 0.5$ with $h = 0.025$ obtained using continuous linear finite elements and strong (top left) and weak (top right) treatment of the boundary Dirichlet condition, while discontinuous elements in space have been used (bottom left). Finally, we show (bottom left) the behavior of $|u_h(0) - u(0)|$ as a function of h for $t = 0.1$, in weak treatment of the Dirichlet condition

which represents the transport of a discontinuity entering the domain. We have considered continuous linear finite elements, with both strong and weak treatment of the boundary conditions, as well as discontinuous linear finite elements. This time, as well, we have used the backward Euler method for the temporal discretization. The grid-size is $h = 0.025$ and the time step is $\Delta t = h$.

The results at time $t = 0.5$ are represented in Fig. 14.9. We can note how the Dirichlet datum is well represented also by schemes with weak boundary treatment. To this end, for the case of continuous finite elements with weak boundary treatment, we have computed the behaviour of $|u_h(0) - u(0)|$ for $t = 0.1$ for several values of h, Δt being constant. We can note a linear convergence to zero with respect to h.

14.4.2 The multi-dimensional case

Let us now consider the multi-dimensional case (14.22). Let W_h be the space of discontinuous piecewise polynomials of degree r on each element $K \in \mathcal{T}_h$, introduced in (14.34). The discontinuous Galerkin (DG) finite element semi-discretization of problem (14.22) becomes: for each $t > 0$ find $u_h(t) \in W_h$ such that

$$\int_{\Omega} \frac{\partial u_h(t)}{\partial t} v_h d\Omega + \sum_{K \in \mathcal{T}_h} \left[a_K(u_h(t), v_h) - \int_{\partial K^{in}} \mathbf{a} \cdot \mathbf{n}_K \, [u_h(t)] v_h^+ d\gamma \right] = \int_{\Omega} f(t) v_h d\Omega \qquad \forall v_h \in W_h, \tag{14.47}$$

with $u_h(0) = u_{0,h}$, where $\mathbf{n}_K$ denotes the outward unit normal vector on ∂K, and

$$\partial K^{in} = \{\mathbf{x} \in \partial K : \ \mathbf{a}(\mathbf{x}) \cdot \mathbf{n}_K(\mathbf{x}) < 0\}.$$

The bilinear form a_K is defined in the following way

$$a_K(u, v) = \int_K \left(\mathbf{a} \cdot \nabla u v + a_0 u v\right) d\mathbf{x},$$

while

$$[u_h(\mathbf{x})] = \begin{cases} u_h^+(\mathbf{x}) - u_h^-(\mathbf{x}), & \mathbf{x} \notin \partial\Omega^{in}, \\ u_h^+(\mathbf{x}) - \varphi_h(\mathbf{x}), & \mathbf{x} \in \partial\Omega^{in}, \end{cases}$$

$\partial\Omega^{in}$ being the inflow boundary (14.23) and with

$$u_h^{\pm}(\mathbf{x}) = \lim_{s \to 0^{\pm}} u_h(\mathbf{x} + s\mathbf{a}), \quad \mathbf{x} \in \partial K.$$

For each $t > 0$, the stability estimate obtained for problem (14.47) is (thanks to the hypothesis (14.26))

$$\|u_h(t)\|^2_{\mathrm{L}^2(\Omega)} + \int_0^t \left(\mu_0 \|u_h(\tau)\|^2_{\mathrm{L}^2(\Omega)} + \sum_{K \in \mathcal{T}_h} \int_{\partial K^{in}} |\mathbf{a} \cdot \mathbf{n}_K| \, [u_h(\tau)]^2 \right) d\tau \leq C \left[\|u_{0,h}\|^2_{\mathrm{L}^2(\Omega)} + \int_0^t \left(\|f(\tau)\|^2_{\mathrm{L}^2(\Omega)} + |\varphi_h|^2_{\mathbf{a},\partial\Omega^{in}} \right) d\tau \right],$$

having introduced, for each subset Γ of $\partial\Omega$ of positive measure, the seminorm

$$|v|_{\mathbf{a},\Gamma} = \left(\int_{\Gamma} |\mathbf{a} \cdot \mathbf{n}| v^2 \, d\gamma \right)^{1/2}.$$

Supposing for simplicity that $f = 0$, $\varphi = 0$, and that $u_0 \in H^{r+1}(\Omega)$, we can prove the following a priori error estimate

$$\max_{t\in[0,T]} \|u(t) - u_h(t)\|_{L^2(\Omega)} + \left(\int_0^T \sum_{K\in\mathcal{T}_h} \int_{\partial K^{in}} |\mathbf{a}\cdot\mathbf{n}_K| \, [u(t) - u_h(t)]^2 \, dt \right)^{\frac{1}{2}}$$
$$\leq C h^{r+1/2} \|u_0\|_{H^{r+1}(\Omega)}. \quad (14.48)$$

For the proofs, we refer to [QV94, Chap. 14], [Joh87], and to the references cited therein.

14.4.3 DG method with jump stabilization

Other formulations are possible, based on different forms of stabilization. Let us consider a diffusion and reaction problem such as (14.22) but written in conservation form

$$\frac{\partial u}{\partial t} + \mathrm{div}(\mathbf{a}u) + a_0 u = f, \quad \mathbf{x} \in \Omega, \, t > 0. \quad (14.49)$$

Having now set

$$a_K(u_h, v_h) = \int_K \big(- u_h(\mathbf{a}\cdot\nabla v_h) + a_0 u_h v_h \big) \, d\mathbf{x},$$

we consider the following approximation: for each $t > 0$, find $u_h(t) \in W_h$ such that, $\forall v_h \in W_h$,

$$\int_\Omega \frac{\partial u_h(t)}{\partial t} v_h \, d\Omega + \sum_{K\in\mathcal{T}_h} a_K(u_h(t), v_h) + \sum_{e\not\subset\partial\Omega^{in}} \int_e \{\!\{\mathbf{a} u_h(t)\}\!\} [v_h] \, d\gamma$$
$$+ \sum_{e\not\subset\partial\Omega} \int_e c_e \, [u_h(t)] \, [v_h] \, d\gamma \quad (14.50)$$
$$= \int_\Omega f(t) v_h \, d\Omega - \sum_{e\subset\partial\Omega^{in}} \int_e (\mathbf{a}\cdot\mathbf{n}) \, \varphi_h(t) \, v_h \, d\gamma \, .$$

The notations are the following: we denote by e any side of the grid $\mathcal{T}_h$ shared by two triangles, say K_1 and K_2. For each scalar function ψ, piecewise regular on the mesh, with $\psi^i = \psi|_{K_i}$, we have defined its jump on e as follows:

$$[\psi] = \psi^1 \mathbf{n}_1 + \psi^2 \mathbf{n}_2,$$

$\mathbf{n}_i$ being the outward unit normal to element K_i. Instead, if σ is a vector function, then its average on e is defined as

$$\{\!\{\sigma\}\!\} = \frac{1}{2}(\sigma^1 + \sigma^2).$$

Note that the jump $[\psi]$ through e of a scalar function ψ is a vector parallel to the normal to e.
These definitions do not depend on the ordering of the elements.
If e is a side belonging to the boundary $\partial\Omega$, then

$$[\psi] = \psi\mathbf{n}, \qquad \{\!\{\sigma\}\!\} = \sigma.$$

Concerning c_e, this is a non-negative function which will typically be chosen to be constant on each side. Choosing, for instance, $c_e = |\mathbf{a}\cdot\mathbf{n}|/2$ on each internal side, $c_e = -\mathbf{a}\cdot\mathbf{n}/2$ on $\partial\Omega^{in}$, $c_e = \mathbf{a}\cdot\mathbf{n}/2$ on $\partial\Omega^{out}$, the formulation in (14.50) is reduced to the standard upwind formulation

$$\begin{aligned}
&\int_\Omega \frac{\partial u_h(t)}{\partial t} v_h\, d\Omega + \sum_{K\in\mathcal{T}_h} a_K(u_h(t), v_h) + \sum_{e\not\subset\partial\Omega^{in}} \int_e \{\!\{\mathbf{a}u_h(t)\}\!\}_{\mathbf{a}}[v_h]\, d\gamma \\
&= \int_\Omega f(t) v_h\, d\Omega - \sum_{e\subset\partial\Omega^{in}} \int_e (\mathbf{a}\cdot\mathbf{n})\,\varphi_h(t)\, v_h\, d\gamma \qquad \forall v_h \in W_h.
\end{aligned} \tag{14.51}$$

Here $\{\!\{\mathbf{a}u_h\}\!\}_{\mathbf{a}}$ denotes the upwind value of $\mathbf{a}u_h$, that coincides with $\mathbf{a}u_h^1$ if $\mathbf{a}\cdot\mathbf{n}_1 > 0$, with $\mathbf{a}u_h^2$ if $\mathbf{a}\cdot\mathbf{n}_1 < 0$, and finally with $\{\!\{\mathbf{a}u_h\}\!\}$ if $\mathbf{a}\cdot\mathbf{n}_1 = 0$ (see definition (12.92)). Finally, if $\mathbf{a}$ is a constant (or divergence-free) vector, then $\mathrm{div}(\mathbf{a}u_h) = \mathbf{a}\cdot\nabla u_h$ and (14.51) coincides with (14.47). Formulation (14.50) is called discontinuous Galerkin method with *jump stabilization*. The latter is stable if $c_e \geq \theta_0\,|\mathbf{a}\cdot\mathbf{n}_e|$ (for a suitable $\theta_0 > 0$) for each internal side e, and also convergent with optimal order. Indeed, in the case of the stationary problem it can be proven that

$$\|u - u_h\|^2_{\mathrm{L}^2(\Omega)} + \sum_{e\in\mathcal{T}_h} \|\sqrt{c_e}\,[u - u_h]\|^2_{\mathrm{L}^2(e)} \leq C h^{2r+1}\,\|u\|^2_{\mathrm{H}^{r+1}(\Omega)}.$$

For the proof and for other formulations with jump stabilization, including the case of advection-diffusion equations, we refer the reader to [BMS04].

14.5 Approximation using spectral methods

In this section we will briefly discuss the approximation of hyperbolic problems with spectral methods. For simplicity, we will limit our discussion to one-dimensional problems. We will first treat the G-NI approximation in a single interval, then the SEM approximation corresponding to a decomposition in sub-intervals where we use discontinuous polynomials when we move from an interval to its neighbors. This provides a generalization of discontinuous finite elements, in the case where we consider polynomials of "high" degree on each element, and the integrals on each element are approximated using the GLL integration formula (10.18).

14.5.1 The G-NI method in a single interval

Let us consider the first-order hyperbolic transport-reaction problem (13.3) and let us suppose that $(\alpha,\beta) = (-1,1)$. Then we approximate in space by a spectral collocation method, with strong imposition of the boundary conditions. Having denoted by $\{x_0 = -1, x_1, \ldots, x_N = 1\}$ the GLL nodes introduced in Sect. 10.2.3, the semi-discretized problem is:
for each $t > 0$, find $u_N(t) \in \mathbb{Q}_N$ (the space of polynomials (10.1)) such that

$$\begin{cases} \left(\dfrac{\partial u_N}{\partial t} + a\dfrac{\partial u_N}{\partial x} + a_0\, u_N\right)(x_j,t) = f(x_j,t), & j = 1,\ldots,N, \\ u_N(-1,t) = \varphi(t), & \\ u_N(x_j,0) = u_0(x_j), & j = 0,\ldots,N. \end{cases} \tag{14.52}$$

Suitably using the discrete GLL scalar product defined in (10.25), the G-NI approximation of problem (14.52) becomes: for each $t > 0$, find $u_N(t) \in \mathbb{Q}_N$ such that

$$\begin{cases} \left(\dfrac{\partial u_N(t)}{\partial t}, v_N\right)_N + \left(a\dfrac{\partial u_N(t)}{\partial x}, v_N\right)_N + \left(a_0\, u_N(t), v_N\right)_N = \left(f(t), v_N\right)_N \quad \forall v_N \in \mathbb{Q}_N^-, \\ u_N(-1,t) = \varphi(t), \\ u_N(x,0) = u_{0,N}, \end{cases} \tag{14.53}$$

where $u_{0,N} \in \mathbb{Q}_N$ is a suitable approximation of u_0, and having set $\mathbb{Q}_N^- = \{v_N \in \mathbb{Q}_N : v_N(-1) = 0\}$. At the inflow, the solution u_N satisfies the imposed condition at each time $t > 0$, while test functions vanish.
In fact, the solutions of problems (14.52) and (14.53) coincide if $u_{0,N}$ in (14.53) is chosen as the interpolated $\Pi_N^{GLL} u_0$. To prove this, it is sufficient to choose in (14.53) v_N coinciding with the characteristic polynomial ψ_j (defined in (10.12), (10.13)) associated to the GLL node x_j, for each $j = 1,\ldots,N$.

Let us now derive a stability estimate for formulation (14.53) in the norm (10.53) induced from the discrete scalar product (10.25). For simplicity, we choose a homogeneous inflow datum, that is $\varphi(t) = 0$, for each t, and a and a_0 constant. Having chosen, for each $t > 0$, $v_N = u_N(t)$, we obtain

$$\frac{1}{2}\frac{\partial}{\partial t}\|u_N(t)\|_N^2 + \frac{a}{2}\int_{-1}^{1}\frac{\partial u_N^2(t)}{\partial x}\,dx + a_0\|u_N(t)\|_N^2 = \left(f(t), u_N(t)\right)_N.$$

Suitably rewriting the convective term, integrating with respect to time and using the Young inequality, we have

$$\begin{aligned}\|u_N(t)\|_N^2 + a\int_0^t (u_N(1,\tau))^2\, d\tau + 2a_0\int_0^t \|u_N(\tau)\|_N^2\, d\tau \\ = \|u_{0,N}\|_N^2 + 2\int_0^t \left(f(\tau), u_N(\tau)\right)_N d\tau \\ \le \|u_{0,N}\|_N^2 + a_0\int_0^t \|u_N(\tau)\|_N^2\, d\tau + \frac{1}{a_0}\int_0^t \|f(\tau)\|_N^2\, d\tau,\end{aligned}$$

that is

$$\begin{aligned}\|u_N(t)\|_N^2 \; + a\int_0^t (u_N(1,\tau))^2\, d\tau + a_0\int_0^t \|u_N(\tau)\|_N^2\, d\tau \\ \le \|u_{0,N}\|_N^2 + \frac{1}{a_0}\int_0^t \|f(\tau)\|_N^2\, d\tau.\end{aligned} \tag{14.54}$$

The norm of the initial data can be bounded as follows

$$\|u_{0,N}\|_N^2 \le \|u_{0,N}\|_{\mathrm{L}^\infty(-1,1)}^2 \left(\sum_{i=0}^N \alpha_i\right) = 2\,\|u_{0,N}\|_{\mathrm{L}^\infty(-1,1)}^2,$$

and a similar bound holds for $\|f(\tau)\|_N^2$ provided that f is a continuous function. Hence, reverting to (14.54) and using inequality (10.54) to bound the norms of the left-hand side, we deduce

$$\begin{aligned}\|u_N(t)\|_{\mathrm{L}^2(-1,1)}^2 \; + a\int_0^t (u_N(1,\tau))^2\, d\tau + a_0\int_0^t \|u_N(\tau)\|_{\mathrm{L}^2(-1,1)}^2\, d\tau \\ \le 2\,\|u_{0,N}\|_{\mathrm{L}^\infty(-1,1)}^2 + \frac{2}{a_0}\int_0^t \|f(\tau)\|_{\mathrm{L}^2(-1,1)}^2\, d\tau.\end{aligned}$$

The reinterpretation of the G-NI method as a collocation method is less immediate in the case where the convective term a is not constant and if we start from a conservative formulation of the differential equation in (14.52), that is when the second term on the left-hand side is replaced by $\partial(au)/\partial x$. In such case, we can show again that the G-NI approximation is equivalent to the collocation approximation where the convective term is replaced by $\partial\left(\Pi_N^{GLL}(au_N)\right)/\partial x$, i.e. by the interpolation derivative (10.40).

Also in the case of a G-NI approximation, we can resort to a weak imposition of the boundary conditions. Such approach is more flexible than the one considered above, and more suitable for the generalization to multi-dimensional problems or systems of equations. As we have seen in the previous section, the starting point for imposing boundary conditions weakly is a suitable integration by parts of the transport terms. Referring to the one-dimensional problem (14.52), we have (if a is constant)

$$\begin{aligned}\int_{-1}^{1} a\frac{\partial u(t)}{\partial x}v\,dx &= -\int_{-1}^{1} a u(t)\frac{\partial v}{\partial x}dx + \left[a u(t)\,v\right]_{-1}^{1}\\ &= -\int_{-1}^{1} a u(t)\frac{\partial v}{\partial x}dx + a u(1,t)\,v(1) - a\,\varphi(t)\,v(-1).\end{aligned}$$

Thanks to the above identity, we can immediately formulate the G-NI approximation of problem (14.52) with a weak treatment of boundary conditions:
for each $t>0$, find $u_N(t)\in\mathbb{Q}_N$ such that

$$\begin{aligned}&\left(\frac{\partial u_N(t)}{\partial t}, v_N\right)_N - \left(a u_N(t), \frac{\partial v_N}{\partial x}\right)_N + \left(a_0\, u_N(t), v_N\right)_N +\\ &a u_N(1,t)\,v_N(1) = \left(f(t), v_N\right)_N + a\,\varphi(t)\,v_N(-1)\quad \forall v_N\in\mathbb{Q}_N,\end{aligned}\tag{14.55}$$

with $u_N(x,0)=u_{0,N}(x)$. We note that both the solution u_N and the test function v_N are free at the boundary.
An equivalent formulation of (14.55) is obtained by suitably counter-integrating the convective term by parts:
for each $t>0$, find $u_N(t)\in\mathbb{Q}_N$ such that

$$\begin{aligned}&\left(\frac{\partial u_N(t)}{\partial t}, v_N\right)_N + \left(a\frac{\partial u_N(t)}{\partial x}, v_N\right)_N + \left(a_0\, u_N(t), v_N\right)_N +\\ &a\left(u_N(-1,t)-\varphi(t)\right)v_N(-1) = \left(f, v_N\right)_N\quad \forall v_N\in\mathbb{Q}_N.\end{aligned}\tag{14.56}$$

It is now possible to reinterpret such weak formulation as a suitable collocation method. To this end, it is sufficient to choose in (14.56) the test function v_N to be the characteristic polynomials (10.12), (10.13) associated to the GLL nodes. Considering first the internal and outflow nodes, and choosing therefore $v_N=\psi_i$, with $i=1,\ldots,N$, we have

$$\left(\frac{\partial u_N}{\partial t} + a\frac{\partial u_N}{\partial x} + a_0\, u_N\right)(x_i,t) = f(x_i,t),\tag{14.57}$$

having previously simplified the weight α_i common to all terms of the equation. On the other hand, by choosing $v_N=\psi_0$ we obtain the following relation at the inflow node

$$\begin{aligned}&\left(\frac{\partial u_N}{\partial t} + a\frac{\partial u_N}{\partial x} + a_0\, u_N\right)(-1,t)\\ &\quad + \frac{1}{\alpha_0}a\left(u_N(-1,t)-\varphi(t)\right) = f(-1,t),\end{aligned}\tag{14.58}$$

$\alpha_0 = 2/(N^2+N)$ being the GLL weight associated to node $x_0=-1$. From equations (14.57) and (14.58) it then follows that a reformulation in terms of collocation is possible at all the GLL nodes except for the inflow node, for which we find the relation

$$a\left(u_N(-1,t)-\varphi(t)\right)=\alpha_0\left(f-\frac{\partial u_N}{\partial t}-a\frac{\partial u_N}{\partial x}-a_0\,u_N\right)(-1,t). \tag{14.59}$$

The latter can be interpreted as the fulfillment of the boundary condition of the differential problem (14.52) up to the residue associated to the u_N approximation. Such condition is therefore satisfied exactly only in the limit, for $N\longrightarrow\infty$ (i.e. in a natural way).

In accordance with what we noted previously, formulation (14.56) would be complicated, for instance, in case of a non-constant convective field a. Indeed,

$$-\left(a\,u_N(t),\frac{\partial v_N}{\partial x}\right)_N=\left(a\frac{\partial u_N(t)}{\partial x},v_N\right)_N-a\,u_N(1,t)\,v_N(1)+a\,\varphi(t)\,v_N(-1),$$

would not be true as, in this case, the product $a\,u_N(t)\frac{\partial v_N}{\partial x}$ no longer identifies a polynomial of degree $2N-1$, so the exactness of the numerical integration formula would not hold. It is therefore necessary to apply the interpolation operator Π_N^{GLL}, introduced in Sect. 10.2.3, before counter-integrating by parts, so that

$$\begin{aligned}-\left(a\,u_N(t),\frac{\partial v_N}{\partial x}\right)_N&=-\left(\Pi_N^{GLL}(a\,u_N(t)),\frac{\partial v_N}{\partial x}\right)_N\\&=-\left(\Pi_N^{GLL}(a\,u_N(t)),\frac{\partial v_N}{\partial x}\right)\\&=\left(\frac{\partial\Pi_N^{GLL}(a\,u_N(t))}{\partial x},v_N\right)-\left[(a\,u_N(t))\,v_N\right]_{-1}^{1}.\end{aligned}$$

In this case, formulation (14.56) then becomes:
for each $t>0$, find $u_N(t)\in\mathbb{Q}_N$ such that

$$\begin{aligned}&\left(\frac{\partial u_N(t)}{\partial t},v_N\right)_N+\left(\frac{\partial\Pi_N^{GLL}(a\,u_N(t))}{\partial x},v_N\right)+\left(a_0\,u_N(t),v_N\right)_N+\\&a(t)\left(u_N(-1,t)-\varphi(t)\right)v_N(-1)=\left(f(t),v_N\right)_N\ \ \forall v_N\in\mathbb{Q}_N,\end{aligned} \tag{14.60}$$

with $u_N(x,0)=u_{0,N}(x)$. Also the collocation reinterpretation of formulation (14.56), represented by relations (14.57) and (14.59), will need to be modified with the introduction of the interpolation operator Π_N^{GLL} (that is by replacing the exact derivative with the interpolation derivative). Precisely, we obtain

$$\left(\frac{\partial u_N}{\partial t}+\frac{\partial\Pi_N^{GLL}(a\,u_N)}{\partial x}+a_0\,u_N\right)(x_i,t)=f(x_i,t),$$

for $i=1,\ldots,N$, and

$$a(-1)\left(u_N(-1,t)-\varphi(t)\right)=\alpha_0\left(f-\frac{\partial u_N}{\partial t}-\frac{\partial\Pi_N^{GLL}(a\,u_N)}{\partial x}-a_0\,u_N\right)(-1,t),$$

at the inflow node $x=-1$.

14.5.2 The DG-SEM-NI method

As anticipated, we will introduce in this section an approximation based on a partition in sub-intervals, in each of which the G-NI method is used. Moreover, the solution will be discontinuous between an interval and its neighbors. This explains the DG (*discontinuous Galerkin*), SEM (*spectral element method*), NI (*numerical integration*) acronym.

Let us reconsider problem (14.52) on the generic interval (α,β). On the latter, we introduce a partition in M subintervals $\Omega_m = (\bar{x}_{m-1},\bar{x}_m)$ with $m = 1,\ldots,M$. Let

$$W_{N,M} = \{v \in \mathrm{L}^2(\alpha,\beta) : v\big|_{\Omega_m} \in \mathbb{Q}_N,\ \forall m = 1,\ldots,M\}$$

be the space of piecewise polynomials of degree $N(\geq 1)$ on each sub-interval. We observe that continuity is not necessarily guaranteed in correspondence of the points $\{\bar{x}_i\}$. Thus, we can formulate the following approximation of problem (14.52): for each $t > 0$, find $u_{N,M}(t) \in W_{N,M}$ such that

$$\sum_{m=1}^{M}\Big[\Big(\frac{\partial u_{N,M}}{\partial t}, v_{N,M}\Big)_{N,\Omega_m} + \Big(a\frac{\partial u_{N,M}}{\partial x}, v_{N,M}\Big)_{N,\Omega_m} + \big(a_0\, u_{N,M}, v_{N,M}\big)_{N,\Omega_m}$$
$$+\, a(\bar{x}_{m-1})\big(u^+_{N,M} - U^-_{N,M}\big)(\bar{x}_{m-1})\, v^+_{N,M}(\bar{x}_{m-1})\Big] = \sum_{m=1}^{M}\big(f, v_{N,M}\big)_{N,\Omega_m} \tag{14.61}$$

for all $v_{N,M} \in W_{N,M}$, with

$$U^-_{N,M}(\bar{x}_i) = \begin{cases} u^-_{N,M}(\bar{x}_i), & i = 1,\ldots,M-1, \\ \varphi(\bar{x}_0), & \text{for } i = 0, \end{cases} \tag{14.62}$$

and where $(\cdot,\cdot)_{N,\Omega_m}$ denotes the approximation via the GLL formula (10.25) of the scalar product L^2 restricted to the element Ω_m. To simplify the notations we have omitted to indicate the dependence on t of $u_{N,M}$ and f explicitly. Given the discontinuous nature of the test functions, we can reformulate equation (14.61) on each of the M sub-intervals, by choosing the test function $v_{N,M}$ so that $v_{N,M}\big|_{[\alpha,\beta]\setminus\Omega_m} = 0$. Proceeding in this way, we obtain

$$\Big(\frac{\partial u_{N,M}}{\partial t}, v_{N,M}\Big)_{N,\Omega_m} + \Big(a\frac{\partial u_{N,M}}{\partial x}, v_{N,M}\Big)_{N,\Omega_m} + \big(a_0\, u_{N,M}, v_{N,M}\big)_{N,\Omega_m}$$
$$+\, a(\bar{x}_{m-1})\big(u^+_{N,M} - U^-_{N,M}\big)(\bar{x}_{m-1})\, v^+_{N,M}(\bar{x}_{m-1}) = \big(f, v_{N,M}\big)_{N,\Omega_m},$$

for each $m = 1,\ldots,M$. We note that, for $m = 1$, the term

$$a(\bar{x}_0)\big(u^+_{N,M} - \varphi\big)(\bar{x}_0)\, v^+_{N,M}(\bar{x}_0)$$

can be regarded as the imposition in weak form of the inflow boundary condition. On the other hand for $m = 2,\ldots,M$, the term

$$a(\bar{x}_{m-1})\big(u^+_{N,M} - U^-_{N,M}\big)(\bar{x}_{m-1})\, v^+_{N,M}(\bar{x}_{m-1})$$

can be interpreted as a penalization term that provides a weak imposition of the continuity of the solution $u_{N,M}$ at the endpoints $\bar{x}_i$, $i = 1,\dots,M-1$.

We now want to interpret formulation (14.61) as a suitable collocation method. To this end, we introduce on each sub-interval Ω_m, the $N+1$ GLL nodes $x_j^{(m)}$, with $j = 0,\dots,N$, and we denote by $\alpha_j^{(m)}$ the corresponding weights (see (10.71)). We now choose the test function $v_{N,M}$ in (14.61) as the characteristic Lagrangian polynomial $\psi_j^{(m)} \in \mathbb{P}^N(\Omega_m)$ associated to node $x_j^{(m)}$ and extended trivially outside the domain Ω_m. Given the presence of the jump term, we will have a non-unique rewriting for equation (14.61). We start by considering the characteristic polynomials associated to the nodes $x_j^{(m)}$, with $j = 1,\dots,N-1$, and $m = 1,\dots,M$. In this case we will have no contribution from the penalization term, yielding

$$\left[\frac{\partial u_{N,M}}{\partial t} + a\frac{\partial u_{N,M}}{\partial x} + a_0\, u_{N,M}\right](x_j^{(m)}) = f(x_j^{(m)}).$$

For this choice of nodes we thus find exactly the collocation of the differential problem (14.52).

Instead, in the case where the function $\psi_j^{(m)}$ is associated to a node of the partition $\{\bar{x}_i\}$, that is $j = 0$, with $m = 1,\dots,M$ we have

$$\begin{aligned}\alpha_0^{(m)}&\left[\frac{\partial u_{N,M}}{\partial t} + a\frac{\partial u_{N,M}}{\partial x} + a_0\, u_{N,M}\right](x_0^{(m)})\\ &+ a(x_0^{(m)})\left(u_{N,M}^+ - U_{N,M}^-\right)(x_0^{(m)}) = \alpha_0^{(m)} f(x_0^{(m)}),\end{aligned} \tag{14.63}$$

recalling that $U_{N,M}^-(x_0^{(1)}) = \varphi(\bar{x}_0)$. We have implicitly adopted the convention that the sub-interval Ω_m should not include $\bar{x}_m$, as the discontinuous nature of the adopted method would make us process twice each node $\bar{x}_i$, with $i = 1,\dots,M-1$. Equation (14.63) can be rewritten as

$$\left[\frac{\partial u_{N,M}}{\partial t} + a\frac{\partial u_{N,M}}{\partial x} + a_0\, u_{N,M} - f\right](x_0^{(m)}) = -\frac{a(x_0^{(m)})}{\alpha_0^{(m)}}\left(u_{N,M}^+ - U_{N,M}^-\right)(x_0^{(m)}).$$

We observe that while the left-hand side represents the residue of the equation at node $x_0^{(m)}$, the right-hand side one is, up to a multiplicative factor, the residue of the weak imposition of the continuity of $u_{N,M}$ at $x_0^{(m)}$.

14.6 Numerical treatment of boundary conditions for hyperbolic systems

We have seen different strategies to impose the inflow boundary conditions for the scalar transport equation. When considering hyperbolic systems, the numerical treatment of boundary conditions requires more attention. We will explain why the point is using a linear system with constant coefficients in one dimension,

$$\begin{cases} \dfrac{\partial \mathbf{u}}{\partial t} + A\dfrac{\partial \mathbf{u}}{\partial x} = \mathbf{0}, & -1 < x < 1,\ t > 0, \\ \mathbf{u}(x,0) = \mathbf{u}_0(x), & -1 < x < 1, \end{cases} \tag{14.64}$$

completed with suitable boundary conditions. Following [CHQZ07], we choose the case of a system made of two hyperbolic equations, and take u to be the vector $(u,v)^T$, while

$$A = \begin{bmatrix} -1/2 & -1 \\ -1 & -1/2 \end{bmatrix},$$

whose eigenvalues are $-3/2$ and $1/2$. We make the choice

$$u(x,0) = \sin(2x) + \cos(2x), \quad v(x,0) = \sin(2x) - \cos(2x)$$

for the initial conditions and

$$\begin{aligned} u(-1,t) &= \sin(-2+3t) + \cos(-2-t) = \varphi(t), \\ v(1,t) &= \sin(2+3t) + \cos(2-t) = \psi(t) \end{aligned} \tag{14.65}$$

for the boundary conditions.
Let us now consider the (right) eigenvector matrix

$$W = \begin{bmatrix} 1/2 & 1/2 \\ 1/2 & -1/2 \end{bmatrix},$$

whose inverse is

$$W^{-1} = \begin{bmatrix} 1 & 1 \\ 1 & -1 \end{bmatrix}.$$

Exploiting the diagonalization

$$\Lambda = W^{-1}AW = \begin{bmatrix} -3/2 & 0 \\ 0 & 1/2 \end{bmatrix},$$

we can rewrite the differential equation in (14.64), in terms of the characteristic variables

$$\mathbf{z} = W^{-1}\mathbf{u} = \begin{bmatrix} u+v \\ u-v \end{bmatrix} = \begin{bmatrix} z_1 \\ z_2 \end{bmatrix}, \tag{14.66}$$

as

$$\frac{\partial \mathbf{z}}{\partial t} + \Lambda\frac{\partial \mathbf{z}}{\partial x} = \mathbf{0}. \tag{14.67}$$

The characteristic variable z_1 propagates towards the left at rate $3/2$, while z_2 propagates towards the right at rate $1/2$.

This suggests to assign a condition for z_1 at $x = 1$ and one for z_2 at $x = -1$. The boundary values of z_1 and z_2 can be generated by using the boundary conditions for u

and v as follows. From relation (14.66) we have

$$\mathbf{u} = W\mathbf{z} = \begin{bmatrix} 1/2 & 1/2 \\ 1/2 & -1/2 \end{bmatrix} \begin{bmatrix} z_1 \\ z_2 \end{bmatrix} = \begin{bmatrix} 1/2\,(z_1 + z_2) \\ 1/2\,(z_1 - z_2) \end{bmatrix},$$

that is, exploiting the boundary values (14.65) assigned for u and v,

$$\frac{1}{2}(z_1 + z_2)(-1,t) = \varphi(t), \quad \frac{1}{2}(z_1 - z_2)(1,t) = \psi(t). \tag{14.68}$$

The conclusion is that, in spite of the diagonal structure of system (14.67), the characteristic variables are in fact coupled by the boundary conditions (14.68).
Hence we have to face the problem of how to handle, from a numerical viewpoint, boundary conditions for systems like (14.64). Indeed, difficulties can arise even from the discretization of the corresponding scalar problem (for a constant > 0)

$$\begin{cases} \dfrac{\partial z}{\partial t} + a\dfrac{\partial z}{\partial x} = 0, & -1 < x < 1, \ t > 0, \\ z(-1,t) = \phi(t), & t > 0, \\ z(x,0) = z_0(x), & -1 < x < 1, \end{cases} \tag{14.69}$$

if we do not use an appropriate discretization scheme. We will illustrate the procedure for a spectral approximation method. As a matter of fact, a correct treatment of the boundary conditions for high-order methods is even more vital than for a finite element or finite difference method, because with spectral methods boundary errors would be propagated inwards with an infinite rate.

Having introduced the partition $x_0 = -1 < x_1 < \ldots < x_{N-1} < x_N = 1$ of the interval $[-1, 1]$, if we decide to use, say, a finite difference scheme, we encounter problems, essentially in determining the value of z at the outflow node x_N, unless we use the first order upwind scheme. As a matter of fact, higher order FD schemes such as the centered finite difference scheme would not be able to provide such an approximation unless additional nodes outside the definition interval $(-1, 1)$ were introduced.
In contrast, a spectral discretization does not involve any boundary problem. For instance, the collocation scheme corresponding to problem (14.69) can be written as follows:
$\forall n \geq 0$, find $z_N^n \in \mathbb{Q}_N$ such that

$$\begin{cases} \dfrac{z_N^{n+1}(x_i) - z_N^n(x_i)}{\Delta t} + a\dfrac{\partial z_N^n}{\partial x}(x_i) = 0, & i = 1, \ldots, N, \\ z_N^{n+1}(x_0) = \phi(t^{n+1}). \end{cases}$$

One equation is associated to each node, whether internal or on the boundary, and the outflow node is treated as any other internal node. However, when moving to system (14.64), two unknowns and two equations are associated with each internal node x_i, with $i = 1, \ldots, N-1$, while at the boundary nodes x_0 and x_N we still have two unknowns but a single equation. Thus, we will need additional conditions for these

points: in general, at the endpoint $x = -1$ we will need as many conditions as the positive eigenvalues, while for $x = 1$ we will need to provide as many additional conditions as the negative eigenvalues.
Let us look for a solution to this problem guided by the spectral Galerkin method. Let us suppose we apply a collocation method to system (14.64); then, we want to find $\mathbf{u}_N = (u_{N,1}, u_{N,2})^T \in (\mathbb{Q}_N)^2$ such that

$$\frac{\partial \mathbf{u}_N}{\partial t}(x_i) + A\frac{\partial \mathbf{u}_N}{\partial x}(x_i) = \mathbf{0}, \quad i = 1,\ldots,N-1, \tag{14.70}$$

and with

$$u_{N,1}(x_0,t) = \varphi(t), \quad u_{N,2}(x_N,t) = \psi(t). \tag{14.71}$$

The simplest idea to obtain the two missing equations for $u_{N,1}$ and $u_{N,2}$ at x_N resp. x_0, is to exploit the vector equation (14.70) together with the known vectors $\varphi(t)$ and $\psi(t)$ in (14.71). The solution computed in this way is, however, strongly unstable.
To seek an alternative approach, the idea is to add to the $2(N-1)$ collocation relations (14.70) and to the "physical" boundary conditions (14.71), the equations of the outgoing characteristics at points x_0 and x_N. More in detail, the characteristic outgoing from the domain at point $x_0 = -1$ is the one associated to the negative eigenvalue of the matrix A, and has equation

$$\frac{\partial z_1}{\partial t}(x_0) - \frac{3}{2}\frac{\partial z_1}{\partial x}(x_0) = 0, \tag{14.72}$$

while the one associated with the point $x_N = 1$ is highlighted by the positive eigenvalue $1/2$ and is given by

$$\frac{\partial z_2}{\partial t}(x_N) + \frac{1}{2}\frac{\partial z_2}{\partial x}(x_N) = 0. \tag{14.73}$$

The choice of the outgoing characteristic is motivated by the fact that the latter carries information from the inside of the domain to the corresponding outflow point, where it makes sense to impose the differential equation.
Equations (14.72) and (14.73) allow us to have a closed system of $2N+2$ equations in the $2N+2$ unknowns $u_{N,1}(x_i,t) = u_N(x_i,t)$, $u_{N,2}(x_i,t) = v_N(x_i,t)$, with $i = 0,\ldots,N$.
For completeness, we can rewrite the characteristic equations (14.72) and (14.73) in terms of the unknowns u_N and v_N, as

$$\frac{\partial(u_N + v_N)}{\partial t}(x_0) - \frac{3}{2}\frac{\partial(u_N + v_N)}{\partial x}(x_0) = 0$$

and

$$\frac{\partial(u_N - v_N)}{\partial t}(x_N) + \frac{1}{2}\frac{\partial(u_N - v_N)}{\partial x}(x_N) = 0,$$

respectively, or in matrix terms as

$$\begin{aligned} \left[W_{11}^{-1} \; W_{12}^{-1}\right]\left[\frac{\partial \mathbf{u}_N}{\partial t}(x_0) + A\frac{\partial \mathbf{u}_N}{\partial x}(x_0)\right] &= 0, \\ \left[W_{21}^{-1} \; W_{22}^{-1}\right]\left[\frac{\partial \mathbf{u}_N}{\partial t}(x_N) + A\frac{\partial \mathbf{u}_N}{\partial x}(x_N)\right] &= 0. \end{aligned} \tag{14.74}$$

Such additional equations are called *compatibility* equations: they represent a linear combination of the differential equations of the problem at the boundary points with coefficients given by the components of the matrix W^{-1}.

Remark 14.4. Due to their global nature, spectral methods (either collocation, Galerkin, or G-NI) propagate immediately, and on the whole domain, every possible numerical perturbation introduced at the boundary. As such, spectral methods represent a good testbed for understanding how suitable numerical strategies are for the boundary treatment of hyperbolic systems. •

14.6.1 Weak treatment of boundary conditions

Now we want to generalize the approach based on compatibility equations, and move from pointwise relations, such as (14.74), to integral relations, more suitable for numerical approximations such as, e.g., finite elements or G-NI.
Let us, once again, consider the constant coefficient system (14.64) and the notations used in Sect. 14.6. Let A be a real, symmetric and non-singular matrix of order d, Λ the diagonal matrix whose diagonal entries are the real eigenvalues of A, and W the square matrix whose columns are the (right) eigenvectors of A. Let us suppose that W is orthogonal, which guarantees that $\Lambda = W^T AW$. The characteristic variables, defined as $\mathbf{z} = W^T \mathbf{u}$, satisfy the diagonal system (14.67). We introduce the splitting $\Lambda = \text{diag}(\Lambda^+, \Lambda^-)$ of the eigenvalue matrix, arising by grouping the positive eigenvalues (Λ^+) and the negative ones (Λ^-). Both sub-matrices are diagonal, Λ^+ positive definite of order p, Λ^- negative definite of order $n = d - p$.
Analogously, we can rewrite $\mathbf{z}$ as $\mathbf{z} = (\mathbf{z}^+, \mathbf{z}^-)^T$, having denoted by $\mathbf{z}^+$ ($\mathbf{z}^-$) the characteristic variables that are constant along the characteristic lines with positive (negative) slope, in other words, lines moving rightwards (leftwards) on the (x,t) reference frame. In correspondence of the right extremum $x = 1$, $\mathbf{z}^+$ is associated to the outgoing characteristic variables while $\mathbf{z}^-$ to the incoming ones. Clearly, the roles are switched at the left boundary point $x = -1$.
A simple case occurs when we assign, as boundary conditions, the values of the incoming characteristics at both domain extrema, that is p conditions at $x = -1$ and n conditions at $x = 1$. In this case, (14.67) represents a fully-fledged decoupled system. Much more frequently, however, the values of suitable linear combinations of the physical variables are assigned at both boundary points. Reading them in terms of the $\mathbf{z}$ variables, these yield linear combinations of the characteristic variables. None of the outgoing characteristics will in principle be determined by these combinations, as the resulting values will generally be incompatible with the ones propagated inwards by the hyperbolic system. In contrast, the boundary conditions should allow to determine the incoming characteristic variables as a function of the outgoing ones and of the problem data.
For the sake of clarity, let us consider the following boundary conditions

$$B_L \mathbf{u}(-1,t) = \mathbf{g}_L(t), \quad B_R \mathbf{u}(1,t) = \mathbf{g}_R(t), \quad t > 0, \tag{14.75}$$

where $\mathbf{g}_L$ and $\mathbf{g}_R$ are assigned vectors and B_L, B_R are suitable matrices. At the left extremum $x = -1$, we have p incoming characteristics, and B_L will have dimension

$p \times d$. Setting $C_L = B_L W$ and using the splitting $\mathbf{z} = (\mathbf{z}^+, \mathbf{z}^-)^T$ introduced for $\mathbf{z}$ and the corresponding splitting $W = (W^+, W^-)^T$ for the eigenvector matrix, we have

$$C_L \mathbf{z}(-1,t) = C_L^+ \mathbf{z}^+(-1,t) + C_L^- \mathbf{z}^-(-1,t) = \mathbf{g}_L(t),$$

where $C_L^+ = B_L W^+$ is a $p \times p$ matrix, while $C_L^- = B_L W^-$ has dimension $p \times n$. We demand that C_L^+ is non-singular. Then the incoming characteristic at $x = -1$ can be obtained by

$$\mathbf{z}^+(-1,t) = S_L \mathbf{z}^-(-1,t) + \mathbf{z}_L(t), \tag{14.76}$$

$S_L = -(C_L^+)^{-1} C_L^-$ being a $p \times n$ matrix and $\mathbf{z}_L(t) = (C_L^+)^{-1} \mathbf{g}_L(t)$. In a similar way, we can assign at the right extremum $x = 1$ the incoming characteristic variable as

$$\mathbf{z}^-(1,t) = S_R \mathbf{z}^+(1,t) + \mathbf{z}_R(t), \tag{14.77}$$

S_R being a $n \times p$ matrix.
The matrices S_L and S_R are called *reflection matrices*.
The hyperbolic system (14.64) will thus be completed by the boundary conditions (14.75) or, equivalently, by conditions (14.76)–(14.77).

Let us see which advantages can be brought by such a choice for boundary conditions. We start from the weak formulation of problem (14.64), integrating by parts the term containing the space derivative

$$\int_{-1}^{1} \mathbf{v}^T \frac{\partial \mathbf{u}}{\partial t} dx - \int_{-1}^{1} \left(\frac{\partial \mathbf{v}}{\partial x}\right)^T A\mathbf{u}\, dx + \left[\mathbf{v}^T A\mathbf{u}\right]_{-1}^{1} = 0,$$

for each $t > 0$, $\mathbf{v}$ being an arbitrary, differentiable test function. We want to rewrite the boundary term $\left[\mathbf{v}^T A\mathbf{u}\right]_{-1}^{1}$ by exploiting the boundary equations (14.76) - (14.77). Introducing the characteristic variable $W^T \mathbf{v} = \mathbf{y} = (\mathbf{y}^+, \mathbf{y}^-)^T$ associated to the test function $\mathbf{v}$, we will have

$$\mathbf{v}^T A\mathbf{u} = \mathbf{y}^T \Lambda \mathbf{z} = (\mathbf{y}^+)^T \Lambda^+ \mathbf{z}^+ + (\mathbf{y}^-)^T \Lambda^- \mathbf{z}^-.$$

Using relations (14.76)–(14.77), it then follows that

$$\begin{aligned}
&\int_{-1}^{1} \mathbf{v}^T \frac{\partial \mathbf{u}}{\partial t} dx - \int_{-1}^{1} \left(\frac{\partial \mathbf{v}}{\partial x}\right)^T A\mathbf{u}\, dx \\
&- (\mathbf{y}^+)^T(-1,t)\Lambda^+ S_L \mathbf{z}^-(-1,t) - (\mathbf{y}^-)^T(-1,t)\Lambda^- \mathbf{z}^-(-1,t) \\
&+ (\mathbf{y}^+)^T(1,t)\Lambda^+ \mathbf{z}^+(1,t) + (\mathbf{y}^-)^T(1,t)\Lambda^- S_R \mathbf{z}^+(1,t) \\
&= (\mathbf{y}^+)^T(-1,t)\Lambda^+ \mathbf{z}_L(t) - (\mathbf{y}^-)^T(1,t)\Lambda^- \mathbf{z}_R(t).
\end{aligned} \tag{14.78}$$

We observe that the boundary conditions (14.76)–(14.77) are naturally incorporated in the right-hand side of the system. Moreover, integrating again by parts, it is possible to obtain an equivalent formulation to (14.78) where the boundary conditions are imposed in a weak way

$$\begin{aligned}
&\int_{-1}^{1} \mathbf{v}^T \frac{\partial \mathbf{u}}{\partial t}\, dx + \int_{-1}^{1} \mathbf{v}^T A \frac{\partial \mathbf{u}}{\partial x}\, dx \\
&+ (\mathbf{y}^+)^T(-1,t)\Lambda^+(\mathbf{z}^+(-1,t) - S_L\mathbf{z}^-(-1,t)) \qquad (14.79)\\
&- (\mathbf{y}^-)^T(1,t)\Lambda^-(\mathbf{z}^-(1,t) - S_R\mathbf{z}^+(1,t)) \\
&= (\mathbf{y}^+)^T(-1,t)\Lambda^+\mathbf{z}_L(t) - (\mathbf{y}^-)^T(1,t)\Lambda^-\mathbf{z}_R(t).
\end{aligned}$$

Finally, we recall that the following assumption, called *dissipation hypothesis*, is usually made on the reflection matrices S_L and S_R

$$\|S_L\|\,\|S_R\| < 1. \qquad (14.80)$$

The matrix norm in (14.80) must be understood as the Euclidean norm of a rectangular matrix, that is the square root of the maximum eigenvalue of $S_L^T S_L$ and $S_R^T S_R$. This assumption is sufficient to guarantee the stability of the previous scheme in the L^2 norm. Formulation (14.78) (or (14.79)) is suitable for Galerkin approximations such as Galerkin finite elements, the spectral Galerkin method, the spectral method with Gaussian numerical integration in a single domain (G-NI), the spectral element version, whether continuous (SEM-NI) or discontinuous (DG-SEM-NI) spectral elements.

14.7 Exercises

1. Prove that the discretization with continuous linear finite elements (14.13) coincides with the finite difference one (13.22) in case the mass matrix is diagonalized using the mass lumping technique.
   [*Solution*: use the partition of unity property (12.35) as in Sect. 12.5.]
2. Prove the stability inequalities provided in Sect. (14.4) for the semi-discretization based on finite elements.
3. Prove relation (14.13).
4. Discretize system (14.78) using the continuous spectral element method, SEM-NI, and the discontinuous one, DG-SEM-NI.

15

Nonlinear hyperbolic problems

In this chapter we introduce some examples of nonlinear hyperbolic problems. We will point out some characteristic properties of such problems, most notably their ability to generate discontinuous solutions also in the case of continuous initial and boundary data. The numerical approximation of these problems is far from easy. Here we will simply limit ourselves to point out how finite difference and finite element schemes can be applied in the case of one-dimensional equations.

For a more complete discussion, we refer to [LeV07, GR96, Bre00, Tor99, Kro97].

15.1 Scalar equations

Let us consider the following equation

$$\frac{\partial u}{\partial t} + \frac{\partial}{\partial x} F(u) = 0, \quad x \in \mathbb{R}, \quad t > 0, \tag{15.1}$$

where F is a nonlinear function of u called *flux* of u, because on each interval (α, β) of $\mathbb{R}$, it satisfies the following relation

$$\frac{d}{dt} \int_{\alpha}^{\beta} u(x,t) dx = F(u(t,\alpha)) - F(u(t,\beta)).$$

For this reason, (15.1) expresses a *conservation law*. A typical example is Burgers' equation

$$\frac{\partial u}{\partial t} + u \frac{\partial u}{\partial x} = 0. \tag{15.2}$$

This equation was already considered in Example 1.3, and corresponds to (15.1) when the flux is $F(u) = u^2/2$. Its characteristic curves are obtained by solving $x'(t) = u$. However, since u is constant on the characteristics, we obtain $x'(t) = constant$, so the characteristics are straight lines. The latter are defined in the plane (x,t) by the map $t \rightarrow (x + tu_0(x), t)$, and the solution to (15.2) is implicitly defined by $u(x + tu_0(x)) =$

A. Quarteroni: *Numerical Models for Differential Problems*, 2nd Ed.
MS&A – Modeling, Simulation & Applications 8
DOI 10.1007/978-88-470-5522-3_15, © Springer-Verlag Italia 2014

$u_0(x), t < t_c, t_c$ being the first instant where such characteristics intersect. For instance, if $u_0(x) = (1+x^2)^{-1}$, then $t_c = 8/\sqrt{27}$.
Indeed, if $u_0'(x)$ is negative at some point, having set

$$t_c = -\frac{1}{\min\ u_0'(x)},$$

for $t > t_c$ there can be no classical solution (i.e. of class C^1), as

$$\lim_{t \to t_c^-} \left(\inf_{x \in \mathbb{R}} \frac{\partial u}{\partial x}(x,t) \right) = -\infty$$

Let us consider Fig. 15.1: note how for $t = t_c$ the solution has a discontinuity.
To account for this loss of uniqueness, we introduce the notion of *weak solution* of a hyperbolic equation: we say that u is a weak solution of (15.1) if it satisfies the differential relation (15.1) at all points $x \in \mathbb{R}$ except for those where it is discontinuous. In the latter, we no longer expect (15.1) to hold (it would make no sense to differentiate a discontinuous function). Rather we require the following *Rankine-Hugoniot* condition to be verified

$$F(u_r) - F(u_l) = \sigma(u_r - u_l), \tag{15.3}$$

where u_r and u_l respectively denote the right and left limit of u at the discontinuity point, and σ is the speed of propagation of the discontinuity. Condition (15.3) therefore expresses the fact that the jump of the flux is proportional to the jump of the solution.

Weak solutions are not necessarily unique: among them, the physically correct one is the so-called *entropic solution*. As we will see at the end of this section, in the case of Burgers' equation, the entropic solution is obtained as the limit, for $\varepsilon \to 0$, of the solution $u^\varepsilon(x,t)$ of the equation having a viscous perturbation term

$$\frac{\partial u^\varepsilon}{\partial t} + \frac{\partial}{\partial x} F(u^\varepsilon) = \varepsilon \frac{\partial^2 u^\varepsilon}{\partial x^2}\ ,\ x \in \mathbb{R},\ t > 0,$$

with $u^\varepsilon(x,0) = u_0(x)$.

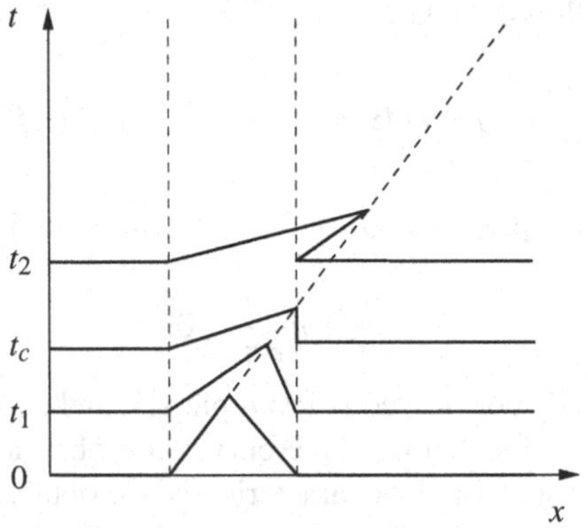

Fig. 15.1. Development of the singularity at the critical time t_c

In general, we can say that:

- *if $F(u)$ is differentiable, a discontinuity that propagates at rate σ given by* (15.3) *satisfies the entropy condition if*

$$F'(u_l) \geq \sigma \geq F'(u_r);$$

- *if $F(u)$ is not differentiable, a discontinuity that propagates at rate σ given by* (15.3) *satisfies the entropy condition if*

$$\frac{F(u)-F(u_l)}{u-u_l} \geq \sigma \geq \frac{F(u)-F(u_r)}{u-u_r},$$

for each u between u_l and u_r.

Example 15.1. Let us consider Burgers' equation with the following initial condition

$$u_0(x) = \begin{cases} u_l & \text{if } x<0, \\ u_r & \text{if } x>0, \end{cases}$$

where u_r and u_l are two constants. If $u_l > u_r$, then there exists a unique weak solution (which is also entropic)

$$u(x,t) = \begin{cases} u_l, & x<\sigma t, \\ u_r, & x>\sigma t, \end{cases} \tag{15.4}$$

where $\sigma = (u_l+u_r)/2$ is the propagation rate of the discontinuity (also called *shock*). In this case the characteristics "enter" the shock (see Fig. 15.2).
If $u_l < u_r$, there are infinitely many weak solutions: one still has (15.4), but in this case the characteristics *exit* the discontinuity (see Fig. 15.3). Such solution is unstable, i.e. small perturbations on the data substantially change the solution itself. Another weak solution is

$$u(x,t) = \begin{cases} u_l & \text{if } x<u_l t, \\ \dfrac{x}{t} & \text{if } u_l t \leq x \leq u_r t, \\ u_r & \text{if } x>u_r t. \end{cases}$$

Such solution, describing a rarefaction wave, is entropic in contrast to the previous one (see Fig. 15.4). ∎

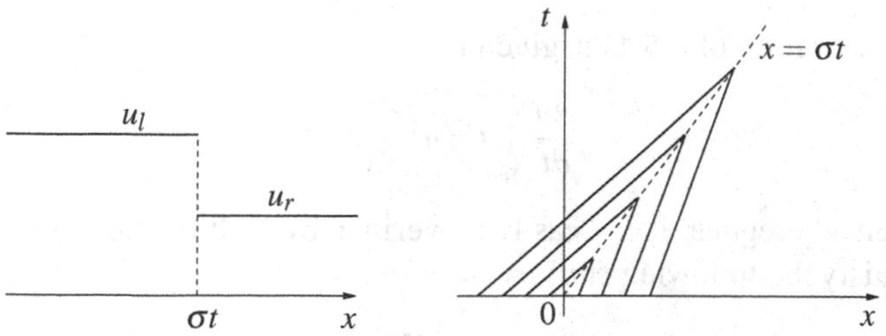

Fig. 15.2. Entropic solution to Burgers' equation

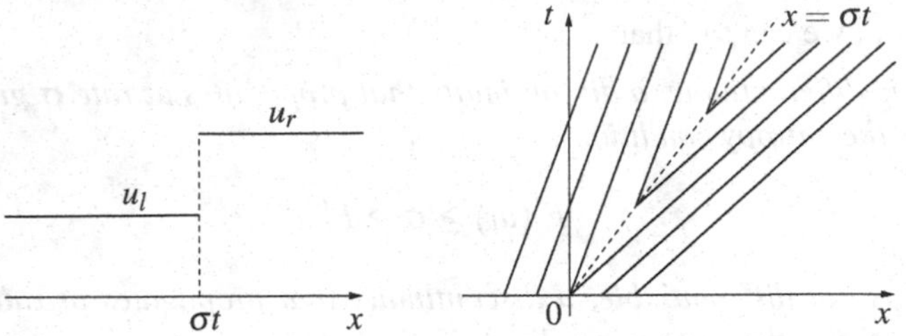

Fig. 15.3. Non-entropic solution to Burgers' equation

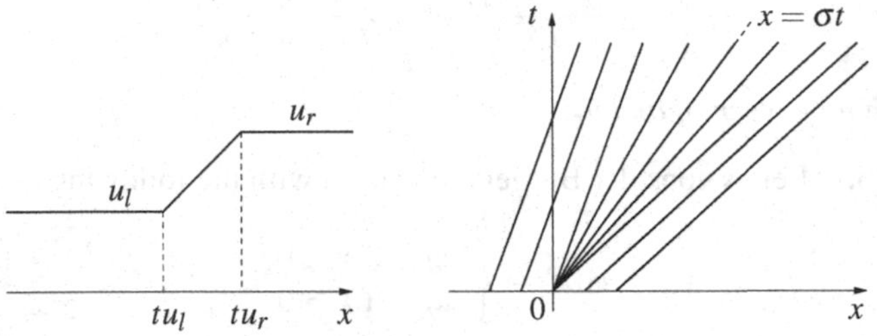

Fig. 15.4. Rarefaction wave

We say that a hyperbolic problem (15.1) has an *entropy function* if there exist a strictly convex function $\eta = \eta(u)$ and a function $\Psi = \Psi(u)$ such that

$$\Psi'(u) = \eta'(u)F'(u), \tag{15.5}$$

where the 'prime' $'$ denotes the derivative with respect to the argument u. The function η is called *entropy* and Ψ is called *entropy flux*. We recall that a function η is said to be convex if for each distinct u and w and for each $\theta \in (0,1)$, we have

$$\eta(u + \theta(w-u)) < (1-\theta)\eta(u) + \theta\eta(w).$$

If η has a continuous second derivative, this is equivalent to requiring that $\eta'' > 0$.

Remark 15.1. The one presented here is a "mathematical" definition of entropy. In the case where (15.1) governs a physical phenomenon, it is often possible to define a "thermodynamic entropy". The latter turns out to be an entropy according to the definition given above. •

The quasi-linear form of (15.1) is given by

$$\frac{\partial u}{\partial t} + F'(u)\frac{\partial u}{\partial x} = 0. \tag{15.6}$$

If u is sufficiently regular, it can easily be verified by multiplying (15.6) by $\eta'(u)$ that η and Ψ satisfy the following *conservation law*

$$\frac{\partial \eta}{\partial t}(u) + \frac{\partial \Psi}{\partial x}(u) = 0. \tag{15.7}$$

For a scalar equation it is generally possible to find different pairs of functions η and Ψ that satisfy the given conditions.

The operations carried out to derive (15.7) make sense only if u is regular, in particular if there are no discontinuities in the solution. However, we can find the conditions to be verified by the entropy variable at a discontinuity in the solution of (15.1) when such equation represents the limit for $\varepsilon \to 0^+$ of the following regularized equation (called *viscosity equation*)

$$\frac{\partial u}{\partial t} + \frac{\partial F}{\partial x}(u) = \varepsilon \frac{\partial^2 u}{\partial x^2}. \tag{15.8}$$

The solution of (15.8) is regular for each $\varepsilon > 0$, and performing the same manipulations used previously we can write

$$\frac{\partial \eta}{\partial t}(u) + \frac{\partial \Psi}{\partial x}(u) = \varepsilon \eta'(u) \frac{\partial^2 u}{\partial x^2} = \varepsilon \frac{\partial}{\partial x}\left[\eta'(u)\frac{\partial u}{\partial x}\right] - \varepsilon \eta''(u)\left(\frac{\partial u}{\partial x}\right)^2.$$

By integrating on a generic rectangle $[x_1, x_2] \times [t_1, t_2]$ we obtain

$$\int_{t_1}^{t_2}\int_{x_1}^{x_2}\left[\frac{\partial \eta}{\partial t}(u) + \frac{\partial \Psi(u)}{\partial x}\right] dxdt = \varepsilon \int_{t_1}^{t_2}\left[\eta'(u(x_2,t))\frac{\partial u}{\partial x}(x_2,t)\right.$$
$$\left. - \eta'(u(x_1,t))\frac{\partial u}{\partial x}(x_1,t)\right] dt - \varepsilon \int_{t_1}^{t_2}\int_{x_1}^{x_2} \eta''(u)\left(\frac{\partial u}{\partial x}\right)^2 dxdt = R_1(\varepsilon) + R_2(\varepsilon),$$

where we have set

$$R_1(\varepsilon) = \varepsilon \int_{t_1}^{t_2}\left[\eta'(u(x_2,t))\frac{\partial u}{\partial x}(x_2,t) - \eta'(u(x_1,t))\frac{\partial u}{\partial x}(x_1,t)\right] dt,$$

$$R_2(\varepsilon) = -\varepsilon \int_{t_1}^{t_2}\int_{x_1}^{x_2} \eta''(u)\left(\frac{\partial u}{\partial x}\right)^2 dxdt.$$

We have

$$\lim_{\varepsilon \to 0^+} R_1(\varepsilon) = 0,$$

while if the solution for $\varepsilon \to 0^+$ of the modified problem denotes a discontinuity along a curve of the (x,t) plane, we have

$$\lim_{\varepsilon \to 0^+} R_2(\varepsilon) \neq 0,$$

as the integral containing the term $(\frac{\partial u}{\partial x})^2$ is, in general, unbounded.

On the other hand $R_2(\varepsilon) \leq 0$ for each $\varepsilon > 0$, with $\partial^2\eta/\partial u^2 > 0$, hence the weak boundary solution for $\varepsilon \to 0^+$ satisfies

$$\int_{t_1}^{t_2}\int_{x_1}^{x_2}\left[\frac{\partial \eta}{\partial t}(u) + \frac{\partial \Psi}{\partial x}(u)\right] dxdt \leq 0 \quad \forall x_1, x_2, t_1, t_2. \tag{15.9}$$

In other words

$$\frac{\partial \eta}{\partial t}(u) + \frac{\partial \Psi}{\partial x}(u) \leq 0, \quad x \in \mathbb{R}, \quad t > 0$$

in a weak sense.

There is obviously a relationship between what we have just seen and the notion of entropic solution. If the differential equation admits an entropy function η, then a weak solution is an entropic solution if and only if η satisfies (15.9). In other words, entropic solutions are limits, as $\varepsilon \to 0^+$, of solutions of the regularized problem (15.8).

15.2 Finite difference approximation

Let us return to the nonlinear hyperbolic equation (15.1), with initial condition

$$u(x,0) = u_0(x),\ x \in \mathbb{R} \ .$$

We denote by $a(u) = F'(u)$ its characteristic rate.
Also for this problem, we can use an explicit finite difference scheme of the form (13.13). The functional interpretation of $H^n_{j+1/2} = H(u^n_j, u^n_{j+1})$ is

$$H^n_{j+1/2} \simeq \frac{1}{\Delta t}\int_{t^n}^{t^{n+1}} F(u(x_{j+1/2},t))dt,$$

so $H^n_{j+1/2}$ approximates the mean flux through $x_{j+1/2}$ in the time interval $[t^n, t^{n+1}]$. To have consistency, the numerical flux $H(\cdot,\cdot)$ must verify

$$H(\overline{u},\overline{u}) = F(\overline{u}), \qquad (15.10)$$

in the case where $\overline{u}$ is a constant. Under hypothesis (15.10), thanks to a classical result by Lax and Wendroff, the functions u such that

$$u(x_j,t^n) = \lim_{\Delta t, h \to 0} u^n_j$$

are weak solutions of the original problem (15.1).
Unfortunately, however, solutions obtained in this manner do not necessarily satisfy the entropy condition (i.e. weak solutions may not be entropic).
In order to "recover" the entropic solutions, numerical schemes must introduce a suitable numerical diffusion, as suggested by the analysis of Sect. 15.1. To this end, we rewrite (13.13) in the form

$$u^{n+1}_j = G(u^n_{j-1}, u^n_j, u^n_{j+1}) \qquad (15.11)$$

and we introduce some definitions. The numerical scheme (15.11) is called:

- *monotone* if G is a monotonically increasing function in each of its arguments;
- *bounded* if there exists $C > 0$ such that $\sup_{j,n}|u^n_j| \leq C$;
- *stable* if $\forall\, h > 0$, $\exists\, \delta_0 > 0$ (possibly dependent on h) such that for each $0 < \Delta t < \delta_0$, if $\mathbf{u}^n$ and $\mathbf{v}^n$ are the finite difference solutions obtained starting from the two initial data $\mathbf{u}^0$ and $\mathbf{v}^0$, then

$$\|\mathbf{u}^n - \mathbf{v}^n\|_\Delta \leq C_T \|\mathbf{u}^0 - \mathbf{v}^0\|_\Delta \ , \qquad (15.12)$$

for each $n \geq 0$ such that $n\Delta t \leq T$ and for any choice of the initial data $\mathbf{u}^0$ and $\mathbf{v}^0$. The constant $C_T > 0$ is independent of Δt and h, and $\|\cdot\|_\Delta$ is a suitable discrete norm, such as those introduced in (13.26). Note that for linear problems, this definition is equivalent to (13.25). We say that the numerical scheme is *strongly stable* when in (15.12) we can take $C_T = 1$ for each $T > 0$.

For example, using $F_j = F(u_j)$ for simplicity of notation, the Lax-Friedrichs scheme for problem (15.1) is realized through the general scheme (13.13) where we take

$$H_{j+1/2} = \frac{1}{2}\left[F_{j+1} + F_j - \frac{1}{\lambda}(u_{j+1} - u_j)\right].$$

This method is consistent, stable and monotone provided that the following condition (analogous to the CFL condition seen previously in the linear case) holds

$$|F'(u_j^n)|\frac{\Delta t}{h} \leq 1 \quad \forall j \in \mathbb{Z}, \ \forall n \in \mathbb{N}. \tag{15.13}$$

A classical result due to N.N. Kuznetsov establishes that monotone schemes of the type (15.11) are bounded, stable, convergent to the entropic solution, and are accurate to order one, at most, with respect to both time and space, that is there exists a constant $C > 0$ such that

$$\max_{j,n} |u_j^n - u(x_j, t^n)| \leq C(\Delta t + h).$$

These schemes are generally too dissipative and do not generate accurate solutions except when using very fine grids.

Higher order schemes (called *high order shock capturing schemes)* can be developed using techniques that allow to calibrate the numerical dissipation as a function of the local regularity of the solution. By doing so one can solve the discontinuities correctly (ensuring the convergence of entropic solution and avoiding spurious oscillations) by using a minimal numerical dissipation. This is a complex topic and cannot be sorted out within a few lines. For an in-depth analysis, we refer to [LeV02b, LeV07, GR96, Hir88].

15.3 Approximation by discontinuous finite elements

For the discretization of problem (15.1) we now consider the space approximation based on discontinuous Galerkin (DG) finite elements. Using the same notations introduced in Sect. 14.4, we seek, for each $t > 0$, $u_h(t) \in W_h$ such that we have $\forall j = 0, \ldots, m-1$ and $\forall v_h \in \mathbb{P}_r(I_j)$,

$$\int_{I_j} \frac{\partial u_h}{\partial t} v_h \, dx - \int_{I_j} F(u_h)\frac{\partial v_h}{\partial x} dx + H_{j+1}(u_h)v_h^-(x_{j+1}) - H_j(u_h)v_h^+(x_j) = 0, \tag{15.14}$$

with $I_j = [x_j, x_{j+1}]$. The initial datum u_h^0 is provided by the relations

$$\int_{I_j} u_h^0 v_h dx = \int_{I_j} u_0 v_h dx, \quad j = 0, \ldots, m-1.$$

The function H_j now denotes the nonlinear flux at node x_j and depends on the values of u_h at x_j, that is

$$H_j(u_h(t)) = H(u_h^-(x_j,t), u_h^+(x_j,t)), \tag{15.15}$$

for a suitable numerical flux $H(\cdot,\cdot)$. If $j=0$ we will have to set $u_h^-(x_0,t) = \phi(t)$, which is the boundary datum at the left extremum (assuming of course that this is the inflow point).

We note that there exist various options for the choice of H. The first requirement is that the numerical flux H has to be consistent with the flux F, i.e. it must satisfy property (15.10) for any constant value $\overline{u}$. Moreover, we want (15.14) to be perturbations of *monotone* schemes. Indeed, as noted in the previous section, the latter are stable and convergent to the entropic solution albeit being only first-order accurate. More precisely, we require (15.14) to be a monotone scheme when $r=0$. In this case, having denoted by $u_h^{(j)}$ the *constant value* of u_h on I_j, (15.14) becomes

$$h_j \frac{\partial}{\partial t} u_h^{(j)}(t) + H(u_h^{(j)}(t), u_h^{(j+1)}(t)) - H(u_h^{(j-1)}(t), u_h^{(j)}(t)) = 0, \tag{15.16}$$

with initial datum $u_h^{0,(j)} = h_j^{-1} \int_{x_j}^{x_{j+1}} u_0\, dx$ in the interval I_j, $j = 0, \ldots, m-1$. We have denoted by $h_j = x_{j+1} - x_j$ the length of I_j.

In order for scheme (15.16) to be monotone, the flux H must be monotone, which is equivalent to saying that $H(v,w)$ is a Lipschitz function of its arguments, not decreasing in v and not increasing in w, that is $v \rightarrow H(v,\cdot)$ is a non-decreasing function while $w \rightarrow H(\cdot,w)$ is non-increasing. In symbols, $H(\uparrow,\downarrow)$.

Three classical examples of monotone fluxes are the following:

1. *Godunov Flux*

$$H(v,w) = \begin{cases} \min\limits_{v \le u \le w} F(u) & \text{if } v \le w, \\ \max\limits_{w \le u \le v} F(u) & \text{if } v > w; \end{cases}$$

2. *Engquist-Osher Flux*

$$H(v,w) = \int_0^v \max(F'(u),0) du + \int_0^w \min(F'(u),0) du + F(0);$$

3. *Lax-Friedrichs Flux*

$$H(v,w) = \frac{1}{2}[F(v) + F(w) - \delta(w-v)], \quad \delta = \max_{\inf_x u_0(x) \le u \le \sup_x u_0(x)} |F'(u)|.$$

The Godunov flux is the one yielding the least amount of numerical dissipation, the Lax-Friedrichs is the cheapest to evaluate. However, numerical experience suggests that if the degree r increases, the choice of the flux H has no significant consequences on the quality of the approximation.

In the linear case, where $F(u) = au$, all previous fluxes coincide and are equal to the upwind flux

$$H(v,w) = a\frac{v+w}{2} - \frac{|a|}{2}(w-v). \tag{15.17}$$

In this case we observe that the scheme (15.14) exactly coincides with the one introduced in (14.42) when $a > 0$. Indeed, having set $a_0 = 0$ and $f = 0$ in (14.42) and integrating by parts we obtain, for each $j = 1, \dots, m-1$

$$\begin{aligned} &\int_{I_j} \frac{\partial u_h}{\partial t} v_h\,dx - \int_{I_j} (au_h)\frac{\partial v_h}{\partial x}\,dx \\ &+ (au_h)^-(x_{j+1})v_h^-(x_{j+1}) - (au_h)^-(x_j)v_h^+(x_j) = 0, \end{aligned} \tag{15.18}$$

i.e. (15.14), keeping in mind that in the case under exam $au_h = F(u_h)$ and, $\forall j = 1, \dots, m-1$,

$$(au_h)^-(x_j) = a\frac{u_h^-(x_j) + u_h^+(x_j)}{2} - \frac{a}{2}(u_h^+(x_j) - u_h^-(x_j)) = H_j(u_h).$$

Verifying the case $j = 0$ is obvious.

We have the following stability result

$$\|u_h(t)\|^2_{L^2(\alpha,\beta)} + \theta(u_h(t)) \leq \|u_h^0\|^2_{L^2(\alpha,\beta)}$$

having set $[u_h]_j = u_h^+(x_j) - u_h^-(x_j)$, and

$$\theta(u_h(t)) = |a| \int_0^t \sum_{j=1}^{m-1} [u_h(t)]_j^2\,dt.$$

Note how jumps are also controlled by the initial datum. The convergence analysis provides the following result (under the assumption that $u_0 \in \mathrm{H}^{r+1}(\alpha,\beta)$)

$$\|u(t) - u_h(t)\|_{L^2(\alpha,\beta)} \leq Ch^{r+1/2}|u_0|_{\mathrm{H}^{r+1}(\alpha,\beta)}, \tag{15.19}$$

hence a convergence order ($= r + 1/2$) larger than the one ($= r$) we would have using continuous finite elements, as previously encountered in the linear case (see (14.48)). In the nonlinear case and for $r = 0$, defining the seminorm

$$|v|_{TV(\alpha,\beta)} = \sum_{j=0}^{m-1} |v_{j+1} - v_j|, \quad v \in W_h,$$

and taking the Engquist-Osher numerical flux in (15.16), we have the following result (due to N.N. Kuznestov)

$$\|u(t)-u_h(t)\|_{L^1(\alpha,\beta)} \leq \|u_0-u_h^0\|_{L^1(\alpha,\beta)} + C|u_0|_{TV(\alpha,\beta)}\sqrt{th}.$$

Moreover, $|u_h(t)|_{TV(\alpha,\beta)} \leq |u_h^0|_{TV(\alpha,\beta)} \leq |u_0|_{TV(\alpha,\beta)}$.

15.3.1 Temporal discretization of DG methods

For the temporal discretization, we first write scheme (15.14) in the algebraic form

$$\begin{aligned}
&\mathrm{M}_h\dot{\mathbf{u}}_h(t) = L_h(\mathbf{u}_h(t),t), \quad t \in (0,T),\\
&\mathbf{u}_h(0) = \mathbf{u}_h^0,
\end{aligned}$$

$\mathbf{u}_h(t)$ being vector of degrees of freedom, $L_h(\mathbf{u}_h(t),t)$ the vector resulting from the discretization of the flux term $-\frac{\partial F}{\partial x}$ and M_h the mass matrix. M_h is a block diagonal matrix whose j-th block is the mass matrix corresponding to the element I_j. As previously observed, the latter is diagonal if we resort to the Legendre polynomial basis, which is orthogonal.

For the temporal discretization, in addition to the previously discussed Euler schemes, we can resort to the following second-order Runge-Kutta method:

$$\begin{aligned}
&\mathrm{M}_h(\mathbf{u}_h^* - \mathbf{u}_h^n) = \Delta t L_h(\mathbf{u}_h^n, t^n),\\
&\mathrm{M}_h(\mathbf{u}_h^{**} - \mathbf{u}_h^*) = \Delta t L_h(\mathbf{u}_h^*, t^{n+1}),\\
&\mathbf{u}_h^{n+1} = \tfrac{1}{2}(\mathbf{u}_h^n + \mathbf{u}_h^{**}).
\end{aligned}$$

In the case of the linear problem (where $F(u) = au$), using $r = 1$ this scheme is stable in the norm $\|\cdot\|_{L^2(\alpha,\beta)}$ provided that the condition

$$\Delta t \leq \frac{1}{3}\frac{h}{|a|}$$

is satisfied. For an arbitrary r, numerical evidence shows that a scheme of order $2r+1$ must be used, in which case we have stability under the condition

$$\Delta t \leq \frac{1}{2r+1}\frac{h}{|a|}.$$

We report the third order Runge-Kutta scheme, to be used preferably when $r = 1$:

$$\begin{aligned}
&\mathrm{M}_h(\mathbf{u}_h^* - \mathbf{u}_h^n) = \Delta t L_h(\mathbf{u}_h^n, t^n),\\
&\mathrm{M}_h(\mathbf{u}_h^{**} - (\tfrac{3}{4}\mathbf{u}_h^n + \tfrac{1}{4}\mathbf{u}_h^*)) = \tfrac{1}{4}\Delta t L_h(\mathbf{u}_h^*, t^{n+1}),\\
&\mathrm{M}_h(u_h^{n+1} - (\tfrac{1}{3}\mathbf{u}_h^n + \tfrac{2}{3}\mathbf{u}_h^{**})) = \tfrac{2}{3}\Delta t L_h(\mathbf{u}_h^{**}, t^{n+1/2}).
\end{aligned} \qquad (15.20)$$

More in general, the following family of Runge-Kutta methods was proposed by [Shu88] and [SO88, SO89]. Let us set for notational convenience $K_h = M_h^{-1} L_h$. Then the new value $\mathbf{u}_h^{n+1}$ is obtained from $\mathbf{u}_h^n$ as follows:

1. Set $\mathbf{u}_h^{(0)} = \mathbf{u}_h^n$;
2. For $i = 1, \ldots, I$ compute the intermediate vectors

$$\text{2a)}\ \mathbf{u}_h^i = \sum_{p=0}^{i-1} \alpha_{ip} \mathbf{w}_h^{ip},$$

$$\text{2b)}\ \mathbf{w}_h^{ip} = \mathbf{u}_h^{(p)} + \frac{\beta_{ip}}{\alpha_{ip}} \Delta t^n K_h(\mathbf{u}_h^{(p)});$$

3. Set $\mathbf{u}_h^{n+1} = \mathbf{u}_h^{(I)}$

The coefficients are requested to satisfy the following conditions:

i) If $\beta_{ip} \neq 0$ then $\alpha_{ip} \neq 0$;
ii) $\alpha_{ip} \geq 0$ (positivity);
iii) $\sum_{p=0}^{i-1} \alpha_{ip} = 1$ (consistency).

Let us make the stability assumption $|\mathbf{w}_h^{ip}| \leq |\mathbf{u}_h^{(p)}|$ for an arbitrary semi-norm $|\cdot|$. Then ii) and iii) give

$$|\mathbf{u}_h^{(i)}| = |\sum_{p=0}^{i-1} \alpha_{ip} \mathbf{w}_h^{ip}| \leq \sum_{p=0}^{i-1} \alpha_{ip} |\mathbf{w}_h^{ip}|$$
$$\leq \sum_{p=0}^{i-1} \alpha_{ip} |\mathbf{u}_h^{(p)}| \leq \max_{0 \leq p \leq i-1} |\mathbf{u}_h^{(p)}|,$$

whence, in particular, $|\mathbf{u}_h^n| \leq |\mathbf{u}_h^0| \quad \forall n \geq 0$.
The RK-DG schemes are analyzed in [SGT01], where they were named strong stability preserving schemes.

Example 15.2. Let us reconsider the problem of Example 14.2, which we solve with the discontinuous finite element method, using the third-order Runge-Kutta scheme for the temporal discretization. Our scope is to verify (15.19) experimentally. To this end, we use a very small time step, $\Delta t = 5 \times 10^{-4}$, and 5 decreasing values for step h obtained by repeatedly halving the initial value $h = 12.5 \times 10^{-3}$. We have compared the error in $L^2(0,1)$ norm at time $t = 0.01$ for elements of degree r equal to 0, 1, 2 and 3. The result is reported in logarithmic scale in Fig. 15.5. This is in accordance with the theory by which the error tends to zero as $h^{r+1/2}$. Indeed, for $r = 1$ in this particular case convergence is faster than what was predicted in theory: the numerical data provides an order of convergence very close to 2. In the case $r > 1$ we have not reported the results for values smaller than h, as for such values (and for the selected Δt) the problem is numerically unstable. ■

Example 15.3. Let us consider the same linear transport problem of the previous example, now using initial datum the square wave illustrated in Fig. 15.6 left. As the initial datum is discontinuous, the use of high-degree elements does not improve the

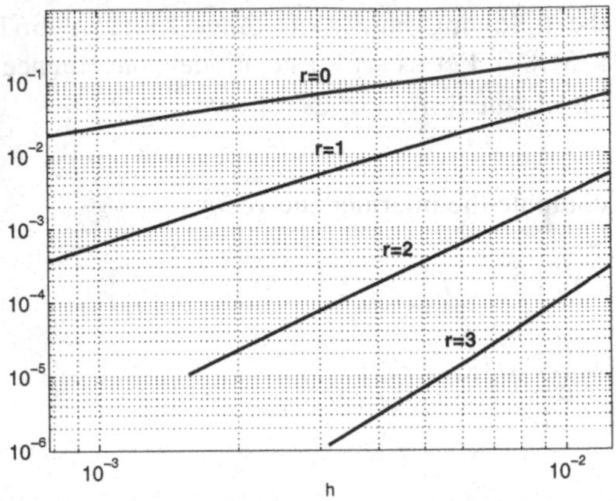

Fig. 15.5. Error $\|u - u_h\|_{L^2(0,1)}$ obtained by solving a linear transport problem with regular initial datum using discontinuous finite elements of degree $r = 0, 1, 2, 3$. The error has been computed at time $t = 0.01$. The time-advancing scheme is the third-order Runge-Kutta scheme with time step $\Delta t = 5 \times 10^{-4}$

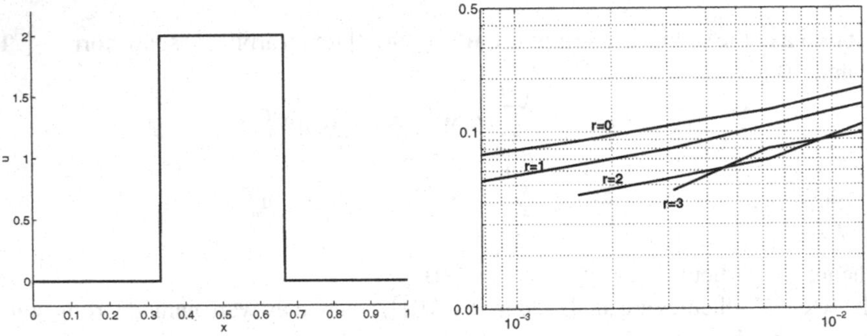

Fig. 15.6. Error $\|u - u_h\|_{L^2(0,1)}$ obtained by solving a linear transport problem with initial datum illustrated in the left figure. We have used discontinuous finite elements of degree r equal 0, 1, 2 and 3. The error has been computed at time $t = 0.01$. The temporal progression scheme is the third-order Runge-Kutta scheme with $\Delta t = 5 \times 10^{-4}$

convergence order, which results to be very close to the theoretical value of $1/2$ for all values of r considered. In Fig. 15.7 we show the oscillations in proximity of the discontinuity of the solution in the case $r = 2$, responsible of the convergence degradation, while the solution for $r = 0$ denotes no oscillation. ■

In the case of the nonlinear problem, using the second-order Runge-Kutta scheme with $r = 0$, under the condition (15.13) we obtain

$$|u_h^n|_{TV(\alpha,\beta)} \leq |u_0|_{TV(\alpha,\beta)},$$

i.e. strong stability in the norm $|\cdot|_{TV(\alpha,\beta)}$.

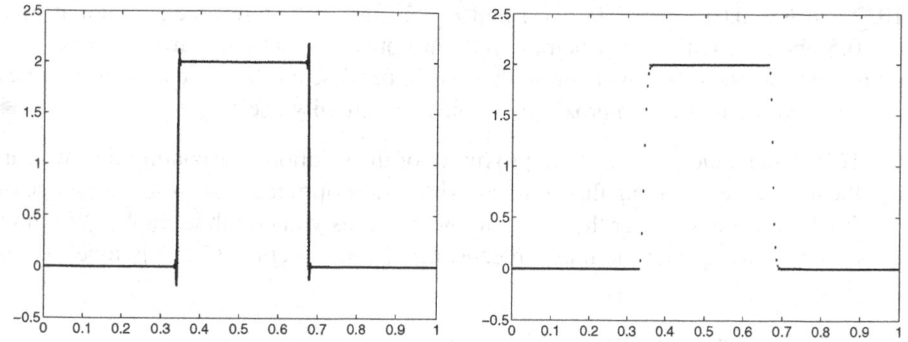

Fig. 15.7. Solution at time $t = 0.01$ and for $h = 3.125 \times 10^{-3}$ for the test case of Fig. 15.6. On the left-hand side, the case $r = 3$: note the presence of oscillations at the discontinuities, while elsewhere the solution is accurate. On the right-hand side, we show the solution obtained when using the same spatial and temporal discretization for $r = 0$

When we do not resort to monotone schemes, it is much more difficult to obtain strong stability. In this case, we can limit ourselves to guaranteeing that the total variation of the *local averages* is uniformly bounded. (See [Coc98].)

Example 15.4. This example illustrates a typical feature of nonlinear problems, that is how discontinuities can show up even if we start from a regular initial datum. To this end, we consider the Burgers equation (15.2) in the $(0,1)$ interval, with initial datum (see Fig. 15.8)

$$u_0(x) = \begin{cases} 1, & 0 \leq x \leq \frac{5}{12}, \\ 54(2x - \frac{5}{6})^3 - 27(2x - \frac{5}{6})^2 + 1, & \frac{5}{12} < x < \frac{7}{12}, \\ 0, & \frac{7}{12} \leq x \leq 1. \end{cases}$$

It can be easily verified that u_0, illustrated in Fig. 15.8, is of class $C^1(0,1)$.

We have then considered the numerical solution obtained with the discontinuous Galerkin method, using the third-order Runge-Kutta scheme with a time step of $\Delta t =$

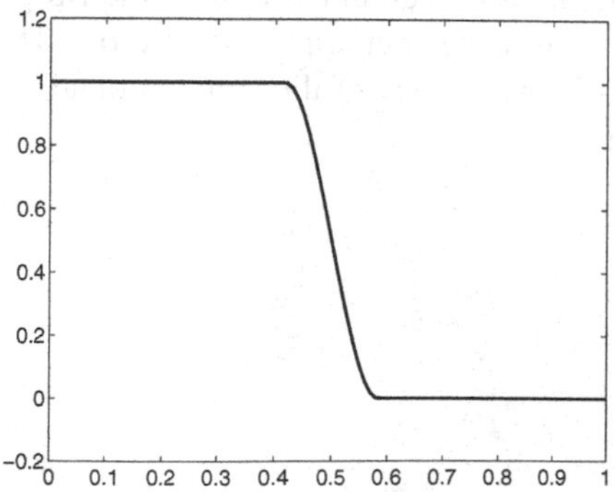

Fig. 15.8. Initial solution u_0 of the first test case of the Burgers problem

10^{-3} and $h = 0.01$, for $r = 0$, $r = 1$ and $r = 2$. Fig. 15.9 shows the solution at time $t = 0.5$ obtained with such schemes. We can note a discontinuity arising, which the numerical scheme solves without oscillations in the case $r = 0$, while for larger values of r we have oscillations in proximity of the discontinuity itself ■

To eliminate the oscillations in proximity of the solution's discontinuities, we can use the technique involving flux limiters, whose description goes beyond the scope of this book. For this we refer the reader to the previously cited bibliography. We limit ourselves to saying that the third-order Runge-Kutta scheme (15.20) is modified as follows

$$\mathbf{u}_h^* = \Lambda_h \left(\mathbf{u}_h^n + \Delta t \mathrm{M}_h^{-1} L_h(\mathbf{u}_h^n, t^n)\right),$$

$$\mathbf{u}_h^{**} = \Lambda_h \left(\tfrac{3}{4}\mathbf{u}_h^n + \tfrac{1}{4}\mathbf{u}_h^* + \tfrac{1}{4}\Delta t \mathrm{M}_h^{-1} L_h(\mathbf{u}_h^*, t^{n+1})\right),$$

$$\mathbf{u}_h^{n+1} = \Lambda_h \left(\tfrac{1}{3}\mathbf{u}_h^n + \tfrac{2}{3}\mathbf{u}_h^{**} + \tfrac{2}{3}\Delta t \mathrm{M}_h^{-1} L_h(\mathbf{u}_h^{**}, t^{n+1/2})\right),$$

Λ_h being the flux limiter, that is a function depending also on the variations of the computed solutions, i.e. the difference between the values of two adjacent nodes. This is equal to the identity operator where the solution is regular, while it limits its variations if these cause high-frequency oscillations in the numerical solution. Clearly Λ_h must be constructed in a suitable way, in particular it has to maintain the properties of consistency and conservation of the scheme, and must differ as little as possible from the identity operator so as to prevent accuracy degradation.

For the sake of an example, we report in Fig. 15.10 the result obtained with linear discontinuous finite elements ($r = 1$) for the same test case of Fig. 15.9 applying flux limiters. The obtained numerical solution is more regular, although slightly more diffusive than that of Fig. 15.9.

Example 15.5. Let us now consider a second problem, where the initial datum is that of Fig. 15.11, obtained by reflecting with respect to the line $x = 0.5$ the datum of the previous test case. By keeping all the remaining parameters of the numerical simulation unchanged, we once again examine the solution at $t = 0.5$. The latter is illustrated in Fig. 15.12. In this case, the solution remains continuous; in fact, with this initial condition, the characteristic lines (which in the case of the Burgers equation are straight lines in the plane (x,t) with slope $\arctan u^{-1}$) never cross. The zoom allows to appreciate qualitatively the better accuracy of the solution obtained for $r = 2$ with respect to the one obtained for $r = 1$. ■

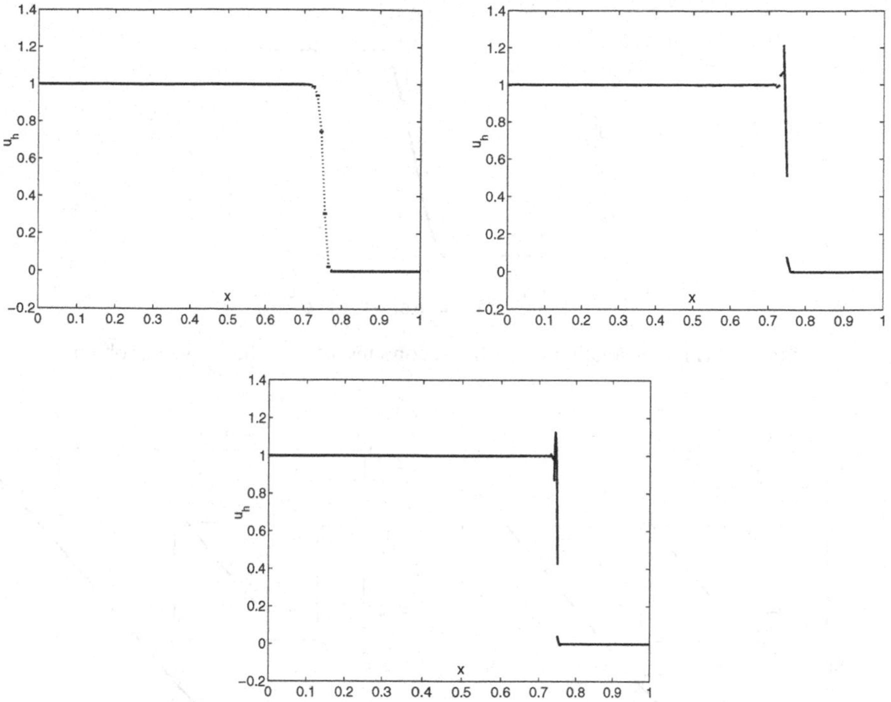

Fig. 15.9. Solution at time $t = 0.5$ of the first test case of the Burgers problem. Comparison between the numerical solution for $r = 0$ (top left), $r = 1$ (top right) and $r = 2$ (bottom). For the case $r = 0$, the piecewise constant discrete solution has been highlighted by connecting with a dashed line the values at the midpoint of each element

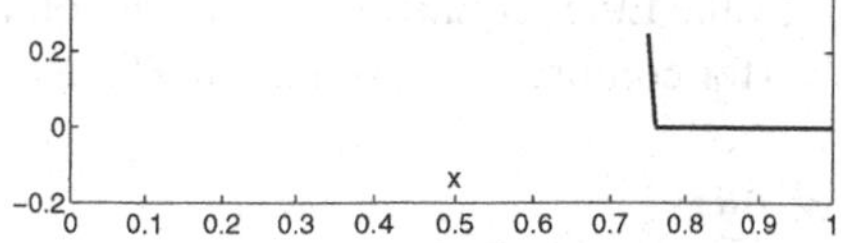

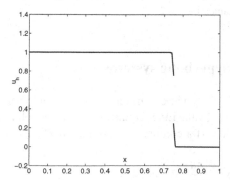

Fig. 15.10. Solution at time $t = 0.5$ for the first test case of the Burgers problem. It has been obtained for $r = 1$ and by applying the flux limiters technique to regularize the numerical solution near the discontinuities

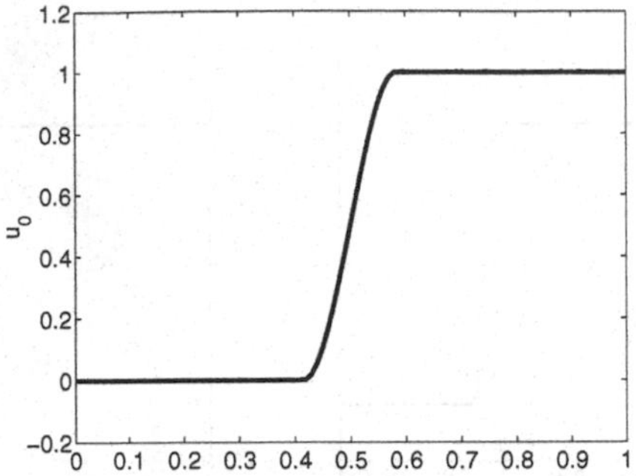

Fig. 15.11. Initial solution u_0 for the second test case of the Burgers problem

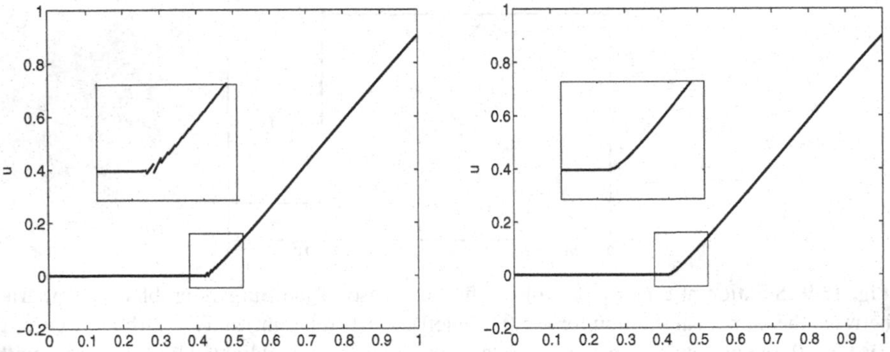

Fig. 15.12. Solution at time $t = 0.5$ for the second test case of the Burgers problem. Comparison between the solution obtained for $r = 1$ (left) and the one obtained for $r = 2$ (right). In the box we illustrate an enlargement of the numerical solution which allows to qualitatively grasp the improved accuracy obtained for $r = 2$

15.4 Nonlinear hyperbolic systems

In this last section we briefly address the case of nonlinear hyperbolic systems. A classical example is provided by the Euler equations, which are obtained from the following Navier-Stokes equations (for compressible fluids) in $\mathbb{R}^d$, $d = 1,2,3$:

$$\begin{aligned}
&\frac{\partial \rho}{\partial t} + \sum_{j=1}^{d} \frac{\partial(\rho u_j)}{\partial x_j} = 0, \\
&\frac{\partial(\rho u_i)}{\partial t} + \sum_{j=1}^{d} \left[\frac{\partial(\rho u_i u_j + \delta_{ij} p)}{\partial x_j} - \frac{\partial \tau_{ij}}{\partial x_j} \right] = 0, \;\; i = 1,\ldots,d\,, \\
&\frac{\partial \rho e}{\partial t} + \sum_{j=1}^{d} \left[\frac{\partial(\rho h u_j)}{\partial x_j} - \frac{\partial(\sum_{i=1}^{d} u_i \tau_{ij} + q_j)}{\partial x_j} \right] = 0.
\end{aligned} \tag{15.21}$$

The variables have the following meaning: $\mathbf{u} = (u_1, \ldots, u_d)^T$ is the vector of velocities, ρ is the density, p the pressure, $e_i + \frac{1}{2}|\mathbf{u}|^2$ the total energy per unit of mass, equal to the sum of the internal energy e_i and of the kinetic energy of the fluid, $h = e + p/\rho$ the total entalpy per mass unit, $\mathbf{q}$ the thermal flux and finally

$$\tau_{ij} = \mu \left[\left(\frac{\partial u_j}{\partial x_i} + \frac{\partial u_i}{\partial x_j} \right) - \frac{2}{3} \delta_{ij} \mathrm{div}\mathbf{u} \right]$$

the stress tensor (μ being the molecular viscosity of the fluid).
The equations in the above system describe the conservation of mass, momentum and energy, respectively. To complete the system, it is necessary to put e in relationship with the variables $\rho, p, \mathbf{u}$, by defining a law

$$e = \Phi(\rho, p, \mathbf{u}).$$

The latter is normally derived from the state equations of the fluid under exam. In particular, the state equations of the ideal gas

$$p = \rho RT, \qquad e_i = C_v T,$$

where $R = C_p - C_v$ is the gas constant and T is its temperature, provide

$$e = \frac{p}{\rho(\gamma - 1)} + \frac{1}{2}|\mathbf{u}|^2,$$

where $\gamma = C_p/C_v$ is the ratio between the specific heats at constant pressure and volume, respectively. The thermal flux $\mathbf{q}$ is usually related to the temperature gradient via the Fick law

$$\mathbf{q} = -\kappa \nabla T = -\frac{\kappa}{C_v} \nabla (e - \frac{1}{2}|\mathbf{u}|^2),$$

κ being the conductibility of the fluid under exam.

If $\mu = 0$ and $\kappa = 0$, we obtain the Euler equations for non-viscous fluids. The interested reader can find them in specialized fluid dynamics textbooks, or in textbooks on nonlinear hyperbolic systems, such as for instance [Hir88] or [GR96]. Such equations can be written in compact form in the following way

$$\frac{\partial \mathbf{w}}{\partial t} + \mathrm{Div} F(\mathbf{w}) = \mathbf{0}, \tag{15.22}$$

with $\mathbf{w} = (\rho, \rho\mathbf{u}, \rho e)^T$ being the vector of the so-called *conservative variables*. The flux matrix $F(\mathbf{w})$, a nonlinear function of $\mathbf{w}$, can be obtained from (15.21). For instance, if $d = 2$, we have

$$F(\mathbf{w}) = \begin{bmatrix} \rho u_1 & \rho u_2 \\ \rho u_1^2 + p & \rho u_1 u_2 \\ \rho u_1 u_2 & \rho u_2^2 + p \\ \rho h u_1 & \rho h u_2 \end{bmatrix}.$$

Finally, in (15.22) Div denotes the divergence operator of a tensor: if $\boldsymbol{\tau}$ is a tensor with components (τ_{ij}), its divergence is a vector with components

$$(\mathrm{Div}(\boldsymbol{\tau}))_k = \sum_{j=1}^{d} \frac{\partial}{\partial x_j}(\tau_{kj}), \quad k = 1, ..., d. \tag{15.23}$$

The form (15.22) is called *conservation form of the Euler equations*. Indeed, by integrating it on any region $\Omega \subset \mathbb{R}^d$ and using the Gauss theorem, we obtain

$$\frac{d}{dt}\int_{\Omega} \mathbf{w}\, d\Omega + \int_{\partial\Omega} F(\mathbf{w}) \cdot \mathbf{n}\, d\gamma = 0.$$

This is interpreted by saying that *the variation in time of* $\mathbf{w}$ *in* Ω *is compensated by the variation of the fluxes through the boundary of* Ω; (15.22) is thus a conservation law.

The Navier-Stokes equations can also be written in conservative form as follows

$$\frac{\partial \mathbf{w}}{\partial t} + \mathrm{Div} F(\mathbf{w}) = \mathrm{Div} G(\mathbf{w}),$$

where $G(\mathbf{w})$ are the so-called *viscous fluxes*. Remaining in the $d = 2$ case, these are given by

$$G(\mathbf{w}) = \begin{bmatrix} 0 & 0 \\ \tau_{11} & \tau_{12} \\ \tau_{21} & \tau_{22} \\ \rho h u_1 + \mathbf{u} \cdot \boldsymbol{\tau}_1 + q_1 & \rho h u_2 + \mathbf{u} \cdot \boldsymbol{\tau}_2 + q_2 \end{bmatrix}$$

where $\boldsymbol{\tau}_1 = (\tau_{11}, \tau_{21})^T$ and $\boldsymbol{\tau}_2 = (\tau_{12}, \tau_{22})^T$.

We now rewrite system (15.22) in the form

$$\frac{\partial \mathbf{w}}{\partial t} + \sum_{i=1}^{d} \frac{\partial F_i(\mathbf{w})}{\partial \mathbf{w}} \frac{\partial \mathbf{w}}{\partial x_i} = 0. \tag{15.24}$$

This is a particular case of the following *quasi-linear form*

$$\frac{\partial \mathbf{w}}{\partial t} + \sum_{i=1}^{d} A_i(\mathbf{w}) \frac{\partial \mathbf{w}}{\partial x_i} = \mathbf{0}. \tag{15.25}$$

If the matrix $A^{\alpha}(\mathbf{w}) = \sum_{i=1}^{d} \alpha_i A_i(\mathbf{w})$ can be diagonalized for all real values of $\{\alpha_1, \ldots, \alpha_d\}$ and its eigenvalues are real and distinct, then system (15.25) is said to be strictly hyperbolic.

Example 15.6. A simple example of a strictly hyperbolic problem is provided by the so-called *p-system*:

$$\frac{\partial v}{\partial t} - \frac{\partial u}{\partial x} = 0,$$

$$\frac{\partial u}{\partial t} + \frac{\partial}{\partial x} p(v) = 0.$$

If $p'(v) < 0$, the two eigenvalues of the Jacobian matrix

$$A(\mathbf{w}) = \begin{pmatrix} 0 & -1 \\ p'(v) & 0 \end{pmatrix}$$

are

$$\lambda_1(v) = -\sqrt{-p'(v)} < 0 < \lambda_2(v) = +\sqrt{-p'(v)}\,. \qquad \blacksquare$$

Example 15.7. For the one-dimensional Euler system (i.e. with $d = 1$) we have: $\mathbf{w} = (\rho, \rho u, e)^T$ and $F(\mathbf{w}) = (\rho u, \rho u^2 + p, u(e+p))^T$. The eigenvalues of the matrix $A_1(\mathbf{w})$ of the system are $u-c, u, u+c$ where $c = \sqrt{\gamma\frac{p}{\rho}}$ is the speed of sound. As $u, c \in \mathbb{R}$, the eigenvalues are real and distinct and therefore the one-dimensional Euler system is strictly hyperbolic. ■

The remarks made on the discontinuities of the solution in the scalar case can be extended to the case of systems, by introducing the notion of weak solution also here. The entropy condition can be extended to the case of systems, using for instance the condition proposed by Lax. We observe that in the case of the one-dimensional system (15.22) the Rankine-Hugoniot jump conditions are written in the form

$$F(\mathbf{w}_+) - F(\mathbf{w}_-) = \sigma(\mathbf{w}_+ - \mathbf{w}_-),$$

$\mathbf{w}_\pm$ being the two constant states that represent the values of the unknowns through the discontinuity, and σ representing once again the rate at which the discontinuity propagates. Using the fundamental theorem of calculus, such relation is written in the form

$$\begin{aligned} \sigma(\mathbf{w}_+ - \mathbf{w}_-) &= \int_0^1 DF(\theta\mathbf{w}_+ + (1-\theta)\mathbf{w}_-)\cdot(\mathbf{w}_+ - \mathbf{w}_-)d\theta \\ &= A(\mathbf{w}_-, \mathbf{w}_+)\cdot(\mathbf{w}_+ - \mathbf{w}_-), \end{aligned} \tag{15.26}$$

where the matrix

$$A(\mathbf{w}_-, \mathbf{w}_+) = \int_0^1 DF(\theta\mathbf{w}_+ + (1-\theta)\mathbf{w}_-)d\theta$$

represents the mean value of the Jacobian of F (denoted by DF) along the segment connecting $\mathbf{w}_-$ with $\mathbf{w}_+$. Relation (15.26) shows that at each point where a discontinuity occurs the difference between the right and left state $\mathbf{w}_+ - \mathbf{w}_-$ is an eigenvector of the matrix $A(\mathbf{w}_-, \mathbf{w}_+)$, while the rate of the jump σ coincides with the corresponding eigenvalue $\lambda = \lambda(\mathbf{w}_-, \mathbf{w}_+)$. Calling $\lambda_i(\mathbf{w})$ the i–th eigenvalue of

$$A(\mathbf{w}) = DF(\mathbf{w}),$$

the *admissibility condition* of Lax requires that

$$\lambda_i(\mathbf{w}_+) \le \sigma \le \lambda_i(\mathbf{w}_-) \quad \text{for each i.}$$

Intuitively, this means that the velocity at which a shock of the i-th family travels must exceed the velocity $\lambda_i(\mathbf{w}_+)$ of the waves that are immediately ahead of the shock, and must be less than the velocity $\lambda_i(\mathbf{w}_-)$ of the waves behind the shock.

In the case of hyperbolic systems of m equations ($m > 1$), the entropy η and its corresponding flux Ψ are still scalar functions, and relation (15.5) becomes

$$\nabla\Psi(\mathbf{u}) = \frac{d\mathbf{F}}{d\mathbf{u}}(\mathbf{u}) \cdot \nabla\eta(\mathbf{u}), \tag{15.27}$$

which represents a system of m equations and 2 unknowns (η and Ψ). If $m > 2$, such system may have no solutions.

Remark 15.2. In the case of the Euler equations, the entropy function exists also in the case $m = 3$. •

16

Navier-Stokes equations

Navier-Stokes equations describe the motion of a fluid with constant density ρ in a domain $\Omega \subset \mathbb{R}^d$ (with $d = 2,3$). They read as follows

$$\begin{cases} \dfrac{\partial \mathbf{u}}{\partial t} - \mathrm{div}\left[\nu\left(\nabla \mathbf{u} + \nabla \mathbf{u}^T\right)\right] + (\mathbf{u} \cdot \nabla)\mathbf{u} + \nabla p = \mathbf{f}, & \mathbf{x} \in \Omega,\ t > 0, \\ \mathrm{div}\mathbf{u} = 0, & \mathbf{x} \in \Omega,\ t > 0, \end{cases} \tag{16.1}$$

$\mathbf{u}$ being the fluid's velocity, p the pressure divided by the density (which will simply be called "pressure"), $\nu = \frac{\mu}{\rho}$ the kinematic viscosity, μ the dynamic viscosity, and $\mathbf{f}$ a forcing term per unit of mass that we suppose belongs in $\mathrm{L}^2(\mathbb{R}^+; [\mathrm{L}^2(\Omega)]^d)$ (see Sect. 5.2). The first equation is that of conservation of linear momentum, the second one that of conservation of mass, which is also called the continuity equation. The term $(\mathbf{u} \cdot \nabla)\mathbf{u}$ describes the process of convective transport, while $-\mathrm{div}\left[\nu(\nabla \mathbf{u} + \nabla \mathbf{u}^T)\right]$ the process of molecular diffusion. System (16.1) can be derived by the analogous system for compressible flows introduced in Chapter 15 by assuming ρ constant, using the continuity equation (which, under current assumptions, takes the simplified form $\mathrm{div}\mathbf{u} = 0$) to simplify the various terms, and finally dividing the equation by ρ. Note that in the incompressible case (16.2) the energy equation has disappeared. Indeed, even though such an equation can still be written for incompressible flows, its solution can be found independently once the velocity field is obtained from the solution of (16.1).

When ν is constant, from the continuity equation we obtain

$$\mathrm{div}\left[\nu(\nabla \mathbf{u} + \nabla \mathbf{u}^T)\right] = \nu\left(\Delta \mathbf{u} + \nabla \mathrm{div}\mathbf{u}\right) = \nu \Delta \mathbf{u}$$

whence system (16.1) can be written in the equivalent form

$$\begin{cases} \dfrac{\partial \mathbf{u}}{\partial t} - \nu \Delta \mathbf{u} + (\mathbf{u} \cdot \nabla)\mathbf{u} + \nabla p = \mathbf{f}, & \mathbf{x} \in \Omega,\ t > 0, \\ \mathrm{div}\mathbf{u} = 0, & \mathbf{x} \in \Omega,\ t > 0, \end{cases} \tag{16.2}$$

which is the one that we will consider in this chapter.

A. Quarteroni: *Numerical Models for Differential Problems*, 2nd Ed.
MS&A – Modeling, Simulation & Applications 8
DOI 10.1007/978-88-470-5522-3_16, © Springer-Verlag Italia 2014

Equations (16.2) are often called incompressible Navier-Stokes equations. More in general, fluids satisfying the *incompressibility condition* $\text{div}\mathbf{u} = 0$ are said to be incompressible. Constant density fluids necessarily satisfy this condition, however there exist incompressible fluids featuring variable density (e.g., stratified fluids) that are governed by a different system of equations in which the density ρ explicitly shows up. This case will be addressed in Sect. 16.9.

In order for problem (16.2) to be well posed it is necessary to assign the initial condition

$$\mathbf{u}(\mathbf{x},0) = \mathbf{u}_0(\mathbf{x}) \quad \forall \mathbf{x} \in \Omega, \tag{16.3}$$

where $\mathbf{u}_0$ is a given divergence-free vector field, together with suitable boundary conditions, such as, e.g., $\forall t > 0$,

$$\begin{cases} \mathbf{u}(\mathbf{x},t) = \varphi(\mathbf{x},t) \quad \forall \mathbf{x} \in \Gamma_D, \\ \left(\nu \dfrac{\partial \mathbf{u}}{\partial \mathbf{n}} - p\mathbf{n}\right)(\mathbf{x},t) = \psi(\mathbf{x},t) \quad \forall \mathbf{x} \in \Gamma_N, \end{cases} \tag{16.4}$$

where φ and ψ are given vector functions, while Γ_D and Γ_N provide a partition of the domain boundary $\partial\Omega$, that is $\Gamma_D \cup \Gamma_N = \partial\Omega$, $\overset{\circ}{\Gamma}_D \cap \overset{\circ}{\Gamma}_N = \emptyset$. Finally, as usual $\mathbf{n}$ is the outward unit normal vector to $\partial\Omega$. If we use the alternative formulation (16.1), the second equation in (16.4) must be replaced by

$$\left[\nu\left(\nabla\mathbf{u} + \nabla\mathbf{u}^T\right)\mathbf{n} - p\mathbf{n}\right](\mathbf{x},t) = \psi(\mathbf{x},t) \quad \forall \mathbf{x} \in \Gamma_N.$$

Further considerations on boundary conditions will follow in Sect. 16.9.2.

Denoting with u_i, $i = 1,\ldots,d$ the components of the vector $\mathbf{u}$ with respect to a Cartesian frame, and with f_i the components of $\mathbf{f}$, system (16.2) can be written componentwise as

$$\begin{cases} \dfrac{\partial u_i}{\partial t} - \nu\Delta u_i + \displaystyle\sum_{j=1}^{d} u_j \dfrac{\partial u_i}{\partial x_j} + \dfrac{\partial p}{\partial x_i} = f_i, \quad i = 1,\ldots,d, \\ \displaystyle\sum_{j=1}^{d} \dfrac{\partial u_j}{\partial x_j} = 0. \end{cases}$$

In the two-dimensional case the Navier-Stokes equations with the boundary conditions previously indicated yield well-posed problems. This means that if all data (initial condition, forcing term, boundary data) are smooth enough, then the solution is continuous together with its derivatives and does not develop singularities in time. Things may go differently in three dimensions, where existence and uniqueness of classical solutions have been proven only locally in time (that is for a sufficiently small time interval). In the following section we will introduce the weak formulation of the Navier-Stokes equations, for which existence of a solution has been proven for all times. However, the issue of uniqueness (which is related to that of regularity) is still open, and is actually the central issue of Navier-Stokes theory.

Remark 16.1. The Navier-Stokes equations have been written in terms of the *primitive variables* $\mathbf{u}$ and p, but other sets of variables may be used, too. For instance, in the two-dimensional case it is common to see the vorticity ω and the streamfunction ψ, that are related to the velocity as follows

$$\omega = \mathrm{rot}\mathbf{u} = \frac{\partial u_2}{\partial x_1} - \frac{\partial u_1}{\partial x_2}, \quad \mathbf{u} = \begin{bmatrix} \dfrac{\partial \psi}{\partial x_2} \\ -\dfrac{\partial \psi}{\partial x_1} \end{bmatrix}.$$

The various formulations are in fact equivalent from a mathematical standpoint, although they give rise to different numerical methods. See, e.g., [Qua93]. •

16.1 Weak formulation of Navier-Stokes equations

A weak formulation of problem (16.2)–(16.4) can be obtained by proceeding formally, as follows. Let us multiply the first equation of (16.2) by a test function $\mathbf{v}$ belonging to a suitable space V that will be specified later on, and integrate Ω

$$\int_\Omega \frac{\partial \mathbf{u}}{\partial t} \cdot \mathbf{v}\, d\Omega - \int_\Omega \nu \Delta \mathbf{u} \cdot \mathbf{v}\, d\Omega + \int_\Omega [(\mathbf{u} \cdot \nabla)\mathbf{u}] \cdot \mathbf{v}\, d\Omega + \int_\Omega \nabla p \cdot \mathbf{v} d\Omega = \int_\Omega \mathbf{f} \cdot \mathbf{v} d\Omega.$$

Using Green's formulae (3.16) and (3.17) we find:

$$-\int_\Omega \nu \Delta \mathbf{u} \cdot \mathbf{v}\, d\Omega = \int_\Omega \nu \nabla \mathbf{u} \cdot \nabla \mathbf{v}\, d\Omega - \int_{\partial\Omega} \nu \frac{\partial \mathbf{u}}{\partial \mathbf{n}} \cdot \mathbf{v}\, d\gamma,$$

$$\int_\Omega \nabla p \cdot \mathbf{v}\, d\Omega = -\int_\Omega p \mathrm{div}\mathbf{v}\, d\Omega + \int_{\partial\Omega} p\mathbf{v} \cdot \mathbf{n}\, d\gamma.$$

Using these relations in the first of (16.2), we obtain

$$\begin{aligned} &\int_\Omega \frac{\partial \mathbf{u}}{\partial t} \cdot \mathbf{v}\, d\Omega + \int_\Omega \nu \nabla \mathbf{u} \cdot \nabla \mathbf{v}\, d\Omega + \int_\Omega [(\mathbf{u} \cdot \nabla)\mathbf{u}] \cdot \mathbf{v}\, d\Omega - \int_\Omega p \mathrm{div}\mathbf{v}\, d\Omega \\ &\quad = \int_\Omega \mathbf{f} \cdot \mathbf{v}\, d\Omega + \int_{\partial\Omega} \left(\nu \frac{\partial \mathbf{u}}{\partial \mathbf{n}} - p\mathbf{n} \right) \cdot \mathbf{v}\, d\gamma \qquad \forall \mathbf{v} \in V. \end{aligned} \tag{16.5}$$

(All boundary integrals should indeed be regarded as duality pairings.)
Similarly, by multiplying the second equation of (16.2) by a test function q, belonging to a suitable space Q to be specified, then integrating on Ω it follows

$$\int_\Omega q \mathrm{div}\mathbf{u}\, d\Omega = 0 \quad \forall q \in Q. \tag{16.6}$$

Customarily V is chosen so that the test functions vanish on the boundary portion where a Dirichlet data is prescribed on $\mathbf{u}$, that is

$$V = [\mathrm{H}^1_{\Gamma_D}(\Omega)]^d = \{\mathbf{v} \in [\mathrm{H}^1(\Omega)]^d : \mathbf{v}|_{\Gamma_D} = \mathbf{0}\}. \tag{16.7}$$

It will coincide with $[\mathrm{H}_0^1(\Omega)]^d$ if $\Gamma_D = \partial\Omega$. If $\Gamma_N \neq \emptyset$, we can choose $Q = \mathrm{L}^2(\Omega)$. Moreover, if $t > 0$, then $\mathbf{u}(t) \in [\mathrm{H}^1(\Omega)]^d$, with $\mathbf{u}(t) = \varphi(t)$ on Γ_D, $\mathbf{u}(0) = \mathbf{u}_0$ and $p(t) \in Q$.

Having chosen these functional spaces, we can note first of all that

$$\int_{\partial\Omega} (\nu \frac{\partial \mathbf{u}}{\partial \mathbf{n}} - p\mathbf{n}) \cdot \mathbf{v} d\gamma = \int_{\Gamma_N} \psi \cdot \mathbf{v} d\gamma \quad \forall \mathbf{v} \in V.$$

Moreover, all the integrals involving bilinear terms are finite. More precisely, by using the vector notation $\mathbf{H}^k(\Omega) = [\mathrm{H}^k(\Omega)]^d, \mathbf{L}^p(\Omega) = [\mathrm{L}^p(\Omega)]^d, k \geq 1, 1 \leq p < \infty$, we find:

$$\left| \nu \int_\Omega \nabla \mathbf{u} \cdot \nabla \mathbf{v} \right| \leq \nu |\mathbf{u}|_{\mathbf{H}^1(\Omega)} |\mathbf{v}|_{\mathbf{H}^1(\Omega)},$$

$$\left| \int_\Omega p \mathrm{div} \mathbf{v} d\Omega \right| \leq \|p\|_{\mathrm{L}^2(\Omega)} |\mathbf{v}|_{\mathbf{H}^1(\Omega)},$$

$$\left| \int_\Omega q \nabla \mathbf{u} d\Omega \right| \leq \|q\|_{\mathrm{L}^2(\Omega)} |\mathbf{u}|_{\mathbf{H}^1(\Omega)}.$$

For every function $\mathbf{v} \in \mathbf{H}^1(\Omega)$, we denote by

$$\|\mathbf{v}\|_{\mathbf{H}^1(\Omega)} = \Big(\sum_{k=1}^d \|v_k\|^2_{\mathrm{H}^1(\Omega)}\Big)^{1/2}$$

its norm and by

$$|\mathbf{v}|_{\mathbf{H}^1(\Omega)} = \Big(\sum_{k=1}^d |v_k|^2_{\mathrm{H}^1(\Omega)}\Big)^{1/2}$$

its seminorm. The notation $\|\mathbf{v}\|_{\mathbf{L}^p(\Omega)}, 1 \leq p < \infty$, has a similar meaning. The same symbols will be used in case of tensor functions. Thanks to Poincaré's inequality, $|\mathbf{v}|_{\mathbf{H}^1(\Omega)}$ is equivalent to the norm $\|\mathbf{v}\|_{\mathbf{H}^1(\Omega)}$ for all functions belonging to V.

Also the integral involving the trilinear term is finite. To see how, let us start by recalling the following result (see (2.19); for its proof, see [Ada75]): if $d \leq 3$, $\exists C > 0$ such that

$$\forall \mathbf{v} \in \mathbf{H}^1(\Omega) \text{ we have } \mathbf{v} \in \mathbf{L}^4(\Omega) \text{ and moreover } \|\mathbf{v}\|_{\mathbf{L}^4(\Omega)} \leq C \|\mathbf{v}\|_{\mathbf{H}^1(\Omega)}.$$

Using the following three-term Hölder inequality

$$\left| \int_\Omega fgh \, d\Omega \right| \leq \|f\|_{\mathrm{L}^p(\Omega)} \|g\|_{\mathrm{L}^q(\Omega)} \|h\|_{\mathrm{L}^r(\Omega)},$$

valid for all $p, q, r > 1$ such that $p^{-1} + q^{-1} + r^{-1} = 1$, we conclude that

$$\left| \int_\Omega [(\mathbf{u} \cdot \nabla)\mathbf{u}] \cdot \mathbf{v} \, d\Omega \right| \leq \|\nabla \mathbf{u}\|_{\mathbf{L}^2(\Omega)} \|\mathbf{u}\|_{\mathbf{L}^4(\Omega)} \|\mathbf{v}\|_{\mathbf{L}^4(\Omega)} \leq C^2 \|\mathbf{u}\|^2_{\mathbf{H}^1(\Omega)} \|\mathbf{v}\|_{\mathbf{H}^1(\Omega)}.$$

As for the solution's uniqueness, let us consider again the Navier-Stokes equations in strong form (16.2) (similar considerations can be made on the weak form (16.5), (16.6)). If $\Gamma_D = \partial\Omega$, when only boundary conditions of Dirichlet type are imposed, the pressure appears merely in terms of its gradient; in such a case, if we call $(\mathbf{u}, p)$ a solution of (16.2), for any possible constant c the couple $(\mathbf{u}, p+c)$ is a solution too, since $\nabla(p+c) = \nabla p$. To avoid such indeterminacy one can fix a priori the value of p at one given point $\mathbf{x}_0$ of the domain Ω, that is set $p(\mathbf{x}_0) = p_0$, or, alternatively, require the pressure to have null average, i.e., $\int_\Omega p\, d\Omega = 0$. The former condition requires to prescribe a pointwise value for the pressure, but this is inconsistent with our Ansatz that $p \in \mathrm{L}^2(\Omega)$. (We anticipate, however, that this is admissible at the numerical level when we look for a continuous finite-dimensional pressure.) For this reason we assume from now on that the pressure is average-free. More specifically, we will consider the following pressure space

$$\mathrm{Q} = \mathrm{L}^2_0(\Omega) = \{p \in \mathrm{L}^2(\Omega) : \int_\Omega p\, d\Omega = 0\}.$$

Further, we observe that if $\Gamma_D = \partial\Omega$, the prescribed Dirichlet data φ must be compatible with the incompressibility constraint; indeed,

$$\int_{\partial\Omega} \varphi \cdot \mathbf{n}\, d\gamma = \int_\Omega \mathrm{div}\mathbf{u}\; \mathrm{d}\Omega = 0.$$

If Γ_N is not empty, i.e. in presence of either Neumann or mixed Dirichlet-Neumann boundary conditions, the problem of pressure indeterminacy (up to an additive constant) no longer exists, as on Γ_N the unknown p appears without derivatives. In this case we can take $\mathrm{Q} = \mathrm{L}^2(\Omega)$. In conclusion, from now on we shall implicitly assume

$$Q = \mathrm{L}^2(\Omega) \quad \text{if} \quad \Gamma_N \neq \emptyset, \quad Q = \mathrm{L}^2_0(\Omega) \quad \text{if } \Gamma_N = \emptyset. \tag{16.8}$$

The weak formulation of the system (16.2), (16.3), (16.4) is therefore:

find $\mathbf{u} \in \mathrm{L}^2(\mathbb{R}^+; [\mathrm{H}^1(\Omega)]^d) \cap C^0(\mathbb{R}^+; [\mathrm{L}^2(\Omega)]^d)$, $p \in \mathrm{L}^2(\mathbb{R}^+; Q)$ such that

$$\begin{cases} \displaystyle\int_\Omega \frac{\partial \mathbf{u}}{\partial t} \cdot \mathbf{v}\, d\Omega + \nu \int_\Omega \nabla\mathbf{u} \cdot \nabla\mathbf{v}\, d\Omega + \int_\Omega [(\mathbf{u}\cdot\nabla)\mathbf{u}] \cdot \mathbf{v}\, d\Omega - \int_\Omega p\, \mathrm{div}\mathbf{v}\, d\Omega \\ \qquad\qquad = \displaystyle\int_\Omega \mathbf{f}\cdot\mathbf{v}\, d\Omega + \int_{\Gamma_N} \psi \cdot \mathbf{v} d\gamma \quad \forall \mathbf{v} \in V, \\ \displaystyle\int_\Omega q\, \mathrm{div}\mathbf{u} d\Omega = 0 \quad \forall q \in Q, \end{cases} \tag{16.9}$$

with $\mathbf{u}|_{\Gamma_D} = \varphi_D$ and $\mathbf{u}|_{t=0} = \mathbf{u}_0$. The space V is the one in (16.7) while Q is the space introduced in (16.8). The spaces of time-dependent functions for $\mathbf{u}$ and p have been introduced in Sect. 2.7.

As we have already anticipated, existence of solutions can be proven for this problem for both dimensions $d = 2$ and $d = 3$, whereas uniqueness has been proven only in the case $d = 2$ for sufficiently small data (see, e.g., [Tem01] and [Sal08]).

Let us define the Reynolds number

$$Re = \frac{|\mathbf{U}|L}{\nu},$$

where L is a representative length of the domain Ω (e.g. the length of a channel where the fluid's flow is studied) and $\mathbf{U}$ a representative fluid velocity.
The Reynolds number measures the extent to which convection dominates over diffusion. When $Re \ll 1$ the convective term $(\mathbf{u} \cdot \nabla)\mathbf{u}$ can be omitted, and the Navier-Stokes equations reduce to the so-called Stokes equations, that will be investigated later in this chapter. On the other hand, if Re is large, problems may arise concerning uniqueness of the solution, the existence of stationary and stable solutions, the possible existence of strange attractors, the transition towards turbulent flows. See [Tem01, FMRT01].

When fluctuations of flow velocity occur at very small spatial and temporal scales, their numerical approximation becomes very difficult if not impossible. In those cases one typically resorts to the so-called *turbulence models*: the latter allow the approximate description of this flow behaviour through either algebraic or differential equations. This topic will not be addressed in this monograph. The interested readers may consult, e.g., [Wil98] for a description of the physical aspects of turbulent flows, [HYR08] for multiscale analysis of incompressible flows, [Le 05] for modelling aspects of multiscale systems, [MP94] for the analysis of one of the most widely used turbulence models, the so-called $\kappa - \varepsilon$ model. [Sag06] and [BIL06] provide the analysis of the so-called *Large Eddy* model, which is more computationally expensive but in principle better suited to provide a more realistic description of turbulent flow fields.

Finally, let us mention the Euler equations introduced in (15.21), which are used for both compressible or incompressible flows in those cases in which the viscosity can be neglected. Formally speaking, this corresponds to taking the Reynolds number equal to infinity.

By eliminating the pressure, the Navier-Stokes equations can be rewritten in *reduced form* in the sole variable $\mathbf{u}$. This can be shown by starting from the weak formulation (16.9) and using the following subspaces of the functional space $[\mathrm{H}^1(\Omega)]^d$,

$$V_{\mathrm{div}} = \{\mathbf{v} \in \left[\mathrm{H}^1(\Omega)\right]^d : \mathrm{div}\mathbf{v} = 0\}, \quad \mathrm{V}^0_{\mathrm{div}} = \{\mathbf{v} \in \mathrm{V}_{\mathrm{div}} : \mathbf{v} = \mathbf{0} \text{ on } \Gamma_{\mathrm{D}}\}.$$

Upon requiring the test function $\mathbf{v}$ in the momentum equation in (16.9) to belong to the space V_{div}, the term associated to the pressure gradient vanishes, whence we find the following reduced problem for the velocity

find $\mathbf{u} \in \mathrm{L}^2(\mathbb{R}^+; V_{\mathrm{div}}) \cap C^0(\mathbb{R}^+; [\mathrm{L}^2(\Omega)]^d)$ such that

$$\begin{aligned} \int_\Omega \frac{\partial \mathbf{u}}{\partial t} \cdot \mathbf{v} \, d\Omega \quad &+ \nu \int_\Omega \nabla \mathbf{u} \cdot \nabla \mathbf{v} \, d\Omega + \int_\Omega [(\mathbf{u} \cdot \nabla)\mathbf{u}] \cdot \mathbf{v} \, d\Omega \\ &= \int_\Omega \mathbf{f} \cdot \mathbf{v} \, d\Omega + \int_{\Gamma_N} \boldsymbol{\psi} \cdot \mathbf{v} \, d\gamma \quad \forall \mathbf{v} \in V^0_{\mathrm{div}}, \end{aligned} \tag{16.10}$$

with $\mathbf{u}|_{\Gamma_D} = \varphi_D$ and $\mathbf{u}|_{t=0} = \mathbf{u}_0$. Since this problem is (nonlinear) parabolic, its analysis can be carried out using techniques similar to those applied in Chapter 5. (See, e.g., [Sal08].)
Obviously, if $\mathbf{u}$ is a solution of (16.9), then it also solves (16.10). Conversely, the following theorem holds. For its proof, see, e.g., [QV94].

Theorem 16.1. *Let $\Omega \subset \mathbb{R}^d$ be a domain with Lipschitz-continuous boundary $\partial\Omega$. Let $\mathbf{u}$ be a solution to the reduced problem* (16.10). *Then there exists a unique function $p \in \mathrm{L}^2(\mathbb{R}^+; Q)$ such that $(\mathbf{u}, p)$ is a solution to* (16.9).

Once the reduced problem is solved, there exists a unique way to recover the pressure p, and hence the complete solution of the original Navier-Stokes problem (16.9).

In practice, however, this approach can be quite unsuitable from a numerical viewpoint. Indeed, in a Galerkin spatial approximation framework, it would require the construction of finite dimensional subspace, say $V_{\mathrm{div},h}$, of *divergence-free* velocity functions. In this regard, see, e.g., [BF91a] for finite element approximations, and [CHQZ06] for spectral approximations. Moreover, the result of Theorem 16.1 is not constructive, as it does not provide a way to build the solution pressure p. For these reasons one usually prefers to approximate the complete coupled problem (16.9) directly.

16.2 Stokes equations and their approximation

In this section we will consider the following *generalized Stokes problem* with homogeneous Dirichlet boundary conditions

$$\begin{cases} \alpha\mathbf{u} - \nu\Delta\mathbf{u} + \nabla p = \mathbf{f} & \text{in } \Omega, \\ \mathrm{div}\mathbf{u} = 0 & \text{in } \Omega, \\ \mathbf{u} = \mathbf{0} & \text{on } \partial\Omega, \end{cases} \tag{16.11}$$

for a given coefficient $\alpha \geq 0$. This problem describes the motion of an incompressible viscous flow in which the (quadratic) convective term has been neglected, a simplification that is acceptable when $Re \ll 1$. However, one can generate a problem like (16.11) also while using an implicit temporal discretization of the Navier-Stokes equations, as we will see in Sect. 16.7.

From an analytical standpoint, Navier-Stokes equations can be regarded as a *compact perturbation* of Stokes' equations, as they differ from the latter solely because of the presence of the convective term, which is of first order, whereas the diffusive term is of second order. On the other hand, it is fair to say that this term can have a fundamental impact on the solution's behaviour when the Reynolds number Re is very large, as already pointed out.

The weak formulation of problem (16.11) reads:
find $\mathbf{u} \in V$ and $p \in Q$ such that

$$\begin{cases} \int_\Omega (\alpha \mathbf{u} \cdot \mathbf{v} + \nu \nabla \mathbf{u} \cdot \nabla \mathbf{v}) \, d\Omega - \int_\Omega p \mathrm{div} \mathbf{v} \, d\Omega = \int_\Omega \mathbf{f} \cdot \mathbf{v} \, d\Omega & \forall \mathbf{v} \in V, \\ \int_\Omega q \mathrm{div} \mathbf{u} \, d\Omega = 0 & \forall q \in Q, \end{cases} \tag{16.12}$$

where $V = [\mathrm{H}_0^1(\Omega)]^d$ and $Q = \mathrm{L}_0^2(\Omega)$. (We recall that we are addressing a homogeneous Dirichlet problem for the velocity field $\mathbf{u}$ and that we are looking for a pressure having null average.) Now define the bilinear forms $a : V \times V \mapsto \mathbb{R}$ and $b : V \times Q \mapsto \mathbb{R}$ as follows:

$$\begin{aligned} a(\mathbf{u}, \mathbf{v}) &= \int_\Omega (\alpha \mathbf{u} \cdot \mathbf{v} + \nu \nabla \mathbf{u} \cdot \nabla \mathbf{v}) \, d\Omega, \\ b(\mathbf{u}, q) &= -\int_\Omega q \mathrm{div} \mathbf{u} \, d\Omega. \end{aligned} \tag{16.13}$$

With these notations, problem (16.12) becomes

find $(\mathbf{u}, p) \in V \times Q$ such that

$$\begin{cases} a(\mathbf{u}, \mathbf{v}) + b(\mathbf{v}, p) = (\mathbf{f}, \mathbf{v}) & \forall \mathbf{v} \in V, \\ b(\mathbf{u}, q) = 0 & \forall q \in Q, \end{cases} \tag{16.14}$$

where $(\mathbf{f}, \mathbf{v}) = \sum_{i=1}^d \int_\Omega f_i v_i \, d\Omega$.

If we consider non-homogeneous boundary conditions, as indicated in (16.4), the weak formulation of the Stokes problem becomes:

find $(\overset{\circ}{\mathbf{u}}, p) \in V \times Q$ such that

$$\begin{cases} a(\overset{\circ}{\mathbf{u}}, \mathbf{v}) + b(\mathbf{v}, p) = \mathbf{F}(\mathbf{v}) & \forall \mathbf{v} \in V, \\ b(\overset{\circ}{\mathbf{u}}, q) = G(q) & \forall q \in Q, \end{cases} \tag{16.15}$$

where V and Q are the spaces introduced in (16.7) and (16.8), respectively. Having denoted with $\mathbf{R}\varphi \in [\mathrm{H}^1(\Omega)]^d$ a lifting of the boundary datum φ, we have set $\overset{\circ}{\mathbf{u}} = \mathbf{u} - \mathbf{R}\varphi$, while the new terms on the right-hand side have the following expression:

$$\mathbf{F}(\mathbf{v}) = (\mathbf{f}, \mathbf{v}) + \int_{\Gamma_N} \psi \mathbf{v} \, d\gamma - a(\mathbf{R}\varphi, \mathbf{v}), \qquad G(q) = -b(\mathbf{R}\varphi, q). \tag{16.16}$$

The following result holds:

Theorem 16.2. *The couple* $(\mathbf{u}, p)$ *solves the Stokes problem* (16.14) *iff it is a saddle point of the Lagrangian functional*

$$\mathcal{L}(\mathbf{v}, q) = \frac{1}{2} a(\mathbf{v}, \mathbf{v}) + b(\mathbf{v}, q) - (\mathbf{f}, \mathbf{v}),$$

or equivalently,

$$\mathcal{L}(\mathbf{u}, p) = \min_{\mathbf{v} \in V} \max_{q \in Q} \mathcal{L}(\mathbf{v}, q).$$

The pressure q hence plays the role of Lagrange multiplier associated to the divergence-free constraint. As we saw in Sect. 16.1 for the Navier-Stokes equations, it is possible to eliminate, formally, the variable p from the Stokes equations, thus obtaining the following reduced Stokes problem (in weak form)

$$\text{find } \mathbf{u} \in V_{\text{div}}^0 \quad \text{such that} \quad a(\mathbf{u}, \mathbf{v}) = (\mathbf{f}, \mathbf{v}) \quad \forall \mathbf{v} \in V_{\text{div}}^0. \tag{16.17}$$

This is an elliptic problem for the vector variable $\mathbf{u}$. Existence and uniqueness can be proven using the Lax-Milgram Lemma 3.1. As a matter of fact, V_{div}^0 is a Hilbert space with respect to the norm $\|\nabla \mathbf{v}\|_{\mathrm{L}^2(\Omega)}$, because the divergence operator is continuous from V into $\mathrm{L}^2(\Omega)$, thus V_{div}^0 is a closed subspace of the space V. Moreover, the bilinear form $a(\cdot,\cdot)$ is continuous and coercive in V_{div}^0, and $\mathbf{f} \in V_{\text{div}}'$. Using the Cauchy-Schwarz and Poincaré inequalities, the following estimates can be obtained by taking $\mathbf{v} = \mathbf{u}$ in (16.17)):

$$\frac{\alpha}{2}\|\mathbf{u}\|^2_{\mathrm{L}^2(\Omega)} + \nu \|\nabla \mathbf{u}\|^2_{\mathrm{L}^2(\Omega)} \leq \frac{1}{2\alpha}\|\mathbf{f}\|^2_{\mathrm{L}^2(\Omega)}, \quad \text{if } \alpha \neq 0,$$

$$\|\nabla \mathbf{u}\|_{L^2(\Omega)} \leq \frac{C_\Omega}{\nu}\|f\|_{L^2(\Omega)}, \quad \text{if } \alpha = 0,$$

where C_Ω is the constant of the Poincaré inequality (2.13). Note that the pressure has disappeared from (16.17). However, still from (16.17) we can infer that the vector $\mathbf{w} = \alpha\mathbf{u} - \nu\Delta\mathbf{u} - \mathbf{f}$, regarded as a linear functional of $\mathrm{H}^{-1}(\Omega)$, vanishes when applied to any vector function of V_{div}^0. Thanks to this property, there exists a unique function $p \in Q$ such that $\mathbf{w} = \nabla p$, that is p satisfies the first equation of (16.11) in distributional sense (see, e.g., [QV94]). The couple $(\mathbf{u}, p)$ is therefore the unique solution of the weak problem (16.14).

The Galerkin approximation of problem (16.14) has the following form: find $(\mathbf{u}_h, p_h) \in V_h \times Q_h$ such that

$$\begin{cases} a(\mathbf{u}_h, \mathbf{v}_h) + b(\mathbf{v}_h, p_h) = (\mathbf{f}, \mathbf{v}_h) & \forall \mathbf{v}_h \in V_h, \\ b(\mathbf{u}_h, q_h) = 0 & \forall q_h \in Q_h, \end{cases} \tag{16.18}$$

where $\{V_h \subset V\}$ and $\{Q_h \subset Q\}$ represent two families of finite-dimensional subspaces depending on a real discretization parameter h. If, instead, we consider problem

(16.15)–(16.16) corresponding to non-homogeneous boundary data (16.4), the above formulation needs to be modified by using $\mathbf{F}(\mathbf{v}_h)$ on the right-hand side of the first equation and $G(q_h)$ on that of the second equation. These new functionals can be obtained from (16.16) by replacing $\mathbf{R}\varphi$ with the interpolant of φ at the nodes of Γ_D (and vanishing at all other nodes), and replacing ψ with its interpolant at the nodes sitting on Γ_N. The algebraic formulation of problem (16.18) will be addressed in Sect. 16.4.

The following theorem is due to F. Brezzi [Bre74], and guarantees uniqueness and existence for problem (16.18):

Theorem 16.3. *The Galerkin approximation* (16.18) *admits one and only one solution if the following conditions hold:*

1. *The bilinear form* $a(\cdot,\cdot)$ *is:*
   a) *coercive, that is* $\exists\alpha > 0$ *(possibly depending on h) such that*
   $$a(\mathbf{v}_h,\mathbf{v}_h) \geq \alpha\|\mathbf{v}_h\|_V^2 \quad \forall \mathbf{v}_h \in V_h^*,$$
   where $V_h^* = \{\mathbf{v}_h \in V_h : b(\mathbf{v}_h, q_h) = 0\ \forall q_h \in Q_h\}$;
   b) *continuous, that is* $\exists\gamma > 0$ *such that*
   $$|a(\mathbf{u}_h,\mathbf{v}_h)| \leq \gamma\|\mathbf{u}_h\|_V\|\mathbf{v}_h\|_V \quad \forall \mathbf{u}_h,\mathbf{v}_h \in V_h.$$
2. *The bilinear form* $b(\cdot,\cdot)$ *is continuous, that is* $\exists\delta > 0$ *such that*
   $$|b(\mathbf{v}_h,q_h)| \leq \delta\|\mathbf{v}_h\|_V\|q_h\|_Q \quad \forall \mathbf{v}_h \in V_h, q_h \in Q_h.$$
3. *Finally, there exists a positive constant* β *(possibly depending on h) such that*
   $$\forall q_h \in Q_h,\ \exists \mathbf{v}_h \in V_h : \ b(\mathbf{v}_h,q_h) \geq \beta\|\mathbf{v}_h\|_{\mathbf{H}^1(\Omega)}\|q_h\|_{\mathrm{L}^2(\Omega)}. \qquad (16.19)$$

Under the previous assumptions the discrete solution fulfills the following a-priori estimates:

$$\|\mathbf{u}_h\|_V \leq \frac{1}{\alpha}\|\mathbf{f}\|_{V'},$$
$$\|p_h\|_Q \leq \frac{1}{\beta}\left(1+\frac{\gamma}{\alpha}\right)\|\mathbf{f}\|_{V'},$$

where, as usual, V' *denotes the dual space of* V. *Moreover, the following convergence results hold:*

$$\begin{aligned}
\|\mathbf{u}-\mathbf{u}_h\|_V &\leq \left(1+\frac{\delta}{\beta}\right)\left(1+\frac{\gamma}{\alpha}\right)\inf_{\mathbf{v}_h\in V_h}\|\mathbf{u}-\mathbf{v}_h\|_V + \frac{\delta}{\alpha}\inf_{q_h\in Q_h}\|p-q_h\|_Q,\\
\|p-p_h\|_Q &\leq \frac{\gamma}{\beta}\left(1+\frac{\gamma}{\alpha}\right)\left(1+\frac{\delta}{\beta}\right)\inf_{\mathbf{v}_h\in V_h}\|\mathbf{u}-\mathbf{v}_h\|_V\\
&\quad + \left(1+\frac{\delta}{\beta}+\frac{\delta\gamma}{\alpha\beta}\right)\inf_{q_h\in Q_h}\|p-q_h\|_Q.
\end{aligned}$$

It is worth noticing that condition (16.19) is equivalent the existence of a positive constant β such that

$$\inf_{q_h \in Q_h, q_h \neq 0} \sup_{\mathbf{v}_h \in V_h, \mathbf{v}_h \neq \mathbf{0}} \frac{b(\mathbf{v}_h, q_h)}{\|\mathbf{v}_h\|_{\mathbf{H}^1(\Omega)} \|q_h\|_{\mathrm{L}^2(\Omega)}} \geq \beta. \tag{16.20}$$

For such a reason it is often called the *inf-sup condition.*
The proof of this theorem requires non-elementary tools of functional analysis. It will be given in Sect. 16.3 for a saddle-point problem that is more general than Stokes' problem. In this perspective, Theorem 16.3 can be regarded as a special case of Theorems 16.5 and 16.6. The reader not interested in the theoretical aspects can skip the next section and jump directly to Sect. 16.4.

16.3 Saddle-point problems

The scope of this section is to study problems (16.14) and (16.18) and show how the latter's solutions converge to the former's. With this aim we will recast those formulations within a more abstract framework, that will eventually allow the use of the theory proposed in [Bre74, Bab71].

16.3.1 Problem formulation

Let X and M be two Hilbert spaces endowed with norms $\|\cdot\|_X$ and $\|\cdot\|_M$. Denoting with X' and M' the corresponding dual spaces (that is the spaces of linear and bounded functionals defined on X and M), we introduce the bilinear forms $a(\cdot,\cdot) : X \times X \longrightarrow \mathbb{R}$ and $b(\cdot,\cdot) : X \times M \longrightarrow \mathbb{R}$ that we suppose to be continuous, meaning there exist two constants $\gamma, \delta > 0$ such that

$$|a(w,v)| \leq \gamma \|w\|_X \|v\|_X, \qquad |b(w,\mu)| \leq \delta \|w\|_X \|\mu\|_M, \tag{16.21}$$

for all $w, v \in X$ and $\mu \in M$.
Consider now the following constrained problem: find $(u,\eta) \in X \times M$ such that

$$\begin{cases} a(u,v) + b(v,\eta) = \langle l, v\rangle & \forall v \in X, \\ b(u,\mu) = \langle \sigma, \mu\rangle & \forall \mu \in M, \end{cases} \tag{16.22}$$

where $l \in X'$ and $\sigma \in M'$ are two assigned linear functionals, while $\langle \cdot,\cdot\rangle$ denotes the pairing between X and X' or M and M'.
Formulation (16.22) is general enough to include the formulation (16.14) of the Stokes problem, that of a generic constrained problem with respect to the bilinear form $a(\cdot,\cdot)$ (with η representing the constraint), or again the formulation which is obtained when mixed finite element approximations are used for various kind of boundary-value problems, for instance those of linear elasticity (see, e.g., [BF91a, QV94]).

Problem (16.22) can be conveniently restated in operator form. For this we associate the bilinear forms $a(\cdot,\cdot)$ and $b(\cdot,\cdot)$ with the operators $A \in \mathcal{L}(X,X')$ and $B \in \mathcal{L}(X,M')$, defined through the following relations:

$$\langle Aw,v\rangle = a(w,v) \qquad \forall w,v \in X,$$
$$\langle Bv,\mu\rangle = b(v,\mu) \qquad \forall v \in X,\ \mu \in M.$$

The relations pair X' to X and M' to M. In accordance with the notations introduced in Sect. 4.5.3, we denote by $\mathcal{L}(V,W)$ the space of linear and bounded functionals between V and W.
Let $B^T \in \mathcal{L}(M,X')$ be the adjoint operator of B defined by

$$\langle B^T\mu,v\rangle = \langle Bv,\mu\rangle = b(v,\mu) \quad \forall v \in X,\ \mu \in M. \tag{16.23}$$

The former duality holds between X' and X, the latter between M' and M. (This operator was denoted by the symbol B' in Sect. 2.6, see the general definition (2.20). Here, however, it is denoted by B^T for consistency with the classical notation used in the framework of saddle-point problems.)

In operator form, the saddle-point problem (16.22) can be restated as follows: find $(u,\eta) \in X \times M$ such that

$$\begin{cases} Au + B^T\eta = l & \text{in } X', \\ Bu = \sigma & \text{in } M'. \end{cases} \tag{16.24}$$

16.3.2 Analysis of the problem

In order to analyze problem (16.24), we introduce the affine manifold

$$X^\sigma = \{v \in X \ :\ b(v,\mu) = \langle\sigma,\mu\rangle\ \forall\mu \in M\}. \tag{16.25}$$

The space X^0 denotes the kernel of B, that is

$$X^0 = \{v \in X \ :\ b(v,\mu) = 0\ \forall\mu \in M\} = \ker(B).$$

This is a closed subspace of X. We can therefore associate (16.22) with the following reduced problem

$$\text{find} \quad u \in X^\sigma \quad \text{such that} \quad a(u,v) = \langle l,v\rangle \quad \forall v \in X^0. \tag{16.26}$$

If (u,η) is a solution of (16.22), then u is a solution to (16.26). In the following we will introduce suitable conditions that allow the converse to hold, too. Moreover, we would like to prove uniqueness for the solution of (16.26). This would allow us to obtain an existence and uniqueness result for the original saddle-point problem (16.22).

We denote by X^0_{polar} the polar set of X^0, that is

$$X^0_{polar} = \{g \in X' \ :\ \langle g,v\rangle = 0\ \forall v \in X^0\}.$$

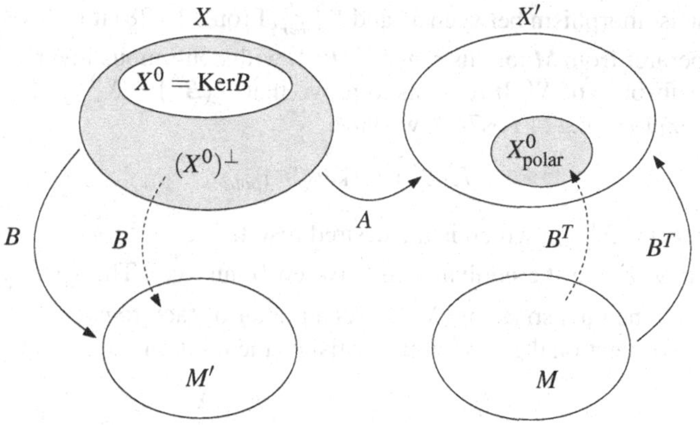

Fig. 16.1. The spaces X and M, the dual spaces X' and M', the subspaces X^0, $(X^0)^\perp$ and X^0_{polar}, and the operators A, B and B^T. Dashed lines indicate isomorphisms

Since $X^0 = \ker(B)$, we have $X^0_{polar} = (\ker(B))_{polar}$. The space X is a direct sum of X^0 and its orthogonal space $(X^0)^\perp$,

$$X = X^0 \oplus (X^0)^\perp.$$

See Fig. 16.1 for a picture. Since, in general, $\ker(B)$ is not empty, we cannot expect B to be an isomorphism between X and M'. The aim is therefore to find a condition which guarantees that B is an isomorphism between $(X^0)^\perp$ and M' (and, similarly, that B^T is an isomorphism between M and X^0_{polar}).

Lemma 16.1. *The three following statements are equivalent:*

a. *there exists a constant $\beta^* > 0$ such that the following compatibility condition holds*

$$\forall\, \mu \in M\ \exists v \in X,\ \text{with } v \neq 0\ :\ b(v,\mu) \geq \beta^* \|v\|_X \|\mu\|_M; \tag{16.27}$$

b. *the operator B^T is an isomorphism from M onto X^0_{polar}; moreover,*

$$\|B^T\mu\|_{X'} = \sup_{v \in V, v \neq 0} \frac{\langle B^T\mu, v\rangle}{\|v\|_X} \geq \beta^* \|\mu\|_M \quad \forall \mu \in M; \tag{16.28}$$

c. *the operator B is an isomorphism from $(X^0)^\perp$ onto M'; moreover,*

$$\|Bv\|_{M'} = \sup_{\mu \in M, \mu \neq 0} \frac{\langle Bv, \mu\rangle}{\|\mu\|_M} \geq \beta^* \|v\|_X \quad \forall v \in (X^0)^\perp. \tag{16.29}$$

Proof. First of all we prove the equivalence between a. and b. Owing to definition (16.23) of B^T, the two inequalities (16.27) and (16.28) do coincide. Let us now prove that B^T is an isomorphism between M and X^0_{polar}. From (16.28) it follows that B^T is an injective operator from M into its range $\mathcal{R}(B^T)$, with continuous inverse. Then $\mathcal{R}(B^T)$ is a closed subspace of X'. It remains to prove that $\mathcal{R}(B^T) = X^0_{polar}$. From the *closed range theorem* (see, e.g., [Yos74]), we have

$$\mathcal{R}(B^T) = (\ker(B))_{polar},$$

whence $\mathcal{R}(B^T) = X^0_{polar}$, which is the desired result.

Let us now prove the equivalence between b. and c. . The space X^0_{polar} can be identified with the dual space of $(X^0)^\perp$. As a matter of fact, to every $g \in ((X^0)^\perp)'$ we can associate a functional $\hat{g} \in X'$ which satisfies the relation

$$\langle \hat{g}, v \rangle = \langle g, P^\perp v \rangle \quad \forall v \in X,$$

where $P^\perp$ denotes the orthogonal projection of X onto $(X^0)^\perp$, that is

$$\forall v \in X, \quad P^\perp v \in (X^0)^\perp : (P^\perp v - v, w)_X = 0 \quad \forall w \in (X^0)^\perp.$$

Clearly, $\hat{g} \in X^0_{polar}$ and it can be verified that $g \longrightarrow \hat{g}$ is an isometric bijection between $((X^0)^\perp)'$ and X^0_{polar}. Consequently, B^T is an isomorphism from M onto $((X^0)^\perp)'$ satisfying the relation

$$\|(B^T)^{-1}\|_{\mathcal{L}(X^0_{polar}, M)} \leq \frac{1}{\beta^*}$$

if and only if B is an isomorphism from $(X^0)^\perp$ onto M' satisfying the relation

$$\|B^{-1}\|_{\mathcal{L}(M', (X^0)^\perp)} \leq \frac{1}{\beta^*}.$$

This completes our proof. ◇

At this point we can prove that problem (16.22) is well posed.

Theorem 16.4. *Let the bilinear form $a(\cdot,\cdot)$ satisfy the continuity condition* (16.21) *and be coercive on the space X^0, that is*

$$\exists \alpha > 0 : a(v, v) \geq \alpha \|v\|^2_X \quad \forall v \in X^0. \tag{16.30}$$

Suppose moreover that the bilinear form $b(\cdot,\cdot)$ satisfies the continuity condition (16.21) *as well as the compatibility condition* (16.27).
Then for every $l \in X'$ and $\sigma \in M'$, there exists a unique solution u of problem (16.26)*; furthermore, there exists a unique function $\eta \in M$ such that (u, η) is the unique solution to the original saddle-point problem* (16.22).
Moreover, the map $(l, \sigma) \longrightarrow (u, \eta)$ is an isomorphism from $X' \times M'$ onto $X \times M$ and the following a priori estimates hold:

$$\|u\|_X \leq \frac{1}{\alpha}\left[\|l\|_{X'} + \frac{\alpha + \gamma}{\beta^*}\|\sigma\|_{M'}\right], \tag{16.31}$$

$$\|\eta\|_M \le \frac{1}{\beta^*}\left[\left(1+\frac{\gamma}{\alpha}\right)\|l\|_{X'} + \frac{\gamma(\alpha+\gamma)}{\alpha\beta^*}\|\sigma\|_{M'}\right]. \tag{16.32}$$

The constants α, β^* and γ are defined in (16.30), (16.27) and (16.21), respectively. The symbols $\|\cdot\|_{X'}$ and $\|\cdot\|_{M'}$ indicate the norms of the dual spaces, and are defined as in (16.28) and (16.29), respectively.

Proof. The uniqueness of the solution to (16.26) directly follows from the coercivity property (16.30). Let us now prove existence. From assumption (16.27) and the equivalence result stated in c. of Lemma 16.1, we can infer that there exists a unique function $u^\sigma \in (X^0)^\perp$ such that $Bu^\sigma = \sigma$, and, moreover,

$$\|u^\sigma\|_X \le \frac{1}{\beta^*}\|\sigma\|_{M'}. \tag{16.33}$$

The saddle-point problem (16.26) can be restated as follows

$$\text{find} \quad \tilde{u} \in X^0 \quad \text{such that} \quad a(\tilde{u}, v) = \langle l, v\rangle - a(u^\sigma, v) \quad \forall v \in X^0. \tag{16.34}$$

The solution u to problem (16.26) is identified by the relation $u = \tilde{u} + u^\sigma$. At this point, existence and uniqueness of the solution $\tilde{u}$ of problem (16.34) follow by the Lax-Milgram Lemma, together with the a priori estimate

$$\|\tilde{u}\|_X \le \frac{1}{\alpha}\left(\|l\|_{X'} + \gamma\|u^\sigma\|_X\right),$$

that is, thanks to (16.33),

$$\|\tilde{u}\|_X \le \frac{1}{\alpha}\left(\|l\|_{X'} + \frac{\gamma}{\beta^*}\|\sigma\|_{M'}\right). \tag{16.35}$$

The uniqueness of the u component of the solution to problem (16.22) is therefore a direct consequence of the uniqueness of $\tilde{u} \in X^0$ and $u^\sigma \in (X^0)^\perp$, while the stability estimate (16.31) follows again from the combination of (16.35) with (16.33).
We focus now on the component η of the solution. Since (16.34) can be restated as

$$\langle Au - l, v\rangle = 0 \quad \forall v \in X^0,$$

it follows that $(Au - l) \in X^0_{polar}$, so we can exploit point b. of Lemma 16.1 and conclude that there exists a unique $\eta \in M$ such that $Au - l = -B^T\eta$, that is (u, η) is a solution of problem (16.22) and η satisfies the inequality

$$\|\eta\|_M \le \frac{1}{\beta^*}\|Au - l\|_{X'}. \tag{16.36}$$

We have already noticed that every solution (u, η) to (16.22) yields a solution u to the reduced problem (16.26), whence the uniqueness of the solution of (16.22). Finally, the a priori estimate (16.32) follows from (16.36), by noting that

$$\|\eta\|_M \le \frac{1}{\beta^*}\left[\|A\|_{\mathcal{L}(X,X')}\|u\|_X + \|l\|_{X'}\right]$$

and using the already proven a priori estimate (16.31) on u. ◇

16.3.3 Galerkin approximation, stability and convergence analysis

To introduce a Galerkin approximation of the abstract saddle-point problem (16.22), we consider two families of finite-dimensional subspaces X_h and M_h of the spaces X and M, respectively. They can be either finite element piecewise polynomial spaces, or global polynomial (spectral) spaces, or spectral element subspaces.
We look for the solution to the following problem:
given $l \in X'$ and $\sigma \in M'$, find $(u_h, \eta_h) \in X_h \times M_h$ such that

$$\begin{cases} a(u_h, v_h) + b(v_h, \eta_h) = \langle l, v_h \rangle & \forall v_h \in X_h, \\ b(u_h, \mu_h) = \langle \sigma, \mu_h \rangle & \forall \mu_h \in M_h. \end{cases} \tag{16.37}$$

By following what we did for the continuous problem, we can introduce the subspace

$$X_h^\sigma = \{ v_h \in X_h \,:\, b(v_h, \mu_h) = \langle \sigma, \mu_h \rangle \; \forall \mu_h \in M_h \} \tag{16.38}$$

which allows us to introduce the following finite dimensional counterpart of the reduced formulation (16.26)

$$\text{find} \quad u_h \in X_h^\sigma \quad \text{such that} \quad a(u_h, v_h) = \langle l, v_h \rangle \quad \forall v_h \in X_h^0. \tag{16.39}$$

Since, in general, M_h is different from M, the space (16.38) is not necessarily a subspace of X^σ.

Clearly, every solution (u_h, η_h) of (16.37) yields a solution u_h for the reduced problem (16.39). In this section we look for conditions that allow us to prove that the converse statement is also true, together with a result of stability and convergence for the solution of problem (16.37).

We start by proving the discrete counterpart of Theorem 16.4.

Theorem 16.5 (Existence, uniqueness and stability). *Suppose that the bilinear form $a(\cdot,\cdot)$ satisfies the continuity property* (16.21) *and that it is coercive on the space X_h^0, that is there exists a constant $\alpha_h > 0$ such that*

$$a(v_h, v_h) \geq \alpha_h \|v_h\|_X^2 \quad \forall v_h \in X_h^0. \tag{16.40}$$

Moreover, suppose that the bilinear form $b(\cdot,\cdot)$ satisfies the continuity condition (16.21) *and that the following discrete compatibility condition holds: there exists a constant $\beta_h > 0$ such that*

$$\forall \mu_h \in M_h \; \exists v_h \in X_h, \; v_h \neq 0 \,:\, b(v_h, \mu_h) \geq \beta_h \|v_h\|_X \|\mu_h\|_M. \tag{16.41}$$

Then, for every $l \in X'$ and $\sigma \in M'$, there exists a unique solution (u_h, η_h) of problem (16.37) *which satisfies the following stability conditions:*

$$\|u_h\|_X \leq \frac{1}{\alpha_h}\left[\|l\|_{X'} + \frac{\alpha_h + \gamma}{\beta_h}\|\sigma\|_{M'}\right], \tag{16.42}$$

$$\|\eta_h\|_M \leq \frac{1}{\beta_h}\left[\left(1 + \frac{\gamma}{\alpha_h}\right)\|l\|_{X'} + \frac{\gamma(\alpha_h + \gamma)}{\alpha_h \beta_h}\|\sigma\|_{M'}\right]. \tag{16.43}$$

Proof. The proof can be obtained by repeating that of Theorem 16.4, considering X_h instead of X, M_h instead of M, and simply noting that

$$\|l\|_{X_h'} \leq \|l\|_{X'}, \quad \|\sigma\|_{M_h'} \leq \|\sigma\|_{M'}. \qquad \diamond$$

The coercivity condition (16.30) does not necessarily guarantee (16.40), as $X_h^0 \not\subset X^0$, nor does the compatibility condition (16.27) in general imply the discrete compatibility condition (16.41), due to the fact that X_h is a proper subspace of X. Moreover, in the case in which the constants α_h and β_h in (16.40) and (16.41) are independent of h, inequalities (16.42) and (16.43) provide the desired stability result.
Condition (16.41) represents the well known *inf-sup* or *LBB* condition (see [BF91a]). (The condition (16.19) (or (16.20)) is just a special case.)

We move now to the convergence result.

Theorem 16.6 (Convergence). *Let the assumptions of Theorems* 16.4 *and* 16.5 *be satisfied. Then the solutions* (u,η) *and* (u_h,η_h) *of problems* (16.22) *and* (16.37), *respectively, satisfy the following error estimates:*

$$\|u-u_h\|_X \leq \left(1+\frac{\gamma}{\alpha_h}\right) \inf_{v_h^*\in X_h^\sigma} \|u-v_h^*\|_X + \frac{\delta}{\alpha_h} \inf_{\mu_h\in M_h} \|\eta-\mu_h\|_M, \tag{16.44}$$

$$\begin{aligned} \|\eta-\eta_h\|_M \; &\leq \frac{\gamma}{\beta_h}\left(1+\frac{\gamma}{\alpha_h}\right) \inf_{v_h^*\in X_h^\sigma} \|u-v_h^*\|_X \\ &+\left(1+\frac{\delta}{\beta_h}+\frac{\gamma\delta}{\alpha_h\beta_h}\right) \inf_{\mu_h\in M_h} \|\eta-\mu_h\|_M, \end{aligned} \tag{16.45}$$

where γ, δ, α_h *and* β_h *are respectively defined by the relations* (16.21), (16.40) *and* (16.41). *Moreover, the following error estimate holds*

$$\inf_{v_h^*\in X_h^\sigma} \|u-v_h^*\|_X \leq \left(1+\frac{\delta}{\beta_h}\right) \inf_{v_h\in X_h} \|u-v_h\|_X. \tag{16.46}$$

Proof. Consider $v_h \in X_h$, $v_h^* \in X_h^\sigma$ and $\mu_h \in M_h$. By subtracting $(16.37)_1$ from $(16.22)_1$, then adding and subtracting the quantities $a(v_h^*, v_h)$ and $b(v_h, \mu_h)$, we find

$$a(u_h - v_h^*, v_h) + b(v_h, \eta_h - \mu_h) = a(u - v_h^*, v_h) + b(v_h, \eta - \mu_h).$$

Let us now choose $v_h = u_h - v_h^* \in X_h^0$. From the definition of the space X_h^0 and using (16.40) and (16.21), we find the bound

$$\|u_h - v_h^*\|_X \leq \frac{1}{\alpha_h}\Big(\gamma\|u-v_h^*\|_X + \delta\,\|\eta-\mu_h\|_M\Big)$$

from which the estimate (16.44) immediately follows, as

$$\|u-u_h\|_X \leq \|u-v_h^*\|_X + \|u_h - v_h^*\|_X.$$

Let us prove now the estimate (16.45). Owing to the compatibility condition (16.41), for every $\mu_h \in M_h$ we can write

$$\|\eta_h - \mu_h\|_M \le \frac{1}{\beta_h} \sup_{v_h \in X_h,\ v_h \neq 0} \frac{b(v_h, \eta_h - \mu_h)}{\|v_h\|_X}. \tag{16.47}$$

On the other hand, by subtracting side by side $(16.37)_1$ from $(16.22)_1$, then adding and subtracting the quantity $b(v_h, \mu_h)$, we obtain

$$b(v_h, \eta_h - \mu_h) = a(u - u_h, v_h) + b(v_h, \eta - \mu_h).$$

Using this identity in (16.47) as well as the continuity inequalities (16.21), it follows that

$$\|\eta_h - \mu_h\|_M \le \frac{1}{\beta_h}\Big(\gamma\|u - u_h\|_X + \delta\|\eta - \mu_h\|_M\Big).$$

This yields the desired result, provided we use the error estimate (16.44) that was previously derived for the variable u.
Finally, let us prove (16.46). Property (16.41) allows us to use the discrete version of Lemma 16.1 (now applied in the finite-dimensional subspaces). Then, owing to the discrete counterpart of (16.29), for every $v_h \in X_h$ we can find a unique function $z_h \in (X_h^0)^\perp$ such that

$$b(z_h, \mu_h) = b(u - v_h, \mu_h) \quad \forall \mu_h \in M_h$$

and, moreover,

$$\|z_h\|_X \le \frac{\delta}{\beta_h}\|u - v_h\|_X.$$

The function $v_h^* = z_h + v_h$ belongs to X_h^σ, as $b(u, \mu_h) = \langle \sigma, \mu_h \rangle$ for all $\mu_h \in M_h$. Moreover,

$$\|u - v_h^*\|_X \le \|u - v_h\|_X + \|z_h\|_X \le \left(1 + \frac{\delta}{\beta_h}\right)\|u - v_h\|_X,$$

whence the estimate (16.46) follows. ◇

The inequalities (16.44) and (16.45) yield error estimates with optimal convergence rate, provided that the constants α_h and β_h in (16.40) and (16.41) are bounded from below by two constants α and β independent of h. Let us also remark that inequality (16.44) holds even if the compatibility conditions (16.27) and (16.41) are not satisfied.

Remark 16.2 (Spurious pressure modes). The compatibility condition (16.41) is essential to guarantee the uniqueness of the η_h-component of the solution. Indeed, if (16.41) does not hold, then one can find functions $\mu_h^* \in M_h$, $\mu_h^* \neq 0$, such that

$$b(v_h, \mu_h^*) = 0 \quad \forall v_h \in X_h.$$

Consequently, if (u_h, η_h) is a solution to problem (16.37), then $(u_h, \eta_h + \tau\mu_h^*)$, for all $\tau \in \mathbb{R}$, is a solution, too.

Any such function μ_h^* is called *spurious mode*, or, more specifically, *pressure* spurious mode when it refers to the Stokes problem (16.18) in which functions μ_h represent discrete pressures. Numerical instabilities can arise since the discrete problem (16.37) is unable to detect such spurious modes. •

For a given couple of finite dimensional spaces X_h and M_h, proving that the discrete compatibility condition (16.41) holds with a constant β_h independent of h is not always easy. Several practical criteria are available, among which we mention those due to Fortin ([For77]), Boland and Nicolaides ([BN83]), and Verfürth ([Ver84]). (See [BF91b].)

16.4 Algebraic formulation of the Stokes problem

Let us investigate the structure of the algebraic system associated to the Galerkin approximation (16.18) to the Stokes problem (or, more generally, to a discrete saddle-point problem like (16.37)). Denote with

$$\{\varphi_j \in V_h\}, \quad \{\phi_k \in Q_h\},$$

the basis functions of the spaces V_h and Q_h, respectively. Le us expand the discrete solutions $\mathbf{u}_h$ and p_h with respect to such bases,

$$\mathbf{u}_h(\mathbf{x}) = \sum_{j=1}^{N} u_j \varphi_j(\mathbf{x}), \quad p_h(\mathbf{x}) = \sum_{k=1}^{M} p_k \phi_k(\mathbf{x}), \tag{16.48}$$

having set $N = \dim V_h$ and $M = \dim Q_h$. By choosing as test functions in (16.18) the same basis functions we obtain the following block linear system

$$\begin{cases} \mathrm{A}\mathbf{U} + \mathrm{B}^T\mathbf{P} = \mathbf{F}, \\ \mathrm{B}\mathbf{U} = \mathbf{0}, \end{cases} \tag{16.49}$$

where $\mathrm{A} \in \mathbb{R}^{N\times N}$ and $\mathrm{B} \in \mathbb{R}^{M\times N}$ are the matrices related respectively to the bilinear forms $a(\cdot,\cdot)$ and $b(\cdot,\cdot)$, whose elements are given by

$$\mathrm{A} = [a_{ij}] = [a(\varphi_j, \varphi_i)], \qquad \mathrm{B} = [b_{km}] = [b(\varphi_m, \phi_k)],$$

while $\mathbf{U}$ and $\mathbf{P}$ are the vectors of the unknowns,

$$\mathbf{U} = [u_j], \quad \mathbf{P} = [p_j].$$

The $(N+M)\times(N+M)$ matrix

$$\mathrm{S} = \begin{bmatrix} \mathrm{A} & \mathrm{B}^T \\ \mathrm{B} & 0 \end{bmatrix} \tag{16.50}$$

is *block symmetric* (as A is symmetric) and *indefinite*, featuring real eigenvalues with variable sign (either positive and negative). S is non-singular iff no eigenvalue is null, a property that follows from the *inf-sup* condition (16.20). To prove the latter statement we proceed as follows.

Since A is non-singular – it is associated to the coercive bilinear form $a(\cdot,\cdot)$ – from the first of (16.49) we can formally obtain $\mathbf{U}$ as

$$\mathbf{U} = \mathrm{A}^{-1}(\mathbf{F} - \mathrm{B}^T\mathbf{P}). \tag{16.51}$$

Using (16.51) in the second equation of (16.49) yields

$$\mathrm{R}\mathbf{P} = \mathrm{B}\mathrm{A}^{-1}\mathbf{F}, \quad \text{where} \quad \mathrm{R} = \mathrm{B}\mathrm{A}^{-1}\mathrm{B}^T. \tag{16.52}$$

This corresponds to having carried out a block Gaussian elimination on system (16.50).

This way we obtain a reduced system for the sole unknown $\mathbf{P}$ (the pressure), which admits a unique solution in case R is non-singular. Since A is non-singular and positive definite, the latter condition is satisfied iff B^T has a null kernel, that is

$$\mathrm{ker}\mathrm{B}^T = \{\mathbf{0}\}, \tag{16.53}$$

where $\mathrm{ker}\mathrm{B}^T = \{\mathbf{x} \in \mathbb{R}^N : B^T\mathbf{x} = \mathbf{0}\}$. The latter algebraic condition is in fact equivalent to the *inf-sup* condition (16.20) (see Exercise 1).

On the other hand, since A is non-singular, from the existence and uniqueness of $\mathbf{P}$ we infer that there exists a unique vector $\mathbf{U}$ which satisfies (16.51).

In conclusion, system (16.49) admits a unique solution $(\mathbf{U},\mathbf{P})$ if and only if condition (16.53) holds.

Remark 16.3. Condition (16.53) is equivalent to asking that B^T (and consequently B) has *full rank*, i.e. that $\mathrm{rank}(B^T) = \min(N,M)$, because $\mathrm{rank}(B^T)$ is the maximum number of linearly independent row vectors (or, equivalently, column vectors) of B^T. Indeed, $\mathrm{rank}(B^T) + \dim \ker(B^T) = M$. •

Let us consider again Remark 16.2 concerning the general saddle-point problem and suppose that the *inf-sup* condition (16.20) does not hold. In this case we can state that

$$\exists q_h^* \in Q_h : \quad b(\mathbf{v}_h, q_h^*) = 0 \qquad \forall \mathbf{v}_h \in V_h. \tag{16.54}$$

Consequently, if $(\mathbf{u}_h, p_h)$ is a solution to the Stokes problem (16.18), then $(\mathbf{u}_h, p_h + q_h^*)$ is a solution too, as

$$\begin{aligned} a(\mathbf{u}_h,\mathbf{v}_h) + b(\mathbf{v}_h, p_h + q_h^*) &= a(\mathbf{u}_h,\mathbf{v}_h) + b(\mathbf{v}_h, p_h) + b(\mathbf{v}_h, q_h^*) \\ &= a(\mathbf{u}_h,\mathbf{v}_h) + b(\mathbf{v}_h, p_h) = (\mathbf{f},\mathbf{v}_h) \qquad \forall \mathbf{v}_h \in V_h. \end{aligned}$$

Functions q_h^* which fail to satisfy the *inf-sup* condition are invisible to the Galerkin problem(16.18). For this reason, as already observed, they are called spurious pressure modes, or even *parasitic modes*. Their presence inhibits the pressure solution from being unique, yielding numerical instabilities. For this reason, those finite-dimensional subspaces that violate the compatibility condition (16.20) are said to be *unstable*, or *incompatible*.

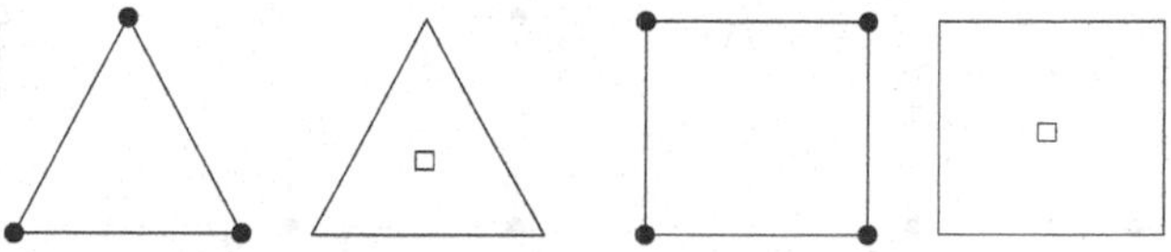

Fig. 16.2. Case of discontinuous pressure: choices that do not satisfy the *inf-sup* condition, on triangles (left), and on quadrilaterals (right)

Two strategies are generally adopted:

- choose spaces V_h and Q_h that satisfy the *inf-sup* condition;
- stabilize (either a priori or a posteriori) the finite dimensional problem by eliminating the spurious modes.

Let us analyze the first type of strategy. To start with, we will consider the case of finite element spaces. To characterize Q_h and V_h it suffices to choose on every element of the triangulation the degrees of freedom for velocity and pressure. The weak formulation does not require a continuous pressure. We will therefore start considering the case of *discontinuous pressures*.

As Stokes equations are of order one in p and order two in $\mathbf{u}$, generally speaking it makes sense to use piecewise polynomials of degree $k \geq 1$ for the velocity space V_h and of degree $k-1$ for the space Q_h.

In particular, we might want to use piecewise linear finite elements $\mathbb{P}_1$ for each velocity component, and piecewise constant finite elements $\mathbb{P}_0$ for the pressure (see Fig. 16.2 in which, as in all those that will follow, by means of the symbol □ we indicate the degrees of freedom for the pressure, whereas the symbol • identifies those for each velocity component). In fact, this choice, although being quite natural, does not pass the *inf-sup* test (16.20) (see Exercise 3).
When looking for a compatible couple of spaces, the larger the velocity space V_h, the higher the probability that the *inf-sup* condition is satisfied. Otherwise said, the space V_h should be "rich" enough compared to the space Q_h. In Fig. 16.3 we report three different choices of spaces that fulfill the *inf-sup* condition, still in the case of continuous velocity and discontinuous pressure. Choice (a) is made by $\mathbb{P}_2 - \mathbb{P}_0$ elements, (b) by $\mathbb{Q}_2 - \mathbb{P}_0$ elements, while choice (c) by piecewise linear discontinuous elements for the pressure, while the velocity components are made by piecewise quadratic continuous

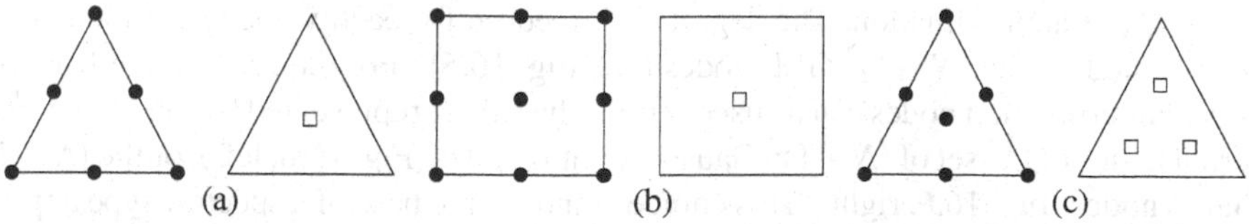

Fig. 16.3. Case of discontinuous pressure: choices that do satisfy the *inf-sup* condition: on triangles, (a), and on quadrilaterals, (b). Also the couple (c), known as Crouzeix-Raviart elements, satisfies the *inf-sup* condition

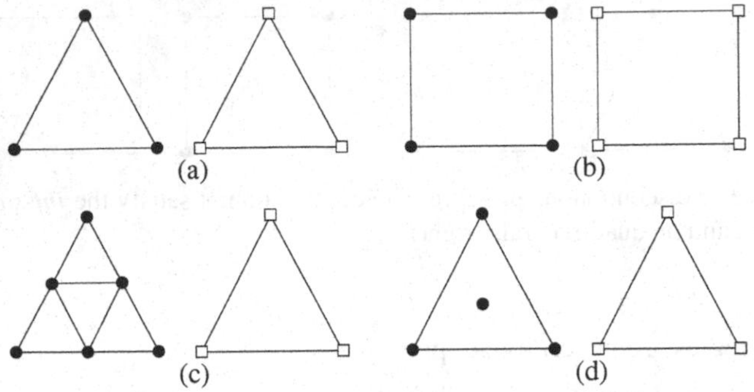

Fig. 16.4. Case of continuous pressure: the couples (a) and (b) do not satisfy the *inf-sup* condition. The elements used for the velocity components in (c) are known as $\mathbb{P}_1$-*iso*$\mathbb{P}_2$ finite elements, whereas couple (d) is called *mini-element*

elements enriched by a cubic bubble function on each triangle – these are the so-called Crouzeix-Raviart elements.

In Fig. 16.4(a), (b) we report two choices of incompatible finite elements in the case of continuous pressure. They consist of piecewise linear elements on triangles (resp. bilinear on quadrilaterals) for both velocity and pressure. More in general, finite elements of the same polynomial degree $k \geq 1$ for both velocity and pressures are unstable. In the same figure, the elements displayed in (c) and (d) are instead stable. In both cases, pressure is a piecewise linear continuous function, whereas velocities are piecewise linear polynomials on each of the four sub-triangles (case (c)), or piecewise linear polynomials enriched by a cubic bubble function (case (d)). The pair $\mathbb{P}_2 - \mathbb{P}_1$ (continuous piecewise quadratic velocities and continuous piecewise linear pressure) is stable. This is the smallest degree representative of the family of the so-called Taylor-Hood elements $\mathbb{P}_k - \mathbb{P}_{k-1}$, $k \geq 2$ (continuous velocities and continuous pressure), that are *inf-sup* stable. For the proof of the stability results mentioned here, as well for the convergence analysis, the reader can refer to [BF91a].

If we use spectral methods, using equal-order polynomial spaces for both velocity and pressure yields subspaces that violate the *inf-sup* condition. Compatible spectral spaces can instead be obtained by using, e.g., polynomials of degree N (≥ 2) for each velocity component, and degree $N-2$ for the pressure, yielding the so-called $\mathbb{Q}_N - \mathbb{Q}_{N-2}$ approximation. The degrees of freedom for each velocity component are represented by the $(N+1)^2$ GLL nodes (see Fig. 16.5). For the pressure, at least two sets of interpolation nodes can be used: either the subset represented by the $(N-1)^2$ internal nodes of the set of $(N+1)^2$ Gauss-Lobatto nodes (Fig. 16.6, left), or the $(N-1)^2$ Gauss nodes (Fig. 16.6, right). This choice stands at the base of a spectral-type approximation, such as collocation, G-NI (Galerkin with numerical integration), or SEM-NI (spectral element with numerical integration) (see [CHQZ07]).

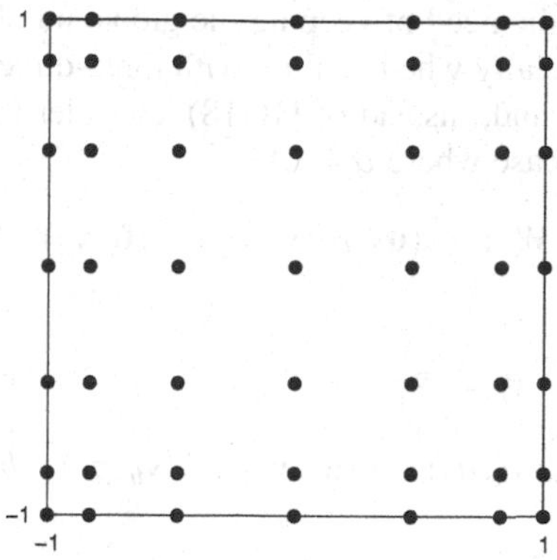

Fig. 16.5. The $(N+1)^2$ Gauss-Legendre-Lobatto (GLL) nodes (here $N=6$), hosting the degrees of freedom of the velocity components

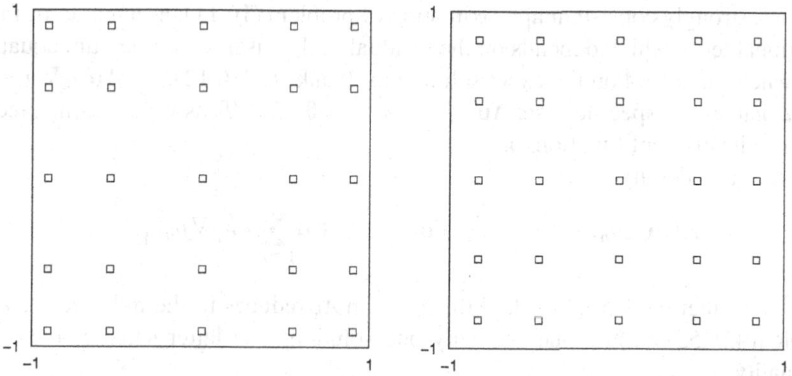

Fig. 16.6. The $(N-1)^2$ internal Gauss-Legendre-Lobatto (GLL) nodes (left) and the $(N-1)^2$ Gauss-Legendre (GL) nodes (right) (here for $N=6$), hosting the degrees of freedom of the pressure

16.5 An example of stabilized problem

We have seen that finite element or spectral methods that make use of equal-degree polynomials for both velocity and pressure do not fulfill the *inf-sup* condition and are therefore "unstable". However, stabilizing them is possible by SUPG or GLS techniques like those encountered in Chapter 12 in the framework of the numerical approximation of advection-diffusion equations.

For a general discussion on stabilization techniques for Stokes equations, the reader can refer e.g. to [BF91a]. Here we limit ourselves to show how the GLS stabilization can be applied to problem (16.18) in case piecewise continuous linear finite elements are used for velocity components as well as for the pressure

$$V_h = [\overset{\circ}{X}{}_h^1]^2, \quad Q_h = \{q_h \in X_h^1 : \int_\Omega q_h \, d\Omega = 0\}.$$

This choice is urged by the need of keeping the global number of degrees of freedom as low as possible, especially when dealing with three-dimensional problems. We set therefore $W_h = V_h \times Q_h$ and, instead of (16.18), consider the following problem (we restrict ourselves to the case where $\alpha = 0$):

$$\text{find } (\mathbf{u}_h, p_h) \in W_h : \ A_h(\mathbf{u}_h, p_h; \mathbf{v}_h, q_h) = (\mathbf{f}_h, \mathbf{v}_h) \quad \forall (\mathbf{v}_h, q_h) \in W_h. \tag{16.55}$$

We have set

$$A_h : W_h \times W_h \to \mathbb{R},$$

$$A_h(\mathbf{u}_h, p_h; \mathbf{v}_h, q_h) = a(\mathbf{u}_h, \mathbf{v}_h) + b(\mathbf{v}_h, p_h) - b(\mathbf{u}_h, q_h)$$

$$+\delta \sum_{K \in \mathcal{T}_h} h_K^2 \int_K (-\nu \Delta \mathbf{u}_h + \nabla p_h - \mathbf{f})(-\nu \Delta \mathbf{v}_h + \nabla q_h)\, dK,$$

and we have denoted with δ a positive parameter that must be chosen conveniently. This is a strongly consistent approximation of problem (16.11): as a matter of fact, the additional term, which depends on the residual of the discrete momentum equation, is null when calculated on the exact solution as, thanks to (16.12), $-\nu \Delta \mathbf{u} + \nabla p - \mathbf{f} = \mathbf{0}$. (Note that, in this specific case, $\Delta \mathbf{u}_{h|K} = \Delta \mathbf{v}_{h|K} = \mathbf{0}\ \forall K \in \mathcal{T}_h$ as we are using piecewise linear finite element functions.).

From the identity

$$A_h(\mathbf{u}_h, p_h; \mathbf{u}_h, p_h) = \nu \|\nabla \mathbf{u}_h\|^2_{\mathbf{L}^2(\Omega)} + \delta \sum_{k \in \mathcal{T}_h} h_K^2 \|\nabla p_h\|^2_{\mathbf{L}^2(K)}, \tag{16.56}$$

we deduce that the kernel of the bilinear form A_h reduces to the null vector, whence problem (16.55) admits one and only one solution. The latter satisfies the stability inequality

$$\nu \|\nabla \mathbf{u}_h\|^2_{\mathbf{L}^2(\Omega)} + \delta \sum_{K \in \mathcal{T}_h} h_K^2 \|\nabla p_h\|^2_{\mathbf{L}^2(K)} \le C \|\mathbf{f}\|^2_{\mathbf{L}^2(\Omega)}, \tag{16.57}$$

C being a constant that depends on ν but not on h (see Exercise 7).

By applying Strang's Lemma 10.1 we can now show that the solution to the generalized Galerkin problem (16.55) satisfies the following error estimate

$$\|\mathbf{u} - \mathbf{u}_h\|_{\mathbf{H}^1(\Omega)} + \left(\delta \sum_{K \in \mathcal{T}_h} h_K^2 \|\nabla p - \nabla p_h\|^2_{\mathbf{L}^2(K)} \right)^{1/2} \le Ch.$$

Still using the notations of Sect. 16.2, we can show that (16.55) admits the following matrix form

$$\begin{bmatrix} \mathrm{A} & \mathrm{B}^T \\ \mathrm{B} & -\mathrm{C} \end{bmatrix} \begin{bmatrix} \mathbf{U} \\ \mathbf{P} \end{bmatrix} = \begin{bmatrix} \mathbf{F} \\ \mathbf{G} \end{bmatrix}. \tag{16.58}$$

This system differs from (16.49) without stabilization because of the presence of the non-null block occupying the position (2,2), which is associated to the stabilization term. More precisely,

$$C = (c_{km}) \ , \ c_{km} = \delta \sum_{K \in \mathcal{T}_h} h_K^2 \int_K \nabla \phi_m \cdot \nabla \phi_k \, dK, \qquad k,m = 1,\dots,M,$$

while the components of the right-hand side $\mathbf{G}$ are

$$g_k = -\delta \sum_{K \in \mathcal{T}_h} h_K^2 \int_K \mathbf{f} \cdot \nabla \phi_k \, dK, \qquad k = 1,\dots,M.$$

In this case, the reduced system for the pressure unknown reads

$$\mathbf{RP} = \mathrm{BA}^{-1}\mathbf{F} - \mathbf{G}.$$

In contrast to (16.52), this time $\mathrm{R} = \mathrm{BA}^{-1}\mathrm{B}^T + \mathrm{C}$. The matrix R is non-singular as C is a positive definite matrix.

16.6 A numerical example

We want to solve the stationary Navier-Stokes equations in the square domain $\Omega = (0,1) \times (0,1)$ with the following Dirichlet conditions

$$\begin{aligned} \mathbf{u} &= \mathbf{0}, & \mathbf{x} &\in \partial\Omega \backslash \Gamma, \\ \mathbf{u} &= (1,0)^T, & \mathbf{x} &\in \Gamma, \end{aligned} \qquad (16.59)$$

where $\Gamma = \{\mathbf{x} = (x_1,x_2)^T \in \partial\Omega : x_2 = 1\}$. This problem is known as flow in a lid-driven cavity. We will use continuous piecewise bilinear $\mathbb{Q}_1 - \mathbb{Q}_1$ polynomials on rectangular finite elements. As we know, these spaces do not fulfill the compatibility condition approximation; in Fig. 16.7, left, we display the spurious pressure modes that are generated by this Galerkin approximation. In the same figure, right, we have drawn the pressure isolines obtained using a GLS stabilization (addressed in the previous section) on the same kind of finite elements. The pressure is now free of numerical oscillations. Still for the stabilized problem, in Fig. 16.8 we display the streamlines for two different values of the Reynolds number, $Re = 1000$ and $Re = 5000$. The stabilization term amends simultaneously pressure instabilities (by getting rid of the spurious modes) and potential instabilities of the pure Galerkin method that develop when diffusion is dominated by convection, an issue that we have extensively addressed in Chapter 12.

For the same problem we consider, as well, a spectral G-NI approximation in which the pressure and each velocity component are polynomials of $\mathbb{Q}_N$ (with $N = 32$). As previously observed, this choice of spaces does not fulfill the *inf-sup* condition, and so it generates spurious pressure modes that are clearly visible in Fig. 16.9, left. A GLS stabilization, similar to that previously used for finite elements, can be set up for

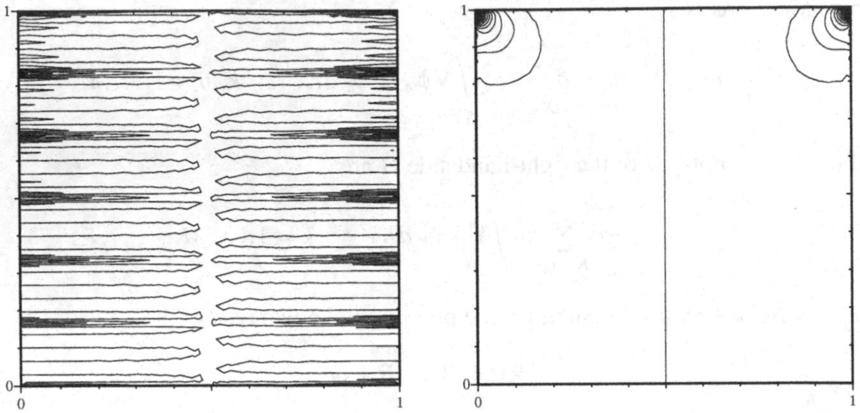

Fig. 16.7. Pressure isolines for the numerical approximation of the lid-driven cavity problem. Stabilized GLS approximation (on the right); the vertical line corresponds to the null value of the pressure. Non-stabilized approximation (on the left); the presence of a spurious numerical pressure is evident

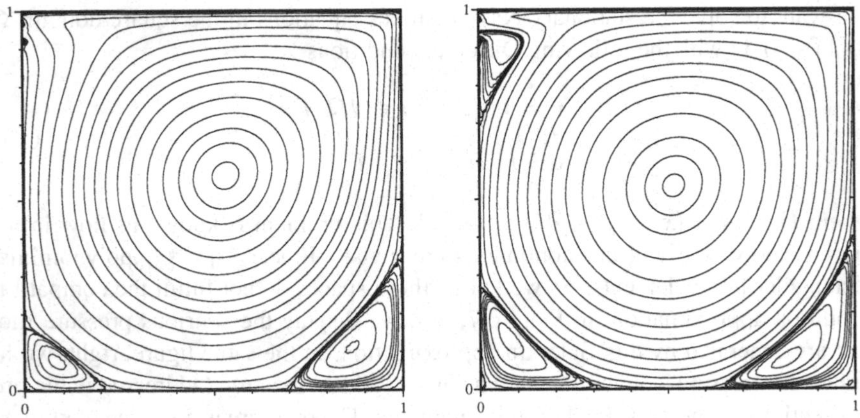

Fig. 16.8. Streamlines of the numerical solution of the lid-driven cavity problem corresponding to two different values of the Reynolds number: $Re = 1000$, left, and $Re = 5000$, right

the G-NI method, too. The corresponding solution is now stable and free of spurious pressure modes, as the pressure isolines displayed on the right hand of the same figure show.

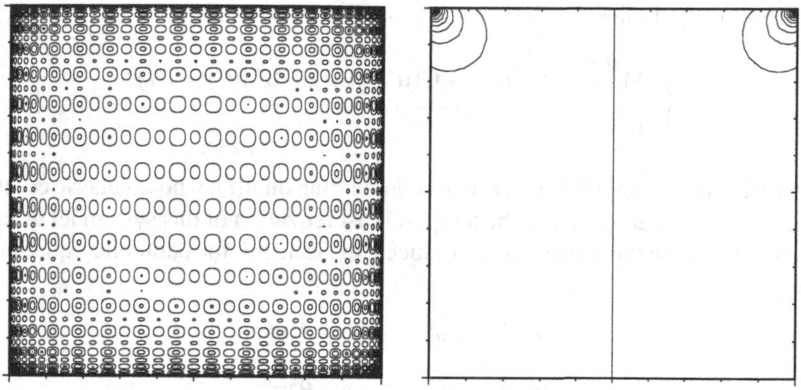

Fig. 16.9. Pressure isolines obtained by the pure spectral G-NI method (on the left), and by the GLS stabilized spectral G-NI method (on the right). In either case, polynomials of the same degree, $N = 32$, are used for both pressure and velocity. As expected, the pure G-NI method yields spurious pressure solutions. The test case is the same lid-driven cavity problem previously approximated by bilinear finite elements

16.7 Time discretization of Navier-Stokes equations

Let us now return to the Navier-Stokes equations (16.2) and focus on the issue of time discretization. To avoid unnecessary cumbersome notation, from now on we will assume that $\Gamma_D = \partial\Omega$ and $\boldsymbol{\varphi} = \mathbf{0}$ in (16.3), whence the velocity space becomes $V = [\mathrm{H}_0^1(\Omega)]^d$. The space discretization of the Navier-Stokes equations yields the following problem:

for every $t > 0$, find $(\mathbf{u}_h(t), p_h(t)) \in V_h \times Q_h$ such that

$$\begin{cases} \left(\dfrac{\partial \mathbf{u}_h(t)}{\partial t}, \mathbf{v}_h\right) + a(\mathbf{u}_h(t), \mathbf{v}_h) + c(\mathbf{u}_h(t), \mathbf{u}_h(t), \mathbf{v}_h) + b(\mathbf{v}_h, p_h(t)) \\ \qquad = (\mathbf{f}_h(t), \mathbf{v}_h) \; \forall \mathbf{v}_h \in V_h, \\ b(\mathbf{u}_h(t), q_h) \quad = 0 \quad \forall q_h \in Q_h, \end{cases} \tag{16.60}$$

where, as usual, $\{V_h \subset V\}$ and $\{Q_h \subset Q\}$ are two families of finite dimensional subspaces of the velocity and pressure functional spaces, respectively. The trilinear form $c(\cdot,\cdot,\cdot)$, defined by

$$c(\mathbf{w}, \mathbf{z}, \mathbf{v}) = \int_\Omega [(\mathbf{w} \cdot \nabla)\mathbf{z}] \cdot \mathbf{v} \, d\Omega \quad \forall \mathbf{w}, \mathbf{z}, \mathbf{v} \in V,$$

is associated to the nonlinear convective term, while $a(\cdot,\cdot)$ and $b(\cdot,\cdot)$ are the same as in (16.13) (setting however $\alpha = 0$).

Problem (16.60) is in fact a system of nonlinear differential algebraic equations. By using notations already employed in the previous sections, it can be restated in compact form as follows

$$\begin{cases} \mathrm{M}\dfrac{d\mathbf{u}(t)}{dt} + \mathrm{A}\mathbf{u}(t) + \mathrm{C}(\mathbf{u}(t))\mathbf{u}(t) + \mathrm{B}^T\mathbf{p}(t) = \mathbf{f}(t), \\ \mathrm{B}\mathbf{u}(t) = \mathbf{0}, \end{cases} \tag{16.61}$$

with $\mathbf{u}(0) = \mathbf{u}_0$. $\mathrm{C}(\mathbf{u}(t))$ is in fact a matrix depending on $\mathbf{u}(t)$, whose generic coefficient is $c_{mi}(t) = c(\mathbf{u}(t), \varphi_i, \varphi_m)$. For the temporal discretization of this system let us use, for instance, the θ-method, that was introduced in Sect. 5.1 for parabolic equations. By setting

$$\mathbf{u}_\theta^{n+1} = \theta\mathbf{u}^{n+1} + (1-\theta)\mathbf{u}^n,$$

$$\mathbf{p}_\theta^{n+1} = \theta\mathbf{p}^{n+1} + (1-\theta)\mathbf{p}^n,$$

$$\mathbf{f}_\theta^{n+1} = \mathbf{f}(\theta t^{n+1} + (1-\theta)t^n),$$

we obtain the following system of algebraic equations

$$\begin{cases} \mathrm{M}\dfrac{\mathbf{u}^{n+1} - \mathbf{u}^n}{\Delta t} + \mathrm{A}\mathbf{u}_\theta^{n+1} + \mathrm{C}(\mathbf{u}_\theta^{n+1})\mathbf{u}_\theta^{n+1} + \mathrm{B}^T\mathbf{p}_\theta^{n+1} = \mathbf{f}_\theta^{n+1}, \\ \mathrm{B}\mathbf{u}^{n+1} = \mathbf{0}. \end{cases} \tag{16.62}$$

Except for the special case $\theta = 0$, which corresponds to the forward Euler method, the solution of this system is quite involved. A possible alternative is to use a *semi-implicit* scheme, in which the linear part of the equation is advanced implicitly, while nonlinear terms explicitly. By doing so, if $\theta \geq 1/2$, the resulting scheme is unconditionally stable, whereas it must obey a stability restriction on the time step Δt (depending on h and ν) in all other cases. We further elaborate on this issue in the next section. Later, in Sects. 16.7.2 and 16.7.3 we will address other temporal discretization schemes. For more details, results and bibliographical references, see, e.g., [QV94, Chap. 13].

16.7.1 Finite difference methods

We consider at first an explicit temporal discretization of the first equation in (16.61), corresponding to the choice $\theta = 0$ in (16.62). If we suppose that all quantities are known at the time t^n, we can write the associated problem at time $t^n + 1$ as follows

$$\begin{cases} \mathrm{M}\mathbf{u}^{n+1} = H(\mathbf{u}^n, \mathbf{p}^n, \mathbf{f}^n), \\ \mathrm{B}\mathbf{u}^{n+1} = \mathbf{0}, \end{cases}$$

where M is the *mass matrix* whose entries are

$$m_{ij} = \int_\Omega \varphi_i \varphi_j \, d\Omega.$$

This system is overdetermined for the unknown vector $\mathbf{u}^{n+1}$, whereas it does not allow the determination of the pressure $\mathbf{p}^{n+1}$. However, if we replace $\mathbf{p}^n$ by $\mathbf{p}^{n+1}$ in the momentum equation, we obtain the new linear system

$$\begin{cases} \frac{1}{\Delta t}\mathrm{M}\mathbf{u}^{n+1} + \mathrm{B}^T\mathbf{p}^{n+1} = \mathbf{G}, \\ \mathrm{B}\mathbf{u}^{n+1} = \mathbf{0}, \end{cases} \tag{16.63}$$

$\mathbf{G}$ being a suitable known vector. This system corresponds to a *semi-explicit* discretization of (16.60). Since M is symmetric and positive definite, if condition (16.53) is satisfied, then the reduced system $\mathrm{BM}^{-1}\mathrm{B}^T\mathbf{p}^{n+1} = \mathrm{BM}^{-1}\mathbf{G}$ is non-singular. Once solved, the velocity vector $\mathbf{u}^{n+1}$ can be recovered from the first equation of (16.63). This discretization method is temporally stable provided the time step satisfies the following limitation

$$\Delta t \leq C\min\left(\frac{h^2}{\nu}, \frac{h}{\max_{\mathbf{x}\in\Omega}|\mathbf{u}^n(\mathbf{x})|}\right).$$

Let us now consider an *implicit* discretization of (16.60), for instance the backward Euler method, which corresponds to choosing $\theta = 1$ in (16.62). As already observed, this scheme is unconditionally stable. It yields a nonlinear algebraic system which can be regarded as the finite element space approximation to the steady Navier-Stokes problem

$$\begin{cases} -\nu\Delta\mathbf{u}^{n+1} + (\mathbf{u}^{n+1}\cdot\nabla)\mathbf{u}^{n+1} + \nabla p^{n+1} + \frac{\mathbf{u}^{n+1}}{\Delta t} = \tilde{\mathbf{f}}, \\ \mathrm{div}\mathbf{u}^{n+1} = 0. \end{cases}$$

The solution of such nonlinear algebraic system can be achieved by Newton-Krylov techniques, that is by using a Krylov method (e.g. GMRES or BiCGStab) for the solution of the linear system that is obtained at each Newton iteration step (see, e.g., [Saa96] or [QV94, Chap. 2]). We recall that Newton's method is based on the full linearization of the convective term, $\mathbf{u}_k^{n+1}\cdot\nabla\mathbf{u}_{k+1}^{n+1} + \mathbf{u}_{k+1}^{n+1}\cdot\nabla\mathbf{u}_k^{n+1}$. A popular approach consists in starting Newton iterations after few Piccard iterations in which the convective term is evaluated as follows: $\mathbf{u}_k^{n+1}\cdot\nabla\mathbf{u}_{k+1}^{n+1}$.
This approach entails three nested cycles:

- temporal iteration: $t^n \to t^{n+1}$;
- Newton iteration: $\mathbf{x}_k^{n+1} \to \mathbf{x}_{k+1}^{n+1}$;
- Krylov iteration: $[\mathbf{x}_k^{n+1}]_j \to [\mathbf{x}_k^{n+1}]_{j+1}$;

for simplicity we have called $\mathbf{x}^n$ the couple $(\mathbf{u}^n, \mathbf{p}^n)$. Obviously, the goal is the following convergence result:

$$\lim_{k\to\infty}\lim_{j\to\infty}[\mathbf{x}_k^{n+1}]_j = \begin{bmatrix} \mathbf{u}^{n+1} \\ \mathbf{p}^{n+1} \end{bmatrix}.$$

Finally, let us operate a *semi-implicit*, temporal discretization, consisting in treating explicitly the nonlinear convective term. The following algebraic linear system, whose form is similar to (16.49), is obtained in this case

$$\begin{cases} \frac{1}{\Delta t}\mathbf{M}\mathbf{u}^{n+1} + \mathbf{A}\mathbf{u}^{n+1} + \mathbf{B}^T\mathbf{p}^{n+1} = \mathbf{G}, \\ \mathbf{B}\mathbf{u}^{n+1} = \mathbf{0}, \end{cases} \tag{16.64}$$

where $\mathbf{G}$ is a suitable known vector. In this case the stability restriction on the time step takes the following form

$$\Delta t \le C \frac{h}{\max\limits_{\mathbf{x}\in\Omega} |\mathbf{u}^n(\mathbf{x})|}. \tag{16.65}$$

In all cases, optimal error estimates can be proven.

16.7.2 Characteristics (or Lagrangian) methods

The *material derivative* (also called Lagrangian derivative) of the velocity vector field is defined as

$$\frac{D\mathbf{u}}{Dt} = \frac{\partial \mathbf{u}}{\partial t} + (\mathbf{u}\cdot\nabla)\mathbf{u}.$$

Characteristics methods are based on approximating the material derivative, e.g. by the backward Euler method

$$\frac{D\mathbf{u}}{Dt}(\mathbf{x}) \approx \frac{\mathbf{u}^{n+1}(\mathbf{x}) - \mathbf{u}^n(\mathbf{x}_p)}{\Delta t},$$

where $\mathbf{x}_p$ is the *foot* (at time t^n) of the characteristic issuing from $\mathbf{x}$ at time t^{n+1}. A system of ordinary differential equations has to be solved to follow backwards the characteristic line $\mathbf{X}$ issuing from the point $\mathbf{x}$

$$\begin{cases} \frac{d\mathbf{X}}{dt}(t;s,\mathbf{x}) = \mathbf{u}(t,\mathbf{X}(t;s,\mathbf{x})), \quad t \in (t^n, t^{n+1}), \\ \mathbf{X}(s;s,\mathbf{x}) = \mathbf{x}, \end{cases}$$

having set $s = t^{n+1}$.
The main difficulty lies in determining the characteristic lines. The first problem is how to suitably approximate the velocity field $\mathbf{u}(\mathbf{t})$ for $t \in (t^n, t^{n+1})$, as $\mathbf{u}^{n+1}$ is unknown. The simplest way to do so consists in using a forward Euler scheme for the discretization of the material derivative. The second difficulty stems from the fact that a characteristic line may cross several elements of the computational grid. An algorithm is therefore necessary to locate the element in which the characteristic foot falls, or to detect those cases in which the latter hits a boundary edge. With the previous discretization of the material derivative, at every time level t^{n+1} the momentum equation becomes (formally)

$$\frac{\mathbf{u}^{n+1}(\mathbf{x}) - \mathbf{u}^n(\mathbf{x}_p)}{\Delta t} - \nu\Delta\mathbf{u}^{n+1}(\mathbf{x}) + \nabla p^{n+1}(\mathbf{x}) = \mathbf{f}^{n+1}(\mathbf{x}).$$

If used in the framework of piecewise linear finite elements in space, this scheme is unconditionally stable. Moreover, it satisfies the error estimate

$$\|\mathbf{u}(t^n) - \mathbf{u}^n\|_{L^2(\Omega)} \leq C(h + \Delta t + h^2/\Delta t) \qquad \forall n \geq 1,$$

for a positive constant C independent of ν. Characteristic-based time discretization strategies for spectral methods are reviewed in [CHQZ07, Chap. 3].

16.7.3 Fractional step methods

Let us consider an abstract time dependent problem,

$$\frac{\partial w}{\partial t} + Lw = f,$$

where L is a differential operator that splits into the sum of two operators, L_1 and L_2, that is

$$Lv = L_1 v + L_2 v.$$

Fractional step methods allow the temporal advancement from time t^n to t^{n+1} in two steps (or more). At first only the operator L_1 is advanced in time implicitly, then the solution so obtained is corrected by performing a second step in which only the other operator, L_2, is in action. This is why these kind of methods are also named *operator splitting*.

In principle, by separating the two operators L_1 and L_2, a complex problem is split into two simpler problems, each one with its own feature. In this respect, the operators L_1 and L_2 can be chosen on the ground of physical considerations: diffusion can be split from transport, for instance. In fact, also the solution of Navier-Stokes equations by the characteristic method can be regarded as a fractional step method whose first step operator is expressed by the Lagrangian derivative.

A simple, albeit not optimal fractional step scheme, is the following, known as *Yanenko splitting*:

1. compute the solution $\tilde{w}$ of the equation

$$\frac{\tilde{w} - w^n}{\Delta t} + L_1 \tilde{w} = 0;$$

2. compute the solution w^{n+1} of the equation

$$\frac{w^{n+1} - \tilde{w}}{\Delta t} + L_2 w^{n+1} = f^n.$$

By eliminating $\tilde{w}$, the following problem is found for w^{n+1}

$$\frac{w^{n+1} - w^n}{\Delta t} + Lw^{n+1} = f^n + \Delta t L_1 (f^n - L_2 w^{n+1}).$$

In the case where both L_1 and L_2 are elliptic operators, this scheme is unconditionally stable with respect to Δt.

This strategy can be applied to the Navier-Stokes equations (16.2), choosing L_1 as $L_1(\mathbf{w}) = -\nu\Delta\mathbf{w} + (\mathbf{w}\cdot\nabla)\mathbf{w}$ whereas L_2 is the operator associated to the remaining terms of the Navier-Stokes problem. In this way we have split the main difficulties arising when treating Navier-Stokes equations, the nonlinear part from that imposing the incompressibility constraint. The corresponding fractional step scheme reads:

1. solve the diffusion-transport equation for the velocity $\tilde{\mathbf{u}}^{n+1}$

$$\begin{cases} \dfrac{\tilde{\mathbf{u}}^{n+1} - \mathbf{u}^n}{\Delta t} - \nu\Delta\tilde{\mathbf{u}}^{n+1} + (\mathbf{u}^*\cdot\nabla)\mathbf{u}^{**} = \mathbf{f}^{n+1} & \text{in } \Omega, \\ \tilde{\mathbf{u}}^{n+1} = \mathbf{0} & \text{on } \partial\Omega; \end{cases} \tag{16.66}$$

2. solve the following coupled problem for the two unknowns $\mathbf{u}^{n+1}$ and p^{n+1}

$$\begin{cases} \dfrac{\mathbf{u}^{n+1} - \tilde{\mathbf{u}}^{n+1}}{\Delta t} + \nabla p^{n+1} = \mathbf{0} & \text{in } \Omega, \\ \mathrm{div}\mathbf{u}^{n+1} = 0 & \text{in } \Omega, \\ \mathbf{u}^{n+1}\cdot\mathbf{n} = 0 & \text{on } \partial\Omega, \end{cases} \tag{16.67}$$

where $\mathbf{u}^*$ and $\mathbf{u}^{**}$ can be either $\tilde{\mathbf{u}}^{n+1}$ or $\mathbf{u}^n$ depending on whether the nonlinear convective terms are treated explicitly, implicitly or semi-implicitly. In such a way, in the first step an intermediate velocity $\tilde{\mathbf{u}}^{n+1}$ is calculated, then it is corrected in the second step in order to satisfy the incompressibility constraint. The diffusion-transport problem of the first step can be successfully addressed by using the approximation techniques investigated in Chapter 12.

More involved is the numerical treatment of the problem associated with the second step. By formally applying the divergence operator to the first equation, we obtain

$$\mathrm{div}\frac{\mathbf{u}^{n+1}}{\Delta t} - \mathrm{div}\frac{\tilde{\mathbf{u}}^{n+1}}{\Delta t} + \Delta p^{n+1} = 0,$$

that is an elliptic boundary-value problem with Neumann boundary conditions

$$\begin{cases} -\Delta p^{n+1} = -\mathrm{div}\dfrac{\tilde{\mathbf{u}}^{n+1}}{\Delta t} & \text{in } \Omega, \\ \dfrac{\partial p^{n+1}}{\partial n} = 0 & \text{on } \partial\Omega. \end{cases} \tag{16.68}$$

The Neumann condition follows from the condition $\mathbf{u}^{n+1}\cdot\mathbf{n} = 0$ on $\partial\Omega$, see (16.67). From the solution of (16.68) we obtain p^{n+1}, and thus $\mathbf{u}^{n+1}$ by using the first equation of (16.67),

$$\mathbf{u}^{n+1} = \tilde{\mathbf{u}}^{n+1} - \Delta t\nabla p^{n+1} \quad \text{in } \Omega. \tag{16.69}$$

This is precisely the correction to operate on the velocity field in order to fulfill the divergence-free constraint.

In conclusion, at first we solve the scalar elliptic problem (16.66) to obtain the intermediate velocity $\tilde{\mathbf{u}}^{n+1}$, then the elliptic problem (16.68) yields the pressure unknown p^{n+1}, and finally we obtain the new velocity field $\mathbf{u}^{n+1}$ through the explicit correction equation (16.69).
Let us now investigate the main features of this method.
Assume that we take $\mathbf{u}^* = \mathbf{u}^{**} = \mathbf{u}^n$ in the first step; after space discretization, we arrive at a linear system as

$$\left(\frac{1}{\Delta t}\mathrm{M} + \mathrm{A}\right)\tilde{\mathbf{u}}^{n+1} = \tilde{\mathbf{f}}^{n+1}.$$

The main limitation of this approach consists in the fact that, having treated explicitly the convective term, the solution undergoes a stability restriction on the time step like (16.65). On the other hand, because of this explicit treatment, this linear system naturally splits into d independent systems of smaller size, one for each spatial component of the velocity field.
If, instead, we use an implicit time advancing scheme, like the one that we would get by setting $\mathbf{u}^* = \mathbf{u}^{**} = \tilde{\mathbf{u}}^{n+1}$, we obtain an unconditionally stable scheme, however with a more involved coupling of all the spatial components due to the nonlinear convective term. This nonlinear algebraic system can be solved by, e.g., a Newton-Krylov method, similar to the one that we have introduced in Sect. 16.7.1. In the second step of the method, we enforce a boundary condition only on the normal component of the velocity field. Yet, we lack any control on the behaviour of the tangential component of the same velocity at the boundary. This generates a so-called *splitting error*: although the solution is divergence-free, the failure to satisfy the physical boundary condition on the tangential velocity component yields the onset of a pressure boundary layer of width $\sqrt{\nu \Delta t}$.

The method just described is due to Chorin and Temam, and is also called *projection method*. The reason can be found in the celebrated Helmholtz-Weyl decomposition theorem:

Theorem 16.7. *Let $\Omega \subset \mathbb{R}^d$, $d = 2,3$, be a domain with Lipschitz boundary. Then, for every $\mathbf{v} \in [\mathrm{L}^2(\Omega)]^d$, there exist two (uniquely-defined) functions* $\mathbf{w}, \mathbf{z}$,

$$\mathbf{w} \in \mathrm{H}^0_{\mathrm{div}} = \{\mathbf{v} \in \left[\mathrm{L}^2(\Omega)\right]^d : \ \mathrm{div}\mathbf{v} = 0 \textit{ in } \Omega,\ \mathbf{v}\cdot\mathbf{n} = 0 \textit{ on } \partial\Omega\},$$

$$\mathbf{z} \in [\mathrm{L}^2(\Omega)]^d, \quad \mathit{rot}\mathbf{z} = \mathbf{0} \quad (\textit{so } \ \mathbf{z} = \nabla\psi, \textit{ for a suitable } \ \psi \in \mathrm{H}^1(\Omega))$$

such that

$$\mathbf{v} = \mathbf{w} + \mathbf{z}.$$

Owing to this result, any function $\mathbf{v} \in [\mathrm{L}^2(\Omega)]^d$ can be univocally represented as being the sum of a solenoidal (that is, divergence-free) field and of an irrotational field (that is, the gradient of a suitable scalar function).

As a matter of fact, after the first step (16.66) in which the preliminary velocity $\tilde{\mathbf{u}}^{n+1}$ is obtained from $\mathbf{u}^n$ by solving the momentum equation, in the course of the second step a solenoidal field $\mathbf{u}^{n+1}$ is constructed in (16.69), with $\mathbf{u}^{n+1} \cdot \mathbf{n} = 0$ on $\partial\Omega$. This solenoidal field is the projection of $\tilde{\mathbf{u}}^{n+1}$, and is obtained by applying the decomposition theorem with the following identifications: $\mathbf{v} = \tilde{\mathbf{w}}^{n+1}$, $\mathbf{v} = \mathbf{u}^{n+1}$, $\psi = +\Delta t p^{n+1}$.

The name projection method is due to the fact that

$$\int_\Omega \mathbf{u}^{n+1} \cdot \boldsymbol{\psi} \, d\Omega = \int_\Omega \tilde{\mathbf{u}}^{n+1} \cdot \boldsymbol{\psi} \, d\Omega \quad \forall \boldsymbol{\psi} \in \mathrm{H}^0_{\mathrm{div}},$$

that is $\mathbf{u}^{n+1}$ is the projection, with respect to the scalar product of $\mathrm{L}^2(\Omega)$, of $\tilde{\mathbf{u}}^{n+1}$ on the space $\mathrm{H}^0_{\mathrm{div}}$.

Remark 16.4. Several variants of the projection method have been proposed with the aim of reducing the splitting error on the pressure, not only for the finite element method but also for higher order spectral or spectral element space approximations. The interested reader can refer to, e.g., [QV94, Qua93, Pro97, KS05] and [CHQZ07, Chap. 3]. •

Example 16.1. In Fig. 16.10 we display the isolines of the modulus of velocity corresponding to the solution of Navier-Stokes equations in a two-dimensional domain $\Omega = (0,17) \times (0,10)$ with five round holes. This can be regarded as the orthogonal section of a three dimensional domain with 5 cylinders. A non-homogeneous Dirichlet condition, $\mathbf{u} = [\arctan(20(5-|5-y|)), 0]^T$, is assigned at the inflow, a homogeneous Dirichlet condition is prescribed on the horizontal side as well as on the border of the cylinders, while at the outflow the normal component of the stress tensor is set to zero. For the space discretization the stabilized spectral element method was used, with 114 spectral elements, and polynomials of degree 7 for both the pressure and the velocity components on every element, plus a second-order BDF2 scheme for temporal discretization (see Sect. 8.5 and also [QSS07]). ■

16.8 Algebraic factorization methods and preconditioners for saddle-point systems

An alternative approach to the solution of systems like (16.49) is the one based on the use of inexact (or incomplete) factorizations of the system matrix (16.50). We remind that these systems can be obtained by the approximate solution of Stokes equations, Navier-Stokes equations (after using one of the linearization approaches described in Sects. 16.7.1, 16.7.2 or 16.7.3), or, more generally, from the approximation of saddle-point problems, as shown in Sect. 16.3. Let us also point out that, in all these cases, the space discretization method can be based on any one of the methods discussed thus far (finite elements, finite differences, finite volumes, spectral methods, etc.).

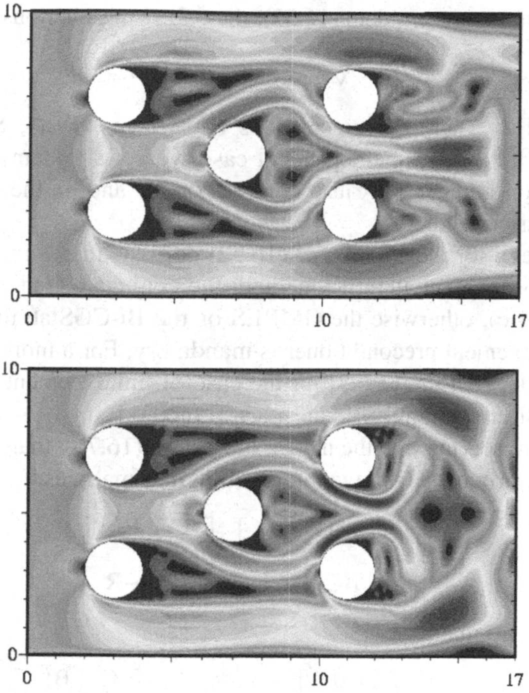

Fig. 16.10. Isolines of the modulus of the velocity vector for the test case of Example 16.1 at the time levels $t = 10.5$ (above) and $t = 11.4$ (below)

Generally speaking, we will suppose to deal with an algebraic system of the following form

$$\begin{bmatrix} \mathrm{C} & \mathrm{B}^T \\ \\ \mathrm{B} & 0 \end{bmatrix} \begin{bmatrix} \mathbf{U} \\ \\ \mathbf{P} \end{bmatrix} = \begin{bmatrix} \mathbf{F} \\ \\ \mathbf{0} \end{bmatrix} \tag{16.70}$$

where C coincides with A in the case of system (16.49), with $\frac{1}{\Delta t}\mathrm{M}+\mathrm{A}$ in case of system (16.64), while more in general it could be given by $\frac{\alpha}{\Delta t}\mathrm{M}+\mathrm{A}+\delta\mathrm{D}$, with D being the matrix associated to the pressure gradient operator, in case a linearization or a semi-implicit treatment are applied to the convective term. In the latter case the coefficients α and δ would depend on the specific linearization or semi-implicit method adopted. Also in this case we can associate (16.70) with the Schur complement

$$\mathrm{R}\mathbf{P} = \mathrm{B}\mathrm{C}^{-1}\mathbf{F}, \quad \text{with } \mathrm{R} = \mathrm{B}\mathrm{C}^{-1}\mathrm{B}^T, \tag{16.71}$$

which reduces to (16.52) if we start from the Stokes system (16.49) instead of (16.70). We start noticing that the condition number of R depends on the *inf-sup* constant β_h, see (16.41), as well as on the continuity constant δ (see (16.21)). More precisely, in the case of the stationary problem (16.52), the following relations hold

$$\beta_h = \sqrt{\lambda_{\min}}, \qquad \delta \geq \sqrt{\lambda_{\max}}$$

where $\lambda_{\min}$ and $\lambda_{\max}$ are the eigenvalues of R (see [QV94, Sect. 9.2.1]). Thus, $\mathrm{cond(R)} \leq \delta^2/\beta_h^2$. In the time-dependent case we get a system like (16.70); in this case the condition number of R also depends on Δt and on the way the convective term has been discretized.
A possible strategy for the solution of (16.52) consists in solving the Schur complement system (16.71) by an iterative method: the conjugate gradient method if C = A (as A is symmetric), otherwise the GMRES or the Bi-CGStab method when $\delta \neq 0$. The use of a convenient preconditioner is mandatory. For a more general discussion, see, e.g., [ESW05, BGL05, QV94] for the case of finite element discretizations, and [CHQZ07] for discretization based on spectral methods.

We start by observing that the matrix of system (16.70), that we denote by S, can be written as the product LU of two block triangular matrices,

$$\mathrm{S} = \begin{bmatrix} \mathrm{I} & 0 \\ \mathrm{BC}^{-1} & \mathrm{I} \end{bmatrix} \begin{bmatrix} \mathrm{C} & \mathrm{B}^T \\ 0 & -\mathrm{R} \end{bmatrix}.$$

Each one of the two matrices

$$\mathrm{P_D} = \begin{bmatrix} \mathrm{C} & 0 \\ 0 & -\mathrm{R} \end{bmatrix} \quad \text{or} \quad \mathrm{P_T} = \begin{bmatrix} \mathrm{C} & \mathrm{B}^T \\ 0 & -\mathrm{R} \end{bmatrix}$$

provides an optimal preconditioner for S, a *block diagonal* preconditioner (P_D), and a *block triangular* preconditioner (P_T). Unfortunately, they are both computationally expensive because of the presence on the diagonal of the Schur complement R, which, in turn, contains the inverse of matrix C. Alternatively, we can use their approximants

$$\widehat{\mathrm{P}}_\mathrm{D} = \begin{bmatrix} \widehat{\mathrm{C}} & 0 \\ 0 & -\widehat{\mathrm{R}} \end{bmatrix} \quad \text{or} \quad \widehat{\mathrm{P}}_\mathrm{T} = \begin{bmatrix} \widehat{\mathrm{C}} & \mathrm{B}^T \\ 0 & -\widehat{\mathrm{R}} \end{bmatrix}$$

where $\widehat{\mathrm{C}}$ and $\widehat{\mathrm{R}}$ are two inexpensive approximations of C and R, respectively. $\widehat{\mathrm{C}}$ can be built from optimal preconditioners of the stiffness matrix, like those that will be introduced in Chapter 18.

The *pressure correction diffusion* preconditioner (PCD) makes use of the following approximation of R

$$\widehat{\mathrm{R}}_{PCD} = \mathrm{A}_P\mathrm{C}_P^{-1}\mathrm{M}_P,$$

where M_P is the pressure mass matrix, A_P the pressure Laplacian matrix, C_P the convection-diffusion pressure matrix. The term "pressure" here means that these matrices are generated by using the basis functions $\{\varphi_k, k = 1, \ldots, M\}$ of the finite dimensional pressure subspace Q_h. This preconditioner is spectrally equivalent to $\mathrm{BM}^{-1}\mathrm{B}^T$, where M is the velocity mass matrix. See [ESW05]. The application of this preconditioner requires the action of one Poisson pressure solve, a mass matrix solve, and a matrix-vector product with F_P. Boundary conditions should be taken into account while constructing A_P and C_P.

The *least-squares commutator* preconditioner (LSC) is

$$\widehat{R}_{LSC} = (B\widehat{M}_V^{-1}B^T)(B\widehat{M}_V^{-1}C\widehat{M}_V^{-1}B^T)^{-1}(B\widehat{M}_V^{-1}B^T),$$

where $\widehat{M}_V$ is the diagonal matrix obtained from the velocity mass matrix M by disregarding the extra-diagonal terms. Using this preconditioner entails two Poisson solves. The convergence of Krylov iterations with the LSC preconditioner is independent of the grid-size and mildly dependent on the Reynolds number. See [EHS+06].

The *augmented Lagrangian* preconditioner (AL), introduced in [BO06], reads

$$\widehat{R}_{AL} = (\nu\widehat{M}_P^{-1} + \gamma W^{-1})^{-1}$$

where $\widehat{M}_P$ is a diagonal matrix that approximates M_P, W is a suitably chosen matrix that, in the simplest case, is also given by $\widehat{M}_P$, ν is the flow viscosity and γ is a positive parameter (usually taken to be 1). This preconditioner requires the original system (16.70) to be modified by replacing the $(1,1)$ block by $C + \gamma B^T W^{-1} B$, which is consistent because $B\mathbf{u} = \mathbf{0}$. The new term $\gamma B^T W^{-1} B$ introduces a coupling between the velocity vector components. Convergence, however, is independent of both the grid-size and the Reynolds number.

Finally, let us remark that direct algebraic preconditioners based on *incomplete LU factorization* (ILU) of the global matrix S can be used, in combination with suitable reordering of the unknowns. An in-depth discussion is found in [RVS08].

A different LU factorization of S,

$$S = \begin{bmatrix} C & 0 \\ B & -R \end{bmatrix} \begin{bmatrix} I & C^{-1}B^T \\ 0 & I \end{bmatrix} \qquad (16.72)$$

stands at the base of the so-called *SIMPLE* preconditioner introduced in [Pat80], and obtained by replacing C^{-1} in both factors L and U by a triangular matrix D^{-1} (for instance, D could be the diagonal of C). More precisely,

$$P_{SIMPLE} = \begin{bmatrix} C & 0 \\ B & -\widehat{R} \end{bmatrix} \begin{bmatrix} I & D^{-1}B^T \\ 0 & I \end{bmatrix} = \widehat{L}\widehat{U},$$

with $\widehat{R} = BD^{-1}B^T$.

Convergence of preconditioned iterative methods deteriorates when the grid-size h decreases and/or the Reynolds number increases.

Note that using P_{SIMPLE} once, say

$$P_{SIMPLE}\mathbf{w} = \mathbf{r} \qquad (16.73)$$

with $\mathbf{r} = [\mathbf{r}_u, \mathbf{r}_p]$ and $\mathbf{w} = [\mathbf{u}, \mathbf{p}]$, yields $\widehat{\mathrm{L}}\mathbf{w}^* = \mathbf{r}$, and so $\widehat{\mathrm{U}}\mathbf{w} = \mathbf{w}^*$, that is, setting $\mathbf{w}^* = [\mathbf{u}^*, \mathbf{p}^*]$:

$$\mathrm{C}\mathbf{u}^* = \mathbf{r}_u, \tag{16.74}$$

$$\widehat{\mathrm{R}} = \mathrm{B}\mathbf{u}^* - \mathbf{r}_p, \tag{16.75}$$

$$\mathbf{u} = \mathbf{u}^* - \mathrm{D}^{-1}\mathrm{B}^T\mathbf{p}^*. \tag{16.76}$$

This requires a C-solve for the velocity and a pressure Poisson solve (for $\mathrm{BD}^{-1}\mathrm{B}^T$).

Several generalizations of the *SIMPLE* preconditioner have been proposed, in particular *SIMPLER*, *h-SIMPLE* and *MSIMPLER*. Using $\mathrm{P}_{SIMPLER}$ instead of P_{SIMPLE} in (16.73) involves the following steps:

$$\widehat{\mathrm{R}}\mathbf{p}^0 = \mathrm{BD}^{-1}\mathbf{r}_u - \mathbf{r}_p, \tag{16.77}$$

$$\mathrm{C}\mathbf{u}^* = \mathbf{r}_u - \mathrm{B}^T\mathbf{p}^0, \tag{16.78}$$

$$\widehat{\mathrm{R}}\mathbf{p}^* = \mathrm{B}\mathbf{u}^* - \mathbf{r}_p, \tag{16.79}$$

$$\mathbf{u} = \mathbf{u}^* - \mathrm{D}^{-1}\mathrm{B}^T\mathbf{p}^*, \tag{16.80}$$

$$\mathbf{p} = \mathbf{p}^* + \omega\mathbf{p}^0, \tag{16.81}$$

with $\omega \in]0,1]$ being a possible relaxation parameter ($\omega = 1$ in *SIMPLER*, $\omega \neq 1$ in *SIMPLER*(ω)). It therefore involves two pressure Poisson solves and one C-velocity solve; however, in general it enjoys faster convergence than *SIMPLE*.

The preconditioner *hSIMPLE* (h = hybrid) is based on a combined application of *SIMPLE* and *SIMPLER* preconditioners. Finally, the preconditioner *MSIMPLER* makes use of the same steps (16.77)–(16.81) as *SIMPLER*, but the approximate Schur complement $\widehat{\mathrm{R}} = \mathrm{BD}^{-1}\mathrm{B}^T$ is replaced by the least-squares commutator $\widehat{\mathrm{R}}_{LSC}$. The convergence is better than with other variants of *SIMPLE*.

For more discussion and a comparative analysis see [RVS09] and also [Wes01].

The $\widehat{\mathrm{L}}\widehat{\mathrm{U}}$ factorization used in P_{SIMPLE} can be regarded as a special case of a more general family of inexact or algebraic factorizations that read as follows

$$\widehat{\mathrm{S}} = \widehat{\mathrm{L}}\widehat{\mathrm{U}} = \begin{bmatrix} \mathrm{C} & 0 \\ \mathrm{B} & -\mathrm{B}\mathcal{L}\mathrm{B}^T \end{bmatrix} \begin{bmatrix} \mathrm{I} & \mathcal{U}\mathrm{B}^T \\ 0 & \mathrm{I} \end{bmatrix} \tag{16.82}$$

Here $\mathcal{L}$ and $\mathcal{U}$ represent two (not necessarily coincident) approximations of C^{-1}. Using this inexact factorization, the solution of the linear system

$$\widehat{\mathrm{S}} \begin{bmatrix} \widehat{\mathbf{u}} \\ \widehat{\mathbf{p}} \end{bmatrix} = \begin{bmatrix} \mathbf{F} \\ \mathbf{0} \end{bmatrix}$$

can be found through the following steps:

$$\text{step } \widehat{\mathrm{L}} : \quad \begin{cases} \mathrm{C}\mathbf{u}^* = \mathbf{F} & \text{(intermediate velocity)} \\ -\mathrm{B}\mathcal{L}\mathrm{B}^T\widehat{\mathbf{p}} = -\mathrm{B}\mathbf{u}^* & \text{(pressure)} \end{cases}$$

$$\text{step } \widehat{\mathrm{U}} : \quad \widehat{\mathbf{u}} = \mathbf{u}^* - \mathcal{U}\mathrm{B}^T\widehat{\mathbf{p}} \qquad \text{(final velocity).}$$

When used in connection with time-dependent (either Stokes or Navier-Stokes) problems, e.g. (16.64), two different possibilities stand out [QSV00]:

$$\mathcal{L} = \mathcal{U} = \left(\frac{1}{\Delta t}\mathrm{M}\right)^{-1}, \tag{16.83}$$

$$\mathcal{L} = \left(\frac{1}{\Delta t}\mathrm{M}\right)^{-1} \quad \text{and} \quad \mathcal{U} = \mathrm{C}^{-1}. \tag{16.84}$$

The former (16.83) is named *Chorin-Temam algebraic approximation* because the steps $\widehat{\mathrm{L}}$ and $\widehat{\mathrm{U}}$ can be regarded as the algebraic counterpart of the Chorin-Temam fractional step method described previously (see Sect. 16.7.3).

The second choice, (16.84), is called a *Yosida approximation* as it can be interpreted as a Yosida regularization of the Schur complement ([Ven98]).
The potential advantage of this strategy with respect to the one based on differential fractional step methods is that it does not require any special care about boundary conditions. The latter are implicitly accounted for in the algebraic formulation (16.70) and no further requirement is needed in the course of the $\widehat{\mathrm{L}}$ and $\widehat{\mathrm{U}}$ steps.
Several generalizations of the inexact factorization technique (16.82) are possible, based on different choices of the factors $\mathcal{L}$ and $\mathcal{U}$. In case the time dependent Navier-Stokes equations are discretized in time by high-order (≥ 2) temporal schemes, inexact factors are chosen so that the time discretization order is maintained. See [GSV06, SV05, Ger08].
In Fig. 16.11 we display the error behaviour corresponding to the approximation of the time dependent Navier-Stokes equations on the domain $\Omega = (0,1)^2$ using the spectral element method (SEM) with 4×4 square elements with side-length $H = 0.25$, and polynomials of degree $N = 8$ for the velocity components and $N = 6$ for the pressure. The exact solution is $\mathbf{u}(x,y,t) = (\sin(x)\sin(y+t), \cos(x)\cos(y+t))^T$, $p(x,y,t) = \cos(x)\sin(y+t)$. The temporal discretization is based on implicit backward differentiation formulae of order 2 (BDF2), 3 (BDF3), and 4 (BDF4) (see [QSS07]), then on inexact (Yosida) algebraic factorizations of order 2, 3, and 4, respectively. Denoting by $(\mathbf{u}_N^n, p_N^n)$ the numerical solution at the time level t^n, the errors on velocity and pressure

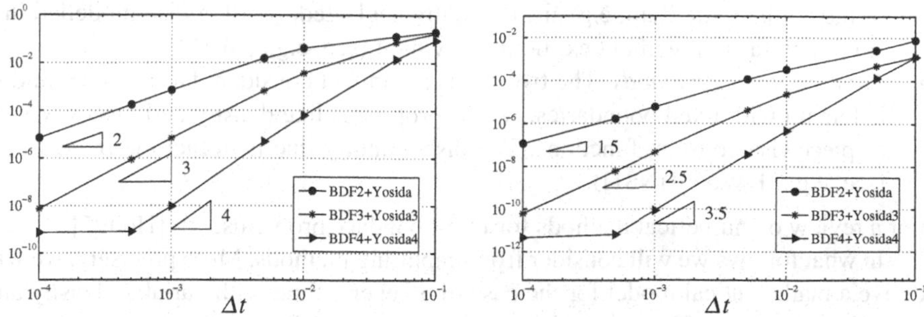

Fig. 16.11. Velocity errors $E_{\mathbf{u}}$ on the left; pressure errors E_p on the right

are defined as

$$E_{\mathbf{u}} = \left(\Delta t \sum_{n=0}^{N_T} \|\mathbf{u}(t^n) - \mathbf{u}_N^n\|_{H^1(\Omega)}^2 \right)^{1/2} \quad \text{and} \quad E_p = \left(\Delta t \sum_{n=0}^{N_T} \|p(t^n) - p_N^n\|_{L^2(\Omega)}^2 \right)^{1/2}$$

Errors on velocity are infinitesimal with respect to Δt of order 2, 3, and 4, respectively, whereas errors on pressure are of order 3/2, 5/2 and 7/2, respectively.

16.9 Free surface flow problems

Free surface flows can manifest under various situations and different shapes. A free surface is generated every time that two immiscible fluids get in contact. They can give rise to jets, [LR98], bubbles [HB76, TF88], droplets [Max76] and films. This kind of fluids are encountered in a variety of different applications, such as waves in rivers, lakes and oceans [Bla02, Qu02], the interaction between waves with solid media (boats, coasts, etc.) [Wya00, KMI$^+$83], injection, moulding and extrusion of polymers and liquid metals [Cab03], chemical reactors or bioreactors, etc. Depending upon the spatial and temporal scales involved, processes like heat transfer, surface tension, laminar to turbulent transition, compressibility and chemical reactions, interaction with solids, might have a relevant impact on the flow behaviour. In what follows we will focus on laminar flows for viscous Newtonian fluids subject to surface tension; in these circumstances, the flow can be described by the incompressible Navier-Stokes equations.

When modeling this kind of fluids, two different approaches can be adopted:

- *Front-tracking methods.* These methods consider the free surface as being the boundary of a moving domain on which suitable boundary conditions are specified. At the interior of the domain, a conventional fluid model is used; special attention however should be paid to the fact that the domain is not fixed. On the other side of the domain, the fluid, e.g. air, is usually neglected, or otherwise modelled in a simplified fashion without explicitly solving it (see, e.g., [MP97]).
- *Front-capturing methods.* The two fluids are in fact considered as a single fluid in a domain with fixed boundaries, whose properties like density and viscosity vary as piecewise constant functions. The discontinuity line is in fact the free surface (see, e.g., [HW65, HN81]).

For a review on numerical methods for free-boundary problems, see [Hou95].

In what follows we will consider front-capturing methods. More precisely, we will derive a mathematical model for the case of a general fluid with variable density and viscosity, which would therefore be appropriate to model the flow of two fluids separated by a free surface.

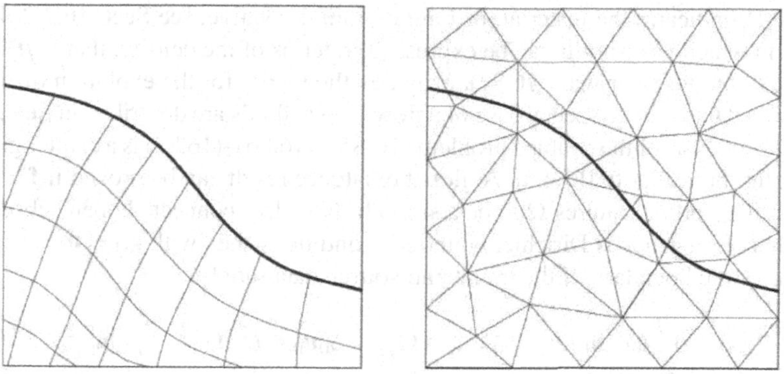

Fig. 16.12. Two typical grids in two dimensions for front-tracking methods (left) and front-capturing methods (right). The thick line represents the free surface

16.9.1 Navier-Stokes equations with variable density and viscosity

We consider the general case of a viscous incompressible flow whose density ρ and dynamical viscosity μ vary both in space and in time. Within a given spatial domain $\Omega \subset \mathbb{R}^d$, the evolution of the fluid's velocity $\mathbf{u} = \mathbf{u}(\mathbf{x},t)$ and pressure $p = p(\mathbf{x},t)$ are modelled by the following equations

$$\rho\partial_t\mathbf{u} + \rho(\mathbf{u}\cdot\nabla)\mathbf{u} - \mathrm{div}\,(2\mu\mathbf{D}(\mathbf{u})) + \nabla p = \mathbf{f}, \qquad \mathbf{x}\in\Omega,\ t>0, \qquad (16.85)$$

$$\mathrm{div}\mathbf{u} = 0, \qquad \mathbf{x}\in\Omega,\ t>0, \qquad (16.86)$$

in which $(\mathbf{D}(\mathbf{v})) = \frac{\nabla\mathbf{v}+\nabla\mathbf{v}^T}{2}$ is the *symmetric gradient* of $\mathbf{v}$, a tensor that is also called *rate of deformation*, while $\mathbf{f}$ denotes a volumetric force, for instance gravity. (Here ∂_t stands for $\partial/\partial t$.)

These equations must be supplemented by suitable initial and boundary conditions. In case ρ is constant we retrieve the form (16.1). Note that incompressibility is not in contradiction with variable density. Incompressibility means that one single fluid parcel does not change volume and thus density, whereas variable density means that different fluid parcels may have different densities. The last two terms of the left-hand side (16.85) can be rewritten as $-\mathrm{div}\mathbf{T}(\mathbf{u},p)$, where

$$\mathbf{T}(\mathbf{u},p) = 2\mu\mathbf{D}(\mathbf{u}) - \mathbf{I}p$$

is the *stress tensor* while $\mathbf{I}$ is the $d\times d$ identity tensor. The divergence of a tensor was introduced in (15.23). A complete derivation of this model can be found, e.g., in [LL59].

An equation for the density function ρ can be obtained from the mass balance equation

$$\mathrm{D}_t\rho = \partial_t\rho + \mathbf{u}\cdot\nabla\rho = 0, \qquad \mathbf{x}\in\Omega,\ t>0, \qquad (16.87)$$

$$\rho|_{t=0} = \rho_0, \qquad \mathbf{x}\in\Omega,$$

where D_t indicates the material, or Lagrangian, derivative, see Sect. 16.7.2. In those cases in which viscosity μ can be expressed in terms of the density, that is $\mu = \mu(\rho)$, this relation, together with (16.87), provides the model for the evolution of ρ and μ. Models adapted to the special case of a flow of two fluids are described in Sect. 16.9.3.

The analysis of the coupled problem (16.85)–(16.86)–(16.87) is a challenging task. We refer the reader to [Lio96]. A global existence result can be proved if $\mathbf{f} = \rho\mathbf{g}$ and $\sigma = 0$. This proof requires Ω to be a smooth, bounded, connected open subset of $\mathbb{R}^d$, and that homogeneous Dirichlet boundary conditions (i.e., with $\mathbf{g}_D = \mathbf{0}$) are imposed on the whole boundary. If the initial and source data satisfy

$$\rho_0 \geq 0 \quad \text{a.e. in } \Omega, \quad \rho_0 \in L^\infty(\Omega), \quad \rho_0\mathbf{u}_0 \in L^2(\Omega)^d, \quad \rho_0|\mathbf{u}_0|^2 \in L^1(\Omega),$$
$$\text{and} \quad \mathbf{g} \in L^2(\Omega \times (0,T))^d,$$

then there exist global weak solutions which satisfy

$$\begin{aligned}
&\rho \in L^\infty(\Omega \times (0,T)), \quad \rho \in C([0,\infty); L^p(\Omega)) \quad \forall p \in [1,\infty);\\
&\mathbf{u} \in [L^2(0,T; H_0^1(\Omega))]^d, \quad \nabla\mathbf{u} \in [L^2(\Omega \times (0,T))]^{d\times d};\\
&\rho|\mathbf{u}|^2 \in L^\infty(0,T; L^1(\Omega)).
\end{aligned}$$

Another result by Tanaka [Tan93] treats the case where the surface tension coefficient σ is different from zero but constant. Under some (stronger) regularity assumptions on the initial data, it has been proved that a global solution exists for sufficiently small initial data and external forces. Moreover, local uniqueness (in time) is proved.

16.9.2 Boundary conditions

Let us generalize the discussion on boundary conditions of the beginning of this chapter to the case of the more general formulation (16.85), (16.86) of the Navier-Stokes equations. We still consider a splitting of the boundary $\partial\Omega$ of the domain Ω into a finite number of components and impose on them appropriate boundary conditions. Several kind of conditions are admissible: for a general discussion see, e.g., [QV94] and the references therein. In the following we just describe the most commonly used conditions for free surface flows.

The Dirichlet boundary conditions prescribe the value of the velocity vector on a boundary subset Γ_D

$$\mathbf{u} = \varphi \quad \text{on} \quad \Gamma_D \subset \partial\Omega. \tag{16.88}$$

They are used either for imposing a velocity profile on the *inflow* boundary, or to model a solid boundary moving with a prescribed velocity. In the latter case they are said to be *no-slip* boundary condition, as they force the fluid not to slip but to stick to the wall.

As we have already noted, when Dirichlet boundary conditions are specified on the entire boundary $\partial\Omega$, the pressure is not uniquely defined. In this case, if $(\mathbf{u}, p)$ is a solution of (16.85), (16.86) and (16.88), then $(\mathbf{u}, p+c)$, $c \in \mathbb{R}$ is also a solution of the same set of equations. Using the Gauss theorem, from equation (16.86) it follows that

$\mathbf{g}_D$ has to satisfy the compatibility condition

$$\int_{\partial\Omega} \mathbf{g}_D \cdot \mathbf{n} \, d\gamma = 0.$$

Neumann boundary conditions prescribe a force ψ_N per unit area as the normal component of the stress tensor

$$\mathbf{T}(\mathbf{u}, p)\mathbf{n} = 2\mu\mathbf{D}(\mathbf{u})\mathbf{n} - p\mathbf{n} = \psi \quad \text{on } \Gamma_N \subset \partial\Omega, \tag{16.89}$$

where $\mathbf{n}$ is the outer unit normal on Γ_N. When $\psi_N = \mathbf{0}$ the subset Γ_N is called a *free outflow*. For vanishing velocity gradients, the force ψ_N corresponds to the pressure on the boundary. See also [HRT96] for more details about the interpretation and implications of this type of boundary conditions. Neumann boundary conditions are used to model a given force per unit area $\mathbf{g}_N$ on the boundary.

Mixed boundary conditions prescribe values of the normal component of the velocity field, as well as on the tangential component of the normal stresses, that is:

$$\begin{aligned} &\mathbf{u} \cdot \mathbf{n} = \varphi \cdot \mathbf{n} && \text{on } \Gamma_D, \\ &(\mathbf{T}(\mathbf{u}, p)\mathbf{n}) \cdot \tau = (2\mu\mathbf{D}(\mathbf{u})\mathbf{n}) \cdot \tau = 0 && \text{on } \Gamma_N, \qquad \forall \tau : \tau \cdot \mathbf{n} = 0. \end{aligned}$$

The choice $\varphi = \mathbf{0}$ models the symmetry of the solution along Γ_D, but also a free slip on Γ_D without penetration. In this case we talk about *free-slip* boundary conditions.

In some situations, a smooth transition from slip to no-slip boundary conditions is desired. This can be realized by imposing Dirichlet boundary conditions in the normal direction, in analogy to the free slip boundary conditions, and to replace the boundary condition in the tangential direction by Robin boundary conditions, a linear combination of Dirichlet and Neumann boundary conditions:

$$\begin{aligned} &\mathbf{u} \cdot \mathbf{n} = \varphi \cdot \mathbf{n} && \text{on } \Gamma_D, \\ &(\omega C_\tau \mathbf{u} + (1-\omega)(\mathbf{T}(\mathbf{u}, p)\mathbf{n})) \cdot \tau = && \\ &(\omega C_\tau \mathbf{u} + (1-\omega)(2\mu\mathbf{D}(\mathbf{u})\mathbf{n})) \cdot \tau = \omega C_\tau \mathbf{g}_D \cdot \tau && \text{on } \Gamma_N, \qquad \forall \tau : \tau \cdot \mathbf{n} = 0. \end{aligned}$$

Here, $\omega \in [0,1]$ determines the regime. For $\omega = 0$ we have free-slip boundary conditions, whereas for $\omega = 1$ we have no-slip boundary conditions. In practice, ω can be a smooth function of space and time, with values in $[0,1]$, allowing thus a smooth transition between the two cases. This holds for $\varphi = \mathbf{0}$, but transition boundary conditions cover also the general Dirichlet case for $\varphi \neq \mathbf{0}$ and $\omega = 1$. The weight C_τ can be seen as a conversion factor between velocities and force per unit area. This type of boundary conditions has been studied in detail in [Joe05].

16.9.3 Application to free surface flows

A free surface flow can be modeled by (16.85)–(16.86). In this perspective the free surface is an interface, denoted by $\Gamma(t)$, cutting the domain Ω into two open subdomains $\Omega^+(t)$ and $\Omega^-(t)$. The initial position of the interface is known, $\Gamma(0) = \Gamma_0$, and the interface moves with fluid velocity $\mathbf{u}$. On each subdomain, we have constant densities and viscosities denoted by ρ^+, ρ^-, μ^+ and μ^-. We require $\rho^\pm > 0$ and $\mu^\pm > 0$.

Density and viscosity are then globally defined as follows:

$$\rho(\mathbf{x},t) = \begin{cases} \rho^- & \mathbf{x} \in \Omega^-(t) \\ \rho^+ & \mathbf{x} \in \Omega^+(t), \end{cases} \qquad \mu(\mathbf{x},t) = \begin{cases} \mu^- & \mathbf{x} \in \Omega^-(t) \\ \mu^+ & \mathbf{x} \in \Omega^+(t). \end{cases}$$

In order to model buoyancy effects, the gravitational force $\mathbf{f} = \rho\mathbf{g}$, where $\mathbf{g}$ is the vector of gravity acceleration, has to be inserted in the right-hand side.

As the viscosity is discontinuous across the interface, equation (16.85) can hold strongly only on the interior of the two subdomains. The latter must therefore be coupled with suitable interface conditions (see, e.g., [Smo01]).

We denote by $\mathbf{n}_\Gamma$ the interface unit normal pointing from Ω^- into Ω^+ and by κ the interface curvature, defined as

$$\kappa = \sum_{i=1}^{d-1} \frac{1}{R_{\tau_i}}, \tag{16.90}$$

where R_{τ_i} are the radii of curvature along the principal vectors τ_i which span the tangent space to the interface Γ. The sign of R_{τ_i} is such that $R_{\tau_i}\mathbf{n}_\Gamma$ points from Γ to the center of the circle approximating Γ locally.

The jump of a quantity v across the interface is denoted by $[v]_\Gamma$ and defined as

$$\begin{aligned} [v]_\Gamma(\mathbf{x},t) &= \lim_{\varepsilon \to 0^+} (v(\mathbf{x}+\varepsilon\mathbf{n}_\Gamma,t) - v(\mathbf{x}-\varepsilon\mathbf{n}_\Gamma,t)) \\ &= v|_{\Omega^+(t)}(\mathbf{x},t) - v|_{\Omega^-(t)}(\mathbf{x},t) \qquad \forall \mathbf{x} \in \Gamma(t). \end{aligned}$$

The interface conditions then read

$$[\mathbf{u}]_\Gamma = \mathbf{0}, \tag{16.91}$$

$$[\mathbf{T}(\mathbf{u},p)\mathbf{n}_\Gamma]_\Gamma = [2\mu\mathbf{D}(\mathbf{u})\mathbf{n}_\Gamma - p\mathbf{n}_\Gamma]_\Gamma = \sigma\kappa\mathbf{n}_\Gamma. \tag{16.92}$$

Equation (16.91) is called the *kinematic interface condition*. It expresses the property that all components of the velocity are continuous. In fact the normal component has to be continuous because there is no flow through the interface, whereas the tangential component(s) have to be continuous because both fluids are assumed viscous ($\mu^+ > 0$ and $\mu^- > 0$).

Equation (16.92) is refered to as the *dynamic interface condition*. It expresses the property that the normal stress jumps by the same amount given by the surface tension force. This force is proportional to the interface curvature and points in the same direction of the interface normal. The surface tension coefficient σ depends on the fluid pairing, and in general also on temperature. We will assume it to be constant, as all heat transfer effects are neglected.

Note that the evolution of the interface has to be compatible with the mass conservation equation (16.87). Mathematically, this equation has to be understood in the weak sense, i.e, in the sense of distributions, as the density is discontinuous across the interface and, consequently, its derivatives can only be interpreted weakly.

As this form of the mass conservation equation is often not convenient for numerical simulations, other equivalent models that describe the evolution of the interface $\Gamma(t)$ have been introduced. A short overview is presented in Sect. 16.10.

16.10 Interface evolution modelling

We give here a short overview of different approaches for modelling the evolution of an interface $\Gamma(t)$ in a fixed domain Ω.

16.10.1 Explicit interface descriptions

An interface can be represented explicitly by a set of marker points or line segments (in 2D, surface segments in 3D) on the interface, that are transported by the fluid velocity.

In the case of marker points, introduced in [HW65], the connectivity of the interface between the points is not known and has to be reconstructed whenever needed. In order to simplify this task, additional markers are usually placed near the interface, marking Ω^+ or Ω^-. The advection of the markers is simple, and connectivity can change easily. However it is still somewhat cumbersome to reconstruct the interface from the marker distribution. Typically, it is also necessary to redistribute the markers, introduce new ones or discard existing ones.

Several markers can be connected to define a line or surface, either straight (plane) or curved, e.g. by NURBS. A set of such geometrical objects can now define the surface. Its evolution is modeled by the evolution of the constituting objects, and thus by the markers defining them. The connectivity of the interface is thereby conserved. This solves the difficulty of pure marker methods, and brings a new drawback in turn: topological changes of the interface are allowed by the underlying physics but not by this description. Sophisticated procedures have to be applied to detect and handle interface breakup correctly.

16.10.2 Implicit interface descriptions

In front-capturing methods, the interface is represented implicitly by the value of a scalar function $\phi : \Omega \times (0,T) \to \mathbb{R}$ that tells to which subset any point $\mathbf{x}$ belongs: $\Omega^+(t)$ or $\Omega^-(t)$. A transport equation solved for ϕ then describes the evolution of the interface. By this feature, all implicit interface models share the advantage that topology changes of the interface are possible in the model, and that these happen without special intervention.

Volume-of-fluid methods

The *volume of fluid methods* (VOF) were originally introduced by Hirt and Nichols [HN81]. Let ϕ be a piecewise constant function such that

$$\phi(\mathbf{x},t) = \begin{cases} 1, & \mathbf{x} \in \Omega^+(t), \\ 0, & \mathbf{x} \in \Omega^-(t); \end{cases}$$

the interface $\Gamma(t)$ is thus located at the discontinuity of the function ϕ, while density and viscosity are simply defined as

$$\begin{aligned} \rho &= \rho^- + (\rho^+ - \rho^-)\phi, \\ \mu &= \mu^- + (\mu^+ - \mu^-)\phi. \end{aligned} \tag{16.93}$$

The transport equation is usually discretized with cell-centered finite volume methods, approximating ϕ by a constant value in each grid cell (see Sects. 9.1 and 16.12). Due to discretization errors and diffusive transport schemes, the approximation ϕ will take values between 0 and 1, which by virtue of equation (16.93) can be (and usually are) interpreted as the volume fraction of the fluid occupying Ω^+. This explains the name *volume of fluid*. Volume fractions between 0 and 1 actually represent a mixture of the two fluids. As the fluids are assumed immiscible, this behaviour is not desired, especially because mixing effects may not stay concentrated near the interface but spread over the whole domain Ω. When this happens, the supposedly sharp interface becomes more and more diffuse. Several techniques exist to limit this problem. Elaborate procedures have been developed for the reconstruction of normals and curvature of a diffuse interface.

Volume of fluid methods have the advantage that applying a conservative discretization of the transport equation ensures mass conservation of the fluid, because the relation (16.93) between ϕ and ρ is linear.

Level-set methods

In order to circumvent the problems with volume of fluid methods, Dervieux and Thomasset [DT80] proposed in 1980 to define the interface as the zero level set of a continuous *pseudo-density* function and to apply this method to flow problems. Their approach was then studied more systematically in [OS88] and subsequent publications, where the term *level-set method* was coined. The first application to flow problems was by Mulder, Osher and Sethian in 1992 [MOS92]. In constrast with volume of fluid approaches, these methods allow to keep the interface sharp, as ϕ is defined as a *continuous* function such that

$$\begin{aligned} \phi(\mathbf{x},t) &> 0, \quad \mathbf{x} \in \Omega^+(t), \\ \phi(\mathbf{x},t) &< 0, \quad \mathbf{x} \in \Omega^-(t), \\ \phi(\mathbf{x},t) &= 0, \quad \mathbf{x} \in \Gamma(t). \end{aligned}$$

The function ϕ is called *level-set function*, because the interface $\Gamma(t)$ is its zero level set, its isoline or isosurface associated to the value zero

$$\Gamma(t) = \{\mathbf{x} \in \Omega : \ \phi(\mathbf{x},t) = 0\}. \tag{16.94}$$

The density and the viscosity can now be expressed in function of ϕ as

$$\rho = \rho^- + (\rho^+ - \rho^-)H(\phi), \tag{16.95}$$
$$\mu = \mu^- + (\mu^+ - \mu^-)H(\phi), \tag{16.96}$$

where $H(\cdot)$ is the Heaviside function

$$H(\xi) = \begin{cases} 0, & \xi < 0 \\ 1, & \xi > 0. \end{cases}$$

By construction, the interface stays sharp in a level-set model, and the immiscible fluids do not start to mix. In addition, the determination of the normals and the curvature of the interface are more straightforward and very natural. In turn, as the relation (16.95) is not linear, applying a conservative discretization of the transport equation for ϕ does not ensure mass conservation of the fluid after discretization. This is not a big problem, however, as the mass error still disappears with grid refinement and is outweighed by advantages of the level-set formulation.

The evolution of the free surface is described by an advection equation for the level-set function

$$\begin{aligned} &\partial_t \phi + \mathbf{u} \cdot \nabla \phi = 0 && \text{in } \Omega \times (0,T), \\ &\phi = \phi_0 && \text{in } \Omega \text{ at } t = 0, \\ &\phi = \phi_{in} && \text{on } \partial\Sigma_{in} \times (0,T), \end{aligned} \tag{16.97}$$

where Σ_{in} is the inflow boundary

$$\Sigma_{in} = \{(\mathbf{x},t) \in \partial\Omega \times (0,T) : \mathbf{u}(\mathbf{x},t) \cdot \mathbf{n} < 0\}.$$

The flow equations (16.85)–(16.86) and the level-set equation (16.97) are therefore coupled. Equation (16.97) can be derived as follows [MOS92]: let $\bar{\mathbf{x}}(t)$ be the path of a point on the interface $\Gamma(t)$. This point moves with the fluid, thus $\mathrm{D}_t\bar{\mathbf{x}}(t) = \mathbf{u}(\bar{\mathbf{x}}(t),t)$. Since the function ϕ is always zero on the moving interface, we must have

$$\phi(\bar{\mathbf{x}}(t),t) = 0.$$

Deriving with respect to time and applying the chain rule, we obtain

$$\partial_t \phi + \nabla\phi \cdot \mathbf{u} = 0 \quad \text{on } \Gamma(t) \quad \forall t \in (0,T). \tag{16.98}$$

If we consider instead a path of a point in $\Omega^{\pm}$, we may require $\phi(\bar{\mathbf{x}}(t),t) = \pm c$, $c > 0$, in order to ensure that the sign of $\phi(\bar{\mathbf{x}},t)$ does not change and that $\bar{\mathbf{x}}(t) \in \Omega^{\pm}(t)$ for all t hereby.

In this way, equation (16.98) generalizes to the whole domain Ω, which gives us equation (16.97).

We can now verify that mass conservation is satisfied: using (16.95), we obtain formally

$$\begin{aligned} \partial_t \rho + \mathbf{u} \cdot \nabla \rho &= (\rho^+ - \rho^-)(\partial_t H(\phi) + \mathbf{u} \cdot \nabla H(\phi)) \\ &= (\rho^+ - \rho^-)\delta(\phi)(\partial_t \phi + \mathbf{u} \cdot \nabla \phi) \end{aligned} \tag{16.99}$$

where $\delta(\cdot)$ denotes the Dirac delta function. By equation (16.97), the third factor in (16.99) is zero. Hence equation (16.87) holds and the mass conservation is satisfied by the level-set interface evolution model.

Interface-related quantities

In the context of two fluid flow, the interface normal and curvature are of particular interest. Namely the surface tension is proportional to the curvature and acting in the normal direction.

We shall now explain how the quantitites depend on ϕ, without going into the details of the differential geometry involved. See, e.g., [Spi99] for a detailed and rigorous derivation.

The unit normal $\mathbf{n}_\Gamma$ is orthogonal to all tangent directions τ, which in turn are characterized by the fact that the directional derivative of ϕ in any tangent direction must vanish:

$$0 = \partial_\tau \phi = \nabla\phi \cdot \tau \quad \text{on } \Gamma.$$

The gradient of ϕ is thus orthogonal to all tangent directions, and we can define the interface unit normal by normalizing it

$$\mathbf{n}_\Gamma = \frac{\nabla\phi}{|\nabla\phi|}. \tag{16.100}$$

Note that by this definition, $\mathbf{n}_\Gamma$ points from Ω^- into Ω^+. Moreover, as ϕ is defined not only on the interface but in the whole domain, the expression for the normal generalizes naturally to the entire domain, too.

In order to derive the expression for the curvature, we need to consider the principal tangent direction(s) τ_i, $i = 1\ldots d-1$. These are the directions in which the interface is approximated by a circle (cylinder), i.e., the directional derivative of $\mathbf{n}_\Gamma$ in the direction τ_i is parallel to τ_i

$$\partial_{\tau_i}\mathbf{n}_\Gamma = \nabla\mathbf{n}_\Gamma\,\tau_i = -\kappa_i\tau_i, \quad \kappa_i \in \mathbb{R}, \quad i = 1\ldots d-1 \tag{16.101}$$

The bigger $|\kappa_i|$, the more curved the surface in this direction, and the κ_i are in fact called *principal curvatures*. It follows from straightforward computations that $\kappa_i = (R_{\tau_i})^{-1}$, where the values R_{τ_i} are the radii of the approximating circles (cylinders) as of equation (16.90).

We can see from equation (16.101) that the $d-1$ values $-\kappa_i$ are eigenvalues of the $d \times d$-tensor $\nabla\mathbf{n}_\Gamma$. By (16.100), $\mathbf{n}_\Gamma$ is (essentially) a gradient field which is smooth near the interface. The rank-two tensor $\nabla\mathbf{n}_\Gamma$ is thus (essentially) a tensor of second derivatives of a smooth function, and hence symmetric. So it has one more real eigenvalue, whose associated eigenvector must be $\mathbf{n}_\Gamma$, because the eigenvectors of a symmetric tensor are orthogonal. It is easy to see that the respective eigenvalue is zero

$$(\nabla\mathbf{n}_\Gamma\,\mathbf{n}_\Gamma)_i = \sum_{j=1}^{d}(\partial_{x_i}n_j)n_j = \sum_{j=1}^{d}\frac{1}{2}\partial_{x_i}(n_j^2) = \frac{1}{2}\partial_{x_i}|\mathbf{n}_\Gamma|^2 = 0,$$

as $|\mathbf{n}_\Gamma| = 1$ by construction (16.100).

Starting from equation (16.90), we obtain for the curvature

$$\kappa = \sum_{i=1}^{d-1}\frac{1}{R_{\tau_i}} = \sum_{i=1}^{d-1}\kappa_i = -\operatorname{tr}(\nabla\mathbf{n}_\Gamma) = -\nabla\cdot\mathbf{n}_\Gamma,$$

and using equation (16.100), we get

$$\kappa = -\nabla \cdot \left(\frac{\nabla \phi}{|\nabla \phi|} \right). \qquad (16.102)$$

Initial Condition

Although we know the position Γ_0 of the interface at $t = 0$, the associated level-set function ϕ_0 is not uniquely defined. The freedom of choice can be used to simplify further subsequent tasks. We notice that steep gradients of ϕ make the numerical solution of equation (16.97) more difficult (see e.g. [QV94]), whereas flat gradients decrease the numerical stability when determining Γ from ϕ. A good compromise is thus the further constraint $|\nabla \phi| = 1$.

A function which fulfills this constraint is the distance function

$$\text{dist}(\mathbf{x};\Gamma) = \min_{\mathbf{y}\in\Gamma} |\mathbf{x}-\mathbf{y}|,$$

which at each point $\mathbf{x}$ takes the value of the minimum Euclidean distance from $\mathbf{x}$ to Γ. Multiplying this function by -1 on Ω^-, we obtain the *signed distance function*

$$\text{sdist}(\mathbf{x};\Gamma) = \begin{cases} \text{dist}(\mathbf{x};\Gamma), & \mathbf{x} \in \Omega^+, \\ 0, & \mathbf{x} \in \Gamma, \\ -\text{dist}(\mathbf{x};\Gamma), & \mathbf{x} \in \Omega^-. \end{cases}$$

It is thus usual and reasonable to choose ϕ_0, representing an initial interface Γ_0, as $\phi_0(\mathbf{x}) = \text{sdist}(\mathbf{x};\Gamma_0)$.

Since $|\nabla \phi| = 1$, the expressions of the interface normal and curvature simplify further to

$$\mathbf{n}_\Gamma = \nabla \phi \quad \text{and} \quad \kappa = -\nabla \cdot \nabla \phi = -\Delta \phi.$$

Reinitialization

Unfortunately, the property $|\nabla \phi| = 1$ is not preserved under advection of ϕ with the fluid velocity $\mathbf{u}$. This is not a problem as long as $|\nabla \phi|$ does not stay too far from 1, which however cannot be guaranteed in general. Two different strategies can be followed to cope with this issue.

One approach is to determine an advection velocity field that gives the same interface motion as the fluid velocity field, while preserving the distance property. Actually such a velocity field exists and is called *extension velocity*, as it is constructed by extending the velocity prescribed on the interface to the whole domain; efficient algorithms are described in [AS99].

Alternatively, we can still use the fluid velocity $\mathbf{u}$ for advecting the level-set function ϕ, and intervene when $|\nabla \phi|$ becomes too large or too small. The action to be taken in this case is known as *reinitialization*, as the procedure is partially the same as for initialization with the initial condition. Suppose we decide to reinitialize at time $t = t_r$:

1. Given $\phi(\cdot,t_r)$, find $\Gamma(t_r) = \{\mathbf{x} : \phi(\mathbf{x},t_r) = 0\}$;
2. Replace $\phi(\cdot,t_r)$ by $\text{sdist}(\cdot,\Gamma(t_r))$.

Interestingly, it turns out that the problem of finding the extension velocity is closely related to the problem of reinitializing ϕ with a signed distance function. The same algorithms can be used and the same computational cost has to be expected. Two conceptual differences favour the reinitialization approach, though: firstly, the extension velocities have to be computed at every time step, whereas reinitialization can be performed only when necessary, which results in a global reduction of the computational costs. Secondly, the approximated extension velocities will only approximately preserve the distance property and may not guarantee that reinitialization is unnecessary.

Algorithmic details about the efficient construction of an approximation to the signed distance function, especially for the three-dimensional case, can be found in [Win07].

16.11 Finite volume approximation

The finite volume approach is widely used for the solution of problems described by differential equations, with applications in different engineering fields. In particular, the most frequently used commercial codes in the field of fluid dynamics adopt finite volume schemes for the solution of the Navier-Stokes equations. The latter are often coupled with models of turbulence, transition, combustion, transport and reaction of chemical species.

When applied to incompressible Navier-Stokes equations, the saddle-point nature of the problem makes the choice of control volumes critical. The most natural choice, with coinciding velocity and pressure nodes, can generate spurious pressure modes. The reason is similar to what was previously noticed for Galerkin finite element approximations: discrete spaces which implicitly underlie the choice of control volumes must satisfy a compatibility condition if we want the problem to be well-posed.

For this reason, it is commonplace to adopt different control volumes, and henceforth nodes, for velocity and pressure. An example is illustrated in Fig. 16.13, where we display a possible choice of nodes for the velocity components and for pressure (on the staggered grid), as well as the corresponding control volumes. The control volumes corresponding to the velocity are used for the discretization of momentum equations,

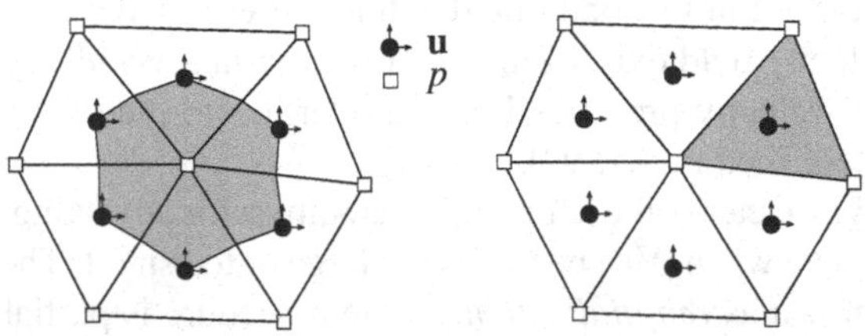

Fig. 16.13. A staggered finite volume grid for velocity and pressure. On the left-hand side we sketch the control volumes for the continuity equation, on the right-hand side the ones used for momentum equations

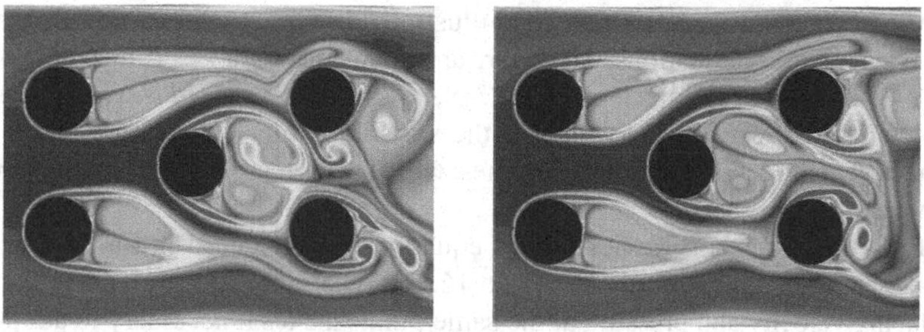

Fig. 16.14. Vorticity field of an incompressible flow around 5 cylinders at time instants $t = 100$ (left) and $t = 102$ (right), $Re = 200$

while the pressure ones are used for the continuity equation. We recall that the latter does not contain the temporal derivative term.

Alternatively, we can adopt stabilization techniques similar to the ones seen in Sect. 16.5, that allow to place the velocity and pressure nodes on the same grid. The interested reader can consult the monographs [FP02], [Kro97], [Pat80] and [VM96] for further details.

In Fig. 16.14 we display the vorticity field of an incompressible flow around 5 cylinders at two time instants (this is the same problem described in Example 16.1, see Fig. 16.10) with a Reynolds number of 200. In this case the Navier-Stokes equations are solved by a cell-centered finite volume discretization. The computational grid used here features 103932 elements and a time step $\Delta t = 0.001$.

Let us also report the simulation of the hydrodynamic flow around an America's Cup sailing boat in upwind regime, targeted at studying the efficiency of its appendages (bulb, keel and winglets) (see Fig. 16.15, left). The computational grid used in this case is *hybrid*, with surface elements of triangular and quadrangular shape, and volume elements of tetrahedral, hexahedral, prismatic and pyramidal shape (see Fig. 16.15, right).

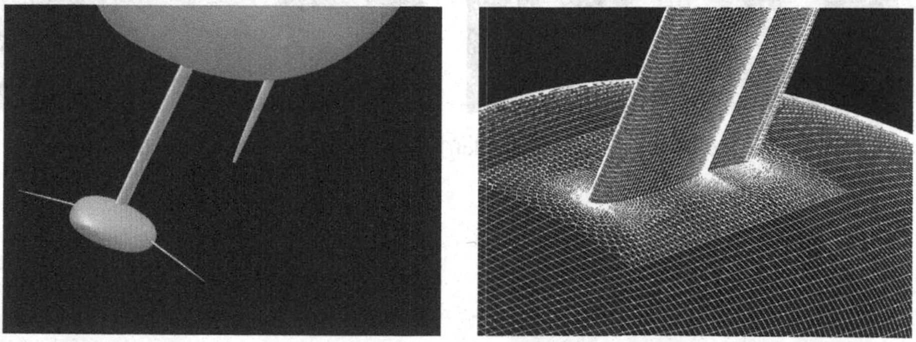

Fig. 16.15. Geometry of the hull and the appendages (left) and detail of the surface grid at the bulb-keel intersection (right)

The mathematical model is the one illustrated in Sect. 16.9 for free-surface fluids. The Navier-Stokes equations, however, are coupled with a $k-\varepsilon$ turbulence model [MP94], through an approach of type RANS (*Reynolds Averaged Navier-Stokes*). The problem's unknowns are the values of the variables (velocity, pressure and turbulent quantities) at the center of control volumes, which in this case correspond to the volume elements of the grid.

The time-dependent Navier-Stokes equations are advanced in time using a fractional-step scheme, as described in Sect. 16.7.3. As previously pointed out, the choice of placing velocity and pressure at the same points makes it necessary to adopt a suitable stabilization of the equations [RC83]. For the computation of the free surface, we have used both the *volume-of-fluid* method and the one based on the *level-set* technique, described in Sect. 16.10.2, the latter being more costly from a computational viewpoint but less dissipative.

Simulations of this kind can require grids with very many elements, in the cases where one wants to reproduce complex fluid dynamics phenomena such as the turbulent flow around complex geometries, or the presence of regions of flow separation. The grid used in this case is composed of 5 million cells and generates an algebraic system with more than 30 million unknowns. Problems of this size are generally solved by resorting to parallel computation techniques based on domain decomposition methods (see Chap. 18) in order to distribute the computation over several processors.

The analysis of pressure distributions and of wall shear stresses, as well as the visualization of the 3D flow through streamlines (see Figs. 16.16 and 16.17) are indeed very useful during the hydrodynamic-project phase, whose aim is the optimization of the boat's performances (see e.g. [PQ05, PQ07, DPQ08]).

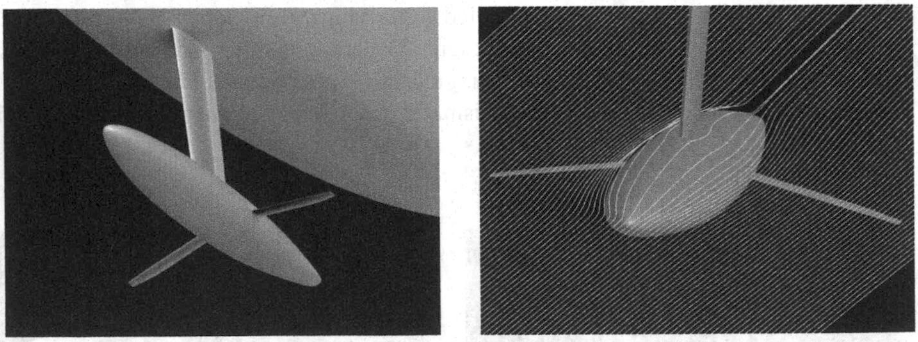

Fig. 16.16. Surface pressure distribution (left) and streamlines around the hull appendages (right)

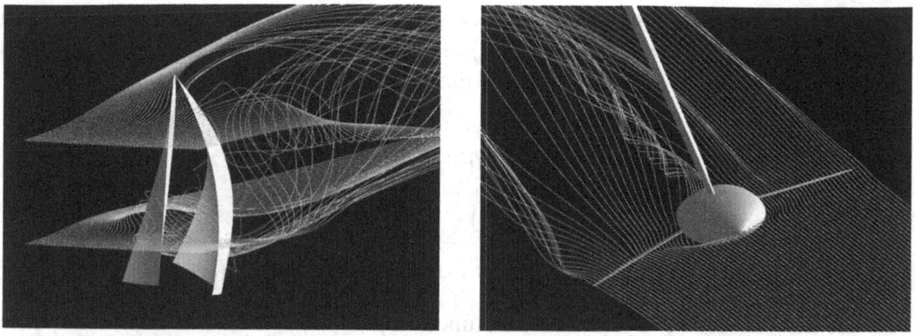

Fig. 16.17. Current lines around the sails during downwind navigation (left) and streamlines around the hull appendages (right)

16.12 Exercises

1. Prove that condition (16.53) is equivalent to the *inf-sup* condition (16.20).
[*Solution*: note that condition (16.53) is violated iff $\exists \mathbf{p}^* \neq \mathbf{0}$ with $\mathbf{p}^* \in \mathbb{R}^M$ such that $\mathrm{B}^T \mathbf{p}^* = \mathbf{0}$ or, equivalently, $\exists p_h^* \in \mathbb{Q}_h$ such that $b(\varphi_n, p_h^*) = 0\ \forall n = 1,\ldots,N$. This is equivalent to $b(\mathbf{v}_h, p_h^*) = 0 \forall \mathbf{v}_h \in V_h$, which in turn is equivalent to violating (16.20).]

2. Prove that a necessary condition in order that (16.53) be satisfied is that $2N \geq M$.
[*Solution*: we have $N = \mathrm{rank}(\mathrm{B}) + \dim(\ker \mathrm{B})$, while $M = \mathrm{rank}(\mathrm{B}^T) + \dim(\ker \mathrm{B}^T) = \mathrm{rank}(\mathrm{B}^T) = \mathrm{rank}(\mathrm{B})$. Consequently, we have $N - M = \dim(\ker \mathrm{B}) \geq 0$, thus the condition $N \geq M$ is necessary for the solution to be unique.]

3. Show that the finite element couple $\mathbb{P}_1 - \mathbb{P}_0$ for velocity and pressure does not satisfy the *inf-sup* condition.
[*Solution*: we restrict ourselves to a two-dimensional Dirichlet problem, and consider a simple uniform triangulation made of $2n^2$ triangles, $n \geq 2$, like the one displayed in Fig. 16.18, left. This triangulation carries $M = 2n^2 - 1$ degrees of freedom for the pressure (one value for every triangle except for one, as our pressure we must have null average), $N = 2(n-1)^2$ for the velocity field (which correspond to the values of two components at each internal vertex). Thus the necessary condition $N \geq M$ proven in Exercise 2 is not fulfilled in the current case.]

4. Show that on a grid made by rectangles, the finite element couple $\mathbb{Q}_1 - \mathbb{Q}_0$ of bilinear polynomials for the velocity components and constant pressure on each rectangle does not satisfy the *inf-sup* condition.
[*Solution*: consider a square computational domain and a uniform Cartesian grid made of $n \times n$ squares as in Fig. 16.18 right). There are $(n-1)^2$ internal nodes carrying $N = 2(n-1)^2$ degrees of freedom for the velocity and $M = n^2 - 1$ for the pressure. The necessary condition is therefore satisfied provided $n \geq 3$. We therefore verify directly that the *inf-sup* condition does not hold. Let h be the uni-

form size of the element edges and denote by $q_{i\pm1/2,j\pm1/2}$ the value at the midpoint $(x_{i\pm1/2}, y_{j\pm1/2}) = (x_i \pm h/2, y_i \pm h/2)$ of a given function q. Let K_{ij} be a square of the grid. A simple calculation shows that

$$\int_\Omega q_h \mathrm{div}\mathbf{u}_h \, d\Omega = \frac{h}{2} \sum_{i,j=1}^{n-1} u_{ij}(q_{i-1/2,j-1/2} + q_{i-1/2,j+1/2}$$
$$-q_{i+1/2,j-1/2} - q_{i+1/2,j+1/2})$$
$$+v_{ij}(q_{i-1/2,j-1/2} - q_{i-1/2,j+1/2} + q_{i+1/2,j-1/2} - q_{i+1/2,j+1/2}).$$

Clearly, any element-wise constant function p^* whose value is 1 on the black elements and -1 on the white ones of Fig. 16.18, right, is a spurious pressure.]

5. Consider the steady Stokes problem with non-homogeneous Dirichlet boundary conditions

$$\begin{cases} -\nu\Delta\mathbf{u} + \nabla p = \mathbf{f} & \text{in } \Omega \subset \mathbb{R}^2, \\ \mathrm{div}\mathbf{u} = 0 & \text{in } \Omega, \\ \mathbf{u} = \mathbf{g} & \text{on } \Gamma = \partial\Omega, \end{cases}$$

where $\mathbf{g}$ is a given vector function. Show that $\int_\Gamma \mathbf{g} \cdot \mathbf{n} = 0$ is a necessary condition for the existence of a weak solution. Show that the right-hand side of the weak form of the momentum equation identifies an element of V', the dual of the space $V = [\mathrm{H}_0^1(\Omega)]^d$.

6. Do the same (as in Exercise 5) for the non-homogeneous Navier-Stokes problem

$$\begin{cases} (\mathbf{u} \cdot \nabla)\mathbf{u} - \nu\Delta\mathbf{u} + \nabla p = \mathbf{f} & \text{in } \Omega \subset \mathbb{R}^2, \\ \mathrm{div}\mathbf{u} = 0 & \text{in } \Omega, \\ \mathbf{u} = \mathbf{g} & \text{on } \Gamma = \partial\Omega. \end{cases}$$

7. Prove the a priori estimate (16.57).
[*Solution*: choose $\mathbf{v}_h = \mathbf{u}_h$ and $q_h = p_h$ as test functions in (16.55). Then apply the Cauchy-Schwarz, Young and Poincaré inequalities to bound the right-hand side.]

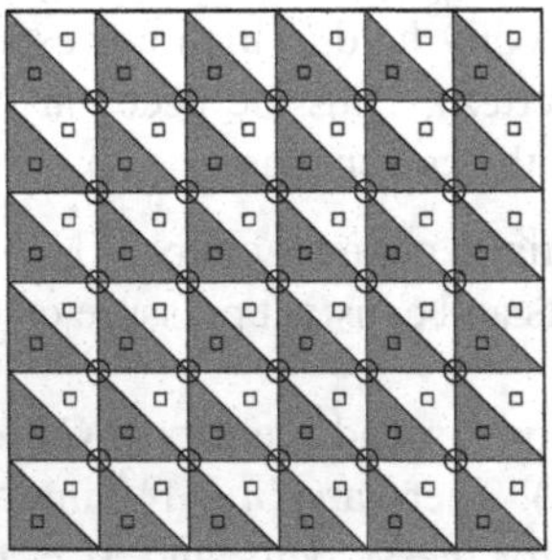

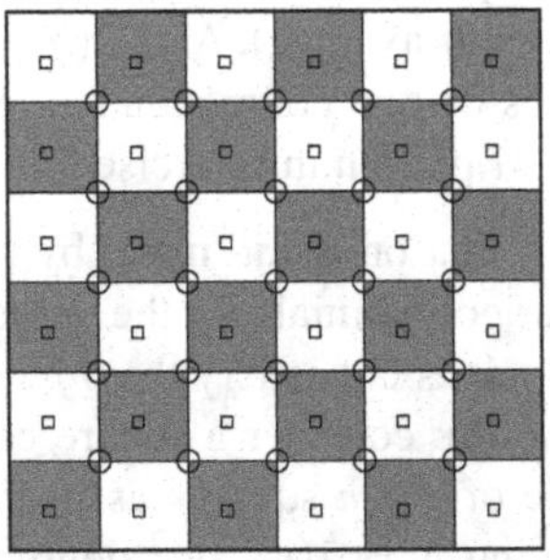

Fig. 16.18. Uniform grid for a finite element discretization using $\mathbb{P}_1 - \mathbb{P}_0$ (left) and $\mathbb{Q}_1 - \mathbb{Q}_0$ (right) finite elements with spurious pressure modes

17

Optimal control of partial differential equations

In this chapter we will introduce the basic concepts of optimal control for linear elliptic partial differential equations. At first we present the classical theory in functional spaces "à la J.L.Lions", see [Lio71, Lio72]; then we will address the methodology based on the use of the Lagrangian functional (see, e.g., [Mau81, BKR00, Jam88]). Finally, we will show two different numerical approaches for control problems, based on the Galerkin finite element method.

This is intended to be an elementary introduction to this fascinating and complex subject. The interested reader is advised to consult more specialized monographs such as, e.g., [Lio71, AWB71, ATF87, Ago03, BKR00, Gun03, Jam88, APV98, MP01, FCZ04, DZ06, Zua05]. For the basic concepts of functional analysis here used, see Chapter 2 and also [Ada75, BG87, Bre86, Rud91, Sal08, TL58].

17.1 Definition of optimal control problems

In abstract terms, a control problem can be expressed by the paradigm illustrated in Fig. 17.1. There is a system expressed by a *state* problem that can be either an algebraic problem, an initial-value problem for ordinary differential equations, or a boundary-value problem for partial differential equations. Its solution, that will generically be denoted by y, depends on a variable u representing the *control* that can be exerted on the system. The goal of a control problem is to find the control u in such a way that a suitable output variable, denoted by z and called observed variable (which is a function of u through y), takes a desired "value" z_d, the so-called *observation*, or target.
The problem is said to be *controllable* if a control u exists such that the observed variable z matches *exactly* the desired value z_d. Not all systems are controllable (see, in this respect, the review paper [Zua06]): take for instance the simple case in which the state problem is the linear algebraic system $A\mathbf{y} = \mathbf{b}$, where A is a given $n \times n$ non-singular matrix and $\mathbf{b}$ a given vector of $\mathbb{R}^n$. Assume moreover that the observation is represented by one component σ, say the first one, and suppose that the control is one

A. Quarteroni: *Numerical Models for Differential Problems*, 2nd Ed.
MS&A – Modeling, Simulation & Applications 8
DOI 10.1007/978-88-470-5522-3_17, © Springer-Verlag Italia 2014

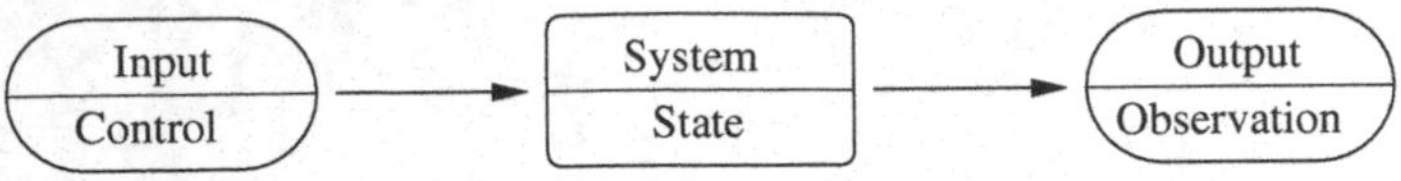

Fig. 17.1. The essential ingredients of a control problem

of the components of the right-hand side, say the last one. The question therefore reads: "Find $u \in \mathbb{R}$ such that the solution of the linear system $A\mathbf{y} = \mathbf{b} + [0, ..., 0, u]^T$ satisfies $y_1 = y_1^*$", y_1^* being a given value. In general, this problem admits no solution.
For this reason it is often preferred to replace the problem of controllability by one of *optimization*: by doing so one does not expect the output variable z to be exactly equal to the observation z_d, but that the difference between z and z_d (in a suitable sense) be the smallest possible. Therefore, control and optimization are two intimately related concepts, as we will see later on in this chapter.
As already noted, we will only consider systems governed by elliptic PDEs. With this aim, we start by introducing the mathematical entities that enter in the control problem.

- The *control function* u. It belongs to a functional space $\mathcal{U}_{ad}$, called the space of *admissible controls*. In general, $\mathcal{U}_{ad} \subseteq \mathcal{U}$, where $\mathcal{U}$ is a functional space apt to describe the role assumed by u in the given state equation. If $\mathcal{U}_{ad} = \mathcal{U}$ the control problem is *unconstrained*; if $\mathcal{U}_{ad} \subset \mathcal{U}$ the control problem is said to be *constrained*.
- The *state* of the system $y(u) \in \mathcal{V}$ (a suitable functional space), a function depending on the control u that satisfies the *equation of state*

$$Ay(u) = f, \tag{17.1}$$

  where $A : \mathcal{V} \mapsto \mathcal{V}'$ is a differential operator (linear or not). This problem describes a physical problem subject to suitable boundary conditions. As we will see, the control function can enter in the right-hand side, in the boundary data, or in the coefficients of the differential operator.
- The *observation function* in $z(u)$, also depending on the control u through y and on a suitable operator $C : \mathcal{V} \to \mathcal{Z}$,

$$z(u) = Cy(u).$$

  This function belongs to the space $\mathcal{Z}$ of the observed functions and must "approach" the observation function z_d. As a matter of fact, optimizing system (17.1) means finding the control function u such that the function $z(u)$ is "as close as possible" to the observation function z_d. This goal will be achieved through a minimization process that we are going to describe.
- Define a *cost functional* $J(u)$, defined on the space $\mathcal{U}_{ad}$

$$u \in \mathcal{U}_{ad} \mapsto J(u) \in \mathbb{R} \qquad \text{with } J(u) \geq 0.$$

  In general, J will depend on u (also) through $z(u)$, that is $J(u) = \tilde{J}(u, z(u))$, for a suitable functional $\tilde{J} : \mathcal{U}_{ad} \times \mathcal{Z} \to \mathbb{R}$.

The optimal control problem can be formulated in either following way:

i) find $u \in \mathcal{U}_{ad}$ such that

$$J(u) = \inf J(v) \qquad \forall v \in \mathcal{U}_{ad}; \tag{17.2}$$

ii) find $u \in \mathcal{U}_{ad}$ such that the following inequality holds

$$J(u) \leq J(v) \qquad \forall v \in \mathcal{U}_{ad}. \tag{17.3}$$

The function u that satisfies (17.2) (or (17.3)) is called *optimal control* of system (17.1). Before analyzing the existence and uniqueness properties of the control problem and characterizing the condition of optimality, let us consider a simple finite-dimensional example.

17.2 A control problem for linear systems

Let A be a $n \times n$ non-singular matrix and B a $n \times q$ matrix. Moreover, let $\mathbf{f}$ be a vector of $\mathbb{R}^n$, $\mathbf{u}$ a vector of $\mathbb{R}^q$ representing the control. The vector $\mathbf{y} = \mathbf{y}(\mathbf{u}) \in \mathbb{R}^n$ which represents the state satisfies the following linear system

$$A\mathbf{y} = \mathbf{f} + B\mathbf{u}. \tag{17.4}$$

We look for a control $\mathbf{u}$ that minimizes the following linear functional:

$$J(\mathbf{u}) = \|\mathbf{z}(\mathbf{u}) - \mathbf{z}_d\|^2_{\mathbb{R}^m} + \|\mathbf{u}\|^2_N . \tag{17.5}$$

In this equation, $\mathbf{z}_d$ is a given vector (the target) of $\mathbb{R}^m$, $\mathbf{z}(\mathbf{u}) = C\mathbf{y}(\mathbf{u})$ is the vector to be observed, where C is a $m \times n$ matrix, $\|\mathbf{u}\|_N = (N\mathbf{u}, \mathbf{u})^{1/2}_{\mathbb{R}^q}$ is the *N-norm* of $\mathbf{u}$, N being a given symmetric and positive definite matrix of dimension $q \times q$.

Upon interpreting the term $\|\mathbf{u}\|^2_N$ as the energy associated to the control, the problem is therefore: choose the control so that the observation $\mathbf{z}(\mathbf{u})$ is close to the target $\mathbf{z}_d$ and its energy is small.

Note that

$$J(\mathbf{u}) = (CA^{-1}(\mathbf{f} + B\mathbf{u}) - \mathbf{z}_d, CA^{-1}(\mathbf{f} + B\mathbf{u}) - \mathbf{z}_d)_{\mathbb{R}^m} + (N\mathbf{u}, \mathbf{u})_{\mathbb{R}^q}. \tag{17.6}$$

The cost functional $J(\mathbf{u})$ is therefore a quadratic function of $\mathbf{u}$ which has on $\mathbb{R}^q$ a global minimum. The latter is characterized by the condition

$$J'(\mathbf{u})\mathbf{h} = 0 \qquad \forall \mathbf{h} \in \mathbb{R}^q \tag{17.7}$$

where $J'(\mathbf{u})\mathbf{h}$ is the directional derivative along the direction $\mathbf{h}$ computed at the "point" $\mathbf{u}$, that is (see Definition 2.6 of Chap. 2)

$$J'(\mathbf{u})\mathbf{h} = \lim_{t \to 0} \frac{J(\mathbf{u} + t\mathbf{h}) - J(\mathbf{u})}{t}.$$

Since

$$Ay'(\mathbf{u})\mathbf{h} = B\mathbf{h} \quad \text{and} \quad \mathbf{z}'(\mathbf{u})\mathbf{h} = Cy'(\mathbf{u})\mathbf{h}$$

for all **u** and **h**, from (17.6) we obtain

$$\begin{aligned} J'(\mathbf{u})\mathbf{h} &= 2[(\mathbf{z}'(\mathbf{u})\mathbf{h}, \mathbf{z}(\mathbf{u}) - \mathbf{z}_d)_{\mathbb{R}^m} + (N\mathbf{u}, \mathbf{h})_{\mathbb{R}^q}] \\ &= 2[(CA^{-1}B\mathbf{h}, C\mathbf{y}(\mathbf{u}) - \mathbf{z}_d)_{\mathbb{R}^m} + (N\mathbf{u}, \mathbf{h})_{\mathbb{R}^q}] \,. \end{aligned} \tag{17.8}$$

Let us introduce the solution $\mathbf{p} = \mathbf{p}(\mathbf{u}) \in \mathbb{R}^n$ of the following system, that is called the *adjoint state* of (17.4)

$$A^T \mathbf{p}(\mathbf{u}) = C^T (C\mathbf{y}(\mathbf{u}) - \mathbf{z}_d) \,. \tag{17.9}$$

From (17.8) we deduce

$$J'(\mathbf{u})\mathbf{h} = 2[(B\mathbf{h}, \mathbf{p}(\mathbf{u}))_{\mathbb{R}^n} + (N\mathbf{u}, \mathbf{h})_{\mathbb{R}^q}],$$

that is

$$J'(\mathbf{u}) = 2[B^T \mathbf{p}(\mathbf{u}) + N\mathbf{u}]. \tag{17.10}$$

Since J attains its minimum at the point **u** for which $J'(\mathbf{u}) = \mathbf{0}$, we can conclude that the three-field system

$$\begin{cases} A\mathbf{y} = \mathbf{f} + B\mathbf{u}, \\ A^T \mathbf{p} = C^T (C\mathbf{y} - \mathbf{z}_d), \\ B^T \mathbf{p} + N\mathbf{u} = \mathbf{0}, \end{cases} \tag{17.11}$$

admits a unique solution $(\mathbf{u}, \mathbf{y}, \mathbf{p}) \in \mathbb{R}^q \times \mathbb{R}^n \times \mathbb{R}^n$, and that **u** is the unique optimal control of the original system.

In the next section we will introduce several examples of optimal control problems for the Laplace equation.

17.3 Some examples of optimal control problems for the Laplace equation

Consider for simplicity the case where the elliptic operator A is the Laplacian. We define two different families of optimal control problems: the distributed control and the boundary control.

- *Distributed control.* Let us introduce the *state* problem

$$\begin{cases} -\Delta y = f + u & \text{in } \Omega, \\ y = 0 & \text{on } \Gamma = \partial\Omega, \end{cases} \tag{17.12}$$

  where Ω is a domain in $\mathbb{R}^n$, $y \in \mathcal{V} = H_0^1(\Omega)$ is the state variable, $f \in L^2(\Omega)$ is a given source term, and $u \in \mathcal{U}_{ad} = L^2(\Omega)$ is the control function. We can consider two different kinds of cost functional:

  - *on the domain*, for instance

$$J(u) = \int_\Omega (y(u) - z_d)^2 d\Omega, \tag{17.13}$$

- *on the boundary*, for instance (provided $y(u)$ is sufficiently regular)

$$J(u) = \int_{\Gamma} (\frac{\partial y(u)}{\partial n} - z_{d_\Gamma})^2 d\gamma.$$

The functions z_d and z_{d_Γ} are two prescribed observation (or target) functions.

- *Boundary control.* Consider now the following *state* problem

$$\begin{cases} -\Delta y = f & \text{in } \Omega, \\ y = u & \text{on } \Gamma_D, \\ \dfrac{\partial y}{\partial n} = 0 & \text{on } \Gamma_N, \end{cases} \tag{17.14}$$

with $\Gamma_D \cup \Gamma_N = \partial\Omega$ and $\overset{\circ}{\Gamma}_D \cap \overset{\circ}{\Gamma}_N = \emptyset$. The *control* $u \in H^{\frac{1}{2}}(\Gamma_D)$ is defined on the Dirichlet boundary. Two different kinds of cost functional can be considered:

- *on the domain*, as in (17.13);
- *on the boundary*, for instance

$$J(u) = \int_{\Gamma_N} (y(u) - z_{d_{\Gamma_N}})^2 d\gamma.$$

Here, too, $z_{d_{\Gamma_N}}$ represents a given observation function.

17.4 On the minimization of linear functionals

In this section we recall some results about the existence and uniqueness of extrema of linear functionals, with focus on those associated to control problems addressed in this chapter. For more results see, e.g., [Lio71, BG87, Bre86, TL58].
We consider a Hilbert space $\mathcal{U}$, endowed with a scalar product $(\cdot,\cdot)$, and a bilinear form π

$$u, v \mapsto \pi(u, v) \qquad \forall u, v \in \mathcal{U}, \tag{17.15}$$

that we assume to be symmetric, continuous and coercive. The norm induced on $\mathcal{U}$ by the scalar product will be denoted by $\|w\| = \sqrt{(w,w)}$. Let

$$v \mapsto F(v) \qquad \forall v \in \mathcal{U}, \tag{17.16}$$

be a linear and bounded functional on $\mathcal{U}$. Finally, let $\mathcal{U}_{ad}$ be the closed subspace of $\mathcal{U}$ of admissible control functions, and consider the following cost functional

$$J(v) = \pi(v, v) - 2F(v) \qquad \forall v \in \mathcal{U}_{ad}. \tag{17.17}$$

The following result holds:

Theorem 17.1. *Under the previous assumptions, there exists a unique* $u \in \mathcal{U}_{ad}$ *such that*

$$J(u) = \inf J(v) \qquad \forall v \in \mathcal{U}_{ad}, \tag{17.18}$$

where $J(v)$ *is defined in (17.17); u is called* optimal control.
Moreover:

(i) *The function* $u \in \mathcal{U}_{ad}$ *satisfies the variational inequality*

$$\pi(u, v-u) \geq F(v-u) \qquad \forall v \in \mathcal{U}_{ad}. \tag{17.19}$$

(ii) *If* $\mathcal{U}_{ad} = \mathcal{U}$ *(that is we consider a* non-constrained *optimization problem), owing to the Lax–Milgram Lemma 3.1, u satisfies the following* Euler equation *associated to (17.18):*

$$\pi(u, w) = F(w) \qquad \forall w \in \mathcal{U}. \tag{17.20}$$

(iii) *If* $\mathcal{U}_{ad}$ *is a closed convex cone with vertex at the origin 0*[1], *u satisfies*

$$\pi(u, v) \geq F(v) \qquad \forall v \in \mathcal{U}_{ad} \quad \text{and} \quad \pi(u, u) = F(u). \tag{17.21}$$

(iv) *Suppose the map* $v \mapsto F(v)$ *is strictly convex and differentiable, J (not necessarily quadratic) satisfies:* $J(v) \to \infty$ *when* $||v||_{\mathcal{U}} \to \infty$ $\forall v \in \mathcal{U}_{ad}$*. Then the unique function* $u \in \mathcal{U}_{ad}$ *which satisfies condition (17.18) is characterized by the variational inequality*

$$J'(u)(v-u) \geq 0 \qquad \forall v \in \mathcal{U}_{ad} \tag{17.22}$$

or, equivalently,

$$J'(v)(v-u) \geq 0 \qquad \forall u \in \mathcal{U}_{ad}. \tag{17.23}$$

(The symbol J' *denotes the Gâteaux derivative of J, see Definition 2.5 of Chap. 2.)*

Proof. For a complete proof see, e.g., [Lio71, Chap. 1, Theorem 1.1]. Here we prove (17.19). If u minimizes (17.18), then for all $v \in \mathcal{U}_{ad}$ and every $0 < \vartheta < 1$, $J(u) \leq J((1-\vartheta)u + \vartheta v)$, thus $\frac{1}{\vartheta}[J(u + \vartheta(v-u)) - J(u)] \geq 0$. This inequality still holds when $\vartheta \to 0$ (provided this limit exists), whence

$$J'(u)(v-u) \geq 0 \qquad \forall v \in \mathcal{U}_{ad}. \tag{17.24}$$

Inequality (17.19) follows by recalling the definition (17.17) of J.

[1] A linear metric space W is a *closed convex cone with vertex in the origin* 0 if: (1) $0 \in W$, (2) $x \in W \Rightarrow kx \in W \ \ \forall k \geq 0$, (3) $x, y \in W \Rightarrow x + y \in W$, (4) W is closed.

The converse holds as well (whence (17.18) and (17.19) are in fact equivalent). Indeed, should u satisfy (17.19), and so (17.24), thanks to the convexity of the map $v \mapsto J(v)$ for every $0 < \vartheta < 1$ one has

$$J(v) - J(w) \geq \frac{1}{\vartheta}[J((1-\vartheta)w + v) - J(w)] \qquad \forall v, w \in \mathcal{U}_{ad}.$$

Taking the limit as $\vartheta \to 0$ we obtain

$$J(v) - J(w) \geq J'(w)(v-w).$$

Taking $w = u$ and using (17.24) we obtain that $J(v) \geq J(u)$, that is (17.18).

To prove (17.20) it is sufficient to choose $v = u \pm w \in \mathcal{U}$ in (17.19).

Let us now prove (17.21). The first inequality can be obtained by replacing v with $v+u$ in (17.19). Setting now $v = 0$ in (17.19) we obtain $\pi(u,u) \leq F(u)$. By combining the latter with the first inequality in (17.21) we obtain the second equation in (17.21). The converse (that is, (17.21) $\Rightarrow$ (17.19)) is obvious.

For the proof of (17.22) and (17.23), see [Lio71, Chap. 1, Theorem 1.4]. ◇

Remark 17.1. If $J(v)$ is differentiable in v, $\forall v \in \mathcal{U}$, then for every minimizing function $u \in \mathcal{U}$ of J (provided it does exist) we have $J'(u) = 0$. Moreover, under the assumptions of Theorem 17.1 (step (iv)), there exists at least one minimizing element $u \in \mathcal{U}$. •

We can summarize by saying that the solution $u \in \mathcal{U}_{ad}$ of the minimization problem satisfies the following (equivalent) conditions:

i) $J(u) = \inf J(v)$ $\quad \forall v \in \mathcal{U}_{ad}$,
ii) $J(u) \leq J(v)$ $\quad \forall v \in \mathcal{U}_{ad}$,
iii) $J'(u)(v-u) \geq 0$ $\quad \forall v \in \mathcal{U}_{ad}$,
iv) $J'(v)(v-u) \geq 0$ $\quad \forall u \in \mathcal{U}_{ad}$.

Before closing this section, consider the abstract problem of finding $u \in \mathcal{U}_{ad}$ satisfying the variational inequality (17.19) (when $\pi(\cdot,\cdot)$ is not symmetric, this problem does not correspond to a problem of minimization in the calculus of variations).

Theorem 17.2. *If there exists a constant $c > 0$ such that*

$$\pi(v_1 - v_2, v_1 - v_2) \geq c\|v_1 - v_2\|^2 \qquad \forall v_1, v_2 \in \mathcal{U}_{ad}, \tag{17.25}$$

then there exists a unique function $u \in \mathcal{U}_{ad}$ which satisfies (17.19).

For the proof, see [Lio71, Chap. 1, Theorem 2.1].

17.5 The theory of optimal control for elliptic problems

In this section we illustrate some existence and uniqueness results for the solution of a control problem governed by a linear elliptic equation (the equation of state). For the sake of simplicity we restrict to *distributed control* (see Sect. 17.3); however, similar results hold for boundary control problems as well, see [Lio71].

Let $\mathcal{V}$ and $\mathcal{H}$ be two Hilbert spaces, $\mathcal{V}'$ the dual of $\mathcal{V}$ and $\mathcal{H}'$ that of $\mathcal{H}$, and assume that $\mathcal{V}$ is dense in $\mathcal{H}$ with continuous injection. We recall that in this case, property (2.10) of Chapter 2 holds. In addition, denote by $(\cdot,\cdot)$ the scalar product of $\mathcal{H}$ and suppose that $a(u,v)$ is a bilinear, continuous and coercive form on $\mathcal{V}$ (but not necessarily symmetric). Under these assumptions, the Lax–Milgram lemma guarantees that there exists a unique solution $y \in \mathcal{V}$ of problem

$$a(y,\varphi) = (f,\varphi) \qquad \forall \varphi \in \mathcal{V}. \tag{17.26}$$

By introducing the operator A associated with the bilinear form $a(\cdot,\cdot)$ (see (3.39))

$$A \in \mathcal{L}(\mathcal{V},\mathcal{V}') \;:\; {}_{\mathcal{V}'}\langle A\varphi,\psi\rangle_{\mathcal{V}} = a(\varphi,\psi) \qquad \forall \varphi,\psi \in \mathcal{V},$$

problem (17.26) becomes (in operator form)

$$Ay = f \qquad \text{in } \mathcal{V}'. \tag{17.27}$$

This is the equation of state of a physical system governed by the operator A, and will be supplemented by a distributed control term.

For that, let $\mathcal{U}$ be a Hilbert space of control functions, and B an operator belonging to the space $\mathcal{L}(\mathcal{U},\mathcal{V}')$. For every control function u the *equation of state* of the system is

$$Ay(u) = f + Bu \qquad \text{in } \mathcal{V}', \tag{17.28}$$

or, in weak form,

$$y(u) \in \mathcal{V} \;:\; a(y(u),\varphi) = (f,\varphi) + b(u,\varphi) \qquad \forall \varphi \in \mathcal{V}, \tag{17.29}$$

where $b(\cdot,\cdot)$ is the bilinear form associated with the operator B, that is

$$b(u,\varphi) = {}_{\mathcal{V}'}\langle Bu,\varphi\rangle_{\mathcal{V}} \qquad \forall u \in \mathcal{U},\; \forall \varphi \in \mathcal{V}. \tag{17.30}$$

Let us denote with $\mathcal{Z}$ the Hilbert space of observation functions, and introduce the *equation of observation*

$$z(u) = Cy(u), \tag{17.31}$$

for a suitable operator $C \in \mathcal{L}(\mathcal{V},\mathcal{Z})$. At last, let us define the *cost functional*

$$J(y(u),u) = \| Cy(u) - z_d \|^2_{\mathcal{Z}} + (Nu,u)_{\mathcal{U}}, \tag{17.32}$$

that we will indicate with the shorthand notation $J(u)$. Here $N \in \mathcal{L}(\mathcal{U},\mathcal{U})$ is a symmetric positive definite form such that

$$(Nu,u)_{\mathcal{U}} \geq \nu \|u\|^2_{\mathcal{U}} \qquad \forall u \in \mathcal{U}, \tag{17.33}$$

where $\nu > 0$ and $z_d \in \mathcal{Z}$ is the desired (target) observation.
The optimal control problem consists in finding $u \in \mathcal{U}_{ad} \subseteq \mathcal{U}$ such that

$$J(u) = \inf J(v) \qquad \forall v \in \mathcal{U}_{ad}. \tag{17.34}$$

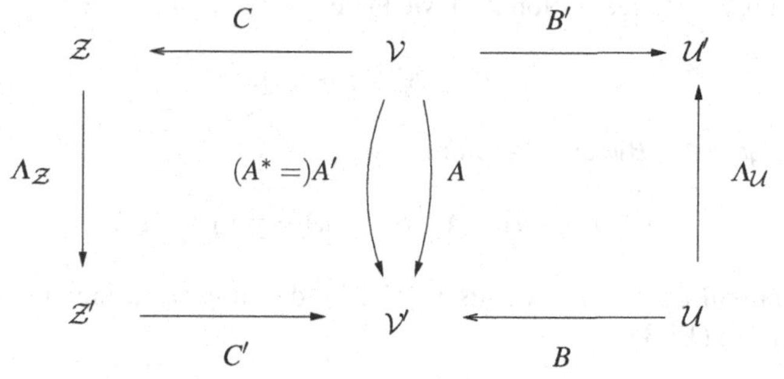

Fig. 17.2. Functional spaces and operators involved in the statement of the control problem

A reminder of the spaces and operators involved in the above definition of optimal control problem is depicted in Fig. 17.2.

Remark 17.2. When minimizing (17.32), one actually minimizes a combination between two terms. The former enforces that the observation $z(u)$ is closed to the desired value (the target) z_d. The latter penalizes the use of a control u that is "too expensive". Heuristically speaking, we are trying to make $z(u)$ go towards z_d by a reduced effort. Note that this theory applies also if the form N is null, but in this case one can only prove the existence of an optimal control, not its uniqueness. •

In order to apply the abstract theoretical results stated in Sect. 17.4, by noticing that the map $u \mapsto y(u)$ from $\mathcal{U}$ in $\mathcal{V}$ is affine we can rewrite (17.32) as follows

$$J(u) = \|C[y(u) - y(0)] + Cy(0) - z_d\|^2_{\mathcal{Z}} + (Nu, u)_{\mathcal{U}}. \tag{17.35}$$

Let us now define a bilinear form π that is continuous in $\mathcal{U}$ and a functional F as follows, for all $u, v \in \mathcal{U}$:

$$\pi(u,v) = (C[y(u) - y(0)],\ C[y(v) - y(0)])_{\mathcal{Z}} + (Nu, v)_{\mathcal{U}},$$
$$F(v) = (z_d - Cy(0),\ C[y(v) - y(0)])_{\mathcal{Z}}.$$

Then

$$J(v) = \pi(v,v) - 2F(v) + \|z_d - Cy(0)\|^2_{\mathcal{Z}}.$$

Since $\|C[y(v) - y(0)]\|^2_{\mathcal{Z}} \geq 0$, owing to (17.33) we obtain

$$\pi(v,v) \geq \nu\|v\|^2_{\mathcal{U}} \qquad \forall v \in \mathcal{U}.$$

By doing so we have cast the control problem in the general formulation addressed in Sect. 17.4. Then Theorem 17.1 guarantees existence and uniqueness of the control function $u \in \mathcal{U}_{ad}$.

At this stage we would like to study the structure of the equations useful to the *solution* of the control problem. Thanks to Theorem 17.1, and since A is an isomorphism between $\mathcal{V}$ and $\mathcal{V}'$ (see Definition 2.4), we have

$$y(u) = A^{-1}(f + Bu),$$

whence $y'(u)\psi = A^{-1}B\psi$ and therefore

$$y'(u)(v-u) = A^{-1}B(v-u) = y(v) - y(u).$$

Since the optimal control must satisfy (17.22), dividing by 2 inequality (17.22) we obtain, thanks to (17.35)

$$(Cy(u) - z_d,\, C[y(v) - y(u)])_{\mathcal{Z}} + (Nu, v-u)_{\mathcal{U}} \geq 0 \qquad \forall v \in \mathcal{U}_{ad}. \tag{17.36}$$

Let now $C' \in \mathcal{L}(\mathcal{Z}', \mathcal{V}')$ be the adjoint of the operator $C \in \mathcal{L}(\mathcal{V}, \mathcal{Z})$ (see (2.20)). Then

$$_{\mathcal{Z}}\langle Cy, v\rangle_{\mathcal{Z}'} = {}_{\mathcal{V}}\langle y, C'v\rangle_{\mathcal{V}'} \qquad \forall y \in \mathcal{V},\ \forall v \in \mathcal{Z}',$$

hence (17.36) becomes

$$_{\mathcal{V}'}\langle C'\Lambda_{\mathcal{Z}}(Cy(u) - z_d),\, y(v) - y(u)\rangle_{\mathcal{V}} + (Nu, v-u)_{\mathcal{U}} \geq 0 \qquad \forall v \in \mathcal{U}_{ad}, \tag{17.37}$$

where $\Lambda_{\mathcal{Z}}$ denotes the canonical Riesz isomorphism from $\mathcal{Z}$ to $\mathcal{Z}'$ (see (2.5)). Let us now introduce the adjoint operator $A' \in \mathcal{L}(\mathcal{V}, \mathcal{V}')$ of A

$$_{\mathcal{V}'}\langle A'\varphi, \psi\rangle_{\mathcal{V}} = {}_{\mathcal{V}}\langle \varphi, A\psi\rangle_{\mathcal{V}'} \qquad \forall \varphi, \psi \in \mathcal{V}.$$

This operator was denoted with the symbol A^* in Sect. 3.5 (see the Lagrange identity (3.42)). Owing to (3.40) and (3.41) we obtain

$$_{\mathcal{V}'}\langle A'\varphi, \psi\rangle_{\mathcal{V}} = a(\psi, \varphi) \quad \forall \varphi, \psi \in \mathcal{V}. \tag{17.38}$$

We define *adjoint state* (or *adjoint variable*) $p(u) \in \mathcal{V}$ the solution of the *adjoint equation*

$$A'p(u) = C'\Lambda_{\mathcal{Z}}(Cy(u) - z_d), \tag{17.39}$$

with $u \in \mathcal{U}$. Thanks to (17.39), we obtain

$$\begin{aligned} &_{\mathcal{V}'}\langle C'\Lambda_{\mathcal{Z}}(Cy(u) - z_d),\, y(v) - y(u)\rangle_{\mathcal{V}} = \\ &_{\mathcal{V}'}\langle A'p(u), y(v) - y(u)\rangle_{\mathcal{V}} = \\ \text{(thanks to the definition of } A') \quad &_{\mathcal{V}}\langle p(u), A(y(v) - y(u))\rangle_{\mathcal{V}'} = \\ \text{(thanks to (17.28))} \quad &_{\mathcal{V}}\langle p(u), B(v-u)\rangle_{\mathcal{V}'} = {}_{\mathcal{U}'}\langle B'p(u), v-u\rangle_{\mathcal{U}}, \end{aligned}$$

where $B' \in \mathcal{L}(\mathcal{V}, \mathcal{U}')$ is the adjoint operator of B (see (2.20)). It follows that, with the help of the Riesz canonical isomorphism $\Lambda_{\mathcal{U}}$ of $\mathcal{U}$ and $\mathcal{U}'$ (see again (2.5)), inequality (17.22) can be rewritten as

$$\frac{1}{2}J'(u)(v-u) = (\Lambda_{\mathcal{U}}^{-1}B'p(u) + Nu, v-u)_{\mathcal{U}} \geq 0 \qquad \forall v \in \mathcal{U}_{ad}. \tag{17.40}$$

In the case of unconstrained control, that is when $\mathcal{U}_{ad} = \mathcal{U}$, the last inequality becomes in fact an equality, that is

$$B'p(u) + \Lambda_{\mathcal{U}} N u = 0. \tag{17.41}$$

This follows by taking $v = u - (\Lambda_{\mathcal{U}}^{-1} B'p(u) + Nu)$ in (17.40). In the case where $\mathcal{V} \subset \mathcal{U}$ the previous equation implies that

$$b(v,p) + n(u,v) = 0 \quad \forall v \in \mathcal{V}.$$

The final result is recapped in the following Theorem ([Lio71, Chap. 2, Thm. 1.4]).

Theorem 17.3. *A necessary and sufficient condition for the existence of an optimal control $u \in \mathcal{U}_{ad}$ is that the following equations and inequalities hold (see (17.28), (17.39), (17.40)):*

$$\begin{cases} y = y(u) \in \mathcal{V}, & Ay(u) = f + Bu, \\ p = p(u) \in \mathcal{V}, & A'p(u) = C'\Lambda(Cy(u) - z_d), \\ u \in \mathcal{U}_{ad}, & (\Lambda_{\mathcal{U}}^{-1} B'p(u) + Nu, v - u)_{\mathcal{U}} \geq 0 \quad \forall v \in \mathcal{U}_{ad}, \end{cases} \tag{17.42}$$

or, in weak form:

$$\begin{cases} y = y(u) \in \mathcal{V}, & a(y(u), \varphi) = (f, \varphi) + b(u, \varphi) \quad \forall \varphi \in \mathcal{V}, \\ p = p(u) \in \mathcal{V}, & a(\psi, p(u)) = (Cy(u) - z_d, C\psi)_{\mathcal{Z}} \quad \forall \psi \in \mathcal{V}, \\ u \in \mathcal{U}_{ad}, & (\Lambda_{\mathcal{U}}^{-1} B'p(u) + Nu, v - u)_{\mathcal{U}} \geq 0 \quad \forall v \in \mathcal{U}_{ad}. \end{cases} \tag{17.43}$$

This is called the optimality system.
If N is symmetric and positive definite, then the control u is unique*; on the other hand, if $N = 0$ and $\mathcal{U}_{ad}$ is bounded, then there exists at least one solution, and the family of optimal controls forms a closed and convex subset $\mathcal{X}$ of $\mathcal{U}_{ad}$.*
The third condition of (17.42) can be expressed as follows

$$(\Lambda_{\mathcal{U}}^{-1} B'p(u) + Nu, u)_{\mathcal{U}} = \inf_{v \in \mathcal{U}_{ad}} (\Lambda_{\mathcal{U}}^{-1} B'p(u) + Nu, v)_{\mathcal{U}}. \tag{17.44}$$

The u-derivative of the cost functional can be expressed in terms of the adjoint state $p(u)$ as follows

$$\frac{1}{2} J'(u) = B'p(u) + \Lambda_{\mathcal{U}} N u. \tag{17.45}$$

Remark 17.3. Apart from the term depending on the form N, J' can be obtained from the adjoint variable p through the operator B'. This result will stand at the base of the *numerical methods* which are useful to determine an approximate control function. If

$\mathcal{U}_{ad} = \mathcal{U}$, then the optimal control satisfies

$$Nu = -\Lambda_{\mathcal{U}}^{-1} B' p(u) \,. \tag{17.46}$$

Thanks to identity (2.24) we have

$$\Lambda_{\mathcal{U}}^{-1} B' = B^T \Lambda_{\mathcal{V}}, \tag{17.47}$$

where $\Lambda_{\mathcal{V}}$ is the Riesz canonical isomorphism from $\mathcal{V}$ to $\mathcal{V}'$ and $B^T : \mathcal{V}' \to \mathcal{U}$ is the transpose operator of B introduced in (2.22). •

17.6 Some examples of optimal control problems

In this section we introduce three examples of optimal control problems.

17.6.1 A Dirichlet problem with distributed control

Let us recover the example of distributed control (17.12) and consider the following cost functional (to be minimized)

$$J(v) = \int_{\Omega} (y(v) - z_d)^2 \, d\Omega + (Nv, v), \tag{17.48}$$

in which, for instance, we can set $N = \nu I$, $\nu > 0$. In this case $\mathcal{V} = H_0^1(\Omega)$, $\mathcal{H} = L^2(\Omega)$, $\mathcal{U} = \mathcal{H}$ (then $(Nv, v) = (Nv, v)_{\mathcal{U}}$) therefore $\Lambda_{\mathcal{U}}$ is the identity operator. Moreover, B is the identity operator, C is the injection operator of $\mathcal{V}$ in $\mathcal{H}$, $\mathcal{Z} = \mathcal{H}$ and therefore $\Lambda_{\mathcal{Z}}$ is the identity operator. Finally, $a(u,v) = \int_{\Omega} \nabla u \cdot \nabla v \, d\Omega$. Owing to Theorem 17.3 we obtain, with $A = -\Delta$, the following *optimality system*:

$$\begin{cases} y(u) \in H_0^1(\Omega) & : \quad Ay(u) = f + u \quad \text{in } \Omega, \\ p(u) \in H_0^1(\Omega) & : \quad A'p(u) = y(u) - z_d \quad \text{in } \Omega, \\ u \in \mathcal{U}_{ad} & : \quad \displaystyle\int_{\Omega} (p(u) + Nu)(v - u) \, d\Omega \geq 0 \qquad \forall v \in \mathcal{U}_{ad}. \end{cases} \tag{17.49}$$

In the unconstrained case in which $\mathcal{U}_{ad} = \mathcal{U}\ (= L^2(\Omega))$, the last inequality reduces to the equation

$$p(u) + Nu = 0,$$

as we can see by taking $v = u - (p(u) + Nu)$.
The first two equations of (17.49) provide a system for the two variables y and p

$$\begin{cases} Ay + N^{-1}p = f \quad \text{in } \Omega, & y = 0 \quad \text{on } \partial\Omega, \\ A'p - y = -z_d \quad \text{in } \Omega, & p = 0 \quad \text{on } \partial\Omega, \end{cases}$$

whose solution provides the optimal control $u = -N^{-1}p$.

If Ω has a smooth boundary, by the elliptic regularity property both y and p belong to the space $H^2(\Omega)$. Since N^{-1} maps $H^2(\Omega)$ to itself, the optimal control u belongs to $H^2(\Omega)$, too.

17.6.2 A Neumann problem with distributed control

Consider now the problem

$$\begin{cases} Ay(u) = f + u & \text{in } \Omega, \\ \dfrac{\partial y(u)}{\partial n_A} = g & \text{on } \partial\Omega, \end{cases} \tag{17.50}$$

where A is an elliptic operator and $\frac{\partial}{\partial n_A}$ is the conormal derivative associated with A (see (3.34)). The cost functional to be minimized is the one introduced in (17.48). In this case $\mathcal{V} = H^1(\Omega)$, $\mathcal{H} = L^2(\Omega)$, $\mathcal{U} = \mathcal{H}$, B is the identity operator, C is the injection map of $\mathcal{V}$ into $\mathcal{H}$,

$$a(\psi,\varphi) = {}_{\mathcal{V}'}\langle A\psi,\varphi\rangle_{\mathcal{V}}, \qquad F(\varphi) = \int_\Omega f\varphi\, d\Omega + \int_{\partial\Omega} g\varphi\, d\gamma,$$

for given $f \in L^2(\Omega)$ and $g \in H^{-1/2}(\partial\Omega)$.
If $A\varphi = -\Delta\varphi + \beta\varphi$, with $\beta > 0$, then

$$a(\psi,\varphi) = \int_\Omega \nabla\psi \cdot \nabla\varphi\, d\Omega + \int_\Omega \beta\psi\varphi\, d\Omega.$$

The variational formulation of the state problem (17.50) is

$$\text{find } y(u) \in H^1(\Omega) \ : \ a(y(u),\varphi) = F(\varphi) \qquad \forall\varphi \in H^1(\Omega). \tag{17.51}$$

The adjoint problem is the following Neumann problem

$$\begin{cases} A'p(u) = y(u) - z_d & \text{in } \Omega, \\ \dfrac{\partial p(u)}{\partial n_{A'}} = 0 & \text{on } \partial\Omega. \end{cases} \tag{17.52}$$

The optimal control can be obtained by solving the system formed by (17.50), (17.52), and

$$u \in \mathcal{U}_{ad} \ : \ \int_\Omega (p(u) + Nu)(v - u)\, d\Omega \geq 0 \qquad \forall v \in \mathcal{U}_{ad}. \tag{17.53}$$

17.6.3 A Neumann problem with boundary control

Consider now the problem

$$\begin{cases} Ay(u) = f & \text{in } \Omega, \\ \dfrac{\partial y(u)}{\partial n_A} = g + u & \text{on } \partial\Omega, \end{cases} \tag{17.54}$$

where the operator is the same as before and the cost functional is still that of (17.48). In this case,

$$\mathcal{V}=H^1(\Omega), \qquad \mathcal{H}=L^2(\Omega), \qquad \mathcal{U}=H^{-1/2}(\partial\Omega).$$

For all $u \in \mathcal{U}$, $Bu \in \mathcal{V}'$ is given by ${}_{\mathcal{V}'}\langle Bu,\varphi\rangle_{\mathcal{V}} = \int_{\partial\Omega} u\varphi \, d\gamma$, and C is the injection map of $\mathcal{V}$ in $\mathcal{H}$.
The weak formulation of (17.54) is

$$\text{find } y(u) \in \mathcal{H}^1(\Omega) \; : \; a(y(u),\varphi) = \int_\Omega f\varphi \, d\Omega + \int_{\partial\Omega} (g+u)\varphi \, d\gamma \qquad \forall \varphi \in \mathcal{H}^1(\Omega).$$

The adjoint problem is still given by (17.52), while the variational inequality yielding the optimal control is the third of (17.42). The interpretation of this inequality is far from trivial. If we choose

$$(u,v)_{\mathcal{U}} = \int_{\partial\Omega} (-\Delta_{\partial\Omega})^{-1/4} u \, (-\Delta_{\partial\Omega})^{-1/4} v \, d\gamma = \int_{\partial\Omega} (-\Delta_{\partial\Omega})^{-1/2} u \, v \, d\gamma$$

as scalar product in $\mathcal{U}$, where $\Delta_{\partial\Omega}$ is the Laplace–Beltrami operator (see, e.g., [QV94]), it can be proven that the third inequality of (17.42) is equivalent to (see [Lio71, Chap. 1, Sect. 2.4])

$$\int_{\partial\Omega} (p(u)_{|\partial\Omega} + (-\Delta_{\partial\Omega})^{-1/2} Nu)(v-u) \, d\gamma \geq 0 \qquad \forall v \in \mathcal{U}_{ad}.$$

In Tables 17.1 and 17.2 we summarize the main conclusions that were drawn for the problems just considered.

17.7 Numerical tests

In this section we present some numerical tests for the solution of 1D optimal control problems similar to those summarized in Tables 17.1 and 17.2.
For all numerical simulations we consider the domain $\Omega = (0,1)$, a simple diffusion-reaction operator A

$$Ay = -\mu y'' + \gamma y,$$

and the very same cost functional considered in the tables, with a regularization coefficient $\nu = 10^{-2}$ (unless otherwise specified). We discretize both state and adjoint problems by means of piecewise-linear finite elements, with grid-size $h = 10^{-2}$; for solving the minimization problem we use the conjugate gradient method with an acceleration parameter τ^k initialized with $\tau^0 = \overline{\tau}$ and then, when necessary for the convergence, reduce it by 2 at every subsequent step. This satisfies the Armijo rule (see Sect. 17.9). The tolerance tol for the iterative method is fixed to 10^{-3}, with the following stopping criterium $\|J'(u^k)\| < Tol\|J'(u^0)\|$.

Table 17.1. Summary of Dirichlet control problems

Dirichlet conditions	Distributed Observation	Boundary Observation
Distributed Control	$\begin{cases} Ay = f + u & \text{in } \Omega \\ y = 0 & \text{on } \partial\Omega \end{cases}$ $J(y,u) = \int_{\Omega} (y - z_d)^2 d\Omega + \nu \int_{\Omega} u^2 d\Omega$ $\begin{cases} A'p = y - z_d & \text{in } \Omega \\ p = 0 & \text{on } \partial\Omega \end{cases}$ $\frac{1}{2} J'(u) = p(u) + \nu u \quad \text{in } \Omega$	$\begin{cases} Ay = f + u & \text{in } \Omega \\ y = 0 & \text{on } \partial\Omega \end{cases}$ $J(y,u) = \int_{\partial\Omega} (\frac{\partial y}{\partial n_A} - z_d)^2 d\gamma + \nu \int_{\Omega} u^2 d\Omega$ $\begin{cases} A'p = 0 & \text{in } \Omega \\ p = -(\frac{\partial y}{\partial n_A} - z_d) & \text{on } \partial\Omega \end{cases}$ $\frac{1}{2} J'(u) = p(u) + \nu u \quad \text{in } \Omega$
Boundary Control	$\begin{cases} Ay = f & \text{in } \Omega \\ y = u & \text{on } \partial\Omega \end{cases}$ $J(y,u) = \int_{\Omega} (y - z_d)^2 d\Omega + \nu \int_{\partial\Omega} u^2 d\gamma$ $\begin{cases} A'p = y - z_d & \text{in } \Omega \\ p = 0 & \text{on } \partial\Omega \end{cases}$ $\frac{1}{2} J'(u) = -\frac{\partial p}{\partial n_{A'}} + \nu u \quad \text{on } \partial\Omega$	$\begin{cases} Ay = f & \text{in } \Omega \\ y = u & \text{on } \partial\Omega \end{cases}$ $J = \int_{\partial\Omega} (\frac{\partial y}{\partial n_A} - z_d)^2 d\gamma + \nu \int_{\partial\Omega} u^2 d\gamma$ $\begin{cases} A'p = 0 & \text{in } \Omega \\ p = -(\frac{\partial y}{\partial n_A} - z_d) & \text{on } \partial\Omega \end{cases}$ $\frac{1}{2} J'(u) = -\frac{\partial p}{\partial n_{A'}} + \nu u \quad \text{on } \partial\Omega$

- Case D1 (Table 17.1 top left): distributed control and observation, with Dirichlet boundary conditions. We assume

$$\mu = 1, \ \ \gamma = 0, \ \ f = 1, \ \ u^0 = 0, \ \ z_d = \begin{cases} x & x \leq 0.5 \\ 1 - x & x > 0.5 \end{cases}, \quad \overline{\tau} = 10.$$

The value of the cost functional for u^0 is $J^0 = 0.0396$. Ater 11 iterations we obtain the optimal cost functional $J = 0.0202$. In Fig. 17.3 we report the state variable for the initial and optimal control u and the desired function z_d (left), and the optimal control function (right).
As displayed in Table 17.3, the number of iterations increases as ν decreases. In the same Table, we also report, for the sake of comparison, the values of the cost functional J corresponding to the optimal value of u for different values of ν.
In Fig. 17.4 we report the optimal state (left) and the control functions (right) obtained for different values of ν.

Table 17.2. Summary of Neumann control problems

Neumann conditions	Distributed Observation	Boundary Observation
Distributed Control	$\begin{cases} Ay = f+u & \text{in } \Omega \\ \dfrac{\partial y}{\partial n_A} = g & \text{on } \partial\Omega \end{cases}$ $J(y,u) = \int_{\Omega} (y-z_d)^2 d\Omega + \nu \int_{\Omega} u^2 d\Omega$ $\begin{cases} A'p = y - z_d & \text{in } \Omega \\ \dfrac{\partial p}{\partial n_{A'}} = 0 & \text{on } \partial\Omega \end{cases}$ $\frac{1}{2}J'(u) = p + \nu u \quad \text{in}\Omega$	$\begin{cases} Ay = f+u & \text{in } \Omega \\ \dfrac{\partial y}{\partial n_A} = g & \text{on } \partial\Omega \end{cases}$ $J(y,u) = \int_{\partial\Omega} (y-z_d)^2 d\gamma + \nu \int_{\Omega} u^2 d\Omega$ $\begin{cases} A'p = 0 & \text{in } \Omega \\ \dfrac{\partial p}{\partial n_{A'}} = y - z_d & \text{on } \partial\Omega \end{cases}$ $\frac{1}{2}J'(u) = p + \nu u \quad \text{in } \Omega$
Boundary Control	$\begin{cases} Ay = f & \text{in } \Omega \\ \dfrac{\partial y}{\partial n_A} = g+u & \text{on } \partial\Omega \end{cases}$ $J(y,u) = \int_{\Omega} (y-z_d)^2 d\Omega + \nu \int_{\partial\Omega} u^2 d\gamma$ $\begin{cases} A'p = y - z_d & \text{in } \Omega \\ \dfrac{\partial p}{\partial n_{A'}} = 0 & \text{on } \partial\Omega \end{cases}$ $\frac{1}{2}J'(u) = p + \nu u \quad \text{on } \partial\Omega$	$\begin{cases} Ay = f & \text{in } \Omega \\ \dfrac{\partial y}{\partial n_A} = g+u & \text{on } \partial\Omega \end{cases}$ $J(y,u) = \int_{\partial\Omega} (y-z_d)^2 d\Omega + \nu \int_{\partial\Omega} u^2 d\gamma$ $\begin{cases} A'p = 0 & \text{in } \Omega \\ \dfrac{\partial p}{\partial n_{A'}} = y - z_d & \text{on } \partial\Omega \end{cases}$ $\frac{1}{2}J'(u) = p + \nu u \quad \text{on } \partial\Omega$

Table 17.3. Case D1. Number of iterations and optimal cost functional corresponding to different values of ν

ν	*it*	*J*	ν	*it*	*J*	ν	*it*	*J*
10^{-2}	11	0.0202	10^{-3}	71	0.0047	10^{-4}	349	0.0011

- Case D2 (Table 17.1 top right): distributed control and boundary observation, with Dirichlet boundary condition. We consider $\mu = 1$, $\gamma = 0$, $f = 1$, $u^0 = 0$, while the target function z_d is such that $z_d(0) = -1$ and $z_d(1) = -4$; finally, $\overline{\tau} = 0.1$. At the initial step we have $J = 12.5401$ and after 89 iterations $J = 0.04305$; we can observe how the normal derivative of the state variable is "near" the desired value $z_d \left[\mu \frac{\partial y}{\partial n}(0), \mu \frac{\partial y}{\partial n}(1)\right] = [-1.0511, -3.8695]$. In Fig. 17.5 we report the initial and optimal state (left) and the corresponding optimal control function (right).

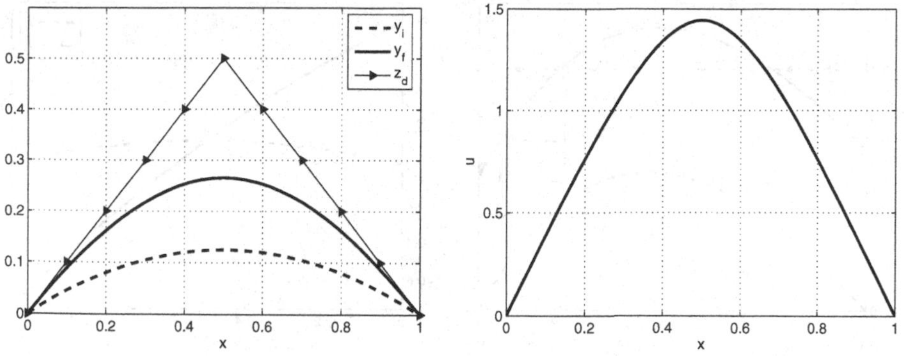

Fig. 17.3. Case D1. Initial and optimal state variables and the desired function (left); optimal control function (right)

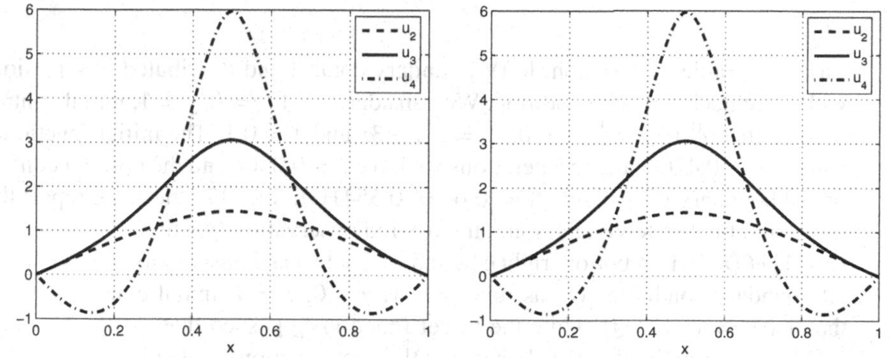

Fig. 17.4. Case D1. Optimal state variables y_2 (for $\nu = 1e-2$), y_3 (for $\nu = 1e-3$), y_4 (for $\nu = 1e-4$) and desired function z_d (left); optimal control u_2 (for $\nu = 1e-2$), u_3 (for $\nu = 1e-3$), u_4 (for $\nu = 1e-4$) (right)

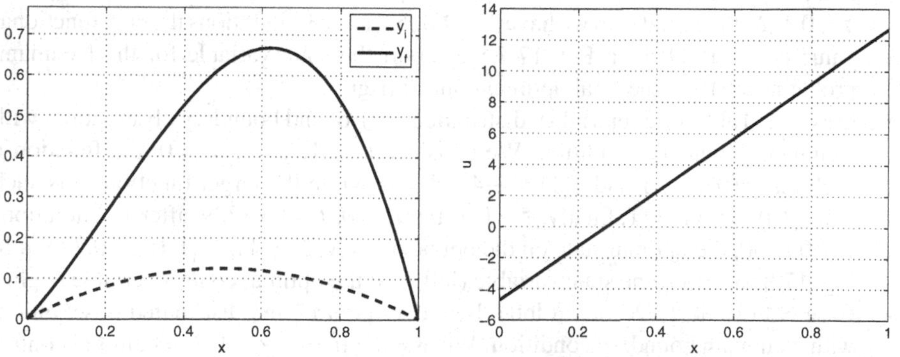

Fig. 17.5. Case D2. Initial and optimal state variables (left); optimal control variable (right)

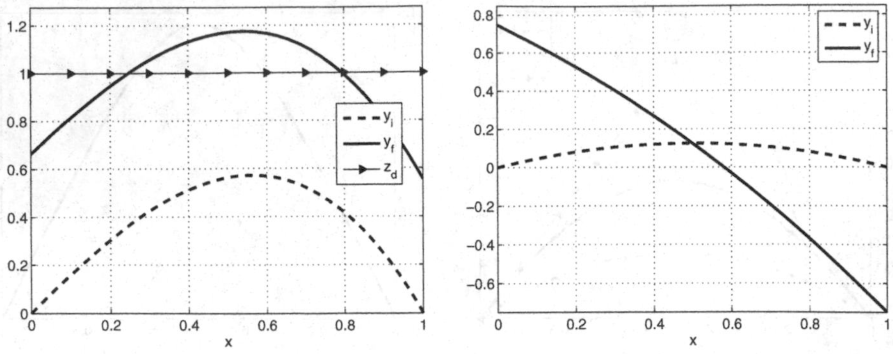

Fig. 17.6. Left: Case D3. Initial and optimal state variables and desired observation function. Right: Case D4. Initial and optimal state variables

- Case D3 (Table 17.1 bottom left): boundary control and distributed observation, with Dirichlet boundary condition. We consider $\mu = 1$, $\gamma = 0$, $f = 1$, initial control u^0 such that $u^0(0) = u^0(1) = 0$, $z_d = -1 - 3x$ and $\overline{\tau} = 0.1$. The initial functional value is $J = 0.4204$, after 55 iterations we have $J = 0.0364$ and the optimal control on the boundary is $[u(0), u(1)] = [0.6638, 0.5541]$. In Fig. 17.6, left, we report the initial and final state variables and the desired observation function.
- Case D4 (Table 17.1 bottom right): boundary control and observation, with Dirichlet boundary condition. We assume $\mu = 1$, $\gamma = 0$, $f = 1$, initial control u^0 such that $u^0(0) = u^0(1) = 0$, while the target function z_d is such that $z_d(0) = -1$ and $z_d(1) = -4$; finally, $\overline{\tau} = 0.1$. For $it = 0$ the cost functional value is $J = 12.5401$, after only 4 iterations $J = 8.0513$ and the optimal control on the boundary is $[u(0), u(1)] = [0.7481, -0.7481]$. In Fig. 17.6, right, we report the state variable.
- Case N1 (Table 17.2 top left): distributed control and observation, with Neumann boundary condition. We consider $\mu = 1$, $\gamma = 1$, $f = 0$, $g = -1$, $u^0 = 0$, $z_d = 1$, $\overline{\tau} = 0.1$. At the initial step we have $J = 9.0053$, after 18 iterations the cost functional value is $J = 0.0944$. In Fig. 17.7 we report the state variable for the Neumann problem (left) and the final optimal control (right).
- Case N2 (Table 17.2 top right): distributed control and boundary observation, with Neumann boundary condition. We consider $\mu = 1$, $\gamma = 1$ $f = 0$, the function g such that $g(0) = -1$ and $g(1) = -4$, $u^0 = 0$, while the target function z_d is such that $z_d(0) = z_d(1) = 1$; finally, $\overline{\tau} = 0.1$. For $it = 0$, $J = 83.1329$, after 153 iterations $J = 0.6280$, the optimal state on the boundary is $[y(0), y(1)] = [1.1613, 0.7750]$. In Fig. 17.8 we report the state variable (left) and the optimal control variable (right).
- Case N3 (Table 17.2 bottom left): boundary control and distributed observation, with Neumann boundary condition. We assume $\mu = 1$, $\gamma = 1$, $f = 0$, initial control u^0 such that $u^0(0) = u^0(1) = 0$, $z_d = -1 - 3x$, $\overline{\tau} = 0.1$. The initial cost functional value is $J = 9.0053$, after 9 iterations we have $J = 0.0461$, and the optimal control is $[u(0), u(1)] = [1.4910, 1.4910]$. In Fig. 17.9, left, we report the state variable for the initial and optimal control u, together with the desired observation function.

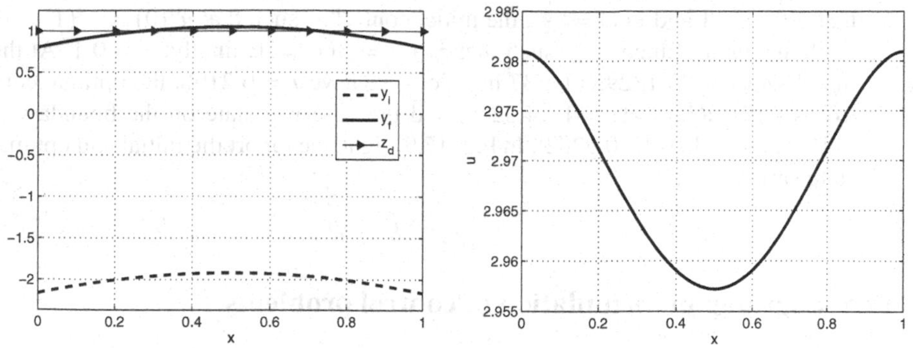

Fig. 17.7. Case N1. Initial and optimal state variables and the desired function (left); optimal control variable (right)

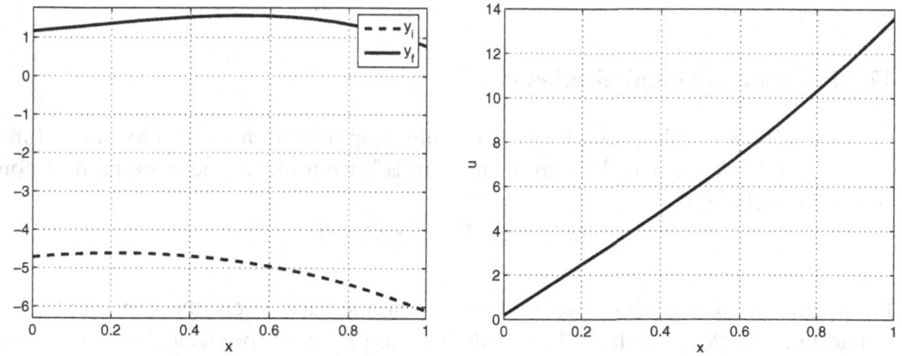

Fig. 17.8. Case N2. Initial and optimal state variables (left); optimal control variable (right)

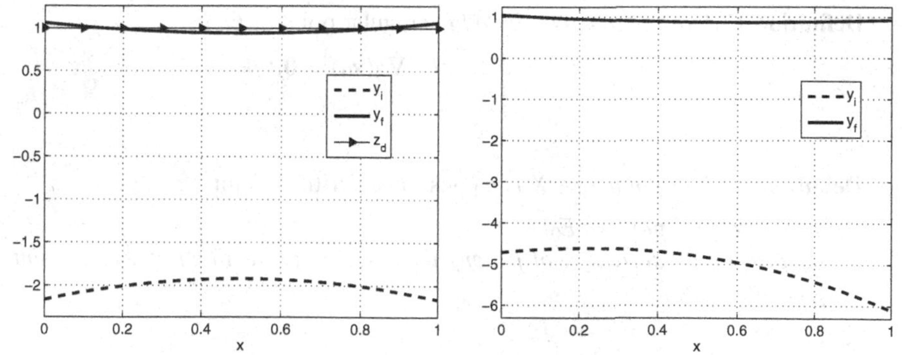

Fig. 17.9. Left: Case N3. Initial and optimal state variables and desired observation function. Right: Case N4. Initial and optimal state variables.

- Case N4 (Table 17.2 bottom right): boundary control and observation, with Neumann boundary condition. We consider $\mu = 1$, $\gamma = 1$, $f = 0$, the function g such that $g(0) = -1$ and $g(1) = -4$, the initial control u^0 such that $u^0(0) = u^0(1) = 0$, while the target function z_d such that $z_d(0) = z_d(1) = 1$; finally, $\overline{\tau} = 0.1$. At the initial step $J = 83.1329$, after 37 iterations we have $J = 0.2196$, the optimal control is $[u(0), u(1)] = [1.5817, 4.3299]$ and the observed state on the boundary is $[y(0), y(1)] = [1.0445, 0.9282]$. In Fig. 17.9, right, we report the initial and optimal state variables.

17.8 Lagrangian formulation of control problems

In this section we present another methodological approach for the solution of optimal control problems, based on Lagrange multipliers; this approach better highlights the role played by the adjoint variable.

17.8.1 Constrained optimization in $\mathbb{R}^n$

Let us start by a simple example, the constrained optimization in $\mathbb{R}^n$. Given two functions $f, g \in C^1(X)$, where X is an open set in $\mathbb{R}^n$, we look for the extrema of f constrained to belong to

$$E_0 = \{\mathbf{x} \in \mathbb{R}^n \; : \; g(\mathbf{x}) = 0\}.$$

For simplicity of exposition we are considering the case where the constraint g is a scalar function. We give the following definitions of *regular* points and of *constrained critical* points:

Definition 17.1. *A point* $\mathbf{x}_0$ *is said to be a* regular point *of* E_0 *if*

$$g(\mathbf{x}_0) = 0 \quad and \quad \nabla g(\mathbf{x}_0) \neq \mathbf{0}.$$

Definition 17.2. *A point* $\mathbf{x}_0 \in X$ *is a* constrained critical point *if:*

i) $\mathbf{x}_0$ *is a regular point of* E_0*;*
ii) *the directional derivative of* f *along the tangential direction to the constraint* g *is null in* $\mathbf{x}_0$.

On the basis of these definitions the following result holds:

Theorem 17.4. *A regular point* $\mathbf{x}_0$ *of* E_0 *is a constrained critical point if and only if there exists* $\lambda_0 \in \mathbb{R}$ *such that*

$$\nabla f(\mathbf{x}_0) = \lambda_0 \nabla g(\mathbf{x}_0).$$

We introduce the function $\mathcal{L} : X \times \mathbb{R} \mapsto \mathbb{R}$, called the *Lagrangian* (or Lagrangian functional),

$$\mathcal{L}(\mathbf{x}, \lambda) = f(\mathbf{x}) - \lambda g(\mathbf{x}).$$

From Theorem 17.4 we deduce that $\mathbf{x}_0$ is a constrained critical point if and only if $(\mathbf{x}_0, \lambda_0)$ is a (free) critical point for $\mathcal{L}$. The number λ_0 is called *Lagrange multiplier* and can be obtained, together with $\mathbf{x}_0$, by solving the system

$$\nabla \mathcal{L}(\mathbf{x}, \lambda) = \mathbf{0},$$

that is

$$\begin{cases} \mathcal{L}_{\mathbf{x}} = \nabla f - \lambda \nabla g = \mathbf{0}, \\ \mathcal{L}_{\lambda} = -g = 0. \end{cases}$$

17.8.2 The solution approach based on the Lagrangian

In this section we will extend the theory developed in Sect. 17.8.1 to optimal control problems. Also in this context, the Lagrangian approach can be used as an alternative to the approach "*à la Lions*" (see, for example, [BKR00]).
It is actually used to integrate the techniques of grid adaptivity based on a posteriori error estimates with optimal control problems (see [BKR00, Ded04]).
The approach based on Lagrange multipliers is also widely used to solve shape optimization problems in which the control is represented by the shape of the computational domain. The optimal control function u is therefore a function defined on the boundary (or on a subset of it) which describes the optimal displacement from the original position. The interested reader can consult, e.g., [Jam88, MP01, SZ91].
In Sect. 17.8.1 the approach based on the Lagrangian allows the determination of the extrema of a function f with constraint g. Instead, in control problems we look for a *function* $u \in \mathcal{U}_{ad} \subset \mathcal{U}$ satisfying the minimization problem (17.34), $y(u)$ being the solution of the state equation (17.28). As usual, A is an elliptic differential operator applied to the state variable, while B is an operator applied to the control function in the state equation. This problem can be regarded as a constrained minimization problem, provided we state a suitable correspondence between the cost functional J and the function f of Sect. 17.8.1), between the equation of state and the constraint equation $g = 0$, and, finally, between the control function u and the extremum $\mathbf{x}$.
We assume, for the sake of simplicity, that $\mathcal{U}_{ad} = \mathcal{U}$. For a more general treatment, see, e.g., [Gun03].
The solution of the optimal control problem can therefore be regarded as the search for the (unconstrained) critical "points" of the following Lagrangian functional

$$\mathcal{L}(y, p, u) = J(y(u), u) \; + \; {}_{\mathcal{V}'}\langle \, p, \; f + Bu - Ay(u) \, \rangle_{\mathcal{V}}, \qquad (17.55)$$

where p is the *Lagrange multiplier*. In this framework the (unconstrained) critical points are represented by the functions y, p and u. The problem therefore becomes

$$\text{find } (y, p, u) \in \mathcal{V} \times \mathcal{V} \times \mathcal{U} \; : \; \nabla \mathcal{L}(y, p, u)[(\psi, \varphi, \phi)] = 0 \qquad \forall (\psi, \varphi, \phi) \in \mathcal{V} \times \mathcal{V} \times \mathcal{U}$$

that is

$$\begin{cases} \mathcal{L}_p(y,p,u)[\psi] = 0 & \forall \psi \in \mathcal{V}, \\ \mathcal{L}_y(y,p,u)[\varphi] = 0 & \forall \varphi \in \mathcal{V}, \\ \mathcal{L}_u(y,p,u)[\phi] = 0 & \forall \phi \in \mathcal{U}. \end{cases} \tag{17.56}$$

We have used the abridged notation $\mathcal{L}_y$ to indicate the Gâteaux derivative of $\mathcal{L}$ with respect to y (see Definition 2.6). The notations $\mathcal{L}_p$ and $\mathcal{L}_u$ have a similar meaning. Consider now as an example a state equation of elliptic type with two linear operators A and B that we rewrite in the weak form (17.29); given $u \in \mathcal{U}$ and $f \in \mathcal{H}$,

$$\text{find } y = y(u) \in \mathcal{V} \,:\, a(y,\varphi) = (f,\varphi) + b(u,\varphi) \qquad \forall \varphi \in \mathcal{V}. \tag{17.57}$$

The bilinear form $a(\cdot,\cdot)$ is associated with the operator A, $b(\cdot,\cdot)$ with B. The cost functional to be minimized can be expressed as follows

$$J(y(u),u) = ||Cy(u) - z_d||^2 + n(u,u), \tag{17.58}$$

where C is the operator that maps the state variable into the space $\mathcal{Z}$ of the observed functions, z_d is the target observation and $n(\cdot,\cdot)$ is a symmetric bilinear form. Thus far, no assumption was made on the boundary conditions, the kind of control (either distributed or on the boundary) nor the kind of observation. This was done on purpose in order to consider a very general framework. In weak form, (17.55) becomes

$$\mathcal{L}(y,p,u) = J(y(u),u) + b(u,p) + (f,p) - a(y,p).$$

Since

$$\begin{aligned} \mathcal{L}_p(y,p,u)[\varphi] &= (f,\varphi) + b(u,\varphi) - a(y,\varphi) & \forall \varphi \in \mathcal{V}, \\ \mathcal{L}_y(y,p,u)[\psi] &= 2(Cy - z_d, C\psi) - a(\psi,p) & \forall \psi \in \mathcal{V}, \\ \mathcal{L}_u(y,p,u)[\phi] &= b(\phi,p) + 2n(u,\phi) & \forall \phi \in \mathcal{U}, \end{aligned} \tag{17.59}$$

(17.56) yields the *optimality system*

$$\begin{cases} y \in \mathcal{V} \,:\, a(y,\varphi) = b(u,\varphi) + (f,\varphi) & \forall \varphi \in \mathcal{V}, \\ p \in \mathcal{V} \,:\, a(\psi,p) = 2(Cy - z_d, C\psi) & \forall \psi \in \mathcal{V}, \\ u \in \mathcal{U} \,:\, b(\phi,p) + 2n(u,\phi) = 0 & \forall \phi \in \mathcal{U}. \end{cases} \tag{17.60}$$

It is worth noticing that, upon rescaling p as $\tilde{p} = p/2$, the variables $(y,\tilde{p},u)$ obtained from (17.60) actually satisfy the system (17.43) obtained in the framework of Lions' approach. Note that the vanishing of $\mathcal{L}_p$ yields the state equation (in weak form), that of $\mathcal{L}_y$ generates the equation for the Lagrange multiplier (which can be identified with the adjoint equation), and that of $\mathcal{L}_u$ yields the so-called *sensitivity* equation expressing the condition that the optimum is achieved. The adjoint variable, that is the Lagrange multiplier, is associated to the *sensitivity* of the cost functional J with respect to the variation of the observation function, and therefore to the variation of the control u.

It turns out to be very useful to express the Gâteaux derivative of the Lagrangian $\mathcal{L}_u$ in terms of the derivative of the cost functional J with respect to u, (17.60), according to what we have seen in Sect. 2.2. This correspondence is guaranteed by the Riesz representation theorem (see Theorem 2.1). Indeed, since $\mathcal{L}_u[\phi]$ is a linear and bounded functional, and ϕ belongs to the Hilbert space $\mathcal{U}$, we can compute J' case by case, that is, from the third of (17.59), and from (17.45)

$$\mathcal{L}_u[\phi] = (J'(u), \phi)_{\mathcal{U}} = b(\phi, p) + 2n(u, \phi).$$

It is worth noticing how the adjoint equation is generated. According to Lions' theory the adjoint equation is based on the use of the adjoint operator (see equation (17.39)), whereas when using the approach based on the Lagrangian we obtain it by differentiating $\mathcal{L}$ with respect to the state variable. The adjoint variable $p(u)$, when computed on the optimal control u, corresponds to the Lagrange multiplier. In general, Lions' method and the method based on the use of the Lagrangian do not lead to the same definition of adjoint problem, so they give rise to numerical methods which can behave differently. For a correct solution of the optimal control problem it is therefore essential to be coherent with the kind of approach that we are considering.
Another crucial issue is the derivation of the boundary conditions for the adjoint problem; the two different approaches may lead to different types of boundary conditions. In particular, the approach based on the Lagrangian yields the boundary conditions for the adjoint problem automatically, while this is not the case for the other approach.

17.9 Iterative solution of the optimal control problem

For the numerical solution of optimal control problems, two different paradigms can be adopted:

- (*optimize-then-discretize*) we first apply the iterative method, then discretize the various steps of the algorithm, or
- (*discretize-then-optimize*) we first discretize our optimal control problem and then apply an iterative algorithm to solve its discrete version.

This discussion is deferred until Sect. 17.12. In this section we illustrate the way an iterative algorithm can be used to generate a sequence that hopefully converges to the optimal control function u.
As previously discussed, an optimal control problem can be formulated according to the Lions approach, yielding the set of equations (17.43), or by means of the Lagrangian, in which case the equations to be solved are (17.60). In either case we end up with an optimality system made by:

i) the state equation,
ii) the adjoint equation,
iii) the equation expressing the optimality condition.

In particular, the third equation is related to the variation of the cost functional, either explicitly in the Lions approach, or implicitly (through the Riesz representation theorem) when using the Lagrangian approach. Indeed, in the case of linear elliptic equations previously examined we obtain, respectively:

- $\frac{1}{2}J'(u) = B'p(u) + \Lambda_{\mathcal{U}}Nu;$
- $\mathcal{L}_u[\phi] \;=\; b(\phi,p) + 2n(u,\phi) \qquad \forall\phi \in \mathcal{U}.$

In order to simplify our notation, in the remainder of this section we will use the symbol J' not only to indicate the derivative of the cost functional but also that of the Lagrangian, $\mathcal{L}_u[\phi]$. The evaluation of J' at a given point of the control region (Ω, Γ, or one of their subsets) provides an indication of the sensitivity of the cost functional J (at that very point) with respect to the variations of the control function u. Otherwise said, an infinitesimal variation δu of the control about a given value u, entails, up to infinitesimals of higher order, a variation δJ of the cost functional that is proportional to $J'(u)$. This suggests the use of the following *steepest descent* iterative algorithm. If u^k denotes the value of the control function at step k, the control function at the next step, $k+1$, can be obtained as follows:

$$u^{k+1} = u^k - \tau^k J'(u^k), \tag{17.61}$$

where J' rappresents the *descent direction*, and τ^k the *acceleration parameter*. Although not necessarily the most efficient, the choice (17.61) is however pedagogically useful to understand the role played by J' and therefore that of the adjoint variable p. A method for the search of an optimal control can therefore be devised in terms of the following iterative algorithm:

1. Find the expression of the adjoint equation and of the derivative J' by either one of the two approaches (Lions or Lagrangian);
2. Provide an initial guess u^0 of the control function u;
3. Solve the *equation of state* in y using the above guess;
4. Knowing the state variable and the observation target z_d, evaluate the *cost functional* J;
5. Solve the *adjoint equation* for p, knowing y and z_d;
6. Knowing the adjoint variable, compute J';
7. If the chosen stopping test is fulfilled (up to a given *tolerance*) then exit (jump to point 10), otherwise continue;
8. Compute the parameter(s) for the acceleration of the iterative algorithm (for instance τ^k);
9. Compute the new *control function*, e.g. through equation (17.61), and return to point 3.;
10. Take the last computed variable to generate the "converged" unknowns u, y and p.

In Fig. 17.10 we display a flow chart that illustrates the above algorithm.

Remark 17.4. A convenient stopping test can be built on the measure of the distance, in a suitable norm, between the observed variable z and the (desired) target observation z_d, say

$$||z^k - z_d||_{\mathcal{Z}} \leq Tol.$$

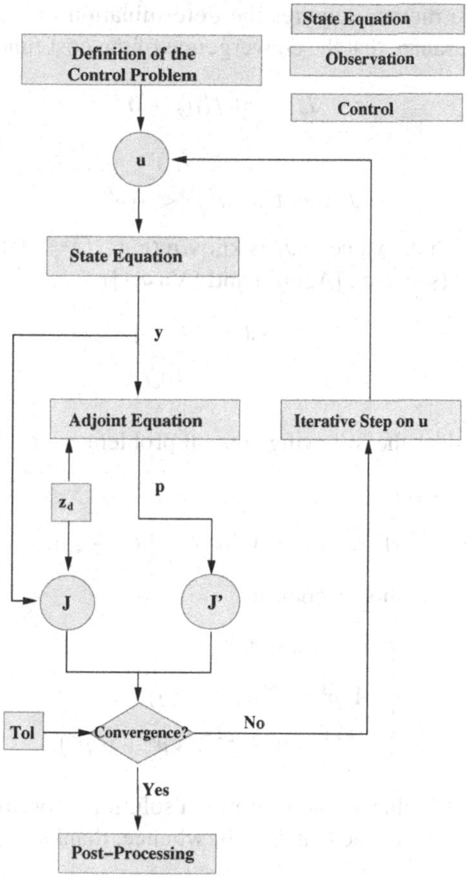

Fig. 17.10. Flow chart of a possible iterative algorithm for the solution of an optimal control problem

However, in general this does not guarantee that $J(u^k) \to 0$ as $k \to \infty$. That is to say that J might not converge to 0. A different stopping criterion is based on the evaluation of the derivative of the cost functional

$$||J'(u^k)||_{\mathcal{U}'} \leq Tol.$$

The value of the tolerance must be sufficiently small with respect to both the value of $||J'(u^0)||$ on the initial control and the proximity to the target observation that we want to achieve. •

Remark 17.5. The adjoint variable is defined on the whole computational domain. For the evaluation of $J'(u)$ it would be necessary to *restrict* the adjoint variable p on that portion of the domain, or of its boundary, on which the control function u is defined. See Tables 17.1 and 17.2 for several examples. •

The descent iterative method requires the determination of a suitable parameter τ^k. The latter should guarantee that the convergence of the cost functional to its minimum

$$J_* = \inf_{u \in \mathcal{U}} J(u) \geq 0$$

is monotone, that is

$$J(u^k - \tau^k J'(u^k)) < J(u^k).$$

In those cases in which the value of J_* is known (e.g., $J_* = 0$), then the parameter can be chosen as follows (see, e.g., [Ago03] and [Vas81])

$$\tau^k = \frac{(J(u^k) - J_*)}{||J'(u^k)||^2_{\mathcal{U}'}}. \tag{17.62}$$

As an example, consider the following control problem

$$\begin{cases} Ay = f + Bu, \\ \inf J(u), \quad \text{where } J(u) = \nu ||u||^2_{\mathcal{U}} + ||Cy - z_d||^2_{\mathcal{Z}}, \qquad \nu \geq 0. \end{cases}$$

The previous iterative method becomes:

$$\begin{cases} Ay^k = f + Bu^k, \\ A^* p^k = C'(Cy^k - z_d), \\ u^{k+1} = u^k - \tau^k 2(\nu u^k + B' p^k). \end{cases}$$

If $Ker(B'A^{*-1}C') = \{0\}$ this problem admits a solution, moreover $J(u) \to 0$ if $\nu \to 0$. Thus, if $\nu \simeq 0^+$ we can assume that $J_* \simeq 0$, whence, thanks to (17.62),

$$\tau^k = \frac{J(u^k)}{||J'(u^k)||^2_{\mathcal{U}'}} = \frac{\nu ||u^k||^2_{\mathcal{U}} + ||Cy^k - z_d||^2_{\mathcal{Z}}}{||2\nu u^k + B' p^k||^2_{\mathcal{U}'}}. \tag{17.63}$$

When considering a *discretized* optimal control problem, for instance using the Galerkin–finite element method, as we will see in Sect. 17.12, instead of looking for the minimum of $J(u)$, with $J : \mathcal{U} \mapsto \mathbb{R}$, one looks for the minimum of $J(\mathbf{u})$, where $J : \mathbb{R}^n \mapsto \mathbb{R}$ and $\mathbf{u} \in \mathbb{R}^n$ is the vector whose components are the nodal values of the control $u \in \mathcal{U}$. We will make this assumption in the remainder of this section.
As previously noted, the steepest descent method (17.61) is one among several iterative algorithms that could be used for the solution of an optimal control problem. As a matter of fact, this method is a special case of *gradient method*

$$\mathbf{u}^{k+1} = \mathbf{u}^k + \tau^k \mathbf{d}^k, \tag{17.64}$$

where $\mathbf{d}^k$ represents a descent direction, that is a vector that satisfies

$$\mathbf{d}^{k^T} \cdot J'(\mathbf{u}^k) < 0 \quad \text{if } \nabla J(\mathbf{u}^k) \neq \mathbf{0}.$$

Depending upon the criterion that is followed to choose $\mathbf{d}^k$, we obtain several special cases:

- *Newton method*, for which
$$\mathbf{d}^k = -H(\mathbf{u}^k)^{-1}\nabla J(\mathbf{u}^k),$$
where $H(\mathbf{u}^k)$ is the Hessian matrix of $J(\mathbf{u})$ computed at $\mathbf{u} = \mathbf{u}^k$;
- *quasi–Newton methods*, for which
$$\mathbf{d}^k = -B_k \nabla J(\mathbf{u}^k),$$
where B_k is an approximation of the inverse of $H(\mathbf{u}^k)$;
- *conjugate gradient method*, for which
$$\mathbf{d}^k = -\nabla J(\mathbf{u}^k) + \beta_k \mathbf{d}^{k-1},$$
where β_k is a scalar to be chosen in such a way that $\mathbf{d}^{k^T}\mathbf{d}^{k-1} = 0$. (See also Chap. 7.)

Once $\mathbf{d}^k$ is computed, the parameter τ^k should be chosen in such a way to guarantee the monotonicity property
$$J(\mathbf{u}^k + \tau^k \mathbf{d}^k) < J(\mathbf{u}^k). \tag{17.65}$$
A more stringent requirement is that the following scalar minimization problem is solved
$$\text{find } \tau^k \ : \ \phi(\tau^k) = J(\mathbf{u}^k + \tau^k \mathbf{d}^k) \text{ minimum;}$$
this would guarantee the following orthogonality property
$$\mathbf{d}^{k^T} \cdot \nabla J(\mathbf{u}^k) = 0.$$
Often, the computation of τ^k is based on heuristic methods. One way is to start from a relatively large value of the parameter τ^k, which is then halved until (17.65) is verified. However, this approach is not always successful. The idea is therefore to adopt more stringent criteria than (17.65) when choosing τ^k, with the aim of achieving on one hand a high convergence rate, and on the other avoiding too small steps. The first goal is achieved by requiring that
$$J(\mathbf{u}^k) - J(\mathbf{u}^k + \tau^k \mathbf{d}^k) \geq -\sigma \tau^k \mathbf{d}^{k^T} \cdot \nabla J(\mathbf{u}^k), \tag{17.66}$$
for a suitable $\sigma \in (0, 1/2)$. This inequality ensures that the average rate of decrease of J at $\mathbf{u}^{k+1}$ along the direction $\mathbf{d}^k$ is at least equal to a given fraction of the initial decrease rate at $\mathbf{u}^k$. On the other hand, too small steps are avoided by requiring that
$$|\mathbf{d}^{k^T} \cdot \nabla J(\mathbf{u}^k + \tau^k \mathbf{d}^k)| \leq \beta |\mathbf{d}^{k^T} \cdot \nabla J(\mathbf{u}^k)|, \tag{17.67}$$
for a suitable $\beta \in (\sigma, 1)$, so to guarantee that (17.66) is satisfied, too. In practice, $\sigma \in [10^{-5}, 10^{-1}]$ and $\beta \in [10^{-1}, 1/2]$. Several strategies can be chosen for the choice of τ^k that are compatible with conditions (17.66) and (17.67). A popular one is based on

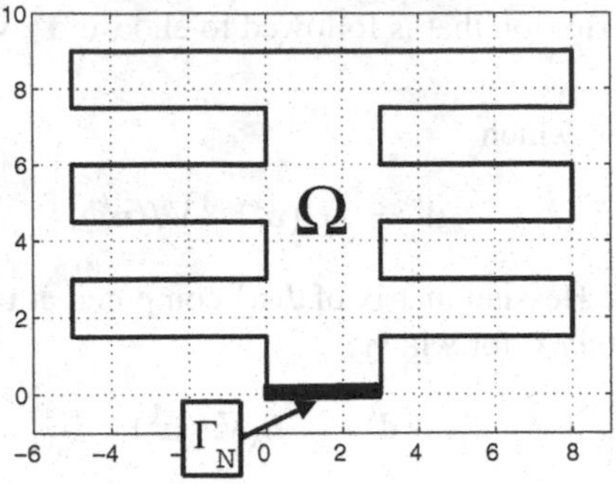

Fig. 17.11. Thermal fin: computational domain; unit measure in *mm*

the *Armijo formulae* (see, e.g. [MP01]). For fixed $\sigma \in (0,1/2)$, $\beta \in (0,1)$ and $\bar{\tau} > 0$, one chooses $\tau^k = \beta^{m_k}\bar{\tau}$, m_k being the first non-negative integer for which (17.66) is satisfied. One can even take $\tau^k = \bar{\tau}$ for all k, at least in those cases in which the evaluation of the cost functional J is very involved.
For a more comprehensive discussion on this issue see, e.g., [GMSW89, KPTZ00, MP01, NW06].

17.10 Numerical examples

In this section we illustrate two examples of control problems inspired by real life applications. Both problems are analyzed by means of the Lagrangian approach outlined in Sect. 17.8.2; for simplicity the optimal control function is in fact a scalar value.

17.10.1 Heat dissipation by a thermal fin

Thermal fins are used to dissipate the heat produced by some devices with the goal of maintaining their temperature below some limit values. Typically, they are used for electronic devices such as transistors; when active, and depending on the electrical power, the latter could incur in failure with a higher frequency when the operational temperature increases. This represents a major issue when designing the dissipator, which is often used in combination with a fan able to improve considerably the thermal dissipation via forced convection, thus controlling the temperature of the device. For further details we refer the reader e.g. to [Ç07]; for another example in the field of parametrized problems see [OP07].
In our example, we aim at regulating the intensity of the forced convection associated with the fan in order to keep the temperature of the transistor as close as possible to a desired value. The control is represented by the scalar coefficient of forced convection, while the observation is the temperature on the boundary of the thermal fin which is in contact with the transistor.

Let us consider the following state problem, whose solution y (in Kelvin degrees $[K]$) represents the temperature in the thermal fin

$$\begin{cases} -\nabla\cdot(k\nabla y)=0 & \text{in } \Omega, \\ -k\dfrac{\partial y}{\partial n}=-q & \text{on } \Gamma_N, \\ -k\dfrac{\partial y}{\partial n}=(h+U)\,(y-y_\infty) & \text{on } \Gamma_R=\partial\Omega\backslash\Gamma_N, \end{cases} \tag{17.68}$$

where the domain Ω and its boundary are reported in Fig. 17.11. The coefficient k ($[W/(mm\ K)]$) represents the thermal conductivity (aluminium is considered in this case), while h and U (our control variable) are the natural and forced convection coefficients ($[W/(mm^2\ K)]$), respectively. Let us remark that when the fan is active, the value of U is greater than zero; if $U=0$ heat dissipation is due only to natural convection. The temperature y_∞ corresponds to the temperature of the air far away from the dissipator, while q ($[W/mm^2]$) is the heat per unit of area emitted by the transistor and entering in the thermal fin through the boundary Γ_N.
The weak form of problem (17.68) reads, for a given $U\in\mathcal{U}=\mathbb{R}$

$$\text{find } y\in\mathcal{V}\ :\ a(y,\varphi;U)=b(U,\varphi) \qquad \forall\varphi\in\mathcal{V}, \tag{17.69}$$

where $\mathcal{V}=H^1(\Omega)$, $a(\varphi,\psi;\phi)=\int_\Omega k\nabla\varphi\cdot\nabla\psi\,d\Omega+\int_{\Gamma_R}(h+\phi)\varphi\psi\,d\gamma$ and $b(\phi,\psi)=\int_{\Gamma_R}(h+\phi)y_\infty\psi\,d\gamma+\int_{\Gamma_N}q\psi\,d\gamma$. Existence and uniqueness of the solution of the Robin-Neumann problem (17.69) are ensured by the Peetre-Tartar lemma (see Remark 3.5). The optimal control problem consists in finding the value of the forced convection coefficient U such that the following cost functional $J(y,U)$ is smallest, $y\in\mathcal{V}$ being the solution of (17.69)

$$J(y,U)=\nu_1\int_{\Gamma_N}(y-z_d)^2\,d\gamma+\nu_2U^2. \tag{17.70}$$

This leads to keeping the temperature of the transistor as close as possible to the desired value z_d ($[K]$) and the forced convection coefficient close to zero depending on the value of the coefficient $\nu_2>0$; in particular we assume $\nu_1=1/\int_{\Gamma_N}z_d^2\,d\gamma$ and $\nu_2=\nu_2^0/h^2$, for a suitable ν_2^0.
The analysis of the problem is carried out by means of the Lagrangian functional $\mathcal{L}(y,p,U)=J(y,U)+b(p;U)-a(y,p;U)$. In particular, we obtain via differentiation of $\mathcal{L}(\cdot)$ the following adjoint equation for a given $U\in\mathbb{R}$ and the corresponding $y=y(U)\in\mathcal{V}$

$$\text{find } p\in\mathcal{V}\ :\ a(\psi,p;U)=c(y,\psi) \qquad \forall\psi\in\mathcal{V}, \tag{17.71}$$

where $c(\varphi,\psi)=2\nu_1\int_{\Gamma_N}(\varphi-z_d)\psi\,d\gamma$. Similarly, from the optimality condition we deduce that

$$J'(U)=2\nu_2U-\int_{\Gamma_R}(y(U)-y_\infty)p(U)\,d\gamma. \tag{17.72}$$

We assume now $k=2.20\,W/(mm\,K)$, $h=15.0\cdot10^{-6}\,W/(mm^2\,K)$, $y_\infty=298.15\,K$ $(=25\,°C)$, $z_d=353.15\,K\,(=80\,°C)$ and $\nu_2^0=10^{-3}$. The problem is solved by means of

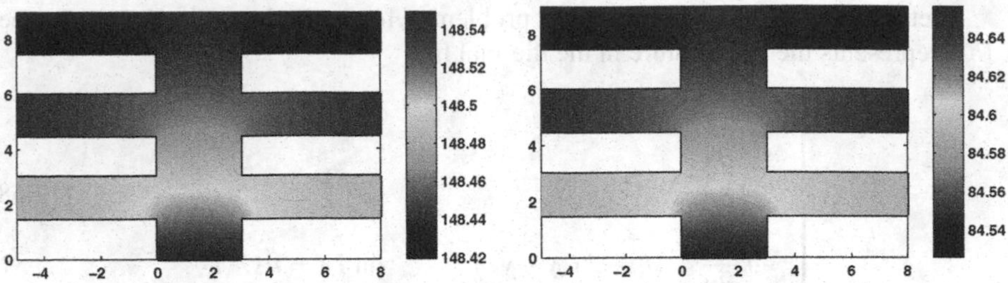

Fig. 17.12. Thermal fin: state solution (temperature [°C]) at the initial step (natural convection) (left) and at the optimum (right)

the finite element method with piecewise quadratic basis functions on a triangular mesh with 1608 elements and 934 degrees of freedom. The steepest descent iterative method is used for the optimization with $\tau^k = \tau = 10^{-9}$ (see (17.61)); the iterative procedure is stopped when $|J'(U^k)|/|J'(U^0)| < tol = 10^{-6}$. At the initial step we consider natural convection for the dissipation of the heat such that $U = 0.0$, to which corresponds a cost functional $J = 0.0377$. The optimum is reached after 132 iterations yielding the optimal cost functional $J = 0.00132$ for the optimal value of the forced convection coefficient $U = 16.1 \cdot 10^{-6}\ W/(mm^2\ K)$. Ideally, the fan should be designed in order to warrant this value of the forced convection coefficient. In Fig. 17.12 we display the state solution at the initial step and that at the optimum; we observe that the temperature on Γ_N is not equal to z_d, because $v_2^0 \neq 0$.

17.10.2 Thermal pollution in a river

Industrial activities are often related with pollution phenomena which need to be properly taken into account while designing a new plant or planning its operations. In this field, thermal pollution could often affect a river or a channel used for cooling the hot liquids produced by industrial plants, thus affecting the vital flora and fauna.
In this case the goal could consist in regulating the heat emission in a branch of a river in order to maintain the temperature of the river close to a desired threshold without considerably affecting the ideal heat emission rate of the plant.

We introduce the following state problem, whose solution y represents the temperature in the channels and branches of the river considered

$$\begin{cases} \nabla\cdot(-k\nabla y + \mathbf{V}y) = f\chi_1 + U\chi_2 & \text{in } \Omega, \\ y = 0 & \text{on } \Gamma_{IN}, \\ (-k\nabla y + \mathbf{V}y)\cdot\mathbf{n} = 0 & \text{on } \Gamma_N. \end{cases} \tag{17.73}$$

The domain Ω and the boundary Γ_{IN} are indicated in Fig. 17.13, while $\Gamma_N = \partial\Omega \backslash \Gamma_{IN}$ (note that the outflow boundary Γ_{OUT} displayed in Fig. 17.13 is part of Γ_N). χ_1, χ_2 and χ_{OBS} represent the characteristic functions of the subdomains Ω_1, Ω_2 and Ω_{OBS}, respectively. Dimensionless quantities are considered for this test case. The coefficient k is the thermal diffusivity coefficient, which also accounts for the contribution to the

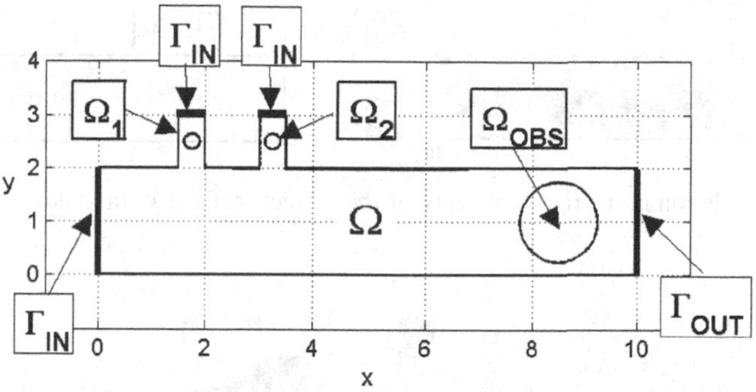

Fig. 17.13. Pollution in the river: computational domain

diffusion of turbulence phenomena, while $\mathbf{V}$ is the advection field which describes the motion of the water in the domain Ω (we comment later on the way to find it). The source term $f \in \mathbb{R}$ and the control $U \in \mathcal{U} = \mathbb{R}$ represent the heat emission rates from two industrial plants; f is given, whereas U has to be determined on the basis of the solution of the optimal control problem. In particular, we want the following cost functional to be minimized

$$J(y,U) = \int_{\Omega_{OBS}} (y - z_d)^2 \, d\Omega + \nu(U - U_d)^2, \tag{17.74}$$

where z_d is the desired temperature in Ω_{OBS}, U_d the ideal heat emission rate and $\nu > 0$ is chosen conveniently.
The optimal control problem is set up by means of the Lagrangian approach. With this aim, (17.73) is rewritten in weak form, for a given U, as

$$\text{find } y \in \mathcal{V} \ : \ a(y,\varphi) = b(U,\varphi) \qquad \forall \varphi \in \mathcal{V}, \tag{17.75}$$

where $\mathcal{V} = H^1_{\Gamma_{IN}}(\Omega)$, $a(\varphi,\psi) = \int_\Omega k \nabla\varphi \cdot \nabla\psi \, d\Omega$ and $b(U,\psi) = f \int_{\Omega_1} \psi \, d\Omega + U \int_{\Omega_2} \psi \, d\Omega$. Existence and uniqueness of the solution of problem (17.75) are proved as indicated in Sect. 3.4.
The Lagrangian functional is $\mathcal{L}(y,p,U) = J(y,U) + b(U,p) - a(y,p)$. Differentiating $\mathcal{L}(\cdot)$ in $y \in \mathcal{V}$ we obtain the adjoint equation

$$\text{find } p \in \mathcal{V} \ : \ a(\psi,p) = c(y,\psi) \qquad \forall \psi \in \mathcal{V}, \tag{17.76}$$

where $c(\varphi,\psi) = 2\int_{\Omega_{OBS}} (\varphi - z_d)\psi \, d\Omega$. Similarly, we deduce the following derivative of the cost functional

$$J'(U) = 2\nu(U - U_d) + \int_{\Omega_2} p(U) \, d\Omega. \tag{17.77}$$

We assume now $k = 0.01$, $f = 10.0$, $z_d = 0$ and $U_d = f$. The advection field $\mathbf{V}$ is deduced by solving the Navier–Stokes equations (see Chap. 16) in the domain Ω, with the following boundary conditions: on Γ_{IN} a parabolic profile is prescribed for

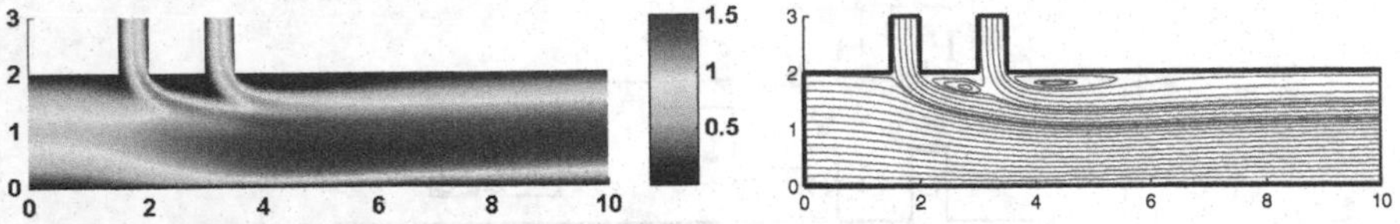

Fig. 17.14. Pollution in the river: intensity of the advection field **V**, modulus (left) and streamlines (right)

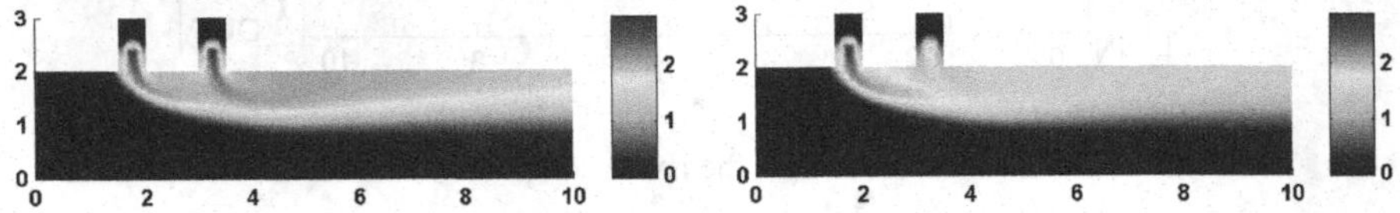

Fig. 17.15. Pollution in the river: state solution (temperature), at the initial step ($U = U_d$) (left) and at the optimum (right)

the velocity, with a maximum velocity equal to 1; on Γ_{OUT} no stress conditions are assumed in the normal direction, together with the slip condition $\mathbf{V} \cdot \mathbf{n} = 0$; finally, no-slip conditions are prescribed on $\partial\Omega \setminus (\Gamma_{IN} \cup \Gamma_{OUT})$. The notations are those displayed in Fig. 17.13. The Reynolds number is equal to $\mathbb{R}e = 500$. The Navier–Stokes problem is solved by means of the Taylor–Hood $\mathbb{P}^2$–$\mathbb{P}^1$ (see Sect. 16.4) pairs of finite elements on a mesh composed by 32248 triangles and 15989 nodes. In Fig. 17.14 we report the intensity of the advection field **V** and the corresponding streamlines.
The optimal control problem is solved by means of the finite element method with $\mathbb{P}^2$ basis functions on a triangular mesh with 32812 elements and 16771 degrees of freedom, using the steepest descent method for the functional optimization; we select $\tau^k = \tau = 5$ (see (17.61)) and the stopping criterium is $|J'(U^k)|/|J'(U^0)| < tol = 10^{-6}$. The advection field **V** obtained by solving the Navier–Stokes equations is interpolated on this new mesh. At the initial step we assume that $U = U_d$, thus obtaining a cost functional $J = 1.884$. The optimal solution is obtained after 15 iterations, the corresponding optimal cost functional is $J = 1.817$ obtained for an optimal heat emission rate of $U = 6.685$. In practice the heat from the plant in Ω_2 should be reduced in order to maintain the temperature in Ω_{OBS} low. In Fig. 17.15 we report the state solutions y (temperature) before and after optimization.

17.11 A few considerations about observability and controllability

A few considerations can be made on the behaviour of iterative methods with respect to the kind of optimal control problem that we are solving, particularly on which kind of variable z we are observing, and which kind of control function u we are using. Briefly, on the relationship between *observability* and *controllability*. We warn the reader that the conclusions that we are going to draw are not supported by a general theory, nor they apply to any kind of numerical discretization method.

- *Where we observe.* In general, optimal control problems based on an observation variable distributed in the domain enjoy higher convergence rate than those for which the observation variable is concentrated on the domain boundary. Within the same discretization error tolerance, in the former case the number of iterations can be up to one order of magnitude lower than in the latter.
- *Where we control.* In general, the optimization process is more robust if also the control function is distributed in the domain (as a source term to the state equation, or as coefficient of the differential operator governing the state equation), rather than being concentrated on the domain boundary. More precisely, the convergence rate is higher and its sensitivity to the choice of the acceleration parameter lower for distributed control problems than for boundary control ones, provided of course all other parameters are the same.
- *What we observe.* Also the kind of variable that we observe affects the convergence behaviour of the iterative scheme. For instance, observing the state variable is less critical than observing either its gradients or some of its higher-order derivatives. The latter circumstance occurs quite commonly. e.g., in fluid dynamics problems, when for potential problems one observes the velocity field, or for Navier-Stokes equations one observes the fluid vorticity or its stresses.
- *Shape optimization.* Shape optimization problems are a special class of optimal control problems: as a matter of fact, in this case the control function is not only *on* the boundary, it is *the* boundary itself. The cost functional to be minimized is called *shape functional* as it depends on the domain itself. One looks for

$$J(\Omega_{opt}) \leq J(\Omega) \qquad \forall \Omega \in \mathcal{D}_{ad},$$

  where $\mathcal{D}_{ad}$ is a set of admissible domains. Shape optimization problems are difficult to analyze theoretically and hard to solve numerically. The numerical grid needs to be changed at every iteration. Besides, non-admissible boundary shapes might be generated in the course of the iterations, unless additional geometrical constraints are imposed. Moreover, special stabilization and regularization techniques might be necessary to prevent numerical oscillations in the case of especially complex problems. More in general, shape optimization problems are more sensitive to the variation of the various parameters that characterize the control problem.
- *Adjoint problem and state problem.* For steady elliptic problems like those considered in this chapter, the use of the adjoint problem provides the gradient of the cost functional at the same computational cost of the state problem. This approach can be considered as an alternative to those based on inexact or automatic differentiation of the cost functional. In the case of shape optimization problems the use of the adjoint problem allows a computational saving with respect to the method based on the shape sensitivity analysis, as the latter depends on the (often prohibitive) number of parameters that characterize the shape (the control points). See, e.g., [KAJ02].

17.12 Two alternative paradigms for numerical approximation

Let us start by considering a simple example that illustrates some additional difficulties that arise when solving an optimal control problem numerically. For a more insightful analysis of the numerical discretization of optimal control problems, see, e.g., [FCZ03, Gun03].
Consider again the state equation (17.1) and assume that the optimal control problem reads

"find $u \in \mathcal{U}_{ad}$ such that

$$J(u) \le J(v) \qquad \forall v \in \mathcal{U}_{ad}, \tag{17.78}$$

where J is a given cost functional".

Now the question is

"How can this problem be conveniently approximated?"

As already anticipated at the beginning of Sect. 17.8, at least two alternative strategies can be pursued:

1) **Discretize–then–optimize**
According to this strategy, we discretize first the control space $\mathcal{U}_{ad}$ by a finite dimensional space $\mathcal{U}_{ad,h}$ and the state equation (17.1) by a discrete equation written for short, as

$$A_h y_h(u_h) = f_h. \tag{17.79}$$

In a finite element context, the parameter h would denote the finite element gridsize. We assume that the choice of the discrete control space and the discretized state equation are such that a "discrete state" $y_h(v_h)$ exists for every "admissible" discrete control $v_h \in \mathcal{U}_{ad,h}$.
At this stage we look for a discrete optimal control, that is a function $u_h \in \mathcal{U}_{ad,h}$ such that

$$J(u_h) \le J(v_h) \qquad \forall v_h \in \mathcal{U}_{ad,h}, \tag{17.80}$$

or, more precisely,

$$J(y_h(u_h), u_h) \le J(y_h(v_h), v_h) \qquad \forall v_h \in \mathcal{U}_{ad,h}. \tag{17.81}$$

This corresponds to the following scheme

MODEL $\longrightarrow$ DISCRETIZATION $\longrightarrow$ CONTROL,

for which the "*discretize–then–optimize*" expression was coined.

2) **Optimize–then–discretize**
Alternatively we could proceed as follows. We start from the control problem (17.1), (17.78) and we write down the corresponding *optimality system* based on the Euler–Lagrange equations :

$$\begin{aligned} Ay(u) &= f, \\ A'p &= G(y(u)), \end{aligned} \tag{17.82}$$

for a suitable G which depends on the state $y(u)$ and represents the right-hand side of the adjoint problem, plus an additional equation (formally corresponding to the third equation of (17.56)) relating the three variables y, p and u

$$Q(y,p,u) = 0. \tag{17.83}$$

At this stage we discretize system (17.82), (17.83) and solve it numerically. This corresponds to the following procedure:

$$\text{MODEL} \longrightarrow \text{CONTROL} \longrightarrow \text{DISCRETIZATION},$$

for which the expression "*optimize–then–discretize*" is used. With respect to the former approach, here we have swapped the last two steps.

The two strategies do not necessarily yield the same results. For instance, in [IZ99] it is shown that if the state equation is a dynamic problem that describes the vibrations of an elastic structure and a finite element approximation is used, then the first strategy yields wrong results. This can be attributed to the lack of accuracy of the finite element method for high frequency solutions of the wave equation (see [Zie00]).
At the same time it was also observed that for several shape optimization problems in optimal design, the former strategy should be preferred; see. e.g., [MP01, Pir84].

The strategy of choice certainly depends on the nature of the differential problem at hand. In this respect, control problems governed by elliptic or parabolic PDEs are less problematic than those governed by hyperbolic equations because of their intrinsic dissipative nature. See for a discussion [Zua03]. The reader should however keep abreast of the many important developments expected in this field in the coming years.

17.13 A numerical approximation of an optimal control problem for advection–diffusion equations

In this section we consider an optimal control problem for an advection–diffusion equation formulated with the Lagrangian approach. For its numerical discretization the two different strategies: "discretize–then–optimize" and "optimize–then–discretize" will be considered. The numerical approximation will be based on stabilized finite element methods, as seen in Chapter 12. Besides, an a posteriori error analysis will be carried out, according to the guidelines illustrated in Chapter 4. For more details we refer to [QRDQ06, DQ05], and the references therein.

We consider the following advection-diffusion boundary-value problem

$$\begin{cases} L(y) = -\nabla\cdot(\mu\nabla y) + \mathbf{V}\cdot\nabla y = u & \text{in } \Omega, \\ y = 0 & \text{on } \Gamma_D, \\ \mu\dfrac{\partial y}{\partial n} = 0 & \text{on } \Gamma_N, \end{cases} \tag{17.84}$$

where Ω is a two-dimensional domain, Γ_D and Γ_N provide a disjoint partition of the domain boundary $\partial\Omega$, $u \in L^2(\Omega)$ is the control variable while μ and $\mathbf{V}$ are two given functions (the former being a positive viscosity). Here $\Gamma_D = \{\mathbf{x} \in \partial\Omega : \mathbf{V}(\mathbf{x}) \cdot \mathbf{n}(\mathbf{x}) < 0\}$ is the inflow boundary, $\mathbf{n}(\mathbf{x})$ is the outward unit normal, while $\Gamma_N = \partial\Omega \setminus \Gamma_D$ is the outflow boundary.
We assume that the observation function is restricted to a subdomain $D \subseteq \Omega$ and that the optimal control problem reads

$$\text{find } u \ : \ J(u) = \int_D (gy(u) - z_d)^2 \, dD \quad \text{minimum}, \tag{17.85}$$

where $g \in C^\infty(\Omega)$ maps the variabile y into the space of observations, and z_d is the desired observation (the target). By setting

$$\mathcal{V} = H^1_{\Gamma_D} = \{v \in H^1(\Omega) : v_{|\Gamma_D} = 0\} \ \text{and} \ \mathcal{U} = L^2(\Omega),$$

the Lagrangian functional introduced in Sect. 17.8 becomes

$$\mathcal{L}(y, p, u) = J(u) + F(p; u) - a(y, p), \tag{17.86}$$

where:

$$a(\varphi, \psi) = \int_\Omega \mu \nabla\varphi \cdot \nabla\psi \, d\Omega + \int_\Omega \mathbf{V} \cdot \nabla\varphi \, \psi \, d\Omega, \tag{17.87}$$

$$F(\varphi; u) = \int_\Omega u\varphi \, d\Omega. \tag{17.88}$$

By differentiating $\mathcal{L}$ with respect to the state variable y, we obtain the adjoint equation (in weak form)

$$\text{find } p \in \mathcal{V} \ : \ a^{ad}(p, \psi) = F^{ad}(\psi; \varphi) \qquad \forall \psi \in \mathcal{V}, \tag{17.89}$$

where:

$$a^{ad}(p, \psi) = \int_\Omega \mu \nabla p \cdot \nabla\psi \, d\Omega + \int_\Omega \mathbf{V} \cdot \nabla\psi \, p \, d\Omega, \tag{17.90}$$

$$F^{ad}(\psi; y) = \int_D 2(g\, y - z_d)\, g\, \psi \, dD. \tag{17.91}$$

Its differential (distributional) counterpart reads

$$\begin{cases} L^{ad}(p) = -\nabla \cdot (\mu \nabla p + \mathbf{V} p) = \chi_D g\, (g\, y - z_d) & \text{in } \Omega, \\ p = 0 & \text{on } \Gamma_D, \\ \mu \dfrac{\partial p}{\partial n} + \mathbf{V} \cdot \mathbf{n}\, p = 0 & \text{on } \Gamma_N, \end{cases} \tag{17.92}$$

where χ_D denotes the characteristic function of the region D. By differentiating $\mathcal{L}$ with respect to the control function u, we obtain the optimality equation (see the third equation of (17.56)), that is

$$\int_\Omega \phi \, p \, d\Omega = 0 \qquad \forall \phi \in L^2(\Omega). \tag{17.93}$$

This equation provides the sensitivity $J'(u)$ of the cost functional with respect to the control variable. Denoting for simplicity this sensitivity by δu, in this case we obtain $\delta u = p(u) = p$. Finally, by differentiating $\mathcal{L}$ with respect to the adjoint variable p, as usual we obtain the state equation (in weak form)

$$\text{find } y \in \mathcal{V} \;:\; a(y,\varphi) = F(\varphi;u) \qquad \forall \varphi \in \mathcal{V}. \tag{17.94}$$

17.13.1 The strategies "optimize–then–discretize" and "discretize–then–optimize"

From a numerical viewpoint the minimization algorithm introduced in Sect. 17.9 requires, at every step, the numerical approximation of both the state and the adjoint boundary-value problems. For both problems we can use the stabilized Galerkin-Least-Squares finite element formulations introduced in Sect. 12.8.6. The corresponding discretized equations respectively read:

$$\text{find } y_h \in \mathcal{V}_h \;:\; a(y_h,\varphi_h) + \bar{s}_h(y_h,\varphi_h) = F(\varphi_h;u_h) \qquad \forall \varphi_h \in \mathcal{V}_h, \tag{17.95}$$

$$\bar{s}_h(y_h,\varphi_h) = \sum_{K\in\mathcal{T}_h} \int_K \delta_K R(y_h;u_h)\, L(\varphi_h)\, dK, \tag{17.96}$$

$$\text{find } p_h \in \mathcal{V}_h \;:\; a^{ad}(p_h,\psi_h) + \bar{s}_h^{ad}(p_h,\psi_h) = F^{ad}(\psi_h;y_h) \qquad \forall \psi_h \in \mathcal{V}_h, \tag{17.97}$$

$$\bar{s}_h^{ad}(p_h,\psi_h) = \sum_{K\in\mathcal{T}_h} \int_K \delta_K R^{ad}(p_h;y_h)\, L^{ad}(\psi_h)\, dK, \tag{17.98}$$

where δ_K is a stabilization parameter, $R(y;u) = L(y) - u$, $R^{ad}(p;y) = L^{ad}(p) - G(y)$, with $G(y) = 2\chi_D g\,(g\,y - z_d)$. This is the paradigm "*optimize–then–discretize*"; see Sect. 17.12 and, for the specific problem at hand, [Bec01, CH01, Gun03].
In the paradigm "*discretize–then–optimize*" , the one that we will adopt in the following, we first discretize (by the same stabilized GLS formulation (Eq. (17.95) and (17.96)), and then introduce the discrete Lagrangian functional

$$\mathcal{L}_h(y_h,p_h,u_h) = J(y_h,u_h) + F(p_h;u_h) - a(y_h,p_h) - \bar{s}_h(y_h,p_h). \tag{17.99}$$

At this stage, by differentiating with respect to y_h, we obtain the discrete adjoint equation (17.97), however this time the stabilization term is different, precisely

$$\bar{\bar{s}}_h^{ad}(p_h,\psi_h) = \sum_{K\in\mathcal{T}_h} \int_K \delta_K L(\psi_h)\, L(p_h)\, dK. \tag{17.100}$$

Now, by differentiating $\mathcal{L}_h$ with respect to u_h and using the Riesz representation theorem (Theorem 2.1), we obtain, noting that $u_h \in \mathcal{V}_h$,

$$\delta u_h = p_h + \sum_{K\in\mathcal{T}_h} \int_K \delta_K L(p_h)\, dK.$$

In particular, the new stabilized Lagrangian reads [DQ05]

$$\mathcal{L}_h^s(y_h, p_h, u_h) = \mathcal{L}(y_h, p_h, u_h) + S_h(y_h, p_h, u_h), \tag{17.101}$$

where we have set

$$S_h(y, p, u) = \sum_{K \in \mathcal{T}_h} \int_K \delta_K R(y; u)\, R^{ad}(p; y)\, dK. \tag{17.102}$$

By differentiating $\mathcal{L}_h^s$ we obtain the new discretized state and adjoint problems, which can still be written as in (17.95) and (17.97), however this time the stabilization terms read, respectively, as follows:

$$s_h(y_h, \varphi_h; u_h) = -\sum_{K \in \mathcal{T}_h} \int_K \delta_K R(y_h; u_h)\, L^{ad}(\varphi_h)\, dK, \tag{17.103}$$

$$s_h^{ad}(p_h, \psi_h; y_h) = -\sum_{K \in \mathcal{T}_h} \int_K \delta_K \left(R^{ad}(p_h; y_h)\, L(\psi_h) - R(y_h; u_h)\, G'(\psi_h) \right) dK, \tag{17.104}$$

having set $G'(\psi) = 2\chi_D g^2 \psi$. Finally, the sensitivity of the cost functional becomes now

$$\delta u_h(p_h, y_h) = p_h - \sum_{K \in \mathcal{T}_h} \delta_K\, R^{ad}(p_h; y_h). \tag{17.105}$$

17.13.2 A posteriori error estimates

With the aim of obtaining a suitable a posteriori error estimate for the optimal control problem we shall use as error indicator the error on the cost functional, as done in [BKR00]. Moreover, we will split this error into two terms, that we will identify as *iteration error* and *discretization error*. In particular, for the discretization error we will make use of duality principle advocated in [BKR00] for the grid adaptivity.

Iteration error and discretization error

At every step k of the iterative algorithm for the minimization of the cost functional we consider the error

$$\varepsilon^{(k)} = J(y^*, u^*) - J(y_h^k, u_h^k), \tag{17.106}$$

where the symbol $*$ identifies the variables corresponding to the optimal value of the control, while y_h^k denotes the discrete state variable at step k. (The variables y_h^k and u_h^k have a similar meaning.) We call *discretization error* $\varepsilon_D^{(k)}$ [DQ05] the component of the total error $\varepsilon^{(k)}$ arising from step k, and *iteration error* $\varepsilon_{IT}^{(k)}$ [DQ05] the component of $\varepsilon^{(k)}$ that represents the difference between the value of the cost functional computed on the exact variables at step k and the value $J^* = J(y^*, u^*)$ of the cost functional at the optimum. In conclusion, the total error $\varepsilon^{(k)}$ (17.106) can be written as

$$\varepsilon^{(k)} = \left(J(y^*, u^*) - J(y^k, u^k) \right) + \left(J(y^k, u^k) - J(y_h^k, u_h^k) \right) = \varepsilon_{IT}^{(k)} + \varepsilon_D^{(k)}. \tag{17.107}$$

In the following we will apply an a posteriori error estimate only on $\varepsilon_D^{(k)}$, that is the only part of $\varepsilon^{(k)}$ that can be reduced by a grid refinement procedure. Since the gradient of $\mathcal{L}(\mathbf{x}), \mathbf{x}=(y,p,u)$, is linear in $\mathbf{x}$, when using algorithm (17.61) with $\tau^k=\tau=$ constant, the iteration error $\varepsilon_{IT}^{(k)}$ becomes $\varepsilon_{IT}^{(k)}=\frac{1}{2}\,(\,\delta u(p^k,u^k)\;,\;u^*-u^k\,)$, which in the current case becomes ([DQ05])

$$\varepsilon_{IT}^{(k)}=-\frac{1}{2}\tau\|p^k\|^2_{L^2(\Omega)}-\frac{1}{2}\tau\sum_{r=k+1}^{\infty}(\,p^k,p^r\,)_{L^2(\Omega)}. \tag{17.108}$$

Since the iteration error cannot be exactly defined by this formula, we will approximate $\varepsilon_{IT}^{(k)}$ as

$$|\varepsilon_{IT}^{(k)}|\approx\frac{1}{2}\tau\|p^k\|^2_{L^2(\Omega)},$$

or, more simply,

$$|\varepsilon_{IT}^{(k)}|\approx\|p^k\|^2_{L^2(\Omega)},$$

which yields the usual stopping criterium

$$|\varepsilon_{IT}^{(k)}|\approx\|\delta u(p^k)\|_{L^2(\Omega)}. \tag{17.109}$$

In practice, $\varepsilon_{IT}^{(k)}$ is computed on the discrete variables, that is $|\varepsilon_{IT}^{(k)}|\approx\|\delta u_h(p_h^k)\|$. Suppose that at a given iteration k the grid is adaptively refined, and denote with $\mathbf{x}_h=(y_h,p_h,u_h)$ the variables computed on the old grid (before the refinement) $\mathcal{T}_h$, and with $\mathbf{x}_{h,ref}=(y_h^{ref},p_h^{ref},u_h^{ref})$ those of the refined grid $\mathcal{T}_{h,ref}$. Then at step k the discretization error associated with the grid $\mathcal{T}_{h,ref}$ is lower than the one associated to $\mathcal{T}_h$. However, the discretization error $\varepsilon_{IT}^{(k)}$ computed on $\mathbf{x}_{h,ref}$, is lower than the iteration error computed on $\mathbf{x}_h$.

A posteriori error estimate and adaptive strategy

The a posteriori error estimate for the discretization error $\varepsilon_D^{(k)}$ can be characterized as follows ([DQ05]).

Theorem 17.5. *For the linear advection-diffusion control problem under exam, with stabilized Lagrangian $\mathcal{L}_h^s$ (Eq. (17.101) and Eq. (17.102)), the discretization error at step k of the iterative optimization algorithm can be written as*

$$\varepsilon_D^{(k)}=\frac{1}{2}(\,\delta u(p^k,u^k),u^k-u_h^k\,)+\frac{1}{2}\nabla\mathcal{L}_h^s(\mathbf{x}_h^k)\cdot(\mathbf{x}^k-\mathbf{x}_h^k)+\Lambda_h(\mathbf{x}_h^k), \tag{17.110}$$

where $\mathbf{x}_h^k=(y_h^k,p_h^k,u_h^k)$ represents the GLS linear finite element solution, $\Lambda_h(\mathbf{x}_h^k)=S_h(\mathbf{x}_h^k)+s_h(y_h^k,p_h^k;u_h^k)$, and $s_h(w_h^k,p_h^k;u_h^k)$ is the stabilization term (17.103).

By adapting (17.110) to the specific problem at hand and expressing the contributions on the different finite elements $K \in \mathcal{T}_h$ ([BKR00]), the following estimate can be obtained

$$|\varepsilon_D^{(k)}| \leq \eta_D^{(k)} = \frac{1}{2} \sum_{K \in \mathcal{T}_h} \{ \, (\omega_K^p \rho_K^y + \omega_K^y \rho_K^p + \omega_K^u \rho_K^u) + \lambda_K \, \}, \tag{17.111}$$

where:

$$\begin{aligned}
&\rho_K^y = \|R(y_h^k; u_h^k)\|_K + h_K^{-\frac{1}{2}} \|r(y_h^k)\|_{\partial K},\\
&\omega_K^p = \|(p^k - p_h^k) - \delta_K L^{ad}(p^k - p_h^k) + \delta_K G'(y^k - y_h^k)\|_K + h_K^{\frac{1}{2}} \|p^k - p_h^k\|_{\partial K},\\
&\rho_K^p = \|R^{ad}(p_h^k; y_h^k)\|_K + h_K^{-\frac{1}{2}} \|r^{ad}(p_h^k)\|_{\partial K},\\
&\omega_K^y = \|(y^k - y_h^k) - \delta_K L(y^k - y_h^k)\|_K + h_K^{\frac{1}{2}} \|y^k - y_h^k\|_{\partial K},\\
&\rho_K^u = \|\delta u_h(p_h^k, y_h^k) + \delta u(p^k)\|_K = \|p^k + p_h^k - \delta_K R^{ad}(p_h^k; y_h^k)\|_K,\\
&\omega_K^u = \|u^k - u_h^k\|_K,\\
&\lambda_K = 2\delta_K \|R(y_h^k; u_h^k)\|_K \, \|G(y_h^k)\|_K,\\
&r(y_h^k) \; = \begin{cases} -\dfrac{1}{2}\left[\mu \dfrac{\partial y_h^k}{\partial n}\right], & \text{on } \partial K \backslash \partial \Omega,\\[2ex] -\mu \dfrac{\partial y_h^k}{\partial n}, & \text{on } \partial K \in \Gamma_N, \end{cases}\\
&r^{ad}(p_h^k) = \begin{cases} -\dfrac{1}{2}\left[\mu \dfrac{\partial p_h^k}{\partial n} + \mathbf{V}\cdot\mathbf{n}\; p_h^k\right], & \text{on } \partial K \backslash \partial \Omega,\\[2ex] -\left(\mu \dfrac{\partial p_h^k}{\partial n} + \mathbf{V}\cdot\mathbf{n}\; p_h^k\right), & \text{on } \partial K \in \Gamma_N. \end{cases}
\end{aligned} \tag{17.112}$$

As usual, ∂K denotes the boundary of $K \in \mathcal{T}_h$, and $[\cdot]$ the jump operator across ∂K.

For a practical use of estimate (17.111) it is necessary to evaluate y^k, p^k and u^k. With this in mind we replace y^k and p^k by their quadratic reconstructions $(y_h^k)^q$ and $(p_h^k)^q$, while u^k is replaced by $(u_h^k)^q = u_h^k - \tau(\delta u_h((p_h^k)^q, (y_h^k)^q) - \delta u_h(p_h^k, y_h^k))$, according to the steepest descent method with constant acceleration parameter $\tau^k = \tau$. Consider the following adaptive strategy for the iterative optimization algorithm:

1. use a coarse grid and iterate until the tolerance on the iterative error Tol_{IT} is achieved;
2. adapt the grid, by equi-distributing the error on the different elements $K \in \mathcal{T}_h$, according to the estimate $\eta_D^{(k)}$ (17.111), until convergence on the discretization error within a tolerance Tol_D;
3. re-evaluate of the variables as well as $\varepsilon_{IT}^{(k)}$ on the refined grid: return to point 1 and repeat the procedure if $\varepsilon_{IT}^{(k)} \geq Tol_{IT}$, stop the algorithm if $\varepsilon_{IT}^{(k)} < Tol_{IT}$.

17.13.3 A test problem on control of pollutant emission

As an example we are going to apply the a posteriori estimate (17.111) on the discretization error $\eta_D^{(k)}$ and the strategy illustrated in Sect. 17.13.2 to a test case on emission of pollutants into the atmosphere. The specific problem is how to regulate the emission from industrial chimneys so to keep the pollutant's concentration in a certain critical area below a prescribed admissible threshold.

With this aim we consider a state equation that is given by a quasi–3D advection–diffusion boundary–value problem [DQ05]. The pollutant concentration y at the source (the emission height) $\mathrm{z} = H$ is described by (17.84), while the concentration at ground level is obtained by applying the projection function $g(\mathrm{x},\mathrm{y})$. The form of the diffusion coefficients $\mu(\mathrm{x},\mathrm{y})$ and $g(\mathrm{x},\mathrm{y})$ depends both on the distance H of the source from the ground, and on the class of atmospheric stability (stable, neutral or unstable). We will refer to a neutral atmosphere and, with reference to the domain illustrated in Fig. 17.16, we assume that the wind field is $\mathbf{V} = V_{\mathrm{x}}\hat{\mathrm{x}} + V_{\mathrm{y}}\hat{\mathrm{y}}$, with $V_{\mathrm{x}} = V\cos(\frac{\pi}{30})$ and $V_{\mathrm{y}} = V\sin(\frac{\pi}{30})$, and with wind intensity $V = 2.5\ m/s$. Moreover, we assume that there are three chimneys (represented by the three aligned small circles in Fig.17.16), all at the same height $H = 100\ m$, and that the maximum discharge allowed from every chimney is $u_{\max} = 800\ g/s$. We assume that the pollutant emitted be SO_2 and we fix at $z_d = 100\ \mu g/m^3$ (the target observation in our control problem) the desired concentration in the region of observation, a circular region of the computational domain that we indicate by D in Fig.17.16. In (17.84) we have considered the case of a distributed control u on the whole computational domain Ω, whereas in the current example $u = \sum_{i=1}^{N} u_i\chi_i$, where χ_i is the characteristic function of the (tiny) region U_i which represents the location of the i–th chimney.

In Fig.17.17(a) we display the concentration at ground level corresponding to the highest possible discharge ($u_{max} = 800\ g/s$) from each chimney, while in Fig.17.17(b) we display the concentration that we have after having applied the optimal control procedure (the cost functional being the square of the $L^2(D)$ norm of the distance from the target concentration z_d). We observe that the "optimal" emission rates become $u_1 = 0.0837 \cdot u_{\max}$, $u_2 = 0.0908 \cdot u_{\max}$ and $u_3 = 1.00 \cdot u_{\max}$.

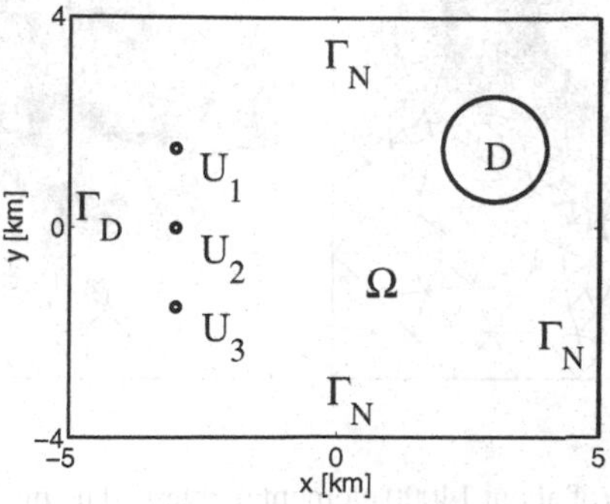

Fig. 17.16. Computational domain for the test problem on pollutant control

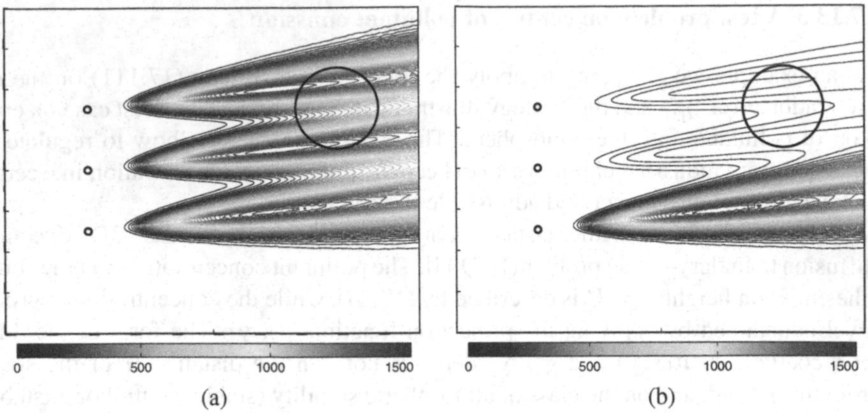

Fig. 17.17. Pollutant concentration measured in $[\mu g/m^3]$ at the ground level: (a) before and (b) after regulating the emissions from the chimneys

In Fig.17.18(a) we report the grid obtained by the a posteriori estimate on $\eta_D^{(k)}$, whereas in Fig.17.18(b) the one obtained by the following *energy norm* error indicator $(\eta_E)^{(k)} = \sum_{K\in\mathcal{T}_h} h_K\, \rho_K^y$. The symbols adopted are those of (17.112).

These results are then compared with those that are obtained with a very fine grid of about 80000 elements. The grid adaptivity driven by the error indicator $\eta_D^{(k)}$ tends to concentrate nodes in those areas that are more relevant for the optimal control. This is confirmed by comparing the errors on the cost functional using the same number of gridpoints. For instance, the indicator $\eta_D^{(k)}$ yields an error of about 20%, against the 55% that would be obtained using the indicator $(\eta_E)^{(k)}$ on a grid of about 4000 elements, while on a grid of about 14000 elements it would be 6% against 15%.

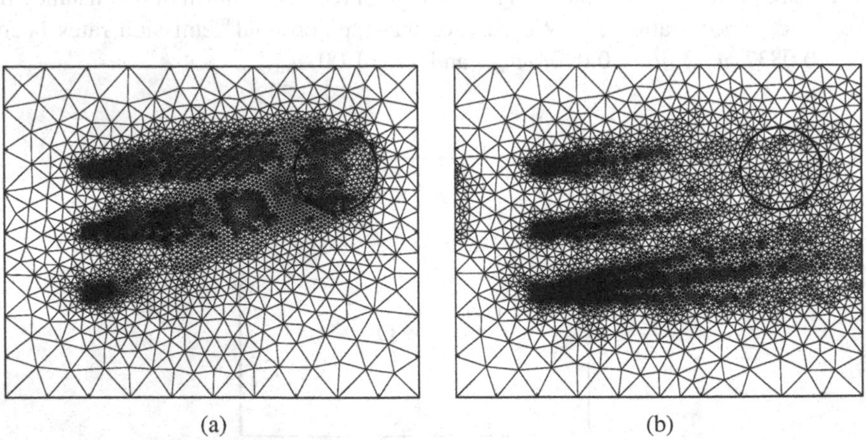

Fig. 17.18. Adapted grids (of about 14000 elements) obtained using the error indicator $\eta_D^{(j)}$ (see (17.111)) (a) and $(\eta_E)^{(j)}$)

17.14 Exercises

1. Consider the optimal control problem with boundary control

$$\begin{cases} -\nabla\cdot(\alpha\nabla y)+\beta\cdot\nabla y+\gamma y=f & \text{in } \Omega=(0,1)^2, \\ \dfrac{\partial y}{\partial n}=u & \text{on } \partial\Omega, \end{cases} \tag{17.113}$$

$u\in L^2(\Omega)$ being the control function and $f\in L^2(\Omega)$ a given function. Consider the cost functional

$$J(u)=\frac{1}{2}\|\eta y-z_d\|^2_{L^2(\Omega)}+\nu\|u\|^2_{L^2(\partial\Omega)}, \tag{17.114}$$

with $\eta\in L^\infty(\Omega)$.
Write the equations (equation of state, adjoint equation and equation of optimality) of the optimal control problem (17.113)-(17.114) based on the Lagrangian approach, and those based on Lions' approach.

2. Consider the optimal control problem

$$\begin{cases} -\nabla\cdot(\alpha\nabla y)+\beta\cdot\nabla y+\gamma y=f+cu & \text{in } \Omega=(0,1)^2, \\ \dfrac{\partial y}{\partial n}=g & \text{on } \partial\Omega, \end{cases} \tag{17.115}$$

where $u\in L^2(\Omega)$ is a distributed control, c a given constant, $f\in L^2(\Omega)$ and $g\in H^{-1/2}(\partial\Omega)$ two given functions. Consider the cost functional

$$J(u)=\frac{1}{2}\|\eta y-z_d\|^2_{L^2(\Omega)}+\nu\|u\|^2_{L^2(\Omega)}, \tag{17.116}$$

with $\eta\in L^\infty(\Omega)$.
Find the formulation of the optimal control problem (17.115)–(17.116) by the Lagrangian-based approach, then the one based on Lions' formulation.

18

Domain decomposition methods

In this chapter we will introduce the domain decomposition method (DD, in short). In its most common version, DD can be used in the framework of any discretization method for partial differential equations (such as, e.g. finite elements, finite volumes, finite differences, or spectral element methods) to make their algebraic solution more efficient on parallel computer platforms. In addition, DD methods allow the reformulation of any given boundary-value problem on a partition of the computational domain into subdomains. As such, it provides a very convenient framework for the solution of *heterogeneous* or multiphysics problems, i.e. those that are governed by differential equations of different kinds in different subregions of the computational domain.

The basic idea behind DD methods consists in subdividing the computational domain Ω, on which a boundary-value problem is set, into two or more subdomains on which discretized problems of smaller dimension are to be solved, with the further potential advantage of using parallel solution algorithms. More in particular, there are two ways of subdividing the computational domain into subdomains: one with disjoint subdomains, the others with overlapping subdomains (for an example, see Fig. 18.1). Correspondingly, different DD algorithms will be set up.
For reference lectures on DD methods we refer to [BGS96, QV99, TW05].

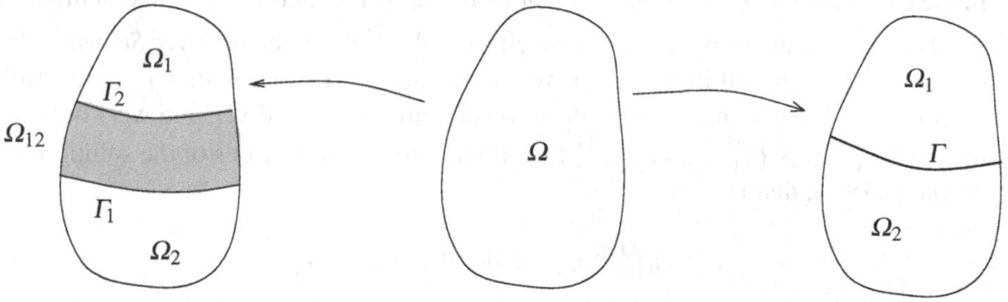

Fig. 18.1. Two examples of subdivision of the domain Ω, with and without overlap

A. Quarteroni: *Numerical Models for Differential Problems*, 2nd Ed.
MS&A – Modeling, Simulation & Applications 8
DOI 10.1007/978-88-470-5522-3_18, © Springer-Verlag Italia 2014

18.1 Some classical iterative DD methods

In this section we introduce four different iterative schemes starting from the model problem: find $u : \Omega \to \mathbb{R}$ such that

$$\begin{cases} Lu = f & \text{in } \Omega, \\ u = 0 & \text{on } \partial\Omega, \end{cases} \tag{18.1}$$

L being a generic second order elliptic operator, whose weak formulation reads

$$\text{find } u \in V = H_0^1(\Omega) \; : \; a(u,v) = (f,v) \quad \forall v \in V, \tag{18.2}$$

being $a(\cdot,\cdot)$ the bilinear form associated with L.

18.1.1 Schwarz method

Consider a decomposition of the domain Ω in two subdomains Ω_1 and Ω_2 such that $\overline{\Omega} = \overline{\Omega}_1 \cup \overline{\Omega}_2$, $\Omega_1 \cap \Omega_2 = \Omega_{12} \neq \emptyset$ (see Fig. 18.1) and let $\Gamma_i = \partial\Omega_i \setminus (\partial\Omega \cap \partial\Omega_i)$.

Consider the following iterative method. Given $u_2^{(0)}$ on Γ_1, solve the following problems for $k \geq 1$:

$$\begin{cases} Lu_1^{(k)} = f & \text{in } \Omega_1, \\ u_1^{(k)} = u_2^{(k-1)} & \text{on } \Gamma_1, \\ u_1^{(k)} = 0 & \text{on } \partial\Omega_1 \setminus \Gamma_1, \end{cases} \tag{18.3}$$

$$\begin{cases} Lu_2^{(k)} = f & \text{in } \Omega_2, \\ u_2^{(k)} = \begin{cases} u_1^{(k)} \\ u_1^{(k-1)} \end{cases} & \text{on } \Gamma_2, \\ u_2^{(k)} = 0 & \text{on } \partial\Omega_2 \setminus \Gamma_2. \end{cases} \tag{18.4}$$

In the case in which one chooses $u_1^{(k)}$ on Γ_2 in (18.4) the method is named *multiplicative Schwarz*, whereas that in which we choose $u_1^{(k-1)}$ is named *additive Schwarz*. The reason will be clarified in Sect.18.6. We have thus two elliptic boundary-value problems with Dirichlet conditions for the two subdomains Ω_1 and Ω_2, and we would like the two sequences $\{u_1^{(k)}\}$ and $\{u_2^{(k)}\}$ to converge to the restrictions of the solution u of problem (18.1), that is

$$\lim_{k\to\infty} u_1^{(k)} = u_{|\Omega_1} \text{ and } \lim_{k\to\infty} u_2^{(k)} = u_{|\Omega_2}.$$

It can be proven that the Schwarz method applied to problem (18.1) always converges, with a rate that increases as the measure $|\Omega_{12}|$ of the overlapping region Ω_{12} increases. Let us show this result on a simple one-dimensional case.

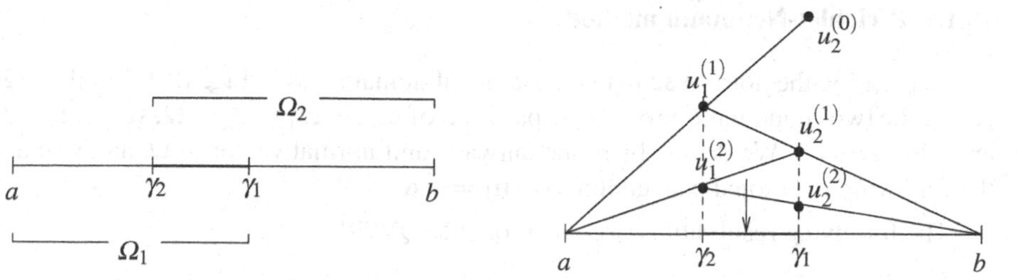

Fig. 18.2. Example of a decomposition with overlap in dimension 1 (left). A few iterations of the multiplicative Schwarz method for problem (18.7) (right)

Example 18.1. Let $\Omega = (a,b)$ and let $\gamma_1, \gamma_2 \in (a,b)$ be such that $a < \gamma_2 < \gamma_1 < b$ (see Fig. 18.2). The two problems (18.3) and (18.4) become:

$$\begin{cases} Lu_1^{(k)} = f, & a < x < \gamma_1, \\ u_1^{(k)} = u_2^{(k-1)}, & x = \gamma_1, \\ u_1^{(k)} = 0, & x = a, \end{cases} \tag{18.5}$$

$$\begin{cases} Lu_2^{(k)} = f, & \gamma_2 < x < b, \\ u_2^{(k)} = u_1^{(k)}, & x = \gamma_2, \\ u_2^{(k)} = 0, & x = b. \end{cases} \tag{18.6}$$

To show that this scheme converges, let us bound ourselves to the simpler problem

$$\begin{cases} -u''(x) = 0, & a < x < b, \\ u(a) = u(b) = 0, \end{cases} \tag{18.7}$$

that is the model problem (18.1) with $L = -d^2/dx^2$ and $f = 0$, whose solution clearly is $u = 0$ in (a,b). This is not restrictive since at every step the error: $u - u_1^{(k)}$ in Ω_1, $u - u_2^{(k)}$ in Ω_2, satisfies a problem like (18.5)-(18.6) with null forcing term.
Let $k = 1$; since $(u_1^{(1)})'' = 0$, $u_1^{(1)}(x)$ is a linear function; moreover, it vanishes at $x = a$ and takes the value $u_2^{(0)}$ at $x = \gamma_1$. As we know the value of $u_1^{(1)}$ at γ_2, we can solve the problem (18.6) which, in its turn, features a linear solution. Then we proceed in a similar manner. In Fig. 18.2 we show a few iterations: we clearly see that the method converges, moreover the convergence rate reduces as the length of the interval (γ_2, γ_1) gets smaller. ■

At each iteration the Schwarz iterative method (18.3)–(18.4) requires the solution of two subproblems with boundary conditions of the same kind as those of the original problem: indeed, by starting with a Dirichlet boundary-value problem in Ω we end up with two subproblems with Dirichlet conditions on the boundary of Ω_1 and Ω_2.
Should the differential problem (18.1) had been completed by a Neumann boundary condition on the whole boundary $\partial\Omega$, we would have been led to the solution of a mixed Dirichlet-Neumann boundary-value problem on either subdomain Ω_1 and Ω_2.

18.1.2 Dirichlet-Neumann method

Let us partition the domain Ω in two disjoint subdomains (as in Fig. 18.1): let then Ω_1 and Ω_2 be two subdomains providing a partition of Ω, i.e. $\overline{\Omega}_1 \cup \overline{\Omega}_2 = \overline{\Omega}$, $\overline{\Omega}_1 \cap \overline{\Omega}_2 = \Gamma$ and $\Omega_1 \cap \Omega_2 = \emptyset$. We denote by $\mathbf{n}_i$ the outward unit normal vector to Ω_i and will use the following notational convention: $\mathbf{n} = \mathbf{n}_1 = -\mathbf{n}_2$.

The following result holds (for its proof see [QV99]):

Theorem 18.1 (of equivalence). *The solution u of problem (18.1) is such that* $u_{|\Omega_i} = u_i$ *for* $i = 1, 2$, *where* u_i *is the solution to the problem*

$$\begin{cases} Lu_i = f & \text{in } \Omega_i, \\ u_i = 0 & \text{on } \partial\Omega_i \setminus \Gamma, \end{cases} \tag{18.8}$$

with interface conditions

$$u_1 = u_2 \tag{18.9}$$

and

$$\frac{\partial u_1}{\partial n_L} = \frac{\partial u_2}{\partial n_L} \tag{18.10}$$

on Γ, *having denoted with* $\partial/\partial n_L$ *the conormal derivative (see (3.34)).*

Thanks to this result we could split problem (18.1) by assigning the interface conditions (18.9)-(18.10) the role of "boundary conditions" for the two subproblems on the interface Γ. In particular, we can set up the following *Dirichlet-Neumann* (DN) iterative algorithm : given $u_2^{(0)}$ on Γ, for $k \geq 1$ solve the problems:

$$\begin{cases} Lu_1^{(k)} = f & \text{in } \Omega_1, \\ u_1^{(k)} = u_2^{(k-1)} & \text{on } \Gamma, \\ u_1^{(k)} = 0 & \text{on } \partial\Omega_1 \setminus \Gamma, \end{cases} \tag{18.11}$$

$$\begin{cases} Lu_2^{(k)} = f & \text{in } \Omega_2, \\ \dfrac{\partial u_2^{(k)}}{\partial n} = \dfrac{\partial u_1^{(k)}}{\partial n} & \text{on } \Gamma, \\ u_2^{(k)} = 0 & \text{on } \partial\Omega_2 \setminus \Gamma. \end{cases} \tag{18.12}$$

Condition (18.9) has generated a Dirichlet boundary condition on Γ for the subproblem in Ω_1 whereas (18.10) has generated a Neumann boundary condition on Γ for the subproblem in Ω_2.

Differently than Schwarz's method, the DN algorithm yields a Neumann boundary-value problem on the subdomain Ω_2. Theorem 18.1 guarantees that when the two sequences $\{u_1^{(k)}\}$ and $\{u_2^{(k)}\}$ converge, then their limit will be perforce the solution to the exact problem (18.1). The DN algorithm is therefore *consistent.* Its convergence however is not always guaranteed, as we can see on the following simple example.

Example 18.2. Let $\Omega = (a,b)$, $\gamma \in (a,b)$, $L = -d^2/dx^2$ and $f = 0$. At every $k \geq 1$ the DN algorithm generates the two subproblems:

$$\begin{cases} -(u_1^{(k)})'' = 0, & a < x < \gamma, \\ u_1^{(k)} = 0, & x = a, \\ u_1^{(k)} = u_2^{(k-1)}, & x = \gamma, \end{cases} \tag{18.13}$$

$$\begin{cases} -(u_2^{(k)})'' = 0, & \gamma < x < b, \\ (u_2^{(k)})' = (u_1^{(k)})', & x = \gamma, \\ u_2^{(k)} = 0, & x = b. \end{cases} \tag{18.14}$$

Proceeding as done in Example 18.1, we can prove that the two sequences converge only if $\gamma > (a+b)/2$, as shown graphycally in Fig. 18.3.

■

In general, for a problem in arbitrary dimension $d > 1$, the measure of the "Dirichlet" subdomain Ω_1 must be larger than that of the "Neumann" one Ω_2 in order to guarantee the convergence of (18.11)-(18.12). This however yields a severe constraint to fulfill, especially if several subdomains will be used.
To overcome such limitation, a variant of the DN algorithm can be set up by replacing the Dirichlet condition $(18.11)_2$ in the first subdomain by

$$u_1^{(k)} = \theta u_2^{(k-1)} + (1-\theta) u_1^{(k-1)} \quad \text{on } \Gamma, \tag{18.15}$$

that is by introducing a *relaxation* which depends on a positive parameter θ. In such a way it is always possible to reduce the error between two subsequent iterates.

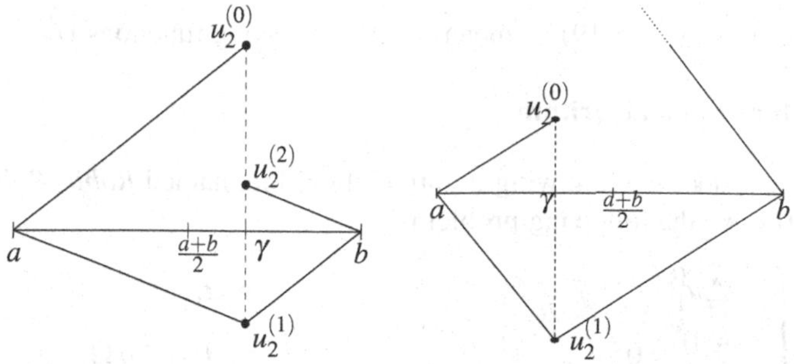

Fig. 18.3. Example of converging (left) and diverging (right) iterations for the DN method in 1D

In the case displayed in Fig. 18.3 we can easily verify that, by choosing

$$\theta_{opt} = -\frac{u_1^{(k-1)}}{u_2^{(k-1)} - u_1^{(k-1)}}, \tag{18.16}$$

the algorithm converges to the exact solution in a single iteration.

More in general, it can be proven that in any dimension $d \geq 1$, there exists a suitable value $\theta_{\max} < 1$ such that the DN algorithm converges for any possible choice of the relaxation parameter θ in the interval $(0, \theta_{\max})$.

18.1.3 Neumann-Neumann algorithm

Consider again a partition of Ω into two disjoint subdomains and denote by λ the (unknown) value of the solution u at their interface Γ. Consider the following iterative algorithm: for any given $\lambda^{(0)}$ on Γ, for $k \geq 0$ and $i = 1,2$ solve the following problems:

$$\begin{cases} -\Delta u_i^{(k+1)} = f & \text{in } \Omega_i, \\ u_i^{(k+1)} = \lambda^{(k)} & \text{on } \Gamma, \\ u_i^{(k+1)} = 0 & \text{on } \partial\Omega_i \setminus \Gamma, \end{cases} \tag{18.17}$$

$$\begin{cases} -\Delta \psi_i^{(k+1)} = 0 & \text{in } \Omega_i, \\ \dfrac{\partial \psi_i^{(k+1)}}{\partial n} = \sigma_i \left(\dfrac{\partial u_1^{(k+1)}}{\partial n} - \dfrac{\partial u_2^{(k+1)}}{\partial n} \right) & \text{on } \Gamma, \\ \psi_i^{(k+1)} = 0 & \text{on } \partial\Omega_i \setminus \Gamma, \end{cases} \tag{18.18}$$

with

$$\lambda^{(k+1)} = \lambda^{(k)} - \theta \left(\sigma_1 \psi_{1|\Gamma}^{(k+1)} - \sigma_2 \psi_{2|\Gamma}^{(k+1)} \right), \tag{18.19}$$

where θ is a positive acceleration parameter, while σ_1 and σ_2 are two positive coefficients such that $\sigma_1 + \sigma_2 = 1$. This iterative algorithm is named *Neumann-Neumann* (NN). Note that in the first stage (18.17) we care about the continuity on Γ of the functions $u_1^{(k+1)}$ and $u_2^{(k+1)}$ but not that of their derivatives. The latter are addressed in the second stage (18.18), (18.19) by means of the correcting functions $\psi_1^{(k+1)}$ and $\psi_2^{(k+1)}$.

18.1.4 Robin-Robin algorithm

At last, we consider the following iterative algorithm, named *Robin-Robin* (RR). For every $k \geq 0$ solve the following problems:

$$\begin{cases} -\Delta u_1^{(k+1)} = f & \text{in } \Omega_1, \\ u_1^{(k+1)} = 0 & \text{on } \partial\Omega_1 \cap \partial\Omega, \\ \dfrac{\partial u_1^{(k+1)}}{\partial n} + \gamma_1 u_1^{(k+1)} = \dfrac{\partial u_2^{(k)}}{\partial n} + \gamma_1 u_2^{(k)} & \text{on } \Gamma, \end{cases} \tag{18.20}$$

then

$$\begin{cases} -\Delta u_2^{(k+1)} = f & \text{in } \Omega_2, \\ u_2^{(k+1)} = 0 & \text{on } \partial\Omega_2 \cap \partial\Omega, \\ \dfrac{\partial u_2^{(k+1)}}{\partial n} + \gamma_2 u_2^{(k+1)} = \dfrac{\partial u_1^{(k+1)}}{\partial n} + \gamma_2 u_1^{(k+1)} & \text{on } \Gamma, \end{cases} \tag{18.21}$$

where u_0 is assigned and γ_1, γ_2 are non-negative acceleration parameters that satisfy $\gamma_1 + \gamma_2 > 0$. Aiming at the algorithm parallelization, in (18.21) we could use $u_1^{(k)}$ instead of $u_1^{(k+1)}$, provided in such a case an initial value for u_1^0 is assigned as well.

18.2 Multi-domain formulation of Poisson problem and interface conditions

In this section, for the sake of exposition, we choose $L = -\Delta$ and consider the Poisson problem with homogeneous Dirichlet boundary conditions (3.13). Generalization to an arbitrary second order elliptic operator with different boundary conditions is in order.

In the case addressed in Sect. 18.1.2 of a domain partitioned into two disjoint subdomains, the equivalence Theorem 18.1 allows the following *multidomain formulation* of problem (18.1), in which $u_i = u|_{\Omega_i}$, $i = 1,2$:

$$\begin{cases} -\Delta u_1 = f & \text{in } \Omega_1, \\ u_1 = 0 & \text{on } \partial\Omega_1 \setminus \Gamma, \\ -\Delta u_2 = f & \text{in } \Omega_2, \\ u_2 = 0 & \text{on } \partial\Omega_2 \setminus \Gamma, \\ u_1 = u_2 & \text{on } \Gamma, \\ \dfrac{\partial u_1}{\partial n} = \dfrac{\partial u_2}{\partial n} & \text{on } \Gamma. \end{cases} \tag{18.22}$$

18.2.1 The Steklov-Poincaré operator

We denote again by λ the unknown value of the solution u of problem (3.13) on the interface Γ, that is $\lambda = u_{|\Gamma}$. Should we know a priori the value λ on Γ, we could solve the following two independent boundary-value problems with Dirichlet condition on Γ $(i = 1,2)$:

$$\begin{cases} -\Delta w_i = f & \text{in } \Omega_i, \\ w_i = 0 & \text{on } \partial\Omega_i \setminus \Gamma, \\ w_i = \lambda & \text{on } \Gamma. \end{cases} \tag{18.23}$$

With the aim of obtaining the value λ on Γ, let us split w_i as follows

$$w_i = w_i^* + u_i^0,$$

where w_i^* and u_i^0 represent the solutions of the following problems ($i = 1,2$):

$$\begin{cases} -\triangle w_i^* = f & \text{in } \Omega_i\,, \\ w_i^* = 0 & \text{on } \partial\Omega_i \cap \partial\Omega, \\ w_i^* = 0 & \text{on } \Gamma, \end{cases} \tag{18.24}$$

and

$$\begin{cases} -\triangle u_i^0 = 0 & \text{in } \Omega_i\,, \\ u_i^0 = 0 & \text{on } \partial\Omega_i \cap \partial\Omega, \\ u_i^0 = \lambda & \text{on } \Gamma, \end{cases} \tag{18.25}$$

respectively. Note that the functions w_i^* depend solely on the source data f, while u_i^0 solely on the value λ on Γ, henceforth we can write $w_i^* = G_i f$ and $u_i^0 = H_i\lambda$. Both operators G_i and H_i are linear; H_i is the so-called harmonic extension operator of λ on the domain Ω_i.
By a formal comparison of problem (18.22) with problem (18.23), we infer that the equality

$$u_i = w_i^* + u_i^0\ ,\ i = 1,2\,,$$

holds iff the condition $(18.22)_6$ on the normal derivatives on Γ is satisfied, that is iff

$$\frac{\partial w_1}{\partial n} = \frac{\partial w_2}{\partial n} \qquad \text{on } \Gamma.$$

By using the previously introduced notations the latter condition can be reformulated as

$$\frac{\partial}{\partial n}(G_1 f + H_1\lambda) = \frac{\partial}{\partial n}(G_2 f + H_2\lambda)$$

and therefore

$$\left(\frac{\partial H_1}{\partial n} - \frac{\partial H_2}{\partial n}\right)\lambda = \left(\frac{\partial G_2}{\partial n} - \frac{\partial G_1}{\partial n}\right) f \qquad \text{on } \Gamma.$$

In this way we have obtained an equation for the unknown λ on the interface Γ, named *Steklov-Poincaré equation*, that can be rewritten in compact form as

$$S\lambda = \chi \qquad \text{on } \Gamma. \tag{18.26}$$

S is the *Steklov-Poincaré* pseudo-differential operator; its formal definition is

$$S\mu = \frac{\partial}{\partial n} H_1\mu - \frac{\partial}{\partial n} H_2\mu = \sum_{i=1}^{2} \frac{\partial}{\partial n_i} H_i\mu = \sum_{i=1}^{2} S_i\mu, \tag{18.27}$$

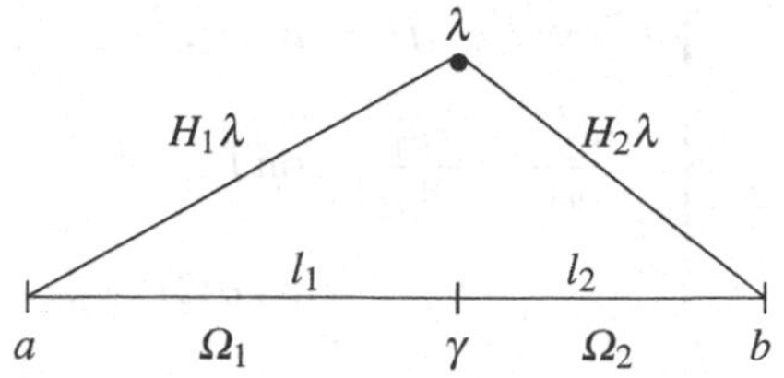

Fig. 18.4. Harmonic extensions in one dimension

while χ is a linear functional which depends on f

$$\chi = \frac{\partial}{\partial n} G_2 f - \frac{\partial}{\partial n} G_1 f = -\sum_{i=1}^{2} \frac{\partial}{\partial n_i} G_i f. \tag{18.28}$$

The operator

$$S_i : \mu \to S_i \mu = \frac{\partial}{\partial n_i} (H_i \mu)\Big|_{\Gamma}, \quad i = 1,2, \tag{18.29}$$

is called local Steklov-Poincaré operator. Note that S, S_1 and S_2 operate between the trace space

$$\Lambda = \{\mu \mid \exists v \in V \, : \, \mu = v|_{\Gamma}\} \tag{18.30}$$

(that is $H_{00}^{1/2}(\Gamma)$, see [QV99]), and its dual Λ', whereas $\chi \in \Lambda'$.

Example 18.3. With the aim of providing a practical (elementary) example of operator S, let us consider a simple one-dimensional problem. Let $\Omega = (a,b) \subset \mathbb{R}$ as shown in Fig. 18.4 and $Lu = -u''$. By subdividing Ω in two disjoint subdomains, the interface Γ reduces to a single point $\gamma \in (a,b)$, and the Steklov-Poincaré operator S becomes

$$S\lambda = \left(\frac{dH_1}{dx} - \frac{dH_2}{dx} \right) \lambda = \left(\frac{1}{l_1} + \frac{1}{l_2} \right) \lambda,$$

with $l_1 = \gamma - a$ and $l_2 = b - \gamma$. ■

18.2.2 Equivalence between Dirichlet-Neumann and Richardson methods

The Dirichlet-Neumann (DN) method introduced in Sect. 18.1.2 can be reinterpreted as a (preconditioned) Richardson method for the solution of the Steklov-Poincaré interface equation. To check this statement, consider again, for the sake of simplicity, a domain Ω partitioned into two disjoint subdomains Ω_1 and Ω_2 with interface Γ. Then we re-write the DN algorithm (18.11), (18.12), (18.15) in the case of the operator $L = -\Delta$: for a given λ^0, for $k \geq 1$ solve:

$$\begin{cases} -\Delta u_1^{(k)} = f_1 & \text{in } \Omega_1, \\ u_1^{(k)} = \lambda^{(k-1)} & \text{on } \Gamma, \\ u_1^{(k)} = 0 & \text{on } \partial\Omega_1 \setminus \Gamma, \end{cases} \tag{18.31}$$

$$\begin{cases} -\Delta u_2^{(k)} = f_2 & \text{in } \Omega_2, \\ \dfrac{\partial u_2^{(k)}}{\partial n_2} = \dfrac{\partial u_1^{(k)}}{\partial n_2} & \text{on } \Gamma, \\ u_2^{(k)} = 0 & \text{on } \partial\Omega_2 \setminus \Gamma, \end{cases} \tag{18.32}$$

$$\lambda^{(k)} = \theta u_2^{(k)}{}_{|\Gamma} + (1-\theta)\lambda^{(k-1)}. \tag{18.33}$$

The following result holds:

Theorem 18.2. *The Dirichlet-Neumann iterative algorithm (18.31)–(18.33) is equivalent to the preconditioned Richardson algorithm*

$$P_{DN}(\lambda^{(k)} - \lambda^{(k-1)}) = \theta(\chi - S\lambda^{(k-1)}). \tag{18.34}$$

The preconditioning operator is $P_{DN} = S_2$.

Proof. The solution $u_1^{(k)}$ of (18.31) can be written as

$$u_1^{(k)} = H_1\lambda^{(k-1)} + G_1 f_1. \tag{18.35}$$

Since $G_2 f_2$ satisfies the differential problem

$$\begin{cases} -\Delta(G_2 f_2) = f_2 & \text{in } \Omega_2, \\ G_2 f_2 = 0 & \text{on } \partial\Omega_2, \end{cases}$$

thanks to (18.32) the function $u_2^{(k)} - G_2 f_2$ satisfies the differential problem

$$\begin{cases} -\Delta(u_2^{(k)} - G_2 f_2) = 0 & \text{in } \Omega_2, \\ \dfrac{\partial}{\partial n_2}(u_2^{(k)} - G_2 f_2) = -\dfrac{\partial u_1^{(k)}}{\partial n} + \dfrac{\partial}{\partial n}(G_2 f_2) & \text{on } \Gamma, \\ u_2^{(k)} - G_2 f_2 = 0 & \text{on } \partial\Omega_2 \setminus \Gamma. \end{cases} \tag{18.36}$$

In particular $u_2^{(k)}{}_{|\Gamma} = (u_2^{(k)} - G_2 f_2)_{|\Gamma}$. Since the operator S_i (18.29) maps a Dirichlet data to a Neumann data on Γ, its inverse S_i^{-1} transforms a Neumann data in a Dirichlet one on Γ.
Otherwise said, $S_2^{-1}\eta = w_{2|\Gamma}$, where w_2 is the solution of

$$\begin{cases} -\Delta w_2 = 0 & \text{in } \Omega_2, \\ \dfrac{\partial w_2}{\partial n} = \eta & \text{on } \Gamma, \\ w_2 = 0 & \text{on } \partial\Omega_2 \setminus \Gamma. \end{cases} \tag{18.37}$$

Setting now

$$\eta = -\frac{\partial u_1^{(k)}}{\partial n} + \frac{\partial}{\partial n}(G_2 f_2),$$

and comparing (18.36) with (18.37), we conclude that

$$u_2^{(k)}{}_{|\Gamma} = (u_2^{(k)} - G_2 f_2)_{|\Gamma} = S_2^{-1}\left(-\frac{\partial u_1^{(k)}}{\partial n} + \frac{\partial}{\partial n}(G_2 f_2)\right).$$

On the other hand, owing to (18.35) and to the definition (18.28) of χ, we obtain

$$\begin{aligned} u_2^{(k)}{}_{|\Gamma} &= S_2^{-1}\left(-\frac{\partial}{\partial n}(H_1\lambda^{(k-1)}) - \frac{\partial}{\partial n}(G_1 f_1) + \frac{\partial}{\partial n}(G_2 f_2)\right) \\ &= S_2^{-1}(-S_1\lambda^{(k-1)} + \chi). \end{aligned}$$

Using (18.33) we can therefore write

$$\lambda^{(k)} = \theta\left[S_2^{-1}(-S_1\lambda^{(k-1)} + \chi)\right] + (1-\theta)\lambda^{(k-1)},$$

that is

$$\lambda^{(k)} - \lambda^{(k-1)} = \theta\left[S_2^{-1}(-S_1\lambda^{(k-1)} + \chi) - \lambda^{(k-1)}\right].$$

Since $-S_1 = S_2 - S$, we finally obtain

$$\begin{aligned} \lambda^{(k)} - \lambda^{(k-1)} &= \theta\left[S_2^{-1}((S_2 - S)\lambda^{(k-1)} + \chi) - \lambda^{(k-1)}\right] \\ &= \theta S_2^{-1}(\chi - S\lambda^{(k-1)}), \end{aligned}$$

that is (18.34). The preconditioned *DN* operator is therefore $S_2^{-1}S = I + S_2^{-1}S_1$. ◇

Using an argument similar to that used for the proof of Theorem 18.2, also the Neumann-Neumann (NN) algorithm (18.17)–(18.19) can be interpreted as a preconditioned Richardson algorithm

$$P_{NN}(\lambda^{(k)} - \lambda^{(k-1)}) = \theta(\chi - S\lambda^{(k-1)}),$$

this time the preconditioner being $P_{NN} = (D_1 S_1^{-1} D_1 + D_2 S_2^{-1} D_2)^{-1}$ where D_i is a diagonal matrix whose entries are equal to σ_i. Note that the preconditioned operator becomes (if $D_i = I$) $S_2^{-1}S_1 + 2I + (S_2^{-1}S_1)^{-1}$.

Consider at last the Robin-Robin iterative algorithm (18.20)–(18.21). Denoting by $\mu_i^{(k)} \in \Lambda$ the approximation at step k of the trace of $u_i^{(k)}$ on the interface Γ, $i = 1, 2$, it can be proven that (18.20)–(18.21) is equivalent to the following alternating direction (ADI) algorithm:

$$\begin{aligned} (\gamma_1 i_\Lambda + S_1)\mu_1^{(k)} &= \chi + (\gamma_1 i_\Lambda + S_2)\mu_2^{(k-1)}, \\ (\gamma_2 i_\Lambda + S_2)\mu_2^{(k)} &= \chi + (\gamma_2 i_\Lambda + S_1)\mu_1^{(k-1)}, \end{aligned}$$

where $i_\Lambda : \Lambda \to \Lambda'$ here denotes the Riesz isomorphism between the Hilbert space Λ and its dual Λ' (see (2.5)).
Should, for a convenient choice of the two parameters γ_1 and γ_2, the algorithm converge to two limit functions μ_1 and μ_2, then $\mu_1 = \mu_2 = \lambda$, the latter function being the solution to the Steklov-Poincaré equation (18.26).
The RR preconditioner reads $P_{RR} = (\gamma_1+\gamma_2)^{-1}(\gamma_1 i_\Lambda + S_1)(\gamma_2 i_\Lambda + S_2)$.

Remark 18.1. In the Dirichlet-Neumann algorithm, the value λ of the solution u at the interface Γ is the principal unknown. Once it has been determined, we can use it as Dirichlet data to recover the original solution in the whole domain. Alternatively, one could use the normal derivative $\eta = \frac{\partial u}{\partial n}$ on Γ as principal unknown (or, for a more general partial differential operator, the conormal derivative - or flux). By proceeding as above, we can show that η satisfies the new Steklov-Poincaré equation

$$(S_1^{-1} + S_2^{-1})\eta = T_1 f_1 + T_2 f_2 \quad \text{on}\,\Gamma \tag{18.38}$$

where for $i = 1,2$, $T_i f_i$ is the solution of the following Neumann problem

$$\begin{cases} -\triangle(T_i f_i) = f_i & \text{in}\,\Omega_i, \\ \dfrac{\partial}{\partial n_i}(T_i f_i) = 0 & \text{on}\,\Gamma, \\ T_i f_i = 0 & \text{on}\,\partial\Omega\backslash\Gamma. \end{cases} \tag{18.39}$$

The so-called FETI algorithms (see Sect. 18.5.4) are examples of iterative algorithms designed for the solution of problems like (18.38). The FETI preconditioner is $P_{FETI} = S_1 + S_2$, hence the preconditioned FETI operator is $(S_1+S_2)(S_1^{-1}+S_2^{-1})$.

•

18.3 Multidomain formulation of the finite element approximation of the Poisson problem

What seen thus far can be regarded as propedeutical to numerical solution of boundary-value problems. In this section we will see how the previous ideas can be reshaped in the framework of a numerical discretization method. Although we will only address the case of finite element discretization, this is however not restrictive. We refer, e.g., to [CHQZ07] and [TW05] for the case of spectral or spectral element discretizations and to [Woh01] for discretization based on DG and mortar methods.

Consider the Poisson problem (3.13), its weak formulation (3.18) and its Galerkin finite element approximation (4.40) on a triangulation $\mathcal{T}_h$. Recall that $V_h = \overset{\circ}{X}{}_h^r = \{v_h \in X_h^r : v_h|_{\partial\Omega} = 0\}$ is the space of finite element functions of degree r vanishing on $\partial\Omega$, whose basis is $\{\varphi_j\}_{j=1}^{N_h}$ (see Sect. 4.5.1).

For the finite element nodes in the domain Ω we consider the following partition: let $\{x_j^{(1)}, 1 \le j \le N_1\}$ be the nodes located in the subdomain Ω_1, $\{x_j^{(2)}, 1 \le j \le N_2\}$ those in Ω_2 and, finally, $\{x_j^{(\Gamma)}, 1 \le j \le N_\Gamma\}$ those lying on the interface Γ. Let us split the basis functions accordingly: $\varphi_j^{(1)}$ will denote those associated to the nodes $x_j^{(1)}$, $\varphi_j^{(2)}$ those associated with the nodes $x_j^{(2)}$, and $\varphi_j^{(\Gamma)}$ those associated with the nodes $x_j^{(\Gamma)}$ lying on the interface. This yields

$$\varphi_j^{(\alpha)}(x_j^{(\beta)}) = \delta_{ij}\delta_{\alpha\beta}\,, \quad 1 \leq i \leq N_\alpha, \quad \leq j \leq \mathrm{N}_\beta\,, \tag{18.40}$$

with $\alpha, \beta = 1, 2, \Gamma$; δ_{ij} is the Kronecker symbol.
By letting v_h in (4.40) to coincide with a test function, (4.40) can be given the following equivalent formulation: find $u_h \in V_h$ such that

$$\begin{cases} a(u_h, \varphi_i^{(1)}) = F(\varphi_i^{(1)}) & \forall i = 1, \ldots, N_1\,, \\ a(u_h, \varphi_j^{(2)}) = F(\varphi_j^{(2)}) & \forall j = 1, \ldots, N_2\,, \\ a(u_h, \varphi_k^{(\Gamma)}) = F(\varphi_k^{(\Gamma)}) & \forall k = 1, \ldots, N_\Gamma\,, \end{cases} \tag{18.41}$$

having set $F(v) = \int_\Omega f v \, d\Omega$. Let now

$$a_i(v, w) = \int_{\Omega_i} \nabla v \cdot \nabla w \, d\Omega \qquad \forall v, w \in V,\ i = 1, 2$$

be the restriction of the bilinear form $a(.,.)$ to the subdomain Ω_i and define $V_{i,h} = \{v \in H^1(\Omega_i) \mid v = 0 \text{ on } \partial\Omega_i \setminus \Gamma\}$ $(i = 1, 2)$. Similarly we set $F_i(v) = \int_{\Omega_i} f v \, d\Omega$ and denote by $u_h^{(i)} = u_{h|\Omega_i}$ the restriction of u_h to the subdomain Ω_i, with $i = 1, 2$. Problem (18.41) can be rewritten in the equivalent form: find $u_h^{(1)} \in V_{1,h}$, $u_h^{(2)} \in V_{2,h}$ such that

$$\begin{cases} a_1(u_h^{(1)}, \varphi_i^{(1)}) = F_1(\varphi_i^{(1)}) & \forall i = 1, \ldots, N_1, \\ a_2(u_h^{(2)}, \varphi_j^{(2)}) = F_2(\varphi_j^{(2)}) & \forall j = 1, \ldots, N_2 \\ a_1(u_h^{(1)}, \varphi_k^{(\Gamma)}|_{\Omega_1}) + a_2(u_h^{(2)}, \varphi_k^{(\Gamma)}|_{\Omega_2}) & \\ = F_1(\varphi_k^{(\Gamma)}|_{\Omega_1}) + F_2(\varphi_k^{(\Gamma)}|_{\Omega_2}) & \forall k = 1, \ldots, N_\Gamma. \end{cases} \tag{18.42}$$

The interface continuity condition $(18.22)_5$ is automatically satisfied thanks to the continuity of the functions $u_h^{(i)}$. Moreover, equations $(18.42)_1$-$(18.42)_3$ correspond to the finite element discretization of equations $(18.22)_1$-$(18.22)_6$, respectively. In particular, the third of equations (18.42) must be regarded as the discrete counterpart of condition $(18.22)_6$ expressing the continuity of normal derivatives on Γ.
Let us expand the solution u_h with respect to the basis functions V_h

$$u_h(x) = \sum_{j=1}^{N_1} u_h(x_j^{(1)})\varphi_j^{(1)}(x) + \sum_{j=1}^{N_2} u_h(x_j^{(2)})\varphi_j^{(2)}(x) + \sum_{j=1}^{N_\Gamma} u_h(x_j^{(\Gamma)})\varphi_j^{(\Gamma)}(x). \tag{18.43}$$

From now on, the nodal values $u_h(x_j^{(\alpha)})$, for $\alpha = 1, 2, \Gamma$ and $j = 1, \ldots, N_\alpha$, which are the expansion coefficients, will be indicated with the shorthand notation $u_j^{(\alpha)}$.

Using (18.43), we can rewrite (18.42) as follows:

$$\begin{cases} \sum_{j=1}^{N_1} u_j^{(1)} a_1(\varphi_j^{(1)}, \varphi_i^{(1)}) + \sum_{j=1}^{N_\Gamma} u_j^{(\Gamma)} a_1(\varphi_j^{(\Gamma)}, \varphi_i^{(1)}) = F_1(\varphi_i^{(1)}) \quad \forall i = 1, \dots, N_1, \\ \sum_{j=1}^{N_2} u_j^{(2)} a_2(\varphi_j^{(2)}, \varphi_i^{(2)}) + \sum_{j=1}^{N_\Gamma} u_j^{(\Gamma)} a_2(\varphi_j^{(\Gamma)}, \varphi_i^{(2)}) = F_2(\varphi_i^{(2)}) \quad \forall i = 1, \dots, N_2, \\ \sum_{j=1}^{N_\Gamma} u_j^{(\Gamma)} \left[a_1(\varphi_j^{(\Gamma)}, \varphi_i^{(\Gamma)}) + a_2(\varphi_j^{(\Gamma)}, \varphi_i^{(\Gamma)}) \right] \\ \quad + \sum_{j=1}^{N_1} u_j^{(1)} a_1(\varphi_j^{(1)}, \varphi_i^{(\Gamma)}) + \sum_{j=1}^{N_2} u_j^{(2)} a_2(\varphi_j^{(2)}, \varphi_i^{(\Gamma)}) \\ \quad = F_1(\varphi_i^{(\Gamma)}|_{\Omega_1}) + F_2(\varphi_i^{(\Gamma)}|_{\Omega_2}) \qquad \forall i = 1, \dots, N_\Gamma. \end{cases} \tag{18.44}$$

Let us introduce the following arrays:

$$\begin{aligned} (A_{11})_{ij} &= a_1(\varphi_j^{(1)}, \varphi_i^{(1)}), & (A_{1\Gamma})_{ij} &= a_1(\varphi_j^{(\Gamma)}, \varphi_i^{(1)}), \\ (A_{22})_{ij} &= a_2(\varphi_j^{(2)}, \varphi_i^{(2)}), & (A_{2\Gamma})_{ij} &= a_2(\varphi_j^{(\Gamma)}, \varphi_i^{(2)}), \\ (A_{\Gamma\Gamma}^1)_{ij} &= a_1(\varphi_j^{(\Gamma)}, \varphi_i^{(\Gamma)}), & (A_{\Gamma\Gamma}^2)_{ij} &= a_2(\varphi_j^{(\Gamma)}, \varphi_i^{(\Gamma)}), \\ (A_{\Gamma 1})_{ij} &= a_1(\varphi_j^{(1)}, \varphi_i^{(\Gamma)}), & (A_{\Gamma 2})_{ij} &= a_2(\varphi_j^{(2)}, \varphi_i^{(\Gamma)}), \\ (\mathbf{f}_1)_i &= F_1(\varphi_i^{(1)}), & (\mathbf{f}_2)_i &= F_2(\varphi_i^{(2)}), \\ (\mathbf{f}_1^\Gamma)_i &= F_1(\varphi_i^{(\Gamma)}), & (\mathbf{f}_2^\Gamma)_i &= F_2(\varphi_i^{(\Gamma)}, \varphi_i^{(1)}), \end{aligned}$$

then set

$$\mathbf{u} = (\mathbf{u}_1, \mathbf{u}_2, \boldsymbol{\lambda})^T, \text{with } \mathbf{u}_1 = \left(u_j^{(1)}\right), \ \mathbf{u}_2 = \left(u_j^{(2)}\right) \text{ and } \boldsymbol{\lambda} = \left(u_j^{(\Gamma)}\right). \tag{18.45}$$

Problem (18.44) can be casted in the following algebraic form

$$\begin{cases} A_{11}\mathbf{u}_1 + A_{1\Gamma}\boldsymbol{\lambda} = \mathbf{f}_1, \\ A_{22}\mathbf{u}_2 + A_{2\Gamma}\boldsymbol{\lambda} = \mathbf{f}_2, \\ A_{\Gamma 1}\mathbf{u}_1 + A_{\Gamma 2}\mathbf{u}_2 + \left(A_{\Gamma\Gamma}^{(1)} + A_{\Gamma\Gamma}^{(2)}\right)\boldsymbol{\lambda} = \mathbf{f}_1^\Gamma + \mathbf{f}_2^\Gamma, \end{cases} \tag{18.46}$$

or, equivalently,

$$A\mathbf{u} = \mathbf{f}, \text{ that is } \begin{bmatrix} A_{11} & 0 & A_{1\Gamma} \\ 0 & A_{22} & A_{2\Gamma} \\ A_{\Gamma 1} & A_{\Gamma 2} & A_{\Gamma\Gamma} \end{bmatrix} \begin{bmatrix} \mathbf{u}_1 \\ \mathbf{u}_2 \\ \boldsymbol{\lambda} \end{bmatrix} = \begin{bmatrix} \mathbf{f}_1 \\ \mathbf{f}_2 \\ \mathbf{f}_\Gamma \end{bmatrix}, \tag{18.47}$$

having set $A_{\Gamma\Gamma} = \left(A_{\Gamma\Gamma}^{(1)} + A_{\Gamma\Gamma}^{(2)}\right)$ and $\mathbf{f}_\Gamma = \mathbf{f}_1^\Gamma + \mathbf{f}_2^\Gamma$. (18.47) is nothing but a blockwise representation of the finite element system (4.46), the blocks being determined by the partition (18.45) of the vector of unknowns.

More precisely, the first and second equations of (18.46) are discretizations of the given Poisson problems in Ω_1 and Ω_2, respectively for the interior values $\mathbf{u}_1$ and $\mathbf{u}_2$, with Dirichlet data vanishing on $\partial\Omega_i \setminus \Gamma$ and equal to the common value $\boldsymbol{\lambda}$ on Γ.
Alternatively, by setting (from the third equation of (18.46))

$$A_{\Gamma_1}\mathbf{u}_1 + A_{\Gamma\Gamma}^{(1)}\boldsymbol{\lambda} - \mathbf{f}_\Gamma^1 = -(A_{\Gamma_2}\mathbf{u}_2 + A_{\Gamma\Gamma}^{(2)}\boldsymbol{\lambda} - \mathbf{f}_\Gamma^2) \equiv \eta, \tag{18.48}$$

the first and third equations of (18.46) provide a discretization of the Poisson problem in Ω_1 with vanishing Dirichlet data on $\partial\Omega_1 \setminus \Gamma$ and with Neumann data η on Γ.

Similar considerations apply to the second and third equations of (18.46): they represent the discretization of a Poisson problem in Ω_2 with zero Dirichlet data in $\partial\Omega_2 \setminus \Gamma$ and Neumann data equal to η on Γ.

18.3.1 The Schur complement

Consider now the Steklov-Poincaré interface equation (18.26) and look for its finite element counterpart. Since λ represents the unknown value of u on Γ, its finite element correspondent is the vector $\boldsymbol{\lambda}$ of the values of u_h at the interface nodes.

By gaussian elimination operated on system (18.47), we can obtain a new reduced system on the sole unknown $\boldsymbol{\lambda}$.

Matrices A_{11} and A_{22} are invertible since they are associated with two homogeneous Dirichlet boundary-value problems for the Laplace operator, hence

$$\mathbf{u}_1 = A_{11}^{-1}(\mathbf{f}_1 - A_{1\Gamma}\boldsymbol{\lambda}) \quad \text{and} \quad \mathbf{u}_2 = A_{22}^{-1}(\mathbf{f}_2 - A_{2\Gamma}\boldsymbol{\lambda}). \tag{18.49}$$

From the third equation in (18.46), we obtain

$$\begin{aligned}&\left[\left(A_{\Gamma\Gamma}^{(1)} - A_{\Gamma 1}A_{11}^{-1}A_{1\Gamma}\right) + \left(A_{\Gamma\Gamma}^{(2)} - A_{\Gamma 2}A_{22}^{-1}A_{2\Gamma}\right)\right]\boldsymbol{\lambda}\\ &= \mathbf{f}_\Gamma - A_{\Gamma 1}A_{11}^{-1}\mathbf{f}_1 - A_{\Gamma 2}A_{22}^{-1}\mathbf{f}_2 = (\mathbf{f}_\Gamma^{(1)} - A_{\Gamma_1}A_{11}^{-1}\mathbf{f}_1) + (\mathbf{f}_\Gamma^{(2)} - A_{\Gamma_2}A_{22}^{-1}\mathbf{f}_2).\end{aligned} \tag{18.50}$$

Using the following definitions:

$$\Sigma = \Sigma_1 + \Sigma_2, \qquad \Sigma_i = A_{\Gamma\Gamma}^{(i)} - A_{\Gamma i}A_{ii}^{-1}A_{i\Gamma}, \qquad i = 1,2, \tag{18.51}$$

and

$$\chi_\Gamma = \chi_\Gamma^{(1)} + \chi_\Gamma^{(2)}, \quad \chi_\Gamma^{(i)} = \mathbf{f}_\Gamma^{(i)} - A_{\Gamma_i}A_{ii}\mathbf{f}_i, \tag{18.52}$$

(18.50) becomes

$$\Sigma\boldsymbol{\lambda} = \chi_\Gamma. \tag{18.53}$$

Since Σ and χ_Γ approximate S and χ, respectively, (18.53) can be considered as a finite element approximation to the Steklov-Poincaré equation (18.26). Matrix Σ is the so-called *Schur complement* of A with respect to $\mathbf{u}_1$ and $\mathbf{u}_2$, whereas matrices Σ_i are the Schur complements related to the subdomains Ω_i $(i = 1,2)$.
Once system (18.53) is solved w.r.t the unknown $\boldsymbol{\lambda}$, by virtue of (18.49) we can compute $\mathbf{u}_1$ and $\mathbf{u}_2$. This computation amounts to solve numerically two Poisson problems

on the two subdomains Ω_1 and Ω_2, with Dirichlet boundary conditions $u_h^{(i)}{}_{|\Gamma} = \lambda_h$ ($i = 1,2$) on the interface Γ.

The Schur complement Σ inherits some of the properties of its generating matrix A, as stated by the following result:

Lemma 18.1. *Matrix Σ satisfies the following properties:*
1. *if A is singular, so is Σ;*
2. *if A (respectively, A_{ii}) is symmetric, then Σ (respectively, Σ_i) is symmetric too;*
3. *if A is positive definite, so is Σ.*

Recall that the condition number of the finite element stiffness matrix A satisfies $K_2(A) \simeq Ch^{-2}$ (see (4.50)). As of Σ, it can be proven that

$$K_2(\Sigma) \simeq Ch^{-1}. \tag{18.54}$$

In the specific case under consideration, A (and therefore Σ, thanks to Lemma 18.1) is symmetric and positive definite. It is therefore convenient to use the conjugate gradient method (with a suitable preconditioner) for the solution of system (18.53). At every iteration, the computation of the residue will involve the finite element solution of two independent Dirichlet boundary-value problems on the subdomains Ω_i.

By employing a similar procedure we can derive instead of (18.53) an interface equation for the flux η introduced in (18.48). From (18.47) and (18.48) we derive

$$\begin{bmatrix} A_{11} & A_{1\Gamma} \\ A_{\Gamma 1} & A_{\Gamma\Gamma}^{(1)} \end{bmatrix} \begin{bmatrix} \mathbf{u}_1 \\ \boldsymbol{\lambda} \end{bmatrix} = \begin{bmatrix} \mathbf{f}_1 \\ \mathbf{f}_\Gamma^{(1)} + \eta \end{bmatrix}. \tag{18.55}$$

By eliminating $\mathbf{u}_1$ from the first row and replacing it in the second one we obtain

$$\Sigma_1 \boldsymbol{\lambda} = \chi_\Gamma^{(1)} + \eta, \text{ that is } \boldsymbol{\lambda} = \Sigma_1^{-1}(\chi_\Gamma^{(1)} + \eta). \tag{18.56}$$

Proceeding in a similar way we obtain

$$\Sigma_2 \boldsymbol{\lambda} = \chi_\Gamma^{(2)} - \eta, \text{ that is } \boldsymbol{\lambda} = \Sigma_2^{-1}(\chi_\Gamma^{(2)} - \eta). \tag{18.57}$$

By equating the last two equations (whose common value is $\boldsymbol{\lambda}$) we finally obtain the Schur-complement equation for the flux η:

$$T\eta = \psi_\Gamma, \quad \text{with } T = \Sigma_1^{-1} + \Sigma_2^{-1},\ \psi_\Gamma = \Sigma_2^{-1}\chi_\Gamma^{(2)} - \Sigma_1^{-1}\chi_\Gamma^{(1)}. \tag{18.58}$$

This algebraic equation can be regarded as a direct discretization of the Steklov-Poincaré problem for the flux (18.38).

18.3.2 The discrete Steklov-Poincaré operator

In this section we will find the discrete operator associated with the Schur complement. With this aim, besides the space $V_{i,h}$ previously introduced, we will need the one $V_{i,h}^0$ generated by the functions $\{\varphi_j^{(i)}\}$ exclusively associated to the internal nodes of the subdomain Ω_i, and the space Λ_h generated by the set of functions $\{\varphi_j^{(\Gamma)}{}_{|\Gamma}\}$. We have $\Lambda_h = \{\mu_h \mid \exists v_h \in V_h \,:\, v_h\big|_\Gamma = \mu_h\}$, whence Λ_h represents a finite element subspace of the trace functions space Λ introduced in (18.30).
Consider now the following problem: find $H_{i,h}\eta_h \in V_{i,h}$, with $H_{i,h}\eta_h = \eta_h$ on Γ, such that

$$\int_{\Omega_i} \nabla(H_{i,h}\eta_h)\cdot\nabla v_h \, d\Omega_i = 0 \qquad \forall v_h \in V_{i,h}^0. \tag{18.59}$$

Clearly, $H_{i,h}\eta_h$ represents a finite element approximation of the harmonic extension $H_i\eta_h$, and the operator $H_{i,h} : \eta_h \to H_{i,h}\eta_h$ can be regarded as an approximation of H_i. By expanding $H_{i,h}\eta_h$ in terms of the basis functions

$$H_{i,h}\eta_h = \sum_{j=1}^{N_i} u_j^{(i)}\varphi_j^{(i)} + \sum_{k=1}^{N_\Gamma} \eta_k \varphi_k^{(\Gamma)}{}_{|\Omega_i},$$

we can rewrite (18.59) in matrix form

$$A_{ii}\mathbf{u}^{(i)} = -A_{i\Gamma}\boldsymbol{\eta}. \tag{18.60}$$

The following result, called *the uniform discrete extension theorem*, holds:

Theorem 18.3. *There exist two constants $\hat{C}_1, \hat{C}_2 > 0$, independent of h, such t hat*

$$\hat{C}_1\|\eta_h\|_\Lambda \le \|H_{i,h}\eta_h\|_{H^1(\Omega_i)} \le \hat{C}_2\|\eta_h\|_\Lambda \qquad \forall \eta_h \in \Lambda_h \;\; i = 1,2. \tag{18.61}$$

Consequently, there exist two constants $K_1, K_2 > 0$, independent of h, such that

$$K_1\|H_{1,h}\eta_h\|_{H^1(\Omega_1)} \le \|H_{2,h}\eta_h\|_{H^1(\Omega_2)} \le K_2\|H_{1,h}\eta_h\|_{H^1(\Omega_1)} \quad \forall \eta_h \in \Lambda_h. \tag{18.62}$$

For the proof see, e.g., [QV99].

Now for $i = 1,2$ the (local) discrete Steklov-Poincaré operator is defined as follows: $S_{i,h} : \Lambda_h \to \Lambda_h'$,

$$\langle S_{i,h}\eta_h, \mu_h\rangle = \int_{\Omega_i} \nabla(H_{i,h}\eta_h)\cdot\nabla(H_{i,h}\mu_h)\,d\Omega_i \quad \forall \eta_h, \mu_h \in \Lambda_h, \tag{18.63}$$

then we define the (global) discrete Steklov-Poincaré operator as $S_h = S_{1,h} + S_{2,h}$.

Lemma 18.2. *The local discrete Steklov-Poincaré operator can be expressed in terms of the local Schur complement as*

$$\langle S_{i,h}\eta_h, \mu_h\rangle = \mu^T \Sigma_i \eta \quad \forall \eta_h, \mu_h \in \Lambda_h\ , i = 1,2, \tag{18.64}$$

where

$$\eta_h = \sum_{k=1}^{N_\Gamma} \eta_k \varphi_k^{(\Gamma)}{}_{|\Gamma}, \quad \mu_h = \sum_{k=1}^{N_\Gamma} \mu_k \varphi_k^{(\Gamma)}{}_{|\Gamma}$$

and

$$\eta = (\eta_1, \ldots, \eta_{N_\Gamma})^T, \quad \mu = (\mu_1, \ldots, \mu_{N_\Gamma})^T.$$

Therefore the global discrete Steklov-Poincaré operator $S_h = S_{1,h} + S_{2,h}$ satisfies the relation

$$\langle S_h \eta_h, \mu_h\rangle = \mu^T \Sigma \eta \quad \forall \eta_h, \mu_h \in \Lambda_h. \tag{18.65}$$

Proof. For $i = 1,2$ we have

$$\begin{aligned}
\langle S_{i,h}\eta_h, \mu_h\rangle &= a_i(H_{i,h}\eta_h, H_{i,h}\mu_h) \\
&= a_i\Big(\sum_{j=1}^{N_\Gamma} u_j \varphi_j^{(i)} + \sum_{k=1}^{N_\Gamma} \eta_k \varphi_k^{(\Gamma)}{}_{|\Omega_i}, \sum_{l=1}^{N_\Gamma} w_l \varphi_l^{(i)} + \sum_{m=1}^{N_\Gamma} \mu_m \varphi_m^{(\Gamma)}{}_{|\Omega_i}\Big) \\
&= \sum_{j,l=1}^{N_\Gamma} w_l a_i(\varphi_j^{(i)}, \varphi_l^{(i)}) u_j + \sum_{j,m=1}^{N_\Gamma} \mu_m a_i(\varphi_j^{(i)}, \varphi_m^{(\Gamma)}{}_{|\Omega_i}) u_j \\
&\quad + \sum_{k,l=1}^{N_\Gamma} w_l a_i(\varphi_k^{(\Gamma)}{}_{|\Omega_i}, \varphi_l^{(i)}) \eta_k + \sum_{k,m=1}^{N_\Gamma} \mu_m a_i(\varphi_k^{(\Gamma)}{}_{|\Omega_i}, \varphi_m^{(\Gamma)}{}_{|\Omega_i}) \eta_k \\
&= \mathbf{w}^T A_{ii}\mathbf{u} + \mu^T A_{\Gamma i}\mathbf{u} + \mathbf{w}^T A_{i\Gamma}\eta + \mu^T A_{\Gamma\Gamma}^{(i)}\eta.
\end{aligned}$$

Thanks to (18.60) we obtain

$$\begin{aligned}
\langle S_{i,h}\eta_h, \mu_h\rangle &= -\mathbf{w}^T A_{i\Gamma}\eta - \mu^T A_{\Gamma i} A_{ii}^{-1} A_{i\Gamma}\eta + \mathbf{w}^T A_{i\Gamma}\eta + \mu^T A_{\Gamma\Gamma}^{(i)}\eta \\
&= \mu^T \left(A_{\Gamma\Gamma}^{(i)} - A_{\Gamma i} A_{ii}^{-1} A_{i\Gamma}\right)\eta \\
&= \mu^T \Sigma_i \eta.
\end{aligned}$$

◇

From Theorem 18.3 and thanks to the representation (18.63), we deduce that there exist two constants $\hat{K}_1, \hat{K}_2 > 0$, independent of h, such that

$$\hat{K}_1\langle S_{1,h}\mu_h, \mu_h\rangle \le \langle S_{2,h}\mu_h, \mu_h\rangle \le \hat{K}_2\langle S_{1,h}\mu_h, \mu_h\rangle \qquad \forall \mu_h \in \Lambda_h. \tag{18.66}$$

Thanks to (18.64) we can infer that there exist two constants $\tilde{K}_1, \tilde{K}_2 > 0$, independent of h, such that

$$\tilde{K}_1 \left(\boldsymbol{\mu}^T \Sigma_1 \boldsymbol{\mu}\right) \leq \boldsymbol{\mu}^T \Sigma_2 \boldsymbol{\mu} \leq \tilde{K}_2 \left(\boldsymbol{\mu}^T \Sigma_1 \boldsymbol{\mu}\right) \qquad \forall \boldsymbol{\mu} \in \mathbb{R}^{N_\Gamma}. \tag{18.67}$$

This amounts to say that the two matrices Σ_1 and Σ_2 are spectrally equivalent, that is their spectral condition number features the same asymptotic behaviour w.r.t h. Henceforth, both Σ_1 and Σ_2 provide an optimal preconditioner of the Schur complement Σ, that is there exists a constant C, independent of h, such that

$$K_2(\Sigma_i^{-1}\Sigma) \leq C, \quad i = 1,2. \tag{18.68}$$

As we will see in Sect. 18.3.3, this property allows us to prove that the discrete version of the Dirichlet-Neumann algorithm converges with a rate independent of h. A similar result holds for the discrete Neumann-Neumann and Robin-Robin algorithms.

18.3.3 Equivalence between the Dirichlet-Neumann algorithm and a preconditioned Richardson algorithm in the discrete case

Let us now prove the analogue of the equivalence theorem 18.2 in the algebraic case. The finite element approximation of the Dirichlet problem (18.31) has the following algebraic form

$$A_{11}\mathbf{u}_1^{(k)} = \mathbf{f}_1 - A_{1\Gamma}\boldsymbol{\lambda}^{(k-1)}, \tag{18.69}$$

whereas that of the Neumann problem (18.32) reads

$$\begin{bmatrix} A_{22} & A_{2\Gamma} \\ A_{\Gamma 2} & A_{\Gamma\Gamma}^{(2)} \end{bmatrix} \begin{bmatrix} \mathbf{u}_2^{(k)} \\ \boldsymbol{\lambda}^{(k-1/2)} \end{bmatrix} = \begin{bmatrix} \mathbf{f}_2 \\ \mathbf{f}_\Gamma - A_{\Gamma 1}\mathbf{u}_1^{(k)} - A_{\Gamma\Gamma}^{(1)}\boldsymbol{\lambda}^{(k-1)} \end{bmatrix}. \tag{18.70}$$

In its turn, (18.33) becomes

$$\boldsymbol{\lambda}^{(k)} = \theta\boldsymbol{\lambda}^{(k-1/2)} + (1-\theta)\boldsymbol{\lambda}^{(k-1)}. \tag{18.71}$$

By eliminating $\mathbf{u}_2^{(k)}$ from (18.70) we obtain

$$\left(A_{\Gamma\Gamma}^{(2)} - A_{\Gamma 2}A_{22}^{-1}A_{2\Gamma}\right)\boldsymbol{\lambda}^{(k-1/2)} = \mathbf{f}_\Gamma - A_{\Gamma 1}\mathbf{u}_1^{(k)} - A_{\Gamma\Gamma}^{(1)}\boldsymbol{\lambda}^{(k-1)} - A_{\Gamma 2}A_{22}^{-1}\mathbf{f}_2.$$

By the definition (18.51) of Σ_2 and by (18.69), one has

$$\Sigma_2\boldsymbol{\lambda}^{(k-1/2)} = \mathbf{f}_\Gamma - A_{\Gamma 1}A_{11}^{-1}\mathbf{f}_1 - A_{\Gamma 2}A_{22}^{-1}\mathbf{f}_2 - \left(A_{\Gamma\Gamma}^{(1)} - A_{\Gamma 1}A_{11}^{-1}A_{1\Gamma}\right)\boldsymbol{\lambda}^{(k-1)},$$

that is, owing to the definition (18.51) of Σ_1 and to (18.52),

$$\boldsymbol{\lambda}^{(k-1/2)} = \Sigma_2^{-1}\left(\boldsymbol{\chi}_\Gamma - \Sigma_1\boldsymbol{\lambda}^{(k-1)}\right).$$

Now, by virtue of (18.71) we deduce

$$\boldsymbol{\lambda}^{(k)} = \theta\Sigma_2^{-1}\left(\boldsymbol{\chi}_\Gamma - \Sigma_1\boldsymbol{\lambda}^{(k-1)}\right) + (1-\theta)\boldsymbol{\lambda}^{(k-1)},$$

that is, since $-\Sigma_1 = -\Sigma + \Sigma_2$,

$$\boldsymbol{\lambda}^{(k)} = \theta \Sigma_2^{-1} \left(\chi_\Gamma - \Sigma \boldsymbol{\lambda}^{(k-1)} + \Sigma_2 \boldsymbol{\lambda}^{(k-1)} \right) + (1-\theta) \boldsymbol{\lambda}^{(k-1)}$$

whence

$$\Sigma_2(\boldsymbol{\lambda}^{(k)} - \boldsymbol{\lambda}^{(k-1)}) = \theta(\chi_\Gamma - \Sigma \boldsymbol{\lambda}^{(k-1)}).$$

The latter is nothing but a Richardson iteration on the system (18.53) using the local Schur complement Σ_2 as preconditioner.

Remark 18.2. The Richardson preconditioner induced by the Dirichlet-Neumann algorithm is in fact the local Schur complement associated to that subdomain on which we solve a Neumann problem. So, in the so-called *Neumann-Dirichlet* algorithm, in which at every iteration we solve a Dirichlet problem in Ω_2 and a Neumann one in Ω_1, the preconditioner of the associated Richardson algorithm would be Σ_1 and not Σ_2. •

Remark 18.3. An analogous result can be proven for the discrete version of the Neumann-Neumann algorithm introduced in Sect. 18.1.3. Precisely, the Neumann-Neumann algorithm is equivalent to the Richardson algorithm applied to system (18.53) with a preconditioner whose inverse is given by $P_h^{-1} = \sigma_1 \Sigma_1^{-1} + \sigma_2 \Sigma_2^{-1}$, σ_1 and σ_2 being the coefficients used for the (discrete) interface equation which corresponds to (18.19). Moreover we can prove that there exists a constant $C > 0$, independent of h, such that

$$K_2((\sigma_1 \Sigma_1^{-1} + \sigma_2 \Sigma_2^{-1})\Sigma) \leq C\,.$$

Proceeding in a similar way we can show that the discrete version of the Robin-Robin algorithm (18.20)-(18.21) is also equivalent to a Richardson algorithm for (18.53), using this time as preconditioner the matrix $(\gamma_1+\gamma_2)^{-1}(\gamma_1 I + \Sigma_1)(\gamma_2 I + \Sigma_2)$. •

Let us recall that a matrix P_h is an optimal preconditioner for Σ if the condition number of $P_h^{-1}\Sigma$ is bounded uniformely w.r.t the dimension N of the matrix Σ (and therefore from h in the case in which Σ arises from a finite element discretization).

We can therefore summarize by saying that for the solution of system $\Sigma \boldsymbol{\lambda} = \chi_\Gamma$, we can make use of the following preconditioners, all of them being optimal:

$$P_h = \begin{cases} \Sigma_2 & \text{for the Dirichlet-Neumann algorithm,} \\ \Sigma_1 & \text{for the Neumann-Dirichlet algorithm,} \\ (\sigma_1 \Sigma_1^{-1} + \sigma_2 \Sigma_2^{-1})^{-1} & \text{for the Neumann-Neumann algorithm,} \\ (\gamma_1+\gamma_2)^{-1}(\gamma_1 I + \Sigma_1)(\gamma_2 I + \Sigma_2) & \text{for the Robin-Robin algorithm.} \end{cases} \tag{18.72}$$

When solving the flux equation (18.58), the FETI preconditioner reads $P_h = (\Sigma_1 + \Sigma_2)^{-1}$, yelding the preconditioned matrix $(\Sigma_1+\Sigma_2)(\Sigma_1^{-1}+\Sigma_2^{-1})$. For all these preconditioners, optimality follows from the spectral equivalence (18.67), hence $K_2(P_h^{-1}\Sigma)$ is bounded independently of h.

From the convergence theory of Richardson method we know that if both Σ and P_h are symmetric and positive definite, one has $\|\boldsymbol{\lambda}^n - \boldsymbol{\lambda}\|_\Sigma \leq \rho^n \|\boldsymbol{\lambda}^0 - \boldsymbol{\lambda}\|_\Sigma$, $n \geq 0$, being $\|\boldsymbol{v}\|_\Sigma = (\boldsymbol{v}^T \Sigma \boldsymbol{v})^{1/2}$. The optimal convergence rate is given by

$$\rho = \frac{K_2(P_h^{-1}\Sigma) - 1}{K_2(P_h^{-1}\Sigma) + 1},$$

and is therefore independent of h.

18.4 Generalization to the case of many subdomains

To generalize the previous DD algorithms to the case in which the domain Ω is partitioned into an arbitrary number $M > 2$ of subdomains we proceed as follows.
Let Ω_i, $i = 1, \ldots, M$, denote a family of disjoint subdomains such that $\cup \overline{\Omega}_i = \overline{\Omega}$, and denote $\Gamma_i = \partial\Omega_i \setminus \partial\Omega$ and $\Gamma = \cup \Gamma_i$ (the skeleton).

Let us consider the Poisson problem (3.13). In the current case the equivalence Theorem 18.1 generalizes as follows:

$$\begin{cases} -\Delta u_i = f & \text{in } \Omega_i, \\ u_i = u_k & \text{on } \Gamma_{ik},\ \forall k \in \mathcal{A}(i), \\ \dfrac{\partial u_i}{\partial n_i} = \dfrac{\partial u_k}{\partial n_i} & \text{on } \Gamma_{ik},\ \forall k \in \mathcal{A}(i), \\ u_i = 0 & \text{on } \partial\Omega_i \cap \partial\Omega, \end{cases} \tag{18.73}$$

for $i = 1, \ldots, M$, being $\Gamma_{ik} = \partial\Omega_i \cap \partial\Omega_k \neq \emptyset$, $\mathcal{A}(i)$ the set of indices k such that Ω_k is adjacent to Ω_i; as usual, $\mathbf{n}_i$ denotes the outward unit normal vetor to Ω_i.

Assume now that (3.13) has been approximated by the finite element method. Following the ideas presented in Sect. 18.3 and denoting by $\mathbf{u} = (\mathbf{u}_I, \mathbf{u}_\Gamma)^T$ the vector of unknowns split in two subvectors, the one ($\mathbf{u}_I$) related with the internal nodes, and that ($\mathbf{u}_\Gamma$) related with the nodes lying on the skeleton Γ, the finite element algebraic system can be reformulated in blockwise form as follows

$$\begin{bmatrix} A_{II} & A_{I\Gamma} \\ A_{\Gamma I} & A_{\Gamma\Gamma} \end{bmatrix} \begin{bmatrix} \mathbf{u}_I \\ \mathbf{u}_\Gamma \end{bmatrix} = \begin{bmatrix} \mathbf{f}_I \\ \mathbf{f}_\Gamma \end{bmatrix}, \tag{18.74}$$

being $A_{\Gamma I} = A_{I\Gamma}^T$. Matrix $A_{I\Gamma}$ is banded, while A_{II} has the block diagonal form

$$A_{II} = \begin{bmatrix} A_{\Omega_1,\Omega_1} & 0 & \ldots & 0 \\ 0 & \ddots & & \vdots \\ \vdots & & \ddots & 0 \\ 0 & \ldots & 0 & A_{\Omega_M,\Omega_M} \end{bmatrix}. \tag{18.75}$$

We are using the following notations:

$$(A_{\Omega_i,\Omega_i})_{lj} = a_i(\varphi_j, \varphi_l), \quad 1 \le l, j \le N_i,$$
$$(A^{(i)}_{\Gamma\Gamma})_{sr} = a_i(\psi_r, \psi_s), \quad 1 \le r, s \le N_{\Gamma_i},$$
$$(A_{\Omega_i,\Gamma})_{lr} = a_i(\psi_r, \varphi_l), \quad 1 \le r \le N_{\Gamma_i}, \qquad 1 \le l \le N_i,$$

where N_i is the number of nodes internal to Ω_i, N_{Γ_i} that of the nodes sitting on the interface Γ_i, φ_j and ψ_r the basis functions associated with the internal and interface nodes, respectively.

Let us remark that on every subdomain Ω_i the matrix

$$A_i = \begin{bmatrix} A_{\Omega_i,\Omega_i} & A_{\Omega_i,\Gamma} \\ A_{\Gamma,\Omega_i} & A^{(i)}_{\Gamma\Gamma} \end{bmatrix} \tag{18.76}$$

represents the local finite element stiffness matrix associated to a Neumann problem on Ω_i. Since A_{II} is non-singular, from (18.74) we can formally derive

$$\mathbf{u}_I = A_{II}^{-1}(\mathbf{f}_I - A_{I\Gamma}\mathbf{u}_\Gamma). \tag{18.77}$$

By eliminating the unknown $\mathbf{u}_I$ from system (18.74), it follows

$$A_{\Gamma\Gamma}\mathbf{u}_\Gamma = \mathbf{f}_\Gamma - A_{\Gamma I}A_{II}^{-1}(\mathbf{f}_I - A_{I\Gamma}\mathbf{u}_\Gamma),$$

that is

$$\begin{pmatrix} A_{II} & A_{I\Gamma} \\ 0 & \Sigma \end{pmatrix} \begin{pmatrix} u_I \\ u_\Gamma \end{pmatrix} = \begin{pmatrix} \mathbf{f}_I \\ \chi_\Gamma \end{pmatrix} \tag{18.78}$$

having set

$$\Sigma = A_{\Gamma\Gamma} - A_{\Gamma I}A_{II}^{-1}A_{I\Gamma} \text{ and } \chi_\Gamma = \mathbf{f}_\Gamma - A_{\Gamma I}A_{II}^{-1}\mathbf{f}_I.$$

Denoting, as usual, $\boldsymbol{\lambda} = \mathbf{u}_\Gamma$, (18.78) yields

$$\Sigma\boldsymbol{\lambda} = \chi_\Gamma. \tag{18.79}$$

This is the Schur complement system in the multidomain case. It can be regarded as a finite element approximation of the interface Steklov-Poincaré problem in the case of M subdomains.

The local Schur complements are defined as

$$\Sigma_i = A^{(i)}_{\Gamma\Gamma} - A_{\Gamma,\Omega_i}A^{-1}_{\Omega_i,\Omega_i}A_{\Omega_i,\Gamma}, \quad i = 1, \dots M, \tag{18.80}$$

hence

$$\Sigma = R^T_{\Gamma_1}\Sigma_1 R_{\Gamma_1} + \dots + R^T_{\Gamma_M}\Sigma_M R_{\Gamma_M} \tag{18.81}$$

where R_{Γ_i} is a restriction operator, that is a rectangular matrix of zeros and ones that map values on Γ onto those on Γ_i, $i = 1, \dots, M$. Note that the r.h.s. of (18.79) can be written as a sum of local contributions,

$$\chi_\Gamma = \sum_{i=1}^{M} R_{\Gamma_i}^T \left(\mathbf{f}_\Gamma^{(i)} - A_{\Gamma,\Omega_i} A_{\Omega_i,\Omega_i}^{-1} \mathbf{f}_I^{(i)} \right) \tag{18.82}$$

A general algorithm to solve the finite element Poisson problem in Ω could be formulated as follows:

1. compute the solution of (18.79) to obtain the value of $\boldsymbol{\lambda}$ on the skeleton Γ;
2. then solve (18.77); since A_{II} is block-diagonal, this step yields the solution of M independent subproblems of reduced dimension, $A_{\Omega_i,\Omega_i}\mathbf{u}_I^i = \mathbf{g}^i$, $i = 1,\dots,M$, which can therefore be carried out in parallel.

About the condition number of Σ, the following estimate can be proven: there exists a constant $C > 0$, independent of h and H_{min}, H_{max}, such that

$$K_2(\Sigma) \leq C \frac{H_{max}}{h H_{min}^2}, \tag{18.83}$$

H_{max} being the maximum diameter of the subdomains and H_{min} the minimum one.

Remark 18.4 (Approximation of the inverse of A). The inverse of the block matrix A in (18.74) admits the following LDU factorization

$$A^{-1} = \begin{bmatrix} I & -A_{II}^{-1}A_{I\Gamma} \\ 0 & I \end{bmatrix} \begin{bmatrix} A_{II}^{-1} & 0 \\ 0 & \Sigma^{-1} \end{bmatrix} \begin{bmatrix} I & 0 \\ -A_{\Gamma I}A_{II}^{-1} & I \end{bmatrix} \tag{18.84}$$

Should we have suitable preconditioners B_{II} of A_{II} and P of Σ, an approximation of A^{-1} would be given by

$$P_A^{-1} = \begin{bmatrix} I & -B_{II}^{-1}A_{I\Gamma} \\ 0 & I \end{bmatrix} \begin{bmatrix} B_{II}^{-1} & 0 \\ 0 & P^{-1} \end{bmatrix} \begin{bmatrix} I & 0 \\ -A_{\Gamma I}B_{II}^{-1} & I \end{bmatrix}. \tag{18.85}$$

An application of P_A^{-1} to a given vector involves B_{II}^{-1} in two matrix-vector multiplies and P^{-1} in only one matrix-vector multiply (see [TW05, Sect. 4.3]). •

Remark 18.5 (Saddle-point systems). In case of a saddle-point (block) matrix like the one in (16.58), an LDU factorization can be obtained as follows

$$K = \begin{bmatrix} A & B^T \\ B & -C \end{bmatrix} = \begin{bmatrix} I_A & 0 \\ BA^{-1} & I_C \end{bmatrix} \begin{bmatrix} A & 0 \\ 0 & S \end{bmatrix} \begin{bmatrix} I_A & A^{-1}B^T \\ 0 & I_C \end{bmatrix} \tag{18.86}$$

where $S = -C - BA^{-1}B^T$ is the Schur complement computed with respect to the second variable (e.g. $\mathbf{P}$ in the case of system (16.58)).
An inverse of K is obtained as

$$K^{-1} = \begin{bmatrix} A^{-1} & 0 \\ 0 & 0 \end{bmatrix} + QS^{-1}Q^T, \quad Q = \begin{bmatrix} -A^{-1}B^T \\ I \end{bmatrix}, \tag{18.87}$$

being I_A and I_C two identity matrices having the size of A and C, respectively. A preconditioner for K can be constructed by replacing in (18.87) A^{-1} and S^{-1} by suitable domain decomposition preconditioners of A and S, respectively.

This observation stands at the ground of the design of the so-called FETI-DP and BDDC preconditioners, see Sects. 18.5.5 and 18.5.6. •

18.4.1 Some numerical results

Consider the Poisson problem (3.13) on the domain $\Omega = (0,1)^2$ whose finite element approximation was given in (4.40).
Let us partition Ω into M disjoint squares Ω_i whose sidelength is H, such that $\cup_{i=1}^M \overline{\Omega_i} = \overline{\Omega}$. An example with four subdomains is displayed in Fig. 18.5 (left).

In Table 18.1 we report the numerical values of $K_2(\Sigma)$ for the problem at hand, for several values of the finite element grid-size h; it grows linearly with $1/h$ and with $1/H$, as predicted by the formula (18.83). In Fig. 18.5 (right) we display the *pattern* of the Schur complement matrix Σ in the particular case of $h = 1/8$ and $H = 1/2$. The matrix has a blockwise structure that accounts for the interfaces Γ_1, Γ_2, Γ_3 and Γ_4, plus the contribution arising from the crosspoint Γ_c. Since Σ is a *dense* matrix, when solving the linear system (18.79) the explicit computation of its entries is not convenient. Instead, we can use the following **Algorithm 18.1** to compute the matrix-vector product $\Sigma \mathbf{x}_\Gamma$, for any vector $\mathbf{x}_\Gamma$ (and therefore the residue at every step of an iterative algorithm). We have denoted by R_{Γ_i} the rectangular matrix associated to the restriction operator $R_{\Gamma_i} : \Gamma \to \Gamma_i = \partial\Omega_i \setminus \partial\Omega$, while $\mathbf{x} \leftarrow \mathbf{y}$ indicates the algebraic operation $\mathbf{x} = \mathbf{x} + \mathbf{y}$.

Table 18.1. Condition number of the Schur complement Σ

$K_2(\Sigma)$	$H = 1/2$	$H = 1/4$	$H = 1/8$
$h = 1/8$	9.77	14.83	25.27
$h = 1/16$	21.49	35.25	58.60
$h = 1/32$	44.09	75.10	137.73
$h = 1/64$	91.98	155.19	290.43

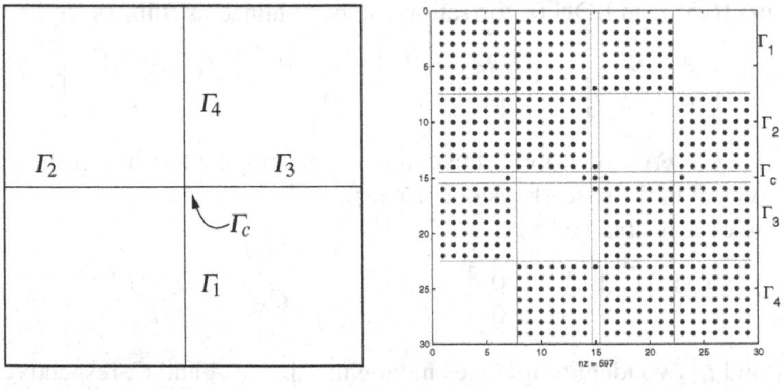

Fig. 18.5. Example of partition of $\Omega = (0,1)^2$ into four squared subdomains (left). Pattern of the Schur complement Σ (right) corresponding to the domain partition displayed on the left

Algorithm 18.1 (Schur complement multiplication by a vector)

Given $\mathbf{x}_\Gamma$, compute $\mathbf{y}_\Gamma = \Sigma\mathbf{x}_\Gamma$ as follows:

a. Set $\mathbf{y}_\Gamma = \mathbf{0}$

b. For $i = 1,\ldots,M$ Do in parallel:

c. $\quad \mathbf{x}_i = R_{\Gamma_i}\mathbf{x}_\Gamma$

d. $\quad \mathbf{z}_i = A_{\Omega_i,\Gamma_i}\mathbf{x}_i$

e. $\quad \mathbf{z}_i \leftarrow A^{-1}_{\Omega_i,\Omega_i}\mathbf{z}_i$

f. $\quad$ sum up in the local vector $\mathbf{y}_{\Gamma_i} \leftarrow A_{\Gamma_i,\Gamma_i}\mathbf{x}_i - A_{\Gamma_i,\Omega_i}\mathbf{z}_i$

g. $\quad$ sum up in the global vector $\mathbf{y}_\Gamma \leftarrow R^T_{\Gamma_i}\mathbf{y}_{\Gamma_i}$

h. EndFor

Since no communication is required among the subdomains, this is a fully parallel algorithm.

Before using for the first time the Schur complement, a start-up phase, described in **Algorithm 18.2**, is requested. Note that this is an *off-line* procedure.

Algorithm 18.2 (Start-up phase for the solution of the Schur complement system)

Given $\mathbf{x}_\Gamma$, compute $\mathbf{y}_\Gamma = \Sigma\mathbf{x}_\Gamma$ as follows:

a. For $i = 1,\ldots,M$ Do in parallel:

b. $\quad$ Compute the entries of A_i

c. $\quad$ Reorder A_i as in (18.76) then extract the submatrices $A_{\Omega_i,\Omega_i}, A_{\Omega_i,\Gamma_i}, A_{\Gamma_i,\Omega_i}$ and $A^{(i)}_{\Gamma,\Gamma}$

d. $\quad$ Compute the (either LU or Cholesky) factorization of A_{Ω_i,Ω_i}

e. EndFor

18.5 DD preconditioners in case of many subdomains

Before introducing the preconditioners for the Schur complement in the case in which Ω is partitioned in many subdomains we recall the following definition:

Definition 18.1. *A preconditioner P_h of Σ is said to be* scalable *if the condition number of the preconditioned matrix $P_h^{-1}\Sigma$ is independent of the number of subdomains.*

Iterative methods using scalable preconditioners allow henceforth to achieve convergence rates independent of the subdomain number. This is a very desirable property in those cases where a large number of subdomains is used.

Let R_i be a *restriction operator* which, to any vector $\mathbf{v}_h$ of nodal values on the global domain Ω, associates its restriction to the subdomain Ω_i

$$R_i : \mathbf{v}_{h|\Omega} \to \mathbf{v}^i_{h|\Omega_i \cup \Gamma_i}.$$

Let moreover

$$R_i^T : \mathbf{v}^i_{h|\Omega_i \cup \Gamma_i} \to \mathbf{v}_{h|\Omega}$$

be the *prolongation (or extension-by-zero) operator*. In algebraic form R_i can be represented by a matrix that coincides with the identity matrix in correspondence with the subdomain Ω_i

$$R_i = \left[\begin{array}{ccc|ccc|ccc} 0 & \dots & 0 & 1 & & & 0 & \dots & 0 \\ \vdots & \ddots & \vdots & & \ddots & & \vdots & \ddots & \vdots \\ 0 & \dots & 0 & & & 1 & 0 & \dots & 0 \end{array}\right]$$
$$\underbrace{\qquad\qquad}_{\Omega_i}$$

Similarly we can define the restriction and prolongation operators R_{Γ_i} and $R^T_{\Gamma_i}$, respectively, that act on the vector of interface nodal values (as done in (18.81)). In order to find a preconditioner for Σ the strategy consists of combining the contributions of local subdomain preconditioners with that of a global contribution referring to a coarse grid whose elements are the subdomains themselves. Without the latter coarse grid term the preconditioner could not be scalable since it would lack any mechanism for global communication of information across the domain in each iteration step. This idea can be formalized through the following relation that provides the inverse of the preconditioner

$$(P_h)^{-1} = \sum_{i=1}^{M} R^T_{\Gamma_i} P^{-1}_{i,h} R_{\Gamma_i} + R^T_{\Gamma} P_H^{-1} R_{\Gamma}\,.$$

We have denoted by H the maximum value of the diameters H_i of the subdomains Ω_i; moreover, $P_{i,h}$ is either the local Schur complement Σ_i, or (more frequently) a suitable preconditioner of Σ_i, while R_Γ and P_H refer to operators that act on the global scale (that of the coarse grid).

Many different choices are possible for the local Schur complement preconditioner $P_{i,h}$; they will give rise to different condition numbers of the preconditioned matrix $P_h^{-1}\Sigma$.

18.5.1 Jacobi preconditioner

Let $\{e_1, \dots, e_m\}$ be the set of edges and $\{v_1, \dots, v_n\}$ that of vertices of a partition of Ω into subdomains (see Fig. 18.6 for an example).

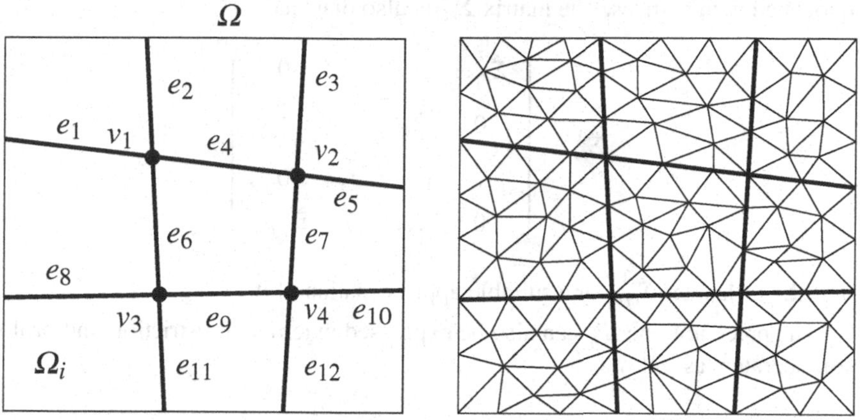

Fig. 18.6. A decomposition into 9 subdomains (left) with a fine triangulation in small triangles and a coarse triangulation in large quadrilaterals (the 9 subdomains) (right)

The Schur complement Σ features the following blockwise representation

$$\Sigma = \left[\begin{array}{c|c} \Sigma_{ee} & \Sigma_{ev} \\ \hline \Sigma_{ev}^T & \Sigma_{vv} \end{array}\right],$$

having set

$$\Sigma_{ee} = \begin{bmatrix} \Sigma_{e_1e_1} & \cdots & \Sigma_{e_1e_m} \\ \vdots & \ddots & \vdots \\ \Sigma_{e_me_1} & \cdots & \Sigma_{e_me_m} \end{bmatrix}, \quad \Sigma_{ev} = \begin{bmatrix} \Sigma_{e_1v_1} & \cdots & \Sigma_{e_1v_n} \\ \vdots & \ddots & \vdots \\ \Sigma_{e_mv_1} & \cdots & \Sigma_{e_mv_n} \end{bmatrix}$$

and

$$\Sigma_{vv} = \begin{bmatrix} \Sigma_{v_1v_1} & 0 & \ldots & 0 \\ 0 & \ddots & & \vdots \\ \vdots & & \ddots & 0 \\ 0 & \ldots & 0 & \Sigma_{v_nv_n} \end{bmatrix}.$$

In 3D there should be a further block row and column due to the presence of faces.

The *Jacobi preconditioner* of the Schur complement Σ is a block diagonal matrix defined by

$$P_h^J = \left[\begin{array}{c|c} \hat{\Sigma}_{ee} & 0 \\ \hline 0 & \Sigma_{vv} \end{array}\right]$$

where $\hat{\Sigma}_{ee}$ is either Σ_{ee} or a suitable approximation of it. This preconditioner does not account for the interaction between the basis functions associated with edges and those associated with vertices. The matrix $\hat{\Sigma}_{ee}$ is also diagonal

$$\hat{\Sigma}_{ee} = \begin{bmatrix} \hat{\Sigma}_{e_1e_1} & 0 & \dots & 0 \\ 0 & \ddots & & \vdots \\ \vdots & & \ddots & 0 \\ 0 & \dots & 0 & \hat{\Sigma}_{e_me_m} \end{bmatrix}.$$

Here $\hat{\Sigma}_{e_ke_k}$ denotes $\Sigma_{e_ke_k}$ or a suitable approximation of it.

The preconditioner P_h^J can also be expressed in terms of restriction and prolongation operators as follows

$$\left(P_h^J\right)^{-1} = \sum_{k=1}^{m} R_{e_k}^T \hat{\Sigma}_{e_ke_k}^{-1} R_{e_k} + R_v^T \Sigma_{vv}^{-1} R_v , \tag{18.88}$$

where R_{e_k} and R_v denote edge and vertices restriction operators, respectively.

Regarding the condition number of the preconditioned Schur complement, there exists a constant $C > 0$, indipendent of both h and H, such that

$$K_2\left((P_h^J)^{-1}\Sigma\right) \leq CH^{-2}\left(1 + \log\frac{H}{h}\right)^2.$$

Should the conjugate gradient method be used to solve the preconditioned Schur complement system (18.79) with preconditioner P_h^J, the number of iterations necessary to converge (within a prescribed tolerance) would be proportional to H^{-1}. The presence of H indicates that the Jacobi preconditioner is not scalable.

Moreover, we notice that the presence of the logarithmic term $\log(H/h)$ introduces a relation between the size of the subdomains and the size of the computational grid $\mathcal{T}_h$. This generates a propagation of information among subdomains characterized by a finite (rather than infinite) speed of propagation. Note that the ratio H/h measures the maximum number of elements across any subdomain.

18.5.2 Bramble-Pasciak-Schatz preconditioner

With the aim of accelerating the speed of propagation of information among subdomains we can devise a mechanism of global coupling among subdomains. As already anticipated, the family of subdomains can be regarded as a *coarse* grid, say $\mathcal{T}_H$, of the original domain. For instance, in Fig. 18.6 $\mathcal{T}_H$ is made of 9 (macro) elements and 4 internal nodes. It identifies a stiffness matrix of piecewise bilinear elements, say A_H, of dimension 4×4 which guarantees a global coupling in Ω. We can now introduce a restriction operator that, for simplicity, we indicate $R_H : \Gamma_h \to \Gamma_H$. More precisely, this operator transforms a vector of nodal values on the skeleton Γ_h into a vector of nodal values on the internal vertices of the coarse grid (4 in the case at hand). Its transpose

R_H^T is an extension operator. The matrix P_h^{BPS}, whose inverse is

$$(P_h^{BPS})^{-1} = \sum_{k=1}^{m} R_{e_k}^T \hat{\Sigma}_{e_k e_k}^{-1} R_{e_k} + R_H^T A_H^{-1} R_H \,, \tag{18.89}$$

is named Bramble-Pasciak-Schatz preconditioner. The main difference with Jacobi preconditioner (18.88) is due to the presence of the global (coarse-grid) stiffness matrix A_H instead of the diagonal vertex matrix Σ_{vv}. The following results hold:

$$K_2\left((P_h^{BPS})^{-1}\Sigma\right) \leq C\left(1+\log\frac{H}{h}\right)^2 \quad \text{in 2D},$$

$$K_2\left((P_h^{BPS})^{-1}\Sigma\right) \leq C\frac{H}{h} \quad \text{in 3D}.$$

Note that the factor H^{-2} does not show up anymore. The number of iterations of the conjugate gradient method with preconditioner P_h^{BPS} is now proportional to $\log(H/h)$ in 2D and to $(H/h)^{1/2}$ in 3D.

18.5.3 Neumann-Neumann preconditioner

Although the Bramble-Pasciak-Schatz preconditioner has better properties than Jacobi's, yet in 3D the condition number of the preconditioned Schur complement still contains a linear dependence on H/h.

In this respect, a further improvement is achievable using the so-called Neumann-Neumann preconditioner, whose inverse has the following expression

$$(P_h^{NN})^{-1} = \sum_{i=1}^{M} R_{\Gamma_i}^T D_i \Sigma_i^* D_i R_{\Gamma_i}. \tag{18.90}$$

As before, R_{Γ_i} denotes the restriction from the nodal values on the whole skeleton Γ to those on the local interface Γ_i, whereas Σ_i^* is either Σ_i^{-1} (should the local inverse exist) or an approximation of Σ_i^{-1}, e.g. the pseudo-inverse Σ_i^+ of Σ_i. The matrix D_i is a diagonal matrix of positive weights $d_j > 0$, for $j = 1,\ldots,n$, n being the number of nodes on Γ_i. For instance, d_j coincides with the inverse of the number of subdomains that share the $j-th$ node. If we still consider the 4 internal vertices of Fig. 18.6, we will have $d_j = 1/4$, for $j = 1,\ldots,4$.

For the preconditioner (18.90) the following estimate (similar to that of Jacobi preconditioner) holds: there exists a constant $C > 0$, indipendent of both h and H, such that

$$K_2\left((P_h^{NN})^{-1}\Sigma\right) \leq CH^{-2}\left(1+\log\frac{H}{h}\right)^2.$$

The last (logarithmic) factor drops out in case the subdomains partition features no cross points.

The presence of D_i and R_{Γ_i} in (18.90) only entails matrix-matrix multiplications. On the other hand, if $\Sigma_i^* = \Sigma_i^{-1}$, applying Σ_i^{-1} to a given vector can be reconducted to the use of local inverses. As a matter of fact, let $\mathbf{q}$ be a vector whose components are the nodal values on the local interface Γ_i; then

$$\Sigma_i^{-1}\mathbf{q} = [0, I]A_i^{-1}[0, I]^T\mathbf{q}.$$

In particular, $[0, I]^T\mathbf{q} = [0, \mathbf{q}]^T$, and the matrix-vector product

$$\underbrace{\left[\begin{array}{c|c} \text{internal} & \\ \text{nodes} & \\ & \\ \hline \text{boundary nodes} & \end{array}\right]}_{A_i^{-1}} \left[\begin{array}{c} 0 \\ \vdots \\ 0 \\ \hline \mathbf{q} \end{array}\right]$$

corresponds to the solution on Ω_i of the Neumann boundary-value problem

$$\begin{cases} -\Delta w_i = 0 & \text{in } \Omega_i, \\ \dfrac{\partial w_i}{\partial n} = q & \text{on } \Gamma_i. \end{cases} \tag{18.91}$$

Algorithm 18.3 (Neumann-Neumann preconditioner)

Given a vector $\mathbf{r}_\Gamma$, compute $\mathbf{z}_\Gamma = (P_h^{NN})^{-1}\mathbf{r}_\Gamma$ as follows:

a. Set $\mathbf{z}_\Gamma = \mathbf{0}$

b. For $i = 1, \ldots, M$ Do in parallel:

c. restrict the residue on Ω_i: $\mathbf{r}_i = R_{\Gamma_i}\mathbf{r}_\Gamma$

d. compute $\mathbf{z}_i = [0, I]A_i^{-1}[0, \mathbf{r}_i]^T$

e. Sum up the global residue: $\mathbf{z}_\Gamma \leftarrow R_{\Gamma_i}^T\mathbf{z}_i$

f. EndFor

Also in this case a start-up phase is required, consisting in the preparation for the solution of linear systems with local stiffness matrices A_i. Note that in the case of the model problem (3.13), A_i is singular if Ω_i is an internal subdomain, that is if $\partial\Omega_i \setminus \partial\Omega = \emptyset$. One of the following strategies should be adopted:

1. compute a (either LU or Cholesky) factorization of $A_i + \varepsilon I$, for a given $\varepsilon > 0$ sufficiently small;
2. compute a factorization of $A_i + \dfrac{1}{H^2}M_i$, where M_i is the mass matrix whose entries are
$$(M_i)_{k,j} = \int_{\Omega_i} \varphi_k\varphi_j\, d\Omega_i;$$
3. compute the singular-value decomposition of A_i.

Table 18.2. Condition number of the preconditioned matrix $(P_h^{NN})^{-1}\Sigma$

$K_2((P_h^{NN})^{-1}\Sigma)$	$H=1/2$	$H=1/4$	$H=1/8$	$H=1/16$
$h=1/16$	2.55	15.20	47.60	—
$h=1/32$	3.45	20.67	76.46	194.65
$h=1/64$	4.53	26.25	105.38	316.54
$h=1/128$	5.79	31.95	134.02	438.02

The matrix Σ_i^* is defined accordingly. In our numerical results we have adopted the third approach.

The convergence history of the preconditioned conjugate gradient method with preconditioner P_h^{NN} in the case $h = 1/32$ is displayed in Fig. 18.7. In Table 18.2 we report the values of the condition number of $(P_h^{NN})^{-1}\Sigma$ for several values of H.

As already pointed out, the Neumann-Neumann preconditioner of the Schur complement matrix is not scalable. A substantial improvement of (18.90) can be achieved by adding a coarse grid correction mechanism, yielding the following new preconditioned Schur complement matrix (see, e.g., [TW05, Sect. 6.2.1])

$$\left(P_h^{BNN}\right)^{-1}\Sigma = P_0 + (I-P_0)((P_h^{NN})^{-1}\Sigma)(I-P_0), \tag{18.92}$$

in which we have used the shorthand notation $P_0 = \bar{R}_0^T\Sigma_0^{-1}\bar{R}_0\Sigma$, $\Sigma_0 = \bar{R}_0\Sigma\bar{R}_0^T$, and $\bar{R}_0$ denotes restriction from Γ onto the coarse level skeleton.
The matrix P_h^{BNN} is called *balanced Neumann-Neumann preconditioner.*

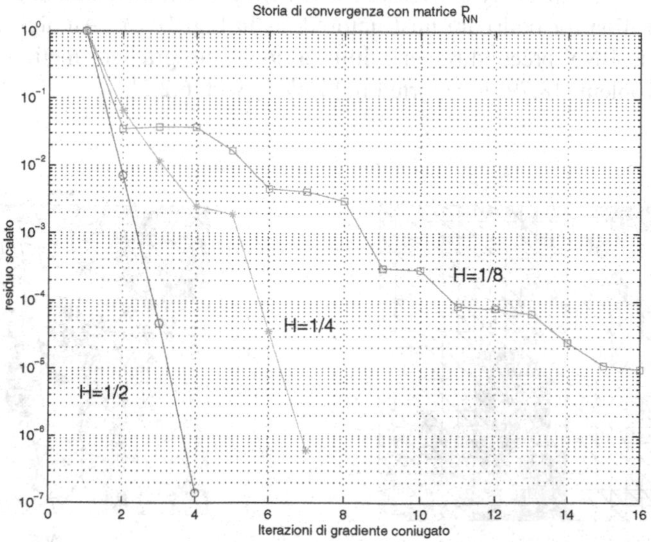

Fig. 18.7. Convergence history for the preconditioned conjugate gradient method with preconditioner P_h^{NN} when $h = 1/32$

It can be proven that there exists a constant $C > 0$, independent of h and H, such that

$$K_2\left((P_h^{BNN})^{-1}\Sigma\right) \leq C\left(1+\log\frac{H}{h}\right)^2$$

both in 2D and 3D. The balanced Neumann-Neumann preconditioner therefore guarantees optimal scalability up to a light logarithmic dependence on H and h.
The coarse grid matrix Σ_0 that is a constituent of Σ_H can be built up using the Algorithm 18.4:

Algorithm 18.4 (construction of the coarse matrix for preconditioner P_h^{BNN})

a. Build the restriction operator $\bar{R}_0$ that returns, for every subdomain, the weighted sum of the values at all the nodes at the boundary of that subdomain
For every node the corresponding weight is given by the inverse of the number of subdomains sharing that node

b. Build up the matrix $\Sigma_0 = \bar{R}_0 \Sigma \bar{R}_0^T$

Step a. of this Algorithm is computationally very cheap, whereas step b. requires several (e.g., ℓ) matrix-vector products involving the Schur complement matrix Σ. Since Σ is never built explicitly, this involves the finite element solution of $\ell \times M$ Dirichlet boundary value problems to generate A_H. Observe moreover that the restriction operator introduced at step a. implicitly defines a coarse space whose functions are piecewise constant on every Γ_i. For this reason the balanced Neumann-Neumann preconditioner is especially convenient when either the finite element grid or the subdomain partition (or both) are unstructured, as in Fig. 18.8). An algorithm that implements the BNN preconditioner within a conjugate gradient method to solve the interface problem (18.79) is reported in [TW05, Sect. 6.2.2].

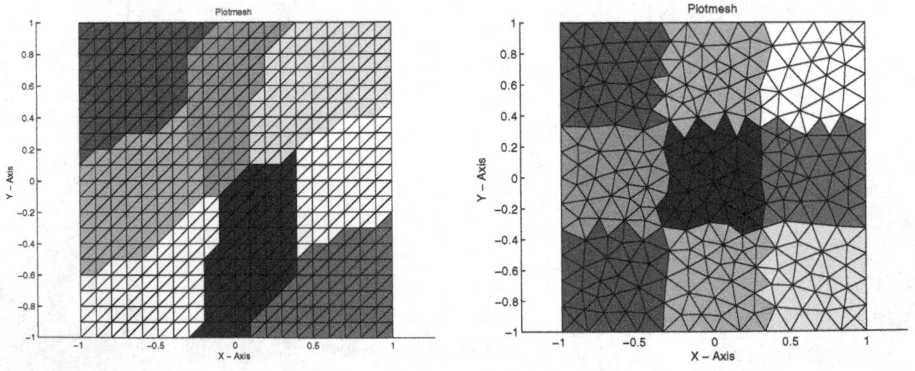

Fig. 18.8. Example of an unstructured subdomain partition in 8 subdomains for a finite element grid which is either structured (left) or unstructured (right)

Table 18.3. Condition number of $(P_h^{BNN})^{-1}\Sigma$ for several values of H

$K_2((P_h^{BNN})^{-1}\Sigma)$	$H=1/2$	$H=1/4$	$H=1/8$	$H=1/16$
$h=1/16$	1.67	1.48	1.27	—
$h=1/32$	2.17	2.03	1.47	1.29
$h=1/64$	2.78	2.76	2.08	1.55
$h=1/128$	3.51	3.67	2.81	2.07

By a comparison of the results obtained using the Neumann-Neumann preconditioner (with and without balancing), the following conclusions can be drawn:

- although featuring a better condition number than A, Σ is still ill-conditioned. The use of a suitable preconditioner is therefore mandatory;
- the Neumann-Neumann preconditioner can be satisfactorily used for partitions featuring a moderate number of subdomains;
- the balancing Neumann-Neumann preconditioner is almost optimally scalable and therefore recommandable for partitions with a large number of subdomains.

18.5.4 FETI (Finite Element Tearing & Interconnecting) methods

In this section we will denote by $H_i = diam(\Omega_i)$, $W_i = W^h(\partial\Omega_i)$ (the space of traces of finite element functions on the boundaries $\partial\Omega_i$), and by $W = \prod_{i=1}^M W_i$ the product space of such trace spaces.

At a later stage we will need two further finite element trace spaces, $\widehat{W} \subset W$ a subspace of *continuous traces* across the skeleton Γ, and $\widetilde{W}$, a possible intermediate space $\widehat{W} \subset \widetilde{W} \subset W$ that will fulfill a smaller number of continuity constraints.

We will consider the variable coefficient elliptic problem

$$\begin{cases} -div(\rho\nabla u) & = f \quad \text{in } \Omega, \\ u & = 0 \quad \text{on } \partial\Omega, \end{cases} \tag{18.93}$$

where ρ is piecewise constant, $\rho = \rho_i \in \mathbb{R}^+$ in Ω_i.
Finally, we will denote by Ω_{ih} the nodes in Ω_i, $\partial\Omega_{ih}$ the nodes on $\partial\Omega_i$, $\partial\Omega_h$ the nodes on $\partial\Omega$, and Γ_h the nodes on Γ. See Fig. 18.9.

Let us introduce the following scaling counting functions: $\forall x \in \Gamma_h \cup \partial\Omega_h$

$$\delta_i(x) = \begin{cases} 1 & x \in \partial\Omega_{ih} \cap (\partial\Omega_h \setminus \Gamma_h), \\ \sum\limits_{j\in N_x} \rho_j^\gamma(x)/\rho_i^\gamma(x) & x \in \partial\Omega_{ih} \cap \Gamma_h, \\ 0 & \text{elsewhere} \end{cases} \tag{18.94}$$

where $\gamma \in [1/2, +\infty)$ and N_x is the set of indices of the subregions having x on their boundary. Then we set

$$\delta_i^\dagger(x) \quad (= \text{pseudo inverses}\,) = \begin{cases} \delta_i^{-1}(x) & \text{if } \delta_i(x) \neq 0, \\ 0 & \text{if } \delta_i(x) = 0. \end{cases} \tag{18.95}$$

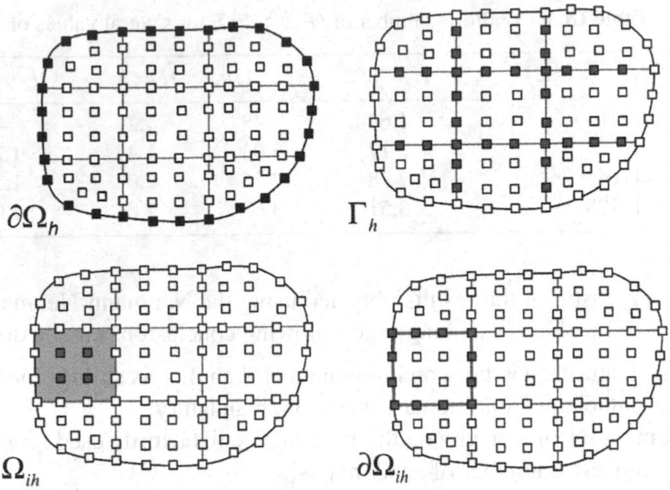

Fig. 18.9. Finite element sets of nodes $\partial\Omega_h$, Γ_h, Ω_{ih}, and $\partial\Omega_{ih}$

Based on the finite element approximation of (18.93), let us consider the local Schur complements (18.80), which are positive semi-definite matrices. In this section we will indicate the interface nodal values on $\partial\Omega_i$ as $\mathbf{u}_i$, and we set $\mathbf{u} = (\mathbf{u}_1, \ldots, \mathbf{u}_M)$, the local load vectors on $\partial\Omega_i$ as χ_i and we set $\chi_\Delta = (\chi_1, \ldots, \chi_M)$. Finally, we set

$$\Sigma_\Delta = diag(\Sigma_1, \ldots, \Sigma_M) = \begin{bmatrix} \Sigma_1 & 0 & \cdots & 0 \\ \vdots & \Sigma_2 & & \vdots \\ & & \ddots & \\ 0 & 0 & \cdots & \Sigma_M \end{bmatrix},$$

a block diagonal matrix.
The original FEM problem, when reduced to the interface Γ, reads

$$\begin{cases} \text{Find } \mathbf{u} \in W \text{ such that} & J(\mathbf{u}) = \frac{1}{2}\langle \Sigma_\Delta \mathbf{u}, \mathbf{u}\rangle - \langle \chi_\Delta, \mathbf{u}\rangle \to min, \\ & B_\Gamma \mathbf{u} = \mathbf{0}. \end{cases} \tag{18.96}$$

B_Γ is not unique, so that we should impose continuity when $\mathbf{u}$ belongs to more than one subdomain; B_Γ is made of $\{0, -1, 1\}$, since it enforces continuity constraints at interfaces' nodes. Here, we are using the same notation W to denote the finite element space trace and that of their nodal values at points of Γ_h.
In 2D, there is a little choice on how to write the constraint of continuity at a point sitting on an edge, there are many options for a vertex point. For the edge node we only need to choose the sign, whereas for a vertex node, e.g. one common to 4 subdomains, a minimum set of three constraints can be chosen in many different ways to assure continuity at the node in question. See, e.g., Fig. 18.10.

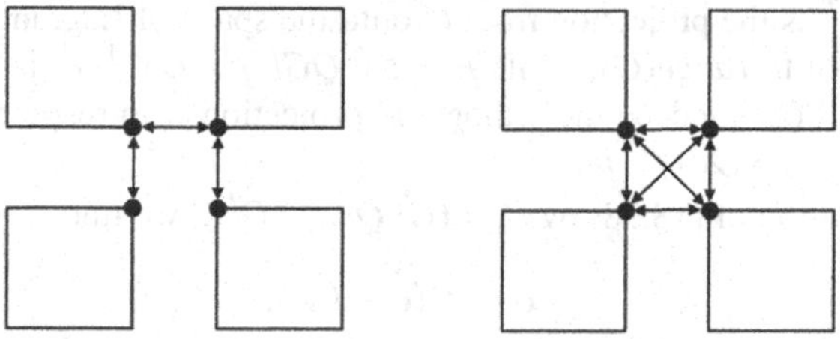

Fig. 18.10. Continuity constraints enforced by 3 (non-redundant) conditions on the left, by 6 (redundant) conditions on the right

Problem (18.96) admits a unique solution iff $Ker\{\Sigma_\Delta\} \cap Ker\{B_\Gamma\} = 0$, that is if Σ_Δ is invertible on $Ker(B_\Gamma)$.

We can reformulate (18.96) using Lagrange multipliers:

$$\begin{cases} \text{Find } (\mathbf{u}, \boldsymbol{\lambda}) \in W \times U \text{ such that} & \Sigma_\Delta \mathbf{u} + B_\Gamma^T \boldsymbol{\lambda} = \chi_\Delta, \\ & B_\Gamma \mathbf{u} = \mathbf{0}. \end{cases} \tag{18.97}$$

Because of the inf-sup (LBB) condition (see Chap. 16), the component $\boldsymbol{\lambda}$ of the solution to (18.97) is unique up to an additive vector of $Ker(B_\Gamma^T)$, so we choose $U = range(B_\Gamma)$. Let $R = diag(R^{(1)}, \ldots, R^{(M)})$ be made of null-space elements of Σ_Δ. (E.g. $R^{(i)}$ corresponds to the rigid body motions of Ω_i, in case of linear elasticity operator.)

R is a full column rank matrix. The solution of the first equation of (18.97) exists iff $\chi_\Delta - B_\Gamma^T \boldsymbol{\lambda} \in range(\Sigma_\Delta)$, a limitation that will be resolved by introducing a suitable projection operator P. Then,

$$\mathbf{u} = \Sigma_\Delta^\dagger (\chi_\Delta - B_\Gamma^T \boldsymbol{\lambda}) + R\alpha \quad \text{if } \chi_\Delta - B_\Gamma^T \boldsymbol{\lambda} \perp Ker(\Sigma_\Delta),$$

where α is an arbitrary vector and $\Sigma_\Delta^\dagger$ is a pseudoinverse of Σ_Δ. (Even though there are several pseudo-inverses of a given matrix, the following algorithm will be invariant to the specific choice.) It is convenient to choose a symmetric $\Sigma_\Delta^\dagger$, e.g. that of Moore-Penrose, see [QSS07].

Substituting $\mathbf{u}$ into the second equation of (18.97) yields

$$B_\Gamma \Sigma_\Delta^\dagger B_\Gamma^T \boldsymbol{\lambda} = B_\Gamma \Sigma_\Delta^\dagger \chi_\Delta + B_\Gamma R\alpha. \tag{18.98}$$

Let us set $F = B_\Gamma \Sigma_\Delta^\dagger B_\Gamma^T$ and $\mathbf{d} = B_\Gamma \Sigma_\Delta^\dagger \chi_\Delta$. Then choose P^T to be a suitable projection matrix, e.g. $P^T = I - G(G^T G)^{-1} G^T$, with $G = B_\Gamma R$. Then

$$\begin{cases} P^T F \boldsymbol{\lambda} = P^T \mathbf{d}, \\ G^T \boldsymbol{\lambda} = \mathbf{e} \, (= R^T \chi_\Delta). \end{cases} \tag{18.99}$$

More in general, one can introduce a s.p.d. matrix Q, and set

$$P^T = I - G(G^T Q G)^{-1} G^T Q.$$

The operator P^T is the projection from U onto the space of Lagrange multipliers that are Q-orthogonal to $range(G)$, while $P = I - QG(G^TQG)^{-1}G^T$ is a projection from U onto $Ker(G^T)$ (it is indeed the orthogonal projection with respect to the Q^{-1}-inner product $(\boldsymbol{\lambda},\boldsymbol{\mu})_{Q^{-1}} = (\boldsymbol{\lambda}, Q^{-1}\boldsymbol{\mu})$).
Upon multiplication of (18.98) by $H = (G^TQG)^{-1}G^TQ$ we find

$$\alpha = H(\mathbf{d} - F\boldsymbol{\lambda}),$$

which fully determines the primal variables in terms of $\boldsymbol{\lambda}$.
If the differential operator has constant coefficients, choosing $Q = I$ suffices. In case of jumps in the coefficients, Q is typically chosen as a scaling diagonal matrix and can be regarded as a scaling from the left of matrix B_Γ by $Q^{1/2}$.

The original one-level FETI method is a CG method in the space V applied to

$$P^TF\boldsymbol{\lambda} = P^T\mathbf{d}, \quad \boldsymbol{\lambda} \in \boldsymbol{\lambda}_0 + V \tag{18.100}$$

with an initial $\boldsymbol{\lambda}_0$ such that $G^T\boldsymbol{\lambda}_0 = \mathbf{e}$. Here

$$V = \{\boldsymbol{\lambda} \in U : \langle \boldsymbol{\lambda}, B\mathbf{z}\rangle = 0, \mathbf{z} = Ker(\Sigma_\Delta)\}$$

is the so-called space of admissible increments, $Ker(G^T) = range(P)$ and

$$V' = \{\boldsymbol{\mu} \in U : \langle \boldsymbol{\mu}, B\mathbf{z}\rangle_Q = 0, \mathbf{z} \in Ker(\Sigma_\Delta)\} = range(P^T).$$

The above simplest version of FETI with no preconditioner (or only a diagonal preconditioner) in the subdomain is scalable with the number of subdomains, but the condition number grows polynomially with the number of elements per subdomain. The original, most basic FETI preconditioner is

$$P_h^{-1} = B_\Gamma \Sigma_\Delta B_\Gamma^T = \sum_{i=1}^{M} B^{(i)} \Sigma_i {B^{(i)}}^T. \tag{18.101}$$

It is called a Dirichlet preconditioner since its application to a given vector involves the solution of M independent Dirichlet problems, one in every subdomain. The coarse space in FETI consists of the nullspace on each substructure.

To keep the search directions of the resulting preconditioned CG method in the space V, the application of P_h^{-1} is followed by an application of the projection P. Thus, the so-called Dirichlet variant of the FETI method is the CG algorithm applied to the modified equation

$$PP_h^{-1}P^TF\boldsymbol{\lambda} = PP_h^{-1}P^T\mathbf{d}, \quad \boldsymbol{\lambda} \in \boldsymbol{\lambda}_0 + V. \tag{18.102}$$

Since, for $\boldsymbol{\lambda} \in V$, $PP_h^{-1}P^TF\boldsymbol{\lambda} = PP_h^{-1}P^TP^TFP\boldsymbol{\lambda}$, the matrix on the left of (18.102) can be regarded as the product of two symmetric matrices. In case B_Γ has full row rank, i.e. the constraints are linearly independent and there are no redundant Lagrange multipliers, a better preconditioner can be defined as follows

$$\widehat{P}_h^{-1} = (B_\Gamma D^{-1} B_\Gamma^T)^{-1} B_\Gamma D^{-1} \Sigma_\Delta D^{-1} B_\Gamma^T (B_\Gamma D^{-1} B_\Gamma^T)^{-1} \tag{18.103}$$

where D is a block diagonal matrix $D = diag(D^{(1)}, \ldots, D^{(M)})$ and each block $D^{(i)}$ is a diagonal matrix whose elements are $\delta_i^\dagger(x)$ (see (18.95)) corresponding to the point x of $\partial\Omega_{i,h}$.
Since $B_\Gamma D^{-1} B_\Gamma^T$ is block-diagonal, its inverse can be easily computed by inverting small blocks whose size is n_x, the number of Lagrange multipliers used to enforce continuity at point x.
The matrix D, that operates on elements of the product space W, can be regarded as a scaling from the right of B_Γ by $D^{-1/2}$. With this choice

$$K_2(P\widehat{P}_h^{-1}P^T F) \leq C(1 + log(H/h))^2, \tag{18.104}$$

where $K_2(\cdot)$ is the spectral condition number and C is a constant independent of h, H, γ and the values of the ρ_i.

18.5.5 FETI-DP (Dual Primal FETI) methods

The FETI-DP method is a domain decomposition method introduced in [FLT+01] that enforces equality of the solution at subdomains interfaces by Lagrange multipliers except at subdomains corners, which remain primal variables. The first mathematical analysis of the method was provided by Mandel and Tezaur [MT01]. The method was further improved by enforcing the equality of averages across the edges or faces on subdomain interfaces [FLP00], [KWD02]. This is important for parallel scalability.

Let us consider a 2D case for simplicity. As anticipated at the beginning of Sect. 18.5.4, this idea is implemented by introducing an additional space $\widetilde{W}$ such that $\widehat{W} \subset \widetilde{W} \subset W$ for which we have continuity of the primal variables at subdomain vertices, and also common values of the averages over all edges of the interface. However, for simplicity we will confine ourselves to the case of primal variables associated to subdomain vertices only. This space can be written as the sum of two subspaces

$$\widetilde{W} = \widehat{W}_\Pi \oplus \widetilde{W}_\Delta \tag{18.105}$$

where $\widehat{W}_\Pi \subset \widehat{W}$ is the space of continuous interface functions that vanish at all nodal points of Γ_h except at the subdomain vertices. $\widehat{W}_\Pi$ is given in terms of the vertex variables and the averages of the values over the individual edges of the set of interface nodes Γ_h. $\widetilde{W}_\Delta$ is the direct sum of local subspaces $\widetilde{W}_{\Delta,i}$:

$$\widetilde{W}_\Delta = \prod_{i=1}^{M} \widetilde{W}_{\Delta,i} \tag{18.106}$$

where $\widetilde{W}_{\Delta,i} \subset W_i$ consists of local functions on $\partial\Omega_i$ that vanish at the vertces of Ω_i and have zero average on each individual edge.

According to this space splitting, the continuous degrees of freedom associated with the subdomain vertices and with the subspace $\widehat{W}_\Pi$ are called *primal* (Π), while those (that are potentially discontinuous across Γ) that are associated with the subspaces $\widetilde{W}_{\Delta,i}$ and with the interior of the subdomain edges are called *dual* (Δ).

The subspace $\widehat{W}_{\Pi}$, together with the interior subspace, defines the subsystem which is fully assembled, factored, and stored in each iteration step.

At this stage, all unknowns of the first subspace as well as the interior variables are eliminated to obtain a new Schur complement $\widetilde{\Sigma}_{\Delta}$. More precisely, we proceed as follows.

Let $\widetilde{A}$ denote the stiffness matrix obtained by restricting $diag(A_1,\ldots,A_M)$ (see (18.76)) from $\prod_{i=1}^{M} W^h(\Omega_i)$ to $\widetilde{W}^h(\Omega)$ (these spaces now refer to subdomains, not to their boundaries). Then $\widetilde{A}$ is no longer block diagonal because of the coupling that now exists between subdomains sharing a common vertex. According to the previous space decomposition, $\widetilde{A}$ can be split as follows

$$\widetilde{A} = \begin{bmatrix} A_{II} & A_{I\Pi} & A_{I\Delta} \\ A_{I\Pi}^T & A_{\Pi\Pi} & A_{\Pi\Delta} \\ A_{I\Delta}^T & A_{\Pi\Delta}^T & A_{\Delta\Delta} \end{bmatrix}.$$

Here the subscript I refers to the internal degrees of freedom of the subdomains, Π to those associated to the subdomains vertices, and Δ to those of the interior of the subdomains edges, see Fig. 18.11, right. The matrices A_{II} and $A_{\Delta\Delta}$ are block diagonal (one block per subdomain). Any non-zero entry of $A_{I\Delta}$ represents a coupling between degrees of freedom associated with the same subdomain. Upon eliminating the variables of the I and Π sets, a Schur complement associated with the variables of the Δ sets (interior and edges) is obtained as follows

$$\widetilde{\Sigma} = A_{\Delta\Delta} - \left[A_{I\Delta}^T A_{\Pi\Delta}^T\right] \begin{bmatrix} A_{II} & A_{I\Pi} \\ A_{I\Pi}^T & A_{\Pi\Pi} \end{bmatrix}^{-1} \begin{bmatrix} A_{I\Delta} \\ A_{\Pi\Delta} \end{bmatrix}. \tag{18.107}$$

Correspondingly we obtain a reduced right hand side $\widetilde{\chi}_{\Delta}$. By indicating with $\mathbf{u}_{\Delta} \in \widetilde{W}_{\Delta}$ the vector of degrees of freedom associated with the edges, similarly to what done in (18.96) for FETI, the finite element problem can be reformulated as a minimization

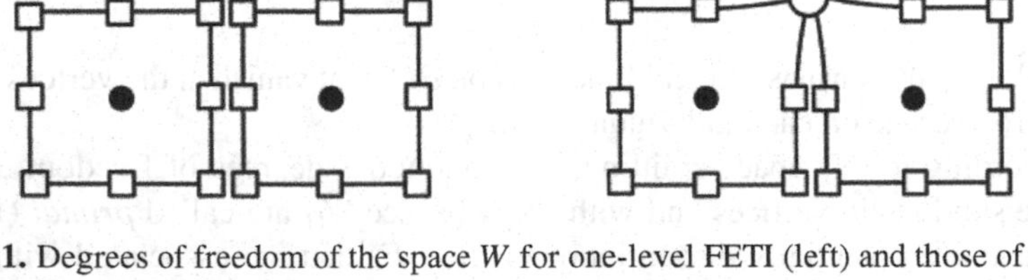

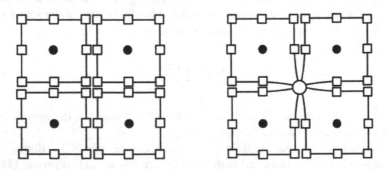

Fig. 18.11. Degrees of freedom of the space W for one-level FETI (left) and those of the space $\widetilde{W}$ for one-level FETI-DP (right) in the case of primal vertices only

problem with constraints given by the requirement of continuity across all of Γ

$$\begin{cases} \text{Find } \mathbf{u}_\Delta \in \widetilde{W}_\Delta : & J(\mathbf{u}_\Delta) = \frac{1}{2}\langle \widetilde{\Sigma}\mathbf{u}_\Delta, \mathbf{u}_\Delta\rangle - \langle \tilde{\chi}_\Delta, \mathbf{u}_\Delta\rangle \rightarrow min, \\ & B_\Delta \mathbf{u}_\Delta = \mathbf{0}. \end{cases} \tag{18.108}$$

The matrix B_Δ is made of $\{0,-1,1\}$ as it was for B_Γ. Note however that this time the constraints associated with the vertex nodes are dropped since they are assigned to the primal set. Note also that since all the constraints refer to edge points, no distinction needs to be made between redundant and non-redundant constraints and Lagrange multipliers.
A saddle point formulation of (18.108), similar to (18.97), can be obtained by introducing a set of Lagrange multipliers $\boldsymbol{\lambda} \in V = range(B_\Delta)$. Indeed, since $\widetilde{A}$ is s.p.d., so is $\widetilde{\Sigma}$: by eliminating the subvectors $\mathbf{u}_\Delta$ we obtain the reduced system

$$F_\Delta \boldsymbol{\lambda} = \mathbf{d}_\Delta, \tag{18.109}$$

where $F_\Delta = B_\Delta \widetilde{\Sigma}^{-1} B_\Delta^T$ and $\mathbf{d}_\Delta = B_\Delta \widetilde{\Sigma}^{-1}\tilde{\chi}_\Delta$.
Note that once $\boldsymbol{\lambda}$ is found, $\mathbf{u}_\Delta = \widetilde{\Sigma}^{-1}(\tilde{\chi}_\Delta - B_\Delta^T \boldsymbol{\lambda}) \in \widetilde{W}_\Delta$, while the interior variables $\mathbf{u}_I$ and the vertex variables $\mathbf{u}_\Pi$ are obtained by back-solving the system associated with $\widetilde{A}$.

A preconditioner for F is introduced as done in (18.103) for FETI (in case of non-redundant Lagrange multipliers)

$$P_\Delta^{-1} = (B_\Delta D_\Delta^{-1} B_\Delta^T)^{-1} B_\Delta D_\Delta^{-1} S_{\Delta\Delta} D_\Delta^{-1} B_\Delta^T (B_\Delta D_\Delta^{-1} B_\Delta^T)^{-1}. \tag{18.110}$$

Here D_Δ is a block diagonal scaling matrix with blocks $D_\Delta^{(i)}$: each of their diagonal elements corresponds to a Lagrange multiplier that enforces continuity between the nodal values of some $w_i \in W_i$ and $w_j \in W_j$ at some point $x \in \Gamma_h$ and it is given by $\delta_j^\dagger(x)$. Moreover, $\Sigma_{\Delta\Delta} = diag(\Sigma_{1,\Delta\Delta}, \ldots, \Sigma_{M,\Delta\Delta})$ with $\Sigma_{i,\Delta\Delta}$ being the restriction of the local Schur complement Σ_i to $\widetilde{W}_{\Delta,i} \subset W_i$.

When using the conjugate gradient method for the preconditioned system

$$P_\Delta^{-1} F_\Delta \boldsymbol{\lambda} = P_\Delta^{-1} \mathbf{d}_\Delta,$$

in contrast with one level FETI methods we can use an arbitrary initial guess $\boldsymbol{\lambda}^0$.
For an efficient implementation of this algorithm see [TW05, Sect. 6.4.1]. Also in this case we have a condition number that scales polylogarithmically, that is

$$K_2(P_\Delta^{-1} F_\Delta) \leq C(1 + log(H/h))^2,$$

where C is independent of h, H, γ and the values of the ρ_i. For a comprehensive presentation and analysis, see [KWD02] and [TW05].
For a conclusive comparative remark between FETI and FETI-DP methods, by following [TW05] we can note that FETI-DP algorithms do not require the characterization of the kernels of local Neumann problems (as required by one-level methods), because the enforcement of the additional constraints in each iteration always makes the local problems nonsingular and at the same time provides an underlying coarse global prob-

lem. FETI-DP methods do not require the introduction of a scaling matrix Q, which enters in the construction of a coarse solver for one-level FETI algorithms.
Finally, it is worth noticing that one-level FETI methods are projected conjugate gradient algorithms that cannot start from an arbitrary initial guess. In contrast, FETI-DP methods are standard preconditioned conjugate algorithms and can therefore employ an arbitrary initial guess λ^0.

18.5.6 BDDC (Balancing Domain Decomposition with Constraints) methods

This method was introduced by Dohrmann [Doh03] as a simpler primal alternative to the FETI-DP domain decomposition method. The name BDDC was coined by Mandel and Dohrmann because it can be understood as further development of the balancing domain decomposition method [Man93] with the coarse, global component of a BDDC algorithm expressed interms of a set of primal constraints.

In contrast to the original Neumann-Neumann and one-levet FETI methods, FETI-DP and BDDC algorithms do not require the solution of any singular linear systems of equations (those associated with a pure Neumann problem). In fact, any given choice of the primal set of variables determines a FETI-DP method and an associated BDDC method. This pair defines a duality, and features the same spectrum of eigenvalues (up to the eigenvalues 0 and 1) (see [LW06]). The choice of the primal constraints is of course a crucial question in order to obtain an efficient FETI-DP or BDDC algorithm.

BDDC is used as a preconditioner for the conjugate gradient method. A specific version of BDDC is characterized by the choice of coarse degrees of freedom, which can be values at the corners of the subdomains, or averages over the edges of the interface between the subdomains. One application of the BDDC preconditioner then combines the solution of local problems on each subdomain with the solution of a global coarse problem with the coarse degrees of freedom as the unknowns. The local problems on different subdomains are completely independent of each other, so the method is suitable for parallel computing.
A BDDC preconditioner reads

$$P_{BDDC}^{-1} = \widetilde{R}_{D\Gamma}^{T} \widetilde{\Sigma}^{-1} \widetilde{R}_{D\Gamma},$$

where $\widetilde{R}_{\Gamma} : \widehat{W} \to \widetilde{W}$ is a restriction matrix, $\widetilde{R}_{D\Gamma}$ is a scaled variant of $\widetilde{R}_{\Gamma}$ with scale factor $\delta_i^{\dagger}$ (featuring the same sparsity pattern of $\widetilde{R}_{\Gamma}$). This scaling is chosen in such a way that $\widetilde{R}_{\Gamma}\widetilde{R}_{D\Gamma}^{T}$ is a projection (then it coincides with its square).

Theoretical analysis of BDDC preconditioner (and its spectral analogy with FETI-DP preconditioner) was first provided in [MDT05] and later in [LW06] and [BS07].

18.6 Schwarz iterative methods

Schwarz method, in its original form described in Sect. 18.1.1, was proposed by H. Schwarz [Sch69] as an iterative scheme to prove existence of solutions to elliptic equations set in domains whose shape inhibits a direct application of Fourier series.

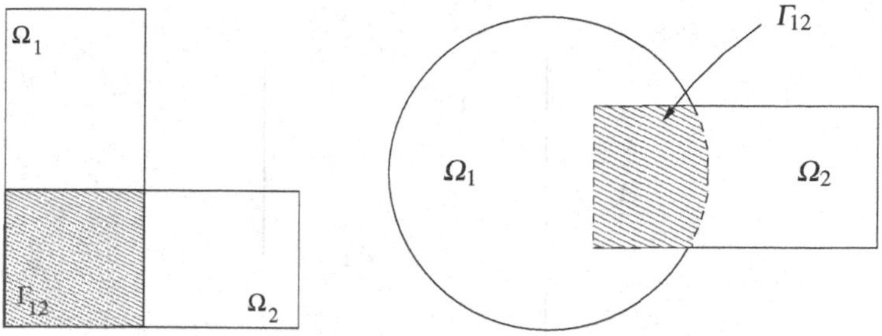

Fig. 18.12. Two examples for which the Schwarz method in its classical form applies

Two elementary examples are displayed in Fig. 18.12. This method is still used in some quarters as solution method for elliptic equations in arbitrarily shaped domains. However, nowadays it is mostly used in a somehow different version, that of DD preconditioner of conjugate gradient (or, more generally, Krylov) iterations for the solution of algebraic systems arising from finite element (or other kind of) discretizations of boundary-value problems.

As seen in Sect. 18.1.1, the distinctive feature of Schwarz method is that it is based on an overlapping subdivision of the original domain. Let us still denote $\{\Omega_m\}$ these subdomains.

To start with, in the following subsection we will show how the Schwarz method can be formulated as an iterative algorithm to solve the algebraic system associated with the finite element discretization of problem (18.1).

18.6.1 Algebraic form of Schwarz method for finite element discretizations

Consider as usual a finite element triangulation $\mathcal{T}_h$ of the domain Ω. Then assume that Ω is decomposed in two overlapping subdomains, Ω_1 and Ω_2, as shown in Fig. 18.1 (left).

Denote with N_h the total number of nodes of the triangulation that are internal to Ω (i.e., they don't sit on its boundary), and with N_1 and N_2, respectively, those internal to Ω_1 and Ω_2, as done in Sect. 18.3. Note that $N_h \leq N_1 + N_2$ and that equality holds only if the overlap reduces to a single layer of elements. Indeed, if we denote with $I = \{1, \ldots, N_h\}$ the set of indices of the nodes of Ω, and with I_1 and I_2 those associated with the internal nodes of Ω_1 and Ω_2, respectively, one has $I = I_1 \cup I_2$, while $I_1 \cap I_2 \neq \emptyset$ unless the overlap consists of a single layer of elements.

Let us order the nodes in such a way that the first block corresponds to those in $\Omega_1 \setminus \Omega_2$, the second to those in $\Omega_1 \cap \Omega_2$, and the third to those in $\Omega_2 \setminus \Omega_1$. The stiffness matrix A of the finite element discretization contains two submatrices, A_1 and A_2, corresponding to the local stiffness matrices in Ω_1 e Ω_2, respectively (see Fig. 18.13). They are related to A as follows

$$A_1 = R_1 A R_1^T \in \mathbb{R}^{N_1 \times N_1} \quad \text{and} \quad A_2 = R_2 A R_2^T \in \mathbb{R}^{N_2 \times N_2}, \tag{18.111}$$

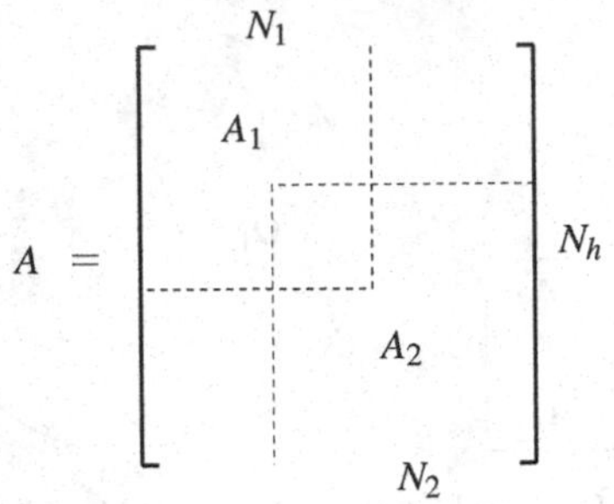

Fig. 18.13. The submatrices A_1 and A_2 of the stiffness matrix A

being R_i and R_i^T, for $i=1,2$, the restriction and prolongation operators, respectively. The matrix representation of the latter is

$$R_1^T = \begin{bmatrix} 1 & \dots & 0 \\ \vdots & \ddots & \vdots \\ 0 & \dots & 1 \\ & \mathbf{0} & \end{bmatrix} \in \mathbb{R}^{N_h \times N_1}, \qquad R_2^T = \begin{bmatrix} & \mathbf{0} & \\ 1 & \dots & 0 \\ \vdots & \ddots & \vdots \\ 0 & \dots & 1 \end{bmatrix} \in \mathbb{R}^{N_h \times N_2}. \quad (18.112)$$

If $\mathbf{v}$ is a vector of $\mathbb{R}^{N_h}$, then $R_1\mathbf{v}$ is a vector of $\mathbb{R}^{N_1}$ whose components coincide with the first N_1 components of $\mathbf{v}$. Should $\mathbf{v}$ instead be a vector of $\mathbb{R}^{N_1}$, then $R_1^T\mathbf{v}$ would be a vector of dimension N_h whose last $N_h - N_1$ components are all zero.

By using these definitions, an iteration of the multiplicative Schwarz method applied to system $A\mathbf{u}=\mathbf{f}$ can be expressed as follows:

$$\mathbf{u}^{(k+1/2)} = \mathbf{u}^{(k)} + R_1^T A_1^{-1} R_1(\mathbf{f} - A\mathbf{u}^{(k)}), \quad (18.113)$$

$$\mathbf{u}^{(k+1)} = \mathbf{u}^{(k+1/2)} + R_2^T A_2^{-1} R_2(\mathbf{f} - A\mathbf{u}^{(k+1/2)}). \quad (18.114)$$

Equivalently, by setting

$$P_i = R_i^T A_i^{-1} R_i A\ , \quad i=1,2, \quad (18.115)$$

we have

$$\mathbf{u}^{(k+1/2)} = (I - P_1)\mathbf{u}^{(k)} + P_1\mathbf{u},$$

$$\mathbf{u}^{(k+1)} = (I - P_2)\mathbf{u}^{(k+1/2)} + P_2\mathbf{u} = (I-P_2)(I-P_1)\mathbf{u}^{(k)} + (P_1 + P_2 - P_2P_1)\mathbf{u}.$$

Similarly, an iteration of the additive Schwarz method reads

$$\mathbf{u}^{(k+1)} = \mathbf{u}^{(k)} + (R_1^T A_1^{-1} R_1 + R_2^T A_2^{-1} R_2)(\mathbf{f} - A\mathbf{u}^{(k)}), \quad (18.116)$$

that is

$$\mathbf{u}^{(k+1)} = (I - P_1 - P_2)\mathbf{u}^{(k)} + (P_1 + P_2)\mathbf{u}. \quad (18.117)$$

Introducing the matrices

$$Q_i = R_i^T A_i^{-1} R_i = P_i A^{-1}, \; i = 1,2,$$

from (18.113) and (18.114) we derive the following recursive formula for the multiplicative Schwarz method

$$\begin{aligned}\mathbf{u}^{(k+1)} &= \mathbf{u}^{(k)} + Q_1(\mathbf{f} - A\mathbf{u}^{(k)}) + Q_2[\mathbf{f} - A(\mathbf{u}^{(k)} + Q_1(\mathbf{f} - A\mathbf{u}^{(k)}))] \\ &= \mathbf{u}^{(k)} + (Q_1 + Q_2 - Q_2 A Q_1)(\mathbf{f} - A\mathbf{u}^{(k)}),\end{aligned}$$

whereas for the additive Schwarz method we obtain from (18.116) that

$$\mathbf{u}^{(k+1)} = \mathbf{u}^{(k)} + (Q_1 + Q_2)(\mathbf{f} - A\mathbf{u}^{(k)}). \tag{18.118}$$

This last formula can easily be extended to the case of a decomposition of Ω into $M \geq 2$ overlapping subdomains $\{\Omega_i\}$ (see Fig. 18.14 for an example). In this case we have

$$\mathbf{u}^{(k+1)} = \mathbf{u}^{(k)} + \left(\sum_{i=1}^{M} Q_i\right)(\mathbf{f} - A\mathbf{u}^{(k)}). \tag{18.119}$$

18.6.2 Schwarz preconditioners

Denoting with

$$P_{as} = \left(\sum_{i=1}^{M} Q_i\right)^{-1}, \tag{18.120}$$

from (18.119) it follows that an iteration of the additive Schwarz method corresponds to an iteration of the preconditioned Richardson method applied to the solution of the linear system $A\mathbf{u} = \mathbf{f}$ using P_{as} as preconditioner. For this reason the matrix P_{as} is named *additive Schwarz preconditioner*.

In case of disjoint subdomains (no overlap), P_{as} coincides with the *block Jacobi preconditioner*

$$P_J = \begin{bmatrix} A_1 & & 0 \\ & \ddots & \\ 0 & & A_M \end{bmatrix}, \quad P_J^{-1} = \begin{bmatrix} A_1^{-1} & & 0 \\ & \ddots & \\ 0 & & A_M^{-1} \end{bmatrix} \tag{18.121}$$

in which we have removed the off-diagonal blocks of A.

Equivalently, one iteration of the additive Schwarz method corresponds to an iteration by the Richardson method on the preconditioned linear system $Q_a\mathbf{u} = \mathbf{g}_a$, with $\mathbf{g}_a = P_{as}^{-1}\mathbf{f}$, and the preconditioned matrix Q_a is

$$Q_a = P_{as}^{-1} A = \sum_{i=1}^{M} P_i.$$

By proceeding similarly, using the multiplicative Schwarz method would yield the following preconditioned matrix

$$Q_M = P_{ms}^{-1} A = I - (I - P_M) \dots (I - P_1).$$

Lemma 18.3. *Matrices P_i defined in (18.115) are symmetric and non-negative w.r.t the following scalar product induced by A*

$$(\mathbf{w}, \mathbf{v})_A = (A\mathbf{w}, \mathbf{v}) \qquad \forall \mathbf{w}, \mathbf{v} \in \mathbb{R}^{N_h}.$$

Proof. For $i = 1, 2$, we have

$$(P_i\mathbf{w}, \mathbf{v})_A = (AP_i\mathbf{w}, \mathbf{v}) = (R_i^T A_i^{-1} R_i A\mathbf{w}, A\mathbf{v}) = (A\mathbf{w}, R_i^T A_i^{-1} R_i A\mathbf{v})$$
$$= (\mathbf{w}, P_i\mathbf{v})_A \quad \forall \mathbf{v}, \mathbf{w} \in \mathbb{R}^{N_h}.$$

Moreover, $\forall \mathbf{v} \in \mathbb{R}^{N_h}$,

$$(P_i\mathbf{v}, \mathbf{v})_A = (AP_i\mathbf{v}, \mathbf{v}) = (R_i^T A_i^{-1} R_i A\mathbf{v}, A\mathbf{v}) = (A_i^{-1} R_i A\mathbf{v}, R_i A\mathbf{v}) \geq 0.$$ ◇

Lemma 18.4. *The preconditioned matrix Q_a of the additive Schwarz method is symmetric and positive definite w.r.t the scalar product induced by A.*

Proof. Let us first prove the symmetry: for all $\mathbf{u}, \mathbf{v} \in \mathbb{R}^{N_h}$, since A and P_i are both symmetric, we obtain

$$(Q_a\mathbf{u}, \mathbf{v})_A = (AQ_a\mathbf{u}, \mathbf{v}) = (Q_a\mathbf{u}, A\mathbf{v}) = \sum_i (P_i\mathbf{u}, A\mathbf{v})$$
$$= \sum_i (P_i\mathbf{u}, \mathbf{v})_A = \sum_i (\mathbf{u}, P_i\mathbf{v})_A = (\mathbf{u}, Q_a\mathbf{v})_A.$$

Concerning the positivity, choosing in the former identities $\mathbf{u} = \mathbf{v}$, we obtain

$$(Q_a\mathbf{v}, \mathbf{v})_A = \sum_i (P_i\mathbf{v}, \mathbf{v})_A = \sum_i (R_i^T A_i^{-1} R_i A\mathbf{v}, A\mathbf{v}) = \sum_i (A_i^{-1}\mathbf{q}_i, \mathbf{q}_i) \geq 0,$$

having set $\mathbf{q}_i = R_i A\mathbf{v}$. It follows that $(Q_a\mathbf{v}, \mathbf{v})_A = 0$ iff $\mathbf{q}_i = \mathbf{0}$ for every i, that is iff $A\mathbf{v} = \mathbf{0}$. Since A is positive definite, this holds iff $\mathbf{v} = \mathbf{0}$. ◇

Owing to the previous properties we can deduce that a more efficient iterative method can be generated by replacing the preconditioned Richardson iterations with the preconditioned conjugate gradient iterations, yet using the same additive Schwarz preconditioner P_{as}. Unfortunately, this preconditioner is not scalable. In fact, the condition number of the preconditioned matrix Q_a can only be bounded as

$$K_2(P_{as}^{-1}A) \leq C\frac{1}{\delta H}, \tag{18.122}$$

being C a constant independent of h, H and δ; here δ is a characteristic linear measure of the overlapping regions and, as usual, $H = \max_{i=1,\dots,M}\{\text{diam}(\Omega_i)\}$. This is due to the fact that the exchange of information only occurs among neighbooring subdomains, as the application of $(P_{as})^{-1}$ involves only local solvers. This limitation can be overcome by introducing, also in the current context, a global coarse solver defined on the whole domain Ω and apt at guaranteing a global communication among all of the subdomains. This leads to devise two-level domain decomposition strategies, see Sect. 18.6.3.

Let us address some algorithmic aspects. Let us subdivide the domain Ω in M subdomains $\{\Omega_i\}_{i=1}^M$ such that $\cup_{i=1}^M \overline{\Omega}_i = \overline{\Omega}$. Neighbooring subdomains share an overlapping region of size at least equal to $\delta = \xi h$, for a suitable $\xi \in \mathbb{N}$. In particular, $\xi = 1$ corresponds to the case of minimum overlap, that is the overlapping strip reduces to a single layer of finite elements. The following algorithm can be used.

Algorithm 18.5 (introduction of overlapping subdomains)

a. Build a triangulation $\mathcal{T}_h$ of the computational domain Ω
b. Subdivide $\mathcal{T}_h$ in M disjoint subdomains $\{\hat{\Omega}_i\}_{i=1}^M$ such that $\cup_{i=1}^M \overline{\hat{\Omega}}_i = \overline{\Omega}$
c. Extend every subdomain $\hat{\Omega}_i$ by adding all the strips of finite elements of $\mathcal{T}_h$ within a distance δ from $\hat{\Omega}_i$. These extended subdomains identify the family of overlapping subdomains Ω_i

In Fig. 18.14 a rectangular two-dimensional domain is subdivided into 9 disjoint subdomains $\hat{\Omega}_i$ (on the left); also shown is one of the extended (overlapping) subdomains (on the right).

To apply the Schwarz preconditioner (18.120) we can proceed as indicated in **Algorithm 18.5**. We recall that N_i is the number of internal nodes of Ω_i, R_i^T and R_i are the prolongation and restriction matrices, respectively, introduced in (18.112) and A_i are

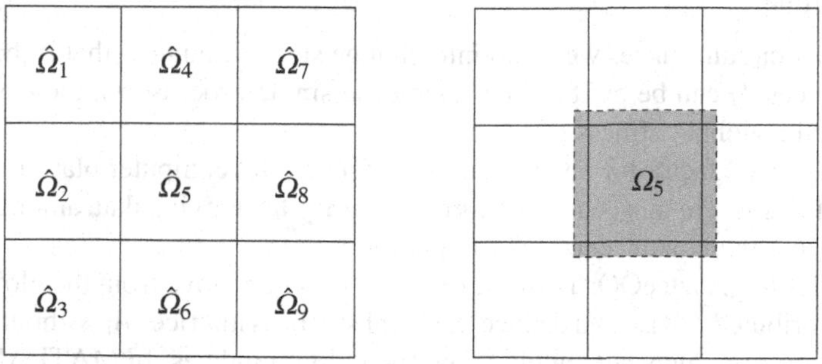

Fig. 18.14. Partition of a rectangular region Ω in 9 disjoint subregions $\hat{\Omega}_i$ (on the left), and an example of an extended subdomain Ω_5 (on the right)

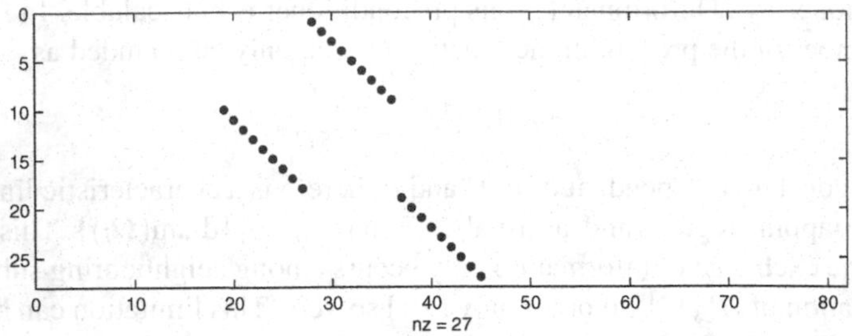

Fig. 18.15. The sparsity pattern of the matrix R_i for a partition of the domain in 4 subdomains

the local stiffness matrices introduced in (18.111). In Fig. 18.15 we display an example of sparsity pattern of R_i.

Algorithm 18.6 (start-up phase for the application of P_{as})

```
a. Build on every subdomanin Ω_i the matrices R_i and R_i^T
b. Build the stiffness matrix A corresponding to the
   finite element discretization on the grid T_h
c. On every Ω_i build the local submatrices A_i = R_i A R_i^T
d. On every Ω_i set up the code for the solution of a linear system with
   matrix A_i.
   For instance, compute a suitable (exact or incomplete) LU or
   Cholesky factorization of A_i
```

A few general comments on **Algorithm 18.5** and **Algorithm 18.6** are in order:

- steps `a.` and `b.` of algorithm 18.5 can be carried out in reverse order, that is we could first subdivide the computational domain into subdomains (based, for instance, on physical considerations), then set up a triangulation;
- depending upon the general code structure, steps `b.` and `c.` of the algorithm 18.6 could be glued together with the scope of optimizing memory requirements and CPU time.

In other circumstances we could interchange steps `b.` and `c.`, that is the local stiffness matrices A_i can be built at first (using the single processors), then assembled to construct the global stiffness matrix A.
Indeed, a crucial factor for an efficient use of a parallel computer platform is keeping data locality since in most cases the time necessary for moving data among processors can be higher than that needed for computation.
Other codes (e.g. AztecOO, Trilinos, IFPACK) instead move from the global stiffness matrix distributed rowise and deduce the local stiffness matrices A_i without performing matrix-matrix products but simply using the column indices. In MATLAB, however, it seems more convenient to build A at first, next the restriction matrices R_i, and finally to carry out matrix multiplications $R_i A R_i^T$ to generate the A_i.

Table 18.4. Condition number of $P_{as}^{-1}A$ for several values of h and H

$K_2(P_{as}^{-1}A)$	$H=1/2$	$H=1/4$	$H=1/8$	$H=1/16$
$h=1/16$	15.95	27.09	52.08	–
$h=1/32$	31.69	54.52	104.85	207.67
$h=1/64$	63.98	109.22	210.07	416.09
$h=1/128$	127.99	218.48	420.04	832.57

In Table 18.4 we analyze the case of a decomposition with minimum overlap ($\delta = h$), considering several values for the number M of subdomains. The subdomains Ω_i are overlapping squares of area H^2. Note that the theoretical estimate (18.122) is satisfied by our results.

18.6.3 Two-level Schwarz preconditioners

As anticipated in Sect. 18.6.2, the main limitation of Schwarz methods is to propagate information only among neighbooring subdomains. As for the Neumann-Neumann method, a possible remedy consists of introducing a coarse grid mechanism that allows for a sudden information diffusion on the whole domain Ω. The idea is still that of considering the subdomains as macro-elements of a new coarse grid $\mathcal{T}_H$ and to build a corresponding stiffness matrix A_H. The matrix

$$Q_H = R_H^T A_H^{-1} R_H,$$

where R_H is the restriction operator from the fine to the coarse grid, represents the coarse level correction for the new two-level preconditioner. More precisely, setting for notational convenience $Q_0 = Q_H$, the two-level preconditioner P_{cas} is defined through its inverse as

$$P_{cas}^{-1} = \sum_{i=0}^{M} Q_i. \qquad (18.123)$$

The following result can be proven in 2D: there exists a constant $C > 0$, independent of both h and H, such that

$$K_2(P_{cas}^{-1}A) \leq C(1 + \frac{H}{\delta}).$$

The ratio H/δ measures the relative overlap between neighboring overlapping subdomains. For "generous" overlap, that is if δ is a fraction of H, the preconditioner P_{cas} is scalable. Consequently, conjugate gradient iterations on the original finite element system using the preconditioner P_{cas} converges with a rate independent of h and H (and therefore of the number of subdomains). Moreover, thanks to the additive structure (18.123), the preconditioning step is fully parallel as it involves the solution of M independent systems, one per each local matrix A_i.

In 3D, we would get a bound with a factor H/h, unless the elliptic differential operator has constant coefficients (or variable coefficients which don't vary too much).

The use of P_{cas} involves the same kind of operations required by P_{as}, plus those of the following algorithm.

Algorithm 18.7 (start-up phase for the use of $\boldsymbol{P_{cas}}$)

a. Execute **Algorithm 18.6**

b. Define a coarse level triangulation $\mathcal{T}_H$ whose elements are of the order of H, then set $n_0 = \dim(V_0)$. Suppose that $\mathcal{T}_h$ be nested in $\mathcal{T}_H$. (See Fig. 18.16 for an example.)

c. Build the restriction matrix $R_0 \in \mathbb{R}^{n_0 \times N_h}$ whose elements are

$$R_0(i,j) = \Phi_i(\mathbf{x}_j),$$

where Φ_i is the basis function associated to the node i of the coarse grid, while by $\mathbf{x}_j$ we indicate the coordinates of the $j-th$ node on the fine grid

d. Build the coarse matrix A_H. This can be done by discretizing the original problem on the coarse grid $\mathcal{T}_H$, that is by computing

$$A_H(i,j) = a(\Phi_j, \Phi_i) = \int_\Omega \sum_{\ell=1}^{d} \frac{\partial \Phi_i}{\partial x_\ell} \frac{\partial \Phi_j}{\partial x_\ell},$$

or, otherwise, by setting

$$A_H = R_H A R_H^T.$$

For a computational domain with a simple shape (like the one we are considering) one typically generates the coarse grid $\mathcal{T}_H$ first, and then, by multiple refinements, the fine grid $\mathcal{T}_h$. In other cases, when the domain has a complex shape and/or a non struc-

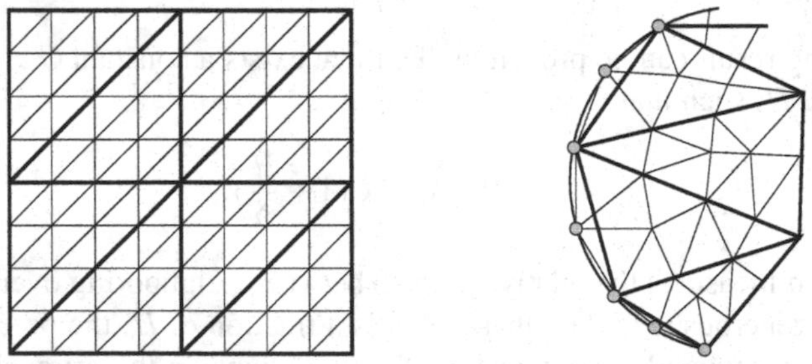

Fig. 18.16. On the left, example of a coarse grid for a 2D domain, based on a structured mesh. The triangles of the fine grid has thin edges; thick edges identify the boundaries of the coarse grid elements. On the right, a similar construction is displayed, this time for an unstructured fine grid

tured fine grid $\mathcal{T}_h$ is already available, the generation of a coarse grid might be difficult or computationally expensive. A first option would be to generate $\mathcal{T}_H$ by successive derefinements of the fine grid, in which case the nodes of the coarse grid will represent a subset of those of the fine grid. This approach, however, might not be very efficient in 3D.

Alternatively, one could generate the two (not necessarily nested) grids $\mathcal{T}_h$ and $\mathcal{T}_H$ independently, then generate the corresponding restriction and prolongation operators from the fine to the coarse grid, R_H and R_H^T.

The final implementation of P_{cas} could therefore be made as follows:

Algorithm 18.8 (P_{cas} solve)

For any given vector $\mathbf{r}$, the computation of $\mathbf{z} = P_{cas}^{-1}\mathbf{r}$ can be carried out as follows:

a. Set $\mathbf{z} = \mathbf{0}$

b. For $i = 1, \ldots, M$ Do in parallel:

c. restrict the residue on Ω_i: $\mathbf{r}_i = R_i\mathbf{r}$

d. compute $\mathbf{z}_i$: $A_i\mathbf{z}_i = \mathbf{r}_i$

e. add to the global residue: $\mathbf{z} \leftarrow R_i^T\mathbf{z}_i$

f. EndFor

g. Compute the coarse grid contribution $\mathbf{z}_H$: $A_H\mathbf{z}_H = R_H\mathbf{r}$

h. Add to the global residue: $\mathbf{z} \leftarrow R_H^T\mathbf{z}_H$

In Table 18.5 we report the condition number of $P_{cas}^{-1}A$ in the case of a minimum overlap $\delta = h$. Note that the condition number is almost the same on each NW-SE diagonal (i.e. for fixed values of the ratio H/δ).

An alternative approach to the coarse grid correction can be devised as follows. Suppose that the coefficients of the restriction matrix be given by

$$\hat{R}_H(i,j) = \begin{cases} 1 & \text{if the } j-th \text{ node is in } \Omega_i, \\ 0 & \text{otherwise,} \end{cases}$$

Table 18.5. Condition number of $P_{cas}^{-1}A$ for several values of h and H

$K_2(P_{cas}^{-1}A)$	$H = 1/4$	$H = 1/8$	$H = 1/16$	$H = 1/32$
$h = 1/32$	7.03	4.94	—	—
$h = 1/64$	12.73	7.59	4.98	—
$h = 1/128$	23.62	13.17	7.66	4.99
$h = 1/256$	45.33	24.34	13.28	—

Table 18.6. Condition number of $P_{aggre}^{-1}A$ for several values of h and H

$P_{aggre}^{-1}A$	$H = 1/4$	$H = 1/8$	$H = 1/16$
$h = 1/16$	13.37	8.87	—
$h = 1/32$	26.93	17.71	9.82
$h = 1/64$	54.33	35.21	19.70
$h = 1/128$	109.39	70.22	39.07

then we set $\hat{A}_H = \hat{R}_H A \hat{R}_H^T$. This procedure is named *aggregation* because the elements of $\hat{A}_H$ are obtained by simply summing up the entries of A. Note that we don't need to construct a coarse grid in this case. The corresponding preconditioner, denoted by P_{aggre}, has an inverse that reads

$$P_{aggre}^{-1} = \hat{R}_H^T \hat{A}_H^{-1} \hat{R}_H + P_{as}.$$

It can be proven that

$$K_2(P_{aggre}^{-1}A) \leq C\left(1 + \frac{H}{\delta}\right).$$

In Table 18.6 we report several numerical values of the condition number for different values of h and H.

If H/δ =constant, this two-level preconditioner is either optimal and scalable, that is the condition number of the preconditioned stiffness matrix is independent of both h and H.

We can conclude this section with the following practical indications:

- for decompositions with a small number of subdomains, the single level Schwarz preconditioner P_{as} is very efficient;
- when the number M of subdomains gets large, using two-level preconditioners becomes crucial; aggregation techniques can be adopted, in alternative to the use of a coarse grid in those cases in which the generation of the latter is difficult.

18.7 An abstract convergence result

The analysis of overlapping and non-overlapping domain decomposition preconditioners is based on the following abstract theory, due to P.L. Lions, J. Bramble, M. Dryja, O. Wildlund.

Let V_h be a Hilbert space of finite dimension. In our applications, V_h is one of the finite element spaces or spectral element spaces. Let V_h be decomposed as follows:

$$V_h = V_0 + V_1 + \cdots + V_M.$$

Let $F \in V'$ and $a : V \times V \rightarrow \mathbb{R}$ be a symmetric, continuous and coercive bilinear form.

Consider the problem

$$\text{find } u_h \in V_h : a(u_h, v_h) = F(v_h) \quad \forall v_h \in V_h. \tag{18.124}$$

Let $P_i : V_h \to V_i$ be a projection operator defined by

$$b_i(P_i u_h, v_h) = a(u_h, v_h) \quad \forall v_h \in V_i$$

with $b_i : V_i \times V_i \to \mathbb{R}$ being a local symmetric, continuous and coercive bilinear form on each subspace V_i. Assume that the following properties hold:

a. stable subspace decomposition:
$\exists C_0 > 0$ such that every $u_h \in V_h$ admits a decomposition $u_h = \sum_{i=0}^{M} u_i$ with $u_i \in V_i$ and

$$\sum_{i=0}^{M} b_i(u_i, u_i) \leq C_0^2 a(u_h, u_h);$$

b. strengthened Cauchy-Schwarz inequality:
$\exists \varepsilon_{ij} \in [0,1], i, j = 0, \ldots, M$ such that

$$a(u_i, u_i) \leq \varepsilon_{ij} \sqrt{a(u_i, u_i)} \sqrt{a(u_j, u_j)} \quad \forall u_i \in V_i, u_j \in V_j;$$

c. local stability:
$\exists \omega \geq 1$ such that $\forall i = 0, \ldots, M$

$$a(u_i, u_i) \leq \omega b_i(u_i, u_i) \quad \forall u_i \in Range(P_i) \subset V_i.$$

Then, $\forall u_h \in V_h$,

$$C_0^{-2} a(u_h, u_h) \leq a(P_{as} u_h, u_h) \leq \omega(\rho(E) + 1) a(u_h, u_h) \tag{18.125}$$

where $\rho(E)$ is the spectral radius of the matrix $E = (\varepsilon_{ij})$, and $P_{as} = P_0 + \cdots + P_M$ is the domain decomposition preconditioner.
From inequality (18.125) the following bound holds for the preconditioned system

$$K(B^{-1}A) \leq C_0^2 \omega(\rho(E) + 1)$$

where $K(\cdot)$ denotes the spectral condition number, A the matrix associated with the original system (18.124), B the matrix associated to the operator P_{as}. For the proof, see e.g. [TW05].

18.8 Interface conditions for other differential problems

Theorem 18.1 in Sect. 18.1.2 allows a second order elliptic problem (18.1) to be reformulated in a DD version thanks to suitable interface conditions (18.9) and (18.10). On the other hand, as we have extensively discussed, such reformulation sets the ground for several iterative algorithms on disjoint DD partitions. They comprise Dirichlet-Neumann, Neumann-Neumann, Robin-Robin algorithms and, more generally, all of

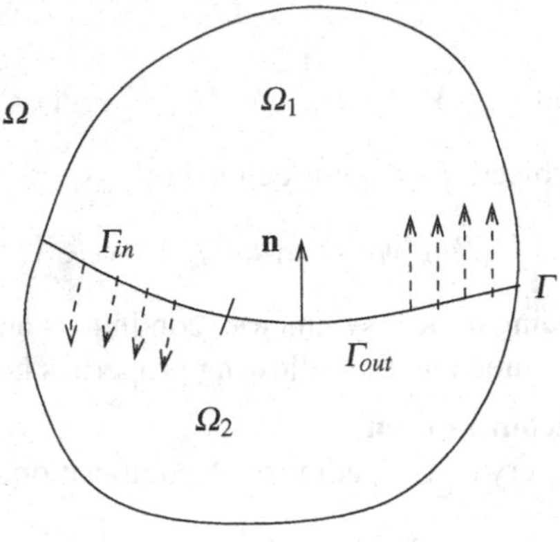

Fig. 18.17. Domain partition and interface splitting for the advection problem (18.126)

the preconditioned iterative algorithms of the Schur complement system (18.53) using suitable DD preconditioners.
In this section we consider other kind of boundary-value problems and formulate the associated interface conditions. Table 18.7 displays the interface conditions for these problems. For more details, analysis and investigation of associated iterative DD algorithms, the interested reader can consult [QV99].

Here we limit ourselves to provide a few additional insights in the case of advection and Stokes equations.

Advection (transport) problems. Consider the differential problem

$$Lu = \nabla \cdot (\mathbf{b}u) + a_0 u = f \qquad \text{in } \Omega, \tag{18.126}$$

supplemented by suitable conditions on the boundary $\partial\Omega$. Consider a partition of the computational domain Ω into two disjoint subdomains whose interface is Γ. Let us partition the latter as follows (see Fig. 18.17): $\Gamma = \Gamma_{in} \cup \Gamma_{out}$, where

$$\Gamma_{in} = \{x \in \Gamma \,|\, \mathbf{b}(x) \cdot \mathbf{n}(x) > 0\} \text{ and } \Gamma_{out} = \Gamma \setminus \Gamma_{in}.$$

Example 18.4. The Dirichlet-Neumann method for the problem at hand could be generalized as follows: being given two functions $u_1^{(0)}$, $u_2^{(0)}$ on Γ, $\forall k \geq 0$ solve:

$$\begin{cases} Lu_1^{(k+1)} = f & \text{in } \Omega_1, \\ (\mathbf{b}\cdot\mathbf{n})u_1^{(k+1)} = (\mathbf{b}\cdot\mathbf{n})u_2^{(k)} & \text{on } \Gamma_{out}, \end{cases}$$

$$\begin{cases} Lu_2^{(k+1)} = f & \text{in } \Omega_2, \\ (\mathbf{b}\cdot\mathbf{n})u_2^{(k+1)} = \theta(\mathbf{b}\cdot\mathbf{n})u_1^{(k)} + (1-\theta)(\mathbf{b}\cdot\mathbf{n})u_2^{(k)} & \text{on } \Gamma_{in}. \end{cases}$$

where $\theta > 0$ denotes a suitable relaxation parameter. The adaptation to the case of a finite element discretization is straightforward. ■

Stokes problem. The Stokes equations (16.11) feature two fields of variables: fluid velocity and fluid pressure. When considering a DD partition, at subdomain interface only the velocity field is requested to be continuous. Pressure needs not necessarily be continuous, since in the weak formulation of the Stokes equations it is "only" requested to be in L^2. Moreover, on the interface Γ the continuity of the normal Cauchy stress $\nu \frac{\partial \mathbf{u}}{\partial n} - p\mathbf{n}$ needs only be satisfied in weak (natural) form.

Example 18.5. A Dirichlet-Neumann algorithm for the Stokes problem would entail at each iteration the solution of the following subproblems (we use the short-hand notation $\mathcal{S}$ to indicate the Stokes operator):

$$\begin{cases} \mathcal{S}(\mathbf{u}_2^{(k+1)}, p_2^{(k+1)}) = \mathbf{f} & \text{in } \Omega_2, \\ \nu \dfrac{\partial \mathbf{u}_2^{(k+1)}}{\partial n} - p_2^{(k+1)} = \nu \dfrac{\partial \mathbf{u}_1^{(k)}}{\partial n} - p_1^{(k)} & \text{on } \Gamma, \\ \mathbf{u}_2^{(k+1)} = \mathbf{0} & \text{on } \partial\Omega_2 \setminus \Gamma, \end{cases} \tag{18.127}$$

$$\begin{cases} \mathcal{S}(\mathbf{u}_1^{(k+1)}, p_1^{(k+1)}) = \mathbf{f} & \text{in } \Omega_1, \\ \mathbf{u}_1^{(k+1)} = \theta \mathbf{u}_2^{(k+1)} + (1-\theta)\mathbf{u}_1^{(k)} & \text{on } \Gamma, \\ \mathbf{u}_1^{(k+1)} = \mathbf{0} & \text{on } \partial\Omega_1 \setminus \Gamma. \end{cases} \tag{18.128}$$

Should the boundary conditions of the original problem be prescribed on the velocity field, e.g. $\mathbf{u} = \mathbf{0}$, pressure p would be defined only up to an additive constant, which could be fixed by, e.g., imposing the constraint $\int_\Omega p\, d\Omega = 0$.
To fulfill this constraint we can proceed as follows. When solving the Neumann problem (18.127) on the subdomain Ω_2, both the velocity $\mathbf{u}_2^{(k+1)}$ and the pressure $p_2^{(k+1)}$ are univocally determined. When solving the Dirichlet problem (18.128) on Ω_1, the pressure is defined only up to an additive constant; we fix it by imposing the additional equation

$$\int_{\Omega_1} p_1^{(k+1)}\, d\Omega_1 = -\int_{\Omega_2} p_2^{(k+1)}\, d\Omega_2.$$

Should the four sequences $\{\mathbf{u}_1^{(k)}\}$, $\{\mathbf{u}_2^{(k)}\}$, $\{p_1^{(k)}\}$ and $\{p_2^{(k)}\}$ converge, the null average condition on the pressure would be automatically verified. ■

Example 18.6. Suppose now that the Schwarz iterative method is used on an overlapping subdomain decomposition of the domain like that on Fig. 18.1, left. At every step we have to solve two Dirichlet problems for the Stokes equations:

$$\begin{cases} \mathcal{S}(\mathbf{u}_1^{(k+1)}, p_1^{(k+1)}) = \mathbf{f} & \text{in } \Omega_1, \\ \mathbf{u}_1^{(k+1)} = \mathbf{u}_2^{(k)} & \text{on } \Gamma_1, \\ \mathbf{u}_1^{(k+1)} = 0 & \text{on } \partial\Omega_1 \setminus \Gamma_1, \end{cases} \tag{18.129}$$

$$\begin{cases} \mathcal{S}(\mathbf{u}_2^{(k+1)}, p_2^{(k+1)}) = \mathbf{f} & \text{in } \Omega_2, \\ \mathbf{u}_2^{(k+1)} = \mathbf{u}_1^{(k+1)} & \text{on } \Gamma_2, \\ \mathbf{u}_2^{(k+1)} = 0 & \text{on } \partial\Omega_2 \setminus \Gamma_2. \end{cases} \tag{18.130}$$

No continuity is required on the pressure field at subdomain boundaries.

The constraint on the fluid velocity to be divergence free on the whole domain Ω requires special care. Indeed, after solving (18.129), we have $\mathrm{div}\mathbf{u}_1^{(k+1)} = 0$ in Ω_1, hence, thanks to the Green formula,

$$\int_{\partial\Omega_1} \mathbf{u}_1^{(k+1)} \cdot \mathbf{n}\, d\gamma = 0.$$

This relation implies a similar relation for $\mathbf{u}_2^{(k)}$ in $(18.129)_2$; indeed

$$0 = \int_{\partial\Omega_1} \mathbf{u}_1^{(k+1)} \cdot \mathbf{n}\, d\gamma = \int_{\Gamma_1} \mathbf{u}_1^{(k+1)} \cdot \mathbf{n}\, d\gamma = \int_{\Gamma_1} \mathbf{u}_2^{(k)} \cdot \mathbf{n}\, d\gamma. \tag{18.131}$$

At the very first iteration we can select $\mathbf{u}_2^{(0)}$ in such a way that the compatibility condition (18.131) be satisfied, however this control is lost, a priori, in the course of the subsequent iterations. For the same reason, the solution of (18.130) yields the compatibility condition

$$\int_{\Gamma_2} \mathbf{u}_1^{(k+1)} \cdot \mathbf{n}\, d\gamma = 0. \tag{18.132}$$

Fortunately, Schwarz method automatically guarantees that this condition holds. Indeed, in $\Gamma_{12} = \Omega_1 \cap \Omega_2$ we have $\mathrm{div}\mathbf{u}_1^{(k+1)} = 0$, moreover on $\Gamma_{12} \setminus (\Gamma_1 \cup \Gamma_2)$, $\mathbf{u}_1^{(k+1)} = \mathbf{0}$ because of the given homogeneous Dirichlet boundary conditions. Thus

$$0 = \int_{\partial\Gamma_{12}} \mathbf{u}_1^{(k+1)} \cdot \mathbf{n}\, d\gamma = \int_{\Gamma_1} \mathbf{u}_1^{(k+1)} \cdot \mathbf{n}\, d\gamma + \int_{\Gamma_2} \mathbf{u}_1^{(k+1)} \cdot \mathbf{n}\, d\gamma.$$

The first integral on the right hand side vanishes because of (18.131), therefore (18.132) is satisfied. ■

18.9 Exercises

1. Consider the one-dimensional advection-transport-reaction problem

$$\begin{cases} -(\alpha\, u_x)_x + (\beta\, u)_x + \gamma u = f & \text{in } \Omega = (a,b) \\ u(a) = 0, \quad \alpha\, u_x(b) - \beta\, u(b) = g, \end{cases} \tag{18.133}$$

with α and $\gamma \in \mathrm{L}^\infty(a,b)$, $\beta \in \mathrm{W}^{1,\infty}(a,b)$ and $f \in L^2(a,b)$.

a) Write the addititive Schwarz iterative method, then the multiplicative one, on the two overlapping intervals $\Omega_1 = (a,\gamma_2)$ and $\Omega_2 = (\gamma_1, b)$, with $a < \gamma_1 < \gamma_2 < b$.

Table 18.7. Interface continuity conditions for several kind of differential operators; D stands for Dirichlet condition, N for Neumann

Operator	*Problem*	*D*	*N*
Laplace	$-\Delta u = f,$	u	$\dfrac{\partial u}{\partial n}$
Elasticity	$-\nabla \cdot (\sigma(\mathbf{u})) = \mathbf{f},$ with $\sigma_{kj} = \hat{\mu}(D_k u_j + D_j u_k) + \hat{\lambda}\,\mathrm{div}\mathbf{u}\,\delta_{kj}\,,$ $\mathbf{u}$ in-plane membrane displacement	$\mathbf{u}$	$\sigma(\mathbf{u}) \cdot \mathbf{n}$
Transport-diffusion	$-\sum_{kj} D_k(A_{kj} D_j u) + \mathrm{div}(\mathbf{b}u) + a_0 u = f$	u	$\dfrac{\partial u}{\partial n_L}$ $= \sum_k a_{kj} D_j u \cdot n_k$
Transport	$\mathrm{div}(\mathbf{b}u) + a_0 u = f$		$\mathbf{b} \cdot \mathbf{n} u$
Incompressible viscous flows	$-\mathrm{div}\mathsf{T}(\mathbf{u}, p) + (\mathbf{u}^* \cdot \nabla)\mathbf{u} = \mathbf{f},$ $\mathrm{div}\mathbf{u} = 0,$ with $\mathsf{T}_{kj} = \nu(D_k u_j + D_j u_k) - p\delta_{kj},$ $\mathbf{u}^* = \begin{cases} 0 & \text{(Stokes equations)} \\ \mathbf{u}_\infty & \text{(Oseen equations)} \\ \mathbf{u} & \text{(Navier-Stokes equations)} \end{cases}$	$\mathbf{u}$	$\mathsf{T}(\mathbf{u}, p) \cdot \mathbf{n}$
Compressible viscous flows	$\alpha \mathbf{u} - \mathrm{div}\hat{\mathsf{T}}(\mathbf{u}, \sigma) = \mathbf{f},$ $\alpha\sigma + \mathrm{div}\mathbf{u} = g,$ with $\hat{\mathsf{T}}_{kj} = \nu(D_k u_j + D_j u_k)$ $-\beta\sigma\delta_{kj} + \left(g - \frac{2\nu}{d}\right)\mathrm{div}\mathbf{u}\,\delta_{kj},$ $\rho =$ fluid density $= \log \sigma$	$\mathbf{u}$	$\hat{\mathsf{T}}(\mathbf{u}, \sigma) \cdot \mathbf{n}$
Compressible inviscid flows	$\alpha \mathbf{u} + \beta \nabla \sigma = \mathbf{f},$ $\alpha\sigma + \mathrm{div}\mathbf{u} = 0$	$\mathbf{u} \cdot \mathbf{n}$	σ
Maxwell (harmonic regime)	$\mathbf{rot}\left(\dfrac{1}{\mu}\mathbf{rot}\mathbf{E}\right)$ $-\alpha^2 \varepsilon \mathbf{E} + i\alpha\sigma\mathbf{E} = \mathbf{f}$	$\mathbf{n} \times \mathbf{E}$	$\mathbf{n} \times \left(\dfrac{1}{\mu}\mathbf{rot}\mathbf{E}\right)$

b) Interpret these methods as suitable Richardson algorithms to solve the given differential problem.
c) In case we approximate (18.133) by the finite element method, write the corresponding additive Schwarz preconditioner, with and without coarse-grid component. Then provide an estimate of the condition number of the preconditioned matrix, in both cases.

2. Consider the one-dimensional diffusion-transport-reaction problem

$$\begin{cases} -(\alpha u_x)_x + (\beta u)_x + \delta u = f & \text{in } \Omega = (a,b) \\ \alpha u_x(a) - \beta u(a) = g, \quad u_x(b) = 0, \end{cases} \tag{18.134}$$

with α and $\gamma \in \mathrm{L}^\infty(a,b)$, $\alpha(x) \geq \alpha_0 > 0$, $\beta \in \mathrm{W}^{1,\infty}(a,b)$, $f \in L^2(a,b)$ and g a given real number.

a) Consider two disjoined subdomains of Ω, $\Omega_1 = (a,\gamma)$ and $\Omega_2 = (\gamma,b)$, with $a < \gamma < b$. Formulate problem (18.134) using the Steklov-Poincaré operator, both in differential and variational form. Analyze the properties of this operator starting from those of the bilinear form associated with problem (18.134).
b) Apply the Dirichlet-Neumann method to problem (18.134) using the same domain partition introduced at point a).
c) In case of finite element approximation, derive the expression of the Dirichlet-Neumann preconditioner of the Schur complement matrix.

3. Consider the one-dimensional Poisson problem

$$\begin{cases} -u_{xx}(x) = f(x) & \text{in } \Omega = (0,1) \\ u(0) = 0, \quad u_x(1) = 0, \end{cases} \tag{18.135}$$

with $f \in L^2(\Omega)$.

a) If $\mathcal{T}_h$ indicates a partition of the interval Ω with step-size h, write the Galerkin-finite element approximation of problem (18.135).
b) Consider now a partition of Ω into the subintervals $\Omega_1 = (0,\gamma)$ and $\Omega_2 = (\gamma,1)$, being $0 < \gamma < 1$ a node of the partition $\mathcal{T}_h$ (See Fig. 18.18). Write the algebraic blockwise form of the Galerkin-finite element stiffness matrix relative to this subdomain partition.
c) Derive the discrete Steklov-Poincaré interface equation which corresponds to the DD formulation at point b). Which is the dimension of the Schur complement?
d) Consider now two overlapping subdomains $\Omega_1 = (0,\gamma_2)$ and $\Omega_2 = (\gamma_1,1)$, with $0 < \gamma_1 < \gamma_2 < 1$, the overlap being reduced to a single finite element of the

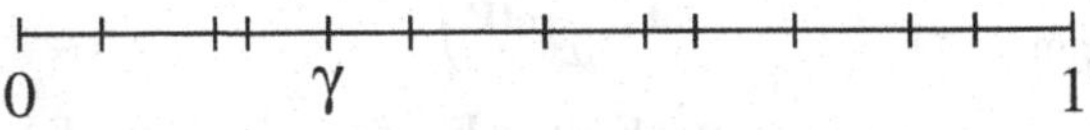

Fig. 18.18. Subdomain partition $\mathcal{T}_h$ of the interval $(0,1)$

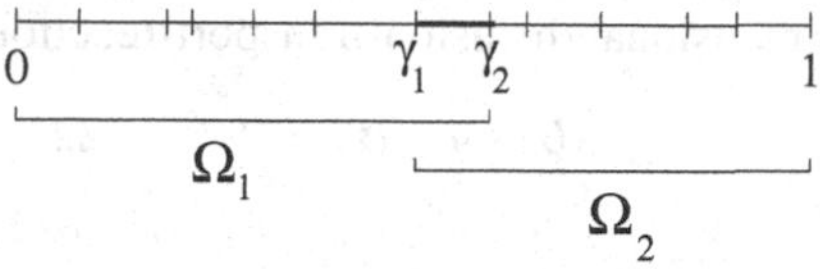

Fig. 18.19. Overlapping decomposition of the interval $(0,1)$

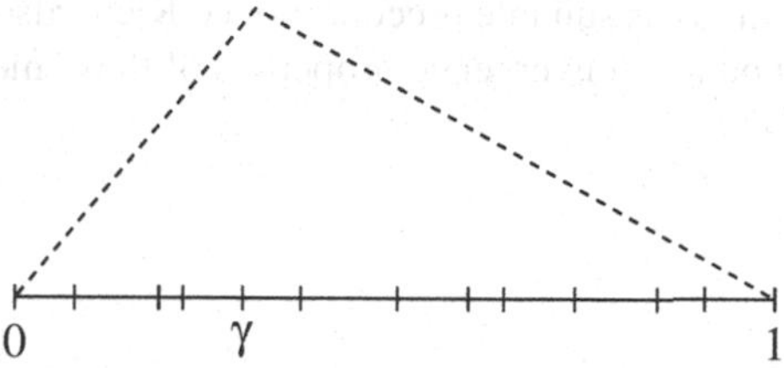

Fig. 18.20. Coarse-grid partition made of two macro elements for the construction of matrix A_H and Lagrangian characteristic function associated with the node γ

partition $\mathcal{T}_h$ (see Fig. 18.19). Provide the algebraic formulation of the additive Schwarz iterative method.

e) Provide the general expression of the two-level additive Shwarz preconditioner, by assuming as coarse matrix A_H that associated with only two elements, as displayed in Fig. 18.20.

4. Consider the diffusion-transport-reaction problem

$$\begin{cases} Lu = -\nabla\cdot(\alpha\nabla u) + \nabla\cdot(\beta u) + \gamma u = f & \text{in } \Omega = (0,2)\times(0,1), \\ u = 0 & \text{on } \Gamma_D, \\ \alpha\dfrac{\partial u}{\partial n} + \delta u = 0 & \text{on } \Gamma_R, \end{cases} \tag{18.136}$$

with $\alpha = \alpha(\mathbf{x})$, $\beta = \beta(\mathbf{x})$, $\gamma = \gamma(\mathbf{x})$, $\delta = \delta(\mathbf{x})$ and $f = f(\mathbf{x})$ being given functions, and $\partial\Omega = \overline{\Gamma}_D \cup \overline{\Gamma}_R$, with $\overset{\circ}{\Gamma}_D \cap \overset{\circ}{\Gamma}_R = \emptyset$.
Let Ω in (18.136) be partitioned into two disjoined subdomains $\Omega_1 = (0,1)\times(0,1)$ and $\Omega_2 = (1,2)\times(0,1)$.

a) Formulate problem (18.136) in terms of the Steklov-Poincaré operator, both in differential and variational form.
b) Apply the Dirichlet-Neumann method to problem (18.136) using the same decomposition introduced before.
c) Prove the equivalence between the Dirichlet-Neumann method at point b) and a suitable preconditioned Richardson operator, after setting $\alpha = 1$, $\beta = 0$, $\gamma = 1$ and $\Gamma_R = \emptyset$ in (18.136). Do the same for the Neumann-Neumann method.

5. Consider the two-dimensional diffusion-transport-reaction problem

$$\begin{cases} Lu = -\nabla \cdot (\mu \nabla u) + \mathbf{b} \cdot \nabla u + \sigma u = f & \text{in } \Omega = (a,c) \times (d,e), \\ u = 0 & \text{on } \partial\Omega. \end{cases} \tag{18.137}$$

Consider a decomposition of Ω made of the overlapping subdomains $\Omega_3 = (a, f) \times (d,e)$ and $\Omega_4 = (g,c) \times (d,e)$, with $g < f$. On such a decomposition, write for problem (18.137) the Schwarz method in both multiplicative and additive versions. Then interpret these methods as suitable preconditioned Richardson iterative algorithms. Finally, comment on the convergence properties of these methods.

19

Reduced basis approximation for parametrized partial differential equations

Reduced basis (RB) methods are computational reduction techniques for the rapid and reliable evaluation of *input-output relationships*: the *output* is expressed as a functional of the solution of a *parametrized* partial differential equation (PDE), the input being the set of parameters.

Parametrized PDEs can model several processes that are relevant in applications, such as, e.g., steady and unsteady heat and mass transfer, acoustics, solid and fluid mechanics, but also electromagnetics or even finance. The input-parameter vector may characterize either the geometric configuration, or some physical properties, or else boundary conditions and source terms. The outputs of interest might be the maximum system temperature, an added mass coefficient, a crack stress intensity factor, an effective constitutive property, an acoustic waveguide transmission loss, or a channel flowrate or pressure drop, just to mention a few. Finally, the field variables that connect the input parameters to the outputs can represent a distribution function, temperature or concentration, displacement, pressure, or velocity, etc.

The goal of an RB approximation is to capture the essential features of the input/output behaviour of a system *(i)* by improving computational performances and *(ii)* by keeping the approximation error between the reduced-order solution and the full-order one (the parametrized PDE) under control. In particular, the aim is to approximate a PDE solution using a handful of degrees of freedom instead of the many more (thousands, or millions, sometimes even billions) that would be needed for a full-order approximation. In fact, the idea at the heart of computational reduction strategies is the assumption (often occurring in real work) that the behaviour of a system can be well described by a small number of dominant modes.

In this way, we need to solve the full-order problem only for few instances of the input through a computationally demanding *Offline* stage, in order to construct a *reduced space* of *basis* solutions. This makes possible to perform many low-cost reduced-order simulations at a very inexpensive *Online* stage for new instances of the input, by expressing the reduced solution as a linear combination of the basis solutions and exploiting a Galerkin projection onto this reduced space.

A. Quarteroni: *Numerical Models for Differential Problems*, 2nd Ed.
MS&A – Modeling, Simulation & Applications 8
DOI 10.1007/978-88-470-5522-3_19, © Springer-Verlag Italia 2014

RB methods do not replace Galerkin methods (or any other method suitable to approximate PDEs). Rather, they build upon, and are measured against (as regards accuracy), a given approximation method, for instance the finite element method (FEM): the RB solution does not approximate directly the exact solution, but rather a "given" finite element discretization of (typically) very large dimension N_h of it. In a nutshell, RB methods establish an algorithmic collaboration (rather than competing) with the finite element method.

In particular, we shall consider in this chapter the case of linear functional outputs of affinely parametrized linear elliptic coercive PDEs. This class of problems – relatively simple, yet relevant to many important applications, such as conduction and convection-diffusion, linear elasticity, etc. – proves to be a convenient expository vehicle for the methodology, which can be applied, up to suitable extension, to more general equations. We refer the reader interested in, say, linear parabolic PDEs to the review presented in [QRM11], for instance.

Although our focus is on the affine linear elliptic coercive case, the reduced basis approximation and a posteriori error estimation we discuss in this chapter are much more general, and their combination is a key factor for reduction techniques to be computationally successful. We also point out that, despite the increasing computer power nowadays makes the numerical solution of problems of very large dimensions and/or modelling complex phenomena essential, a computational reduction is still crucial whenever one is interested in real-time simulations and/or repeated output evaluations for different values of some inputs of interest. Typical cases are, for instance, real time visualization, the sensitivity analysis of PDE solutions with respect to parameters, or optimization problems under PDE constraints (such as optimal control problems, as the ones addressed in Chap. 17).

We outline below the chapter's content. In Sect. 19.1 we introduce the affine linear elliptic coercive setting in the case of so-called *compliant* problems, by considering the most relevant examples of parametrizations in Sect. 19.9. In Sect. 19.2 we illustrate some basic ingredients shared by several computational reduction approaches. Then we describe the main features of the reduced basis method for parametrized problems in Sect. 19.3: reduced spaces, Galerkin projection and an Offline/Online procedure ensuring computational efficiency. Moreover, we provide in Sect. 19.4 both an algebraic and a geometrical interpretation of the reduced basis approximation problem. We address in Sect. 19.5 the most popular strategies to construct snapshots (that is, basis functions) and reduced spaces: greedy algorithms (the core of RB methods for parametrized PDEs) and proper orthogonal decomposition, POD. In Sect. 19.6 we sketch some ideas related to a priori convergence theory, whereas in Sect. 19.7 we present rigorous and relatively sharp a posteriori output error bounds for RB approximations. We extend both the RB approximation and a posteriori error bounds to the case of non-compliant problems in Sect. 19.8. We discuss a simple numerical test case and provide a brief overview for more general classes of problems in Sect. 19.10.

19.1 Elliptic coercive parametric PDEs

Before introducing the main features of computational reduction, let us describe the class of problems we deal with throughout the chapter; more details will be given in Sect. 19.9. We denote by $\mathcal{D} \subset \mathbb{R}^p$, for an integer $p \geq 1$, a set of *input* parameters which may describe physical properties of the system, as well as boundary terms, source terms or the geometry of the computational domain. The class of problems we focus on can be written in the following form:

given $\mu \in \mathcal{D}$, evaluate the output of interest $s(\mu) = J(u(\mu))$ where $u(\mu) \in V = V(\Omega)$ is the solution of the following parametrized PDE

$$L(\mu)u(\mu) = F(\mu). \tag{19.1}$$

Here $\Omega \subset \mathbb{R}^d$, $d = 1,2,3$ is a regular domain, V a suitable Hilbert space, V' its dual, $L(\mu) : V \to V'$ a second-order differential operator and $F(\mu) \in V'$. The weak formulation of problem (19.1) reads: find $u(\mu) \in V = V(\Omega)$ such that

$$a(u(\mu), v; \mu) = f(v; \mu) \quad \forall v \in V, \tag{19.2}$$

where the bilinear form[1] is obtained from $L(\mu)$:

$$a(u, v; \mu) = {}_{V'}\langle L(\mu)u, v\rangle_V \;\; \forall u, v \in V, \tag{19.3}$$

while

$$f(v; \mu) = {}_{V'}\langle F(\mu), v\rangle_V \tag{19.4}$$

is a continuous linear form. We assume, for each $\mu \in \mathcal{D}$, $a(\cdot,\cdot;\mu)$ to be continuous and coercive, i.e. $\exists\, \bar{\gamma} < +\infty, \alpha_0 > 0$:

$$\gamma(\mu) = \sup_{u\in V}\sup_{v\in V} \frac{a(u, v; \mu)}{\|u\|_V \|v\|_V} < \bar{\gamma}, \quad \alpha(\mu) = \inf_{u\in V} \frac{a(u, u; \mu)}{\|u\|_V^2} \geq \alpha_0. \tag{19.5}$$

If the coercivity assumption is not satisfied, we have stability in the more general sense of the *inf-sup* condition. J (the output functional) is a linear and bounded form on V. Under these standard hypotheses on a and f, (19.2) admits a unique solution, due to the Lax-Milgram lemma. We shall exclusively consider second-order elliptic partial differential equations, in which case $V = H^1_{\Gamma_D}(\Omega)$ – see (3.26). Furthermore, we assume that a is symmetric and that $J = f$. The latter is merely a simplifying assumption and it means that we are in the so-called compliant case [PR07], a situation occurring quite frequently in engineering problems (see Sect. 19.1.1). The generalization to the non-compliant case, where a may be non-symmetric and J may be any bounded linear functional over V, is provided in Sect. 19.8.

We make one last assumption, crucial to enhance the computational efficiency: we require both the parametric bilinear form a and the parametric linear form f to be affine

[1] To be rigorous, we should introduce the Riesz identification operator $R : V' \to V$ by which we identify V and its dual, so that, given a third Hilbert space H such that $V \hookrightarrow H'$ and $H' \hookrightarrow V'$, ${}_{V'}\langle L(\mu)u, v\rangle_V = (RL(\mu)u, v)_H$; see also Sect. 2.1. However, the Riesz operator will be omitted for the sake of simplicity.

with respect to the parameter μ, that is

$$a(w,v;\mu) = \sum_{q=1}^{Q_a} \Theta_a^q(\mu)\, a^q(w,v) \quad \forall v,w \in V, \mu \in \mathcal{D}\,, \tag{19.6}$$

$$f(v;\mu) = \sum_{q=1}^{Q_f} \Theta_f^q(\mu)\, f^q(w) \quad \forall w \in V, \mu \in \mathcal{D}\,. \tag{19.7}$$

Here $\Theta_a^q : \mathcal{D} \to \mathbb{R}, q = 1,\ldots,Q_a$ and $\Theta_f^q : \mathcal{D} \to \mathbb{R}, q = 1,\ldots,Q_f$, are μ-dependent functions, whereas $a^q : V \times V \to \mathbb{R}$, $f^q : V \to \mathbb{R}$ are μ-independent. As a general principle, parameter-independent terms will be computed Offline, thus making Online computation much lighter.

Let us also remark that, since a is symmetric, we can define the *energy* inner product and the *energy* norm for elements of V as follows:

$$(w,v)_\mu = a(w,v;\mu) \quad \forall w,v \in V\,, \tag{19.8}$$

$$\|w\|_\mu = (w,w)_\mu^{1/2} \quad \forall w \in V\,. \tag{19.9}$$

Next, for given $\overline{\mu} \in \mathcal{D}$ and non-negative real τ,

$$(w,v)_V = (w,v)_{\overline{\mu}} + \tau (w,v)_{L^2(\Omega)} \qquad \forall w,v \in V\,, \tag{19.10}$$

$$\|w\|_V = (w,w)_V^{1/2} \qquad \forall w \in V\,, \tag{19.11}$$

shall define the inner product and norm on our V, respectively. The role of this scalar product will be clear in Sect.19.5.

19.1.1 Two simple examples

Before describing the main features shared by several computational reduction approaches, we provide two simple examples of parametrized problems which fit the framework and the methodology presented in this chapter. More involved examples, as well as a general formulation of physical and/or geometrical parametrizations fulfilling the key assumptions of RB methods, will be addressed later on, in Sect.19.9.

The simplest elliptic coercive parametrized problem we may think of is a Poisson problem, defined over a domain Ω, modelling the diffusion/reaction of e.g. a pollutant. Here $u = u(\mu)$ denotes its concentration, and the diffusion coefficient μ plays the role of input parameter; the output of interest is the average of the concentration over the domain,

$$s(\mu) = \int_\Omega u(\mu)\, d\Omega.$$

Consider homogeneous Neumann boundary conditions over $\partial\Omega$, a unit source term over the domain Ω and set $V = H^1(\Omega)$. We recover the abstract formulation of (19.2) by defining

$$a(w,v;\mu) = \mu \int_\Omega \nabla w \cdot \nabla v d\Omega + \int_\Omega wv d\Omega, \qquad f(v;\mu) = \int_\Omega v d\Omega\,. \tag{19.12}$$

We can easily observe that the problem is coercive, symmetric, and compliant. In this case we deal with $p=1$ parameters, encoding a physical property; f does not depend on μ, while a is affine in μ: in this case we have $Q_a=2$, $Q_f=1$, $\Theta_a^1(\mu)=\mu$, $\Theta_a^2(\mu)=\Theta_f^1(\mu)=1$, and

$$a^1(w,v)=\int_\Omega \nabla w\cdot\nabla v d\Omega, \quad a^2(w,v)=\int_\Omega wv d\Omega, \quad f^1(v)=\int_\Omega v d\Omega .$$

A slightly more involved case, still dealing with physical parameters only, is given by a heat conduction problem in a square domain Ω which comprises $B_1\times B_2$ blocks, each one representing a subregion with (a priori different) constant thermal conductivity; the geometry is depicted in Fig. 19.1. Here

$$\overline{\Omega}=\bigcup_{i=1}^{P+1}\overline{\mathcal{R}}_i ,$$

where the $\mathcal{R}_i$, $i=1\dots,P+1$, correspond to the subregions featuring a conductivity equal to $\mu_i>0$. Inhomogeneous Neumann (non-zero flux) boundary conditions on Γ_{base}, homogeneous Dirichlet (temperature) conditions on Γ_{top}, and homogeneous (zero flux) Neumann conditions are imposed on the two vertical sides. The output of interest is the average temperature over Γ_{base}.

The parameters $\mu=(\mu_1,\dots,\mu_P)$ are then the conductivities in the first $P=B_1B_2-1$ blocks (with the blocks numbered as shown in Figure 19.1); the conductivity of the last block, which serves for normalization, is one.

By setting $V=\{v\in H^1(\Omega)\mid v|_{\Gamma_{\mathrm{top}}}=0\}$, we recover the abstract formulation (19.2), with

$$a(w,v;\mu)=\sum_{i=1}^{P}\mu_i\int_{\mathcal{R}_i}\nabla w\cdot\nabla v d\Omega+\int_{\mathcal{R}_{P+1}}\nabla w\cdot\nabla v d\Omega , \tag{19.13}$$

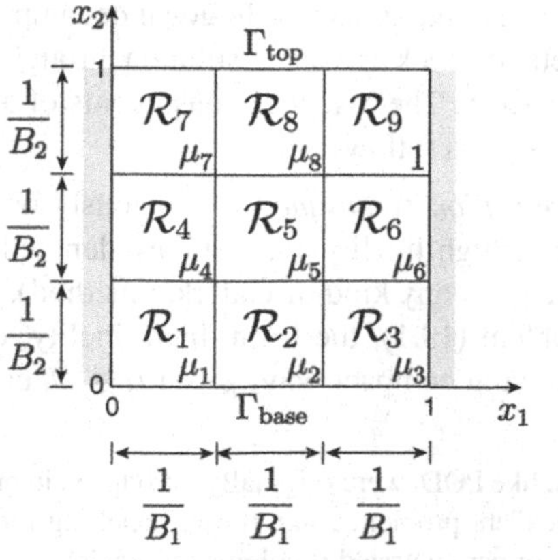

Fig. 19.1. Thermal block problem for $B_1=B_2=3$

which is associated to the Laplace operator with homogeneous Neumann conditions (as well as internal flux continuity conditions), and

$$f(v;\mu) = \int_{\Gamma_{\text{base}}} v d\Gamma \,, \tag{19.14}$$

which imposes the inhomogeneous Neumann conditions. The problem is coercive, symmetric, and compliant (the functional (19.14) yields the mean temperature). Moreover, F is independent of μ. The dependence of (19.13) on the parameters is affine; by direct inspection we note that $Q_a = P+1$, $Q_f = 1$, $\Theta_a^q(\mu) = \mu_q$, $1 \le q \le P$, $\Theta_a^{P+1} = 1$, $\Theta_f^1(\mu) = 1$ and

$$a^q(w,v) = \int_{\mathcal{R}_q} \nabla w \cdot \nabla v d\Omega, \quad 1 \le q \le P+1\,, \qquad f^1(v) = \int_{\Gamma_{\text{base}}} v d\Gamma \,.$$

19.2 Main components of computational reduction techniques

By *computational reduction techniques* (CRT) we denote *problem-dependent* methods which aim at reducing the dimension of the algebraic system arising from the discretization of a given PDE problem, for instance (19.2).

The reduced solution is obtained through a projection onto a small subspace made by *global* basis functions, constructed for the specific problem at hand, rather than onto a large space of generic basis functions (either local, like in finite elements or global, like in spectral methods).

In order to highlight their essential features, in this section we rely on the strong form (19.1) of the PDE problem. In fact, the CRTs we are going to introduce can be built upon any discretization technique described throughout the book, and not necessarily on those based on the weak form of the PDE problem.

The goal of a computational reduction technique for parametrized PDE problems is to compute, in a cheap way, a low-dimensional approximation of the PDE solution. The most common choices, like proper orthogonal decomposition (POD) or (*greedy*) reduced basis (RB) methods, seek a reduced solution through a *projection* onto suitable low-dimensional subspaces[2]. The essential constituents of a computational reduction technique can be described as follows:

- *High-fidelity discretization technique*: as previously observed, a CRT is not intended to replace a high-fidelity (sometimes denoted as *truth*) discretization method (obtained e.g. by any kind of Galerkin method).
  In the case of problem (19.1), the truth, high-fidelity approximation can be expressed in the following compact way: given $\mu \in \mathcal{D}$, evaluate $s_h(\mu) = f(u_h(\mu))$

[2] Indeed, several CRTs, like POD, were originally introduced in order to speed-up the solution of complex time-dependent problems, like those modelling turbulent flows, without being specifically designed for parametrized problems (otherwise said, time was considered as the only parameter).

where $u_h(\mu) \in V^{N_h}$ is such that

$$L_h(\mu)u_h(\mu) = F_h(\mu). \tag{19.15}$$

Here $V^{N_h} \subset V$ is a finite-dimensional space of very large dimension N_h, $L_h(\mu)$ a suitable discrete operator and $F_h(\mu)$ a given term. Recall that compliance means $J = f$.

For instance, assume that the truth approximation is based on the following Galerkin high-fidelity approximation of problem (19.2): find $u_h(\mu) \in V^{N_h}$ such that

$$a(u_h(\mu), v_h; \mu) = f(v_h; \mu) \quad \forall v_h \in V^{N_h}. \tag{19.16}$$

Moreover, let us introduce the injection operator $Q_h : V^{N_h} \to V$, and its adjoint $Q_h' : V' \to (V^{N_h})'$ between the dual spaces. The Galerkin problem (19.15) reads

$$Q_h'(L(\mu)Q_h u_h(\mu) - F(\mu)) = 0, \tag{19.17}$$

which corresponds to (19.15) upon defining

$$L_h(\mu) = Q_h' L(\mu) Q_h, \qquad F_h(\mu) = Q_h' F(\mu). \tag{19.18}$$

Note that $(L_h(\mu))^{-1} = \Pi_h (L(\mu))^{-1} \Pi_h'$, where $\Pi_h : V \to V^{N_h}$ is the L^2-projection operator and $\Pi_h' : (V^{N_h})' \to V'$ its adjoint. It follows directly from our assumptions on a, f, and V^{N_h} that (19.16) admits a unique solution. Let us assume that

$$\|u(\mu) - u_h(\mu)\|_V \le \mathcal{E}(h) \qquad \forall \mu \in \mathcal{D}, \tag{19.19}$$

$\mathcal{E}(h)$ being an estimate of the discretization error, which can be made as small as desired by choosing a suitable discretization space. Moreover, we define the coercivity and continuity constants (related to the subspace V^{N_h}) as

$$\alpha^{N_h}(\mu) = \inf_{w \in V^{N_h}} \frac{a(w,w;\mu)}{\|w\|_V^2}, \qquad \gamma^{N_h}(\mu) = \sup_{w \in V^{N_h}} \sup_{v \in V^{N_h}} \frac{a(w,v;\mu)}{\|w\|_V \|v\|_V}, \tag{19.20}$$

respectively. By (19.5), from the continuity and coercivity of a it follows that

$$\alpha^{N_h}(\mu) \ge \alpha(\mu), \qquad \gamma^{N_h}(\mu) \le \gamma(\mu) \qquad \forall \mu \in \mathcal{D}.$$

- *(Galerkin) projection*: a CRT usually consists in selecting a reduced basis of few high-fidelity PDE solutions $\{u_h(\mu^i)\}_{i=1}^N$ (called *snapshots*) and seeking a reduced approximation $u_N(\mu)$ expressed as a linear combination of these basis functions. The coefficients of this combination are determined through a projection of the equations onto the reduced space

$$V_N = \text{span}\{u_h(\mu^i)\ ,\ i = 1, \dots, N\},$$

with $N = \dim(V_N) \ll N_h$. The reduced problem reads, therefore: given $\mu \in \mathcal{D}$, evaluate $s_N(\mu) = f(u_N(\mu))$, where $u_N(\mu) \in V_N$ solves

$$a(u_N(\mu), v_N; \mu) = f(v_N; \mu) \quad \forall v_N \in V_N. \tag{19.21}$$

The smaller N, the cheaper the reduced problem to solve. We remark that our RB field and RB output approximate, for given N_h, the high-fidelity solution $u_h(\mu)$ and output $s_h(\mu)$ (hence, indirectly, $u(\mu)$ and $s(\mu)$).

As before, we can interpret (19.21) with the aid of suitable operators as

$$L_N(\mu)u_N(\mu) = F_N(\mu). \tag{19.22}$$

Indeed, let us introduce the injection operator $Q_N : V_N \to V^{N_h}$, and its adjoint $Q'_N : (V^{N_h})' \to V'_N$ operating between the dual spaces. Then, since

$$Q'_N(L_h(\mu)Q_N u_N(\mu) - F_h(\mu)) = 0, \tag{19.23}$$

we can obtain (19.22) from (19.23) by identifying

$$L_N(\mu) = Q'_N L_h(\mu) Q_N, \qquad F_N(\mu) = Q'_N F_h(\mu). \tag{19.24}$$

Similarly to what was done before, here $(L_N(\mu))^{-1} = \Pi_N(L_h(\mu))^{-1}\Pi'_N$, where $\Pi_N : V^{N_h} \to V_N$ is the L^2-projection operator and $\Pi'_N : V'_N \to (V^{N_h})'$.

- *Offline/Online procedure*: under suitable assumptions (see Sect. 19.3.3) the extensive generation of the snapshots database can be performed Offline once, and is completely decoupled from each new subsequent input-output Online query. Clearly, during the *Online* stage, the goal is to solve the reduced problem for parameter instances $\mu \in \mathcal{D}$ not selected during the Offline stage. In addition, the expensive Offline computations have to be amortized over the Online stage – in the RB context the break-even point is usually reached with $\mathcal{O}(10^2)$ Online queries.
- *Error estimation procedure*: sharp, inexpensive bounds $\Delta_N(\mu)$ such that

$$\|u_h(\mu) - u_N(\mu)\|_V \le \Delta_N(\mu) \qquad \forall \mu \in \mathcal{D},\ N = 1, \ldots, N_{max}, \tag{19.25}$$

may be available [PR07], as well as output error bounds $\Delta_N^s(\mu)$ such that

$$|s_h(\mu) - s_N(\mu)| \le \Delta_N^s(\mu).$$

These error estimators might also be employed to generate a *clever* parameter sampling during the construction of the reduced space, as we will see in Sect. 19.5.1. Their construction in the case of elliptic coercive problems is describe in detail in Sect. 19.7.

By putting (19.19) and (19.25) together we finally obtain, for all $\mu \in \mathcal{D}$, the error bound:

$$\|u(\mu) - u_N(\mu)\|_V \le \|u(\mu) - u_h(\mu)\|_V + \|u_h(\mu) - u_N(\mu)\|_V \le \mathcal{E}(h) + \Delta_N(\mu).$$

In the following section we provide further particulars on the construction of a reduced basis approximation, which is the main goal of this chapter.

19.3 The reduced basis method

We now characterize the general features presented in Sect. 19.2 in the case of the *reduced basis* method. Reduced Basis (RB) discretization is, in brief, a Galerkin (sometimes, a Petrov-Galerkin) projection on an N-dimensional approximation space V_N that approximates the manifold

$$\mathcal{M}_h = \{u_h(\mu) \in V^{N_h} \,:\, \mu \in \mathcal{D}\}, \tag{19.26}$$

given by the set of fields generated as the input varies over the whole parameter domain $\mathcal{D}$. We assume that this manifold is sufficiently smooth; in the case of a single parameter, the parametrically induced manifold is a one-dimensional filament within the infinite-dimensional space which characterizes all possible solutions to the given PDE. We depict the retained snapshots in Fig. 19.2.

If indeed the manifold is low-dimensional and smooth (a point we will return to later), then we expect any point of the manifold – any solution $u_h(\mu)$ for some μ in $\mathcal{D}$ – to be well approximated in terms of relatively few retained snapshots. However, we must ensure that not only we can choose our retained snapshots optimally (see Sect. 19.5), but also that we can *(i)* select a good combination of the available retained snapshots, *(ii)* represent the retained snapshots in a stable reduced basis and *(iii)* obtain the associated basis coefficients efficiently. These three important points will be discussed in the following section.

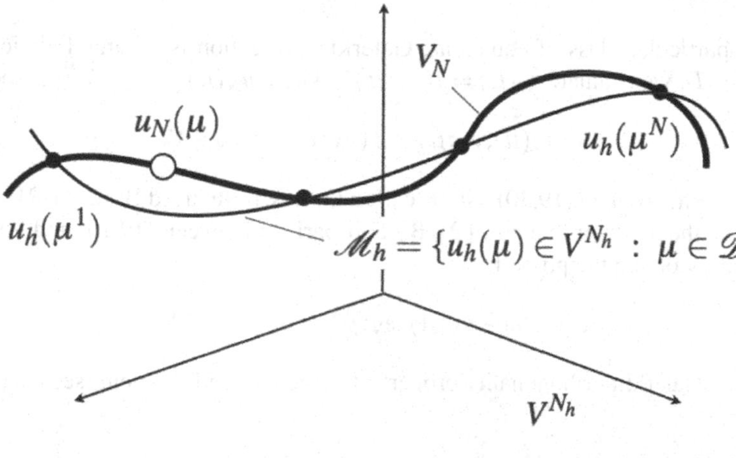

Fig. 19.2. The "snapshots" $u_h(\mu^n), 1 \le n \le N$, on the parametric manifold $\mathcal{M}_h$

19.3.1 RB Spaces

We restrict our attention to reduced spaces arising from[3] "snapshot" high-fidelity solutions of the PDE, and review the construction of the RB approximation in the elliptic case, also from an algebraic standpoint. Given a positive integer $N_{\max}$, we define a (hierarchical) sequence of RB spaces V_N^{RB}, $1 \leq N \leq N_{max}$, such that each V_N^{RB} is an N-dimensional subspace of V^{N_h}. We further suppose that

$$V_1^{RB} \subset V_2^{RB} \subset \cdots V_{N_{\max}}^{RB} \subset V^{N_h}. \tag{19.27}$$

As we shall see, the nesting or *hierarchy* condition (19.27) is important in ensuring memory efficiency of the resulting RB approximation.

In order to define a (hierarchical) sequence of spaces V_N^{RB}, $1 \leq N \leq N_{\max}$, we first introduce, for given $N \in \{1,\ldots,N_{max}\}$, a sample

$$S_N = \{\mu^1,\ldots,\mu^N\} \tag{19.28}$$

of parameter points $\mu^n \in \mathcal{D}$, $1 \leq n \leq N$, to be properly selected (e.g. by means of the *greedy* procedure that will be presented in Sect. 19.5.1). This produces corresponding *snapshots* $u_h(\mu^n) \in V^{N_h}$. The associated greedy-RB spaces are thus given by

$$V_N^{RB} = \text{span}\{u_h(\mu^n),\ 1 \leq n \leq N\}. \tag{19.29}$$

In the rest of the section the superscript ^{RB} will often be omitted for ease of notation. We observe that, by construction, the spaces V_N satisfy (19.27) – i.e. RB spaces (19.29) are hierarchical – and the samples (19.28) are nested, that is $S_1 = \{\mu^1\} \subset S_2 = \{\mu^1,\mu^2\} \subset \cdots \subset S_{N_{\max}}$.

19.3.2 Galerkin projection

For our particular class of equations, Galerkin projection is a natural choice. Thus, given $\mu \in \mathcal{D}$, we evaluate $s_N(\mu) = J(u_N(\mu))$, where $u_N(\mu) \in V_N \subset V^{N_h}$ is such that

$$a(u_N(\mu), v_N; \mu) = f(v_N; \mu) \quad \forall v_N \in V_N. \tag{19.30}$$

From now on, problem (19.30) will be called Galerkin Reduced Basis (G-RB) approximation of the given problem (19.2). By comparison between (19.16) and (19.30) we immediately obtain the property

$$a(u_h(\mu) - u_N(\mu), v_N; \mu) = 0 \quad \forall v_N \in V_N, \tag{19.31}$$

which is a Galerkin orthogonality property for the reduced problem (see Chap. 4).

---

[3] This is the most common way of constructing reduced subspaces, which are also called Lagrange RB spaces. Other known methods are based onTaylor [Por85] and Hermite [IR98] spaces.

Moreover, we obtain from (19.31) and Céa's lemma – see Sect. 4.2 – the classical optimality result in the energy norm (19.9)

$$\|u_h(\mu)-u_N(\mu)\|_\mu \le \inf_{w\in V_N} \|u_h(\mu)-w\|_\mu \,. \tag{19.32}$$

In other words, in the energy norm the Galerkin procedure automatically selects the *best* combination of snapshots. It is also clear that

$$s_h(\mu)-s_N(\mu) = \|u_h(\mu)-u_N(\mu)\|^2_\mu \,, \tag{19.33}$$

i.e. the output converges as the "square" of the energy error. In fact, using the compliance assumption, we can write

$$\begin{aligned} s_h(\mu)-s_N(\mu) &= a(u_h(\mu),u_h(\mu);\mu)-a(u_N(\mu),u_N(\mu);\mu)\\ &= a(u_h(\mu),u_h(\mu)-u_N(\mu);\mu)+a(u_h(\mu),u_N(\mu);\mu)-a(u_N(\mu),u_N(\mu);\mu)\\ &= a(u_h(\mu)-u_N(\mu),u_h(\mu)-u_N(\mu);\mu)+a(u_h(\mu)-u_N(\mu),u_N(\mu);\mu), \end{aligned}$$

where the second term in the last row vanishes by (19.31). Although this result depends critically on the compliance assumption, a generalisation *via* adjoint approximations to the non-compliant case is possible; see Sect. 19.8.

Let us remark that, by choosing the V-norm (19.11) instead of (19.32), we would find

$$\|u_h(\mu)-u_N(\mu)\|_V \le \left(\frac{\bar\gamma}{\alpha_0}\right)^{1/2} \inf_{w\in V_N} \|u_h(\mu)-w\|_V \,, \tag{19.34}$$

$\bar\gamma$ and α_0 being the uniform continuity and coercivity constants defined in (19.5).

We now consider the discrete equations associated with the Galerkin approximation (19.30). We must first choose an appropriate basis for our space; note that an ill-advised choice of the RB basis can lead to very poorly conditioned systems. Moreover, if V_N provides rapid convergence, the snapshots of (19.29) will be increasingly co-linear as N increases, by construction. To avoid this situation (and generate an independent set of snapshots) we apply the Gram-Schmidt process [Mey00, TI97] in the $(\cdot,\cdot)_V$ inner product to the snapshots $u_h(\mu^n)$, $1\le n\le N_{\max}$, to obtain mutually orthonormal functions ζ_n, $1\le n\le N_{\max}$: $(\zeta_n,\zeta_m)_V=\delta_{nm}$, $1\le n,m\le N_{\max}$, where δ_{nm} is the Kronecker delta symbol. We then choose the sets $\{\zeta_n\}_{n=1,\dots,N}$ as our bases for V_N, $1\le N\le N_{\max}$.

We now insert

$$u_N(\mu)=\sum_{m=1}^N u_N^{(m)}(\mu)\zeta_m \tag{19.35}$$

and then $v_N=\zeta_n$, $1\le n\le N$, into (19.30) to obtain the RB algebraic system

$$\sum_{m=1}^N a(\zeta_m,\zeta_n;\mu)\,u_N^{(m)}(\mu)=f(\zeta_n;\mu), \tag{19.36}$$

for the RB coefficients $u_N^{(m)}(\mu)$, $1 \le m,n \le N$. We can subsequently evaluate the RB output prediction as

$$s_N(\mu) = \sum_{m=1}^{N} u_N^{(m)}(\mu) f(\zeta_m;\mu) \,. \tag{19.37}$$

By using the Rayleigh quotient, as explained in (4.51), we can show that the condition number of the matrix $a(\zeta_m,\zeta_n;\mu)$, $1 \le n,m \le N$, is bounded by $\gamma(\mu)/\alpha(\mu)$, independently of N and N_h, owing to the orthogonality of the $\{\zeta_n\}$ and to (19.5); see e.g. [PR07] for further details. For the sake of simplicity, from now on we consider the case where f does not depend on the parameter μ.

19.3.3 Offline-Online computational procedure

System (19.36) is nominally of small size, yet it involves entities ζ_n, $1 \le n \le N$, associated with our N_h-dimensional high-fidelity approximation space. If we have to invoke high-fidelity fields in order to form the RB stiffness matrix *for each new value of* μ the marginal cost per input-output evaluation $\mu \to s_N(\mu)$ will remain unacceptably large. Fortunately, the crucial assumption of *affine parametric dependence* will cause a major increase in computational speed. In particular, thanks to (19.6), system (19.36) can be expressed as

$$\sum_{m=1}^{N} \left(\sum_{q=1}^{Q_a} \Theta_a^q(\mu)\, \mathbb{A}_N^q \right) \mathbf{u}_N(\mu) = \mathbf{f}_N \tag{19.38}$$

and (19.37) reads

$$s_N(\mu) = \mathbf{f}_N \cdot \mathbf{u}_N(\mu), \tag{19.39}$$

where $(\mathbf{u}_N(\mu))_m = u_N^{(m)}(\mu)$, $(\mathbb{A}_N^q)_{mn} = a^q(\zeta_n,\zeta_m)$, $(\mathbf{f}_N)_n = f(\zeta_n)$, for $1 \le m,n \le N$.

The computation thus entails an expensive μ-independent Offline stage, performed only once, and an inexpensive Online stage for any chosen parameter value $\mu \in \mathcal{D}$:

- in the Offline stage, we first compute the $u_h(\mu^n)$, and subsequently the ζ_n by Gram-Schmidt orthonormalization, $1 \le n \le N_{\max}$; we then form and store the terms

$$f(\zeta_n), \quad 1 \le n \le N_{\max} \,, \tag{19.40}$$

$$a^q(\zeta_n,\zeta_m), \quad 1 \le n,m \le N_{\max},\ 1 \le q \le Q_a \,. \tag{19.41}$$

  The Offline operation count depends on $N_{\max}$, Q_a, and N_h;

- in the Online stage, we retrieve (19.41) to form

$$\sum_{q=1}^{Q_a} \Theta_a^q(\mu) a^q(\zeta_n,\zeta_m), \quad 1 \le n,m \le N \,; \tag{19.42}$$

  we solve the resulting $N \times N$ stiffness system (19.38) to obtain the $u_N^{(m)}(\mu)$, $1 \le m \le N$; finally, we access (19.40) to evaluate the output (19.37). The Online operation count is $O(Q_a N^2)$ to perform the sum (19.42), $O(N^3)$ to invert (19.38) – note that the RB stiffness matrix is full – and finally $O(N)$ to evaluate the inner product (19.37). The Online storage (the data archived in the Offline stage) is only $O(Q_a N_{\max}^2) + O(N_{\max})$, because of the hierarchical condition (19.27): for any given N, we extract the necessary $N \times N$ RB matrices (resp. N-vectors) as principal submatrices (resp. subvectors) of the corresponding $N_{\max} \times N_{\max}$ (resp. $N_{\max}$) quantities.

The Online cost (operation count and storage) to evaluate $\mu \to s_N(\mu)$ is thus independent of N_h. The consequence is two-fold: first, *if* N is indeed small, we will achieve very fast response in real-time and many-query contexts; secondly, we may choose N_h large enough, to make sure that the error $\|u(\mu) - u_h(\mu)\|_V$ is very small, without affecting the Online marginal cost.

19.4 Algebraic and geometric interpretations of the RB problem

We now discuss the relationship between the Galerkin Reduced Basis (G-RB) approximation (19.30) and the Galerkin high-fidelity approximation (19.16) from both an algebraic and a geometric point of view.

Let us denote by $\mathbf{u}_h(\mu) \in \mathbb{R}^{N_h}$ and $\mathbf{u}_N(\mu) \in \mathbb{R}^N$ the vectors of degrees of freedom associated to the functions $u_h(\mu) \in V^{N_h}$ and $u_N(\mu) \in V_N$, respectively, which are given by

$$\mathbf{u}_h(\mu) = (u_h^{(1)}(\mu), \dots, u_h^{(N_h)}(\mu))^T, \qquad \mathbf{u}_N(\mu) = (u_N^{(1)}(\mu), \dots, u_N^{(N)}(\mu))^T.$$

Let $\{\tilde{\varphi}^r\}_{r=1}^{N_h}$ denote the standard FE basis, orthogonal with respect to a *discrete* scalar product

$$(u_h, v_h)_h = \sum_{r=1}^{N_h} w_r u_h(\mathbf{x}_r) v_h(\mathbf{x}_r),$$

$\{\mathbf{x}_r\}_{r=1}^{N_h}$ being the set of FE nodes such that $\tilde{\varphi}^r(\mathbf{x}_s) = \delta_{rs}$ and $\{w_r\}_{r=1}^{N_h}$ a set of weights such that $\sum_{r=1}^{N_h} w_r = |\Omega|$, for $r, s = 1, \dots, N_h$; note that other choices of scalar products can be made. It is useful to normalize these basis functions by defining

$$\varphi^r = \frac{1}{\sqrt{w_r}} \tilde{\varphi}^r, \quad (\varphi^r, \varphi^s)_h = \delta_{rs}, \qquad r, s = 1, \dots, N_h. \tag{19.43}$$

As we saw in Chapter 4 – see relation (4.7) – we can consider the following bijection between the spaces $\mathbb{R}^{N_h}$ and V^{N_h}:

$$\begin{cases} \mathbf{v}_h \in \mathbb{R}^{N_h} \quad \leftrightarrow \quad v_h \in V^{N_h}, \\ \mathbf{v}_h = (v_h^{(1)}, \dots, v_h^{(N_h)})^T \quad \leftrightarrow \quad v_h(\mathbf{x}) = \displaystyle\sum_{r=1}^{N_h} v_h^{(r)} \varphi^r(\mathbf{x}). \end{cases} \tag{19.44}$$

For ease of notation, we will express the bijection (19.44) as follows: $\mathbf{v}_h \in \mathbb{R}^{N_h} \to I_h \mathbf{v}_h = v_h \in V_h$, $I_h^T v_h = \mathbf{v}_h$. Similarly, for the reduced basis approximation we use the notation $\mathbf{v}_N \in \mathbb{R}^N \to I_N \mathbf{v}_N = v_N \in V_N$, $I_N^T v_N = \mathbf{v}_N$. Thanks to the ortonormality of the basis functions, $v_h^{(r)} = (v_h, \varphi^r)_h$, $r = 1, \dots, N_h$.

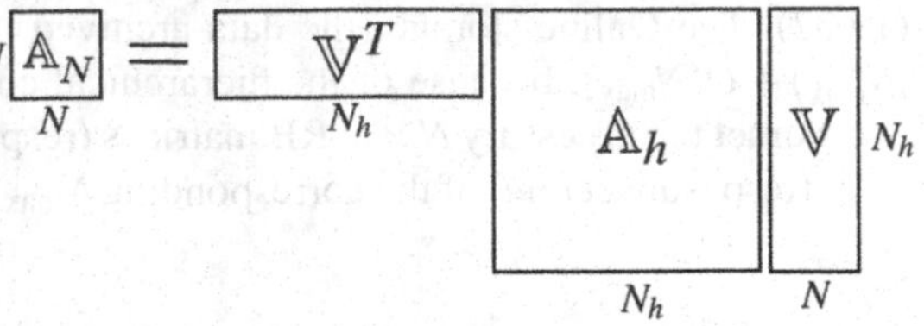

Fig. 19.3. Schematic representation of how to assemble the RB "stiffness" matrix

Using bijection (19.44), the following algebraic relations hold:

$$(\mathbf{u}_h, \mathbf{v}_h)_2 = (u_h, v_h)_h \quad \forall\, \mathbf{u}_h, \mathbf{v}_h \in \mathbb{R}^{N_h} \text{ (equivalently, } \forall\, u_h, v_h \in V^{N_h}). \tag{19.45}$$

Indeed:

$$\begin{aligned}(u_h, v_h)_h &= \left(\sum_{r=1}^{N_h} u_h^{(r)} \varphi^r, \sum_{s=1}^{N_h} v_h^{(s)} \varphi^s \right)_h \\ &= \sum_{r,s=1}^{N_h} u_h^{(r)} v_h^{(s)} (\varphi^r, \varphi^s)_h = \sum_{r,s=1}^{N_h} u_h^{(r)} v_h^{(r)} = (\mathbf{u}_h, \mathbf{v}_h)_2.\end{aligned}$$

19.4.1 Algebraic interpretation of the (G-RB) problem

We first discuss the algebraic connection between the (G-RB) problem (19.30) and the Galerkin high-fidelity approximation (19.16), which has strong consequences on the computational aspects related with RB methods.

In matrix form, the (G-RB) problem (19.36) can be written as

$$\mathbb{A}_N(\mu)\mathbf{u}_N(\mu) = \mathbf{f}_N, \tag{19.46}$$

with $\mathbf{f}_N = (f_N^{(1)}, \dots, f_N^{(N)})^T$, $f_N^{(k)} = f(\zeta_k)$, $(\mathbb{A}_N(\mu))_{km} = a(\zeta_m, \zeta_k; \mu)$, for $k, m = 1, \dots, N$. On the other hand, the Galerkin high-fidelity approximation (19.16) reads in matrix form as

$$\mathbb{A}_h(\mu)\mathbf{u}_h(\mu) = \mathbf{f}_h, \tag{19.47}$$

with $\mathbf{f}_h = (f_h^{(1)}, \dots, f_h^{(N_h)})^T$, $f_h^{(r)} = f(\varphi^r)$, $(\mathbb{A}_h(\mu))_{rs} = a(\varphi^s, \varphi^r; \mu)$, for $r, s = 1, \dots, N_h$.
For the sake of notation, we omit the dependence on μ in the rest of the section.

Let $\mathbb{V} \in \mathbb{R}^{N_h \times N}$ be the *transformation matrix* whose entries are

$$(\mathbb{V})_{rk} = (\zeta_k, \varphi^r)_h, \qquad r = 1, \dots, N_h\,,\; k = 1, \dots, N. \tag{19.48}$$

Using this matrix, we can easily obtain the following algebraic identities:

$$\mathbf{f}_N = \mathbb{V}^T \mathbf{f}_h, \qquad \mathbb{A}_N = \mathbb{V}^T \mathbb{A}_h \mathbb{V}, \tag{19.49}$$

which represent the algebraic counterparts of the operator identities (19.24) (see Fig. 19.3). Indeed,

$$(\mathbb{V}^T \mathbb{A}_h \mathbb{V})_{km} = \sum_{r,s=1}^{N_h} (\mathbb{V})^T_{kr} (\mathbb{A}_h)_{rs} (\mathbb{V})_{sm} = \sum_{r,s=1}^{N_h} (\zeta_k, \varphi^r)_h a(\varphi^s, \varphi^r)(\zeta_m, \varphi^s)_h$$
$$= a\left(\sum_{s=1}^{N_h} (\zeta_m, \varphi^s)_h \varphi^s, \sum_{r=1}^{N_h} (\zeta_k, \varphi^r)_h \varphi^r\right) = a(\zeta_m, \zeta_k) = (\mathbb{A}_N)_{km}$$

and, in the same way,

$$(\mathbb{V}^T \mathbf{f}_h)^{(k)} = \sum_{r=1}^{N_h} (\mathbb{V})^T_{kr} (\mathbf{f}_h)^{(r)} = \sum_{r=1}^{N_h} (\zeta_k, \varphi^r)_h f(\varphi^r)$$
$$= f\left(\sum_{r=1}^{N_h} (\zeta_k, \varphi^r)_h \varphi^r\right) = f(\zeta_k) = (\mathbf{f}_N)_k. \tag{19.50}$$

Thanks to (19.49), each μ-independent RB "stiffness" matrix $\mathbb{A}_N^q$ can be assembled once the corresponding high-fidelity "stiffness" matrix $\mathbb{A}_h^q$ has been computed.

The (vector representation of the) error between the solution of the (G-RB) problem and the Galerkin high-fidelity approximation is

$$\mathbf{e}_N = \mathbf{u}_h - \mathbb{V}\mathbf{u}_N. \tag{19.51}$$

Similarly, the (vector representation of the) high-fidelity residual of the (G-RB) solution reads

$$\mathbf{r}_h(\mathbf{u}_N) = \mathbf{f}_h - \mathbb{A}_h \mathbb{V}\mathbf{u}_N. \tag{19.52}$$

The following lemma provides the main algebraic connection between the (G-RB) problem and the Galerkin high-fidelity approximation:

Lemma 19.1. *The following algebraic relations hold:*

$$\mathbb{A}_h \mathbf{e}_N = \mathbf{r}_h(\mathbf{u}_N), \tag{19.53}$$
$$\mathbb{V}^T \mathbb{A}_h \mathbf{u}_h = \mathbf{f}_N, \tag{19.54}$$
$$\mathbb{V}^T \mathbf{r}_h(\mathbf{u}_N) = \mathbf{0}, \tag{19.55}$$

$\mathbf{e}_N$ *and* $\mathbf{r}_h(\mathbf{u}_N)$ *being defined by* (19.51) *and* (19.52), *respectively.*

Proof. Equation (19.53) follows directly from (19.51) and (19.47).
By left multiplication of (19.47) by $\mathbb{V}^T$, we immediately obtain (19.54), thanks to (19.49).
Finally, (19.55) follows from (19.52) using the identities (19.49) and problem (19.46). ◇

Note that condition (19.53) is the algebraic counterpart of the Galerkin orthogonality property (19.31) valid for the (G-RB) problem.

In summary, for a given matrix $\mathbb{V}$ of reduced bases, the Galerkin Reduced Basis (G-RB) problem (19.46) can be formally obtained as follows:

Galerkin Reduced Basis (G-RB) problem

1. consider the Galerkin high-fidelity problem (19.47);
2. set $\mathbf{u}_h = \mathbb{V}\mathbf{u}_N + \mathbf{e}_N$, where $\mathbf{u}_N \in \mathbb{R}^N$ has to be determined and the error $\mathbf{e}_N$ is the difference between $\mathbf{u}_h$ and $\mathbb{V}\mathbf{u}_N$;
3. left multiply (19.47) by $\mathbb{V}^T$ to obtain $\mathbb{A}_N\mathbf{u}_N - \mathbf{f}_N = -\mathbb{V}^T\mathbb{A}_h\mathbf{e}_N$, that is
$$\mathbb{A}_N\mathbf{u}_N - \mathbf{f}_N = -\mathbb{V}^T\mathbf{r}_h(\mathbf{u}_N);$$
4. require $\mathbf{u}_N$ to satisfy $\mathbb{V}^T\mathbf{r}_h(\mathbf{u}_N) = \mathbf{0}$, or equivalently
$$\mathbb{A}_N\mathbf{u}_N = \mathbf{f}_N.$$

If $\mathbb{A}_h$ is symmetric and positive definite, then the G-RB solution satisfies the following residual minimization property:

$$\mathbf{u}_N = \arg\min_{\tilde{\mathbf{u}}_N \in \mathbb{R}^N} \|\mathbf{r}_h(\tilde{\mathbf{u}}_N)\|^2_{\mathbb{A}_h^{-1}}. \tag{19.56}$$

In fact, by indicating with $\mathbb{K}^{1/2}$ the square root of a (symmetric and positive definite) matrix $\mathbb{K}$, we have

$$\begin{aligned}\|\mathbf{r}_h(\tilde{\mathbf{u}}_N)\|^2_{\mathbb{A}_h^{-1}} &= (\mathbf{f}_h - \mathbb{A}_h\mathbb{V}\mathbf{u}_N, \mathbf{f}_h - \mathbb{A}_h\mathbb{V}\mathbf{u}_N)_{\mathbb{A}_h^{-1}} = \\ &= (\mathbb{A}_h^{-1/2}\mathbf{f}_h - \mathbb{A}_h^{1/2}\mathbb{V}\mathbf{u}_N, \mathbb{A}_h^{-1/2}\mathbf{f}_h - \mathbb{A}_h^{1/2}\mathbb{V}\mathbf{u}_N).\end{aligned}$$

This can be regarded as the least-squares solution of the system $\mathbb{A}_h^{1/2}\mathbb{V}\mathbf{u}_N = \mathbb{A}_h^{-1/2}\mathbf{f}_h$, whose corresponding *normal equations*[4] are

$$\mathbb{V}^T\mathbb{A}_h^{1/2}\mathbb{A}_h^{1/2}\mathbb{V}\mathbf{u}_N = \mathbb{V}^T\mathbb{A}_h^{1/2}\mathbb{A}_h^{-1/2}\mathbf{f}_h = \mathbb{V}^T\mathbf{f}_h.$$

Note that the latter coincide with the (G-RB) problem (19.46).

19.4.2 Geometric interpretation of the (G-RB) problem

We can also characterize from a geometric standpoint the RB approximation obtained by solving the (G-RB) problem, as well as the error $\mathbf{e}_N = \mathbf{u}_h - \mathbb{V}\mathbf{u}_N$ between the solution of the (G-RB) problem and the Galerkin high-fidelity approximation.

[4] We recall that, given $\mathbf{c} \in \mathbb{R}^{N_h}$ and $\mathbb{B} \in \mathbb{R}^{N_h \times N}$, the overdetermined system $\mathbb{B}\tilde{\mathbf{u}} = \mathbf{b}$ can be solved in the least-squares sense, by seeking $\mathbf{u} = \arg\min_{\tilde{\mathbf{u}} \in \mathbb{R}^N} \|\mathbf{c} - \mathbb{B}\tilde{\mathbf{u}}\|_2^2$. The solution is unique provided that the N columns of $\mathbb{B}$ are linearly independent, and can be obtained through the following normal equations:
$$(\mathbb{B}^T\mathbb{B})\mathbf{u} = \mathbb{B}^T\mathbf{c}.$$

To this end, we exploit the fact that the transformation matrix $\mathbb{V}$ defined by (19.48) identifies an orthogonal projection on the reduced subspace $\mathbf{V}_N = \text{span}\{\mathbf{v}_1, \dots, \mathbf{v}_N\}$ of $\mathbb{R}^{N_h}$ generated by the column vectors of the matrix $\mathbb{V}$. Then $\dim(\mathbf{V}_N) = N$ because of the linear independence of the columns of $\mathbb{V}$.

Assume that the basis functions $\{\zeta_k\}_{k=1,\dots,N}$ are orthonormal with respect to the scalar product $(\cdot,\cdot)_h$, that is

$$(\zeta_k, \zeta_m)_h = \sum_{j=1}^{N_h} w_j \zeta_k(x_j)\zeta_m(x_j) = \delta_{km}. \tag{19.57}$$

Then

$$\mathbb{V}^T\mathbb{V} \in \mathbb{R}^{N\times N}, \qquad \mathbb{V}^T\mathbb{V} = \mathbb{I}_N \tag{19.58}$$

where $\mathbb{I}_N$ denotes the identity matrix of dimension N.

As a matter of fact:

$$(\mathbb{V}^T\mathbb{V})_{mk} = (\zeta_m, \zeta_k)_h \qquad \forall k,m = 1,\dots,N. \tag{19.59}$$

Lemma 19.2. *The following results hold:*

1. *The matrix $\boldsymbol{\Pi} = \mathbb{V}\mathbb{V}^T \in \mathbb{R}^{N_h\times N_h}$ is a projection matrix from the whole space $\mathbb{R}^{N_h}$ onto the subspace $\mathbf{V}_N$;*
2. *The matrix $\mathbb{I}_{N_h} - \boldsymbol{\Pi} = \mathbb{I}_{N_h} - \mathbb{V}\mathbb{V}^T \in \mathbb{R}^{N_h\times N_h}$ is a projection matrix from the whole space $\mathbb{R}^{N_h}$ onto the space $\mathbf{V}_N^\perp$, which is the subspace of $\mathbb{R}^{N_h}$ orthogonal to $\mathbf{V}_N$.*
3. *The residual $\mathbf{r}_h(\mathbf{u}_N)$ satisfies*

$$\boldsymbol{\Pi}\,\mathbf{r}_h(\mathbf{u}_N) = \mathbf{0}, \tag{19.60}$$

that is, it belongs to the orthogonal space $\mathbf{V}_N^\perp$.

Proof. Property 1 is a direct consequence of the orthonormality property (19.58). In fact:

$$\forall \mathbf{w}_N \in \mathbf{V}_N \text{ there exists } \mathbf{v}_N \in \mathbb{R}^N \text{ s.t. } \mathbf{w}_N = \mathbb{V}\mathbf{v}_N.$$

Then, $\forall \mathbf{v}_h \in \mathbb{R}^{N_h}$, $\forall \mathbf{w}_N \in \mathbf{V}_N$,

$$(\boldsymbol{\Pi}\mathbf{v}_h, \mathbf{w}_N)_2 = (\boldsymbol{\Pi}\mathbf{v}_h, \mathbb{V}\mathbf{v}_N)_2 = (\mathbb{V}^T\mathbf{v}_h, \mathbb{V}^T\mathbb{V}\mathbf{v}_N)_2 = (\mathbf{v}_h, \mathbb{V}\mathbf{v}_N)_2 = (\mathbf{v}_h, \mathbf{w}_N)_2.$$

Property 2 follows from property 1. Finally, (19.60) follows from (19.55). ◇

If we assume the basis to be orthonormal, the error $\mathbf{e}_N = \mathbf{u}_h - \mathbb{V}\mathbf{u}_N$ can be decomposed into two orthogonal terms:

$$\begin{aligned}\mathbf{e}_N &= \mathbf{u}_h - \mathbb{V}\mathbf{u}_N = (\mathbf{u}_h - \boldsymbol{\Pi}\mathbf{u}_h) + (\boldsymbol{\Pi}\mathbf{u}_h - \mathbb{V}\mathbf{u}_N)\\ &= (\mathbb{I}_{N_h} - \boldsymbol{\Pi})\mathbf{u}_h + \mathbb{V}(\mathbb{V}^T\mathbf{u}_h - \mathbf{u}_N) = \mathbf{e}_{\mathbf{V}_N^\perp} + \mathbf{e}_{\mathbf{V}_N}.\end{aligned}$$

The first term, orthogonal to $\mathbf{V}_N$, accounts for the fact that the high-fidelity solution does not strictly belong to the reduced subspace $\mathbf{V}_N$, whereas the second one, parallel to $\mathbf{V}_N$, accounts for the fact that a (slightly) different problem from the original one is solved; see Fig. 19.4.

We remark that the result (19.58) follows from property (19.57) on the h-orthogonality of the reduced basis functions. Actually, the orthonormalization can be done with respect to a different scalar product that we generically denote $(\cdot,\cdot)_N$, i.e.

$$(\zeta_k,\zeta_m)_N = \delta_{km} \qquad \forall k,m = 1,\ldots,N.$$

In this case, instead of (19.59) we have

$$\delta_{km} = (\zeta_k,\zeta_m)_N = \sum_{r=1}^{N_h}\sum_{s=1}^{N_h}(\mathbb{V}_{rk}\varphi^r,\mathbb{V}_{sm}\varphi^s)_N = \sum_{r=1}^{N_h}\sum_{s=1}^{N_h}\mathbb{V}_{ms}\mathbb{M}_{sr}\mathbb{V}_{rk},$$

where we have used (19.48) and defined the mass matrix for $(\cdot,\cdot)_N$ as

$$\mathbb{M}_{sr} = (\varphi^r,\varphi^s)_N, \qquad 1 \le r,s \le N.$$

Consequently, instead of (19.58) we obtain the new orthonormality relation

$$\mathbb{Y}^T\mathbb{Y} = \mathbb{V}^T\mathbb{M}\mathbb{V} = \mathbb{I}_N,$$

where $\mathbb{Y}^T = \mathbb{V}^T\mathbb{M}^{1/2}$, $\mathbb{Y} = \mathbb{M}^{1/2}\mathbb{V}$. In the same way, the projection matrix is $\Pi = \mathbb{Y}\mathbb{Y}^T$. The results 1. and 2. of Lemma 19.2 proven above will still hold, provided the matrix $\mathbb{V}$ is replaced by $\mathbb{Y}$ and the subspace $\mathbf{V}_N$ by $\mathbf{V}_N = \mathbb{Y}\mathbf{V}_N$.

19.4.3 Alternative formulations: Least-Squares and Petrov-Galerkin RB problems

The Galerkin projection, which leads to the (G-RB) problem discussed so far, is the commonest strategy to build a reduced-order method, since it yields, up to constants,

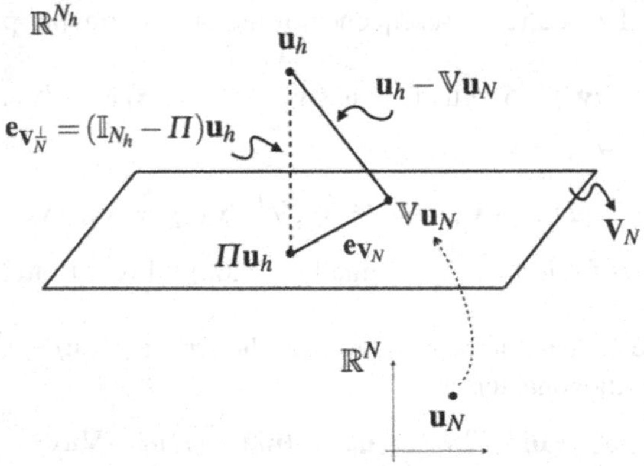

Fig. 19.4. The subspace $\mathbf{V}_N$ of $\mathbb{R}^{N_h}$ and the vectors $\mathbf{u}_N \in \mathbb{R}^N$, $\mathbb{V}\mathbf{u}_N \in \mathbf{V}_N$ and $\mathbf{u}_h \in \mathbb{R}^{N_h}$

the best approximation with respect to the energy norm.

In this case, the trial space (namely, the space where we seek the solution) and the test space are the same; from an algebraic standpoint this is reflected by the identities (19.49), where the matrix by which we pre- and post-multiply the high-fidelity stiffness matrix is the same. However, the trial and the test space may be chosen in a different way, giving rise to what we have called a Petrov-Galerkin formulation. In this section we provide some ideas about this approach.

The Least Squares Reduced Basis method

An alternative approach to (G-RB) is the so-called *Least Squares* Reduced Basis (LS-RB) – sometimes also called *Minimum Residual* – method, where the (LS-RB) solution satisfies

$$\mathbf{u}_N = \arg \min_{\tilde{\mathbf{u}}_N \in \mathbb{R}^N} \|\mathbf{r}_h(\tilde{\mathbf{u}}_N)\|_2^2. \quad (19.61)$$

Note that the minimization criterion (19.61) applies for any matrix $\mathbb{A}_h$, whereas (19.56), relative to the (G-RB) method, requires $\mathbb{A}_h$ to be symmetric and positive definite. The solution to (19.61) coincides with the solution of the *normal equations*

$$(\mathbb{A}_h \mathbb{V})^T \mathbb{A}_h \mathbb{V} \mathbf{u}_N = (\mathbb{A}_h \mathbb{V})^T \mathbf{f}_h,$$

that is

$$(\mathbb{A}_h \mathbb{V})^T \mathbf{r}_h(\mathbf{u}_N) = \mathbf{0}. \quad (19.62)$$

For a given matrix $\mathbb{V}$, the Least Squares (Minimum Residual) RB problem can therefore be obtained as follows:

Least Squares Reduced Basis (LS-RB) problem

1. consider the Galerkin high-fidelity problem (19.47);
2. set $\mathbf{u}_h = \mathbb{V}\mathbf{u}_N + \mathbf{e}_N$, where $\mathbf{u}_N \in \mathbb{R}^N$ has to be determined and the error $\mathbf{e}_N$ is the difference between $\mathbf{u}_h$ and $\mathbb{V}\mathbf{u}_N$;
3. left multiply (19.47) by $(\mathbb{A}_h \mathbb{V})^T$ to obtain
$$(\mathbb{A}_h \mathbb{V})^T \mathbb{A}_h \mathbb{V} \mathbf{u}_N = (\mathbb{A}_h \mathbb{V})^T \mathbf{f}_h - (\mathbb{A}_h \mathbb{V})^T \mathbf{r}_h(\mathbf{u}_N);$$
4. require $\mathbf{u}_N$ to satisfy $(\mathbb{A}_h \mathbb{V})^T \mathbf{r}_h(\mathbf{u}_N) = \mathbf{0}$, that is, equivalently,
$$(\mathbb{A}_h \mathbb{V})^T \mathbb{A}_h \mathbb{V} \mathbf{u}_N = (\mathbb{A}_h \mathbb{V})^T \mathbf{f}_h. \quad (19.63)$$

Note that (19.63) can be rewritten in the form (19.46) provided we set

$$\mathbf{f}_N = (\mathbb{A}_h \mathbb{V})^T \mathbf{f}_h, \qquad \mathbb{A}_N = (\mathbb{A}_h \mathbb{V})^T \mathbb{A}_h \mathbb{V}.$$

The Petrov-Galerkin Reduced Basis method

Problem (19.63) can be regarded as a special instance of the following Petrov-Galerkin (rather than Galerkin) method: find $u_N(\mu) \in V_N$ such that

$$a(u_N(\mu), w_N; \mu) = f(w_N; \mu) \quad \forall w_N \in W_N, \tag{19.64}$$

where $W_N \subset V^{N_h}$ is a subspace of dimension N, different from V_N. If we denote by $\{\eta_k, k = 1, \ldots, N\}$ a basis for W_N, and by $\mathbb{W} \in \mathbb{R}^{N_h \times N}$ the matrix whose entries are

$$(\mathbb{W})_{rk} = (\eta_k, \varphi^r)_h, \qquad r = 1, \ldots, N_h \ , \ k = 1, \ldots, N,$$

we can still express (19.64) in the algebraic form (19.46); this time, however, instead of (19.49) we have

$$\mathbf{f}_N = \mathbb{W}^T \mathbf{f}_h, \qquad \mathbb{A}_N = \mathbb{W}^T \mathbb{A}_h \mathbb{V}. \tag{19.65}$$

For two given matrices $\mathbb{V}$ and $\mathbb{W}$, the Petrov-Galerkin RB (PG-RB) method can be obtained as follows:

Petrov-Galerkin Reduced Basis (PG-RB) problem

1. consider the Galerkin high-fidelity problem (19.47);
2. set $\mathbf{u}_h = \mathbb{V}\mathbf{u}_N + \mathbf{e}_N$, where $\mathbf{u}_N \in \mathbb{R}^N$ has to be determined and the error $\mathbf{e}_N$ is the difference between $\mathbf{u}_h$ and $\mathbb{V}\mathbf{u}_N$;
3. left multiply (19.47) by $\mathbb{W}^T$ to obtain
$$\mathbb{W}^T \mathbb{A}_h \mathbb{V} \mathbf{u}_N = \mathbb{W}^T \mathbf{f}_h - \mathbb{W}^T \mathbf{r}_h(\mathbf{u}_N);$$
4. require $\mathbf{u}_N$ to satisfy $\mathbb{W}^T \mathbf{r}_h(\mathbf{u}_N) = \mathbf{0}$, that is, equivalently,
$$\mathbb{W}^T \mathbb{A}_h \mathbb{V} \mathbf{u}_N = \mathbb{W}^T \mathbf{f}_h. \tag{19.66}$$

As already anticipated, the (LS-RB) problem (19.63) is a special case of the (PG-RB) problem (19.66) corresponding to the choice $\mathbb{W} = \mathbb{A}_h \mathbb{V}$. In fact, we can show (see Exercise 1) the following result:

Property 19.1. *The (LS-RB) formulation corresponds to the following problem in variational form: find $u_N(\mu) \in V_N$ such that*

$$\begin{cases} a(u_N(\mu), y_h^\mu(v_N); \mu) = f(y_h^\mu(v_N); \mu) & \forall v_N \in V_N \\ \text{where } y_h^\mu(v_N) \in V_h \text{ is the solution to } & \text{the following problem:} \\ (y_h^\mu(v_N), z_h)_h = a(v_N, z_h; \mu) & \forall z_h \in V^{N_h}. \end{cases} \tag{19.67}$$

The (LS-RB) method is therefore equivalent to the Petrov-Galerkin method (19.64) provided the test space

$$W_N = \operatorname{span}\{\eta_k^\mu \ , \ k = 1, \ldots, N\} \subset V^{N_h}$$

is defined by taking $\eta_k^\mu = \eta_k^\mu(\zeta_k)$ as the solution to

$$(\eta_k^\mu, z_h)_h = a(\zeta_k, z_h; \mu) \quad \forall z_h \in V^{N_h}. \tag{19.68}$$

Remark 19.1. The choice (19.68) is optimal in the sense that the error in the (PG-RB) problem equals the best approximation error – i.e. the ratio of continuity constant to stability constant is 1 – provided we endow the trial space with a suitable energy norm; the interested reader can refer, for instance, to [DG11].

We also note that in the (LS-RB) case the basis functions of the test space depend on μ because of the parametric dependence of the bilinear form $a(\cdot,\cdot;\mu)$. Several options for their efficient construction and orthonormalization are available in this case; see for instance [RHM13] for further details.

19.5 Construction of reduced spaces

We now describe how to sample the parameter space in order to compute the retained snapshots which actually form the reduced basis. We start by illustrating a sample strategy, the *greedy* algorithm, which was introduced in [PR07, RHP08], and is based on the idea of selecting at each step the locally optimal element. Next we will address an alternative procedure, the so-called *proper orthogonal decomposition.*

19.5.1 Greedy algorithm

We start by formulating this algorithm in an abstract setting, then we characterize the case of the reduced basis method by providing a *computable* version.

A greedy algorithm is a general procedure to approximate each element of a compact set K in a Hilbert space V by a subspace of properly selected elements of K. For a given $N \geq 1$, we seek functions $\{f_1, f_1, \ldots, f_N\}$ such that each $f \in K$ is well approximated by the elements of the subspace $K_N = \text{span}\{f_1, \ldots, f_N\}$. The algorithm can be described as follows:

$f_1 = \arg\max_{f \in K} \|f\|_V;$

assume $f_1, \ldots, f_{N-1}$ are defined

  consider $K_{N-1} = \text{span}\{f_1, \ldots, f_{N-1}\};$

  define $f_N = \arg\max_{f \in K} \|f - \Pi_{K_{N-1}} f\|_V;$

iterate until $\arg\max_{f \in K} \|f - \Pi_{K_{N\max}} f\|_V < \varepsilon_{\text{tol}}^*.$

$\Pi_{K_{N-1}}$ is the orthogonal projection operator on K_{N-1} with respect to the scalar product $(\cdot,\cdot)_V$ and $\varepsilon_{\text{tol}}^*$ is a chosen tolerance; f_N is called the *worst case* element, i.e. the element of K that maximizes the projection error on K_{N-1}.

At each step, the elements provided by the previous algorithm are orthonormalized by the following Gram-Schmidt procedure: we define

$$\zeta_1 = \frac{f_1}{\|f_1\|_V}, \qquad \zeta_N = \frac{f_N - \Pi_{K_{N-1}} f_N}{\|f_N - \Pi_{K_{N-1}} f_N\|_V}, \quad N = 2, \dots, N_{\max}$$

and assume that the construction ends when $N = N_{\max}$. In particular, for any $f \in V$,

$$\Pi_{K_N} f = \sum_{n=1}^{N} \pi_{K_N}^n(f)\zeta_n, \quad \text{with} \quad \pi_{K_N}^n(f) = (f, \zeta_n)_V.$$

We are interested in the case where K is the parametrically induced manifold (19.26); the greedy algorithm for the parameters selection takes the following form:

$$\begin{aligned}
&\mu_1 = \arg\max_{\mu \in \mathcal{D}} \|u_h(\mu)\|_V;\\
&\text{given the samples } \mu^1, \dots, \mu^{N-1},\\
&\qquad \text{consider } V_{N-1} = \text{span}\{u_h(\mu^1), \dots, u_h(\mu^{N-1})\};\\
&\qquad \text{construct } \mu_N = \arg\sup_{\mu \in \mathcal{D}} \|u_h(\mu) - \Pi_{N-1}^{\mu} u_h(\mu)\|_V;\\
&\text{iterate until } \arg\sup_{\mu \in \mathcal{D}} \|u_h(\mu) - \Pi_{N_{\max}}^{\mu} u_h(\mu)\|_V < \varepsilon_{\text{tol}}^*.
\end{aligned} \tag{19.69}$$

Here $\Pi_{N-1}^{\mu} : V^{N_h} \to V_{N-1}$ denotes the elliptic (Galerkin) projection onto V_{N-1}:

$$a(\Pi_{N-1}^{\mu} u_h, v_h; \mu) = a(u_h, v_h; \mu) \qquad \forall v \in V_{N-1}.$$

Note that, thanks to (19.16) and (19.21), $\Pi_N^{\mu} u_h(\mu) = u_N(\mu)$ for all $\mu \in \mathcal{D}$. The set $\{u_h(\mu^1), \dots, u_h(\mu^N)\}$ generated above is then orthonormalized with respect to the scalar product $(\cdot,\cdot)_V$, yielding a new orthonormal basis $\{\zeta_1 \dots, \zeta_N\}$ of V_N.

In the Gram-Schmidt orthonormalization process we cannot use the elliptic (Galerkin) projection Π_N^{μ}, since the latter depends on μ; instead, we employ the orthogonal projection $\Pi_N : V^{N_h} \to V_N$ with respect to the scalar product $(\cdot,\cdot)_V$ corresponding to a specific value of $\bar{\mu}$, see (19.10). The orthonormalization process gives:

$$\zeta_1 = \frac{u_h(\mu^1)}{\|u_h(\mu^1)\|_V}, \qquad \zeta_N = \frac{u_h(\mu^N) - \Pi_{N-1} u_h(\mu^N)}{\|u_h(\mu^N) - \Pi_{N-1} u_h(\mu^N)\|_V}, \quad N = 2, \dots, N_{\max}.$$

In particular,

$$\Pi_N u_h(\mu) = \sum_{n=1}^{N} \pi_N^n(\mu)\zeta_n, \quad \text{with} \quad \pi_N^n(\mu) = (u_h(\mu), \zeta_n)_V.$$

Computationally, the greedy algorithm (19.69) is rather expensive: at each step, seeking the best snapshot entails solving an optimization problem, where the evaluation of the approximation error $\|u_h(\mu) - \Pi^{\mu}_{N-1} u_h(\mu)\|_V$ requires many expensive evaluations of the high-fidelity solution $u_h(\mu)$.

In practice, this cost is alleviated by replacing the sup over $\mathcal{D}$ with a sup over a very fine sample[5] $\Xi_{\text{train}} \subset \mathcal{D}$, of cardinality $|\Xi_{\text{train}}| = n_{\text{train}}$, which shall serve to select our RB space – or train our RB approximation. This nevertheless still requires solving several high-fidelity approximation problems.

A further simplification is then adopted, which consists of replacing the approximation error with an inexpensive a posteriori error estimator $\Delta_{N-1}(\mu)$ such that

$$\|u_h(\mu) - \Pi^{\mu}_{N-1} u_h(\mu)\|_V \le \Delta_{N-1}(\mu) \qquad \forall \mu \in \mathcal{D}.$$

The complete greedy algorithm reads as follows:

$$
\begin{array}{l}
S_1 = \{\mu^1\};\\
\texttt{compute}\ u_h(\mu^1);\\
V_1 = \text{span}\{u_h(\mu^1)\};\\
\texttt{for}\ N = 2, \ldots\\
\quad \mu^N = \arg\max_{\mu \in \Xi_{\text{train}}} \Delta_{N-1}(\mu);\\
\quad \varepsilon_{N-1} = \Delta_{N-1}(\mu^N);\\
\quad \texttt{if}\ \varepsilon_{N-1} \le \varepsilon^*_{\text{tol}}\\
\qquad N_{\max} = N-1;\\
\quad \texttt{end};\\
\quad \texttt{compute}\ u_h(\mu^N);\\
\quad S_N = S_{N-1} \cup \{\mu^N\};\\
\quad V_N = V_{N-1} \cup \text{span}\{u_h(\mu^N)\};\\
\texttt{end}.
\end{array}
\qquad (19.70)
$$

Otherwise said, at the N-th iteration of this algorithm to the *retained* snapshots, over all possible candidate $u_h(\mu)$, $\mu \in \Xi_{\text{train}}$, we append the particular candidate snapshot that the a posteriori error bound (19.96) predicts will be the worst approximated by the RB prediction associated to V_{N-1}.

A similar greedy procedure can also be developed with respect to the energy norm [RHP08]. This is particularly relevant in the compliant case, since the error in the energy norm is directly related to the error in the output (see Sect. 19.3.2).

[5] Typically these samples are chosen by Monte Carlo methods with respect to a uniform or log-uniform density: Ξ is however sufficiently large to ensure that the reported results are insensitive to further refinement of the parameter sample. We usually make a distinction between the test sample Ξ, which serves to assess Online the quality of the RB approximation and the a posteriori error estimators, and the train samples Ξ_{train}, which serves to *generate* the RB approximation; the choice of the latter has important (both Offline and Online) computational implications.

19.5.2 Proper Orthogonal Decomposition

A technique alternative to greedy RB algorithms for the construction of reduced spaces in computational reduction of parametrized systems is *proper orthogonal decomposition* (POD). POD is also popular in multivariate statistical analysis (where it is called *principal component analysis*) or in the theory of stochastic processes (under the name of *Karhunen-Loève decomposition*). The first applications of POD in scientific computing were concerned with the simulation of turbulent flows and date back to the early '90s [Aub91, BHL93]; the interested reader can find further details for instance in [HLB98].

POD techniques[6] reduce the dimensionality of a system by transforming the original variables into a new set of uncorrelated variables (called POD modes, or principal components), the first few modes ideally retaining most of the *energy* present in all of the original variables.

The POD method relies on the use of the singular value decomposition (SVD) algorithm that we briefly describe below. Consider a discrete set of n_{train} snapshot vectors $\{\mathbf{u}_1,\ldots,\mathbf{u}_{n_{\text{train}}}\}$ belonging to $\mathbb{R}^{N_h}$, and form the snapshot matrix $\mathbb{U}\in\mathbb{R}^{N_h\times n_{\text{train}}}$ having them as column vectors:

$$\mathbb{U}=[\mathbf{u}_1\ \mathbf{u}_2\ \ldots\ \mathbf{u}_{n_{\text{train}}}],$$

with $n_{\text{train}}=|\Xi_{\text{train}}|\ll N_h$ (see (19.44)):

$$\mathbf{u}_j=(u_j^{(1)},\ldots,u_j^{(N_h)})\in\mathbb{R}^{N_h},\quad u_j^{(r)}=u_h(x_r;\mu^j)\ \Leftrightarrow\ u_h(x;\mu^j)=\sum_{r=1}^{N_h}u_j^{(r)}\varphi^r(\mathbf{x}). \tag{19.71}$$

The SVD decomposition of $\mathbb{U}$ reads

$$\mathbb{V}^T\mathbb{U}\mathbb{Z}=\begin{pmatrix}\Sigma & 0\\ 0 & 0\end{pmatrix},$$

where $\mathbb{V}=[\zeta_1\ \zeta_2\ \ldots\ \zeta_{N^h}]\in\mathbb{R}^{N_h\times N_h}$ and $\mathbb{Z}=[\psi_1\ \psi_2\ \ldots\ \psi_{n_{\text{train}}}]\in\mathbb{R}^{n_{\text{train}}\times n_{\text{train}}}$ are orthogonal matrices and $\Sigma=\text{diag}(\sigma_1,\ldots,\sigma_r)$ with $\sigma_1\geq\sigma_2\geq\ldots\geq\sigma_r$; here $r\leq n_{\text{train}}$ is the rank of $\mathbb{U}$, which is strictly smaller than n_{train} if the snapshot vectors are not all linearly independent.

Then, we can write

$$\mathbb{U}\psi_i=\sigma_i\zeta_i\quad\text{and}\quad\mathbb{U}^T\zeta_i=\sigma_i\psi_i,\qquad i=1,\ldots,r$$

[6] For a general and concise introduction to POD techniques in view of the reduction of a (time-dependent) dynamical system – which is the first (and most used) application of this strategy – the interested reader may refer to [Pin08, Vol11]. Two additional techniques – indeed quite close to POD – for generating reduced spaces are the Centroidal Voronoi Tessellation (see Chap. 9, [BGL06a, BGL06b]) and the Proper Generalized Decomposition (see for instance [CAC10, CLC11, Nou10]).

or, equivalently,

$$\mathbb{U}^T\mathbb{U}\psi_i = \sigma_i^2\psi_i \quad \text{and} \quad \mathbb{U}\mathbb{U}^T\zeta_i = \sigma_i^2\zeta_i, \qquad i = 1,\dots,r \tag{19.72}$$

i.e. $\sigma_i^2, i = 1,\dots,r$ are the nonzero eigenvalues of the matrix $\mathbb{U}^T\mathbb{U}$ (and also of $\mathbb{U}\mathbb{U}^T$), listed in nondecreasing order; $\mathbb{C} = \mathbb{U}^T\mathbb{U}$ is the *correlation matrix*

$$\mathbb{C}_{ij} = \mathbf{u}_i^T\mathbf{u}_j, \qquad 1 \le i,j \le n_{\text{train}}.$$

For any $N \le n_{\text{train}}$, the POD basis of dimension N is defined as the set of the first N left singular vectors $\zeta_1,\dots,\zeta_N$ of $\mathbb{U}$ or, alternatively, the set of vectors

$$\zeta_j = \frac{1}{\sigma_j}\mathbb{U}\psi_j, \qquad 1 \le j \le N \tag{19.73}$$

obtained from the first N eigenvectors $\psi_1,\dots,\psi_N$ of the correlation matrix $\mathbb{C}$.

By construction, the POD basis is orthonormal. Moreover, if $\{\mathbf{z}_1,\dots,\mathbf{z}_N\}$ is an arbitrary set of N orthonormal vectors in $\mathbb{R}^{N_h}$, and $\Pi_{Z_N}\mathbf{w}$ the projection of a vector $\mathbf{w} \in \mathbb{R}^{N_h}$ onto $Z_N = \text{span}\{\mathbf{z}_1,\dots,\mathbf{z}_N\}$, that is

$$\Pi_{Z_N}\mathbf{u} = \sum_{n=1}^{N} \pi_{Z_N}^n(\mathbf{u})\mathbf{z}_n, \quad \text{with} \quad \pi_{Z_N}^n(\mathbf{u}) = \mathbf{u}^T\mathbf{z}_n,$$

the POD basis (19.73) generated from the set of snapshot vectors $\mathbf{u}_1,\dots,\mathbf{u}_{n_{\text{train}}}$ solves the following minimization problem:

$$\begin{cases} \min\{E(\mathbf{z}_1,\dots,\mathbf{z}_N),\, \mathbf{z}_i \in \mathbb{R}^{N_h},\, \mathbf{z}_i^T\mathbf{z}_j = \delta_{ij},\, \forall 1 \le i,j \le N\} \\ \qquad \text{with } E(\mathbf{z}_1,\dots,\mathbf{z}_N) = \displaystyle\sum_{i=1}^{n_{\text{train}}} \|\mathbf{u}_i - \Pi_{Z_N}\mathbf{u}_i\|_2^2. \end{cases} \tag{19.74}$$

Thus, the POD basis minimizes, over all possible N-dimensional orthonormal sets $\{\mathbf{z}_1,\dots,\mathbf{z}_N\}$ in $\mathbb{R}^{N_h}$, the sum of the squares of the error $E(\mathbf{z}_1,\dots,\mathbf{z}_N)$ between each snapshot vector $\mathbf{u}_i$ and its projection $\Pi_{Z_N}\mathbf{u}_i$ onto the subspace Z_N. $E(\mathbf{z}_1,\dots,\mathbf{z}_N)$ is often referred to as the POD *energy*.

The previous constructive presentation of the POD method was based on the so-called *method of snapshots*, introduced by Sirovich [Sir87]. Alternatively, a POD basis corresponding to a set of snapshot vectors $\mathbf{u}_1,\dots,\mathbf{u}_{n_{\text{train}}}$ can be (and often is) defined by (19.74). In this case, its connection with the SVD of the correlation matrix, as well as relation (19.72), follow by imposing that the POD basis $\{\zeta_1,\dots,\zeta_N\}$ fulfill the first-order necessary optimality conditions (see Exercise 2).

Furthermore, it can be shown that

$$E(\zeta_1,\dots,\zeta_N) = \sum_{i=N+1}^{r} \sigma_i^2, \tag{19.75}$$

so that the error in the POD basis is equal to the squares of the singular values corresponding to the neglected POD modes. In this way, we can select $N_{\max}$ so that $E(\zeta_1,\dots,\zeta_N) \le \varepsilon_{\text{tol}}^*$, for a prescribed tolerance $\varepsilon_{\text{tol}}^*$.

To do this, it is sufficient to choose $N_{\max}$ as the smallest N such that

$$I(N) = \sum_{i=1}^{N} \sigma_i^2 \Big/ \sum_{i=1}^{r} \sigma_i^2 \geq 1 - \delta, \tag{19.76}$$

that is the energy retained by the last $r - N_{\max}$ modes equals to $\delta > 0$, as small as desired; $I(N)$ is referred to as the *relative information content* of the POD basis.

A key feature is that although δ is chosen to be very small, e.g., $\delta = 10^{-\beta}$ with $\beta = 3, 4, \ldots$, in several problems $N_{\max}$ is relatively small (and in particular much smaller than r). This happens because, very often, the singular values of the snapshot matrix decrease very fast (e.g. with exponential rate).

Let us now cast the POD method described so far into the reduced basis context. As already pointed out, see (19.71), the set of snapshot vectors is obtained from a set $\{u_h(\mu^1), \ldots, u_h(\mu^{n_{\text{train}}})\}$ of high-fidelity approximation functions belonging to the space V^{N_h}. In this case the minimization problem (19.74) can be equivalently reformulated as: find the POD basis $\{\zeta_1, \ldots, \zeta_N\}$ that solves the minimization problem:

$$\begin{cases} \min \{E(z_1, \ldots, z_N), z_i \in V^{N_h}, (z_i, z_j)_{L^2(\Omega)} = \delta_{ij}, \forall 1 \leq i, j \leq N\} \\ \text{with } E(z_1, \ldots, z_N) = \displaystyle\sum_{i=1}^{n_{\text{train}}} \|u_h(\mu^i) - \Pi_z u_h(\mu^i)\|^2_{L^2(\Omega)} \end{cases} \tag{19.77}$$

where $\Pi_z : V^{N_h} \to Z_N$ is the $L^2(\Omega)$-projection onto $Z_N = \text{span}\{z_1, \ldots, z_N\}$.

To solve (19.77) we proceed as follows:

- we form the rank r ($\leq n_{\text{train}}$) correlation matrix

$$\mathbb{C}_{ij} = (u_h(\mu^i), u_h(\mu^j))_{L^2(\Omega)}, \qquad 1 \leq i, j \leq n_{\text{train}};$$

- we solve the $n_{\text{train}} \times n_{\text{train}}$ eigenvalue problem: for $i = 1, \ldots, r$,

$$\mathbb{C}\psi_i = \sigma_i^2 \psi_i, \qquad \psi_i^T \psi_j = \delta_{ij}, \quad 1 \leq i, j \leq r;$$

- finally we set

$$\zeta_i = \sum_{j=1}^{n_{\text{train}}} \frac{1}{\sigma_i} \psi_i^{(j)} u_h(\mu^j), \qquad 1 \leq i \leq N, \tag{19.78}$$

$\psi_i^{(j)}$ being the j-th component of the eigenvector ψ_i and $\sigma_i \geq \sigma_{i-1} > 0$.

The correlation matrix factorizes as $\mathbb{C} = \mathbb{U}^T \tilde{\mathbb{M}} \mathbb{U}$, where $\mathbb{U}$ is the snapshot matrix while

$$(\tilde{\mathbb{M}})_{ij} = (\varphi^i, \varphi^j)_{L^2(\Omega)}, \qquad 1 \leq i, j \leq N_h$$

is the mass matrix $\tilde{\mathbb{M}}$ of the high-fidelity approximation space. This allows us to exploit again a SVD to compute the POD basis functions. By using the Cholesky factorization $\tilde{\mathbb{M}} = \mathbb{H}^T \mathbb{H}$, where $\mathbb{H} \in \mathbb{R}^{N_h \times N_h}$ is the Cholesky factor of $\tilde{\mathbb{M}}$, we find $\tilde{\mathbb{U}} = \mathbb{H}\mathbb{U}$, hence $\mathbb{C} = \mathbb{U}^T \tilde{\mathbb{M}} \mathbb{U} = \tilde{\mathbb{U}}^T \tilde{\mathbb{U}}$ and therefore ψ_i, $i = 1, \ldots, N$ represent the first N singular vectors of $\tilde{\mathbb{U}}$.

Typically a POD approach to build an RB space is more expensive than the greedy approach. In the latter, we only need to compute the N – typically very few – high-fidelity *retained* snapshots, whereas in the POD approach we must compute all n_{train} – typically/desirably very many – high-fidelity *candidate* snapshots, as well as the solution of an eigenproblem for the correlation matrix $\mathbb{C} \in \mathbb{R}^{N_h \times N_h}$. Note that (19.76) provides information about the amount of energy neglected by the selected POD modes, that is an indication in the L^2-norm instead than in the V-norm, as it was the case of the a posteriori error estimates used in the greedy algorithm.

19.6 Convergence of RB approximations

In this section we illustrate some convergence results for problems depending on one or several parameters.

19.6.1 A priori convergence theory: a simple case

An *a priori* exponential convergence result, with respect to the number N of basis functions, is known in the case of elliptic PDEs depending on one-dimensional parameters, for instance in [MPT02a, MPT02b]; furthermore, several computational tests shown e.g. in [RHP08] provide a numerical assessment of this behaviour, also for larger dimensions of the parameter space. Here we describe the simplest case, associated with specific *non-hierarchical* spaces V_N^{ln}, $1 \leq N \leq N_{\max}$, given by

$$V_N^{\text{ln}} = \text{span}\{u_h(\mu_N^n), \quad 1 \leq n \leq N\}\,, \tag{19.79}$$

$$\mu_N^n = \mu^{\min} \exp\left\{\frac{n-1}{N-1} \ln\left(\frac{\mu^{\max}}{\mu^{\min}}\right)\right\}, \quad 1 \leq n \leq N,\ 1 \leq N \leq N_{\max}\,. \tag{19.80}$$

We denote the corresponding RB approximation by $u_N^{\text{ln}}(\mu)$. The a priori theory described below suggests that the spaces (19.79) – which we shall call "equi-ln" spaces – do display optimality features; as we will see in the next subsection, a greedy selection would act as just as well. Here we consider the thermal block problem of Sect. 19.1.1 for the case in which $B_1 = 2, B_2 = 1$, as shown in Fig. 19.5.

The governing equations are then given by (19.13) and (19.14) for two blocks/regions $\mathcal{R}_1$ and $\mathcal{R}_2$, the single parameter $\mu = \mu_1$ representing the conductivity of region $\mathcal{R}_1$ (the conductivity of region $\mathcal{R}_2$ is one), and the parameter domain $\mathcal{D} = [\mu^{\min}, \mu^{\max}] = [1/\sqrt{\mu_r}, \sqrt{\mu_r}]$ for $\mu_r = 100$; the associated affine expansion (19.6) now comprises only $Q_a = 2$ terms.

The analysis presented here is, as a matter of fact, relevant to a large class of single-parameter coercive problems. For the RB approximation of this problem, the following result holds:

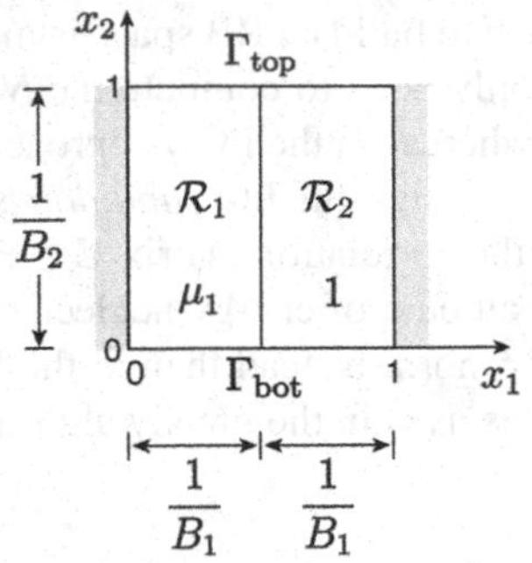

Fig. 19.5. Thermal block problem: $B_1 = 2, B_2 = 1$

Property 19.2. *Given general data F (of which F^{Neu} from (19.14) is a particular example), we obtain that for any $N \geq 1 + C_{\mu_r}$, and every $\mu \in \mathcal{D}$,*

$$\frac{\|u_h(\mu) - u_N^{\text{ln}}(\mu)\|_\mu}{\|u_h(\mu)\|_\mu} \leq \exp\left\{-\frac{N-1}{C_{\mu_r}}\right\}, \tag{19.81}$$

where $C_{\mu_r} = [2e \ln \mu_r]_+$ and $[\,]_+$ denotes the smallest integer greater than or equal to its real argument.

The proof is a "parametric" version of the standard (finite element) variational arguments of Chap. 4. In particular, we first invoke (19.32) and take as our candidate w a high-order polynomial interpolant *in the parameter* μ of $u_h(\mu)$; we next apply the standard Lagrange interpolant remainder formula; finally, we resort to an eigenfunction expansion to bound the parametric (sensitivity) derivatives and optimize the order of the polynomial interpolant.

For the complete proof and more considerations, see [PR07]. We note that the RB convergence estimate (19.81), relative to the model problem we have considered, relies on parameter smoothness and not on the computational grid (through N_h); the exponent in the convergence rate depends on N and logarithmically on μ_r.

19.6.2 A priori convergence theory: greedy algorithms

Several recent results carry out an *a priori* convergence analysis in the more general case of reduced spaces built through the *greedy* algorithm. Below we provide a few preliminary results that explain why approximation spaces V_N built in this way exhibit exponential convergence in N; see e.g. [BMP$^+$12, BCD$^+$11] for further details.

Let us denote, for any $u_h \in \mathcal{M}_h$,

$$\sigma_N(u_h, \mathcal{M}_h) = \inf_{v_N \in V_N} \|u_h - v_N\|_V = \|u_h - \Pi_N^\mu u_h\|_V,$$

where Π_N^μ denotes the projector onto V_N. Then

$$\sigma_N(\mathcal{M}_h) = \sup_{u_h \in \mathcal{M}_h} \sigma_N(u_h, \mathcal{M}_h)$$

is the best approximation error when approximating the set $\mathcal{M}_h$ by V_N. Indeed, this is the quantity which has to be maximized at each step of the greedy algorithm, in order to select the current snapshot. For instance, referring to the case addressed in Sect. 19.6.1, with the specific choice (19.80) we can achieve exponential convergence, that is

$$\sigma_N(\mathcal{M}_h) \leq C\exp(-N^\alpha), \qquad \text{for some } \alpha > 0.$$

A priori convergence analysis aims at providing upper bounds for the sequence $\sigma_N(\mathcal{M}_h)$ in terms of the best N-dimensional subspace, i.e. the one that would minimize the projection error for the whole set $\mathcal{M}_h$ among all N-dimensional subspaces. This minimal error is given by the *Kolmogorov N-width*, which is defined as

$$d_N(\mathcal{M}_h) = \inf_{Z_N \subset V} \sup_{u_h \in \mathcal{M}_h} \text{dist}(u_h, Z_N) \tag{19.82}$$

where

$$\text{dist}(u_h, Z_N) = \min_{w_N \in Z_N} \|u_h - w_N\|_V = \|u_h - \Pi_{Z_N} u_h\|_V$$

and the first infimum is taken over all linear subspaces $Z_N \subset V$ of dimension N. We refer the reader to [LvGM96, Pin85] for a general discussion on Kolmogorov width.

$d_N(\mathcal{M}_h)$ measures the degree to which a subset of the space V can be approximated using finite-dimensional subspaces Z_N. If $\sigma_N(\mathcal{M}_h)$ decayed at a rate comparable to $d_N(\mathcal{M}_h)$, the greedy selection would essentially provide the best possible accuracy attainable by N-dimensional subspaces.

In general the optimal subspace with respect to the Kolmogorov N-width (19.82) is not spanned by elements of the set $\mathcal{M}_h$ being approximated, thus we possibly have that $d_N(\mathcal{M}_h) \ll \sigma_N(\mathcal{M}_h)$. However, if the N-width converges at exponential rate, then also the error of the best approximation in V_N does. This is the meaning of the following result, shown in [BMP+12]; we recall that $\gamma(\mu)$ and $\alpha(\mu)$ are the continuity and the coercivity constants of the bilinear form $a(\cdot,\cdot;\mu)$:

Theorem 19.1. *Assume that the set of all solutions $\mathcal{M}_h = \{u_h(\mu) : \mu \in \mathcal{D}\}$ has an exponentially small Kolmogorov N-width, i.e.*

$$d_N(\mathcal{M}_h) \leq ce^{-\delta N} \quad \textit{with} \quad \delta > \log\left(1 + \left(\frac{\bar{\gamma}}{\alpha_0}\right)^{1/2}\right) = \log \delta_0,$$

where we have assumed that $\gamma(\mu) \leq \bar{\gamma}$ and $\alpha(\mu) \geq \alpha_0$ for all $\mu \in \mathcal{D}$. Then the reduced basis method built using the greedy algorithm (19.69) *converges exponentially, that is, there exists $\eta > 0$ such that*

$$\|u_h(\mu) - u_N(\mu)\|_V \leq Ce^{-\eta N} \quad \forall \mu \in \mathcal{D},$$

where $C = cN\delta_0^N$ is independent of h and μ, and $\delta_0 > 2$. The same result holds if we replace the error $\|u_h(\mu) - \Pi_N^\mu u_h(\mu)\|_V$ with the error estimate $\Delta_N(\mu)$ in (19.69) *provided that*

$$\delta > \log\left(1 + \frac{\bar{\gamma}}{\alpha_0^{N_h}}\left(\frac{\bar{\gamma}}{\alpha_0}\right)^{1/2}\right).$$

Above, $\alpha_0^{N_h} > 0$ is a computable, uniform lower bound of the high-fidelity coercivity constant $\alpha^{N_h}(\mu)$, hence such that $\alpha^{N_h}(\mu) \geq \alpha_0^{N_h}$ for any $\mu \in \mathcal{D}$.

Although interesting from a theoretical viewpoint, the previous comparison is only useful if $d_N(\mathcal{M}_h)$ decays to zero faster than $N^{-1}\delta_0^{-N}$. We remark that the factor $\sqrt{\bar{\gamma}/\alpha_0}$ is related to the V-norm chosen to measure the error in the greedy algorithm. Moreover, the expression of the a posteriori error estimate $\Delta_N(\mu)$, defined in (19.96), involves a computable (μ-dependent) lower bound $\alpha_{\text{LB}}^{N_h}(\mu) \leq \alpha^{N_h}(\mu)$ of the high-fidelity coercivity constant $\alpha^{N_h}(\mu)$; in particular, the parametric lower bound is uniformly bounded below by $\alpha_0^{N_h} > 0$.

In fact, we have exponential convergence of the Kolmogorov N-width when the dependence on the parameter is analytic[7]. This result has been further improved in [BCD$^+$11], where it was shown that if $d_N(\mathcal{M}_h) \leq C\exp\left(-cN^\beta\right)$ for all $N > 0$ and some $C, c > 0$, then, for some $\widetilde{C}, \widetilde{c} > 0$ we have:

$$\sigma_N(\mathcal{M}_h) \leq \widetilde{C}\exp(-\widetilde{c}N^{\beta/(\beta+1)}) \tag{19.83}$$

with $\widetilde{C}$ independent of N. In the case of algebraic convergence, $d_N(\mathcal{M}_h) \leq MN^{-\alpha}$ for all $N > 0$ and some $M, \alpha > 0$, it was proved that

$$\sigma_N(\mathcal{M}_h) \leq CMN^{-\alpha}, \qquad C = C(\gamma, \alpha_0). \tag{19.84}$$

Furthermore, it was shown that

$$\sigma_N(\mathcal{M}_h) \leq \frac{2^{N+1}}{\sqrt{3}} d_N(\mathcal{M}_h),$$

and that this estimate cannot, in general, be improved.

To conclude, the reader will remind that the fast (exponential) convergence of numerical approximations is a distinctive feature of spectral methods (see Chap. 10). The reduced basis method shares indeed some aspects with spectral methods: similarly to the latter, it makes use of basis functions with global support (which are problem-dependent in the RB case, whereas global orthogonal polynomials in the pure spectral case).

[7] However, analytic regularity of the solution manifold $\mathcal{M}_h$ is not necessary in order to apply the reduced basis method.

In general, the reduced basis approximation of solutions of elliptic equations with regular coefficients has indeed been very successful. Further convergence analyses for the greedy algorithm and a numerical proof of these results, extending the ones presented in [MPT02a, MPT02b], can be found e.g. in [LMQR13].

19.7 A posteriori error estimation

Effective a posteriori error bounds for field variables and outputs of interest are crucial for both the efficiency and the reliability of RB approximations. As regards *efficiency*, a posteriori error estimation allows to control the error, and also to minimize the computational effort by controlling the dimension of the RB space. Not only, for the greedy algorithm the application of error bounds (as surrogates for the actual error) allows significantly larger training samples $\Xi_{\text{train}} \subset \mathcal{D}$ and a better parameter space exploration at greatly reduced Offline computational costs. Concerning *reliability*, a posteriori error bounds allow a confident exploitation of the rapid predictive power of the RB approximation and provide an error quantification for each new parameter value μ in the Online stage.

In turn, error bounds must be *rigorous* – valid for all N and for all parameter values in the parameter domain $\mathcal{D}$: non-rigorous error indicators may suffice for adaptivity during basis assembling, but not for reliability. Secondly, the bounds must be reasonably *sharp*: an overly conservative error bound can yield inefficient approximations (N too large) or even dangerous suboptimal engineering results (unnecessary safety margins). In the third place, the bounds must be very *efficient*: the Online operation count and storage to compute the RB error bounds – the marginal average cost – must be independent of N_h (and commensurate with the cost associated with the RB output prediction).

19.7.1 Some preliminary estimates

The central equation for a posteriori error estimates is the error-residual relationship (see Sect. 4.6.2). It follows from the problems statements for $u_h(\mu)$, (19.16), and $u_N(\mu)$, (19.30), by introducing the residual $r(v;\mu) \in (V^{N_h})'$

$$r(v;\mu) = f(v;\mu) - a(u_N(\mu),v;\mu) \quad \forall v \in V^{N_h}\,, \tag{19.85}$$

that the error $e_h(\mu) = u_h(\mu) - u_N(\mu) \in V^{N_h}$ satisfies

$$a(e_h(\mu),v;\mu) = r(v;\mu) \quad \forall v \in V^{N_h}\,. \tag{19.86}$$

The Riesz representation $\hat{e}_h(\mu) \in V^{N_h}$ of $r(\cdot;\mu)$ (see Theorem 2.1) satisfies

$$(\hat{e}_h(\mu),v)_V = r(v;\mu) \quad \forall v \in V^{N_h}\,. \tag{19.87}$$

From the error residual equation (19.86) we obtain

$$a(e_h(\mu),v;\mu) = (\hat{e}_h(\mu),v)_V \quad \forall v \in V^{N_h} \tag{19.88}$$

and therefore

$$\|r(\cdot;\mu)\|_{(V^{N_h})'} = \sup_{v\in V^{N_h}} \frac{r(v;\mu)}{\|v\|_V} = \|\hat{e}_h(\mu)\|_V. \tag{19.89}$$

Computing the dual norm of the residual through the Riesz representation theorem allows to develop a suitable Offline-Online procedure also for the evaluation of the error bounds, as remarked below.

We recall the definition of the exact and FE coercivity constants, respectively (19.5) and (19.20). Moreover, we shall require a (parametric) lower bound $\alpha_{\rm LB}^{N_h}: \mathcal{D} \to \mathbb{R}$ to the (parametric) coercivity constant $\alpha^{N_h}(\mu)$,

$$0 < \alpha_{\rm LB}^{N_h}(\mu) \le \alpha^{N_h}(\mu) \qquad \forall \mu \in \mathcal{D}.$$

19.7.2 Error bounds

We define error estimators for the energy norm and the output respectively as

$$\Delta_N^{\rm en}(\mu) = \frac{\|\hat{e}_h(\mu)\|_V}{(\alpha_{\rm LB}^{N_h}(\mu))^{1/2}} \qquad \text{and} \qquad \Delta_N^{s}(\mu) = \frac{\|\hat{e}_h(\mu)\|_V^2}{\alpha_{\rm LB}^{N_h}(\mu)}. \tag{19.90}$$

We next introduce the effectivity factors associated with these error estimators as

$$\eta_N^{\rm en}(\mu) = \frac{\Delta_N^{\rm en}(\mu)}{\|u_h(\mu) - u_N(\mu)\|_\mu} \qquad \text{and} \qquad \eta_N^{s}(\mu) = \frac{\Delta_N^{s}(\mu)}{(s_h(\mu) - s_N(\mu))},$$

respectively. Clearly, the effectivity factors are a measure of the quality of the proposed estimator: to achieve rigour, we shall insist on effectivities ≥ 1; for sharpness, we desire effectivities as close to one as possible. It has been demonstrated [RHP08, PR07] that

Property 19.3. *For any $N = 1, \ldots, N_{\max}$, the effectivity factors satisfy*

$$1 \le \eta_N^{\rm en}(\mu) \le \sqrt{\frac{\gamma(\mu)}{\alpha_{\rm LB}^{N_h}(\mu)}} \quad \forall \mu \in \mathcal{D}, \tag{19.91}$$

$$1 \le \eta_N^{s}(\mu) \le \frac{\gamma(\mu)}{\alpha_{\rm LB}^{N_h}(\mu)} \quad \forall \mu \in \mathcal{D}. \tag{19.92}$$

Proof. It follows directly from (19.88) for $v = e_h(\mu)$, and from the Cauchy-Schwarz inequality that

$$\|e_h(\mu)\|_\mu^2 \le \|\hat{e}_h(\mu)\|_V \, \|e_h(\mu)\|_V. \tag{19.93}$$

But $(\alpha^{N_h}(\mu))^{\frac{1}{2}} \|e_h(\mu)\|_V \le (a(e_h(\mu), e_h(\mu);\mu))^{\frac{1}{2}} = \|e_h(\mu)\|_\mu$, thus from (19.93) we obtain $\|e_h(\mu)\|_\mu \le \Delta_N^{\rm en}(\mu)$ or $\eta_N^{\rm en}(\mu) \ge 1$. We consider (19.88) again – but now for $v = \hat{e}_h(\mu)$ – and the Cauchy-Schwarz inequality to obtain

$$\|\hat{e}_h(\mu)\|_V^2 \le \|\hat{e}_h(\mu)\|_\mu \, \|e_h(\mu)\|_\mu. \tag{19.94}$$

Thanks to continuity, $\|\hat{e}_h(\mu)\|_\mu \leq (\gamma(\mu))^{\frac{1}{2}} \|\hat{e}_h(\mu)\|_V$, hence from (19.94)

$$\Delta_N^{\text{en}}(\mu) = (\alpha_{\text{LB}}^{N_h}(\mu))^{-\frac{1}{2}} \|\hat{e}_h(\mu)\|_V \leq (\alpha_{\text{LB}}^{N_h}(\mu))^{-\frac{1}{2}}(\gamma(\mu))^{\frac{1}{2}} \|e_h(\mu)\|_\mu,$$

that is $\eta_N^{\text{en}}(\mu) \leq \sqrt{\gamma(\mu)/\alpha_{\text{LB}}^{N_h}(\mu)}$. Next, from (19.33) we have $s_h(\mu) - s_N(\mu) = \|e_h(\mu)\|_\mu^2$, and hence since $\Delta_N^s(\mu) = (\Delta_N^{\text{en}}(\mu))^2$:

$$\eta_N^s(\mu) = \frac{\Delta_N^s(\mu)}{s_h(\mu) - s_N(\mu)} = \frac{(\Delta_N^{\text{en}}(\mu))^2}{\|e(\mu)\|_\mu^2} = (\eta_N^{\text{en}}(\mu))^2 \ ; \tag{19.95}$$

in the end, (19.92) follows directly from (19.91) and (19.95). ◇

A similar result can be obtained for the a posteriori error bound in the V norm, which is defined as

$$\Delta_N(\mu) = \frac{\|\hat{e}_h(\mu)\|_V}{\alpha_{\text{LB}}^{N_h}(\mu)}. \tag{19.96}$$

In fact, by following the same argument as in Proposition 19.3, it can be shown that:

Property 19.4. *For any* $N = 1,\ldots,N_{\max}$,

$$1 \leq \frac{\Delta_N(\mu)}{\|u_h(\mu) - u_N(\mu)\|_V} \leq \frac{\gamma(\mu)}{\alpha_{\text{LB}}^{N_h}(\mu)} \quad \forall \mu \in \mathcal{D}\,. \tag{19.97}$$

The effectivity upper bounds, (19.91), (19.92) and (19.97), are *independent* of N, and hence stable with respect to *RB refinement*.

Finally, we remark that the error bounds of the previous section are of little utility without an accompanying Offline-Online computational approach. The computationally crucial component of all the error bounds of the previous section is $\|\hat{e}_h(\mu)\|_V$, the dual norm of the residual (19.85), whose expression can be expanded according to (19.35) and (19.6), thus yielding the possibility to pre-compute Offline the parameter-independent quantities and then evaluate, for any new value of μ, just some parameter-dependent quantities. The interested reader can find further details for instance in [QRM11, RHP08].

An approach to the construction of lower bounds for coercivity constant – which is a generalized minimum eigenvalue – is the Successive Constraint Method (SCM) introduced in [HRSP07]. The method – based on an Offline-Online strategy – reduces the Online (real-time) calculation to a small linear program for which the operation count is *independent* of N_h. See for instance [RHP08] for more details.

19.8 Non-compliant problems

For the sake of simplicity, we addressed in Sect. 19.7 the RB approximation of affinely parametrized coercive problems in the compliant case. We now consider the elliptic case and the more general non-compliant problem: given $\mu \in \mathcal{D}$, find

$$s(\mu) = J(u(\mu)), \tag{19.98}$$

where $u(\mu) \in V$ satisfies

$$a(u(\mu), v; \mu) = f(v; \mu) \quad \forall v \in V. \tag{19.99}$$

We assume that a is coercive and continuous but not necessarily symmetric; we further suppose that both J and f are bounded linear functionals [8], but we no longer require $J = f$. Moreover, we assume that both a and f are affine, see (19.6)-(19.7).

Following the methodology (and the notation) of Sect. 19.7, we can readily find an a posteriori error bound for $s_N(\mu)$: by standard arguments we obtain

$$|s_h(\mu) - s_N(\mu)| \leq \|J\|_{(V^{N_h})'} \Delta_N^{\text{en}}(\mu),$$

where $\|u_h(\mu) - u_N(\mu)\|_\mu \leq \Delta_N^{\text{en}}(\mu)$ and $\Delta_N^{\text{en}}(\mu)$ is defined in (19.90); see [RHP08, PR07] for further details. We denote this method *"primal-only"*. Although for many outputs primal-only is perhaps the best approach (each additional output, and associated error bound, is a simple "add-on"), this approach has two drawbacks:

1. we lose the "quadratic convergence" effect (19.33) for outputs (unless $J = f$ and a is symmetric);
2. the effectivity factor $\Delta_N^s(\mu)/|s(\mu) - s_N(\mu)|$ may be unbounded: if $J = f$ then we know, from (19.33), that $|s_h(\mu) - s_N(\mu)| \sim \|\hat{e}_h(\mu)\|_V^2$ and hence
$$\frac{\Delta_N^s(\mu)}{|s_h(\mu) - s_N(\mu)|} \sim \frac{1}{\|\hat{e}_h(\mu)\|_V} \to \infty \quad \text{as } N \to \infty,$$
i.e. the effectivity of the output error bound Δ_N^s defined in (19.90) *tends to infinity* as ($N \to \infty$ and) $u_N(\mu) \to u_h(\mu)$.
We may expect a similar behaviour for any J "close" to f: the problem is that $\Delta_N^s(\mu)$ does not reflect the contribution of the test space to the convergence of the output.

The introduction of RB *primal-dual* approximation allows to overcome the previous issues – and ensure a stable limit as $N \to \infty$. Let us introduce the dual problem associated to J, that reads as follows: find $\psi(\mu) \in V$ such that

$$a(v, \psi(\mu); \mu) = -J(v) \quad \forall v \in V;$$

ψ is called the *adjoint* or *dual* field. Let us define the RB spaces for the primal and the dual problem, respectively:

$$V_{N_{pr}} = span\left\{u_h(\mu^{k,pr}), 1 \leq k \leq N_{pr}\right\}, \quad 1 \leq N_{pr} \leq N_{pr,max};$$
$$V_{N_{du}} = span\left\{\Psi_h(\mu^{k,du}), 1 \leq k \leq N_{du}\right\}, \quad 1 \leq N_{du} \leq N_{du,max}.$$

---

[8] Typical output functionals correspond to the "integral" of the field $u(\mu)$ over an area or line (in particular, boundary segment) in $\overline{\Omega}$. However, by appropriate lifting techniques, "integrals" of the *flux* over boundary segments can also be considered.

For our purposes a single FE space suffices for both the primal and dual, even if in the practice the FE primal and dual spaces may be different. The resulting RB approximations $u_{N_{pr}}(\mu) \in V_{N_{pr}}, \Psi_{N_{du}}(\mu) \in V_{N_{du}}$ solve

$$\begin{aligned} a(u_{N_{pr}}(\mu), v; \mu) &= f(v; \mu) \quad \forall v \in V_{N_{pr}}, \\ a(v, \Psi_{N_{du}}(\mu); \mu) &= -J(v) \quad \forall v \in V_{N_{du}}; \end{aligned}$$

then, the RB output can be evaluated as

$$s_{N_{pr},N_{du}}(\mu) = J(u_{N_{pr}}) - r^{pr}(\Psi_{N_{du}}; \mu)$$

where

$$\begin{aligned} r^{pr}(v; \mu) &= f(v; \mu) - a(u_{N_{pr}}, v; \mu), \\ r^{du}(v; \mu) &= -J(v) - a(v, \Psi_{N_{du}}; \mu) \end{aligned}$$

are the primal and the dual residuals, respectively. The term $r^{pr}(\Psi_{N_{du}}; \mu)$ allows to obtain a better convergence to the high-fidelity output $s_h(\mu)$, as remarked in [PG00].

Thus, the output error bound takes the following form:

$$\Delta_N^s(\mu) = \frac{\|r^{\mathrm{pr}}(\cdot\,; \mu)\|_{(V^{N_h})'}}{(\alpha_{\mathrm{LB}}^{N_h}(\mu))^{1/2}} \frac{\|r^{\mathrm{du}}(\cdot\,; \mu)\|_{(V^{N_h})'}}{(\alpha_{\mathrm{LB}}^{N_h}(\mu))^{1/2}} \tag{19.100}$$

in the non-compliant case, so that we are able to recover the "quadratic" output effect. Note that the Offline-Online procedure is very similar to the "primal-only" case, but now we need to do everything both for primal and dual; moreover, we need to evaluate both a primal and a dual residual for the a posteriori error bounds.

19.9 Parametrized geometries and operators

In this section we introduce the general class of scalar problems that fall under the abstract formulation of Sect. 19.1, by extending the simple examples described in Sect. 19.1.1 to the case where the parameters might describe:

- physical properties (material coefficients, source terms, boundary data), possibly varying in different subregions of the computational domain;
- the geometrical configuration of the computational domain.

By following an increasing order of complexity, we first discuss the case of parametrized physical properties; then we describe a generalization to take into account variations of physical properties over different subregions; finally, we present the more difficult case dealing with parametrized geometries.

For the sake of exposition, let us distinguish between d_p physical and d_g geometrical parameter components: we denote the former by μ_p and the latter by μ_g, respectively, so that $\mu = (\mu_p, \mu_g) \in \mathcal{D}_p \times \mathcal{D}_g =: \mathcal{D}$ and $\mathcal{D}_p \subset \mathbb{R}^{d_p}$, $\mathcal{D}_g \subset \mathbb{R}^{d_g}$, with

$p = d_p + d_g$. We focus on the following advection-diffusion-reaction problem:

$$\begin{cases} -\dfrac{\partial}{\partial x_i}\left(\nu_{ij}(\mathbf{x};\mu_p)\dfrac{\partial}{\partial x_j}u(\mu)\right) + \left(b_i(\mathbf{x};\mu_p)\dfrac{\partial}{\partial x_i}\right)u(\mu) & \\ \qquad\qquad +\gamma(\mathbf{x};\mu_p)u(\mu) = F(\mathbf{x};\mu_p), & \mathbf{x} \in \Omega(\mu_g) \\ u(\mu) = 0, & \mathbf{x} \in \Gamma_D \\ \nu_{ij}(\mathbf{x};\mu_p)\dfrac{\partial}{\partial x_j}u(\mu) = g_N(\mathbf{x};\mu_p), & \mathbf{x} \in \Gamma_N, \end{cases}$$

where we suppose that all coefficients (e.g. thermal conductivity, elastic coefficients, and so on) may depend on spatial coordinates and also a vector μ_p of physical parameters. The domain $\Omega(\mu_g) \subset \mathbb{R}^d$, $d = 1,2,3$, may depend instead on a vector μ_g of geometrical parameters. For the sake of simplicity, in the following we consider problems defined over two-dimensional domains ($d = 2$).

Here we assume that only Neumann data imposed on Γ_N are affected by the physical parameters, whereas on $\Gamma_D = \partial\Omega(\mu_g)\setminus\Gamma_N$ we impose homogeneous Dirichlet conditions. Parametrized Dirichlet data, although undergoing a similar treatment, involve suitable (possibly parametrized) lifting functions.

19.9.1 Physical parameters

In the simplest case, input parameters $\mu = \mu_p$ represent physical properties, whereas the computational domain Ω is parameter-independent. Moreover, we assume that the parametric dependence of the physical coefficients does not change on the computational domain. The situation where the domain is split in subregions associated with different physical properties will be covered later on.

The bilinear form and the linear form appearing in (19.2) are given in this case by

$$a(w,v;\mu_p) = \int_\Omega \frac{\partial w}{\partial x_i}\nu_{ij}(\mathbf{x};\mu_p)\frac{\partial v}{\partial x_j}d\Omega + \int_\Omega b_i(\mathbf{x};\mu_p)\frac{\partial w}{\partial x_i}\,v d\Omega + \int_\Omega \gamma(\mathbf{x};\mu_p)wv d\Omega, \tag{19.101}$$

$$f(v;\mu) = \int_\Omega F(\mathbf{x};\mu_p)v d\Omega + \int_{\Gamma_N} g_N(\mathbf{x};\mu_p)v d\Gamma, \tag{19.102}$$

where $V = H^1_{\Gamma_D}(\Omega)$; summation over repeated indices is understood.

In order to ensure an affine expansion of (19.101), we assume that μ_p and $\mathbf{x}$ can be treated as separate variables, i.e. each coefficient shall be expressed as a product of two functions:

$$\nu_{ij}(\mathbf{x};\mu_p) = K_{ij}(\mu_p)\psi_{ij}(\mathbf{x}), \qquad 1 \le i,j \le 2, \tag{19.103}$$

$$b_i(\mathbf{x};\mu_p) = K_{i3}(\mu_p)\psi_{i3}(\mathbf{x}), \qquad i = 1,2, \tag{19.104}$$

$$\gamma(\mathbf{x};\mu_p) = K_{33}(\mu_p)\psi_{33}(\mathbf{x}). \tag{19.105}$$

In this way, by taking the terms depending on the parameters μ_p out of the integrals, we can rewrite (19.101) as follows:

$$a(w,v;\mu) = \sum_{i,j=1}^{3} K_{ij}(\mu_p) \int_\Omega \begin{bmatrix} \frac{\partial w}{\partial x_1} & \frac{\partial w}{\partial x_2} & w \end{bmatrix} \psi_{ij}(\mathbf{x}) \begin{bmatrix} \frac{\partial v}{\partial x_1} \\ \frac{\partial v}{\partial x_2} \\ v \end{bmatrix} d\Omega\,, \tag{19.106}$$

where $\mathbf{x} = (x_1, x_2)$ denotes a point in Ω. Here $K\colon\ \mathcal{D}_p \to \mathbb{R}^{3\times 3}$ and $\Psi\colon\ \Omega \to \mathbb{R}^{3\times 3}$ are given matrices – of components $K_{ij}(\mu_p)$ and $\psi_{ij}(\mathbf{x})$, respectively – which are required to be symmetric and positive definite for any $\mathbf{x} \in \Omega$, $\mu_p \in \mathcal{D}_p$, in order to ensure the coercivity of the bilinear form.

In particular, the upper 2×2 principal submatrices of K and Ψ give the usual conductivity/diffusivity tensor; the $(3,3)$ term allows to consider a reaction (or mass) term. The $(3,1),(3,2)$ (and $(1,3),(2,3)$) elements of K and Ψ are related to first derivative (or convective) terms; in particular, in order to recover (19.101) we assume that $K_{31}(\mu_p) = K_{32}(\mu_p) = 0$.

For the sake of simplicity, in the following we assume that $g_N = 0$. We remark that, in the case of purely physical parametrization, the affinity assumption (19.6) is easy to fulfill; see Exercise 3.

Let us now consider the case in which the domain

$$\bar{\Omega} = \bigcup_{k=1}^{K_{\text{reg}}} \mathcal{R}_k$$

is split into regions[9] $\mathcal{R}_k$, each associated with different physical properties. Instead of (19.101) we might therefore have

$$\begin{aligned} a(w,v;\mu_p) = &\sum_{k=1}^{K_{\text{reg}}} \int_{\mathcal{R}_k} \frac{\partial w}{\partial x_i} v_{ij}^k(\mathbf{x};\mu_p) \frac{\partial v}{\partial x_j} d\Omega \\ &+ \sum_{k=1}^{K_{\text{reg}}} \int_{\mathcal{R}_k} b_i^k(\mathbf{x};\mu_p) \frac{\partial w}{\partial x_i}\, v d\Omega + \sum_{k=1}^{K_{\text{reg}}} \int_{\mathcal{R}_k} \gamma^k(\mathbf{x};\mu_p) w v d\Omega, \end{aligned} \tag{19.107}$$

with

$$v_{ij}^k(\mathbf{x};\mu_p) = K_{ij}^k(\mu_p)\psi_{ij}^k(\mathbf{x}), \qquad 1 \le i,j \le 2, \tag{19.108}$$

$$b_i^k(\mathbf{x};\mu_p) = K_{i3}^k(\mu_p)\psi_{i3}(\mathbf{x}), \quad i = 1,2, \qquad \gamma^k(\mathbf{x};\mu_p) = K_{33}^k(\mu_p)\psi_{33}^k(\mathbf{x}), \tag{19.109}$$

---

[9] We make a distinction between *(i)* a partition of the computational domain accounting for different physical properties over different *subregions* and *(ii)* a domain decomposition in different *subdomains* for the sake of geometrical representation, as shown in subsection 19.9.2.

for any $1 \le k \le K_{\text{reg}}$, and therefore

$$a(w,v;\mu) = \sum_{k=1}^{K_{\text{reg}}} \sum_{i,j=1}^{3} K_{ij}^k(\mu_p) \int_{\mathcal{R}_k} \begin{bmatrix} \dfrac{\partial w}{\partial x_1} & \dfrac{\partial w}{\partial x_2} & w \end{bmatrix} \psi_{ij}^k(x) \begin{bmatrix} \dfrac{\partial v}{\partial x_1} \\ \dfrac{\partial v}{\partial x_2} \\ v \end{bmatrix} d\Omega\,, \tag{19.110}$$

where, for any $k = 1,\ldots,K_{\text{reg}}$, $K^k:\ \mathcal{D}_p \to \mathbb{R}^{3\times 3}$ and $\Psi^k:\ \Omega \to \mathbb{R}^{3\times 3}$ are given matrices. By proceeding as we did before, it is possible to recover the affine expansion also in this case.

19.9.2 Geometrical parameters

We now turn to the more difficult situation of geometrical parametrization. In this case, the original parametrized problem defined over a parametrized domain $\Omega_o(\mu_g)$ needs to be transformed into an equivalent problem defined over a *parameter-independent* domain Ω. This is a requirement of the RB method: if we wish to consider linear combinations of pre-computed solutions (the snapshots), the latter must refer to a common spatial configuration.

Thus, to permit geometric variation, our parameter-independent reference domain Ω must be regarded as the pre-image of $\Omega_o(\mu_g)$, the parameter-dependent "actual" or "original" domain of interest. The geometric transformation will yield variable (parameter-dependent) coefficients of linear and bilinear forms in the reference domain that, under suitable hypotheses to be discussed below, will take the requisite affine form (19.6).

We shall first define an "original" problem (subscript o), over the *parameter-dependent* domain $\Omega_o = \Omega_o(\mu_g)$; we denote V_o a suitable Hilbert space defined on $\Omega_o(\mu_g)$. We shall also take into account possible physical parameters, so that $\mu = (\mu_p, \mu_g) \in \mathcal{D} = \mathcal{D}_p \times \mathcal{D}_g$ and $\mathcal{D}_p \subset \mathbb{R}^{d_p}$, $\mathcal{D}_g \subset \mathbb{R}^{d_g}$, with $p = d_p + d_g$.

In the elliptic case, the original problem reads as follows: given $\mu \in \mathcal{D}$, evaluate

$$s_o(\mu) = J_o(u_o(\mu))\,,$$

where $u_o(\mu) \in V_o$ satisfies

$$a_o(u_o(\mu), v; \mu) = F_o(v;\mu) \quad \forall v \in V_o\,.$$

The reference domain Ω is thus related to the original domain $\Omega_o(\mu_g)$ through a parametric mapping $T(\cdot;\mu_g)$, such that $\Omega_o(\mu_g) = T(\Omega;\mu_g)$; in particular, we may set $\Omega = \Omega_o(\mu_g^{ref})$, for a selected parameter value $\mu_g^{ref} \in \mathcal{D}_g$. In order to build a parametric mapping related to geometrical properties, we shall introduce a conforming domain decomposition of $\Omega_o(\mu_g)$,

$$\overline{\Omega}_o(\mu_g) = \bigcup_{k=1}^{K_{\text{dom}}} \overline{\Omega}_o^k(\mu_g), \tag{19.111}$$

consisting of mutually disjoint open subdomains $\Omega_o^k(\mu_g)$,

$$\Omega_o^k(\mu_g) \cap \Omega_o^{k'}(\mu_g) = \emptyset, \quad 1 \le k < k' \le K_{\text{dom}}.$$

If related to geometrical properties used as input parameters (e.g. lengths, thicknesses, diameters or angles) parametric mappings can be defined in a quite intuitive fashion. We point out that, in the case of a combined physical/geometrical parametrization, the domain decomposition (19.111) can be used not only for algorithmic purposes, to ensure well-behaved mappings, but also to represent different physical coefficients, as in the case described in Sect. 19.9.1.

Hence, the original and reference subdomains are related under a map $T(\cdot;\mu_g)$: $\Omega^k \to \Omega_o^k(\mu_g)$, $1 \le k \le K_{\text{dom}}$,

$$\Omega_o^k(\mu_g) = T^k(\Omega^k;\mu_g), \quad 1 \le k \le K_{\text{dom}}; \tag{19.112}$$

these maps must be individually bijective, collectively continuous, and such that $T^k(\mathbf{x};\mu_g) = T^{k'}(\mathbf{x};\mu_g)\, \forall\, \mathbf{x} \in \Omega^k \cap \Omega^{k'}$, for $1 \le k < k' \le K_{\text{dom}}$. Here we treat the simpler affine case, where the transformation is given, for any $\mu \in \mathcal{D}$, $\mathbf{x} \in \Omega^k$, by

$$T_i^k(\mathbf{x},\mu_g) = c_i^k(\mu_g) + \sum_{j=1}^{d} G_{ij}^k(\mu_g)x_j, \; 1 \le i \le d \tag{19.113}$$

for given translation vectors $\mathbf{c}^k : \mathcal{D}_g \to \mathbb{R}^d$ and linear transformation matrices $\mathbb{G}^k$: $\mathcal{D}_g \to \mathbb{R}^{d\times d}$, also known as "mapping coefficients". The linear transformation matrices can express rotation, scaling and/or shear. We can then define the associated Jacobians

$$J^k(\mu_g) = |\det(\mathbb{G}^k(\mu_g))|, \quad 1 \le k \le K_{\text{dom}}\,. \tag{19.114}$$

By assuming that the geometrical transformation is affine (i.e. given by (19.113)) the Jacobian is constant in space over each subdomain. We further define, for any $\mu_g \in \mathcal{D}_g$,

$$\mathbb{D}^k(\mu_g) = (\mathbb{G}^k(\mu_g))^{-1}, \quad 1 \le k \le K_{\text{dom}}\,; \tag{19.115}$$

this matrix shall prove convenient in subsequent transformations involving derivatives. Under the assumption that the mapping is invertible, we know that the Jacobian $J(\mu_g)$ of (19.114) is strictly positive, and that the derivative transformation matrix, $\mathbb{D}(\mu_g) = (\mathbb{G}(\mu_g))^{-1}$ of (19.115), exists.

We recall that, in two dimensions, an affine transformation maps straight lines to straight lines, parallel lines to parallel lines and indeed parallel lines of equal length to parallel lines of equal length: it follows that a triangle maps to a triangle, a parallelogram to a parallelogram. We also recall that an affine transformation maps ellipses to ellipses. These properties are crucial for the descriptions of domains relevant in engineering contexts, so that piecewise affine mappings, based on a domain decomposition in standard, elliptical, and curvy triangles [RHP08], can be employed to treat a much larger class of geometric variations, still satisfying the affinity assumption crucial for reducing the computational complexity. See for instance [RHP08] for further details.

Let us now place some conditions on a_o and F_o so to ensure, combined with the affine geometry assumption, an affine expansion of the bilinear form. Similarly to what we have done in Sect. 19.9.1 in the case of physical parameters, we require $a_o(\cdot,\cdot;\mu)$: $V_o \times V_o \to \mathbb{R}$ to be expressed as

$$a_o(w,v;\mu) = \sum_{k=1}^{K_{\text{dom}}} \int_{\Omega_o^k(\mu_g)} \begin{bmatrix} \frac{\partial w}{\partial x_{o1}} & \frac{\partial w}{\partial x_{o2}} & w \end{bmatrix} \mathcal{K}_{o,ij}^k(\mathbf{x}_o;\mu_p) \begin{bmatrix} \frac{\partial v}{\partial x_{o1}} \\ \frac{\partial v}{\partial x_{o2}} \\ v \end{bmatrix} d\Omega_o \,, \tag{19.116}$$

where $\mathbf{x}_o = (x_{o1}, x_{o2})$ denotes a point in $\Omega_o(\mu_g)$. Here, for $1 \le k \le K_{\text{dom}}$, $\mathcal{K}_o^k$: $\Omega_o^k \times \mathcal{D}_p \to \mathbb{R}^{3\times 3}$ are given matrices, whose components depend on both spatial coordinates and physical parameters. In case the parametrization is concerned with both physical and geometrical elements, the matrices $\mathcal{K}_o^k$ encode the properties related to the former; the meaning of the terms is the same as in Sect. 19.9.1.

Similarly, we require that $F_o(\cdot;\mu) : V_o \to \mathbb{R}$ can be expressed as

$$F_o(v;\mu) = \sum_{k=1}^{K_{\text{dom}}} \int_{\Omega_o^k(\mu_g)} \mathcal{F}_o^k(\mathbf{x}_o;\mu_p) v \, d\Omega_o \,,$$

where, for $1 \le k \le K_{\text{dom}}$, $\mathcal{F}_o^k$: $\Omega_o^k \times \mathcal{D}_p \to \mathbb{R}$ are given functions, which might encode possible physical parametrizations related to source terms and/or boundary data.

We now transform the problem given on the original domain, into an equivalent problem over the reference domain: given $\mu \in \mathcal{D}$, evaluate

$$s(\mu) = F(u(\mu)) \,,$$

where $u(\mu) \in V$ satisfies

$$a(u(\mu), v; \mu) = F(v;\mu) \quad \forall v \in V.$$

We may then identify $s(\mu) = s_o(\mu)$ and $u(\mu) = u_o(\mu) \circ T(\cdot;\mu_g)$, while

$$a(w,v;\mu) = \sum_{k=1}^{K_{\text{dom}}} \int_{\Omega^k} \left[\frac{\partial w}{\partial x_1} \quad \frac{\partial w}{\partial x_2} \quad w \right] \mathcal{K}_{ij}^k(\mathbf{x};\mu) \begin{bmatrix} \frac{\partial v}{\partial x_1} \\ \frac{\partial v}{\partial x_2} \\ v \end{bmatrix} d\Omega \,, \qquad (19.117)$$

is the transformed bilinear form. Here the $\mathcal{K}^k\colon\ \Omega^k \times \mathcal{D} \to \mathbb{R}^{3\times3}$, $1 \le k \le K_{\text{dom}}$, are symmetric positive definite matrices, whose components depend both on the spatial coordinates $\mathbf{x} = (x_1, x_2)$ and the parameters $\mu = (\mu_p, \mu_g) \in \mathcal{D}$

$$\mathcal{K}^k(\mathbf{x};\mu) = J^k(\mu_g)\,\mathcal{G}^k(\mu_g)\,\mathcal{K}_o^k(T^k(\mathbf{x};\mu_g);\mu_p)(\mathcal{G}^k(\mu_g))^{\mathrm{T}} \,, \qquad (19.118)$$

for $1 \le k \le K_{\text{dom}}$. The matrices $\mathcal{G}^k\colon \mathcal{D}_g \to \mathbb{R}^{3\times3}$, $1 \le k \le K_{\text{dom}}$, are given by

$$\mathcal{G}^k(\mu) = \left(\begin{array}{cc|c} \multicolumn{2}{c|}{\multirow{2}{*}{$\mathbb{D}^k(\mu_g)$}} & 0 \\ & & 0 \\ \hline 0 & 0 & 1 \end{array} \right); \qquad (19.119)$$

$J^k(\mu_g)$ and $\mathbb{D}^k(\mu_g)$, $1 \le k \le K_{\text{dom}}$, are given by (19.114) and (19.115), respectively.

Similarly, the new linear form can be expressed as

$$F(v;\mu) = \sum_{k=1}^{K_{\text{dom}}} \int_{\Omega^k} \mathcal{F}^k(\mathbf{x};\mu) v \, d\Omega,$$

where $\mathcal{F}^k\colon \Omega^k \times \mathcal{D} \to \mathbb{R}$, $1 \le k \le K_{\text{dom}}$, is given by

$$\mathcal{F}^k(\mathbf{x};\mu) = J^k(\mu_g)\mathcal{F}_o^k(T^k(\mathbf{x};\mu_g);\mu_p), \quad 1 \le k \le K_{\text{dom}} \,.$$

We point out that, compared to a purely physical parametrization (19.110), a geometrical parametrization yields to a similar but more involved structure of the tensor appearing in the bilinear form. In order to recover the affine expansion (19.6) for the bilinear form a also in this case, we can proceed as in the case of physical parametrizations, once the original problem is read on the reference domain Ω. In particular, we require that the geometrical mapping $T(\cdot;\mu_g)$ is affine on each subdomain Ω^k, i.e. it is of the form (19.113), and that the parametrized physical coefficients depend affinely on the physical parameters μ_p, i.e. they can be expressed as in (19.108)–(19.109). Under these conditions, the bilinear form satisfies the affine parametric dependence (19.6); F undergoes a similar treatment. An example taking into account both geometrical and physical parameters will be explained thoroughly at the end of the chapter, see Sect. 19.10.

We close by noting that the conditions we provide are sufficient but not necessary. For example, we can permit affine polynomial dependence on $\mathbf{x}_o$ in both $\mathcal{K}_o^k(\mathbf{x}_o;\mu)$ and $\mathcal{F}_o^k(\mathbf{x}_o;\mu)$ and still ensure an affine development under the form (19.6); furthermore, in the absence of geometric variation, $\mathcal{K}_o^k(\mathbf{x}_o;\mu)$ and $\mathcal{F}_o^k(\mathbf{x}_o;\mu)$ can take on *any* "separable" form in $\mathbf{x}_o, \mu$. However, the affine expansion (19.6) is by no means completely general: for more involved data parametric dependencies, non-affine techniques [BNMP04, GMNP07, Roz09] must be invoked.

19.10 A working example: a diffusion-convection problem

In order to illustrate the previous concepts on a real example, in this section[10] we discuss a steady heat diffusion/convection problem. Precisely, we consider heat convection combined with heat diffusion in a straight duct, whose walls can be kept at uniform temperature insulated, or characterized by heat exchange.

This example describes a class of heat transfer problems in fluidic devices with a versatile configuration. In particular, the Péclet number, as a measure of axial transport velocity field (modelling the physics of the problem), and the length of the non-insulated portion of the duct, are only two of the possible parameters that may be varied in order to extract average temperatures. Also discontinuities in Neumann boundary conditions (different heat fluxes) and thermal boundary layers are interesting phenomena to be studied.

We consider the physical domain $\Omega_o(\mu)$ shown in Fig. 19.6; all lengths are non-dimensionalized with respect to a unit length $\tilde{h}$ (dimensional channel width); moreover, let us denote $\tilde{k}$ the dimensional (thermal) conductivity coefficient for the air flowing in the duct, $\tilde{\rho}$ its density and $\tilde{c}_p$ the specific heat capacity under constant pressure. We introduce the (thermal) diffusion coefficient $\tilde{D} = \tilde{k}/\tilde{\rho}\tilde{c}_p$, as well as the Péclet number $\mathbb{P}\mathrm{e} = \tilde{U}\tilde{h}/\tilde{D}$, $\tilde{U}$ being the reference dimensional velocity for the convective field (see Chap. 12). We consider $P = 2$ parameters: μ_1 is the length of the non-insulated bottom portion of the duct (unit heat flux), while μ_2 represents the Péclet number itself; the parameter domain is given by $\mathcal{D} = [1,10] \times [0.1,100]$.

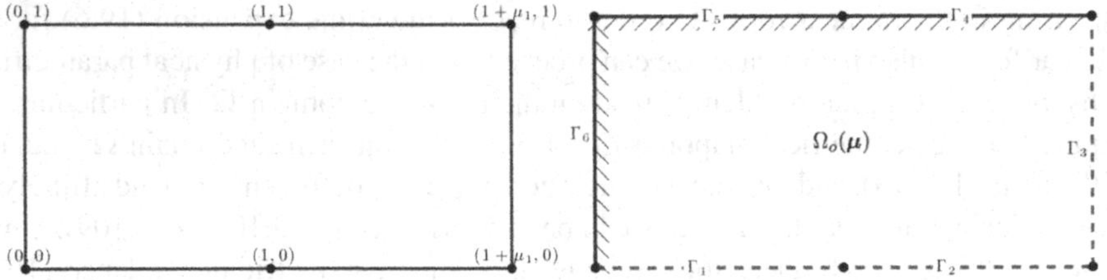

Fig. 19.6. Parametrized geometry and domain boundaries

[10] Throughout the section, $\Omega_o(\mu)$ denotes the original (physical) domain, whose generic point is indicated by (x_1, x_2); for the sake of simplicity, we formulate all problems in the original domain, but remove the subscripts $_o$. Moreover, twiddle $\tilde{}$ denotes dimensional quantities, while its absence signals a non-dimensional quantity.

The solution $u(\mu)$, defined as the non-dimensional temperature $u(\mu) = (\tau(\mu) - \tau_{in})/\tau_{in}$ (where $\tau(\mu)$ is the dimensional temperature, τ_{in} is the dimensional temperature of the air at the inflow and in the first portion of the duct), satisfies the following steady advection-diffusion problem:

$$\begin{cases} -\dfrac{1}{\mu_2}\Delta u(\mu) + x_2 \dfrac{\partial}{\partial x_1} u(\mu) = 0, & \text{in } \Omega_o(\mu) \\ \dfrac{1}{\mu_2}\dfrac{\partial u}{\partial n}(\mu) = 0, & \text{on } \Gamma_1 \cup \Gamma_3 \\ \dfrac{1}{\mu_2}\dfrac{\partial u}{\partial n}(\mu) = 1, & \text{on } \Gamma_2 \\ u(\mu) = 0, & \text{on } \Gamma_4 \cup \Gamma_5 \cup \Gamma_6, \end{cases}$$

with summation $(i, j = 1, 2)$ over repeated indices; hence, we impose the temperature at the top walls and in the "inflow" zone of the duct (Γ_6), while we consider an insulated wall (zero heat flux on Γ_1 and Γ_3) or heat exchange at a fixed rate (i.e. one on Γ_2) on other boundaries. We note that the forced convection field is given by a linear velocity profile $x_2\tilde{U}$ (Couette-type flow, [Pan05]).

The output of interest is the average temperature of the fluid on the non-insulated portion of the bottom wall of the duct, given by

$$s(\mu) = T_{\text{av}}(\mu) = \frac{1}{\mu_1}\int_{\Gamma_2} u(\mu). \tag{19.120}$$

This problem is then mapped to the fixed reference domain Ω and discretized by the Galerkin FEM with piecewise linear elements; the dimension of the corresponding space is $N_h = 5433$. Since we are in a noncompliant case, a further dual problem has to be solved in order to obtain better output evaluations and corresponding error bounds, see Sect. 19.8. In particular, we show in Fig. 19.7 the lower bound of the coercivity constant of the bilinear form associated to our problem.

We plot in Fig. 19.8 the convergence of the greedy algorithm for the primal and the dual problem; with a fixed tolerance $\varepsilon^*_{\text{tol}} = 10^{-2}$, $N_{pr,max} = 21$ and $N_{du,max} = 30$ basis functions have been selected, respectively.

In Fig. 19.9 the selected parameter values $S_{N_{pr}}$ for the primal problem and $S_{N_{du}}$ for the dual problem are shown; in each case Ξ_{train} is a uniform random sample of size $n_{\text{train}} = 1000$.

Moreover, in Fig. 19.10 some representative solutions (computed for $N = N_{max}$) for selected values of parameters are reported.

The thermal boundary layer looks very different in the four cases. In particular, big variations of the temperature, as well as large gradients along the lower wall – are remarkable for high Péclet number, when forced convection dominates steady conduction; moreover, the standard behaviour of boundary layer width – usually given by $O(1/\mathbb{P}\text{e})$ – is captured correctly.

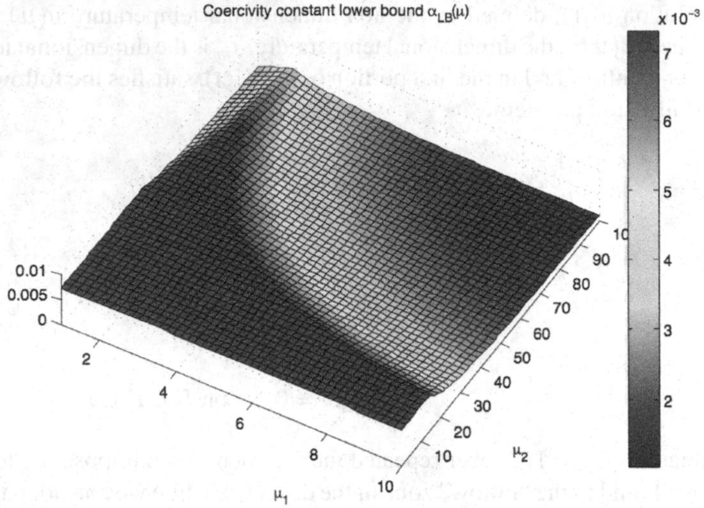

Fig. 19.7. Lower bound of the coercivity constant $\alpha_{\mathrm{LB}}^{\mathcal{N}_h}(\mu)$ as a function of μ

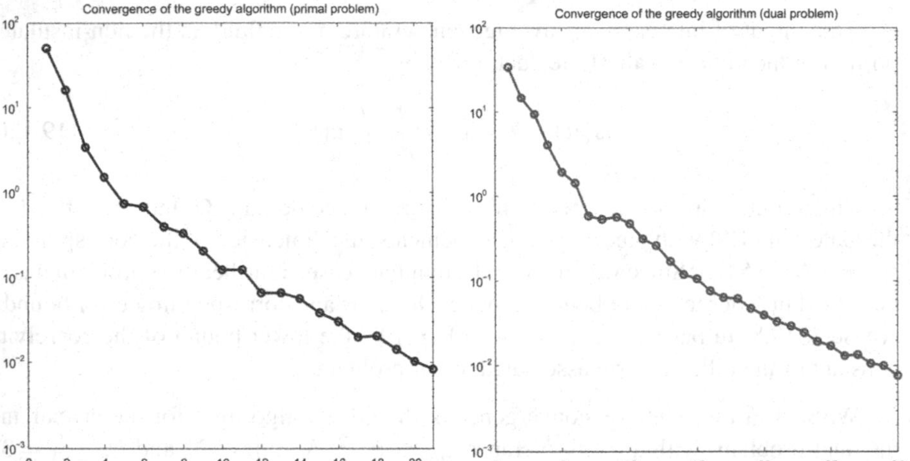

Fig. 19.8. Relative errors $\max_{\mu \in \Xi_{train}}(\Delta_{N_{pr}}(\mu)/\|u_{N_{pr}}(\mu)\|_X)$ and $\max_{\mu \in \Xi_{train}}(\Delta_{N_{du}}(\mu)/\|\psi_{N_{du}}(\mu)\|_X)$ as functions of N^{pr} and N^{du} for the RB approximations computed during the greedy procedure, for the primal (left) and the dual (right) problem, respectively. Here Ξ_{train} is a uniform random sample of size $n_{\mathrm{train}} = 1000$ and the RB tolerance is $\varepsilon_{\mathrm{tol}}^* = 10^{-2}$

In Fig. 19.11 the RB evaluation (for $N = N_{max}$) of the output of interest is reported as a function of the parameters, as well as the corresponding error bound. As we can see, for low values of μ_2 (Péclet number) the dependence of the output on μ_1 (geometrical aspect) is rather modest; for high values of μ_2, instead, the output shows a

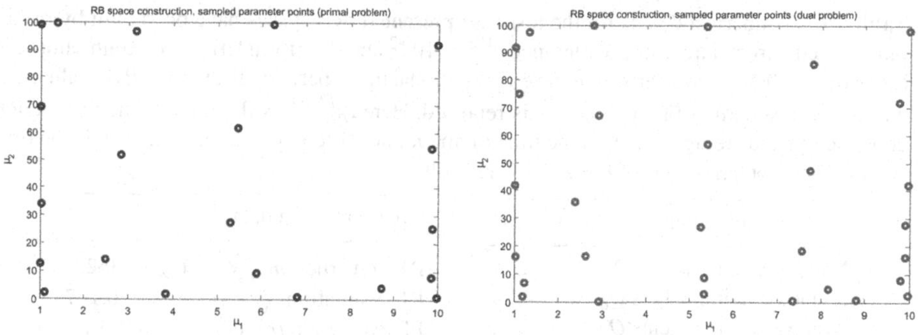

Fig. 19.9. Selected parameter values $S_{N_{pr}}$ for the primal (left) and $S_{N_{du}}$ for the dual (right) in the parameter space

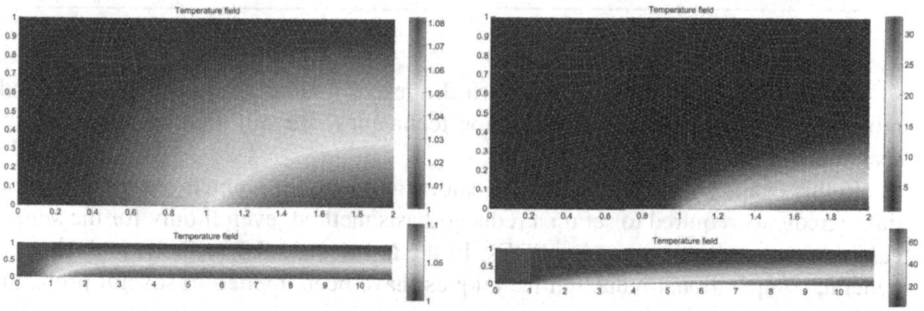

Fig. 19.10. Representative solutions for $\mu = (1, 0.1)$, $\mu = (1, 100)$ (top), $\mu = (10, 0.1)$, $\mu = (10, 100)$ (bottom)

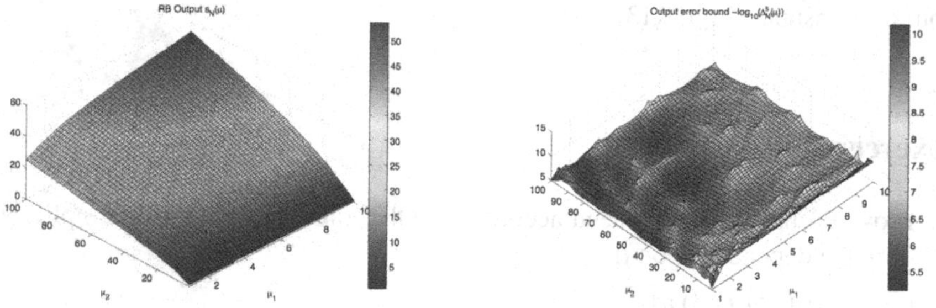

Fig. 19.11. Computed RB output (left) and corresponding error bound (right) as functions of μ in the parameter space

larger variation with respect to μ_1. In the same way, for longer/shorter channels the dependence on the Péclet number is higher/lower.

We conclude by collecting in Table 19.1 all the data relevant to the RB approximation here considered.

Table 19.1. Numerical details for the test case presented. RB spaces have been built by means of the greedy procedure, using a tolerance $\varepsilon_{\text{tol}}^{\text{RB}} = 10^{-2}$ and a uniform RB greedy train sample of size $n_{\text{train}} = 1000$. A comparison of the computational times between the Online RB evaluations and the corresponding FE simulations is reported. Here $t_{RB}^{offline}$ is the time of the Offline RB construction and storage, t_{RB}^{online} is the time of an Online RB computation, while t_{FE} is the time for an FE computation, once FE matrices are built

Approximation data		Approximation data	
Number of parameters P	2	RB construction $t_{RB}^{offline}$ (s)	362.8 s
Affine op. components Q_a	4	RB evaluation t_{RB}^{online} (s)	0.107s
Affine rhs components Q_f	1	FE evaluation t_{FE} (s)	14.3
FE space dim. $\mathcal{N}$	5433	Computational speedup $\mathcal{S}$	133
RB primal space dim. N_{max}^{pr}	21	Break-even point $\mathcal{Q}_{BE}$	26
RB dual space dim. N_{max}^{du}	30		

Before closing this chapter, we warn the reader that reduced basis methods and, more in general, computational reduction techniques, are still a rapidly evolving research area.

In this chapter we have provided a short introduction, pointed out all the essential ingredients required to set up a reduced basis method, even if only for the simple case of steady, elliptic, coercive PDEs. In the last decade RB methods and, more in general, computational reduction techniques, have been applied to several problems modelled by non-affinely parametrized and/or nonlinear, and/or noncoercive, and/or time-dependent problems (see e.g. [AF12, AZF12]), such as Stokes [RHM13] and Navier-Stokes [VP05, QR07] flows, elasticity problems and so on. A recent survey on these classes of problems, and on other computational reduction techniques, can be found for instance in [QR13].

Exercises

1. Prove Property 19.1; proceed according to the following steps (for simplicity, we omit the dependence on μ):

   (i) rewrite (19.63) as

   $$(\mathbb{A}_h \mathbb{V}\mathbf{u}_N, \mathbb{A}_h \mathbb{V}\mathbf{v}_N)_2 = (\mathbf{f}_h, \mathbb{A}_h \mathbb{V}\mathbf{v}_N)_2 \qquad \forall \mathbf{v}_N \in \mathbb{R}^N;$$

   (ii) by using bijection (19.44) and observing that $(\mathbb{V}\mathbf{v}_N)^{(p)} = (v_N, \varphi^p)_h$, show that

   $$(\mathbb{A}_h \mathbb{V}\mathbf{v}_N, \mathbf{z}_h)_2 = a(v_N, z_h) \qquad \forall \mathbf{v}_N \in \mathbb{R}^N, \mathbf{z}_h \in \mathbb{R}^{N_h},$$

   where $z_h = I_h \mathbf{z}_h \in V^{N_h}$, that is, $z_h(\mathbf{x}) = \sum_{r=1}^{N_h} z_h^{(r)} \varphi^r(\mathbf{x})$ and $v_N = I_N \mathbf{v}_N \in V_N$, that is $v_N(\mathbf{x}) = \sum_{k=1}^{N} v_N^{(k)} \zeta_k(\mathbf{x})$;

(iii) define $\mathbf{y}_h = \mathbf{y}_h(\mathbf{v}_N) = \mathbb{A}_h\mathbb{V}\mathbf{v}_N \in \mathbb{R}^{N_h}$; then

$$(\mathbf{f}_h, \mathbb{A}_h\mathbb{V}\mathbf{v}_N)_2 = (\mathbf{f}_h, \mathbf{y}_h(\mathbf{v}_N))_2;$$

note that $y_h(\mathbf{v}_N) = I_h\mathbf{y}_h(\mathbf{v}_N) \in V_h$ is such that

$$(y_h(\mathbf{v}_N), z_h)_h = (\mathbf{y}_h(\mathbf{v}_N), \mathbf{z}_h)_2 = (\mathbb{A}_h\mathbb{V}\mathbf{v}_N, \mathbf{z}_h)_2 \qquad \forall z_h \in V^{N_h}$$

and by using (ii), show that $y_h(\mathbf{v}_N)$ satisfies (19.67);

(iv) show that

$$(\mathbf{f}_h, \mathbf{y}_h(\mathbf{v}_N))_2 = (f_h, y_h(\mathbf{v}_N))_h = f(y_h(\mathbf{v}_N)),$$

where $f_h = I_h\mathbf{f}_h \in V^{N_h}$, that is $f_h(\mathbf{x}) = \sum_{r=1}^{N_h} f_h^{(r)} \varphi^r(\mathbf{x})$, and $f_h^{(r)} = f(\varphi^r)$;

(v) finally, recover the following variational form:

$$a(u_N, y_h(v_N)) = f(y_h(v_N)) \qquad \forall v_N \in V_N.$$

2. By introducing the Lagrangian function

$$L(\mathbf{z}_1, \ldots, \mathbf{z}_N, \tilde{\lambda}_{11}, \ldots, \tilde{\lambda}_{ij}, \ldots, \tilde{\lambda}_{NN}) = E(\mathbf{z}_1, \ldots, \mathbf{z}_N) + \sum_{i,j=1}^{N} \tilde{\lambda}_{ij}(\mathbf{z}_i^T \mathbf{z}_j - \delta_{ij}),$$

and writing the first-order necessary optimality conditions, show that the solution of problem (19.74) satisfies (19.72) and, in particular,

$$\mathbb{U}^T\mathbb{U}\psi_i = \sigma_i^2\psi_i \quad i = 1, \ldots, r.$$

Moreover, by denoting $\lambda_i = \tilde{\lambda}_{ii}$, show that

$$\lambda_i = \sigma_i^2 = \sum_{j=1}^{n_{\text{train}}} (\mathbf{u}_j^T \zeta_i)^2.$$

Furthermore, show that

$$\sum_{i=1}^{n_{\text{train}}} \|\mathbf{u}_i - \sum_{j=1}^{k} (\mathbf{u}_i^T \zeta_j)\zeta_j\|_2^2 = \sum_{i=1}^{n_{\text{train}}} \sum_{j=N+1}^{r} (\mathbf{u}_i^T \zeta_j)^2 = \sum_{i=N+1}^{r} \lambda_i,$$

and thus (19.75).

3. By considering the parametrized bilinear form (19.106)

$$a(w, v; \mu_p) = K_{11}(\mu_p) \int_\Omega \psi_{11}(x) \frac{\partial w}{\partial x_1} \frac{\partial v}{\partial x_1} d\Omega$$

$$+ K_{12}(\mu_p) \int_\Omega \psi_{12}(x) \frac{\partial w}{\partial x_1} \frac{\partial v}{\partial x_2} d\Omega + \cdots + \mathcal{K}_{33}(\mu_p) \int_\Omega \psi_{33}(x) wv d\Omega\,, \tag{19.121}$$

recover the affine expansion (19.6). How many terms (at most) is the affine expansion made of? Now do the same but for the a domain split into K_{reg} regions, each associated with different physical properties.

4. By referring to the problem addressed in Sect. 19.1.1 (the thermal block), consider an example of anisotropic conductivity. The associated bilinear form on each block reads

$$a^q(w,v;\mu) = \sum_{i,j=1}^{2} \mu^q_{x_i x_j} \int_{\mathcal{R}_q} \frac{\partial w}{\partial x_i}\frac{\partial v}{\partial x_j} d\Omega \qquad (19.122)$$

where $\{\mu^q_{x_i x_j}, i,j = 1,2\}$ represent the conductivities modelling an anisotropic heat transfer in block $\mathcal{R}_q$. Provide a weak formulation of the parametrized problem; express $a(w,v;\mu)$ by an affine decomposition like (19.13) and indicate the total number of parameters P and the quantity Q_a for the cases a) $B_1 = B_2 = 3$, b) $B_1 = B_2 = 5$ and c) $B_1 = 3$ and $B_2 = 5$.

5. a) Consider a geometrical parametrization of the thermal block of Sect. 19.1.1 based on the formulation already provided in the text with $P = 8$ physical parameters (i.e. μ_s for $s = 1,\ldots,8$), representing isotropic heat transfer in each sub-block. Consider the case of 3×3 sub-blocks with additional $P_g = 6$ geometrical parameters (i.e. $\mu^i_{x_1}$ and $\mu^i_{x_2}$ for $i = 1,2,3$). The length of the thermal block is $\mu^1_{x_1} + \mu^2_{x_1} + \mu^3_{x_1} = 1$ in the x_1-direction and $\mu^1_{x_2} + \mu^2_{x_2} + \mu^3_{x_2} = 1$ in the x_2-direction; in this way the first sub-block $\mathcal{R}_1$ (see Fig. 19.1) has dimension $\mu^1_{x_1} \times \mu^1_{x_2}$, while the sub-block $\mathcal{R}_9$ has dimension $\mu^3_{x_1} \times \mu^3_{x_2}$.
   Write the complete formulation for $a(w,v;\mu)$ in the form given by (19.13) and indicate the number Q_a of forms $a^q(w,v;\mu)$. Then report the complete formulation for all parameter-dependent functions $\Theta^q(\mu)$, in terms of physical parameters (i.e. μ_s for $s = 1,\ldots,8$) and geometrical parameters (i.e. $\mu^i_{x_1}$ and $\mu^i_{x_2}$ for $i = 1,2,3$).
   b) Propose a range of variation for each geometrical parameter by respecting the given constraints (i.e. $\mu^1_{x_1} + \mu^2_{x_1} + \mu^3_{x_1} = 1$ and $\mu^1_{x_2} + \mu^2_{x_2} + \mu^3_{x_2} = 1$) and the consistency of the thermal block configuration, and by avoiding a geometrical degeneration of some elements $\mathcal{R}_q$.

6. a) Consider the advection-diffusion example of Sect. 19.10 and write the weak formulation of problem (19.120), defined on the original domain.
   b) Propose a suitable domain decomposition of Ω and define a set of affine maps in order to describe the geometric deformation on each subdomain, by considering $\mu_{\text{ref}} = (1,1)$, which in turn defines the reference domain $\Omega = \Omega_o(\mu_{\text{ref}})$. Then write the corresponding weak formulation on the reference domain Ω.
   c) Characterize the expression of the a posteriori estimate $\Delta^{\text{en}}_N(\mu)$ for the error (in the energy norm) on the solution $\|u_h(\mu) - u_N(\mu)\|_\mu$ and the expression of the a posteriori estimate $\Delta^s_N(\mu)$ on the output $|s_h(\mu) - s_N(\mu)|$, $s(\mu)$ being the quantity defined in (19.120).

7. a) Consider the steady Stokes problem of Sect. 16.2 (eq.(16.12)) and the lid driven cavity flow of Sect. 16.6. Take as possible geometrical parametrization for the cavity the aspect ratio $\mu = L/D$, where L is the length and D the height of the cavity, respectively. Write the parametrization and weak formulation of the problem in the reference domain.

b) The reduced basis approximation spaces for a Stokes problem are given by $V_{N,\mathbf{u}} = \{\mathbf{u}_h(\mu^n), \boldsymbol{\sigma}_h(\mu^n), 1 \leq n \leq N\} \subset V_{\mathbf{u}}^{N_h}$ for the velocity and $V_{N,p} = \{p_h(\mu^n), 1 \leq n \leq N\} \subset V_p^{N_h}$ for the pressure, where, for N selected μ^n, $\mathbf{u}_h(\mu^n)$ and $p_h(\mu^n)$ represent the finite elements solutions for velocity and pressure, respectively, and $\boldsymbol{\sigma}_h(\mu^n) \in V_{\mathbf{u}}^{N_h}$ is the solution of an auxiliary problem, called *supremizer* problem, which reads in the original domain

$$\int_{\Omega_o} \nabla \boldsymbol{\sigma}_h \cdot \nabla \mathbf{v} d\Omega = \int_{\Omega_o} p_h(\mu^n) div(\mathbf{v}) d\Omega \qquad \forall \mathbf{v} \in V_{\mathbf{u}}^{N_h}. \tag{19.123}$$

The enrichment of the velocity space by the supremizer guarantees the stability of the RB approximation and the fulfillment of an equivalent *inf-sup* condition (see [RV07, RHM13]).
Write the reduced basis formulation for the Stokes problem; observe that the algebraic system obtained from the RB Galerkin projection features a block structure. Observe that this time the matrices are full, in contrast to what happens with the finite element method.

c) Do the same exercise considering the steady version of the Navier-Stokes equations (Sect. 16.1) by including also the affine transformation and the subsequent parametrization on the trilinear convective term (Sect. 16.7) $c(\mathbf{w}, \mathbf{z}, \mathbf{v}; \mu)$, as considered, for example, in [VP05, QR07].

References

[AAH$^+$98] Achdou Y., Abdoulaev G., Hontand J., Kuznetsov Y., Pironneau O., and Prud'homme C. (1998) Nonmatching grids for fluids. In *Domain decomposition methods, 10 (Boulder, CO, 1997)*, volume 218 of *Contemp. Math.*, pages 3–22. Amer. Math. Soc., Providence, RI.

[ABCM02] Arnold D. N., Brezzi F., Cockburn B., and Marini L. D. (2001/02) Unified analysis of discontinuous Galerkin methods for elliptic problems. *SIAM J. Numer. Anal.* 39(5): 1749–1779.

[ABM09] Antonietti P. F., Brezzi F., and Marini L. D. (2009) Bubble stabilization of discontinuous Galerkin methods. *Comput. Methods Appl. Mech. Engrg.* 198(21-26): 1651–1659.

[Ada75] Adams R. A. (1975) *Sobolev Spaces*. Academic Press, New York.

[AF12] Amsallem D. and Farhat C. (2012) Stabilization of projection-based reduced-order models. *Int. J. Numer. Methods Engr.* 91(4): 358–377.

[AFG$^+$00] Almeida R. C., Feijóo R. A., Galeão A. C., Padra C., and Silva R. S. (2000) Adaptive finite element computational fluid dynamics using an anisotropic error estimator. *Comput. Methods Appl. Mech. Engrg.* 182: 379–400.

[Ago03] Agoshkov V. (2003) *Optimal Control Methods and Adjoint Equations in Mathematical Physics Problems*. Institute of Numerical Mathematics, Russian Academy of Science, Moscow.

[Aki94] Akin J. E. (1994) *Finite Elements for Analysis and Design*. Academic Press, London.

[AM09] Ayuso B. and Marini L. D. (2009) Discontinuous Galerkin methods for advection-diffusion-reaction problems. *SIAM J. Numer. Anal.* 47(2): 1391–1420.

[AMW99] Achdou Y., Maday Y., and Widlund O. (1999) Iterative substructuring preconditioners for mortar element methods in two dimensions. *SIAM J. Numer. Anal.* 36(2): 551–580 (electronic).

[AO00] Ainsworth M. and Oden J. T. (2000) *A posteriori error estimation in finite element analysis*. Pure and Applied Mathematics. John Wiley and Sons, New York.

[Ape99] Apel T. (1999) *Anisotropic Finite Elements: Local Estimates and Applications*. Book Series: Advances in Numerical Mathematics. Teubner, Stuttgart.

[APV98] Achdou Y., Pironneau O., and Valentin F. (1998) *Équations aux dérivées partielles et applications*, chapter Shape control versus boundary control, pages 1–18. Éd. Sci. Méd. Elsevier, Paris.

A. Quarteroni: *Numerical Models for Differential Problems*, 2nd Ed.
MS&A – Modeling, Simulation & Applications 8
DOI 10.1007/978-88-470-5522-3,

[Arn82] Arnold D. N. (1982) An interior penalty finite element method with discontinuous elements. *SIAM J. Numer. Anal.* 19(4): 742–760.

[AS55] Allen D. N. G. and Southwell R. V. (1955) Relaxation methods applied to determine the motion, in two dimensions, of a viscous fluid past a fixed cylinder. *Quart. J. Mech. Appl. Math.* 8: 129–145.

[AS99] Adalsteinsson D. and Sethian J. A. (1999) The fast construction of extension velocities in level set methods. *J. Comput. Phys.* 148(1): 2–22.

[ATF87] Alekseev V., Tikhominov V., and Fomin S. (1987) *Optimal Control*. Consultants Bureau, New York.

[Aub67] Aubin J. P. (1967) Behavior of the error of the approximate solutions of boundary value problems for linear elliptic operators by Galerkin's and finite difference methods. *Ann. Scuola Norm. Sup. Pisa* 21: 599–637.

[Aub91] Aubry N. (1991) On the hidden beauty of the proper orthogonal decomposition. *Theor. Comp. Fluid. Dyn.* 2: 339–352.

[AWB71] Aziz A., Wingate J., and Balas M. (1971) *Control Theory of Systems Governed by Partial Differential Equations*. Academic Press.

[AZF12] Amsallem D., Zahr M., and Farhat C. (2012) Nonlinear model order reduction based on local reduced-order bases. *Int. J. Numer. Methods Engr.* 92(10): 891–916.

[Bab71] Babuška I. (1971) Error bounds for the finite element method. *Numer. Math.* 16: 322–333.

[BCD$^+$11] Binev P., Cohen A., Dahmen W., DeVore R., Petrova G., and Wojtaszczyk P. (2011) Convergence rates for greedy algorithms in reduced basis methods. *SIAM J. Math. Anal.* 43(3): 1457–1472.

[BDR92] Babuška I., Durán R., and Rodríguez R. (1992) Analysis of the efficiency of an a posteriori error estimator for linear triangular finite elements. *SIAM J. Numer. Anal.* 29(4): 947–964.

[BE92] Bern M. and Eppstein D. (1992) Mesh generation and optimal triangulation. In Du D.-Z. and Hwang F. (eds) *Computing in Euclidean Geometry*. World Scientific, Singapore.

[Bec01] Becker R. (2001) Mesh adaptation for stationary flow control. *J. Math. Fluid Mech.* 3: 317–341.

[Bel99] Belgacem F. B. (1999) The mortar finite element method with Lagrange multipliers. *Numer. Math.* 84(2): 173–197.

[BEMS07] Burman E., Ern A., Mozolevski I., and Stamm B. (2007) The symmetric discontinuous Galerkin method does not need stabilization in 1D for polynomial orders $p \geq 2$. *C. R. Math. Acad. Sci. Paris* 345(10): 599–602.

[BF91a] Brezzi F. and Fortin M. (1991) *Mixed and Hybrid Finite Element Methods*. Springer-Verlag, New York.

[BF91b] Brezzi F. and Fortin M. (1991) *Mixed and Hybrid Finite Element Methods*, volume 15 of *Springer Series in Computational Mathematics*. Springer-Verlag, New York.

[BFHR97] Brezzi F., Franca L. P., Hughes T. J. R., and Russo A. (1997) $b = \int g$. *Comput. Methods Appl. Mech. Engrg.* 145: 329–339.

[BG87] Brezzi F. and Gilardi G. (1987) *Functional Analysis and Functional Spaces*. McGraw Hill, New York.

[BG98] Bernardi C. and Girault V. (1998) A local regularisation operator for triangular and quadrilateral finite elements. *SIAM J. Numer. Anal.* 35(5): 1893–1916.

[BGL05] Benzi M., Golub G. H., and Liesen J. (2005) Numerical solution of saddle-point problems. *Acta Numer.* 14: 1–137.

[BGL06a] Burkardt J., Gunzburger M., and Lee H. (2006) Centroidal voronoi tessellation-based reduced-order modeling of complex systems. *SIAM J. Sci. Comput.* 28(2): 459–484.

[BGL06b] Burkardt J., Gunzburger M., and Lee H. (2006) POD and CVT-based reduced-order modeling of Navier-Stokes flows. *Comput. Meth. Appl. Mech. Engrg.* 196(1-3): 337–355.

[BGS96] Bjòrstad P., Gropp P., and Smith B. (1996) *Domain Decomposition, Parallel Multilevel Methods for Elliptic Partial Differential Equations.* Univ. Cambridge Press, Cambridge.

[BHL93] Berkooz G., Holmes P., and Lumley J. (1993) The proper orthogonal decomposition in the analysis of turbulent flows. *Annu. Rev. Fluid Mech.* 25(1): 539–575.

[BHS06] Buffa A., Hughes T. J. R., and Sangalli G. (2006) Analysis of a multiscale discontinuous Galerkin method for convection-diffusion problems. *SIAM J. Numer. Anal.* 44(4): 1420–1440.

[BIL06] Berselli L. C., Iliescu T., and Layton W. J. (2006) *Mathematics of Large Eddy Simulation of Turbulent Flows.* Springer, Berlin Heidelberg.

[BKR00] Becker R., Kapp H., and Rannacher R. (2000) Adaptive finite element methods for optimal control of partial differential equations: Basic concepts. *SIAM, J. Control Opt.* 39(1): 113–132.

[Bla02] Blanckaert K. (2002) *Flow and turbulence in sharp open-channel bends.* PhD thesis, École Polytechnique Fédérale de Lausanne.

[BM92] Bernardi C. and Maday Y. (1992) *Approximations Spectrales de Problèmes aux Limites Elliptiques.* Springer-Verlag, Berlin Heidelberg.

[BM94] Belgacem F. B. and Maday Y. (1994) A spectral element methodology tuned to parallel implementations. *Comput. Methods Appl. Mech. Engrg.* 116(1-4): 59–67. ICOSAHOM'92 (Montpellier, 1992).

[BM06] Brezzi F. and Marini L. D. (2006) Bubble stabilization of discontinuous Galerkin methods. In *Advances in numerical mathematics, Proceedings International Conference on the occasion of the 60th birthday of Y.A. Kuznetsov, September 16-17, 2005*, pages 25–36. Institute of Numerical Mathematics of the Russian Academy of Sciences, Moscow.

[BMMP06] Bottasso C. L., Maisano G., Micheletti S., and Perotto S. (2006) On some new recovery based a posteriori error estimators. *Comput. Methods Appl. Mech. Engrg.* 195(37–40): 4794–4815.

[BMP94] Bernardi C., Maday Y., and Patera A. (1994) A new nonconforming approach to domain decomposition: the mortar element method. In *Nonlinear partial differential equations and their applications. Collège de France Seminar, Vol. XI (Paris, 1989–1991)*, volume 299 of *Pitman Res. Notes Math. Ser.*, pages 13–51. Longman Sci. Tech., Harlow.

[BMP+12] Buffa A., Maday Y., Patera A. T., Prud'homme C., and Turinici G. (2012) A priori convergence of the greedy algorithm for the parametrized reduced basis method. *ESAIM Math. Model. Numer. Anal.* 46(3): 595–603.

[BMS04] Brezzi F., Marini L. D., and Süli E. (2004) Discontinuos Galerkin methods for first-order hyperbolic problems. *Math. Models Methods Appl. Sci.* 14: 1893–1903.

[BN83] Boland J. and Nicolaides R. (1983) Stability of finite elements under divergence constraints. *SIAM J. Numer. Anal.* 20: 722–731.

[BNMP04] Barrault M., Nguyen N. C., Maday Y., and Patera A. T. (2004) An "empirical interpolation" method: Application to efficient reduced-basis discretization of partial differential equations. *C. R. Acad. Sci. Paris, Série I.* 339: 667–672.

[BO06] Benzi M. and Olshanskii M. (2006) An augmented lagrangian-based approach to the oseen problem. *SIAM J. on Scientific Computing* 28 (6): 2095–2113.

[Bre74] Brezzi F. (1974) On the existence, uniqueness and approximation of saddle-point problems arising from Lagrange multipliers. *R.A.I.R.O. Anal. Numér.* 8: 129–151.

[Bre86] Brezis H. (1986) *Analisi Funzionale.* Liguori, Napoli.

[Bre00] Bressan A. (2000) *Hyperbolic Systems of Conservation Laws: The One-dimensional Cauchy Problem.* Oxford Lecture Series in Mathematics and its Applications. The Clarendon Press Oxford University Press, New York.

[BRM$^+$97] Bassi F., Rebay S., Mariotti G., Pedinotti S., and Savini M. (1997) A high-order accurate discontinuous finite element method for inviscid and viscous turbomachinery flows. In Decuypere R. and Dibelius G. (eds) *Proceedings of the 2nd European Conference on Turbomachinery Fluid Dynamics and Thermodynamics*, pages 99–108. Technologisch Instituut, Antwerpen, Belgium.

[BS94] Brenner S. C. and Scott L. R. (1994) *The Mathematical Theory of Finite Element Methods.* Springer-Verlag, New York.

[BS07] Brenner S. and Sung L. (2007) Bddc and feti-dp without matrices or vectors. *Comput. Methods Appl. Mech. Engrg.* 196: 1429–1435.

[BS08] Burman E. and Stamm B. (2008) Symmetric and non-symmetric discontinuous Galerkin methods stabilized using bubble enrichment. *C. R. Math. Acad. Sci. Paris* 346(1-2): 103–106.

[Cab03] Caboussat A. (2003) *Analysis and numerical simulation of free surface flows.* PhD thesis, École Polytechnique Fédérale de Lausanne.

[CAC10] Chinesta F., Ammar A., and Cueto E. (2010) Recent advances and new challenges in the use of the proper generalized decomposition for solving multidimensional models. *Arch. Comput. Methods Engrg* 17: 327–350.

[Ç07] Çengel Y. (2007) *Introduction to Thermodynamics and heat transfer.* McGraw–Hill, New York.

[CH01] Collis S. and Heinkenschloss M. (2001) Analysis of the streamline upwind/petrov galerkin method applied to the solution of optimal control problems. *CAAM report* TR02-01.

[CHQZ06] Canuto C., Hussaini M., Quarteroni A., and Zang T. A. (2006) *Spectral Methods. Fundamentals in Single Domains.* Springer-Verlag, Berlin Heidelberg.

[CHQZ07] Canuto C., Hussaini M. Y., Quarteroni A., and Zang T. A. (2007) *Spectral Methods. Evolution to Complex Geometries and Application to Fluid Dynamics.* Springer-Verlag, Berlin Heidelberg.

[Cia78] Ciarlet P. G. (1978) *The Finite Element Method for Elliptic Problems.* North-Holland, Amsterdam.

[CJRT01] Cohen G., Joly P., Roberts J. E., and Tordjman N. (2001) Higher order triangular finite elements with mass lumping for the wave equation. *SIAM J. Numer. Anal.* 38(6): 2047–2078 (electronic).

[CLC11] Chinesta F., Ladeveze P., and Cueto E. (2011) A short review on model order reduction based on proper generalized decomposition. *Arch. Comput. Methods Engrg.* 18: 395–404.

[Clé75] Clément P. (1975) Approximation by finite element functions using local regularization. *RAIRO, Anal. Numér* 2 pages 77–84.

[Coc98] Cockburn B. (1998) An introduction to the discontinuous Galerkin method for convection-dominated problems. In Quarteroni A. (ed) *Advanced Numerical Approximation of Nonlinesr Hyperbolic Equations*, volume 1697 of *LNM*, pages 151–268. Springer-Verlag, Berlin Heidelberg.

[Coc99] Cockburn B. (1999) Discontinuous Galerkin methods for convection-dominated problems. In *High-order methods for computational physics*, volume 9 of *Lect. Notes Comput. Sci. Eng.*, pages 69–224. Springer, Berlin.

[Ded04] Dedé L. (2004) *Controllo Ottimale e Adattività per Equazioni alle Derivate Parziali e Applicazioni*. Tesi di Laurea, Politecnico di Milano.

[DG11] Demkowicz L. and Gopalakrishnan J. (2011) A class of discontinuous Petrov–Galerkin methods. II. optimal test functions. *Numer. Methods Partial Differential Equations* 27(1): 70–105.

[Doh03] Dohrmann R. (2003) A preconditioner for substructuring based on constrained energy minimization. *SIAM J. Sci. Comput.* 25: 246–258.

[DPQ08] Detomi D., Parolini N., and Quarteroni A. (2008) Mathematics in the wind. *MOX Reports* (25). see the web page `http://mox.polimi.it`.

[DQ05] Dedè L. and Quarteroni A. (2005) Optimal control and numerical adaptivity for advection–diffusion equations. *M2AN Math. Model. Numer. Anal.* 39(5): 1019–1040.

[DSW04] Dawson C., Sun S., and Wheeler M. F. (2004) Compatible algorithms for coupled flow and transport. *Comput. Methods Appl. Mech. Engrg.* 193(23-26): 2565–2580.

[DT80] Dervieux A. and Thomasset F. (1980) *Approximation Methods for Navier–Stokes Problems*, volume 771 of *Lecture Notes in Mathematics*, chapter A finite element method for the simulation of Rayleigh-Taylor instability, pages 145–158. Springer-Verlag, Berlin.

[Dub91] Dubiner M. (1991) Spectral methods on triangles and other domains. *J. Sci. Comput.* 6: 345–390.

[DV02] Darmofal D. L. and Venditti D. A. (2002) Grid adaptation for functional outputs: application to two-dimensional inviscid flows. *J. Comput. Phys.* 176: 40–69.

[DV09] Di Pietro D. A. and Veneziani A. (2009) Expression templates implementation of continuous and discontinuous Galerkin methods. *Computing and Visualization in Science* 12(8): 421–436.

[DZ06] Dáger R. and Zuazua E. (2006) *Wave Propagation, Observation and Control in 1-d Flexible Multi-Structures*. Mathématiques et Applications. Springer, Paris.

[EG04] Ern A. and Guermond J. L. (2004) *Theory and Practice of Finite Elements*, volume 159 of *Applied Mathematics Sciences*. Springer-Verlag, New York.

[EHS+06] Elman H. C., Howte V. E., Shadid J., Shuttleworth R., and Tuminaro R. (2006) Block preconditioners based on approximate commutators. *SIAM J. on Scientific Computing* 27 (5): 1651–1668.

[EJ88] Eriksson E. and Johnson C. (1988) An adaptive method for linear elliptic problems. *Math. Comp.* 50: 361–383.

[Emb99] Embree M. (1999) *Convergence of Krylov subspace methods for non-normal matrices*. PhD thesis, Oxford University Computing Laboratories.

[Emb03] Embree M. (2003) The tortoise and the hare restart GMRES. *SIAM Rev.* 45 (2): 259–266.

[ESW05] Elman H., Silvester D., and Wathen A. (2005) *Finite Elements and Fast Iterative Solvers*. Oxford Science Publications, Oxford.

[Eva98] Evans L. (1998) *Partial differential equations*. American Mathematical Society, Providence.

[FCZ03] Fernández-Cara E. and Zuazua E. (2003) Control theory: History, mathematical achievements and perspectives. *Bol. Soc. Esp. Mat. Apl.* 26: 79–140.

[FCZ04] Fernández-Cara E. and Zuazua E. (2004) On the history and perspectives of control theory. *Matapli* 74: 47–73.

[FLP00] Farhat C., Lesoinne M., and Pierson K. (2000) A scalable dual-primal domain decomposition method. *Numer. Linear Algebra Appl.* 7: 687–714.

[FLT$^+$01] Farhat C., Lesoinne M., Tallec P. L., Pierson K., and Rixen D. (2001) Feti-dp: a dual-primal unified feti method. i. a faster alternative to the two-level feti method. *Intern. J. Numer. Methods Engrg.* 50: 1523–1544.

[FMP04] Formaggia L., Micheletti S., and Perotto S. (2004) Anisotropic mesh adaptation in computational fluid dynamics: application to the advection-diffusion-reaction and the stokes problems. *Appl. Numer. Math.* 51(4): 511–533.

[FMRT01] Foias C., Manley O., Rosa R., and Temam R. (2001) *Navier-Stokes Equations and Turbulence*. Cambridge Univ. Press, Cambridge.

[For77] Fortin M. (1977) An analysis of the convergence of mixed finite element methods. *R.A.I.R.O. Anal. Numér.* 11.

[FP01] Formaggia L. and Perotto S. (2001) New anisotropic a priori error estimates. *Numer. Math.* 89: 641–667.

[FP02] Ferziger J. H. and Peric M. (2002) *Computational Methods for Fluid Dynamics*. Springer, Berlino, III edition.

[FSV05] Formaggia L., Saleri F., and Veneziani A. (2005) *Applicazioni ed esercizi di modellistica numerica per problemi differenziali*. Springer Italia, Milano.

[Fun92] Funaro D. (1992) *Polynomial Approximation of Differential Equations*. Springer-Verlag, Berlin Heidelberg.

[Fun97] Funaro D. (1997) *Spectral Elements for Transport-Dominated Equations*. Springer-Verlag, Berlin Heidelberg.

[Fur97] Furnish G. (May/June 1997) Disambiguated glommable expression templates. *Compuers in Physics* 11(3): 263–269.

[GaAML04] Galeão A., Almeida R., Malta S., and Loula A. (2004) Finite element analysis of connection dominated reaction-diffusion problems. *Applied Numerical Mathematics* 48: 205–222.

[GB98] George P. L. and Borouchaki H. (1998) *Delaunay Triangulation and Meshing*. Editions Hermes, Paris.

[Ger08] Gervasio P. (2008) Convergence analysis of high order algebraic fractional step schemes for timedependent stokes equations. *SINUM* .

[GMNP07] Grepl M. A., Maday Y., Nguyen N. C., and Patera A. T. (2007) Efficient reduced-basis treatment of nonaffine and nonlinear partial differential equations. *M2AN (Math. Model. Numer. Anal.)* 31(3): 575–605 .

[GMSW89] Gill P., Murray W., Saunders M., and Wright M. (1989) *Constrained nonlinear programming*. Elsevier Handbooks In Operations Research And Management Science Optimization. Elsevier North-Holland, Inc., New York.

[GR96] Godlewski E. and Raviart P. A. (1996) *Hyperbolic Systems of Conservations Laws*, volume 118. Springer-Verlag, New York.

[Gri11] Grisvard P. (2011) *Elliptic Problems in Nonsmooth Domains*. SIAM.

[GRS07] Grossmann C., Ross H., and Stynes M. (2007) *Numerical treatment of Partial Differential Equations*. Springer, Heidelberg, Heidelberg.

[GS05] Georgoulis E. H. and Süli E. (2005) Optimal error estimates for the *hp*-version interior penalty discontinuous Galerkin finite element method. *IMA J. Numer. Anal.* 25(1): 205–220.

[GSV06] Gervasio P., Saleri F., and Veneziani A. (2006) Algebraic fractional-step schemes with spectral methods for the incompressible Navier-Stokes equations. *J. Comput. Phys.* 214(1): 347–365.

[Gun03] Gunzburger M. D. (2003) *Perspectives in Flow Control and Optimization*. Advances in Design and Control. SIAM.

[Hac] Hackbush W.

[HB76] Hnat J. and Buckmaster J. (1976) Spherical cap bubbles and skirt formation. *Phys. Fluids* 19: 162–194.

[Hir88] Hirsh C. (1988) *Numerical Computation of Internal and External Flows*, volume 1. John Wiley and Sons, Chichester.

[HLB98] Holmes P., Lumley J., and Berkooz G. (1998) *Turbulence, coherent structures, dynamical systems and symmetry*. Cambridge Univ. Press.

[HN81] Hirt C. W. and Nichols B. D. (1981) Volume of fluid (VOF) method for the dynamics of free boundaries. *J. Comp. Phys.* 39: 201–225.

[Hou95] Hou T. Y. (1995) Numerical solutions to free boundary problems. *ACTA Numerica* 4: 335–415.

[HRSP07] Huynh D. B. P., Rozza G., Sen S., and Patera A. T. (2007) A successive constraint linear optimization method for lower bounds of parametric coercivity and inf-sup stability constants. *C. R. Acad. Sci. Paris, Analyse Numérique*, Series I 345, pp. 473–478.

[HRT96] Heywood J. G., Rannacher R., and Turek S. (1996) Artificial boundaries and flux and pressure conditions for the incompressible Navier-Stokes equations. *Internat. J. Numer. Methods Fluids* 22(5): 325–352.

[HSS02] Houston P., Schwab C., and Süli E. (2002) Discontinuous *hp*-finite element methods for advection-diffusion-reaction problems. *SIAM J. Numer. Anal.* 39(6): 2133–2163 (electronic).

[Hug00] Hughes T. J. R. (2000) *The Finite Element Method. Linear Static and Dynamic Finite Element Analysis*. Dover Publishers, New York.

[HVZ97] Hamacher V. C., Vranesic Z. G., and Zaky S. G. (1997) *Introduzione all'architettura dei calcolatori*. McGraw Hill Italia, Milano.

[HW65] Harlow F. H. and Welch J. E. (1965) Numerical calculation of time-dependent viscous incompressible flow of fluid with free surface. *Physics of Fluids* 8(12): 2182–2189.

[HW08] Hesthaven J. S. and Warburton T. (2008) *Nodal discontinuous Galerkin methods*, volume 54 of *Texts in Applied Mathematics*. Springer, New York. Algorithms, analysis, and applications.

[HYR08] Hou T. Y., Yang D. P., and Ran H. (2008) Multiscale analysis and computation for the 3d incompressible Navier-Stokes equations. *SIAM Multiscale Modeling and Simulation* 6 (4): 1317–1346.

[IR98] Ito K. and Ravindran S. S. (1998) A reduced-order method for simulation and control of fluid flows. *Journal of Computational Physics* 143(2): 403–425.

[IZ99] Infante J. and Zuazua E. (1999) Boundary observability for the space semi–discretizations of the 1–d wave equation. *M2AN Math. Model. Numer. Anal.* 33(2): 407–438.

[Jam88] Jamenson A. (1988) Optimum aerodynamic design using cfd and control theory. *AIAA Paper 95-1729-CP* pages 233–260.

[JK07] John V. and Knobloch P. (2007) On spurious oscillations at layers diminishing (SOLD) methods for convection-diffusion equations. I. A review. *Comput. Methods Appl. Mech. Engrg.* 196(17-20): 2197–2215.

[Joe05] Joerg M. (2005) Numerical investigations of wall boundary conditions for two-fluid flows. Master's thesis, École Polytechnique Fédérale de Lausanne.

[Joh87] Johnson C. (1987) *Numerical Solution of Partial Differential Equations by the Finite Element Method.* Cambridge University Press, Cambridge.

[KA00] Knabner P. and Angermann L. (2000) *Numerical Methods for Elliptic and Parabolic Partial Differential Equations*, volume 44 of *TAM*. Springer-Verlag, New York.

[KAJ02] Kim S., Alonso J., and Jameson A. (2002) Design optimization of hight-lift configurations using a viscous continuos adjoint method. *AIAA paper, 40th AIAA Aerospace Sciences Meeting and Exibit, Jan 14-17 2002* 0844.

[KF89] Kolmogorov A. and Fomin S. (1989) *Elements of the Theory of Functions and Functional Analysis*. V.M. Tikhominov, Nauka - Moscow.

[KMI+83] Kajitani H., Miyata H., Ikehata M., Tanaka H., Adachi H., Namimatzu M., and Ogiwara S. (1983) Summary of the cooperative experiment on Wigley parabolic model in Japan. In *Proc. of the 2nd DTNSRDC Workshop on Ship Wave Resistance Computations (Bethesda, USA)*, pages 5–35.

[KPTZ00] Kawohl B., Pironneau O., Tartar L., and Zolesio J. (2000) *Optimal Shape Design*. Springer-Verlag, Berlin.

[Kro97] Kroener D. (1997) *Numerical Schemes for Conservation Laws*. Wiley-Teubner, Chichester.

[KS05] Karniadakis G. E. and Sherwin S. J. (2005) *Spectral/hp Element Methods for Computational Fluid Dynamics*. Oxford University Press, New York, II edition.

[KWD02] Klawonn A., Widlund O., and Dryja M. (2002) Dual-primal feti methods for three-dimensional elliptic problems with heterogeneous coefficients. *SIAM J. Numer. Anal.* 40: 159–179.

[Le 05] Le Bris C. (2005) *Systèmes multiiéchelles: modélisation et simulation*, volume 47 of *Mathématiques et Applications*. Springer, Paris.

[LeV02a] LeVeque R. J. (2002) *Finite Volume Methods for Hyperbolic Problems*. Cambridge Texts in Applied Mathematics.

[LeV02b] LeVeque R. J. (2002) *Numerical Methods for Conservation Laws*. Birkhäuser Verlag, Basel, II edition.

[LeV07] LeVeque R. J. (2007) *Finite Difference Methods for Ordinary and Partial Differential Equations: Steady-State and Time-Dependent Problems*. SIAM, Philadelphia.

[Lio71] Lions J. (1971) *Optimal Control of Systems Governed by Partial Differential Equations*. Springer-Verlag, New York.

[Lio72] Lions J. (1972) *Some Aspects of the Optimal Control of Distribuited Parameter Systems*. SIAM, Philadelphia.

[Lio96] Lions P.-L. (1996) *Mathematical topics in fluid mechanics. Vol. 1*, volume 3 of *Oxford Lecture Series in Mathematics and its Applications*. The Clarendon Press Oxford University Press, New York. Incompressible models, Oxford Science Publications.

[LL59] Landau L. D. and Lifshitz E. M. (1959) *Fluid mechanics*. Translated from the Russian by J. B. Sykes and W. H. Reid. Course of Theoretical Physics, Vol. 6. Pergamon Press, London.

[LL00] Lippman S. B. and Lajoie J. (2000) *C++ Corso di Programmazione*. Addison Wesley Longman Italia, Milano, III edition.

[LM68] Lions J. L. and Magenes E. (1968) *Quelques Méthodes des Résolution des Problémes aux Limites non Linéaires*. Dunod, Paris.

[LMQR13] Lassila T., Manzoni A., Quarteroni A., and Rozza G. (2013) Generalized reduced basis methods and n-width estimates for the approximation of the solution manifold of parametric PDEs. In Brezzi F., Colli Franzone P., Gianazza U., and Gilardi G. (eds) *Analysis and Numerics of Partial Differential Equations*, volume 4 of *Springer INdAM Series*, pages 307–329. Springer, Milan.

[LR98] Lin S. P. and Reitz R. D. (1998) Drop and spray formation from a liquid jet. *Annu. Rev. Fluid Mech.* 30: 85–105.

[LvGM96] Lorentz G., von Golitschek M., and Makovoz Y. (1996) *Constructive approximation: advanced problems*. Springer-Verlag, New York.

[LW94] Li X. D. and Wiberg N. E. (1994) A posteriori error estimate by element patch post-processing, adaptive analysis in energy and L_2 norms. *Comp. Struct.* 53: 907–919.

[LW06] Li J. and Widlund O. (2006) Feti-dp, bddc, and block cholesky methods. *Internat. J. Numer. Methods Engrg.* 66: 250–271.

[Man93] Mandel J. (1993) Balancing domain decomposition. *Comm. Numer. Methods Engrg.* 9: 233–241.

[Mar95] Marchuk G. I. (1995) *Adjoint Equations and Analysis of Complex Systems*. Kluwer Academic Publishers, Dordrecht.

[Mau81] Maurer H. (1981) First and second order sufficient optimality conditions in mathematical programming and optimal control. *Mathematical Programming Study* 14: 163–177.

[Max76] Maxworthy T. (1976) Experiments on collisions between solitary waves. *Journal of Fluid Mechanics* 76: 177–185.

[MDT05] Mandel J., Dohrmann R., and Tezaur R. (2005) An algebraic theory for primal and dual substructuring methods by contraints. *Appl. Numer. Math.* 54: 167–193.

[Mey00] Meyer C. D. (2000) *Matrix Analysis and Applied Linear Algebra*. SIAM.

[MOS92] Mulder W., Osher S., and Sethian J. (1992) Computing interface motion in compressible gas dynamics. *Journal of Computational Physics* 100(2): 209–228.

[MP94] Mohammadi B. and Pironneau O. (1994) *Analysis of the K-Epsilon Turbulence Model*. John Wiley & Sons, Chichester.

[MP97] Muzaferija S. and Peric M. (1997) Computation of free-surface flows using finite volume method and moving grids. *Numer. Heat Trans., Part B* 32: 369–384.

[MP01] Mohammadi B. and Pironneau O. (2001) *Applied Shape Optimization for Fluids*. Clarendon Press, Oxford.

[MPT02a] Maday Y., Patera A. T., and Turinici G. (2002) Global *a priori* convergence theory for reduced-basis approximation of single-parameter symmetric coercive elliptic partial differential equations. *C. R. Acad. Sci. Paris, Série I* 335(3): 289–294.

[MPT02b] Maday Y., Patera A., and Turinici G. (2002) *A Priori* convergence theory for reduced-basis approximations of single-parameter elliptic partial differential equations. *Journal of Scientific Computing* 17(1-4): 437–446.

[MT01] Mandel J. and Tezaur R. (2001) On the convergence of a dual-primal substructuring method. *Numerische Mathematik* 88: 543–558.

[Nit68] Nitsche J. A. (1968) Ein kriterium für die quasi-optimalitat des Ritzchen Verfahrens. *Numer. Math.* 11: 346–348.

[Nit71] Nitsche J. (1971) Über ein Variationsprinzip zur Lösung von Dirichlet-Problemen bei Verwendung von Teilräumen, die keinen Randbedingungen unterworfen sind. *Abh. Math. Sem. Univ. Hamburg* 36: 9–15. Collection of articles dedicated to Lothar Collatz on his sixtieth birthday.

[Nou10] Nouy A. (2010) Proper generalized decompositions and separated representations for the numerical solution of high dimensional stochastic problems. *Arch. Comput.*

Methods Engrg. 17: 403–434.

[NW06] Nocedal J. and Wright S. (2006) *Numerical Optimization.* Springer Series in Operations Research and Financial Engineering. Springer, New York.

[NZ04] Naga A. and Zhang Z. (2004) A posteriori error estimates based on the polynomial preserving recovery. *SIAM J. Numer. Anal.* 42: 1780–1800.

[OBB98] Oden J. T., Babuška I., and Baumann C. E. (1998) A discontinuous *hp* finite element method for diffusion problems. *J. Comput. Phys.* 146(2): 491–519.

[OP07] Oliveira I. and Patera A. (2007) Reduced-basis techniques for rapid reliable optimization of systems described by affinely parametrized coercive elliptic partial differential equations. *Optimization and Engineering* 8(1): 43–65.

[OS88] Osher S. and Sethian J. A. (1988) Fronts propagating with curvature-dependent speed: algorithms based on Hamilton-Jacobi formulations. *J. Comput. Phys.* 79(1): 12–49.

[Pan05] Panton R. (2005) *Incompressible Flow.* Wiley & Sons, 3rd edition.

[Pat80] Patankar S. V. (1980) *Numerical Heat Transfer and Fluid Flow.* Hemisphere, Washington.

[Pat84] Patera A.T. (1984) A spectral element method for fluid dynamics: laminar flow in a channel expansion. *J. Comput. Phys.* 54: 468–488.

[Pet] `www.mcs.anl.gov/petsc/`.

[PG00] Pierce N. and Giles M. B. (2000) Adjoint recovery of superconvergent functionals from PDE approximations. *SIAM Review* 42(2): 247–264.

[Pin85] Pinkus A. (1985) *n-Widths in Approximation Theory.* Springer-Verlag, Ergebnisse.

[Pin08] Pinnau R. (2008) Model reduction via proper orthogonal decomposition. In Schilder W. and van der Vorst H. (eds) *Model Order Reduction: Theory, Research Aspects and Applications*, pages 96–109. Springer.

[Pir84] Pironneau O. (1984) *Optimal Shape Design for Elliptic Systems.* Springer-Verlag, New York.

[Por85] Porsching T. A. (1985) Estimation of the error in the reduced basis method solution of nonlinear equations. *Mathematics of Computation* 45(172): 487–496.

[PQ05] Parolini N. and Quarteroni A. (2005) Mathematical models and numerical simulations for the America's cup. *Comput. Methods Appl. Mech. Engrg.* 194(9–11): 1001–1026.

[PQ07] Parolini N. and Quarteroni A. (2007) Modelling and numerical simulation for yacht engineering. In *Proceedings of the 26th Symposium on Naval Hydrodynamics.* Strategic Analysis, Inc., Arlington, VA, USA.

[PR07] Patera A. T. and Rozza G. (2006–2007) *Reduced Basis Approximation and A Posteriori Error Estimation for Parametrized Partial Differential Equations.* Copyright MIT. To appear in MIT Pappalardo Monographs in Mechanical Engineering.

[Pro97] Prohl A. (1997) *Projection and Quasi-Compressibility Methods for Solving the Incompressible Navier-Stokes Equations.* Advances in Numerical Mathematics. B.G. Teubner, Stuttgart.

[Pru06] Prud'homme C. (2006) A domain specific embedded language in c++ for automatic differentiation, projection, integration and variational formulations. *Scientific Programming* 14(2): 81–110.

[PS03] Perugia I. and Schötzau D. (2003) The *hp*-local discontinuous Galerkin method for low-frequency time-harmonic Maxwell equations. *Math. Comp.* 72(243): 1179–1214.

[PWY90] Pawlak T. P., Wheeler M. J., and Yunus S. M. (1990) Application of the Zienkiewicz-Zhu error estimator for plate and shell analysis. *Int. J. Numer. Methods Eng.* 29: 1281–1298.

[QR07] Quarteroni A. and Rozza G. (2007) Numerical solution of parametrized Navier-Stokes equations by reduced basis method. *Num. Meth. PDEs* 23: 923–948.

[QRDQ06] Quarteroni A., Rozza G., Dedè L., and Quaini A. (2006) Numerical approximation of a control problem for advection–diffusion processes. System modeling and optimization. *IFIP Int. Fed. Inf. Process.* 199: 261–273.

[QRM11] Quarteroni A., Rozza G., and Manzoni A. (2011) Certified reduced basis approximation for parametrized partial differential equations in industrial applications. *J. Math. Ind.* 1(3).

[QR13] Quarteroni A. and Rozza G. E. (2014, in press) *Reduced Order Methods for Modeling and Computational Reduction.* MS&A Modeling, Simulation and Applications Series. Springer-Verlag Italia, Milano.

[QSS07] Quarteroni A., Sacco R., and Saleri F. (2007) *Numerical Mathematics.* Springer, Berlin and Heidelberg, II edition.

[QSV00] Quarteroni A., Saleri F., and Veneziani A. (2000) Factorization methods for the numerical approximation of Navier-Stokes equations. *Comput. Methods Appl. Mech. Engrg.* 188(1-3): 505–526.

[Qu02] Qu Z. (2002) *Unsteady open-channel flow over a mobile bed.* PhD thesis, École Polytechnique Fédérale de Lausanne.

[Qua93] Quartapelle L. (1993) *Numerical Solution of the Incompressible Navier-Stokes Equations.* Birkhäuser Verlag, Basel.

[QV94] Quarteroni A. and Valli A. (1994) *Numerical Approximation of Partial Differential Equations.* Springer, Berlin Heidelberg.

[QV99] Quarteroni A. and Valli A. (1999) *Domain Decomposition Methods for Partial Differential Equations.* Oxford Science Publications, Oxford.

[Ran99] Rannacher R. (1999) Error control in finite element computations. An introduction to error estimation and mesh-size adaptation. In *Error control and adaptivity in scientific computing (Antalya, 1998)*, pages 247–278. Kluwer Acad. Publ., Dordrecht.

[RC83] Rhie C. M. and Chow W. L. (1983) Numerical study of the turbulent flow past an airfoil with trailing edge separation. *AIAA Journal* 21(11): 1525–1532.

[RHM13] Rozza G., Huynh D.P.B., and Manzoni A. (2013) Reduced basis approximation and a posteriori error estimation for Stokes flows in parametrized geometries: roles of the inf-sup stability constants. *Num. Math.* 125(1), 115–152..

[RHP08] Rozza G., Huynh D.B.P., and Patera A. T. (2008) Reduced basis approximation and a posteriori error estimation for affinely parametrized elliptic coercive partial differential equations: Application to transport and continuum mechanics. *Archives Computational Methods in Engineering* 15(3): 229–275.

[Riv08] Rivière B. (2008) *Discontinuous Galerkin methods for solving elliptic and parabolic equations*, volume 35 of *Frontiers in Applied Mathematics.* Society for Industrial and Applied Mathematics (SIAM), Philadelphia, PA. Theory and implementation.

[Rod94] Rodríguez R. (1994) Some remarks on Zienkiewicz-Zhu estimator. *Numer. Methods Part. Diff. Eq.* 10: 625–635.

[Roz09] Rozza G. (2009) Reduced basis method for Stokes equations in domains with non-affine parametric dependence. *Comp. Vis. Science* 12: 23–35.

[RR04] Renardy M. and Rogers R. C. (2004) *An Introduction to Partial Differential Equations.* Springer-Verlag, New York, II edition.

[RST96] Ross H. G., Stynes M., and Tobiska L. (1996) *Numerical Methods for Singularly Perturbed Differential Equations. Convection-Diffusion and Flow Problems.* Springer-Verlag, Berlin Heidelberg.

[Rud91] Rudin W. (1991) *Analyse Rèelle et Complexe.* Masson, Paris.

[RV07] Rozza G. and Veroy K. (2007) On the stability of reduced basis method for Stokes equations in parametrized domains. *Comp. Meth. Appl. Mech. and Eng.* 196: 1244–1260.

[RVS08] Rehman M., Vuik C., and Segal G. (2008) A comparison of preconditioners for incompressible Navier-Stokes solvers. *Int. J. Numer. Meth. Fluids* 57: 1731–1751.

[RVS09] Rehman M., Vuik C., and Segal G. (2009) Preconditioners for the steady incompressible Navier-Stokes problem. *Int. J. Applied Math.* 38 (4).

[RWG99] Rivière B., Wheeler M. F., and Girault V. (1999) Improved energy estimates for interior penalty, constrained and discontinuous Galerkin methods for elliptic problems. I. *Comput. Geosci.* 3(3-4): 337–360 (2000).

[RWG01] Rivière B., Wheeler M. F., and Girault V. (2001) A priori error estimates for finite element methods based on discontinuous approximation spaces for elliptic problems. *SIAM J. Numer. Anal.* 39(3): 902–931.

[Saa96] Saad Y. (1996) *Iterative Methods for Sparse Linear Systems.* PWS Publishing Company, Boston.

[Sag06] Sagaut P. (2006) *Large Eddy Simulation for Incompressible Flows: an Introduction.* Springer-Verlag, Berlin Heidelberg, III edition.

[Sal08] Salsa S. (2008) *Partial Differential Equations in Action - From Modelling to Theory.* Springer, Milan.

[Sch69] Schwarz H. (1869) Über einige abbildungsdufgaben. *J. Reine Agew. Math.* 70: 105–120.

[Sch98] Schwab C. (1998) *p and hp- Finite Element Methods.* Oxford Science Publication, Oxford.

[SF73] Strang G. and Fix G. J. (1973) *An Analysis of the Finite Element Method.* Wellesley-Cambridge Press, Wellesley, MA.

[SGT01] S. Gottlieb C. S. and Tadmor E. (2001) Strong stability preserving high order time discretization methods. *SIAM review* 43 (1): 89–112.

[She] Shewchuk J. R.`www.cs.cmu.edu/ quake/triangle.html`.

[Shu88] Shu C. (1988) Total-variation-diminishing time discretizations. *SIAM Journal on Scientific and Statistical Computing* 9: 1073–1084.

[Sir87] Sirovich L. (1987) Turbulence and the dynamics of coherent structures, part i: Coherent structures. *Quart. Appl. Math.* 45(3): 561–571.

[Smo01] Smolianski A. (2001) *Numerical Modeling of Two-Fluid Interfacial Flows.* PhD thesis, University of Jyväskylä.

[SO88] Shu C. and Osher S. (1988) Efficient implementation of essentially non-oscillatory shock-capturing schemes. *Journal of Computational Physics* 77: 439–471.

[SO89] Shu C. and Osher S. (1989) Efficient implementation of essentially non-oscillatory shock-capturing schemes ii. *Journal of Computational Physics* 83: 32–78.

[Spi99] Spivak M. (1999) *A comprehensive introduction to differential geometry. Vol. II.* Publish or Perish Inc., Houston, Tex., iii edition.

[Ste98] Stenberg R. (1998) Mortaring by a method of J. A. Nitsche. In *Computational mechanics (Buenos Aires, 1998)*, pages CD–ROM file. Centro Internac. Métodos Numér. Ing., Barcelona.

[Str71] Stroud A. H. (1971) *Approximate calculation of multiple integrals.* Prentice-Hall, Inc., Englewood Cliffs, N.J.

[Str89] Strickwerda J. C. (1989) *Finite Difference Schemes and Partial Differential Equations.* Wadworth & Brooks/Cole, Pacific Grove.

[Str00] Strostroup B. (2000) *C++ Linguaggio, Libreria Standard, Principi di Programmazione.* Addison Welsey Longman Italia, Milano, III edition.

[SV05] Saleri F. and Veneziani A. (2005) Pressure correction algebraic splitting methods for the incompressible Navier-Stokes equations. *SIAM J. Numer. Anal.* 43(1): 174–194.

[SW10] Stamm B. and Wihler T. P. (2010) *hp*-optimal discontinuous Galerkin methods for linear elliptic problems. *Math. Comp.* 79(272): 2117–2133.

[SZ91] Sokolowski J. and Zolesio J. (1991) *Introduction to Shape Optimization (Shape Sensitivity Analysis).* Springer-Verlag, New York.

[Tan93] Tanaka N. (1993) Global existence of two phase nonhomogeneous viscous incompressible fluid flow. *Comm. Partial Differential Equations* 18(1-2): 41–81.

[TE05] Trefethen L. and Embree M. (2005) *Spectra and pseudospectra. The behavior of nonnormal matrices and operators.* Princeton University Press, Princeton.

[Tem01] Temam R. (2001) *Navier Stokes Equations.* North-Holland, Amsterdam.

[TF88] Tsuchiya K. and Fan L.-S. (1988) Near-wake structure of a single gas bubble in a two-dimensional liquid-solid fluidized bed: vortex shedding and wake size variation. *Chem. Engrg. Sci.* 43(5): 1167–1181.

[Tho84] Thomee V. (1984) *Galerkin Finite Element Methods for Parabolic Problems.* Springer, Berlin and Heidelberg.

[TI97] Trefethen L. and III D. B. (1997) *Numerical Linear Algebra.* SIAM.

[TL58] Taylor A. and Lay D. (1958) *Introduction to Functional Analysis.* J.Wiley & Sons, New York.

[Tor99] Toro E. (1999) *Riemann Solvers and Numerical Methods for Fluid Dynamics.* Springer-Verlag, Berlin.

[Tri] `software.sandia.gov/trilinos/`.

[TSW99] Thompson J. F., Soni B. K., and Weatherill N. P. (eds) (1999) *Handook of Grid Generation.* CRC Press, Boca Raton.

[TW05] Toselli A. and Widlund O. (2005) *Domain Decomposition Methods - Algorithms and Theory.* Springer-Verlag, Berlin Heidelberg.

[TWM85] Thompson J. F., Warsi Z. U. A., and Mastin C. W. (1985) *Numerical Grid Generation, Foundations and Applications.* North-Holland, New York.

[UMF] `www.cise.ufl.edu/research/sparse/umfpack/`.

[Vas81] Vasiliev F. (1981) *Methods for Solving the Extremum Problems.* Nauka, Moscow.

[vdV03] van der Vorst H. A. (2003) *Iterative Krylov Methods for Large Linear Systems.* Cambridge University Press, Cambridge.

[Vel95] Veldhuizen T. (1995) Expression templates. *C++ Report Magazine* 7(5): 26–31. see also the web page `http://osl.iu.edu/~tveldhui` .

[Ven98] Veneziani A. (1998) *Mathematical and Numerical Modeling of Blood Flow Problems.* PhD thesis, Università degli Studi di Milano.

[Ver84] Verfürth R. (1984) Error estimates for a mixed finite elements approximation of the stokes equations. *R.A.I.R.O. Anal. Numér.* 18: 175–182.

[Ver96] Verführth R. (1996) *A Review of a Posteriori Error Estimation and Adaptive Mesh Refinement Techniques.* Wiley-Teubner, New York.

[VM96] Versteeg H. and Malalasekra W. (1996) *An Introduction to Computational Fluid Dynamics: the Finite Volume Method Approach.* Prentice-Hall.

[Vol11] Volkwein S. (2011) Model reduction using proper orthogonal decomposition. Lecture Notes, University of Konstanz.

[VP05] Veroy K. and Patera A.T. (2005) Certified real-time solution of the parametrized steady incompressible Navier-Stokes equations: rigorous reduced-basis a posteriori error bounds. *Int. J. Numer. Meth. Fluids* 47(8-9): 773–788.

[Wes01] Wesseling P. (2001) *Principles of Computational Fluid Dynamics.* Springer-Verlag, Berlin Heidelberg New York.

[Whe78] Wheeler M. F. (1978) An elliptic collocation-finite element method with interior penalties. *SIAM J. Numer. Anal.* 15(1): 152–161.

[Wil98] Wilcox D. C. (1998) *Turbulence Modeling in CFD.* DCW Industries, La Cañada, CA, II edition.

[Win07] Winkelmann C. (2007) *Interior penalty finite element approximation of Navier-Stokes equations and application to free surface flows.* PhD thesis, École Polytechnique Fédérale de Lausanne.

[Woh01] Wohlmuth B. (2001) *Discretization Methods and Iterative Solvers Based on Domain Decomposition.* Springer.

[Wya00] Wyatt D. C. (2000) Development and assessment of a nonlinear wave prediction methodology for surface vessels. *Journal of ship research* 44: 96.

[Yos74] Yosida K. (1974) *Functional Analysis.* Springer-Verlag, Berlin Heidelberg.

[Zie00] Zienkiewicz O. (2000) Achievements and some unsolved problems of the finite element method. *Int. J. Numer. Meth. Eng.* 47: 9–28.

[ZT00] Zienkiewicz O. C. and Taylor R. L. (2000) *The Finite Element Method, Vol. 1, The Basis.* Butterworth-Heinemann, Oxford, V edition.

[Zua03] Zuazua E. (2003) Propagation, observation, control and numerical approximation of waves. *Bol. Soc. Esp. Mat. Apl.* 25: 55–126.

[Zua05] Zuazua E. (2005) Propagation, observation, and control of waves approximated by finite difference methods. *SIAM Review* 47 (2): 197–243.

[Zua06] Zuazua E. (2006) Controllability and observability of partial differential equations: Some results and open problems. In Dafermos C. and Feiteisl E. (eds) *Handbook of Differential Equations: Evolutionary Differential Equations*, volume 3, pages 527–621. Elsevier Science.

[ZZ87] Zienkiewicz O. C. and Zhu J. Z. (1987) A simple error estimator and adaptive procedure for practical engineering analysis. *Int. J. Numer. Meth. Engng.* 24: 337–357.

[ZZ92] Zienkiewicz O. C. and Zhu J. Z. (1992) The superconvergent patch recovery and a posteriori error estimates. I: The recovery technique. *Int. J. Numer. Meth. Engng.* 33: 1331–1364.

Index

adaptivity
– a posteriori, 103, 521
– a priori, 100
– goal-oriented, 334
– of type h, 99
– of type p, 70, 99
adjoint
– operator, *see* operator adjoint
– problem, *see* problem adjoint
– state, 486, 492, 493
affine geometry, 623
algorithm
– diagonal exchange, 156
– Laplacian regularization, 157
– Lawson, 157
– steepest descent, 506
– Thomas, 164
analysis
– backward, 368
– Von Neumann, 356, 375, 376
Arnoldi algorithm, 170
assembly, 180
– element-oriented, 189
– node-oriented, 189
Aubin-Nitsche trick, 98, 111

baricentrization, *see* Laplacian regularization algorithm
BDDC, 566
boundary conditions
– Dirichlet, 433
– dual, 54, 56
– essential, 46
– free slip, 471
– natural, 46
– Neumann, 433, 460
– non-slip, 470
– primal, 54
– Robin, 41
– Robin-Dirichlet, 53
– weak treatment, 222
boundary layers, 291
boundary observation, 497, 498
breakdown, 171, 172

CFL
– condition, 352, 353, 415
– number, 352, 361, 362
characteristic
– function, 17
– Lagrangian functions, 67, 79
– lines, 7, 340, 344
  – of the Burgers equation, 409
– rate, 414
– variables, 343
coefficient
– amplification, 356, 376
– dispersion, 376
– dissipation, 361
computational cost, 5

A. Quarteroni: *Numerical Models for Differential Problems*, 2nd Ed.
MS&A – Modeling, Simulation & Applications 8
DOI 10.1007/978-88-470-5522-3, © Springer-Verlag Italia 2014

computational reduction, 590
condition
– CFL, 352, 353, 415
– entropy, 411
– incompressibility, 430
– inf-sup, 438, 439, 449
– Lax admissibility, 427
– Rankine-Hugoniot, 410, 427
conservation law, 213, 409, 426
– of the entropy, 412
consistency, 4, 350
– strong, 4, 314, 316
control
– boundary, 487, 495
– distributed, 486, 490, 494, 495
– function, 506
– optimal, 483, 485, 488
– volume, 214
controllability, 514
convergence, 4, 64, 66, 128, 350
– rate, 5
coordinates
– barycentric, 78, 218

degree
– of a vector, 170
degree of exactness, 185
degrees of freedom, 67, 69
– finite element, 77
derivative
– conormal, 48, 50
– Fréchet, 14, 488
– Gâteaux, 14, 504, 505
– in the sense of distributions, 18
– interpolation, 241, 242, 257, 398, 400
– Lagrangian, 470
– material, 458
– normal, 32
diffusion
– artificial, 305
– numerical, *see* artificial diffusion
Dirac delta, 16, 33
dispersion, 360, 367
dissipation, 360, 367
distributed observation, 497, 498
distributions, 15
domain, 23
domain of dependence, 344
– numerical, 352
Donald
– diagram, 218

element
– finite, 76
  – diameter, 81, 144
  – Lagrangian, 77
  – sphericity, 144
– reference, 77
elliptic regularity, 95, 97
entropy, 412
– flux, 412
equation
– adjoint, 492, 504–506
– Burgers, 2, 409–411, 421, 422
– diffusion-transport, 5, 460
– discriminant, 5
– elliptic, 5
– Euler, 488
– heat, 2, 3, 5, 122
– homogeneous, 1
– hyperbolic, 5
  – nonlinear, 414
– Korteveg-de-Vries, 10
– observation, 490
– parabolic, 5
– Plateau's, 9
– Poisson, 31
– potential, 2, 5
– quasi-linear, 1, 426
– semi-linear, 1
– state, 425, 484, 490, 506
– transport, 1
– transport-reaction, 359
– transportation, 3
– viscosity, 413
– wave, 2, 5, 9, 345
equations
– compatibility, 406
– equivalent, 364
– Euler, 424, 428, 434

– in conservative form, 426
– Euler–Lagrange, 516
– Navier-Stokes, 429, 478
– for compressible fluids, 424
– primitive variables, 431
– reduced form, 434
– weak formulation, 433
– Stokes, 435
– strictly hyperbolic, 426, 427
error
– a posteriori estimate, 103, 107, 111, 112, 520
– goal-oriented, 113
– recovery-based, 112
– residual-based, 107
– a priori estimate, 75, 94, 95, 99, 130, 135, 247, 323, 395, 445
– amplification, 361, 362, 380
– approximation, 65
– discretization, 520
– dispersion, 361, 362, 380
– interpolation
– estimate, 91, 233
– iteration, 520
– truncation, 314, 350
exactness degree, 232, 233, 235

factorization
– Cholesky, 124, 163
– LU, 162
– tridiagonal, 163
FETI
– preconditioner, 538
FETI-DP, 563
Fick law, 425
field of values, 174
finite differences, 241, 301, 346
finite elements, 567
– $\mathbb{P}_1$-*iso*$\mathbb{P}_2$, 450
– compatible, 449
– Crouzeix-Raviart, 449
– discontinuous, 388, 394, 415
– hierarchical, 70
– implementation, 179
– isoparametric, 195
– Lagrangian, 67
– linear, 67, 71
– mini-element, 450
– quadratic, 69
– sphericity, 81
– stabilized, 309
flow
– turbulent, 434
flux
– limiters, 422
– numerical, 220, 222, 347, 349, 414, 416
– Engquist-Osher, 416
– Godunov, 416
– Lax-Friedrichs, 416
– monotone, 416
– upwind, 417
– thermal, 425
– viscous, 426
form, 12
– bilinear, 12
– affine, 587, 627
– eigenfunction, 133
– eigenvalue, 133
– parametric, 587
– coercive, 13
– continuous, 12
– positive, 13
– quadratic, 8
– definite, 8
– degenerate, 8
– indefinite, 8
– quasi-linear, 426
– symmetric, 12
– weakly coercive, 122
formula
– Green, *see* Green formula
formulae
– Armijo, 510
formulation
– conservative, 55
– non-conservative, 55
– strong, 31
– weak, 33
free surface, 468, 480
function

– basis, 68, 70
– Bernoulli, 306
– bubble, 70, 312, 327
– characteristic, 20
– compact support, 15
– control, 484
– Heaviside, 19
– observation, 484, 487
– spacing, 101, 141, 153
– upwind, 319
functional, 11
– bounded, 11
– cost, 484, 490, 504
– Lagrangian, 437
– linear, 11
– norm, 11

Galerkin
– orthogonality, 64
Galerkin orthogonality, 65, 111, 113
Gaussian integration
– Gauss-Legendre, 232
  – exactness, 233
– Gauss-Legendre-Lobatto, 233, 245
Gordon-Hall transformation, 226
greedy
– algorithm, 608
Green formula, 41, 42, 59, 431
grid, *see also* triangulation
– anisotropic, 81, 95
– blockwise structured, 148
– derefinement, 101
– non-structured, 149
– quasi uniform, 86
– refinement, 101
– regular, 81
– structured, 145, 146

hierarchical basis, 70
hyperbolic system, 343, 346, 349

inequality
– Cauchy-Schwarz, 25, 36, 42, 251
  – strengthened, 577
– Hölder, 24, 432
– inverse, 86, 135, 239, 308, 318
– Korn, 60
– Poincaré, 22, 292, 308
– Young, 126
input-parameter, 585
integral
– general, 3
– particular, 3
interpolation, 73, 88
– error estimates, 74, 92
– operator, 73, 88
– transfinite, 226
iterative algorithms
– generalized minimum residual (GMRES)
  – flexible, 174

Kolmogorov N-width, 614
Krylov method, *see* method Krylov
Krylov subspace, 169

Lagrange
– identity, 52, 55
– multiplier, 503
Lagrangian
– basis, 67, 69
– finite elements, 77
– functional, 503
Legendre series, 230
lemma
– Bramble-Hilbert, 90
– Céa, 64, 66
– Deny-Lions, 91
– Gronwall, 28, 130, 343, 383
  – discrete, 28
– Lax-Milgram, 50, 59, 61, 64
– Strang, 247, 250, 311, 452

mass-lumping, 124, 302, 336, 375
matrix
– graph, 187
– interpolation derivative, 242
– iteration, 164
– mass, 123
– preconditioning, 87, 164

– reflection, 407
– sparse, CSR format, 188
– sparsity pattern, 187
– stiffness, 62, 83, 123
  – conditioning, 85
– symmetric positive definite, 167
method
– algebraic factorization, 462
– BDF, 203
– BiCGSTAB, 169
– characteristics, 340, 458
– Chorin-Temam, 461, 467
– collocation, 240
– conjugate gradient
  – convergence, 169
– consistent, 4, 314, 350
– convergent, 4, 351
– Crank-Nicolson, 124
– Dirichlet-Neumann, 530, 545, 546, 579
– Discontinuous Galerkin (DG), 267, 329, 388, 394, 415
  – jump stabilization, 396
  – SEM-NI, 401
– domain decomposition, 192, 480, 527
  – balancing with constraints, 566
  – FETI-DP, 563
– Douglas-Wang (DW), 316, 317
– Euler
  – backward, 124, 372, 385, 392, 457
  – backward/centered, 348
  – explicit, *see* forward Euler method
  – forward, 124, 371, 376, 386, 456
  – forward/centered, 347
  – forward/decentered, *see* upwind method
  – implicit, *see* backward Euler method
– exponential fitting, *see* Scharfetter and Gummel method
  – (FV), 222
– finite volumes, 213
  – a priori estimate, 222
  – cell-centered, 215
  – ghost nodes, 223
  – staggered-grids, 215, 478
  – vertex-centered, 215
– FOM, 171
– fractional step, 459
– front capturing, 468
– front tracking, 468
– G-NI, 236, 240, 243, 246, 255, 308, 397
– Galerkin, 61, 63, 293, 296, 437, 595
  – convergence, 66
  – generalized, 237, 247, 309
  – Least Squares (GLS), 316, 317, 321, 451, 519
  – projection, 594
  – spectral, 225, 236
– GEM, 161
– GMRES, 171, 173
  – convergence, 173
  – flexible, 174
  – with restart, 173
– gradient, 167, 508
  – conjugate, 168, 509
– Incomplete Interior Penalty Galerkin, (IIPG), 269, 271
– Interior Penalty (IP), 269
– Lax-Friedrichs, 348, 354, 368, 415
  – (FEM), 376
– Lax-Wendroff, 348, 349, 354, 368
  – (FEM), 376
– leap-frog, 349
– level set, 480
– MES, 275
– mortar, 267, 273, 277, 282, 285, 330
– Neumann-Dirichlet, 546
– Neumann-Neumann, 532, 537, 546
– Newmark, 349
– Newton, 509
– Non-Symmetric Interior Penalty Galerkin, (NIPG), 269, 271
– numerical
  – bounded, 414
  – monotone, 414, 416
– operator splitting, 459
  – Yanenko, 459
– Petrov-Galerkin, 309, 312
– projection, 461
– quasi–Newton, 509
– reduced basis (RB), 585

– coercivity constant, 591
– continuity constant, 591
– convergence, 612
– hierarchical space, 594
– non-hierarchical space, 612
– residual, 616, 619
– sampling, 606
– snapshot, 593, 608
– space, 594
– relaxation, 531
– Richardson, 165, 170, 536, 545
– Robin-Robin, 532, 537, 546
– Runge-Kutta
– 2nd order, 418
– 3rd order, 418
– Scharfetter and Gummel, 306, 312, 320
– Schwarz, 528, 567, 573
– additive, 528, 569
– multiplicative, 528, 569
– SEM, 226, 228, 273
– SEM-NI, 258
– semi-implicit, 457
– shock capturing, 415
– spectral element, 226
– streamline-diffusion, 313
– Streamline-Upwind Petrov-Galerkin (SUPG), 316, 317, 451
– Symmetric Interior Penalty Galerkin, (SIPG), 269
– Taylor-Galerkin, 378
– upwind, 304, 348, 349, 353, 396
– (FEM), 376, 396
– FV, 222
– volume of fluid, 480
– Yosida, 467
modal
– basis, 263
– boundary-adapted, 263, 264
– representation, 230

nodes, 67, 69, 81, 82
norm
– A, 87
– broken, 278
– discrete, 251
– energy, 65, 75, 88
norma
– broken, 287
number
– CFL, 352, 361, 362, 376
– condition, 85, 542, 545, 549, 554, 575
– Péclet
– global, 295, 299
– grid, *see* local Péclet number
– local, 221, 297, 300, 305, 307
– Reynolds, 434

observability, 514
offline-online computational procedure, 596, 619
operator
– adjoint, 26, 51, 55, 56, 492
– bilaplacian, 57
– dual, *see* adjoint, 54
– extension, 552
– interpolation, 88, 240
– Clément, 105
– Laplace, 2
– Laplace-Beltrami, 496
– lifting, 39, 46, 84
– normal, 52
– primal, 54
– pseudo-spectral, 241
– restriction, 552
– self-adjoint, 52
– skew-symmetric, 315
– Steklov-Poincaré, 533, 534, 544
– symmetric, 315
– trace, 23
– transpose, 26
optimality system, 493, 494, 504, 505
output, 585
– compliant, 587

Parseval identity, 230
partition
– of the domain, 273
partition of unity, 79, 303
phase angle, 361, 376
point

– constrained critical, 502
– regular, 502
polygon, 23
polyhedron, 23
polynomials
– Jacobi, 263
– Legendre, 229, 232, 391
preconditioner, 87, 164
– additive Schwarz, 569
– augmented Lagrangian, 465
– block Jacobi, 569
– Bramble-Pasciak-Schatz, 554
– Jacobi, 553
– MSIMPLER, 466
– Neumann-Neumann, 555, 556
  – balanced, 557
– SIMPLE, 465
– SIMPLER, 466
principal symbol, 8
principle
– discrete maximum, 354
– virtual work, 38
problem
– adjoint, 51, 96, 97, 111
– advection, 578
– collocation, 241
– control, 483
  – constrained, 484
  – discretized, 516
  – unconstrained, 484
– controllable, 483
– diffusion-reaction, 206, 252, 299
– diffusion-transport, 291, 332, 337
– diffusion-transport-reaction, 139, 219, 337
– Dirichlet, 31, 41, 84, 494
– dual, *see* adjoint
– elliptic, 61
– fourth-order, 58, 139
– free surface, 468
– Galerkin, *see* Galerkin method
– generalized eigenvalue, 133
– generalized Stokes, 435
– heat, 266
– heterogeneous, 527
– linear elasticity, 58
– mixed, 40
– Neumann, 32, 39, 40, 495
– optimal design, 517
– optimization, 484
– Poisson, 63, 82, 538
– Robin, 40
– Stokes, 579
  – algebraic formulation, 447
  – Galerkin approximation, 437
  – stabilized finite elements, 451
  – stable finite elements, 449
  – weak form, 436
– transport, 339
– transport-reaction, 341, 382
– variational, 37, 43
product
– discrete scalar, 236
– warped tensor, 262
programming
– object-oriented, 182
proper orthogonal decomposition, 590, 609

quadrature formula
– composite midpoint, 187
– composite trapezoid, 187

random walk, 291
reconstruction
– of the gradient, 100
reduced basis method, *see* method, reduced basis
reference domain, 623, 626
residue, 165
– local, 106
– preconditioned, 165
Riemann invariants, 345
rod
– thin, 114
– vibrating, 9

saddle-point, 437
scheme, *see* method
Schur complement, 541, 551

semi-discretization, 123
seminorm, 22
shape optimization, 515
simplex, 78
solution
– classical, 410
– entropic, 410, 414
  – of the Burgers equation, 411
– weak, 341, 410
space
– dual, 11
– of distributions, 16
– Sobolev, 36
splitting, 164
spurious modes, *see* Stokes problem, spurious solutions
stability, 5, 64, 351, 414
– absolute, 133, 244
– strong, 351, 353, 354, 357, 359, 386, 415, 420
subspace
– Krylov, 169
support
– compact, 15
– of a function, 15, 84

term
– diffusion, 429
– transport, 429
theorem
– closed range, 442
– divergence, 41
– equivalence, 5, 353
– extension, 543
– Helmholtz-Weyl, 461
– Lax-Richtmyer, *see* equivalence theorem
– Riesz, 12, 51
– trace, 23
θ-method, 123, 132
triangulation, 144
– advancing front, 153
– conforming, 144
– Delaunay, 149, 217
  – generalized, 152
– median dual, 218
– regular, 144
turbulence models, 434
– RANS, 480

unisolvent set, 77

variational inequality, 488
vertices, 67
viscosity
– artificial, 310, 313, 314
– numerical, *see* artificial viscosity
– subgrid, 329
Voronoi
– diagram, 216
– polygon, 216
– tessellation, 216
– vertices, 217

Zienkiewicz-Zhu estimator, 112

MS&A – Modeling, Simulation and Applications

Series Editors:

Editor at Springer:

Francesca Bonadei
francesca.bonadei@springer.com

1 L. Formaggia, A. Quarteroni, A. Veneziani (eds.)
Cardiovascular Mathematics
2009, XIV+522 pp, ISBN 978-88-470-1151-9

2. A. Quarteroni
Numerical Models for Differential Problems
2009, XVI+602 pp, ISBN 978-88-470-1070-3

3. M. Emmer, A. Quarteroni (eds.)
MATHKNOW
2009, XII+264 pp, ISBN 978-88-470-1121-2

4. A. Alonso Rodríguez, A. Valli
Eddy Current Approximation of Maxwell Equations
2010, XIV+348 pp, ISBN 978-88-470-1934-8

5. D. Ambrosi, A. Quarteroni, G. Rozza (eds.)
Modeling of Physiological Flows
2012, X+414 pp, ISBN 978-88-470-1934-8

6. W. Liu
Introduction to Modeling Biological Cellular Control Systems
2012, XII+268 pp, ISBN 978-88-470-2489-2

7. B. Maury
The Respiratory System in Equations
2013, XVIII+276 pp, ISBN 978-88-470-5213-0

8. A. Quarteroni
Numerical Models for Differential Problems, 2nd Edition
2014, XX+656pp, ISBN 978-88-470-5521-6

For further information, please visit the following link:
http://www.springer.com/series/8377

Printed in the United States
By Bookmasters